EIGHTH EDITION

ESSENTIALS OF METEOROLOGY

EIGHTH EDITION

Essentials of Meteorology

AN INVITATION TO THE ATMOSPHERE

C. Donald Ahrens
Emeritus, Modesto Junior College

Robert Henson

CENGAGE
Learning

Australia • Brazil • Japan • Korea • Mexico • Singapore •
Spain • United Kingdom • United States

Essentials of Meteorology
An Invitation to the Atmosphere

Eighth Edition

C. Donald Ahrens and Robert Henson

Product Director: Dawn Giovanniello

Senior Content Developer: Lauren Oliveira

Product Assistant: Marina Starkey

Marketing Manager: Ana Albinson

Content Project Manager: Hal Humphrey

Senior Designer: Michael Cook

Manufacturing Planner: Rebecca Cross

Production Service: Janet Bollow Associates

Photo Researcher: Manojkiran Chandler, Lumina Datamatics

Text Researcher: Manjula Subramanian, Lumina Datamatics

Copy Editor: Judith Chaffin

Illustrator: Charles Preppernau

Text Designer: Janet Bollow Associates

Cover Designer: Michael Cook

Cover Image: Ryan McGinnis/Moment/Getty Images

Compositor: Graphic World

For product information and technology assistance, contact us at
Cengage Learning Customer & Sales Support, 1-800-354-9706

For permission to use material from this text or product, submit all requests online at **cengage.com/permissions**
Further permissions questions can be emailed to
permissionrequest@cengage.com

ISBN: 978-1-305-62845-8

Loose-leaf Edition:
ISBN: 978-1-337-27610-8

Cengage Learning
20 Channel Center Street
Boston, MA 02210
USA

Cengage Learning is a leading provider of customized learning solutions with employees residing in nearly 40 different countries and sales in more than 125 countries around the world. Find your local representative at www.cengage.com.

Cengage Learning products are represented in Canada by Nelson Education, Ltd.

To learn more about Cengage Learning Solutions, visit www.cengage.com.

Purchase any of our products at your local college store or at our preferred online store www.cengagebrain.com.

Printed in the United States of America
Print Number: 03 Print Year: 2018

CONTENTS IN BRIEF

CONTENTS

© C. Donald Ahrens

© C. Donald Ahrens

© C. Donald Ahrens

NASA

CHAPTER 10

Thunderstorms and Tornadoes   272

© Robert Henson

CHAPTER 11

Hurricanes   318

© C. Donald Ahrens

CHAPTER 14

Air Pollution   414

© Robert Henson

PREFACE

The world is an ever-changing picture of naturally occurring events. From drought and famine to devastating floods, some of the greatest challenges we face come in the form of natural disasters created by weather. Yet, dealing with weather and climate is an inevitable part of our lives. Sometimes it is as small as deciding what to wear for the day or how to plan a vacation. But it can also have life-shattering consequences, especially for those who are victims of a hurricane or a tornado.

In recent years, weather and climate have become front page news, from record-setting extreme weather events to environmental issues such as global warming and ozone depletion. The dynamic nature of the atmosphere seems to demand our attention and understanding more these days than ever before. Almost daily, there are newspaper articles describing some weather event or impending climate change. For this reason, and the fact that weather influences our daily lives in so many ways, interest in meteorology (the study of the atmosphere) has been growing. This rapidly developing and popular science is giving us more information about the workings of the atmosphere than ever before. One of the reasons that meteorology is such an engaging science to study is that the atmosphere is a universally accessible laboratory for everyone. Although the atmosphere will always provide challenges for us, as research and technology advance, our ability to understand our atmosphere improves as well. The information available to you in this book, therefore, is intended to aid in your own personal understanding and appreciation of our Earth's dynamic atmosphere.

About This Book

Essentials of Meteorology is written for students taking an introductory course on the atmospheric environment. The main purpose of the text is to convey meteorological concepts in a visual, practical, and nonmathematical manner. In addition, the intent of the book is to stimulate curiosity in the reader and to answer questions about weather and climate that arise in our day-to-day lives. Although introductory in nature, this eighth edition maintains scientific integrity and includes up-to-date information on large-scale topics, such as global warming, ozone depletion, and El Niño, as well as discussion of recent high-profile weather events. As in previous editions, no special prerequisites are necessary for understanding.

Written expressly for the student, this book emphasizes the understanding and application of meteorological principles. The text encourages watching the weather so that it becomes "alive," allowing readers to immediately apply textbook material to the world around them. To assist with this endeavor, a color Cloud Chart appears at the back of the text. The Cloud Chart can be separated from the book and used as a learning tool at any place one chooses to observe the sky. To strengthen points and clarify concepts, illustrations are rendered in full color throughout. Color photographs were carefully selected to illustrate features, stimulate interest, and show how exciting the study of weather can be. To enhance the value of the book, several appendices that were only available online in the seventh edition have been reincorporated.

Organized into fifteen chapters, *Essentials of Meteorology* is designed to provide maximum flexibility to instructors of weather and climate courses. Thus, chapters can be covered in any desired order. For example, Chapter 15, "Light, Color, and Atmospheric Optics," is self-contained and can be covered earlier if so desired. Instructors, then, are able to tailor this text to their particular needs. This book basically follows a traditional approach. After an introductory chapter on the origin, composition, and structure of the atmosphere, it covers solar energy, air temperature, humidity, clouds, precipitation, and winds. Then comes a chapter on air masses, fronts, and middle-latitude cyclonic storms. Weather prediction and severe storms are next. A chapter on hurricanes is followed by a chapter on global climate. A chapter on climate change is next. A chapter on air pollution precedes the final chapter on atmospheric optics.

Each chapter contains at least two Focus sections, which either expand on material in the main text or explore a subject closely related to what is being discussed. Focus sections fall into one of three distinct categories: Observations, Special Topics, and Environmental Issues. Some include material that is not always found in introductory meteorology textbooks—subjects such as space weather, the scientific method, and wind energy. Others help to bridge theory and practice. This edition contains several new or rewritten Focus sections, including an updated discussion of nor'easters in Chapter 8 and a new Focus section on tornado damage patterns in Chapter 10.

Set apart as "Did You Know?" features in each chapter is weather information that may not be commonly known, yet pertains to the topic under discussion. Designed to bring the reader into the text, most of these weather highlights relate to some interesting weather fact or astonishing event.

Each chapter incorporates other effective learning aids:

- A major topic outline begins each chapter.
- Interesting introductory pieces draw the reader naturally into the main text.
- Important terms are boldfaced, with their definitions appearing in the glossary or in the text.
- Key phrases are italicized.
- English equivalents of metric units are immediately provided in parentheses.
- A brief review of the main points is placed toward the middle of most chapters.
- Each chapter ends with a summary of the main ideas.
- A list of key terms with page references follows each chapter, allowing students to review and reinforce their knowledge of key concepts.
- Questions for Review act to check how well students assimilate the material.
- Questions for Thought and Exploration encourage students to synthesize learned concepts for deeper understanding.
- References to 19 concept animations are compiled on pp. xx-xxi. These animations convey an immediate appreciation of how a process works and help students visualize the more difficult concepts in meteorology. Animations can be found on the Meteorology Course-Mate, accessed through Cengagebrain.com.
- At the end of each chapter are questions that relate to articles found on the Global Geoscience Watch website, available on its own or via the Meteorology CourseMate.

Eight appendices conclude the book. In addition, at the end of the book, a compilation of supplementary reading material is presented, as is an extensive glossary. On the endsheet at the back of the book is a geophysical map of North America. The map serves as a quick reference for locating states, provinces, and geographical features, such as mountain ranges and large bodies of water.

Supplemental Material and Technology Support

TECHNOLOGY FOR THE INSTRUCTOR

Instructor Companion Website Everything you need for your course in one place! This collection of book-specific lecture and class tools is available online via www.cengage.com/login. Access and download Power-Point presentations, images, instructor's manual, videos, and more.

Cognero Test Bank Cengage Learning Testing Powered by Cognero is a flexible, online system that allows you to:

- author, edit, and manage test bank content from multiple Cengage Learning solutions
- create multiple test versions in an instant
- deliver tests from your LMS, your classroom, or wherever you want

Global Geoscience Watch Updated several times a day, the Global Geoscience Watch is a focused portal into GREENR—our Global Reference on the Environment, Energy, and Natural Resources—an ideal one-stop site for classroom discussion and research projects for all things geoscience! Broken into the four key course areas (Geography, Geology, Meteorology, and Oceanography), you can easily get to the most relevant content available for your course. You and your students will have access to the latest information from trusted academic journals, news outlets, and magazines. You also will receive access to statistics, primary sources, case studies, podcasts, and much more!

TECHNOLOGY FOR THE STUDENT

Earth Science MindTap for Essentials of Meteorology MindTap is well beyond an eBook, a homework solution or digital supplement, a resource center website, a course delivery platform, or a Learning Management System. More than 70 percent of students surveyed said that it was unlike anything they have ever seen before. MindTap is a new personal learning experience that combines all of your digital assets—readings, multimedia, activities, study tools, and assessments—into a singular learning path to improve student outcomes.

Lab Manual Developed by the Oklahoma Climatological Survey (OCS) research and service facility, in concert with the University of Oklahoma, *Explorations in Meteorology* places a strong emphasis on helping students understand weather and climate by using real meteorological data. The activities in this lab manual require that students tap into the OCS archives of meteorological data in order to complete meteorological exercises. Full-color pictures and data graphs help students visually understand weather and severe weather topics. The lab manual also challenges students by providing optional questions intended for honors students, making this lab manual appropriate for both introductory and honors meteorology courses.

Eighth Edition Changes

This edition of *Essentials of Meteorology* includes a coauthor—meteorologist and science journalist Robert Henson (Weather Underground). For more than 20 years, Henson produced publications and websites for the University Corporation for Atmospheric Research, which manages the National Center for Atmospheric Research. He is an expert on severe weather, including tornadoes, thunderstorms, and hurricanes. He has also analyzed how television weathercasters cover major storms and report on climate change. Henson is the author of four trade books on meteorology, including *The Thinking Person's Guide to Climate Change* (previously *The Rough Guide to Climate Change,* the first edition of which was shortlisted for the United Kingdom's Royal Society Prize for Science Books).

The authors have carried out extensive updates and revisions to this eighth edition of *Essentials of Meteorology,* reflecting the ever-changing nature of the field and the atmosphere itself. Dozens of new photos and new or revised color illustrations help students visualize the excitement of the atmosphere.

- Chapter 1, "Earth's Atmosphere," continues to serve as a broad overview of the atmosphere. The text now begins with a discussion of the scientific method and its importance. To help draw students into the material, the introduction to meteorology and the summary of extreme weather types has been placed earlier in the chapter, followed by discussion of the chemistry and vertical structure of Earth's atmosphere. Among

recent events now included are the severe flooding over the Southern Plains and Southeast in 2015 and the Houston flash flood of April 2016.

- Chapter 2, "Warming and Cooling Earth and Its Atmosphere," contains up-to-date statistics and background on greenhouse gases and climate change, topics covered in more detail later in the book. Discussion of the potential impact of clouds on future global warming has been updated.

- In Chapter 3, "Air Temperature," several figures and tables have been updated so that they refer to normals drawn from the most recent reference period (1981–2010).

- Chapter 4, "Humidity, Condensation, and Clouds," includes updated material on satellite observations, including new background and artwork from the Global Precipitation Mission satellite. Also included is a "Did You Know?" box on the high-impact Atlanta snowstorm of January 2014.

- Chapter 5, "Cloud Development and Precipitation," includes a new graphic. New satellite observing techniques are also noted and illustrated in this chapter.

- Chapter 6, "Air Pressure and Winds," includes a substantially enhanced description and revised illustrations of the interplay between the pressure gradient and Coriolis forces in cyclonic and anticyclonic flow. Several other illustrations have been revised for clarity, and the discussion of scatterometers has been updated. A box on wind energy features the most recent data on wind energy adoption.

- Chapter 7, "Atmospheric Circulations," features a major restructuring, update, and expansion of sections dealing with the El Niño/Southern Oscillation, Pacific Decadal Oscillation, North Atlantic Oscillation, and Arctic Oscillation, including several new and updated images. The opening section, which introduces scales of atmospheric motion, has also been revised for clarity.

- Chapter 8, "Air Masses, Fronts, and Middle-Latitude Cyclones," now includes discussion of atmospheric rivers and an illustration of their impacts. In line with recent research, the section on occluded fronts stresses the prevalence of warm-type over cold-type occluded fronts. The discussion of drylines has been expanded, and the Focus box on nor'easters has been reworked to spotlight the record-setting East Coast snowstorm of January 2016.

- "Weather Forecasting" (Chapter 9) has undergone substantial revision, with several updated graphics. Three major types of satellite imagery are introduced near the beginning of the chapter. Explanations of watches, warnings, and forecasts of various durations (including seasonal outlooks) are now incorporated in a new section, "Time Range of Forecasts." The concept of the forecast funnel is also introduced.
- Chapter 10, "Thunderstorms and Tornadoes," includes several new and updated illustrations, depicting low- and high-precipitation supercells, a roll cloud, and a shelf cloud. An expanded section covers both flash flooding and river flooding and their connection to thunderstorms, including examples from Colorado (2013) and Texas and Oklahoma (2015). A new Focus box explores the baffling damage patterns that tornadoes can produce. The effort to accommodate new ways of estimating and reporting tornado wind speed (such as mobile radar reports) is also noted.
- The chapter on "Hurricanes" (Chapter 11) includes a new opening section that introduces students to the terrible impacts of Hurricane Katrina. Several graphics that use satellite imagery to explain basic concepts have been updated with recent tropical cyclones. Charts on hurricane climatology have been brought up to date, and an expanded range of both historical and recent examples are discussed, including the Galveston hurricane of 1900, the New England hurricane of 1938, and Typhoon Haiyan from 2013.
- Chapter 12, "Global Climate," includes a number of updates to climatological charts and discussion, drawing on the most recent set of United States climate normals (1981–2010).
- Chapter 13, "Earth's Changing Climate," has been revised throughout to reflect increasing confidence on a variety of climate change indicators and impacts. Also incorporated are graphics, conclusions, and emission pathways from the Fifth Assessment Report (2013-14) of the Intergovernmental Panel on Climate Change. The extremely quiet solar cycle of the late 2000s and early 2010s is noted, along with a variety of weather extremes from recent years that are relevant to climate change. The Paris Accord is discussed in the context of the Kyoto Protocol

that preceded it. A number of diagrams have been updated.
- The chapter on "Air Pollution" (Chapter 14) has been revised to include the latest air pollution trends across the United States as well as the latest information on ozone depletion in the Arctic and Antarctic. The U.S. Clean Power Plan is introduced, and the devastating impacts of both indoor and outdoor air pollution are discussed, including the effects of tiny particulates on cardiovascular health.
- Several of the photos in Chapter 15, "Light, Color, and Atmospheric Optics," have been replaced with spectacular new examples (e.g., anticrepuscular rays and double rainbows).

Acknowledgments

Many people have contributed to the eighth edition of *Essentials of Meteorology*. A very special and most grateful thank-you goes to Lita Ahrens who proofread each chapter. Special thank you to Charles Preppernau for rendering the beautiful art and to Janet Hansen for careful proofreading.

We are indebted to Janet Alleyn, who not only designed the book but, once again, took the art, photos, and manuscript and turned them into a beautiful book. Thanks goes to Judith Chaffin for her careful and conscientious editing. Special thanks to all the people at Cengage Learning who worked on this edition, including Lauren Oliveira, Morgan Carney, Hal Humphrey, and Dawn Giovanniello.

Thanks to our friends who provided photos and to those reviewers who offered comments and suggestions for this edition, including:

Fidel González Rouco
Universidad Complutense de Madrid

Redina Herman
Western Illinois University

Bette Otto-Bliesner
National Center for Atmospheric Research

David Schultz
University of Manchester

Alex Huang
University of North Carolina at Asheville

Anthony Santorelli
Anne Arundel Community College

Dan Ferandez
Anne Arundel Community College

Dean G Butzow
Western Michigan University

Douglas K. Miller
Purdue University

Edward J. Perantoni
Lindenwood University

Ronald A Dowey
Harrisburg Area Community College

Shaunna L. Donaher
University of Miami

Troy Kimmel
University of Texas

To the Student

Learning about the atmosphere can be a fascinating and enjoyable experience. This book is intended to give you some insight into the workings of the atmosphere. However, for a real appreciation of your atmospheric environment, you must go outside and observe. Although mountains take millions of years to form, a cumulus cloud can develop into a raging thunderstorm in less than an hour. The atmosphere is always producing something new for us to behold. To help with your observations, a color Cloud Chart is at the back of the book for easy reference. Remove it and keep it with you. And, remember, all of the concepts and ideas in this book are manifested out there for you to discover and enjoy. Please take the time to look.

Donald Ahrens and Robert Henson

EXPLORE THE CONCEPT ANIMATIONS

These animations have been carefully created to bring to life key points in the chapters. They are also the perfect tool to help refresh students' memories of previous concepts, so they can keep building on knowledge already acquired. Concept Animations are accessed through the MindTap platform, which can be acquired separately or together with print or looseleaf versions of this book. Some examples of Concept Animations are shown here.

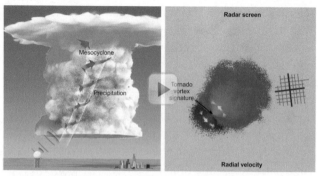

Doppler radar images are used extensively throughout this book. To better understand Doppler radar images, watch all 4 parts of this *Doppler Radar* animation (Chapters 1, 5, and 10).

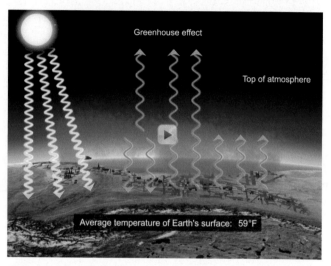

For a visual interpretation of the energy emitted by the earth without and with a greenhouse effect, watch the *Greenhouse* animation (Chapter 2).

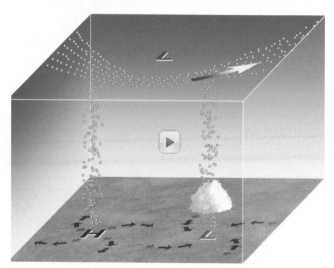

Learn about how air rises above an area of low atmospheric pressure and sinks above an area of high atmospheric pressure. *Converging and Diverging Air* (Chapters 1 and 8).

Additional Animations:

- Ice Crystals (Bergeron) Process (Chapter 5)
- General Circulation of the Atmosphere (Chapter 7)
- Geostrophic Wind (Chapter 6)
- Temperature versus Molecular Movement (Chapter 2)
- Condensation (Chapter 4)
- Air Temperature, Dew Point, and Relative Humidity (Chapter 4)
- Daily Temperature Changes Above the Surface (Chapter 3)

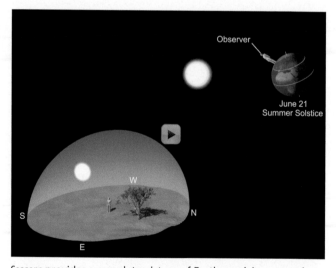

Seasons provides a complete picture of Earth revolving around the sun while it is tilted on its axis. While viewing this animation, look closely at how the sun is viewed by a mid-latitude observer at various times of the year (Chapter 2).

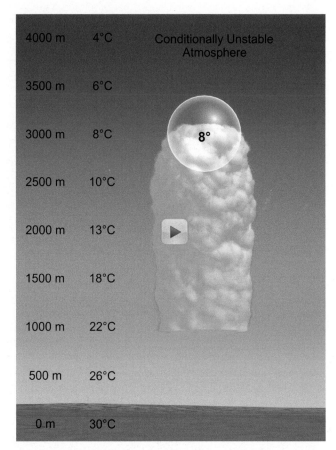

The concept of atmospheric stability can be a bit confusing, especially when comparing the temperature inside a rising air parcel to that of its surroundings. Watch *Stable Atmosphere* (Chapter 5) and the two animations *Unstable Atmosphere* and *Conditionally Unstable Atmosphere* (Chapters 5 and 10).

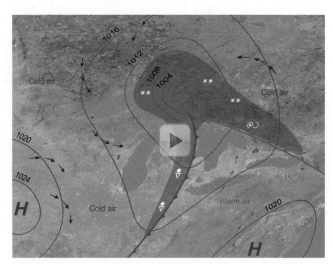

For a visualization of the stages that a wave cyclone goes through from birth to decay, watch the animation entitled *Cyclogenesis* (Chapter 8).

To view air rising over a mountain and the formation of a rain shadow desert, watch *Air Rising Up and Over a Mountain* (Chapters 5 and 7).

For a visual presentation of the Coriolis force, watch *Coriolis Force* (Chapter 6).

For a visualization of a cold front moving across the landscape, watch *Cold Front in Winter* (Chapters 8 and 9), To see a warm front actually move across the surface, watch *Warm Front in Winter* (Chapters 8 and 9).

CHAPTER 1

Earth's Atmosphere

Contents

I remember well a brilliant red balloon which kept me completely happy for a whole afternoon, until, while I was playing, a clumsy movement allowed it to escape. Spellbound, I gazed after it as it drifted silently away, gently swaying, growing smaller and smaller until it was only a red point in a blue sky. At that moment I realized, for the first time, the vastness above us: a huge space without visible limits. It was an apparent void, full of secrets, exerting an inexplicable power over all the Earth's inhabitants. I believe that many people, consciously or unconsciously, have been filled with awe by the immensity of the atmosphere. All our knowledge about the air, gathered over hundreds of years, has not diminished this feeling.

Theo Loebsack, *Our Atmosphere*

Our **atmosphere** is a delicate life-giving blanket of air that surrounds the fragile Earth. In one way or another, it influences everything we see and hear—it is intimately connected to our lives. Air is with us from birth, and we cannot detach ourselves from its presence. In the open air, we can travel for many thousands of kilometers in any horizontal direction, but should we move a mere eight kilometers above the surface, we would suffocate. We may be able to survive without food for a few weeks, or without water for a few days, but, without our atmosphere, we would not survive more than a few minutes. Just as fish are confined to an environment of water, so we are confined to an ocean of air. Anywhere we go, air must go with us.

Earth without an atmosphere would have no lakes or oceans. There would be no sounds, no clouds, no red sunsets. The beautiful pageantry of the sky would be absent. It would be unimaginably cold at night and unbearably hot during the day. All things on Earth would be at the mercy of an intense sun beating down upon a planet utterly parched.

Living on the surface of Earth, we have adapted so completely to our environment of air that we sometimes forget how truly remarkable this substance is. Even though air is tasteless, odorless, and (most of the time) invisible, it protects us from the scorching rays of the sun and provides us with a mixture of gases that allows life to flourish. Because we cannot see, smell, or taste air, it may seem surprising that between your eyes and these words are trillions of air molecules. Some of these may have been in a cloud only yesterday, or over another continent last week, or perhaps part of the life-giving breath of a person who lived hundreds of years ago.

Warmth for our planet is provided primarily by the sun's energy. At an average distance from the sun of nearly 150 million kilometers (km), or 93 million miles (mi), Earth intercepts only a very small fraction of the sun's total energy output. However, it is this radiant energy* that drives the atmosphere into the patterns of everyday wind and weather, and allows life to flourish.

At its surface, Earth maintains an average temperature of about 15°C (59°F).** Although this temperature is mild, Earth experiences a wide range of temperatures, as readings can drop below −85°C (−121°F) during a frigid Antarctic night and climb during the day to above 50°C (122°F) on the oppressively hot, subtropical desert. Not everyone will experience such extremes where they live, but all of us have a chance to observe the day-to-day changes in the atmosphere that we refer to as weather.

In this chapter, we will examine a number of important concepts and ideas about Earth's atmosphere, many of which will be expanded in subsequent chapters.

These concepts and ideas are part of the foundation for understanding the atmosphere and how it produces weather. They are built on knowledge acquired and applied through the *scientific method*, which allows us to make informed predictions about how the natural world will behave.

The Atmosphere and the Scientific Method

For hundreds of years, the scientific method has served as the backbone for advances in medicine, biology, engineering, and many other fields. In the field of atmospheric science, the scientific method has paved the way for the production of weather forecasts that have steadily improved over time.

Investigators use the scientific method by posing a question, putting forth a hypothesis,* predicting what the hypothesis would imply if it were true, and carrying out tests to see if the prediction is accurate. Many common sayings about the weather, such as "red sky at morning, sailor take warning; red sky at night, sailor's delight" (see ● Fig. 1.1) are rooted in careful observation, and there are grains of truth in some of them. However, they are not considered to be products of the scientific method because they are not tested and verified in a standard rigorous way.

To be accepted, a hypothesis has to be shown to be correct through a series of quantitative tests. In many areas of science, such testing is carried out in a laboratory, where it can be replicated again and again. Studying the atmosphere, however, is somewhat different, because our Earth has only one atmosphere. Despite this limitation, scientists have made vast progress by studying the physics and chemistry of air in the laboratory (for instance, studying the way in which molecules absorb energy) and by extending those understandings to the atmosphere as a whole. Observations using weather instruments allow us to quantify how the atmosphere behaves and to determine whether a prediction is accurate. If a particular kind of weather is being studied, such as hurricanes or snowstorms, a field study can gather additional observations to test specific hypotheses.

Over the last fifty years, computers have given atmospheric scientists a tremendous boost. The physical laws that control atmospheric behavior can be represented in software packages known as *numerical models*. Forecasts can be made and tested many times over. The atmosphere described by a model can be used to depict weather conditions from the past and to project them into the future. When a model can accurately simulate past weather conditions and provide confidence in its portrayal of tomorrow's weather, the model can provide valuable information about the weather and climate we may expect decades from now.

---

*Radiant energy, or radiation, is energy transferred in the form of waves that have electrical and magnetic properties. The light that we see is radiation, as is ultraviolet light. More on this important topic is given in Chapter 2.

**The abbreviation °C is used when measuring temperature in degrees Celsius, and °F is the abbreviation for degrees Fahrenheit. More information about temperature scales is given in Appendix A and in Chapter 2.

*A hypothesis is an assertion subject to verification or proof.

• FIGURE 1.1 Observing the natural world is a critical part of the scientific method. Here a vibrant red sky is visible at sunset. One might use scientific method to verify the old proverb: "Red sky at morning, sailor take warning; red sky at night, sailor's delight."

Weather, Climate, and Meteorology

When we talk about the **weather,** we are talking about the condition of the atmosphere at any particular time and place. Weather—which is always changing—is comprised of the elements of:

1. *air temperature*—the degree of hotness or coldness of the air
2. *air pressure*—the force of the air above an area
3. *humidity*—a measure of the amount of water vapor in the air
4. *clouds*—visible masses of tiny water droplets and ice crystals or both that are above Earth's surface
5. *precipitation*—any form of water, either liquid or solid (rain or snow), that falls from clouds and reaches the ground
6. *visibility*—the greatest distance one can see
7. *wind*—the horizontal movement of air

If we measure and observe these **weather elements** over a specified interval of time, say, for many years, we would obtain the "average weather," or the **climate,** of a particular region. Climate, therefore, represents the accumulation of daily and seasonal weather events (the average range of weather) over a long period of time. The concept of climate is much more than this, however, for it also includes the extremes of weather—the heat waves of summer and the cold spells of winter—that occur in a particular region. The *frequency* of these extremes is what helps us distinguish among climates that have similar averages.

If we were able to watch Earth for many thousands of years, even the climate would change. We would see rivers of ice moving down stream-cut valleys and huge glaciers—sheets of moving snow and ice—spreading their icy fingers over large portions of North America. Advancing slowly from Canada, a single glacier might extend as far south as Kansas and Illinois, with ice several thousands of meters thick covering the region now occupied by Chicago. Over an interval of two million years or so, we would see the ice advance and retreat many times. Of course, for this phenomenon to happen, the average temperature of North America would have to decrease and then rise in a cyclic manner.

Suppose we could photograph Earth once every thousand years for many hundreds of millions of years. In time-lapse film sequence, these photos would show that not only is the climate altering, but the whole Earth itself is changing as well: Mountains would rise up only to be torn down by erosion; isolated puffs of smoke and steam would appear as volcanoes spew hot gases and fine dust into the atmosphere; and the entire surface of Earth would undergo a gradual transformation as some ocean basins widen and some of them shrink.*

In summary, Earth and its atmosphere are dynamic systems that are constantly changing. While major transformations of Earth's surface are completed only after long spans of time, the state of the atmosphere can change in a matter of minutes. Hence, a watchful eye turned skyward will be able to observe many of these changes.

Up to this point, we have looked at the concepts of weather and climate without discussing the word *meteorology*. What does this word actually mean, and where did it originate?

METEOROLOGY—THE STUDY OF THE ATMOSPHERE
Meteorology is the study of the atmosphere and its phenomena. The term itself goes back to the Greek

---

*The movement of the ocean floor and continents is explained in the theory of *plate tectonics*.

philosopher Aristotle, who, about 340 B.C., wrote a book on natural philosophy titled *Meteorologica.* This work represented the sum of knowledge on weather and climate at that time, as well as material on astronomy, geography, and chemistry. Some of the topics covered included clouds, rain, snow, wind, hail, thunder, and hurricanes. In those days, all substances that fell from the sky, and anything seen in the air, were called meteors, hence the term meteorology, which actually comes from the Greek word *meteoros,* meaning "high in the air." Today, we differentiate between those meteors that come from extraterrestrial sources outside our atmosphere (meteoroids) and particles of water and ice observed in the atmosphere (hydrometeors).

In *Meteorologica,* Aristotle attempted to explain atmospheric phenomena in a philosophical and speculative manner. Even though many of his ideas were found to be erroneous, Aristotle's work remained a dominant influence in the field of meteorology for almost two thousand years. In fact, the birth of meteorology as a genuine natural science did not take place until the invention of weather instruments, such as the hygrometer in the mid-1400s, the thermometer in the late 1500s, and the barometer (for measuring air pressure) in the mid-1600s. With the newly available observations from instruments, attempts were then made to explain certain weather phenomena employing scientific experimentation and the physical laws that were being developed at the time. As more and better instruments were developed, in the 1800s, the science of meteorology progressed. The invention of the telegraph in 1843 allowed for the transmission of routine weather observations. The understanding of the concepts of wind flow and storm movement became clearer, and in 1869 crude weather maps with *isobars* (lines of equal pressure) were drawn. Around 1920 in Norway, the concepts of air masses and weather fronts were formulated. By the 1940s, daily upper-air balloon observations of temperature, humidity, and pressure gave a three-dimensional view of the atmosphere, and high-flying military aircraft discovered the existence of jet streams.

Meteorology took another step forward in the 1950s, when scientists converted the mathematical equations that describe the behavior of the atmosphere into software called *numerical models* that could be run on new high-speed computers. These calculations were the beginning of *numerical weather prediction.* Today, computers plot the observations, draw the lines on the map, and forecast the state of the atmosphere for some desired time in the future. Meteorologists evaluate the results from different numerical models and use them to issue public forecasts. After World War II, surplus military radars became available, and many were transformed into precipitation-measuring tools. In the mid-1990s, the National Weather Service replaced these conventional radars with the more sophisticated *Doppler radars,* which have the ability to peer into a

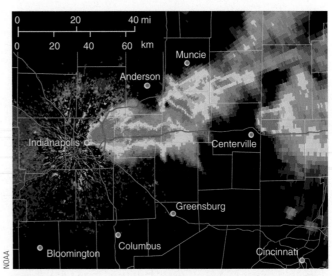

● **FIGURE 1.2** Doppler radar image showing precipitation over portions of Indiana. The areas shaded light green indicate lighter rain, whereas yellow indicates heavier rain. The dark red shaded areas represent the heaviest rain and the possibility of hail and intense thunderstorms.

severe thunderstorm and unveil its wind, as well as show precipitation intensity (see ● Fig. 1.2). Recent upgrades to these Doppler radars make it possible for them to distinguish raindrops, snowflakes, and hailstones.

In 1960, the first weather satellite, *Tiros 1,* was launched, ushering in space-age meteorology. Subsequent satellites provided a wide range of useful information, ranging from day and night time-lapse images of clouds and storms to images that depict swirling ribbons of water vapor flowing around the globe, as shown in ● Fig. 1.3. Over the last several decades, even more sophisticated satellites have been developed. These satellites are supplying computers with a far greater network of data so that more accurate forecasts—perhaps extending up to two weeks or more—will be available in the future.

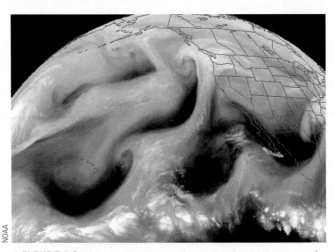

● **FIGURE 1.3** This satellite image shows the dynamic nature of the atmosphere as ribbons of water vapor (gray regions) swirl counterclockwise about huge storms over the North Pacific Ocean.

A Satellite's View of the Weather A good view of the weather can be obtained from a weather satellite. ● Figure 1.4 is a satellite image showing a portion of the Pacific Ocean and the North American continent. The image was obtained from a *geostationary satellite* situated about 36,000 km (22,300 mi) above Earth. At this elevation, the satellite travels at the same rate as Earth spins, which allows it to remain positioned above the same spot so it can continuously monitor what is taking place beneath it.

The thin, solid black lines running from north-to-south on the satellite image are called *meridians*, or lines of longitude. Since the zero meridian (or prime meridian) runs through Greenwich, England, the *longitude* of any place on Earth is simply how far it is in degrees from the prime meridian up to 180° east or west. North America is west of Great Britain and most of the United States lies between 75°W and 125°W longitude.

The thin, solid black lines that parallel the equator are called *parallels of latitude*. The latitude of any place is how far north or south, in degrees, it is from the equator. The latitude of the equator is 0°, whereas the latitude of the North Pole is 90°N and that of the South Pole is 90°S. Most of the United States is located between latitude 30°N and 50°N, a region commonly referred to as the **middle latitudes.**

Storms of All Sizes Probably the most prominent feature in ● Fig. 1.4 is the whirling cloud masses of all shapes and sizes. The clouds appear white because sunlight is reflected back to space from their tops. The largest of the organized cloud masses are the sprawling storms. One such storm shows as an extensive band of clouds, over 2000 km long, west of the Great Lakes. Superimposed on the satellite image is the storm's center (indicated by the large red L) and its adjoining weather fronts in red, blue, and purple. This **middle-latitude cyclonic storm system** (or *extratropical cyclone*) forms outside the tropics and, in the Northern Hemisphere, has winds spinning counterclockwise about its center, which is presently over Minnesota.

A slightly smaller but more vigorous storm is located over the Pacific Ocean near latitude 12°N and longitude 116°W. This tropical storm system, with its swirling band of rotating clouds and sustained surface winds of 65 knots* (74 mi/hr) or more, is known as a **hurricane.** The diameter of the hurricane, as measured by the presence of winds of at least 34 knots (39 mi/hr), is about 800 km (500 mi). The tiny dot at its center is called the *eye*. Near the surface, in the eye, winds are light, skies are generally clear, and the atmospheric pressure is lowest. Around the eye, however, is an extensive region where heavy rain and high surface winds are reaching peak gusts of 100 knots.

Smaller storms are seen as bright spots over the Gulf of Mexico. These spots represent clusters of towering

cumulus clouds that have grown into **thunderstorms,** that is, tall churning clouds accompanied by lightning, thunder, strong gusty winds, and heavy rain. If you look closely at Fig. 1.4, you will see similar cloud forms in many regions. There were probably more than a thousand thunderstorms occurring throughout the world at that very moment. Although they cannot be seen individually, there are even some thunderstorms embedded in the cloud mass west of the Great Lakes. Later in the day on which this image was taken, a few of these storms spawned the most violent disturbance in the atmosphere, **tornadoes.**

A tornado is an intense rotating column of air that usually extends downward from the base of a thunderstorm with a circulation reaching the ground. Sometimes called *twisters*, or *cyclones*, they may appear as ropes or as a large cylinder. They can be more than 2 km (1.2 mi) in diameter, although most are less than a football field wide. Most tornadoes have sustained winds below 100 knots, but some can pack winds exceeding 200 knots (230 mi/hr). Sometimes a visibly rotating funnel cloud dips part of the way down from a thunderstorm, then rises without ever forming a tornado.

A GLIMPSE AT A WEATHER MAP We can obtain a better picture of the middle-latitude storm system by examining a simplified surface weather map for the same day that the satellite image was taken. The weight of the air above different regions varies and, hence, so does the atmospheric pressure. In ● Fig. 1.5, the red letter L on the map indicates a region of low atmospheric pressure, often called a *low*, which marks the center of the middle-latitude cyclonic storm. (Compare the center of the storm in Fig. 1.5 with that in Fig. 1.4.) The two large blue letters H on the map represent regions of high atmospheric pressure, called *highs*, or *anticyclones*. The circles on the map represent other individual weather stations or cities where observations are taken. The **wind** is the horizontal movement of air. The **wind direction**—the direction *from which* the wind is blowing*—is given by *wind barbs*, lines that parallel the wind and extend outward from the center of the station. The **wind speed**—the rate at which the air is moving past a stationary observer—is indicated by *flags*, the short lines that extend off each wind barb.

Notice how the wind blows around the highs and the lows. The horizontal pressure differences create a force

● **FIGURE 1.4** This satellite image (taken in visible reflected light) shows a variety of cloud patterns and storms in Earth's atmosphere.

that starts the air moving from higher pressure toward lower pressure. Because of Earth's rotation, the winds are deflected from their path toward the right in the Northern Hemisphere.* This deflection causes the winds in the Northern Hemisphere to blow *clockwise* and *outward* from the center of the highs, and *counterclockwise* and *inward* toward the center of the low.

As the surface air spins into the low, it flows together and is forced upward, like toothpaste squeezed out of an upward-pointing tube. The rising air cools, and the water vapor in the air condenses into clouds. Notice on the weather map that the area of precipitation (the shaded green area) in the vicinity of the low corresponds to an extensive cloudy region in the satellite image of Fig. 1.4.

Also notice by comparing Figs. 1.4 and 1.5 that, in the regions of high pressure, skies are generally clear. As the surface air flows outward away from the center of a high, air sinking from above must replace the laterally spreading surface air. Since sinking air does not usually produce

clouds, we find generally clear skies and fair weather associated with the regions of high atmospheric pressure.

Areas of high and low pressure, and the swirling air around them, are the major weather producers for the middle latitudes. Look at the middle-latitude storm and the surface temperatures in Fig. 1.5 and notice that, to the southeast of the storm, southerly winds from the Gulf of Mexico are bringing warm, humid air northward over much of the southeastern portion of the nation. On the storm's western side, cool dry northerly breezes combine with sinking air to create generally clear weather over the Rocky Mountains. The boundary that separates the warm and cool air appears as a heavy, colored line on the map—a **front,** across which there is a sharp change in temperature, humidity, and wind direction.

Where the cool air from Canada replaces the warmer air from the Gulf of Mexico, a *cold front* is drawn in blue, with arrowheads showing the front's general direction of movement. Where the warm Gulf air is replacing cooler air to the north, a *warm front* is drawn in red, with half circles showing its general direction of movement. Where the cold front has caught up to the warm front and cold air is now replacing cool air, an *occluded front* is drawn

---

*This deflecting force, known as the *Coriolis force,* is discussed more completely in Chapter 6, as are the winds.

8 CHAPTER 1

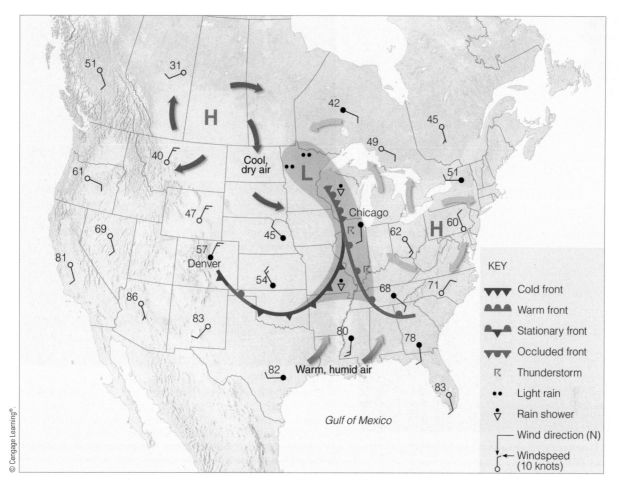

KEY

▼▼▼	Cold front
●●●	Warm front
▼●▼●	Stationary front
▼▼▼	Occluded front
↸	Thunderstorm
••	Light rain
▽	Rain shower
—	Wind direction (N)
↓	Windspeed (10 knots)

● **FIGURE 1.5** Simplified surface weather map that roughly correlates with the satellite image shown in Fig. 1.4. Because this map shows conditions several hours after those in Fig. 1.4, the frontal system across the Midwest is farther east. The shaded green area represents precipitation. The numbers on the map represent air temperatures in °F.

in purple, with alternating arrowheads and half circles to show how it is moving. Along each of the fronts, warm air is rising, producing clouds and precipitation. In the satellite image (Fig. 1.4), the occluded front and the cold front appear as an elongated, curling cloud band that stretches from the low pressure area over Minnesota into the northern part of Texas.

Notice in Fig. 1.5 that the frontal system is to the west of Chicago. As the westerly winds aloft push the front eastward, a person on the outskirts of Chicago might observe the approaching front as a line of towering thunderstorms similar to those shown in ● Fig. 1.6. On a Doppler radar image, the advancing thunderstorms might appear similar to those shown in ● Fig. 1.7. In a few hours, Chicago should experience heavy showers with thunder, lightning, and gusty winds as the front passes. All of this weather, however, should give way to clearing skies and surface winds from the west or northwest after the front has moved on by.

Observing storm systems, we see that not only do they move but they also constantly change. Steered by the upper-level westerly winds, the middle-latitude storm in Fig. 1.5 gradually weakens and moves eastward, carrying its clouds and weather with it. In advance of this system, a sunny day in Ohio will gradually cloud over and yield heavy showers and thunderstorms by nightfall. Behind the storm, cool dry northerly winds rushing into eastern Colorado cause an overcast sky to give way to clearing conditions. Farther south, the thunderstorms presently over the Gulf of Mexico in the satellite image (Fig. 1.4) expand a little, then dissipate as new storms appear over water and land areas. To the west, the hurricane over the Pacific Ocean drifts northwestward and encounters cooler water. Here, away from its warm energy source, it loses its punch; winds taper off, and the storm soon turns into an unorganized mass of clouds and tropical moisture.

WEATHER AND CLIMATE IN OUR LIVES   Weather and climate play a major role in our lives. Weather, for example, often dictates the type of clothing we wear, while climate influences the type of clothing we buy. Climate determines when to plant crops as well as what types of crops can be planted. Weather determines if these same crops will grow to maturity. Although weather and climate affect our lives in many ways, perhaps their most immediate effect is on our comfort. In order to survive

● **FIGURE 1.6** Thunderstorms developing and advancing along an approaching cold front.

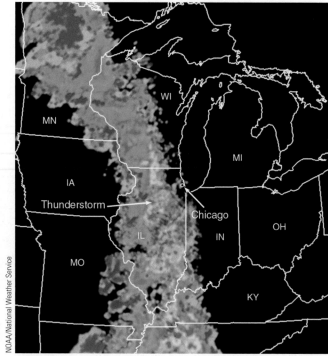

● **FIGURE 1.7** The elongated frontal system in Fig. 1.5 shows up on Doppler radar as a line of different colors. In this composite image, the areas shaded green and blue indicate where light-to-moderate rain is falling. Yellow indicates heavier rainfall. The red-shaded area represents the heaviest rainfall and the possibility of intense thunderstorms. Notice that a thunderstorm is approaching Chicago from the west.

the cold of winter and heat of summer, we build homes, heat them, air condition them, insulate them—only to find that when we leave our shelter, we are at the mercy of the weather elements.

Even when we are dressed correctly for the weather, wind, humidity, and precipitation can change our perception of how cold or warm it feels. On a cold, windy day the effects of *wind chill* tell us that it feels much colder than it really is, and, if we are not correctly dressed, we run the risk of *frostbite* or even *hypothermia* (the rapid, progressive mental and physical collapse that accompanies the lowering of human body temperature). On a hot, humid day we normally feel uncomfortably warm and blame it on the humidity. If we become too warm, our bodies overheat and *heat exhaustion* or *heatstroke* may result. Those most likely to suffer these maladies are the elderly with impaired circulatory systems and infants, whose heat regulatory mechanisms are not yet fully developed.

Weather affects how we feel in other ways, too, not all of them well understood. Arthritic pain is most likely to occur when rising humidity is accompanied by falling pressures. The incidence of heart attacks shows a statistical peak after the passage of warm fronts, when rain and wind are common, and after the passage of cold fronts, when an abrupt change takes place as showery precipitation is accompanied by cold gusty winds. Headaches are common

on days when we are forced to squint, often because of hazy skies or a thin, bright overcast layer of high clouds.

Some people who live near mountainous regions become irritable or depressed when there is a warm, dry wind blowing downslope (a *chinook wind*). Hot, dry downslope *Santa Ana* winds in Southern California can turn dry vegetation into a raging firestorm.

When the weather turns much colder or warmer than normal, it impacts directly on the lives and pocketbooks of many people. For example, the exceptional record warmth observed from January to March 2012 over the United States saved people millions of dollars in heating costs. On the other side of the coin, the colder-than-normal winters of 2013–2014 and 2014–2015 over much of the northeastern United States sent heating costs soaring as demand for heating fuel escalated.

Major cold spells accompanied by heavy snow and ice can play havoc by snarling commuter traffic, curtailing airport services, closing schools, and downing power lines, thereby cutting off electricity to thousands of customers (see ● Fig. 1.8). For example, a huge ice storm during January 1998 in northern New England and Canada left millions of people without power and caused over a billion dollars in damages; and a devastating snow storm during March 1993 buried parts of the East Coast with 14-foot snow drifts and left Syracuse, New York, paralyzed with a

● FIGURE 1.8 Utility workers in Maine clear off broken tree branches from power lines during a major ice storm on December 12, 2008.

snow depth of 36 inches. When the frigid air settles into the Deep South, many millions of dollars worth of temperature-sensitive fruits and vegetables may be ruined, the eventual consequence being higher produce prices for consumers.

Prolonged drought, especially when accompanied by high temperatures, can lead to a shortage of food and, in some places, widespread starvation. Parts of Africa, for example, have periodically suffered through major droughts and famine. During the summer of 2012, much of the United States experienced a severe drought with searing summer temperatures and wilting crops, causing billions of dollars in crop losses. California experienced a destructive multi-year drought beginning in 2011. When the climate turns hot and dry, animals suffer too. In 1986, over 500,000 chickens perished in Georgia during a two-day period at the peak of a summer heat wave. Severe drought also has an effect on water reserves, often forcing communities to ration water and restrict its use. During periods of extended drought, vegetation often becomes tinder-dry and,

sparked by lightning or a careless human, such a dried-up region can quickly become a raging inferno. During the winter of 2005–2006, hundreds of thousands of acres in drought-stricken Oklahoma and northern Texas were ravaged by wildfires.

Every summer, scorching *heat waves* take many lives. During the past twenty years, an annual average of more than 300 deaths in the United States were attributed to excessive heat exposure. In one particularly devastating heat wave that hit Chicago, Illinois, during July 1995, high temperatures coupled with high humidity claimed the lives of more than 700 people. In California, during July 2006, more than 100 people died during a two-week period as air temperatures climbed to over 46°C (115°F). Heat waves have been especially devastating in recent years across Europe, where many cities and buildings are not designed for intense heat. In the summer of 2003, tens of thousands died across Europe, including 14,000 in France alone. A record-breaking heat wave in Russia in 2010 killed nearly 11,000 people in Moscow.

Each year, the violent side of weather influences the lives of millions. Those who live along the Gulf and Atlantic coastlines keep a close watch for hurricanes during the late summer and early autumn. These large tropical systems can be among the nation's most destructive weather events. More than 250,000 people lost their homes when Hurricane Andrew struck the Miami area in 1992, and nearly 2000 people along the central Gulf Coast were killed by Hurricane Katrina in 2005. A late-season hurricane called *Sandy* took a rare path in October 2012, striking the mid-Atlantic coast from the southeast. Although it was no longer classified as a hurricane when it came ashore, Sandy produced wind gusts topping 80 miles per hour in some areas. Because of its vast size and unusual path, Sandy pushed a huge amount of water into the coast, resulting in catastrophic storm-surge flooding and more than 100 deaths across parts of New Jersey, New York, and New England (see ● Fig. 1.9).

● FIGURE 1.9 A resident of Long Beach, New York, digs sand out around his car after Hurricane Sandy, during October 2012, pushed water and sand far inland, destroying homes, commercial businesses, and approximately 10,000 cars in this area alone.

AP Photo/Mark Schiefelbein

● **FIGURE 1.10** Lightning flashes inside a violent tornado that tore through Joplin, Missouri, on May 22, 2011. The tornado ripped through a hospital and destroyed entire neighborhoods. (See tornado damage in Fig. 1.11.)

AP Photo/Mark Schiefelbein

● **FIGURE 1.11** Emergency personnel walk through a neighborhood in Joplin, Missouri, damaged by a violent tornado, with winds exceeding 174 knots (200 mi/hr), on May 22, 2011. The tornado caused hundreds of millions of dollars in damage and took 159 lives, making this single tornado the deadliest in the United States since 1947.

● **FIGURE 1.12** Residents evacuate an apartment complex in the Houston area on April 18, 2016, as torrential rains from severe thunderstorms produced severe flash flooding. More than 1000 high-water rescues took place.

AP Photo/David J Phillip

It is amazing how many people whose family roots are in the Midwest know the story of someone who was severely injured or killed by a tornado. Tornadoes have not only taken many lives, but annually they cause damage to buildings and property totaling in the hundreds of millions of dollars, as a single large tornado can level an entire section of a town (see ● Fig. 1.10 and ● Fig. 1.11).

Although the gentle rains of a typical summer thunderstorm are welcome over much of North America, the heavy downpours, high winds, and large hail of *severe thunderstorms* are not. Cloudbursts from slowly moving, intense thunderstorms can provide too much rain too quickly, creating *flash floods* as small streams become raging rivers composed of mud and sand entangled with uprooted plants and trees. Thunderstorms dumped up to 20 inches of rain in just a few hours over parts of the Houston area in April 2016, leading to severe flash flooding (see ● Fig. 1.12). If heavy rain covers a large area, devastating *river floods* can result. Record rainfall produced both types of flooding over the Southern Plains

in May 2015 and across South Carolina in October 2015. On the average, more people die in the United States from river floods and flash floods than from either lightning strikes or tornadoes. Strong downdrafts originating inside an intense thunderstorm (a *downburst*) create turbulent winds that are capable of destroying crops and inflicting damage upon surface structures. Hundreds of people were killed in airline crashes attributed to turbulent *wind shear* (a rapid change in wind speed, wind direction, or both) from downbursts, before a safety system was implemented in the 1990s, which also has improved since that time. Annually, hail damages crops worth millions of dollars,

DID YOU KNOW?

The folks of Elgin, Manitoba, literally had their "goose cooked" during April 1932, when a lightning bolt killed 52 geese that were flying overhead in formation. As the birds fell to the ground, they were reportedly gathered up and distributed to the townspeople for dinner.

TBD

● **FIGURE 1.13** Estimates are that lightning strikes Earth about 40 to 50 times every second. More than 20 million lightning strikes hit the United States in a typical year. Here, lightning strikes the ground near wind turbines in Texas.

and lightning takes the lives of several dozen people in the United States and starts fires that destroy many thousands of acres of valuable timber (see ● Fig. 1.13).

Up to this point, we have considered the more violent side of weather and its impact on humanity. Weather- and climate-related events can have enormous economic consequences. On average, tens of billions of dollars in property damage occur each year in the United States alone (see ● Fig. 1.14). Yet even the quiet side of weather has its influence. When winds die down and humid air becomes more tranquil, fog may form. Heavy fog can restrict visibility at airports, causing flight delays and cancellations. Every winter, deadly fog-related auto accidents occur along our busy highways and turnpikes. But fog has a positive side, too, especially during a dry spell, as fog moisture collects on tree branches and drips to the ground, where it provides water for the tree's root system.

Weather and climate have become so much a part of our lives that the first thing many of us do in the morning is to listen to the local weather forecast or look it up on our smartphones. For this reason, many radio and most television newscasts have their own "weatherperson" to present weather information and give daily forecasts. More and more of these people are professionally trained in meteorology, and many stations require that the weathercaster be certified by the American Meteorological Society (AMS) or hold a seal of approval from the National Weather Association (NWA). To make their weather presentations as up-to-the-minute as possible, weathercasters draw upon time-lapse satellite images, Doppler radar displays, and other ways of illustrating current weather. Many stations work with private firms that create graphics and customized forecasts, largely based on observations and computer models from the National Weather Service (NWS). Since 1982, a staff of trained professionals at The Weather Channel have provided weather information twenty-four hours a day on cable television. (Many viewers believe the weatherperson they see on TV is a meteorologist and that all meteorologists forecast the weather. If you are interested in learning what a meteorologist or atmospheric scientist is and what he or she might do for a living other than forecast the weather, read Focus section 1.1.)

The National Oceanic and Atmospheric Administration (NOAA), in cooperation with the National Weather Service, sponsors weather radio broadcasts at selected locations across the United States. Known as *NOAA Weather Radio* (and transmitted at VHF-FM frequencies), this service provides continuous weather information and regional forecasts (as well as special weather advisories, including watches and warnings) for over 90 percent of the United States.

Although millions of people rely on weather broadcasts, many use forecasts obtained on their smartphones or on the internet. Websites operated by the NWS and private forecasting companies provide a wealth of local, national, and global data and forecasts. Smartphone applications

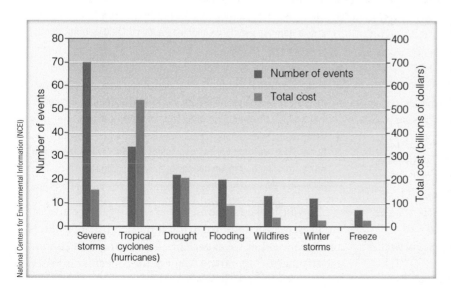

National Centers for Environmental Information (NCEI)

● **FIGURE 1.14** The number of billion-dollar weather and climate events in the United States from 1980 to 2014 (red bar). The total cost (green bar) of the 178 events during this period exceeded $1 trillon. This total is adjusted to the 2014 consumer price index and includes insured and uninsured losses.

What Is a Meteorologist?

Most people associate the term "meteorologist" with the weatherperson they see on television or hear on the radio. Many television and radio weathercasters are in fact professional meteorologists, but some are not. A professional meteorologist is usually considered to be a person who has completed the requirements for a college degree in meteorology or atmospheric science. This individual has strong, fundamental knowledge concerning how the atmosphere behaves, along with a substantial background of coursework in mathematics, physics, and chemistry.

A *meteorologist* uses scientific principles to explain and to forecast atmospheric phenomena. About half of the approximately 9000 meteorologists and atmospheric scientists in the United States work doing weather forecasting for the National Weather Service, the military, or for a television or radio station. The other half work mainly in research, teach atmospheric science courses in colleges and universities, or do meteorological consulting work.

Scientists who do atmospheric research may be investigating how the climate is changing, how snowflakes form, or how pollution impacts temperature patterns. Aided by supercomputers, much of the work of a research meteorologist involves simulating the atmosphere to see how it behaves (see ● Fig. 1). Researchers often work closely with such scientists as chemists, physicists, oceanographers, mathematicians, and environmental scientists to determine how the

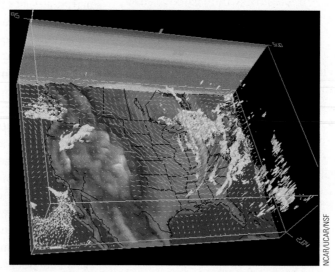

● **FIGURE 1** A model that simulates a three-dimensional view of the atmosphere. This computer model predicts how winds and clouds over the United States will change with time.

atmosphere interacts with the entire ecosystem. Scientists doing work in *physical meteorology* may well study how radiant energy warms the atmosphere; those at work in the field of *dynamic meteorology* might be using the mathematical equations that describe airflow to learn more about jet streams. Scientists working in *operational meteorology* might be preparing a weather forecast by analyzing upper-air information over North America. A *climatologist*, or *climate scientist*, might be studying the interaction of the atmosphere and ocean to see what influence such interchange might have on planet Earth many years from now.

Meteorologists also provide a variety of services not only to the general public in the form of weather forecasts but also to city planners, contractors, farmers, and large corporations. Meteorologists working for private weather firms create the forecasts and graphics that are found in newspapers, on television, and on the internet. Overall, there are many exciting jobs that fall under the heading of "meteorologist"—too many to mention here. However, for more information on this topic, visit this website: http://www.ametsoc.org/ and click on "Students."

can be tailored to provide conditions and forecasts for your hometown or wherever you may be traveling.

BRIEF REVIEW

In the last section, we looked at many ways in which weather and climate impact our lives. A few of the main points described up to now are:

- Our understanding of weather and climate is built on knowledge acquired and applied through the scientific method, which allows us to make informed predictions about the natural world.

- Weather, the state of the atmosphere at any particular time and place, is composed of the weather elements—air temperature, air pressure, humidity, clouds, precipitation, visibility, and wind.

- Climate represents the accumulation of daily and seasonal weather and its extremes over an extended period of time.

- Meteorology is the study of the atmosphere and its phenomena.

- On any given day, a wide variety of storms exist on Earth, ranging in size from the very large middle latitude cyclonic storm to the much smaller tornado.

- Wind direction is the direction *from which* the wind is blowing.

- In the Northern Hemisphere, winds around an area of surface low pressure blow counter clockwise and inward; around an area of a surface high pressure, they blow clockwise and outward.

- Weather and climate impact our lives in many ways. Droughts, floods, heat and cold waves, and violent weather events can cause much suffering and inflict billions of dollars in damage.

Having looked at the many ways that weather can affect our lives, we will now turn to the atmosphere that produces all of the weather we experience and examine its content and structure more closely. Many of the concepts discussed here will be covered in more detail in later chapters.

Components of Earth's Atmosphere

The Earth's *atmosphere* is a relatively thin, gaseous envelope comprised mostly of nitrogen (N_2) and oxygen (O_2), with small amounts of other gases, such as water vapor (H_2O) and carbon dioxide (CO_2). Nestled in the atmosphere are clouds of liquid water and ice crystals.

Although our atmosphere extends upward for many hundreds of kilometers (km), it gets progressively thinner with increasing altitude. Almost 99 percent of the atmosphere lies within a mere 30 km (about 19 mi) of Earth's surface (see ● Fig. 1.15). In fact, if Earth were to shrink to the size of a large beach ball, its inhabitable atmosphere would be thinner than a piece of paper. This thin blanket of air constantly shields the surface and its inhabitants from the sun's dangerous ultraviolet radiant energy, as well as from the onslaught of material from interplanetary space. There is no definite upper limit to the atmosphere; rather, it becomes thinner and thinner, eventually merging with empty space, which surrounds all the planets.

THE EARLY ATMOSPHERE The atmosphere that originally surrounded Earth was probably much different from the air we breathe today. The Earth's first atmosphere (some 4.6 billion years ago) was most likely *hydrogen* and *helium*—the two most abundant gases found in the universe—as well as hydrogen compounds, such as methane (CH_4) and ammonia (NH_3). Most scientists believe that this early atmosphere escaped into space from Earth's hot surface.

A second, more-dense atmosphere, however, gradually enveloped Earth as gases from molten rock within its hot interior escaped through volcanoes and steam vents. We assume that volcanoes spewed out the same gases then as they do today: mostly water vapor (about 80 percent), carbon dioxide (about 10 percent), and up to a few percent nitrogen. These gases probably created Earth's second atmosphere.

As millions of years passed, the constant outpouring of gases from the hot interior—known as **outgassing**—provided a rich supply of water vapor. Moreover, when Earth was very young, some of its water may have originated from numerous collisions with small meteors that pounded Earth, as well as from disintegrating comets. The water vapor condensed into clouds and rain fell upon Earth for many thousands of years, forming the rivers, lakes, and oceans of the world. During this time, large amounts of carbon dioxide (CO_2) were dissolved in the oceans. Through chemical and biological processes, much of the CO_2 became locked up in carbonate sedimentary rocks, such as limestone. With much of the water vapor already condensed and the concentration of CO_2 dwindling, the atmosphere gradually became dominated by molecular nitrogen (N_2), which is usually not chemically active.

It appears that molecular oxygen (O_2), the second most abundant gas in today's atmosphere, probably began an extremely slow increase in concentration as energetic rays from the sun split water vapor (H_2O) into hydrogen and oxygen. The hydrogen, being lighter, probably rose and escaped into space, while the oxygen remained in the atmosphere.

We are uncertain whether this slow increase in oxygen supported the evolution of primitive plants, perhaps two to three billion years ago, or if the plants evolved in an almost oxygen-free (anaerobic) environment. At any rate, plant growth greatly enriched our atmosphere with oxygen. The reason for this enrichment is that, during the process of *photosynthesis*, plants, in the presence of sunlight, combine carbon dioxide and water to produce sugar and oxygen. Hence, after plants evolved, the atmospheric oxygen content increased more rapidly, probably reaching its present composition about several hundred million years ago.

COMPOSITION OF TODAY'S ATMOSPHERE ▼Table 1.1 shows the various gases present in a volume of air near Earth's surface. Notice that molecular **nitrogen** (N_2) occupies about 78 percent and molecular **oxygen** (O_2) about

● **FIGURE 1.15** The Earth's atmosphere as viewed from space. The atmosphere is the thin bluish-white region along the edge of Earth. The photo was taken from the International Space Station on April 12, 2011, over western South America.

NASA/JSC

PERMANENT GASES			VARIABLE GASES			
Gas	Symbol	Percent (by Volume) Dry Air	Gas (and Particles)	Symbol	Percent (by Volume)	Parts per Million (ppm)*
Nitrogen	N_2	78.08	Water vapor	H_2O	0 to 4	
Oxygen	O_2	20.95	Carbon dioxide	CO_2	0.0405	405*
Argon	Ar	0.93	Methane	CH_4	0.00018	1.8
Neon	Ne	0.0018	Nitrous oxide	N_2O	0.00003	0.3
Helium	He	0.0005	Ozone	O_3	0.000004	0.04**
Hydrogen	H_2	0.00006	Particles (dust, soot, etc.)		0.000001	0.01–0.15
Xenon	Xe	0.000009	Chlorofluorocarbons (CFCs) and hydrofluorocarbons (HCFCs)		0.00000001	0.0001

*For CO_2, 405 parts per million means that out of every million air molecules, 405 are CO_2 molecules.
**Stratospheric values at altitudes between 11 km and 50 km are about 5 to 12 ppm.

21 percent of the total volume of dry air. If all the other gases are removed, these percentages for nitrogen and oxygen hold fairly constant up to an elevation of about 80 km (or 50 mi).

At the surface, there is a balance between destruction (output) and production (input) of these gases. For example, nitrogen is removed from the atmosphere primarily by biological processes that involve soil bacteria. Nitrogen is also taken from the air by tiny ocean-dwelling plankton that convert it into nutrients that help fortify the ocean's food chain. It is returned to the atmosphere mainly through the decaying of plant and animal matter. Oxygen, on the other hand, is removed from the atmosphere when organic matter decays and when oxygen combines with other substances, producing oxides. It is also taken from the atmosphere during breathing, as the lungs take in oxygen and release carbon dioxide. The addition of oxygen to the atmosphere occurs during photosynthesis. The concentration of the invisible gas **water vapor,** however, varies greatly from place to place, and from time to time. Close to the surface in warm, steamy, tropical locations, water vapor may account for up to 4 percent of the atmospheric gases, whereas in colder arctic areas, its concentration may dwindle to a mere fraction of a percent. Water vapor molecules are, of course, invisible. They become visible only when they transform into larger liquid or solid particles, such as cloud droplets and ice crystals, which may grow in size and eventually fall to Earth as rain or snow. The changing of water vapor into liquid water is called *condensation,* whereas the process of liquid water becoming water vapor is called *evaporation.* In the lower atmosphere, water is everywhere. It is the only substance that exists as a gas, a liquid, and a solid at those temperatures and pressures normally found near Earth's surface (see ● Fig. 1.16).

Water vapor is an extremely important gas in our atmosphere. Not only does it form into both liquid and solid cloud particles that grow in size and fall to Earth as *precipitation,* but it also releases large amounts of heat—called *latent heat*—when it changes from vapor into liquid water or ice. Latent heat is an important source of atmospheric energy, especially for storms, such as thunderstorms and hurricanes. Moreover, water vapor is a potent *greenhouse gas* because it strongly absorbs a portion of Earth's outgoing radiant energy (somewhat like the glass of a greenhouse prevents the heat inside from escaping

● FIGURE 1.16 Earth's atmosphere is a rich mixture of many gases, with clouds of condensed water vapor and ice crystals. Here, water evaporates from the ocean's surface. Rising air currents then transform the invisible water vapor into many billions of tiny liquid droplets that appear as puffy cumulus clouds. If the rising air in the cloud should extend to greater heights, where air temperatures are quite low, some of the liquid droplets would freeze into minute ice crystals.

© UCAR

and mixing with the outside air). This trapping of heat energy close to Earth's surface, called the *greenhouse effect*, keeps the average air temperature near the surface much warmer than it would be otherwise. Thus, water vapor plays a significant role in Earth's heat-energy balance.*

Carbon dioxide (CO_2), a natural component of the atmosphere, occupies a small but important percent of a volume of air, about 0.04 percent. Carbon dioxide enters the atmosphere mainly from the decay of vegetation, but it also comes from volcanic eruptions, the exhalations of animal life, the burning of fossil fuels (such as coal, gasoline, and natural gas), and deforestation. The removal of CO_2 from the atmosphere takes place during photosynthesis, as plants consume CO_2 to produce green matter. The CO_2 is then stored in roots, branches, and leaves. Rain and snow can react with silicate minerals in rocks and remove CO_2 from the atmosphere through a process known as *chemical weathering*. The oceans act as a huge reservoir for CO_2, as phytoplankton (tiny drifting plants) in surface water fix CO_2 into organic tissues. Carbon dioxide that dissolves directly into surface water mixes downward and circulates through greater depths. Estimates are that the oceans hold more than 50 times the total atmospheric CO_2 content. ● Figure 1.17 illustrates important ways carbon dioxide enters and leaves the atmosphere.

● Figure 1.18 reveals that the atmospheric concentration of CO_2 has risen by almost 30 percent since 1958, when it was first measured at Mauna Loa Observatory in Hawaii. This increase means that CO_2 is entering the atmosphere at a greater rate than it is being removed. The

*A more detailed look at the greenhouse effect is presented in Chapter 2.

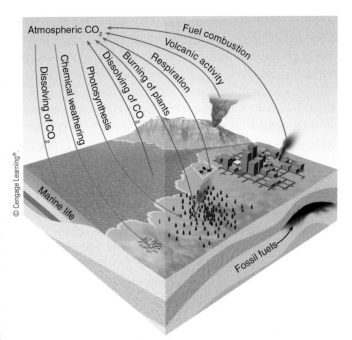

● **FIGURE 1.17** The main components of the atmospheric carbon dioxide cycle. The gray lines show processes that put carbon dioxide into the atmosphere, whereas the red lines show processes that remove carbon dioxide from the atmosphere.

increase appears to be owing mainly to the burning of fossil fuels, such as coal and oil; about half of the CO_2 emitted from fossil fuels remains in the atmosphere, whereas the rest enters the ocean, soil, and plants. Deforestation also plays a role, as cut timber, burned or left to rot, releases CO_2 directly into the air, and the trees can no longer remove CO_2 from the atmosphere. Deforestation accounts for perhaps 10 to 15 percent of the observed CO_2 increase

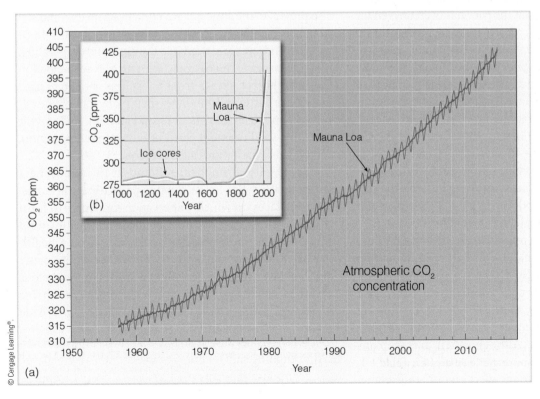

● **FIGURE 1.18** (a) The solid blue line shows the average yearly measurements of CO_2 in parts per million (ppm) at Mauna Loa Observatory, Hawaii, from 1958 through 2015. The jagged dark line illustrates how higher readings occur in winter when plants die and release CO_2 to the atmosphere, and how lower readings occur in summer when more abundant vegetation absorbs CO_2 from the atmosphere. (b) The insert shows CO_2 values in ppm during the past 1000 years from ice cores in Antarctica (orange line) and from Mauna Loa Observatory (blue line). (Mauna Loa data NOAA; Ice Core data courtesy of Carbon Dioxide Information Analysis Center, Oak Ridge National Laboratory)

in recent years. Measurements of CO_2 also come from ice cores. In Greenland and Antarctica, for example, tiny bubbles of air trapped within the ice sheets reveal that before the industrial revolution, CO_2 levels were stable at about 280 parts per million (ppm). (See the insert in Fig. 1.18.) Since the early 1800s, however, CO_2 levels have increased by more than 40 percent. With CO_2 levels presently increasing by approximately 0.5 percent annually (around 2.0 ppm/year), scientists now estimate that the concentration of CO_2 will likely increase from its current value of about 405 ppm to a value perhaps exceeding 750 ppm by the end of this century, assuming that fossil fuel emissions continue at or beyond current levels.

Like water vapor, carbon dioxide is an important greenhouse gas that traps a portion of Earth's outgoing energy. Consequently, with everything else being equal, as the atmospheric concentration of CO_2 increases, so should the average global surface air temperature. In fact, over the last 110 years or so, Earth's average surface temperature has warmed by about 1.0°C (1.8°F). Mathematical climate models, which predict future atmospheric conditions, estimate that if concentrations of CO_2 (and other greenhouse gases) continue to increase at or beyond their present rates, Earth's surface could warm by an additional 3°C (5.4°F) or more by the end of this century. As we will see in Chapter 13, the consequences of this type of *climate change* (such as rising sea levels and the rapid melting of polar ice) will be felt worldwide.

Carbon dioxide and water vapor are not the only greenhouse gases. Others include *methane* (CH_4), *nitrous oxide* (N_2O), and *chlorofluorocarbons* (CFCs). On average, methane concentrations have risen about one-half of one percent per year since the 1990s, but the pace has been uneven for reasons now being studied. Most methane appears to derive from the breakdown of plant material by certain bacteria in rice paddies, wet oxygen-poor soil, the biological activity of termites, and biochemical reactions in the stomachs of cows, although some methane is also leaked into the atmosphere by natural gas operations. Levels of nitrous oxide—commonly known as laughing gas—have been rising annually at the rate of about one-quarter of a percent. As well as being an industrial by-product, nitrous oxide forms in the soil through a chemical process involving bacteria and certain microbes; it is also produced by fossil fuel burning and other activities. Ultraviolet light from the sun destroys it.

Chlorofluorocarbons represent a group of greenhouse gases that, up until the mid-1990s, had been increasing in concentration. At one time, they were the most widely used propellants in spray cans. More recently, they have been used as refrigerants, as propellants for the blowing of plastic-foam insulation, and as solvents for cleaning electronic microcircuits. Although their average concentration in a volume of air is quite small (see Table 1.1, p. 16), they have an important effect on our atmosphere:

They not only trap heat as greenhouse gases but also play a part in destroying the gas ozone in the *stratosphere*, a region in the atmosphere located between about 11 km and 50 km above Earth's surface. By international law, chlorofluorocarbons are gradually being replaced by other compounds, such as hydrochlorofluorocarbons, which are far less harmful to the ozone layer even though they are still greenhouse gases.

On Earth's surface, **ozone** (O_3) is the primary ingredient of *photochemical smog,** which irritates the eyes and throat and damages vegetation. But the majority of atmospheric ozone (about 97 percent) is found in the upper atmosphere—called the stratosphere—where it is formed naturally, as oxygen atoms combine with oxygen molecules. Here, the concentration of ozone averages less than 0.002 percent by volume. This small quantity is important, however, because it shields plants, animals, and humans from the sun's harmful ultraviolet rays. It is ironic that ozone, which damages plant life in a polluted environment, provides a natural protective shield in the upper atmosphere so that plants on the surface may survive. When CFCs enter the stratosphere, ultraviolet rays break them apart, and the CFCs release ozone-destroying *chlorine*. Because of this effect, ozone concentration in the stratosphere has decreased over parts of the Northern and Southern Hemispheres in recent decades. Stratospheric ozone levels over springtime Antarctica plummet each year during September and October, to the point where so little ozone is observed that a seasonal *ozone hole* forms (see ● Fig. 1.19). (We will examine the ozone hole situation, as well as photochemical ozone, in Chapter 14.)

Impurities from both natural and human sources are also present in the atmosphere: Wind picks up dust and soil from Earth's surface and carries it aloft; small saltwater drops from ocean waves are swept into the air (upon evaporating, these drops leave microscopic salt particles suspended in the atmosphere); smoke from forest fires is often carried high above Earth; and volcanoes spew many tons of fine ash particles and gases into the air (see ● Fig. 1.20). Collectively, these tiny solid or liquid particles of various composition, suspended in the air, are called **aerosols.**

Some natural impurities found in the atmosphere are quite beneficial. Small, floating particles, for instance, act as surfaces on which water vapor condenses to form clouds. However, most human-made impurities (and some natural ones) are a nuisance as well as a health hazard. These we call **pollutants.** For example, many automobile engines (especially older ones) emit copious amounts of *nitrogen dioxide* (NO_2), *carbon monoxide* (CO), and *hydrocarbons*. In sunlight, nitrogen dioxide

---

*Originally the word *smog* meant the combining of smoke and fog. Today, however, the word usually refers to the type of smog that forms in large cities, such as Los Angeles, California. Because this type of smog forms when chemical reactions take place in the presence of sunlight, it is termed *photochemical smog*.

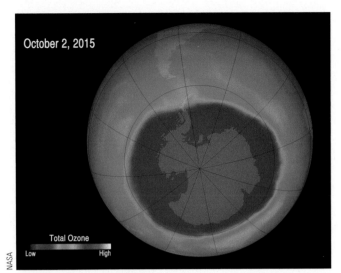

October 2, 2015

Total Ozone
Low   High

NASA

● **FIGURE 1.19** The darkest color represents the area of lowest ozone concentration, or ozone hole, over the Southern Hemisphere on October 2, 2015. Notice that the hole is larger than the continent of Antarctica. A Dobson Unit (DU) is the physical thickness of the ozone layer if it were brought to Earth's surface, where 500 DU equals 5 millimeters.

reacts with hydrocarbons and other gases to produce surface ozone. Carbon monoxide is a major pollutant of city air. Colorless and odorless, this poisonous gas forms during the incomplete combustion of carbon-containing fuel. Hence, more than half of carbon monoxide in urban areas comes from road vehicles.

The burning of sulfur-containing fuels (such as coal and oil) releases sulfur gases into the air. When the atmosphere is sufficiently moist, these gases may transform into tiny dilute drops of sulfuric acid. Rain containing sulfuric acid corrodes metals and painted surfaces and turns freshwater lakes acidic. *Acid rain* (thoroughly discussed in Chapter 14) is a major environmental problem, especially downwind from major industrial areas.

Even the tiniest pollutants are a major concern. *Particulate matter* refers to solid particles and liquid

© David Weintraub/Photo Researchers

● **FIGURE 1.20** Erupting volcanoes can send tons of particles into the atmosphere, along with vast amounts of water vapor, carbon dioxide, and sulfur dioxide.

droplets that are small enough to remain suspended in the air. These particles can obscure visibility and cause respiratory and cardiovascular problems. (More information on these and other pollutants is given in Chapter 14.)

BRIEF REVIEW

Before going on to the next several sections, here is a review of some of the important concepts presented so far:

● The Earth's atmosphere is a mixture of many gases. In a volume of dry air near the surface, nitrogen (N_2) occupies about 78 percent and oxygen (O_2) about 21 percent.

● Water vapor, which normally occupies less than 4 percent of a volume of air near the surface, can condense into liquid cloud droplets or transform into delicate ice crystals. Water is the only substance in our atmosphere that is found naturally as a gas (water vapor), as a liquid (water), and as a solid (ice).

● The majority of water on our planet is believed to have come from its hot interior through outgassing, although some of Earth's water may have come from collisions with meteors and comets.

● Both water vapor and carbon dioxide (CO_2) are important greenhouse gases.

● Ozone (O_3) in the stratosphere protects life from harmful ultraviolet (UV) radiation. At the surface, ozone is the main ingredient of photochemical smog.

Vertical Structure of the Atmosphere

When we examine the atmosphere in the vertical, we see that it can be divided into a series of layers. Each layer can be defined in a number of ways: by the manner in which the air temperature varies through it, by the gases that comprise it, or even by its electrical properties. At any rate, before we examine these various atmospheric layers, we need to look at the vertical profile of two important atmospheric variables: air pressure and air density.

A BRIEF LOOK AT AIR PRESSURE AND AIR DENSITY

Air molecules (as well as everything else) are held near Earth by *gravity*. This strong, invisible force pulling down on the air squeezes (compresses) air molecules closer together, which causes their number in a given volume to increase. The more air above a level, the greater the squeezing effect or compression. Since **air density** is the

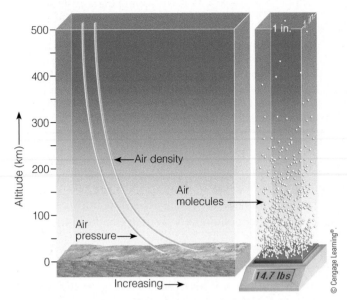

● **FIGURE 1.21** Both air pressure and air density decrease with increasing altitude. The weight of all the air molecules above Earth's surface produces an average pressure near 14.7 lb/in².

number of air molecules in a given space (volume), it follows that air density is greatest at the surface and decreases as we move up into the atmosphere.* Notice in ● Fig. 1.21 that, owing to the fact that the air near the surface is compressed, air density normally decreases rapidly at first, then more slowly as we move farther away from the surface.

Air molecules have weight.** In fact, air is surprisingly heavy. The weight of all the air around Earth is a staggering 5600 trillion tons. The weight of the air molecules acts as a force upon Earth. The amount of force exerted over an area of surface is called *atmospheric pressure* or, simply, **air pressure.**[†] The pressure at any level in the atmosphere can be measured in terms of the total mass of the air above any point. As we climb in elevation, fewer air molecules are above us; hence, *atmospheric pressure always decreases with increasing height.* Like air density, air pressure decreases rapidly at first, then more slowly at higher levels, as illustrated in Fig. 1.21.

If in Fig. 1.21 we weigh a column of air one square inch wide, extending from the average height of the ocean surface (sea level) to the "top" of the atmosphere, it would weigh very nearly 14.7 pounds. Thus, normal atmospheric pressure near sea level is close to 14.7 pounds per square inch (lb/in²). If more molecules are packed into the column,

it becomes more dense, the air weighs more, and the surface pressure goes up. On the other hand, when fewer molecules are in the column, the air weighs less, and the surface pressure goes down. A change in air density can bring about a change in air pressure.

Pounds per square inch is, of course, just one way to express air pressure. Presently, the most common unit for air pressure found on surface weather maps is the *millibar* (mb), although the metric equivalent, the *hectopascal** (hPa), is gradually replacing the millibar as the preferred unit of pressure on surface maps. A more traditional unit of pressure is *inches of mercury* (Hg), which is commonly used both in the field of aviation and in weather reports on television, radio, smartphones, and the Internet. At sea level, the *standard value* for atmospheric pressure is:

1013.25 mb = 1013.25 hPa = 29.92 in. Hg.

Weather conditions may cause the atmospheric pressure to vary from the standard value by 30 millibars or more at a given station.

● Figure 1.22 illustrates how rapidly air pressure decreases with height. Near sea level, atmospheric pressure decreases rapidly, whereas at high levels it decreases more slowly. With a sea-level pressure near 1000 mb, we can see in Fig. 1.22 that, at an altitude of only 5.5 km (or 3.5 mi), the air pressure is about 500 mb, or half of the sea-level pressure. This situation means that, if you were at a mere 5.5 km (which is about 18,000 feet) above the surface, you would be above one-half of all the molecules in the atmosphere.

At an elevation approaching the summit of Mount Everest (about 9 km, or 29,000 ft), the air pressure would

---

*One hectopascal equals 1 millibar.

*Density is defined as the mass of air in a given volume of air. Density = mass/volume.

**The *weight* of an object, including air, is the force acting on the object due to gravity. In fact, weight is defined as the mass of an object times the acceleration of gravity. An object's *mass* is the quantity of matter in the object. Consequently, the mass of air in a rigid container is the same everywhere in the universe. However, if you were to instantly travel to the moon, where the acceleration of gravity is much less than that of Earth, the mass of air in the container would be the same, but its weight would decrease.

[†]Because air pressure is measured with an instrument called a *barometer*, atmospheric pressure is often referred to as *barometric pressure.*

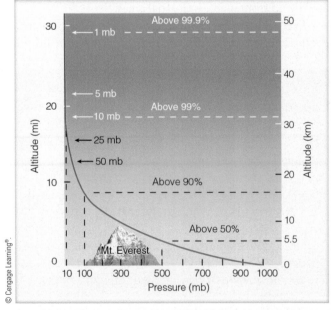

● **FIGURE 1.22** Atmospheric pressure decreases rapidly with height. Climbing to an altitude of only 5.5 km, where the pressure is 500 mb, would put you above one-half of the atmosphere's molecules.

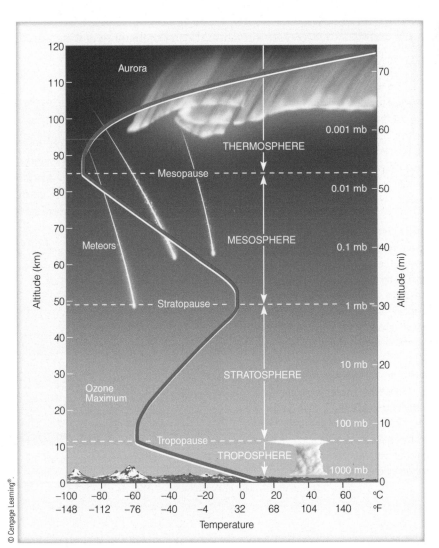

● **FIGURE 1.23** Layers of the atmosphere as related to the average profile of air temperature above Earth's surface. The heavy line illustrates how the average temperature varies in each layer.

© Cengage Learning®.

be about 300 mb. The summit is above nearly 70 percent of all the molecules in the atmosphere. At an altitude of about 50 km (160,000 feet), the air pressure is about 1 mb, which means that 99.9 percent of all the air molecules are below this level. Yet the atmosphere extends upwards for many hundreds of kilometers, gradually becoming thinner and thinner until it ultimately merges with outer space.

LAYERS OF THE ATMOSPHERE   Up to this point, we've looked at how both air pressure and air density decrease with height above Earth—rapidly at first, then more slowly. *Air temperature*, however, has a more complicated vertical profile.*

Look closely at ● Fig. 1.23 and notice that air temperature normally decreases from Earth's surface up to an altitude of about 11 km, which is nearly 36,000 ft, or 7 mi. This decrease in air temperature with increasing height is due primarily to the fact that sunlight warms Earth's surface, and the surface, in turn, warms the air above it (investigated further in Chapter 2). The rate at which the air temperature decreases with height is called the

Air temperature is the degree of hotness or coldness of the air and, as we will see in Chapter 2, it is also a measure of the average speed of the air molecules.

temperature **lapse rate.** The *average* (or *standard*) *lapse rate* in this region of the lower atmosphere is about 6.5 degrees Celsius (°C) for every 1000 meters (m) or about 3.6 degrees Fahrenheit (°F) for every 1000-ft increase in altitude (see ● Fig. 1.24). Keep in mind that these values are only averages. On some days, the air becomes colder more quickly as we move upward, which would increase or steepen the lapse rate. On other days, the air temperature would decrease more slowly with height, and the lapse rate would be less. Occasionally, the air temperature may actually *increase* with height, producing a condition known as a **temperature inversion.** Thus, the lapse rate fluctuates, varying from day to day, season to season, and place to place. The instrument that measures the vertical profile of air temperature in the atmosphere up to an altitude sometimes exceeding 30 km (100,000 ft) is the **radiosonde.** More information on this instrument is given in Focus section 1.2.

Fig. 1.23 shows the region of the atmosphere from the surface up to about 11 km, which contains all of the weather we are familiar with on Earth. Also, this region is kept well stirred by rising and descending air currents, and it is common for air molecules to circulate through a depth of more

The Radiosonde

The vertical distribution of temperature, pressure, and humidity up to an altitude of about 30 km (about 19 mi) can be obtained with an instrument called a *radiosonde*.* The radiosonde is a small, lightweight box equipped with weather instruments and a radio transmitter. It is attached to a cord that has a parachute and a gas-filled balloon tied tightly at the end (see ● Fig. 2). As the balloon rises, the attached radiosonde measures air temperature with a small electrical thermometer— a thermistor— located just outside the box. The radiosonde measures humidity electrically by sending an electric current across a carbon-coated plate. Air pressure is obtained by a small barometer located inside the box. All of this information is transmitted to the surface by radio. Here, a computer rapidly reconverts the various frequencies into values of temperature, pressure, and moisture.

_____
*A radiosonde that is dropped by parachute from an aircraft is called a *dropsonde*.

Special tracking equipment at the surface may be used to provide a vertical profile of winds. Radiosondes are equipped with Global Positioning System (GPS) receivers that lead to highly accurate wind computations. (When winds are added, the observation is called a *rawinsonde*.) When plotted on a graph, the vertical distribution of temperature, humidity, and wind is called a *sounding*. Eventually, the balloon bursts and the radiosonde returns to Earth, its descent being slowed by its parachute.

At most sites, radiosondes are released twice a day, usually at the time that corresponds to midnight and noon in Greenwich, England. Releasing radiosondes is an expensive operation because many of the instruments are never retrieved, and often many of those that are retrieved are often in poor working condition. To complement the radiosonde, modern satellites (using instruments that measure radiant energy) are providing scientists with vertical temperature profiles in inaccessible regions.

NOAA

● FIGURE 2 A radiosonde with parachute and balloon.

than 10 km in just a few days. This region of circulating air extending upward from Earth's surface to where the air stops becoming colder with height is called the **troposphere**— from the Greek *tropein*, meaning "to turn," or "to change."

Notice also in Fig. 1.23 that just above 11 km the air temperature normally stops decreasing with height.

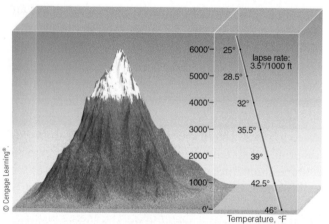

● FIGURE 1.24 Near Earth's surface the air temperature lapse rate is often close to 3.5° F per 1000 ft. If this temperature lapse rate is present and the air temperature at the surface (0 ft) is 46° F, the air temperature about 4000 ft above the surface would be at freezing, and snow and ice might be on the ground.

Here, the lapse rate is zero. This region, where, on average, the air temperature remains constant with height, is referred to as an *isothermal* (equal temperature) *zone*.* The bottom of this zone marks the top of the troposphere and the beginning of another layer, the **stratosphere.** The boundary separating the troposphere from the stratosphere is called the **tropopause.** The height of the tropopause varies. It is normally found at higher elevations over equatorial regions, and it decreases in elevation as we travel poleward. Generally, the tropopause is higher in summer and lower in winter at all latitudes. In some regions, the tropopause "breaks" and is difficult to locate, and here scientists have observed tropospheric air mixing with stratospheric air and vice versa. These breaks mark the position of *jet streams*—high winds that meander in a narrow channel like an old river, often at speeds exceeding 100 knots.**

From Fig. 1.23 we can see that in the stratosphere the air temperature begins to increase with height, producing a *temperature inversion*. The inversion region, along with the lower isothermal layer, tends to keep the vertical currents

_____
*In many instances, the isothermal layer is not present and the air temperature begins to increase with increasing height.

**Recall from p. 7 that one knot equals 1.15 mi/hr.

of the troposphere from spreading into the stratosphere. The inversion also tends to reduce the amount of vertical motion in the stratosphere itself; hence, it is a stratified layer (thus, its name). Even though the air temperature is increasing with height, the air at an altitude of 30 km is extremely cold, averaging less than −46°C (−51°F).

The reason for the inversion in the stratosphere is that the gas ozone plays a major part in heating the air at this altitude. Recall that ozone is important because it absorbs ultraviolet (UV) solar energy. Some of this absorbed energy warms the stratosphere from below, which explains why there is an inversion. If ozone were not present, the air probably would become colder with height, as it does in the troposphere.

Above the stratosphere is the **mesosphere** (middle sphere). The air here is extremely thin and the atmospheric pressure is quite low (again, refer back to Fig. 1.23). Even though the percentage of nitrogen and oxygen in the mesosphere is about the same as it is at Earth's surface, a breath of mesospheric air contains far fewer oxygen molecules than a breath of tropospheric air. At this level, without proper oxygen-breathing equipment, the brain would soon become oxygen-starved—a condition known as *hypoxia*—and suffocation would result. With an average temperature of −90°C (−130°F), the top of the mesosphere represents the coldest part of Earth's atmosphere.

The "hot layer" above the mesosphere is the **thermosphere.** Here, oxygen molecules (O_2) absorb energetic solar rays, warming the air. In the thermosphere, there are relatively few atoms and molecules. Consequently, the absorption of a small amount of energetic solar energy can cause a large increase in air temperature that may exceed 500°C, or 900°F (see ● Fig. 1.25). Moreover, it is in the thermosphere where charged particles from the sun interact with air molecules to produce dazzling aurora displays, which are described in more detail in Chapter 2.

Even though the temperature in the thermosphere is exceedingly high, a person shielded from the sun would not necessarily feel hot. This situation arises because there are too few molecules in this region of the atmosphere to bump against something (exposed skin, for example) and transfer enough heat to it to make it feel warm. The low density of the thermosphere also means that an air molecule will move an average distance of over one kilometer before colliding with another molecule. A similar air molecule at Earth's surface will move an average distance of less than one-millionth of a centimeter before it collides with another molecule. At the top of the thermosphere, about 500 km (300 mi) above Earth's surface, many of the lighter, faster-moving molecules traveling in the right direction actually escape Earth's gravitational pull. The region where atoms and molecules shoot off into space is sometimes referred to as the *exosphere,* which represents the upper limit of our atmosphere.

Up to this point, we have examined the atmospheric layers based on the vertical profile of temperature. The

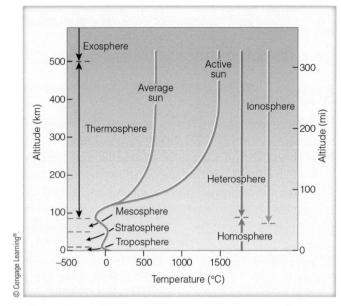

● **FIGURE 1.25** Layers of the atmosphere based on temperature (red line), composition (green line), and electrical properties (blue line). (An active sun is associated with large numbers of solar eruptions.)

atmosphere, however, can also be divided into layers based on its composition. For example, the composition of the atmosphere begins to slowly change in the lower part of the thermosphere. Below the thermosphere, the composition of air remains fairly uniform (78 percent nitrogen, 21 percent oxygen) by turbulent mixing. This lower, well-mixed region is known as the *homosphere* (see Fig. 1.25). In the thermosphere, collisions between atoms and molecules are infrequent, and the air is unable to keep itself stirred. As a result, diffusion takes over as heavier atoms and molecules (such as oxygen and nitrogen) tend to settle to the bottom of the layer, while lighter gases (such as hydrogen and helium) float to the top. The region from about the base of the thermosphere to the top of the atmosphere is often called the *heterosphere.*

The **ionosphere** is not really a layer, but rather an electrified region within the upper atmosphere where fairly large concentrations of ions and free electrons exist. *Ions* are atoms and molecules that have lost (or gained) one or more electrons. Atoms lose electrons and become positively charged when they cannot absorb all of the energy transferred to them by a colliding energetic particle or the sun's energy.

Notice in Fig. 1.25 that the lower region of the ionosphere is usually about 60 km above Earth's surface. From here (60 km), the ionosphere extends upward to the top of the atmosphere. Hence, the bulk of the ionosphere is in the thermosphere. Although the ionosphere allows TV and FM radio waves to pass on through, at night it reflects standard AM radio waves back to Earth. This situation allows AM radio waves to bounce repeatedly off the lower ionosphere and travel great distances.

SUMMARY

This chapter provides an overview of Earth's atmosphere and the many ways weather and climate influence our lives. We looked briefly at the weather map and a satellite image and observed that dispersed throughout the atmosphere are storms and clouds of all sizes and shapes. The movement, intensification, and weakening of these systems, as well as the dynamic nature of air itself, produce a variety of weather events that we described in terms of weather elements. The sum total of weather and its extremes over a long period of time is what we call climate. Although sudden changes in weather may occur in a moment, climatic change takes place gradually over many years. The study of the atmosphere and all of its related phenomena is called *meteorology,* a term whose origin dates back to the days of Aristotle. Weather and climate influence the clothes we wear, the food we eat, and many other parts of our lives. Extreme weather can cause severe damage and major disruption to society.

We learned that our atmosphere is one rich in nitrogen and oxygen as well as smaller amounts of other gases, such as water vapor, carbon dioxide, and other greenhouse gases whose increasing levels may result in additional global warming and climate change. We examined Earth's early atmosphere and found it to be much different from the air we breathe today.

We investigated the various layers of the atmosphere: the troposphere (the lowest layer), where almost all weather events occur, and the stratosphere, where ozone protects us from a portion of the sun's harmful rays. Above the stratosphere lies the mesosphere, where the air temperature drops dramatically with height. Above the mesosphere lies the warmest part of the atmosphere, the thermosphere. At the top of the thermosphere is the exosphere, where collisions between gas molecules and atoms are so infrequent that fast-moving lighter molecules can actually escape Earth's gravitational pull, and shoot off into space. Finally, we looked at the ionosphere, that portion of the upper atmosphere where large numbers of ions and free electrons exist.

KEY TERMS

The following terms are listed (with corresponding page numbers) in the order they appear in the text. Define each. Doing so will aid you in reviewing the material covered in this chapter.

atmosphere, 4
weather, 5
weather elements, 5
climate, 5

meteorology, 5
middle latitudes, 7
middle-latitude cyclonic
  storm, 7

hurricane, 7
thunderstorms, 7
tornadoes, 7
wind, 7
wind direction, 7
wind speed, 7
front, 8
outgassing, 15
nitrogen, 15
oxygen, 15
water vapor, 16
carbon dioxide, 17
ozone, 18

aerosols, 18
pollutants, 18
air density, 19
air pressure, 20
lapse rate, 21
temperature inversion, 21
radiosonde, 21
troposphere, 22
stratosphere, 22
tropopause, 22
mesosphere, 23
thermosphere, 23
ionosphere, 23

QUESTIONS FOR REVIEW

1. What is the primary source of energy for Earth's atmosphere?
2. How could a meteorologist use the scientific method in predicting the weather?
3. List seven common weather elements.
4. How does weather differ from climate?
5. Define *meteorology* and discuss the origin of this word.
6. Rank the following storms in size from largest to smallest: hurricane, tornado, middle-latitude cyclonic storm, thunderstorm.
7. When someone says that "the wind direction today is south," does this mean that the wind is blowing *toward the south* or *from the south*?
8. Weather in the middle latitudes tends to move in what general direction?
9. Describe at least six features observed on a surface weather map.
10. Explain how the wind generally blows around areas of low and high pressure in the Northern Hemisphere.
11. Describe at least six ways weather and climate can influence people's lives.
12. How has Earth's atmosphere changed over time?
13. List the four most abundant gases in today's atmosphere.
14. Of the four most abundant gases in our atmosphere, which one shows the greatest variation from place to place at Earth's surface?
15. Explain how the atmosphere "protects" inhabitants at Earth's surface.
16. What are some of the important roles that water plays in our atmosphere?
17. Briefly explain the production and natural destruction of carbon dioxide near Earth's surface. Give two reasons

for the increase of carbon dioxide over the past 100 plus years.

18. What are some of the aerosols in the atmosphere?
19. What are the two most abundant greenhouse gases in Earth's atmosphere?
20. (a) Explain the concept of air pressure in terms of weight of air above some level.
    (b) Why does air pressure always decrease with increasing height above the surface?
21. What is standard atmospheric pressure at sea level in (a) inches of mercury, (b) millibars, and (c) hectopascals?
22. On the basis of temperature, list the layers of the atmosphere from the lowest layer to the highest.
23. Briefly describe how the air temperature changes from Earth's surface to the lower thermosphere.
24. (a) What atmospheric layer contains all of our weather?
    (b) In what atmospheric layer do we find the highest concentration of ozone? The highest average air temperature?
25. Above what region of the world would you find the ozone hole?
26. Even though the actual concentration of oxygen is close to 21 percent (by volume) in the upper stratosphere, explain why, without proper breathing apparatus, you would not be able to survive there.
27. What is the ionosphere and where is it located?

QUESTIONS FOR THOUGHT AND EXPLORATION

1. Explain how you considered both weather and climate in your choice of the clothing you chose to wear today.
2. Compare a newspaper weather map with a professional weather map obtained from the Internet.

Discuss any differences in the two maps. Look at both maps and see if you can identify a warm front, a cold front, and a middle-latitude cyclonic storm.

3. Which of the following statements relate more to weather and which relate more to climate?
   (a) The summers here are warm and humid.
   (b) Cumulus clouds presently cover the entire sky.
   (c) Our lowest temperature last winter was −29°C (−18°F).
   (d) The air temperature outside is 22°C (72°F).
   (e) December is our foggiest month.
   (f) The highest temperature ever recorded in Phoenixville, Pennsylvania, was 44°C (111°F) on July 10, 1936.
   (g) Snow is falling at the rate of 5 cm (2 in.) per hour.
   (h) The average temperature for the month of January in Chicago, Illinois, is −3°C (26°F).
4. Suppose a friend poses a question about how weather systems generally move in the middle latitudes. He puts forth a hypothesis that these systems generally move from east to west. How would you use the scientific method to prove his hypothesis to be incorrect?
5. Keep track of the weather. On an outline map of North America, mark the daily position of fronts and pressure systems for a period of several weeks or more. (This information can be obtained from newspapers, the TV news, or from the Internet.) Plot the general upper-level flow pattern on the map. Observe how the surface systems move.
6. Compose a one-week journal, including daily newspaper, weather maps and weather forecasts from a newspaper or from the Internet. Provide a commentary for each day regarding the coincidence of actual and predicted weather.

GLOBAL **GEOSCIENCE** WATCH   Go to the News section of the Meteorology portal. Use the search box at left to bring up articles related to the term "storm." Of the first 25 articles, what are the various kinds of weather phenomena mentioned (dust, tornadoes, etc.)? Which ones are mentioned most often? What terms precede the word "storm"?

ONLINE RESOURCES

Visit www.cengagebrain.com to view additional resources, including video exercises, practice quizzes, an interactive eBook, and more.

CHAPTER 2

Warming and Cooling Earth and Its Atmosphere

Contents

The sun doesn't rise or fall: it doesn't move, it just sits there, and we rotate in front of it. Dawn means that we are rotating around into sight of it, while dusk means we have turned another 180 degrees and are being carried into the shadow zone. The sun never "goes away from the sky." It's still there sharing the same sky with us; it's simply that there is a chunk of opaque earth between us and the sun which prevents our seeing it. Everyone knows that, but I really see it now. No longer do I drive down a highway and wish the blinding sun would set; instead I wish we could speed up our rotation a bit and swing around into the shadows more quickly.

Michael Collins, *Carrying the Fire*

As you sit quietly reading this book, you are part of a moving experience. Planet Earth is speeding around the sun at thousands of miles per hour while, at the same time, spinning on its axis. When we look down upon the North Pole, we see that the direction of spin is counterclockwise, meaning that we are moving toward the east at hundreds of miles per hour. We normally don't think of it in that way, but, of course, this is what causes the sun, moon, and stars to rise in the east and set in the west. In fact, it is these motions coupled with energy from the sun, striking a tilted planet, that cause our seasons. But, as we will see later, the sun's energy is not distributed evenly over Earth. Tropical regions receive more energy than polar regions. It is this energy imbalance that drives our atmosphere into the dynamic patterns we experience as wind and weather.

We will begin this chapter by examining the concept of energy and heat transfer. Then we will see how our atmosphere warms and cools. Finally, we will examine how Earth's motions and the sun's energy work together to produce the seasons.

Temperature and Heat Transfer

Temperature is the measurement that tells us how hot or cold something is relative to some set standard value. But we can look at temperature in another way.

We know that air is a mixture of countless billions of atoms and molecules. If they could be seen, they would appear to be moving about in all directions, freely darting, twisting, spinning, and colliding with one another like an angry swarm of bees. Close to Earth's surface, each individual molecule would travel about a thousand times its diameter before colliding with another molecule. Moreover, we would see that all the atoms and molecules are not moving at the same speed, as some are moving faster than others. This is **kinetic energy**, the energy of motion. The temperature of the air (or any substance) is a measure of its average kinetic energy. Simply stated, **temperature** *is a*

measure of the average speed (average motion) of the atoms and molecules, where higher temperatures correspond to faster average speeds.

Suppose we examine a volume of surface air about the size of a large flexible balloon as shown in ●Fig. 2.1. If we warm the air inside, the molecules will move faster, but they also will move slightly farther apart—the air becomes less dense, as illustrated in Fig. 2.1b. Conversely, if we cool the air back to its original temperature, the molecules would slow down, crowd closer together, and the air will become more dense. This molecular behavior is why, in many places throughout the book, we refer to surface air as either *warm, less-dense air* or as *cold, more-dense air.*

Suppose we continue to slowly cool the air. Its atoms and molecules will move more and more slowly until the air reaches a temperature of −273°C (−459°F), which is the lowest temperature possible. At this temperature, called **absolute zero,** the atoms and molecules will possess a minimum amount of energy and theoretically no thermal motion.

Along with temperature, we can also measure internal energy, which is the total energy stored in a group of molecules. **Heat,** on the other hand, *is energy in the process of being transferred from one object to another because of the temperature difference between them.* After heat is transferred, it is stored as internal energy. In the atmosphere, heat is transferred by *conduction, convection,* and *radiation.* We will examine these mechanisms of energy transfer after we look at temperature scales and the important concept of latent heat.

TEMPERATURE SCALES Recall that, theoretically, at a temperature of absolute zero there is no thermal motion. Absolute zero is the starting point for a temperature scale called the *absolute scale,* or **Kelvin scale**, after Lord Kelvin (1824–1907), the British scientist who first introduced it. Since the Kelvin scale contains no negative

●FIGURE 2.1 Air temperature is a measure of the average speed (motion) of the molecules. In the cold volume of air the molecules move more slowly and crowd closer together. In the warm volume, they move faster and farther apart.

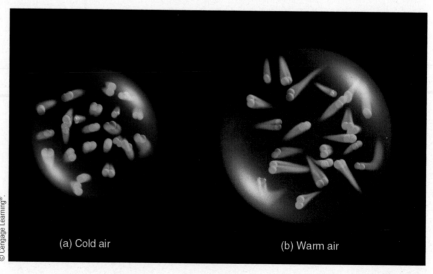

(a) Cold air

(b) Warm air

numbers, it is quite convenient for scientific calculations. Two other temperature scales commonly used today are the Fahrenheit and the Celsius (formerly centigrade). The **Fahrenheit scale** was developed in the early eighteenth century by the physicist G. Daniel Fahrenheit (1686–1736), who assigned the number 32 to the temperature at which water freezes, and the number 212 to the temperature at which water boils. The zero point was simply the lowest temperature that he obtained with a mixture of ice, water, and salt. Between the freezing and boiling points are 180 equal divisions, each of which is called a degree. A thermometer calibrated with this scale is referred to as a Fahrenheit thermometer, for it measures an object's temperature in degrees Fahrenheit (°F).

The **Celsius scale**, named after Swedish astronomer Anders Celsius (1701–1744), was introduced later in the eighteenth century. The number 0 (zero) on this scale is assigned to the temperature at which pure water freezes, and the number 100 to the temperature at which pure water boils at sea level. The space between freezing and boiling is divided into 100 equal degrees. Therefore, each Celsius degree (°C) is 180/100 or 1.8 times bigger than a Fahrenheit degree. Put another way, an increase in temperature of 1°C equals an increase of 1.8°F.

A formula for converting °F to °C is

$$°C = 5/9(°F - 32)$$

On the Kelvin scale, degrees Kelvin are called *kelvins* (abbreviated K). Each degree on the Kelvin scale is exactly the same size as a degree Celsius, and a temperature of 0 K is equal to −273°C. Converting from °C to K can be done by simply adding 273 to the Celsius temperature, as

$$K = °C + 273$$

● Figure 2.2 compares the Kelvin, Celsius, and Fahrenheit scales. Converting a temperature from one scale to another can be done by simply reading the corresponding temperature from the adjacent scale. Thus, 303 K on the Kelvin scale is the equivalent of 30°C and 86°F.*

In most of the world, temperature readings are taken in °C and public weather forecasts use the Celsius scale. In the United States, however, temperatures above the surface are taken in °C, while temperatures at the surface are typically read and reported in °F. Likewise, temperatures on upper-level maps are plotted in °C, while, on surface weather maps, they are in °F. Since both scales are in use, temperature readings in this book will, in most cases, be given in °C followed by their equivalent in °F.

LATENT HEAT—THE HIDDEN WARMTH

We know from Chapter 1 that water vapor is an invisible gas that becomes visible when it changes into larger liquid or solid

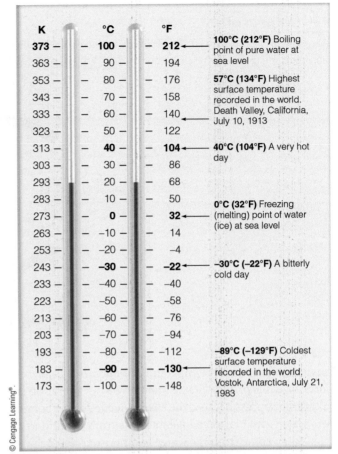

© Cengage Learning®

● **FIGURE 2.2** Comparison of Kelvin, Celsius, and Fahrenheit scales, along with some world temperature extremes.

(ice) particles. This process of transformation is known as a *change of state* or, simply, a *phase change*. The heat energy required to change a substance, such as water, from one state to another is called **latent heat**. But why is this heat referred to as "latent"? To answer this question, we will begin with something familiar to most of us—the cooling produced by evaporating water.

Suppose we microscopically examine a small drop of pure water. At the drop's surface, molecules are constantly escaping (evaporating). Because the more energetic, faster-moving molecules escape most easily, the average motion of all the molecules left behind decreases as each additional molecule evaporates. Since temperature is a measure of average molecular motion, the slower motion

DID YOU KNOW?

We usually think of average human body temperature as being 37°C (98.6°F). This value, established in European studies more than a century ago, may not be quite right. The original research used thermometers that were less accurate than we now have, and the scientists appear to have rounded their data to the nearest degree Celsius (37°C). More recent studies have found that the actual average is closer to 36.8°C (98.2°F). Body temperature varies widely from person to person and from morning to evening.

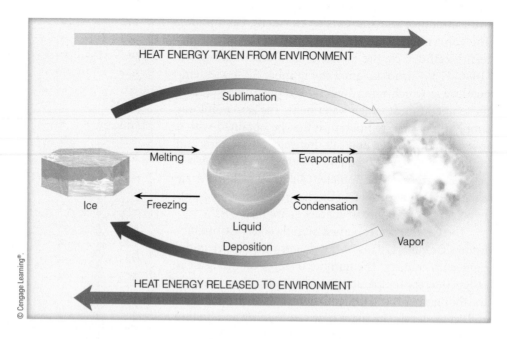

HEAT ENERGY TAKEN FROM ENVIRONMENT

Sublimation

Melting   Evaporation

Ice   Freezing   Condensation

Liquid   Vapor

Deposition

HEAT ENERGY RELEASED TO ENVIRONMENT

© Cengage Learning®.

suggests a lower water temperature. *Evaporation is, therefore, a cooling process.* Stated another way, evaporation is a cooling process because the energy needed to evaporate the water—that is, to change its phase from a liquid to a gas—may come from the water or other sources, including the air.

The energy lost by liquid water during evaporation can be thought of as carried away by, and "locked up" within, the water vapor molecule. The energy is thus in a "stored" or "hidden" condition and is, therefore, called *latent heat*. It is latent (hidden) in that the temperature of the substance changing from liquid to vapor is still the same. However, the heat energy will reappear as **sensible heat** (the heat we can feel and measure with a thermometer) when the vapor condenses back into liquid water. Therefore, *condensation (the opposite of evaporation) is a warming process.*

The heat energy released when water vapor condenses to form liquid droplets is called *latent heat of condensation.* Conversely, the heat energy used to change liquid into vapor at the same temperature is called *latent heat of evaporation* (vaporization). Nearly 600 calories* are required to evaporate a single gram of water at room temperature. With many hundreds of grams of water evaporating from the body, it is no wonder that after a shower we feel cold before drying off. ● Figure 2.3 summarizes the concepts examined so far. When the change of state is from left to right, heat is absorbed by the substance and taken away from the environment. The processes of melting, evaporation, and sublimation (ice to vapor) all cool the environment. When

the change of state is from right to left, heat energy is given up by the substance and added to the environment. The processes of freezing, condensation, and deposition (vapor to ice) all warm their surroundings.

Latent heat is an important source of atmospheric energy. Once vapor molecules become separated from Earth's surface, they are swept away by the wind, like dust before a broom. Rising to high altitudes where the air is cold, the vapor changes into liquid and ice cloud particles. During these processes, a tremendous amount of heat energy is released into the environment (see ● Fig. 2.4).

Water vapor evaporated from warm, tropical water can be carried into polar regions, where it condenses

Robert Henson

● FIGURE 2.4 Every time a cloud forms, it warms the atmosphere. Inside this developing thunderstorm, a vast amount of stored heat energy (latent heat) is given up to the air, as invisible water vapor becomes countless billions of water droplets and ice crystals. In fact, for the duration of this storm alone, more heat energy is released inside this cloud than is unleashed by a small nuclear bomb.

*By definition, a calorie is the amount of heat required to raise the temperature of 1 gram of water from 14.5°C to 15.5°C. In the International System (Système International, SI), the unit of energy is the joule (J), where 1 calorie = 4.186 J. (For pronunciation: joule rhymes with pool.)

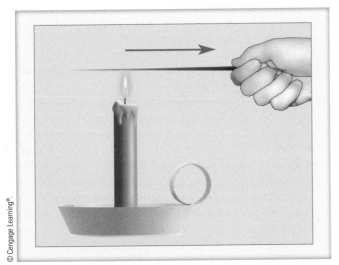

● **FIGURE 2.5** The transfer of heat from the hot end of the metal pin to the cool end by molecular contact is called *conduction*.

▼ **Table 2.1** **Heat Conductivity* of Various Substances**

SUBSTANCE	HEAT CONDUCTIVITY (WATTS** PER METER PER °C)
Still air	0.023 (at 20°C)
Wood	0.08
Dry soil	0.25
Water	0.60 (at 20°C)
Snow	0.63
Wet soil	2.1
Ice	2.1
Sandstone	2.6
Granite	2.7
Iron	80
Silver	427

*Heat (thermal) conductivity describes a substance's ability to conduct heat as a consequence of molecular motion.
**A watt (W) is a unit of power where one watt equals one joule (J) per second (J/s). One joule equals 0.24 calories.

and gives up its heat energy. As we will see, evaporation-transportation-condensation is an extremely important mechanism for the relocation of heat energy (as well as water) in the atmosphere. We are now ready to look at other mechanisms of heat transfer in the atmosphere.

CONDUCTION The transfer of heat from molecule to molecule within a substance is called **conduction**. Hold one end of a metal straight pin between your fingers and place a flaming candle under the other end (see ● Fig. 2.5). Because of the energy they absorb from the flame, the molecules in the pin vibrate faster. The faster-vibrating molecules cause adjoining molecules to vibrate faster. These, in turn, pass vibrational energy on to their neighboring molecules, and so on, until the molecules at the finger-held end of the pin begin to vibrate rapidly. These fast-moving molecules eventually cause the molecules of your finger to vibrate more quickly. Heat is now being transferred from the pin to your finger, and both the pin and your finger feel hot. If enough heat is transferred, your finger will become painful, and you will drop the pin. The transmission of heat from one end of the pin to the other, and from the pin to your finger, occurs by conduction. Heat transferred in this fashion *always flows from warmer to colder* regions. Generally, the greater the temperature difference, the more rapid the heat transfer.

When materials can easily pass energy from one molecule to another, they are considered to be good conductors of heat. How well they conduct heat depends upon how their molecules are structurally bonded together. ▼ Table 2.1 shows that solids, such as metals, are good heat conductors. It is often difficult, therefore, to judge the temperature of metal objects. For example, if you

grab a metal pipe at room temperature, it will seem to be much colder than it actually is because the metal conducts heat away from the hand quite rapidly. Conversely, *air is an extremely poor conductor of heat,* which is why most insulating materials have a large number of air spaces trapped within them. Air is such a poor heat conductor that, in calm weather, the hot ground only warms a shallow layer of air a few centimeters thick by conduction. Yet, air can carry this energy rapidly from one region to another. How, then, does this phenomenon happen?

CONVECTION The transfer of heat by the mass movement of a fluid (such as water and air) is called **convection**. This type of heat transfer takes place in liquids and gases because they can move freely, and it is possible to set up currents within them.

Convection happens naturally in the atmosphere. On a warm, sunny day, certain areas of Earth's surface absorb more heat from the sun than others; as a result, the air near Earth's surface is heated somewhat unevenly. Air molecules adjacent to these hot surfaces bounce against them, thereby gaining some extra energy by conduction. The heated air expands and becomes less dense than the surrounding cooler air. The expanded warm air is buoyed upward and rises. In this manner, large bubbles of warm air rise and transfer heat energy upward. Cooler, heavier air flows toward the surface to replace the rising air. This cooler air becomes heated in turn, rises, and the cycle is repeated. In meteorology, this vertical exchange of heat

Rising Air Cools and Sinking Air Warms

To understand why rising air cools and sinking air warms, we need to examine some air. Suppose we place air in an imaginary thin, elastic wrap about the size of a large balloon (see Fig. 1). This invisible balloonlike "blob" is called an *air parcel*. The air parcel can expand and contract freely, but neither external air nor heat is able to mix with the air inside. By the same token, as the parcel moves, it does not break apart, but remains as a single unit.

At Earth's surface, the parcel has the same temperature and pressure as the air surrounding it. Suppose we lift the parcel. Recall from Chapter 1 that air pressure always decreases as we move up into the atmosphere. Consequently, as the parcel rises, it enters a region where the surrounding air pressure is lower. To equalize the pressure, the parcel molecules inside push the parcel walls outward, expanding it. Because there is no other energy source, the air molecules inside use some of their own energy to expand the parcel. This energy loss shows up as slower molecular speeds, which represent a lower parcel temperature. Hence, *any air that rises always expands and cools.*

If the parcel is lowered to Earth (as shown in Fig. 1), it returns to a region where the air pressure is higher. The higher outside pressure squeezes (compresses) the parcel back to its original (smaller) size. Because air molecules have a faster rebound velocity after striking the sides of a collapsing parcel, the average speed of the molecules inside goes up. (A Ping-Pong ball moves faster after striking a paddle that is moving toward it.) This increase in molecular speed represents a warmer parcel temperature. Therefore, *any air that sinks (subsides) warms by compression.*

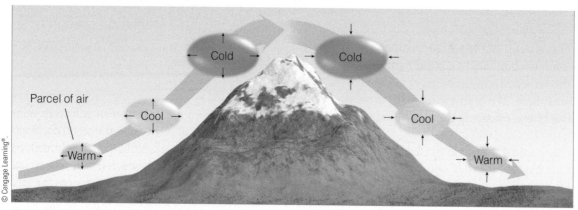

● **FIGURE 1** Rising air expands and cools; sinking air is compressed and warms.

is called *convection,* and the rising air bubbles are known as **thermals** (see Fig. 2.6).

The rising air expands and gradually spreads outward. It then slowly begins to sink. Near the surface, it moves back

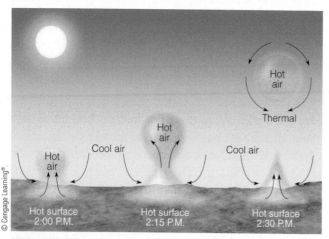

● **FIGURE 2.6** The development of a thermal. A thermal is a rising bubble of air that carries heat energy upward by *convection.*

into the heated region, replacing the rising air. In this way, a *convective circulation,* or thermal "cell," is produced in the atmosphere. In a convective circulation the warm, rising air cools. In our atmosphere, *any air that rises will expand and cool, and any air that sinks is compressed and warms.* To learn more about this important concept, read Focus section 2.1.

Although the entire process of heated air rising, spreading out, sinking, and finally flowing back toward its original location is known as a convective circulation, meteorologists usually restrict the term *convection* to the process of the rising and sinking part of the circulation (see Fig 2.7).

The horizontally moving part of the circulation (called *wind*) carries properties of the air in that particular area with it. The transfer of these properties by horizontally moving air is called **advection.** For example, wind blowing across a body of water will "pick up" water vapor from the evaporating surface and transport it elsewhere in the atmosphere. If the air cools, the water vapor may condense into cloud droplets and release latent heat. In a sense, then,

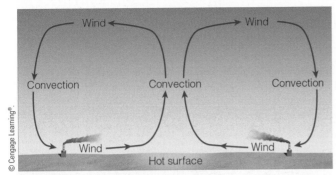

● **FIGURE 2.7** The rising of hot air and the sinking of cool air sets up a convective circulation. Normally, the vertical part of the circulation is called *convection*, whereas the horizontal part is called *wind*. Near the surface the wind is advecting smoke from one region to another.

heat is advected (carried) by the water vapor as it is swept along with the wind. Earlier, we saw that this is an important way to redistribute heat energy in the atmosphere.

BRIEF REVIEW

Before moving on to the next section, here is a summary of some of the important concepts and facts we have covered:

- The temperature of a substance is a measure of the average kinetic energy (average motion) of its atoms and molecules.
- Evaporation (the transformation of liquid into vapor) is a cooling process that can cool the air, whereas condensation (the transformation of vapor into liquid) is a warming process that can warm the air.
- Heat is energy in the process of being transferred from one object to another because of the temperature difference between them.
- In conduction, which is the transfer of heat by molecule-to-molecule contact, heat always flows from warmer to colder regions.
- Air is a poor conductor of heat.
- Convection is an important mechanism of heat transfer, as it represents the vertical movement of warmer air upward and cooler air downward.
- The horizontal transfer of any atmospheric property by the wind (including smoke and warm or cold air) is called *advection*.

There is yet another mechanism for the transfer of energy—radiation, or *radiant energy,* which is what we receive from the sun. In this method, energy can be transferred from one object to another without the space between them necessarily being heated.

DID YOU KNOW?

Some birds are weather savvy. Hawks, for example, seek out rising thermals and ride them up into the air as they scan the landscape for prey. In doing so, these birds conserve a great deal of energy by not having to flap their wings as they circle higher and higher inside the rising air current.

Radiant Energy

On a bright winter day, you may have noticed how warm your face feels as you stand facing the sun. Sunlight travels through the surrounding air with little effect upon the air itself. Your face, however, absorbs this energy and converts it to thermal energy. Thus, sunlight warms your face without actually warming the air. The energy transferred from the sun to your face is called **radiant energy**, or **radiation**. It travels in the form of waves that release energy when they are absorbed by an object. Because these waves have magnetic and electrical properties, we call them **electromagnetic waves**. Electromagnetic waves do not need molecules to propagate them. In a vacuum, they travel at a constant speed of nearly 300,000 km (186,000 mi) per second—the speed of light.

● Figure 2.8 shows some of the different wavelengths of radiation. Notice that the *wavelength* (which is often expressed by the Greek letter lambda, λ) is the distance measured along a wave from one crest to another. Also notice that some of the waves have exceedingly short lengths. For example, radiation that we can see (visible light) has an average wavelength of less than one-millionth of a meter—a distance nearly one-hundredth the diameter of a human hair. To help describe these short lengths, we introduce a new unit of measurement called a **micrometer** (abbreviated μm), which is equal to one-millionth of a meter (m); thus

$$1 \text{ micrometer } (\mu\text{m}) = 0.000001 \text{ m} = 10^{-6} \text{ m}.$$

In Fig. 2.8, we can see that the average wavelength of visible light is about 0.0000005 meters (or 5×10^{-7}), which is the same as 0.5 μm. To give you a common object for comparison, the average height of a letter on this page is about 2000 μm, or 2 millimeters (2 mm).

We can also see in Fig. 2.8 that the longer waves carry less energy than do the shorter waves. When comparing the energy carried by various waves, it is useful to give electromagnetic radiation characteristics of particles in order to explain some of the wave's behavior. We can actually think of radiation as streams of particles, or **photons,** that are discrete packets of energy.*

As shown in Figure 2.8, an ultraviolet (UV) photon carries more energy than a photon of visible light. In fact, certain ultraviolet photons have enough energy to produce sunburns and penetrate skin tissue, sometimes causing skin cancer. (To learn more about radiant energy and its effect on humans, read Focus section 2.2.)

Here are a few important ideas and facts to remember about the concept of radiation:

---

*Packets of photons make up waves, and groups of waves make up a beam of radiation.

● **FIGURE 2.8** Radiation characterized according to wavelength. As the wavelength decreases, the energy carried per wave increases.

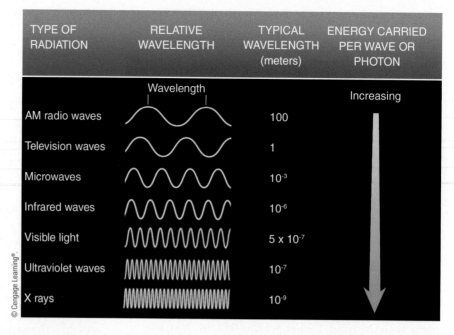

TYPE OF RADIATION	RELATIVE WAVELENGTH	TYPICAL WAVELENGTH (meters)	ENERGY CARRIED PER WAVE OR PHOTON
AM radio waves		100	
Television waves		1	Increasing
Microwaves		10^{-3}	
Infrared waves		10^{-6}	
Visible light		5×10^{-7}	
Ultraviolet waves		10^{-7}	
X rays		10^{-9}	

© Cengage Learning®

1. All things (with a temperature above absolute zero), no matter how big or small, emit radiation. The air, your body, flowers, trees, Earth, and the stars are all radiating a wide range of electromagnetic waves. The energy originates from rapidly vibrating electrons, billions of which exist in every object.

2. The wavelengths of radiation that an object emits depend primarily on the object's temperature. *The higher the object's temperature, the shorter are the wavelengths of emitted radiation.* By the same token, as an object's temperature increases, its peak emission of radiation shifts toward shorter wavelengths. This relationship between temperature and wavelength is called *Wien's law** (or *Wien's displacement law*) after the German physicist Wilhelm Wien (pronounced *Ween, 1864–1928*) who discovered it.

3. Objects that have a high temperature emit radiation at a greater rate or intensity than objects with a lower temperature. Thus, as the temperature of an object increases, more total radiation (over a given surface area) is emitted each second. This relationship between temperature and emitted radiation is known as the *Stefan-Boltzmann law*** after Josef Stefan (1835–1893) and Ludwig Boltzmann (1844–1906), who devised it.

---

*Wien's law:

$$\lambda_{max} = \frac{constant}{T}$$

where λ_{max} is the wavelength at which maximum radiation emission occurs, T is the object's temperature in kelvins (K) and the constant is 2897 μm K. More information on Wien's law is given in Appendix B.

**Stefan-Boltzmann law:

$$E = \sigma T^4$$

where E is the maximum rate of radiation emitted by each square meter of surface of an object, σ (the Greek letter sigma) is a constant, and T is the object's surface temperature in kelvins (K). Additional information on the Stefan-Boltzmann law is given in Appendix B.

Objects at a high temperature (above about 500°C) radiate waves with many lengths, some of which are short enough to stimulate the sensation of color. We actually see these objects glow red. Objects cooler than this radiate at wavelengths that are too long for us to see. The page of this book, for example, is radiating electromagnetic waves. But because its temperature is only about 20°C (68°F), the waves emitted are much too long to stimulate vision. We are able to see the page, however, because light waves from other sources (such as light bulbs or the sun) are being *reflected* (bounced) off the paper. If this book were carried into a completely dark room, it would continue to radiate, but the pages would appear black because there are no visible light waves in the room to reflect off the page.

The sun emits radiation at almost all wavelengths, but because its surface is extremely hot—6000 K (10,500°F)—it radiates the majority of its energy at relatively short wavelengths. If we look at the amount of radiation given off by the sun at each wavelength, we obtain the sun's *electromagnetic spectrum*. A portion of this spectrum is shown in ● Fig. 2.9.

Notice that the sun emits a maximum amount of radiation at wavelengths near 0.5 μm. Since our eyes are sensitive to radiation between 0.4 and 0.7 μm, these waves reach the eye and stimulate the sensation of color. This portion of the spectrum is therefore referred to as the **visible region,** and the radiant energy that reaches our eye is called *visible light.* The color violet is the shortest wavelength of visible light. Wavelengths shorter than violet (0.4 μm) are **ultraviolet (UV).** The longest wavelengths of visible light correspond to the color red. A rainbow spans the spectrum of visible light from violet to red. Wavelengths longer than red (0.7 μm) are called **infrared (IR).**

Sunburning and UV Rays

Earlier, we learned that shorter waves of radiation carry much more energy than longer waves, and that a photon of ultraviolet light carries more energy than a photon of visible light. In fact, ultraviolet (UV) wavelengths in the range of 0.20 and 0.29 μm (known as *UVC radiation*) are harmful to living things, as certain waves can cause chromosome mutations, kill single-celled organisms, and damage the cornea of the eye. Fortunately, virtually all the ultraviolet radiation at wavelengths in the UVC range is absorbed by ozone in the stratosphere.

Ultraviolet wavelengths between about 0.29 and 0.32 μm (known as *UVB radiation*) reach Earth in small amounts. Photons in this wavelength range have enough energy to produce sunburns and penetrate skin tissues, sometimes causing skin cancer. About 90 percent of all skin cancers are linked to sun exposure and UVB radiation. Oddly enough, these same wavelengths activate provitamin D in the skin and convert it into vitamin D, which is essential to health.

Longer ultraviolet waves with lengths of about 0.32 to 0.40 μm (called *UVA radiation*) are less energetic, but they are the main ones that produce skin tanning. Although UVB is the primary wavelength responsible for burning the skin, UVA can cause skin redness. It can also interfere with the skin's immune system and cause long-term skin damage that shows up years later as accelerated aging and skin wrinkling. Moreover, recent studies indicate that the longer UVA exposures needed to create a tan pose about the same cancer risk as a UVB tanning dose.

Upon striking the human body, ultraviolet radiation is absorbed beneath the outer layer of skin. To protect the skin from these harmful rays, the body's defense mechanism kicks in. Certain cells (when exposed to UV radiation) produce a dark pigment (*melanin*) that begins to absorb some of the UV radiation. (It is the production of melanin that produces a tan.) Consequently, a body that produces little melanin—one with pale skin—has little natural protection from UVB.

EXPOSURE CATEGORY	UV INDEX	PROTECTIVE MEASURES
Low	2 or less	Wear sunglasses on bright days; cover up and wear sunscreen
Moderate	3 – 5	Take precautions; stay in shade near midday
High	6 – 7	Apply SPF 15+ sunscreen; wear wide-brim hat and sunscreen; reduce time in sun between 10 a.m. and 4 p.m. (daylight savings time)
Very high	8 – 10	Take extra precautions; apply SPF 15+ sunscreen; wear wide-brim hat and sunscreen; minimize exposure between 10 a.m. and 4 p.m.
Extreme	11+	Take all precautions; apply SPF 15+ sunscreen; wear wide-brim hat and sunscreen; minimize exposure between 10 a.m. and 4 p.m. Unprotected skin can burn in minutes.

Jim Lopes/Shutterstock.com

● FIGURE 2 The UV index.

Additional protection can come from sunscreens that block UV rays from ever reaching the skin. Some contain chemicals (such as zinc oxide) that reflect UV radiation. (These are the white pastes once seen on the noses of lifeguards.) Others consist of a mixture of chemicals that actually absorb ultraviolet radiation. The *Sun Protection Factor (SPF)* number on every container of sunscreen dictates how effective the product is in protecting from UVB—the higher the number, the better the protection. Many "broad-spectrum" sunscreens protect against both UVA and UVB, although only the amount of UVA protection is considered in the SPF rating.

Protecting oneself from excessive exposure to the sun's energetic UV rays is certainly wise. Estimates are that, in a single year, over 70,000 Americans will be diagnosed with malignant melanoma, the most deadly form of skin cancer. And in areas where the protective stratospheric ozone shield has weakened, there is an increased risk of problems associated with UVB. Using a good sunscreen and proper clothing can certainly help. The best way to protect yourself from too much sun, however, is to limit your time in direct sunlight, especially between the hours of 10 a.m. and 4 p.m. daylight saving time when the sun is highest in the sky and its rays are most direct.

Each day the National Weather Service predicts UV radiation levels for selected cities throughout the United States. The forecast, known as the *UV Index*, gives the UV level at its peak, around noon standard time or 1 p.m. daylight saving time. The index corresponds to five exposure categories set by the Environmental Protection Agency (EPA). An index value of 2 or less is considered "low," whereas a value of 11 or greater is deemed "extreme" (see ● Fig. 2). Depending on skin type, a UV index of 10 means that in direct sunlight (without sunscreen protection) a person's skin will likely begin to burn in about 6 to 30 minutes (see ● Fig. 3).

© Cengage Learning®

● FIGURE 3 If this photo was taken around 1 p.m. on a day when the UV index was 10 almost everyone on this beach without sunscreen would experience some degree of sunburning within 30 minutes.

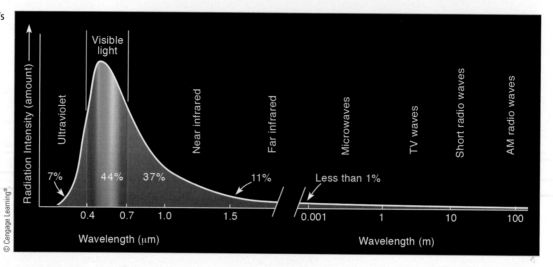

● FIGURE 2.9 The sun's electromagnetic spectrum and some of the descriptive names of each region. The numbers underneath the curve approximate the percent of energy the sun radiates in various regions.

Whereas the hot sun emits only a part of its energy in the infrared portion of the spectrum, the relatively cool earth emits almost all of its energy at infrared wavelengths. In fact, notice in ● Fig. 2.10 that Earth, with an average surface temperature near 288 K (15°C, or 59°F), radiates nearly all its energy between 5 and 20 μm, with a peak intensity (λ_{max}) in the infrared region near 10 μm. The sun, with a much higher surface temperature, radiates with a peak emission near 0.5 μm. Since the sun radiates the majority of its energy at much shorter wavelengths than does Earth, solar radiation is often called **shortwave radiation**, whereas Earth's radiation is referred to as **longwave** (or **terrestrial**) **radiation**.

Radiation—Absorption, Emission, and Equilibrium

If Earth and all things on it are continually radiating energy, why doesn't everything get progressively colder? The answer is that all objects not only radiate energy, they absorb it as well. If an object radiates more energy than it absorbs, it becomes colder; if it absorbs more energy than it emits, it becomes warmer. On a sunny day, Earth's surface warms by absorbing more energy from the sun and the atmosphere than it radiates, whereas at night Earth cools by radiating more energy than it absorbs from its surroundings. When an object emits and absorbs energy at equal rates, its temperature remains constant.

The rate at which something radiates and absorbs energy depends strongly on its surface characteristics, such as color, texture, and moisture, as well as temperature. For example, a black object in direct sunlight is a good absorber of solar radiation. It converts energy from the sun into internal energy, and its temperature ordinarily increases. You need only walk barefoot on a black asphalt road on a summer afternoon to experience this. At night, the blacktop road will cool quickly by emitting

infrared energy and, by early morning, it may be cooler than surrounding surfaces.

Any object that is a perfect absorber (that is, absorbs all the radiation that strikes it) and a perfect emitter (emits the maximum radiation possible at its given temperature) is called a **blackbody**. Blackbodies do not have to be colored black; they simply must absorb and emit all possible radiation. Since Earth's surface and the sun absorb and radiate with nearly 100 percent efficiency for their respective temperatures, they both behave as blackbodies.

When we look at Earth from space, we see that half of it is in sunlight, the other half is in darkness. The outpouring of solar energy constantly bathes Earth with radiation, while Earth, in turn, constantly emits infrared radiation. If we assume that there is no other method of transferring heat, then, when the rate of absorption of solar radiation equals the rate of emission of infrared Earth radiation, a state of *radiative equilibrium* is achieved.

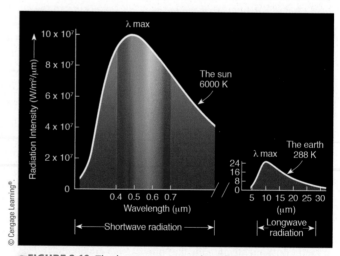

● FIGURE 2.10 The hotter sun not only radiates more energy than that of the cooler Earth (the area under the curve), but it also radiates the majority of its energy at much shorter wavelengths. (The area under the curves is equal to the total energy emitted, and the scales for the two curves differ by a factor of roughly 1 million.)

The average temperature at which this occurs is called the **radiative equilibrium temperature**. At this temperature, Earth (behaving as a blackbody) is absorbing solar radiation and emitting infrared radiation at equal rates, and its average temperature does not change. As Earth is about 150 million km (93 million mi) from the sun, Earth's radiative equilibrium temperature is about 255 K ($-18°C$, 0°F). But this temperature is *much* lower than Earth's observed average surface temperature of 288 K (15°C, 59°F). Why is there such a large difference?

The answer lies in the fact that *Earth's atmosphere absorbs and emits infrared radiation.* Unlike Earth, the atmosphere does *not* behave like a blackbody, as it absorbs some wavelengths of radiation and is transparent to others. Objects that selectively absorb and emit radiation, such as gases in our atmosphere, are known as **selective absorbers**.

SELECTIVE ABSORBERS AND THE ATMOSPHERIC GREENHOUSE EFFECT

There are many selective absorbers in our environment. Snow, for example, is a good absorber of infrared radiation but a poor absorber of sunlight. Objects that selectively absorb radiation usually selectively emit radiation at the same wavelength. Snow is therefore a good emitter of infrared energy. At night, a snow surface usually emits much more infrared energy than it absorbs from its surroundings. This large loss of infrared radiation (coupled with the insulating qualities of snow) enables the air above a snow surface on a clear, calm winter night to become extremely cold.

● Figure 2.11 shows some of the most important selectively absorbing gases in our atmosphere (the purple shaded area represents the absorption characteristics of each gas at various wavelengths). Notice that both water vapor (H_2O) and carbon dioxide (CO_2) are strong absorbers of infrared radiation and poor absorbers of visible solar radiation. Other, less important, selective absorbers include nitrous oxide (N_2O), methane (CH_4), and ozone (O_3), which is most abundant in the stratosphere. As these gases absorb infrared radiation emitted from Earth's surface, they gain kinetic energy (energy of motion). The gas molecules share this energy by colliding with neighboring air molecules, such as oxygen and nitrogen (both of which are poor absorbers of infrared energy). These collisions increase the average kinetic energy of the air, which results in an increase in air temperature. Thus, most of the infrared energy emitted from Earth's surface keeps the lower atmosphere warm.

Besides being selective absorbers, water vapor and CO_2 selectively emit radiation at infrared wavelengths.* This radiation travels away from these gases in all directions.

---

*Nitrous oxide, methane, and ozone also emit infrared radiation, but their concentration in the atmosphere is much smaller than water vapor and carbon dioxide (see Table 1.1, p. 16).

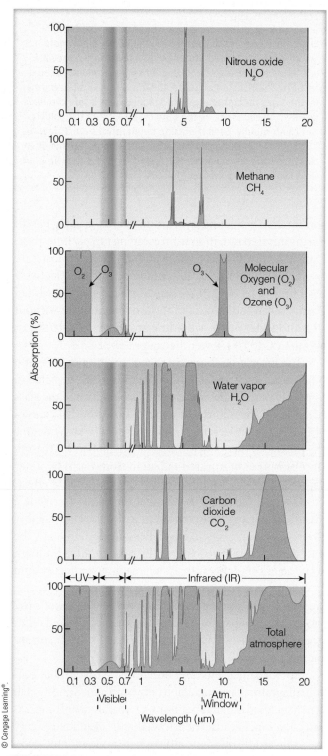

© Cengage Learning®.

● **FIGURE 2.11** Absorption of radiation by gases in the atmosphere. The purple-shaded area represents the percent of radiation absorbed by each gas. The strongest absorbers of infrared radiation are water vapor and carbon dioxide. The bottom figure represents the percent of radiation absorbed by all of the atmospheric gases.

A portion of this energy is radiated toward Earth's surface and absorbed, thus heating the ground.

Earth, in turn, radiates infrared energy upward, where it is absorbed and warms the lower atmosphere. In this way, water vapor and CO_2 absorb and radiate infrared energy and act as an insulating layer around Earth, keeping part of Earth's infrared radiation from escaping rapidly into space. Consequently, Earth's surface and the lower atmosphere are much warmer than they would be if these selectively absorbing gases were not present. In fact, as we saw earlier, Earth's mean radiative equilibrium temperature without CO_2 and water vapor would be around $-18°C$ (0°F), or about 33°C (59°F), lower than at present.

The absorption characteristics of water vapor, CO_2, and other gases (such as methane and nitrous oxide depicted in Fig. 2.11) were at one time thought to be similar to the glass of a florist's greenhouse. In a greenhouse, the glass allows visible radiation to come in, but inhibits to some degree the passage of outgoing infrared radiation. For this reason, the absorption of infrared radiation from Earth by water vapor and CO_2 is popularly called the **greenhouse effect.** However, studies have shown that the warm air inside a greenhouse is probably caused more by the air's inability to circulate and mix with the cooler outside air than by the entrapment of infrared energy. Because of these findings, some scientists suggest that the greenhouse effect should be called the *atmosphere effect.* To accommodate everyone, we will usually use the term *atmospheric greenhouse effect* when describing the role that water vapor, CO_2, and other **greenhouse gases**[*] play in keeping Earth's mean surface temperature higher than it otherwise would be.

Look again at Fig. 2.11 and observe that, in the bottom diagram, there is a region between about 8 and 11 μm where neither water vapor nor CO_2 readily absorbs infrared radiation. Because these wavelengths of emitted energy pass upward through the atmosphere and out into space, the wavelength range (between 8 and 11 μm) is known as the **atmospheric window.** Clouds can enhance the atmospheric greenhouse effect. Tiny liquid cloud droplets are selective absorbers in that they are good absorbers of infrared radiation but poor absorbers of visible solar radiation. Clouds even absorb the wavelengths between 8 and 11 μm, which are otherwise "passed up" by water

vapor and CO_2. Thus, they have the effect of enhancing the atmospheric greenhouse effect by closing the atmospheric window.

Clouds—especially low, thick ones—are excellent emitters of infrared radiation. Their tops radiate infrared energy upward and their bases radiate energy back to Earth's surface where it is absorbed and, in a sense, radiated back to the clouds. This process keeps calm, cloudy nights warmer than calm, clear ones. If the clouds remain into the next day, they prevent much of the sunlight from reaching the ground by reflecting it back to space. Since the ground does not heat up as much as it would in full sunshine, cloudy, calm days are normally cooler than clear, calm days. Hence, the presence of clouds tends to keep nighttime temperatures higher and daytime temperatures lower.

In summary, the atmospheric greenhouse effect occurs because water vapor, CO_2, and other greenhouse gases are selective absorbers. They allow most of the sun's radiation to reach the surface, but they absorb a good portion of Earth's outgoing infrared radiation, preventing it from escaping into space. It is the atmospheric greenhouse effect, then, that keeps the temperature of our planet at a level where life can survive. The greenhouse effect is not just a "good thing"—it is essential to life on Earth, for without it, air at the surface would be extremely cold (see ● Fig. 2.12).

ENHANCEMENT OF THE GREENHOUSE EFFECT Although temperature is not measured perfectly at every spot around the globe, more than enough data exist to allow scientists to estimate the global average. Observations indicate that during the past 110 years or so, Earth's surface air temperature has warmed about 1.0°C (about 1.8°F). Scientific computer models that mathematically simulate the physical processes of the atmosphere, oceans, and ice, predict that should global warming continue unabated, we would be irrevocably committed to major effects from climate change, such as a continuing rise in sea level and a shift in global precipitation patterns.

The main cause of this *climate change* is the greenhouse gas CO_2, whose concentration has been increasing primarily due to the burning of fossil fuels and to deforestation (look back at Fig. 1.18, p. 17). However, increasing concentration of other greenhouse gases, such as methane (CH_4), nitrous oxide (N_2O), and chlorofluorocarbons (CFCs),[*] has collectively been shown to have an effect approaching that of CO_2. In addition, as temperatures warm, more water vapor is added to the air from the world's oceans. Overall, water vapor accounts for about 60 percent of the atmospheric greenhouse effect, CO_2 accounts for about 26 percent, methane about 7 percent, and the remaining greenhouse gases about 7 percent.

---

[*]The term "greenhouse gases" derives from the standard use of "greenhouse effect." Greenhouse gases include, among others, water vapor, carbon dioxide, methane, nitrous oxide, and ozone.

---

[*]To refresh your memory, recall from Chapter 1 that CFCs were once the most widely used propellant in spray cans.

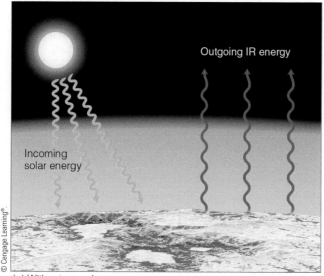 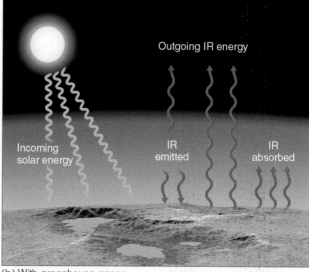

(a) Without greenhouse gases

(b) With greenhouse gases

● **FIGURE 2.12** (a) Near the surface in an atmosphere with little or no greenhouse gases, Earth's surface would constantly emit infrared (IR) radiation upward, both during the day and at night. Incoming energy from the sun would equal outgoing energy from the surface, but the surface would receive virtually no IR radiation from its lower atmosphere. (i.e., there would be no atmospheric greenhouse effect.) Earth's surface air temperature would be quite low, and small amounts of water found on the planet would be in the form of ice. (b) In an atmosphere with greenhouse gases, Earth's surface not only receives energy from the sun but also infrared energy from the atmosphere. Incoming energy still equals outgoing energy, but the added IR energy from the greenhouse gases raises Earth's average surface temperature to a more habitable level.

Presently, the concentration of CO_2 in a volume of air near the surface is just over 0.04 percent, and it is increasing each year. Climate models predict that a continuing increase of CO_2 and other greenhouse gases will cause Earth's current average surface temperature to rise an additional 1°C to 3°C (1.8°F to 5.4°F) or more by the end of this century. How can increasing such a small quantity of CO_2 and adding miniscule amounts of other greenhouse gases bring about such a large temperature increase?

Mathematical climate models predict that rising ocean temperatures will cause an increase in evaporation rates. The added *water vapor*—the primary greenhouse gas—will enhance the atmospheric greenhouse effect and roughly double the temperature rise in what is known as a *positive feedback*. But there are other feedbacks to consider.*

The two potentially largest and least understood feedbacks in the climate system are the clouds and the oceans. Clouds can change area, depth, and radiation properties simultaneously with climatic changes. The net effect of all these changes is not totally clear. Oceans, on the other hand, cover 70 percent of the planet. The response of ocean circulations, ocean temperatures, and sea ice to global warming will determine the global

pattern and speed of climate change. Unfortunately, it is not now known how quickly each of these feedbacks will respond. Satellite data and computer simulations suggest that clouds overall appear to *cool* Earth's climate, as they reflect and radiate away more energy than they retain. (Earth would be about 5°C [9°F] warmer if no clouds were present.) An increase in global cloudiness (if it were to occur) might therefore offset some of the global warming brought on by an enhanced atmospheric greenhouse effect. If clouds were to act on the climate system in this manner, they would provide a *negative feedback* on climate. The actual result would depend on what types of clouds were present, because some clouds are more reflective and have a stronger cooling effect than others. The most recent models tend to show that changes in clouds as a whole would most likely allow for more heat to be retained, thus providing a small positive feedback on the climate system.

Uncertainties unquestionably exist about exactly how much the increasing levels of CO_2 and other greenhouse gases will enhance the atmospheric greenhouse effect. Nonetheless, the most recent studies strongly agree that climate change is presently occurring worldwide owing primarily to increasing levels of greenhouse gases. Evidence for this conclusion comes from increases in global average air and ocean temperatures as well as from widespread melting of snow and ice, rising sea levels, and other conditions. These changes are consistent with the effects one would expect from the increase in greenhouse gases. (We will examine the important topic of climate change in more detail in Chapter 13.)

---

*A feedback is a process whereby an initial change in a process will tend to either reinforce the process (positive feedback) or weaken the process (negative feedback). The *water vapor–greenhouse feedback* is a positive feedback because the initial increase in temperature is reinforced by the addition of more water vapor, which absorbs more of Earth's infrared energy, thus strengthening the greenhouse effect and enhancing the warming.

BRIEF REVIEW

In the last several sections, we have explored examples of some of the ways radiation is absorbed and emitted by various objects. Before we continue, here are a few important facts and principles:

- *All* objects with a temperature above absolute zero emit radiation.
- The higher an object's temperature, the greater the amount of radiation emitted per unit surface area and the shorter the wavelength of maximum emission.
- Earth absorbs solar radiation only during the daylight hours; however, it emits infrared radiation continuously, both during the day and at night.
- Earth's surface behaves as a blackbody, making it a much better absorber and emitter of radiation than the atmosphere.
- Water vapor and carbon dioxide are important greenhouse gases that selectively absorb and emit infrared radiation, thereby keeping Earth's average surface temperature warmer than it otherwise would be.
- Cloudy, calm nights are often warmer than clear, calm nights because clouds strongly emit infrared radiation back to Earth's surface.
- It is *not* the greenhouse effect itself that is of concern, but the *enhancement* of it due to increasing levels of greenhouse gases.
- As greenhouse gases continue to increase in concentration, the average surface air temperature is projected to rise substantially by the end of this century.

With these concepts in mind, we will first examine how the air near the ground warms; then we will consider how Earth and its atmosphere maintain a yearly energy balance.

WARMING THE AIR FROM BELOW If you look back at Fig. 2.11 on p. 37, you'll notice that the atmosphere does not readily absorb radiation with wavelengths between 0.3 and 1.0 μm, the region where the sun emits most of its

energy. Consequently, on a clear day, solar energy passes through the lower atmosphere with little effect upon the air. Ultimately it reaches the surface, warming it (see ●Fig. 2.13). Air molecules in contact with the heated surface bounce against it, gain energy by *conduction*, then shoot upward like freshly popped kernels of corn, carrying their energy with them. Because the air near the ground is very dense, these molecules only travel a short distance before they collide with other molecules. During the collision, these more rapidly moving molecules share their energy with less energetic molecules, raising the average temperature of the air. But air is such a poor heat conductor that this process is only important within a few centimeters of the ground.

As the surface air warms, it actually becomes less dense than the air directly above it. The warmer air rises and the cooler air sinks, setting up thermals, or *free convection cells,* that transfer heat upward and distribute it through a deeper layer of air. The rising air expands and cools, and, if sufficiently moist, the water vapor condenses into cloud droplets, releasing latent heat that warms the air. Meanwhile, Earth constantly emits infrared energy. Some of this energy is absorbed by greenhouse gases (such as water vapor and carbon dioxide) that emit infrared energy upward and downward, back to the surface. Since the concentration of water vapor decreases rapidly above Earth, most of the absorption occurs in a layer near the surface. Hence, the lower atmosphere is mainly heated from the ground upward.

●**FIGURE 2.13** Air in the lower atmosphere is heated from the ground upward. Sunlight warms the ground, and the air above is warmed by conduction, convection, and infrared radiation. Further warming occurs during condensation as latent heat is given up to the air inside the cloud.

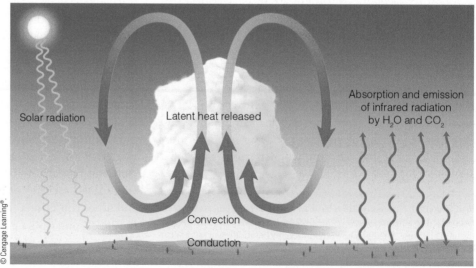

Solar radiation

Latent heat released

Absorption and emission of infrared radiation by H_2O and CO_2

Convection

Conduction

© Cengage Learning®.

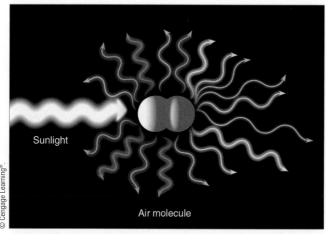

● FIGURE 2.14 The scattering of light by air molecules. Air molecules tend to selectively scatter the shorter (violet, green, and blue) wavelengths of visible white light more effectively than the longer (orange, yellow, and red) wavelengths.

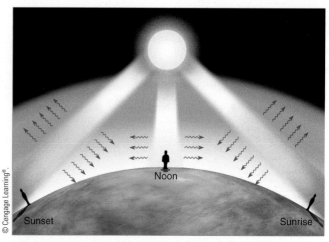

● FIGURE 2.15 At noon, the sun usually appears a bright white. At sunrise and at sunset, sunlight must pass through a thick portion of the atmosphere. Much of the blue light is scattered out of the beam (as illustrated by arrows), causing the sun to appear more red.

SHORTWAVE RADIATION STREAMING FROM THE SUN As the sun's radiant energy travels through space, essentially nothing interferes with it until it reaches the atmosphere. At the top of the atmosphere, solar energy received on a surface perpendicular to the sun's rays appears to remain fairly constant at nearly two calories on each square centimeter each minute, or 1361 W/m²—a value called the **solar constant**.*

When solar radiation enters the atmosphere, a number of interactions take place. For example, some of the energy is absorbed by gases, such as ozone, in the upper atmosphere. Moreover, when sunlight strikes very small objects, such as air molecules and dust particles, the light itself is deflected in all directions—forward, sideways, and backward. The distribution of light in this manner is called **scattering.** (Scattered light is also called *diffuse light.*) Because air molecules are much smaller than the wavelengths of visible light, they are more effective scatterers of the shorter (blue) wavelengths than the longer (red) wavelengths (see ● Fig. 2.14). Hence, when we look away from the direct beam of sunlight, blue light strikes our eyes from all directions, turning the daytime sky blue. At midday, all the wavelengths of visible light from the sun strike our eyes, and the sun is perceived as white (see ● Fig. 2.15). At sunrise and sunset, when the white beam of sunlight must pass through a thick portion of the atmosphere, scattering by air molecules removes the blue light, leaving the longer wavelengths of red, orange, and yellow to pass on through, creating the image of a ruddy or yellowish sun (see ● Fig. 2.16).

Sunlight can be **reflected** from objects. Generally, reflection differs from scattering in that during the process of reflection more light is sent *backward*. **Albedo** is the percent of radiation returning from a given surface compared to the amount of radiation initially striking that surface. Albedo, then, represents the *reflectivity* of the surface. In ▼ Table 2.2, notice that thick clouds have a higher albedo than thin clouds. On the average, the albedo of clouds is near 60 percent. When solar energy strikes a surface covered with snow, up to 95 percent of the sunlight may be reflected. Most of this energy is in the visible and ultraviolet wavelengths. Consequently, reflected radiation, coupled with direct sunlight, can produce severe sunburns on the exposed skin of unwary snow skiers, and unprotected eyes can suffer the agony of snow blindness.

● FIGURE 2.16 A red sunset produced by the process of scattering.

*By definition, the solar constant (which, in actuality, is *not* "constant") is the rate at which radiant energy from the sun is received on a surface at the outer edge of the atmosphere perpendicular to the sun's rays when Earth is at an average distance from the sun. Satellite measurements suggest the solar constant varies slightly as the sun's radiant output varies. The latest measurements and laboratory tests indicate that the solar constant is around 1361 W/m². It rises by about 1 W/m² during peaks of solar activity, which occur about every 11 years.

Table 2.2 Typical Albedo of Various Surfaces

SURFACE	ALBEDO (PERCENT)
Fresh snow	75 to 95
Clouds (thick)	60 to 90
Clouds (thin)	30 to 50
Venus	78
Ice	30 to 40
Sand	15 to 45
Earth and atmosphere	30
Mars	17
Grassy field	10 to 30
Dry, plowed field	5 to 20
Water	10*
Forest	3 to 10
Moon	7

*Daily average.

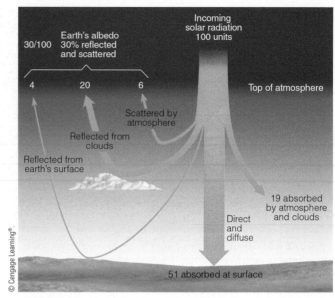

● FIGURE 2.17 On the average, of all the solar energy that reaches Earth's atmosphere annually, about 30 percent ($^{30}/_{100}$) is reflected and scattered back to space, giving Earth and its atmosphere an albedo of 30 percent. Of the remaining solar energy, about 19 percent is absorbed by the atmosphere and clouds, and about 51 percent is absorbed at the surface.

Water surfaces, on the other hand, reflect only a small amount of solar energy. For an entire day, a smooth water surface will have an average albedo of about 10 percent. Averaged for an entire year, Earth and its atmosphere (including its clouds) will redirect about 30 percent of the sun's incoming radiation back to space, which gives Earth and its atmosphere a combined albedo of 30 percent (see ● Fig. 2.17).

EARTH'S ANNUAL ENERGY BALANCE Although the average temperature at any one place may vary considerably from year to year, Earth's overall average equilibrium temperature changes only slightly from one year to the next. Each year, then, Earth and its atmosphere combined must send off into space just as much energy as they receive from the sun. The same type of energy balance must exist between Earth's surface and the atmosphere; that is, each year, Earth's surface must return to the atmosphere the same amount of energy that it absorbs. If this did not occur, Earth's average surface temperature would change. How do Earth and its atmosphere maintain this yearly energy balance?

Suppose 100 units of solar energy reach the top of Earth's atmosphere. We already know from Fig. 2.17 that, on the average, clouds, Earth, and the atmosphere reflect and scatter 30 units back to space, and that the atmosphere and clouds together absorb 19 units, which leaves 51 units of direct and indirect (diffuse) solar radiation to be absorbed at Earth's surface.

● Figure 2.18 shows approximately what happens to the solar radiation that is absorbed by the surface and the atmosphere. Out of 51 units reaching the surface, a large amount (23 units) is used to evaporate water, and about 7 units are lost through conduction and convection, which leaves 21 units to be radiated away as infrared energy. Look closely at Fig. 2.18 and notice that Earth's surface actually radiates upward a whopping 117 units. It does so because, although it receives solar radiation only during the day, it constantly emits infrared energy both during the day and at night. Additionally, the atmosphere above only allows a small fraction of this energy (6 units) to pass through into space. The majority of it (111 units) is absorbed mainly by the greenhouse gases water vapor and CO_2, and by clouds. Much of this energy (96 units) is then radiated back to Earth, producing the atmospheric greenhouse effect. Earth's surface receives nearly twice as much longwave infrared energy from the atmosphere as it does shortwave radiation from the sun. In all these exchanges, notice that the energy lost at Earth's surface (147 units) is exactly balanced by the energy gained there (147 units).

A similar balance exists between Earth's surface and its atmosphere. Again, observe in Fig. 2.18 that the energy gained by the atmosphere (160 units) balances the energy lost. Moreover, averaged for an entire year, the solar energy received at Earth's surface (51 units) and that absorbed by Earth's atmosphere (19 units) balances the infrared energy lost to space by Earth's surface (6 units) and its atmosphere (64 units).

We can see the effect that conduction, convection, and latent heat have in the warming of the atmosphere if we look at the energy balance only in radiative terms. Earth's surface receives 147 units of radiant energy from the sun

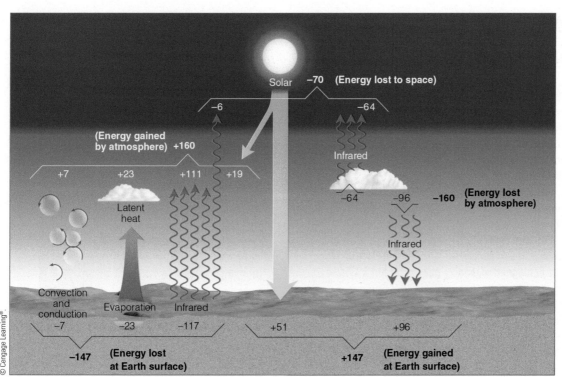

● FIGURE 2.18
Earth-atmosphere
energy balance.
Numbers represent
approximations
based on surface
observations and
satellite data.
While the actual
value of each pro-
cess may vary by
several percent, it
is the relative size
of the numbers
that is important.

and its own atmosphere, while it radiates away 117 units, producing a *surplus* of 30 units. The atmosphere, on the other hand, receives 130 units (19 units from the sun and 111 from Earth), while it loses 160 units, producing a *deficit* of 30 units. The balance (30 units) is the warming of the atmosphere produced by the heat transfer processes of conduction and convection (7 units) and by the release of latent heat (23 units).

So, Earth and the atmosphere absorb energy from the sun as well as from each other. In all of the energy exchanges, a delicate balance is maintained. Essentially, there is no yearly gain or loss of total energy, and the average temperature of Earth and the atmosphere remains fairly constant from one year to the next. This equilibrium does not imply that Earth's average temperature does not change, but rather that the changes are small from year to year (usually less than one-tenth of a degree Celsius) and become significant only when measured over many years. One example is the temperature increase of 0.6°C (1.1°F) during the last century produced by the addition of greenhouse gases to the atmosphere. The added gases led to a radiative imbalance as more infrared radiation is trapped by the atmosphere. However, as Earth warms, the balance is restored.

Even though Earth and the atmosphere together maintain an annual energy balance, such a balance is not maintained at each latitude. High latitudes tend to lose more energy to space each year than they receive from the sun, while low latitudes tend to gain more energy during the course of a year than they lose. From ● Fig. 2.19 we can see that only at middle latitudes near 38° does the amount

of energy received each year balance the amount lost. We might conclude that polar regions are growing colder each year, while tropical regions are becoming warmer. But they are not. To compensate for these gains and losses of energy, winds in the atmosphere and currents in the oceans circulate warm air and water toward the poles, and cold air and water toward the equator. The transfer of heat energy by atmospheric and oceanic circulations prevents low latitudes from steadily becoming warmer and high latitudes from steadily growing colder. These circulations are extremely important to weather and climate, and will be treated more completely in Chapter 7.

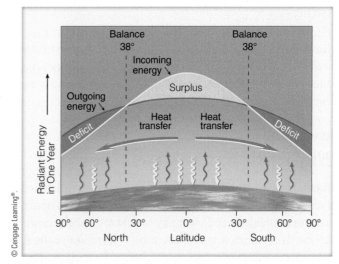

● FIGURE 2.19 The average annual incoming solar radiation (yellow arrows) absorbed by Earth and the atmosphere along with the average annual infrared radiation (red arrows) emitted by Earth and the atmosphere.

Space Weather and Its Impact on Earth

The sun is our nearest star, located some 150 million kilometers (93 million miles) from earth. Even at that distance, it provides enough energy for life to flourish. The sun is a giant celestial furnace with a core temperature estimated to be near 15 million degrees Celsius. Here, thermonuclear processes generate an enormous amount of energy, which gradually works its way outward. At its surface, the sun emits vast quantities of radiant energy. It also generates very large amounts of magnetism, and a constant stream of charged particles that stream outward in all directions from its surface. Sometimes these affect our atmosphere, producing a variety of effects known as *space weather*.

The eerie yet beautiful light show called the *aurora* is one of the most commonly observed aspects of space weather. At high latitudes after darkness has fallen, a faint, white glow may appear in the sky. Lasting from a few minutes to a few hours, the light may move across the sky as a yellow-green arc much wider than a rainbow; or, it may faintly decorate the sky with flickering draperies of blue, green and purple light that constantly changes in form and location, as if blown by a gentle breeze. This impressive light show is the aurora (see ● Fig. 4).

The aurora is caused by charged particles from the sun interacting with our

© Lindsey P. Martin Photography

● **FIGURE 4** The aurora borealis is a phenomenon that forms as energetic particles from the sun interact with Earth's atmosphere.

atmosphere, almost always in the thermosphere, more than 80 kilometers (50 miles) above Earth's surface. From the sun and its tenuous atmosphere comes a continuous discharge of particles. This discharge happens because, at extremely high temperatures, gases become stripped of electrons by violent collisions and acquire enough speed to escape the gravitational pull of the sun. As these charged particles (ions and electrons) travel through space, they are known as the *solar wind*. When the solar wind moves close enough to Earth, it interacts with Earth's magnetic field, disturbing it (see ● Fig. 5). This disturbance causes energetic solar wind particles to enter the upper atmosphere, where they collide with atmospheric gases. These gases then become excited and emit visible radiation (light), which causes the sky to glow like a neon light, thus producing the aurora.

In the Northern Hemisphere, the aurora is called the *aurora borealis*, or northern lights; its counterpart in the Southern

We now turn our attention to how solar energy produces Earth's seasons. Before doing so, you may wish to read Focus section 2.3 to learn how the sun's energy produces *space weather* and a dazzling light show known as the *aurora*.

Why Earth Has Seasons

Earth revolves completely around the sun in an elliptical path (not quite a circle) in about 365 days and six hours (one year, plus a Leap Day every four years in February). As Earth revolves around the sun, it spins on its own axis, completing one spin in 24 hours (one day). The average distance from Earth to the sun is 150 million km (93 million mi). Because Earth's orbit is an ellipse instead of a circle, and is slightly off-center from the sun, the

actual distance from Earth to the sun varies during the year. Earth comes closer to the sun in January (147 million km) than it does in July (152 million km).* (See ● Fig. 2.20.) We might conclude from this fact that our warmest weather should occur in January and our coldest weather in July. But, in the Northern Hemisphere, we normally experience cold weather in January when we are closer to the sun and warm weather in July when we are farther away. If nearness to the sun were the primary cause of the seasons then, indeed, January would be warmer than July. However, nearness to the sun is only a small part of the story.

Our seasons are regulated by the amount of solar energy received at Earth's surface. This amount is determined

---

*The time around January 3, when Earth is closest to the sun, is called *perihelion* (from the Greek *peri*, meaning "near" and *helios*, meaning "sun"). The time when Earth is farthest from the sun (around July 4) is called *aphelion* (from the Greek *ap*, meaning "away from").

Hemisphere is the *aurora australis*, or southern lights. The aurora is most frequently seen in the polar regions, where Earth's magnetic field lines emerge from Earth.

Auroral displays are most frequent and intense during the active part of the sun's cyclic activity, which peaks about every 11 years. During active solar periods, there are numerous sunspots (huge cooler regions on the sun's surface) and giant flares (solar eruptions) that send large quantities of solar wind particles traveling away from the sun at high speeds (hundreds of kilometers a second). When one of these eruptions is pointed toward Earth, the energetic particles may be able to penetrate unusually deep into Earth's magnetic field, creating conditions referred to as *solar storms*. During these conditions in North America, we see the aurora much farther south than usual.

Along with the beauty of the aurora, solar storms can produce many negative effects. In 1859, a huge solar storm disrupted telegraph operations around the world. Another solar storm in March 1989 caused millions of people across Quebec to lose electrical power for several hours. Airlines sometimes reroute planes away from polar regions during the most intense solar storms, which reduces the risk of interference with radio communications.

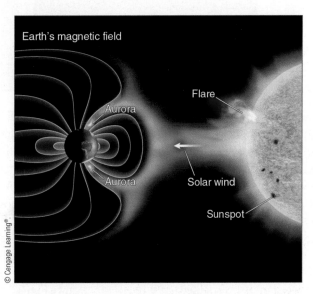

● **FIGURE 5** The stream of charged particles from the sun—called the *solar wind*—distorts Earth's magnetic field. These particles, which spiral in along magnetic field lines, interact with atmospheric gases, and produce the aurora.

Solar storms can also affect satellite operations, which is a growing concern because of the many types of satellites now used for telecommunications, navigation, and other purposes. A $630 million research satellite from Japan failed during an intense solar storm in October 2003. Because solar storms heat the outer atmosphere, they can increase the drag on satellites and reduce their lifespan in orbit.

Space weather was on the quiet side during the solar minimum of 2008–2009, when the number of sunspots observed was at its lowest level in nearly a century.

Activity was also on the low side during the solar maximum that peaked in 2013–2014, which was the weakest maximum in more than a century. Scientists cannot say how long this tendency toward lessened activity during both maximum and minimum might continue, because techniques for predicting the strength of solar cycles are still being researched and tested.

primarily by the angle at which sunlight strikes the surface and by how long the sun shines on any latitude (daylight hours). Let's look more closely at these factors.

Solar energy that strikes Earth's surface perpendicularly (directly) is much more intense than solar energy that strikes the same surface at an angle. Think of shining

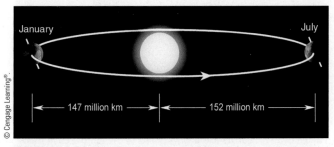

● **FIGURE 2.20** The elliptical path (highly exaggerated) of Earth about the sun brings Earth slightly closer to the sun in January than in July.

a flashlight straight at a wall—you get a small, circular spot of light (see ● Fig. 2.21). Now tip the flashlight and notice how the spot of light spreads over a larger area. The same principle holds for sunlight. Sunlight striking Earth at an angle spreads out and must heat a larger region than sunlight impinging directly on Earth. Everything else being equal, an area experiencing more direct solar rays will receive more heat than the same size area being struck by sunlight at an angle. In addition, the more the sun's rays are slanted from the perpendicular, the more atmosphere they must penetrate. And the more atmosphere they penetrate, the more they can be scattered and absorbed. As a consequence, when the sun is high in the sky, it can heat the ground to a much higher temperature than when it is low on the horizon.

The second important factor determining how warm Earth's surface becomes is the length of time the sun shines

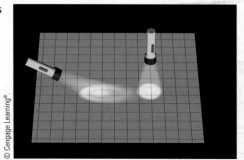

● **FIGURE 2.21** Sunlight that strikes a surface at an angle is spread over a larger area than sunlight that strikes the surface directly. Oblique sun rays deliver less energy (are less intense) to a surface than direct sun rays.

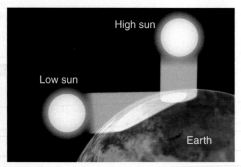

each day, the number of daylight hours. Longer daylight hours, of course, mean that more energy is available from sunlight. In a given location, more solar energy reaches Earth's surface on a clear, long day than on a day that is clear but much shorter. The longer the day, the more surface heating takes place.

From a casual observation, we know that summer days have more daylight hours than winter days. Also, the noontime summer sun is higher in the sky than is the noontime winter sun. Both of these events occur because our spinning planet is inclined on its axis (tilted) as it revolves around the sun. As ● Fig. 2.22 illustrates, the angle of tilt is 23½° from the perpendicular drawn to the plane of Earth's orbit. Earth's axis points to the same direction in space all year long; thus, on one side of Earth's orbit the Northern Hemisphere is tilted toward the sun in summer (June), and on the other side of Earth's orbit it is tilted away from the sun in winter (December).

SEASONS IN THE NORTHERN HEMISPHERE   Notice in Fig. 2.22 that on June 21, the northern half of the world is directed toward the sun. At noon on this day, solar rays beat down upon the Northern Hemisphere more directly than during any other time of year. The sun is at its highest position in the noonday sky, directly above 23½° north (N) latitude (Tropic of Cancer). If you were standing at this latitude on June 21, the sun at noon would be directly overhead. This day, called the **summer solstice**, is the astronomical first day of summer in the Northern Hemisphere.*

Study Fig. 2.22 closely and notice that, as Earth spins on its axis, the side facing the sun is in sunshine and the other side is in darkness. Thus, half of the globe is always illuminated. If Earth's axis were not tilted, the noonday sun would always be directly overhead at the equator,

---

*As we will see later in this chapter, the seasons are reversed in the Southern Hemisphere. Hence, in the Southern Hemisphere, this same day is the winter solstice, or the astronomical first day of winter.

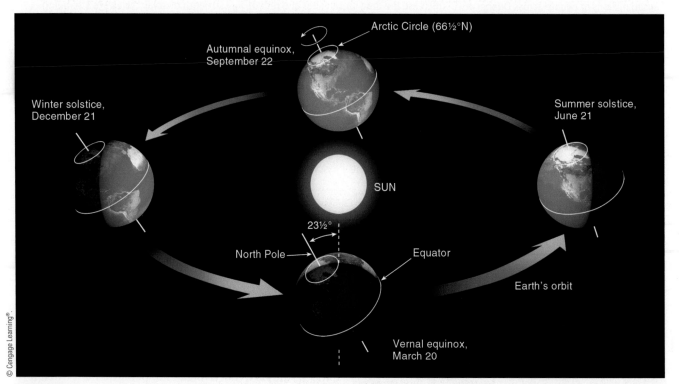

● **FIGURE 2.22** As Earth revolves about the sun, it is tilted on its axis by an angle of 23½°. Earth's axis always points to the same area in space (as viewed from a distant star). Thus, in June, when the Northern Hemisphere is tipped toward the sun, more direct sunlight and long hours of daylight cause warmer weather than in December, when the Northern Hemisphere is tipped away from the sun. (Diagram, of course, is not to scale. The dates of each solstice and equinox may be a day earlier or later than shown here, depending on the year and on your time zone.)

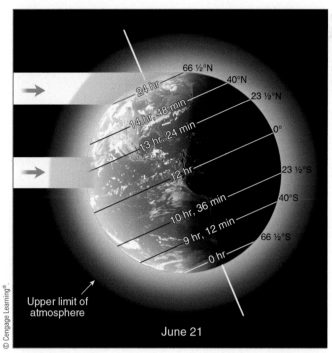

66 ½°N
40°N
24 hr
23 ½°N
14 hr, 48 min
13 hr, 24 min
0°
12 hr
23 ½°S
10 hr, 36 min
40°S
9 hr, 12 min
66 ½°S
0 hr

Upper limit of
atmosphere

June 21

• **FIGURE 2.23** The number of hours of daylight (hours and minutes) for different latitudes on June 21. Notice that daylight hours range from 24 at 66 ½°N to zero at 66 ½°S. Also notice that sunlight that reaches Earth's surface in far northern latitudes has passed through a thicker layer of absorbing, scattering, and reflecting atmosphere than sunlight that reaches Earth's surface farther south. Sunlight is lost through both the thickness of the pure atmosphere and by impurities in the atmosphere. As the sun's rays become more oblique, these effects become more pronounced.

and there would be 12 hours of daylight and 12 hours of darkness at each latitude every day of the year. However, Earth is tilted. Since the Northern Hemisphere faces toward the sun on June 21, each latitude in the Northern Hemisphere will have more than 12 hours of daylight. The farther north we go, the longer are the daylight hours. When we reach the Arctic Circle (66½°N), daylight lasts for 24 hours, as the sun does not set. Notice in Fig. 2.22 and • Fig. 2.23 how the region above 66½°N never gets into the "shadow" zone as Earth spins. At the North Pole,

the sun actually rises above the horizon on March 20 and has six months until it sets on September 22. No wonder this region is called the "Land of the Midnight Sun"! (See • Fig. 2.24.)

Even though in the far north the sun is above the horizon for many hours during the summer (see ▼ Table 2.3, p. 48), the surface air there is not warmer than the air farther south, where days are appreciably shorter. Why this is so is shown in Fig. 2.23. When *incoming solar radiation* (called *insolation*) enters the atmosphere, fine dust, air molecules, and clouds reflect and scatter it, and some of it is absorbed by atmospheric gases. Generally, the greater the thickness of atmosphere that sunlight must penetrate, the greater the chances that it will be either reflected or absorbed by the atmosphere. During the summer in far northern latitudes, the sun is never very high above the horizon, so its radiant energy must pass through a thick portion of atmosphere before it reaches Earth's surface. Some of the solar energy that does reach the surface melts frozen soil or is reflected by snow or ice. That which is absorbed is spread over a large area. So, even though northern cities may experience long hours of sunlight, they are cooler than cities farther south, because overall, they receive less radiation at the surface. What radiation they do receive does not heat the surface as effectively.

Look at Fig. 2.22 again and notice that, by September 22, Earth will have moved so that the sun is directly above the equator. Except at the poles, the days and nights throughout the world are of equal length. This day is called the **autumnal** (fall) **equinox,** and it marks the astronomical beginning of fall in the Northern Hemisphere. At the North Pole, the sun appears on the horizon for 24 hours, due to the bending of light by the atmosphere. The following day (or at least within several days), the sun disappears from view, not to rise again for a long, cold six months. Throughout the northern half of the world on each successive day, there are fewer hours of daylight, and the noon sun is slightly lower in the sky. Less direct

• **FIGURE 2.24** Land of the Midnight Sun. A series of exposures of the sun taken before, during, and after midnight in northern Alaska during July.

Length of Time from Sunrise to Sunset for Various Latitudes on Different Dates in the Northern Hemisphere

LATITUDE	MARCH 20	JUNE 21	SEPT. 22	DEC. 21
0°	12 hr	12.0 hr	12 hr	12.0 hr
10°	12 hr	12.6 hr	12 hr	11.4 hr
20°	12 hr	13.2 hr	12 hr	10.8 hr
30°	12 hr	13.9 hr	12 hr	10.1 hr
40°	12 hr	14.9 hr	12 hr	9.1 hr
50°	12 hr	16.3 hr	12 hr	7.7 hr
60°	12 hr	18.4 hr	12 hr	5.6 hr
70°	12 hr	2 months	12 hr	0 hr
80°	12 hr	4 months	12 hr	0 hr
90°	12 hr	6 months	12 hr	0 hr

sunlight and shorter hours of daylight spell cooler weather for the Northern Hemisphere. Reduced sunlight, lower air temperatures, and cooling breezes stimulate the beautiful pageantry of fall colors (see ● Fig. 2.25).

In some years around the middle of autumn, there is an unseasonably warm spell, especially in the eastern two-thirds of the United States. This warm period, referred to as **Indian summer,*** may last from several days up to a week or more. It usually occurs when a large high-pressure area stalls near the southeast coast. The clockwise flow of air around this system moves warm air from the Gulf of Mexico into the central or eastern half of the nation. The warm, gentle breezes and smoke from a variety of sources respectively make for mild, hazy days. The warm weather ends abruptly when an outbreak of polar air reminds us that winter is not far away.

On December 21 (three months after the autumnal equinox), the Northern Hemisphere is tilted as far away from the sun as it will be all year (see Fig. 2.22, p. 46). Nights are long and days are short. Notice in Table 2.3 that daylight decreases from 12 hours at the equator to 0 (zero) at latitudes above $66\frac{1}{2}$°N. This is the shortest day of the year, called the **winter solstice**—the astronomical beginning of winter in the northern world. On this day, the sun shines directly above latitude $23\frac{1}{2}$°S (Tropic of Capricorn). In the northern half of the world, the sun is at its lowest position in the noon sky. Its rays pass through a

---

*The origin of the term is uncertain, but it dates back to the eighteenth century. It may have originally referred to the good weather that allowed Native Americans time to harvest their crops. Normally, a period of cool autumn weather must precede the warm weather period for the latter to be called Indian summer.

● **FIGURE 2.25** The pageantry of fall colors in New England. The weather most suitable for an impressive display of fall colors is warm, sunny days followed by clear, cool nights with temperatures dropping below 7°C (45°F), but remaining above freezing. Contrary to popular belief, it is not the first frost that causes the leaves of deciduous trees to change color. The yellow and orange colors, which are actually in the leaves, typically begin to show through several weeks before the first frost, as shorter days and cooler nights cause a decrease in the production of the green pigment chlorophyll.

Ron and Patty Thomas/Photographer's Choice/Getty Images

Is December 21 Really the First Day of Winter?

On December 21 (or 22, depending on the year) after nearly a month of cold weather, and perhaps a snowstorm or two (see ● Fig. 6), someone on the radio or television has the audacity to proclaim that "today is the first official day of winter." If during the last several weeks it was not winter, then what season was it?

Actually, December 21 marks the *astronomical* first day of winter in the Northern Hemisphere (NH), just as June 21 marks the *astronomical* first day of summer (NH). Earth is tilted on its axis by $23\frac{1}{2}°$ as it revolves around the sun. This fact causes the sun (as we view it from Earth) to move in the sky from a point where it is directly above $23\frac{1}{2}°$ South latitude on December 21 to a point where it is directly above $23\frac{1}{2}°$ North latitude on June 21. The astronomical first day of spring (NH) occurs around March 20 as the sun crosses the equator moving northward and, likewise, the astronomical first day of autumn (NH) occurs around September 22 as the sun crosses the equator moving southward.

Therefore the "official" beginning of any season is simply the day on which the sun passes over a particular latitude, and has nothing to do with how cold or warm the following day will be. In fact, a period of colder or warmer than normal weather before or after a

● **FIGURE 6** Snow covers Central Park in New York City on December 17, 2013. Since the snowstorm occurred before the winter solstice, is this a late fall storm or an early winter storm?

solstice or equinox is caused mainly by the upper-level winds directing cold or warm air into a region.

In the middle latitudes, summer is defined as the warmest season and winter the coldest season. If the year is divided into four seasons with each season consisting of three months, then the meteorological (or *climatological*) definition of summer over much of the Northern Hemisphere would be the three warmest months of June, July, and August. Winter would be the three coldest months of December, January, and

February. Autumn would be September, October, and November—the transition between summer and winter. And spring would be March, April, and May—the transition between winter and summer.

So, the next time you hear someone remark on December 21 that "winter officially begins today," remember that this is the astronomical definition of the first day of winter. According to the climatological definition, winter has been around for several weeks.

thick section of atmosphere and spread over a large area on the surface.

With so little incident sunlight, Earth's surface cools quickly. A blanket of clean snow covering the ground aids in the cooling. In northern Canada and Alaska, arctic air rapidly becomes extremely cold as it lies poised, ready to do battle with the milder air to the south. Periodically, this cold arctic air pushes down into the northern United States, producing a rapid drop in temperature called a *cold wave,* which occasionally reaches far into the south. Sometimes these cold spells arrive well before the winter solstice—the "official" first day of winter—bringing with them heavy snow and blustery winds. (To learn more about this "official" first day of winter, read Focus section 2.4.)

Three months past the winter solstice marks the astronomical arrival of spring, which is called the **vernal** (spring) **equinox.** The date is March 20 and, once again,

the noonday sun is shining directly on the equator, days and nights throughout the world are of equal length, and, at the North Pole, the sun rises above the horizon after a long six-month absence.

At this point it is interesting to note that although sunlight is most intense in the Northern Hemisphere on June 21, the warmest weather in middle latitudes normally occurs weeks later, usually in July or August. This situation (called the *lag in seasonal temperature*) arises because although incoming energy from the sun is greatest in June, it takes time for oceans and landmasses to release the large amounts of incoming energy they have absorbed. As a result, incoming energy still exceeds outgoing energy from Earth for a period of at least several weeks. Once the incoming solar energy and outgoing earth energy are in balance, the highest average temperature is attained. When outgoing energy exceeds incoming energy, the average temperature drops. As in the summer, there is also a seasonal temperature lag in winter. Because outgoing Earth energy exceeds incoming solar energy well past the winter solstice (December 21), we normally find our coldest weather occurring in January or February.

Up to now, we have seen that the seasons are controlled by the amount of solar energy striking our tilted planet as it makes its annual voyage around the sun. And we know that the tilt of Earth causes a seasonal variation in both the length of daylight and the intensity of sunlight that reaches the surface. These facts are summarized in ● Fig. 2.26, which shows how the sun would appear in the sky to an observer at various latitudes at different times of the year. Earlier we learned that at the North Pole the sun rises above the horizon in March and stays above the horizon for six months, until September. Notice in Fig. 2.26a that at the North Pole even when the sun is at its highest point in June, it is low in the sky—only 23½° above the horizon. Farther south, at the Arctic Circle (Fig. 2.26b), the sun is always fairly low in the sky, even in June, when the sun stays above the horizon for 24 hours.

In the middle latitudes (Fig. 2.26c), notice that in December the sun rises in the southeast, reaches its highest point at noon (only about 26° above the southern horizon), and sets in the southwest.* This apparent path produces little intense sunlight and short daylight hours. On the other hand, in June, the sun rises in the northeast, reaches a much higher position in the sky at noon (about 74° above the southern horizon) and sets in the northwest. This apparent path across the sky produces more intense solar heating, longer daylight hours, and, of course, warmer weather. Figure 2.26d illustrates how the tilt of Earth influences the sun's apparent path across the sky at the Tropic of Cancer (23½°). Figure 2.26e gives the same information for an observer at the equator.

SEASONS IN THE SOUTHERN HEMISPHERE On June 21, the Southern Hemisphere is adjusting to an entirely different season. Again, look back at Fig. 2.22 (p. 46), and notice that this part of the world is now tilted away from the sun. Nights are long, days are short, and solar rays come in at a low angle (see Fig. 2.26f). All of these factors keep air temperatures fairly low. The June solstice marks the astronomical beginning of winter in the Southern Hemisphere. In this part of the world, summer will not "officially" begin until the sun is over

---

*Calculating the noon angle of the sun for any latitude is easy. First, determine the number of degrees between your latitude and the latitude where the sun is directly overhead. Then subtract this number from 90°. The result gives you the elevation of the sun above the southern horizon at noon at your latitude.

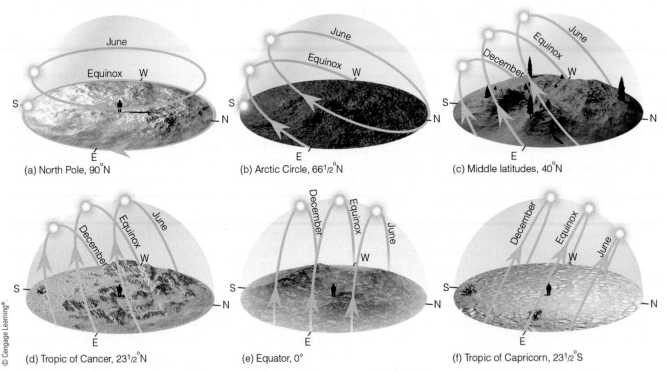

(a) North Pole, 90°N
(b) Arctic Circle, 66½°N
(c) Middle latitudes, 40°N
(d) Tropic of Cancer, 23½°N
(e) Equator, 0°
(f) Tropic of Capricorn, 23½°S

● **FIGURE 2.26** The apparent path of the sun across the sky as observed at different latitudes on the June solstice (June 21), the December solstice (December 21), and the equinoxes (March 20 and September 22).

© Cengage Learning®.

the Tropic of Capricorn (23½°S)—remember that this occurs on December 21. So, when it is winter and June in the Southern Hemisphere, it is summer and June in the Northern Hemisphere. Conversely, when it is summer and December in the Southern Hemisphere, it is winter and December in the Northern Hemisphere. If you are tired of the cold, December weather in your Northern Hemisphere city, travel to the summer half of the world and enjoy the warmer weather. The tilt of Earth as it revolves around the sun makes all this possible.

We know Earth comes nearer to the sun in January than in July. Even though this difference in distance amounts to only about 3 percent, the energy that strikes the top of Earth's atmosphere is almost 7 percent greater on January 3 than on July 4. These numbers might lead us to believe that summer should be warmer in the Southern Hemisphere than in the Northern Hemisphere. However, this is not so. A close examination of the Southern Hemisphere reveals that nearly 81 percent of the surface is water compared to 61 percent in the Northern Hemisphere. The added solar energy due to the closeness of the sun is absorbed by large bodies of water, becoming well mixed and circulated within them. This process keeps the average summer (January) temperatures in the Southern Hemisphere cooler than the average summer (July) temperatures in the Northern Hemisphere. Because of water's large heat capacity, it also tends to keep winters in the Southern Hemisphere warmer than we might expect.*

LOCAL SEASONAL VARIATIONS Look at Fig. 2.26c and observe that in the middle latitudes of the Northern Hemisphere objects facing south will receive more sunlight during a year than those facing north. This fact becomes strikingly apparent in hilly or mountainous country.

Hills that face south receive more sunshine and, hence, become warmer than the partially shielded north-facing hills. Higher temperatures usually mean greater rates of evaporation and slightly drier soil conditions. Thus, south-facing hillsides are usually warmer and drier as compared to north-facing slopes at the same elevation. In many areas of the far western United States, only sparse vegetation grows on south-facing slopes, while, on the same hill, dense vegetation grows on the cool, moist slopes that face north (see ● Fig. 2.27).

In the mountains, snow usually lingers on the ground for a longer time on north slopes than on the warmer south slopes. For this reason, ski runs are built facing

*For a comparison of January and July temperatures see Figs. 3.14 and 3.15, p. 66.

● **FIGURE 2.27** In the middle latitudes of the Northern Hemisphere, where small temperature changes can cause major changes in soil moisture, sparse vegetation on the south-facing slopes will often contrast with lush vegetation on the north-facing slopes.

north wherever possible. Also, homes and cabins built on the north side of a hill usually have a steep pitched roof, as well as a reinforced deck to withstand the added weight of snow from successive winter storms.

The seasonal change in the sun's position during the year can have an effect on the vegetation around the home. In winter, a large two-story home can shade its own north side, keeping it much cooler than its south side. Trees that require warm, sunny weather should be planted on the south side, where sunlight reflected from the house can even add to the warmth.

The design of a home can be important in reducing heating and cooling costs. Large windows should face south, allowing sunshine to penetrate the home in winter. To block out excess sunlight during the summer, a small eave or overhang should be built. A kitchen with windows facing east will let in enough warm morning sunlight to help heat this area. Because the west side warms rapidly in the afternoon, rooms having small windows (such as garages) can be placed here to act as a thermal buffer. Deciduous trees planted on the west side of a home provide shade in the summer. In winter, they drop their leaves, allowing the winter sunshine to warm the house. If you like the bedroom slightly cooler than the rest of the home, face it toward the north. Let nature help with the heating and air conditioning. Proper house design, orientation, and landscaping can help cut the demand for electricity, as well as for natural gas and fossil fuels, which are rapidly being depleted.

SUMMARY

In this chapter, we looked at the concepts of heat and temperature and learned that latent heat is an important source of atmospheric heat energy. We also learned that the transfer of heat can take place by conduction, convection, and radiation—the transfer of energy by means of electromagnetic waves.

The hot sun emits most of its radiation as shortwave radiation. A portion of this energy heats Earth, and Earth, in turn, warms the air above. The cool Earth emits most of its radiation as longwave infrared energy. Selective absorbing greenhouse gases in the atmosphere, such as water vapor and carbon dioxide, absorb some of Earth's infrared radiation and radiate a portion of it back to the surface, where it warms the surface, producing the atmospheric greenhouse effect. The average equilibrium temperature of Earth and the atmosphere remains fairly constant from one year to the next because the amount of energy they absorb each year is equal to the amount of energy they lose.

We examined the seasons and found that Earth has seasons because it is tilted on its axis as it revolves around the sun. The tilt of Earth causes a seasonal variation in both the length of daylight and the intensity of sunlight that reaches the surface. Finally, on a more local setting, we saw that Earth's inclination influences the amount of solar energy received on the north and south side of a hill, as well as around a home.

KEY TERMS

The following terms are listed (with corresponding page numbers) in the order they appear in the text. Define each. Doing so will aid you in reviewing the material covered in this chapter.

kinetic energy, 28	radiant energy
temperature, 28	(radiation), 33
absolute zero, 28	electromagnetic waves, 33
heat, 28	micrometer, 33
Kelvin scale, 28	photons, 33
Fahrenheit scale, 29	visible region, 34
Celsius scale, 29	ultraviolet radiation (UV), 34
latent heat, 29	infrared radiation (IR), 34
sensible heat, 30	shortwave radiation, 36
conduction, 31	longwave (terrestrial)
convection, 31	radiation, 36
thermals, 32	blackbody, 36
advection, 32	selective absorbers, 37

radiative equilibrium	reflected (light), 41
temperature, 37	albedo, 41
greenhouse effect, 38	summer solstice, 46
greenhouse gases, 38	autumnal equinox, 47
atmospheric window, 38	Indian summer, 48
solar constant, 41	winter solstice, 48
scattering, 41	vernal equinox, 49

QUESTIONS FOR REVIEW

1. Distinguish between temperature and heat.
2. How does the average speed (motion) of air molecules relate to the air temperature?
3. Explain how heat is transferred in our atmosphere by: (a) conduction (b) convection (c) radiation.
4. What is latent heat? How is latent heat an important source of atmospheric energy?
5. How does the Kelvin temperature scale differ from the Celsius scale?
6. How does the amount of radiation emitted by Earth differ from that emitted by the sun?
7. How does the temperature of an object influence the radiation it emits?
8. How do the wavelengths of most of the radiation emitted by the sun differ from those emitted by the surface of Earth?
9. When a body reaches a radiative equilibrium temperature, what is taking place?
10. Why are carbon dioxide and water vapor called selective absorbing greenhouse gases?
11. List four important greenhouse gases in Earth's atmosphere.
12. Explain how Earth's atmospheric greenhouse effect works.
13. What greenhouse gases appear to be responsible for the enhancement of Earth's greenhouse effect?
14. Why does the albedo of Earth and its atmosphere average about 30 percent?
15. How is the lower atmosphere warmed from the surface upward?
16. Explain how Earth and its atmosphere balance incoming energy with outgoing energy.
17. In the Northern Hemisphere, why are summers warmer than winters even though Earth is actually closer to the sun in January?
18. What are the main factors that determine seasonal temperature variations?

19. If it is winter and January in New York City, what is the season and month in Sydney, Australia?

20. During the Northern Hemisphere's summer, the daylight hours in northern latitudes are longer than in middle latitudes. Explain why northern latitudes are not warmer.

21. During July, daylight hours in far northern latitudes of the Northern Hemisphere are longer than daylight hours in the middle latitudes. Explain why far northern latitudes are not warmer than middle latitudes.

22. Explain why the vegetation on the north-facing side of a hill is frequently different from the vegetation on the south-facing side of the same hill.

QUESTIONS FOR THOUGHT AND EXPLORATION

1. Explain why the bridge in ● Fig. 2.28 is the first to become icy.

© Cengage Learning®

BRIDGE FREEZES BEFORE ROAD

● **FIGURE 2.28**

2. If the surface of a puddle freezes, is heat energy *released to* or *taken from* the air above the puddle? Explain.

3. In houses and apartments with forced-air furnaces, heat registers are usually placed near the floor rather than near the ceiling. Explain why.

4. How is heat transferred away from the surface of the moon? (Hint: The moon has no atmosphere.)

5. Which do you feel would have the greatest effect on Earth's greenhouse effect: removing all of the CO_2 from the atmosphere or removing all of the water vapor? Explain your answer.

6. How would the seasons be affected where you live if the tilt of Earth's axis *increased* from 23½° to 40°?

7. Explain why an increase in cloud cover surrounding Earth would increase Earth's albedo, yet not necessarily lead to a lower Earth surface temperature.

8. Why does the surface temperature often increase on a clear, calm night as a low cloud moves overhead?

9. Would you expect Earth's surface temperature to continue to rise if CO_2 levels continue to increase but levels of atmospheric water vapor begin to decrease? Explain your reasoning.

10. Explain (with the aid of a diagram) why the morning sun in the Northern Hemisphere shines brightly through a south-facing bedroom window in December, but not in June.

11. In New York City, the intensity of sunlight and the number of daylight hours are almost identical on October 21 and February 21. Why, then, in New York City is it normally much colder on February 21?

GLOBAL **GEOSCIENCE** WATCH   Go to the Arctic Circle portal. Under Websites and Blogs, access "Arctic Change: A Near-Realtime Arctic Change Indicator Website." On the left-hand index, click on "Clouds." What is the trend in springtime cloudiness (March, April, May) over the Arctic? Would this trend act to reduce or increase the amount of solar radiation reaching the Arctic? What if the same trend were observed in December?

ONLINE RESOURCES

Visit www.cengagebrain.com to view additional resources, including video exercises, practice quizzes, an interactive eBook, and more.

Air Temperature

Contents

We threw a dish of water high into the air, just to see what would happen. Before it hit the ground, it made a hissing noise, froze, and fell as tiny round pellets of ice the size of wheat kernels. Ice became so hard the ax rebounded from it. At such temperatures, metal snapped, wood became petrified, and rubber was just like cement. The dog's leather harness could not bend or it would break . . . Becoming lost was of no concern. As an observer walked along, each breath remained as a tiny motionless mist behind him at head level. These patches of human breath fog remained in the still air for three or four minutes before fading away. One person even found such a trail still marking his path when he returned 15 minutes later. It was easy to freeze your nose without even knowing it.

David Phillips, *Blame it on the Weather*

Our opening vignette is an actual account made by two weather observers on February 3, 1947, at Snag Airport in the Yukon of Canada when the air temperature fell to –63°C (–81°F), the lowest temperature ever measured in North America. It illustrates the profound effect air temperature can have on a variety of things, especially when it drops to extremely low readings.

Air temperature is a critical weather element. The careful recording and application of temperature data are tremendously important to us all. Without accurate information of this type, the work of farmers, weather analysts, power company engineers, and many others would be a great deal more difficult, not to mention knowing how to dress for the day. Therefore, we begin this chapter by examining the daily variation in air temperature. We will answer such questions as why the warmest time of the day is normally in the afternoon, why the coldest is usually in the early morning, and why calm, clear nights are usually colder than windy, clear nights. After we examine the factors that cause temperatures to vary from one place to another, we will look at daily, monthly, and yearly temperature averages and ranges with an eye toward practical applications for everyday living. Near the end of the chapter, we will see how air temperature is measured and how the wind can change our perception of air temperature.

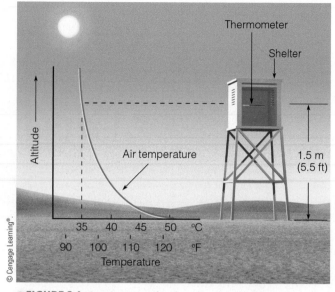

● FIGURE 3.1 On a sunny, calm day, the air near the surface can be substantially warmer than the air a meter or so above the surface, where thermometers are typically located.

Daily Warming and Cooling Air Near the Surface

In Chapter 2, we learned how the sun's energy coupled with the motions of Earth produce the seasons. In a way, each sunny day is like a tiny season as the air goes through a daily cycle of warming and cooling. The air warms during the morning hours as the sun gradually rises higher in the sky, spreading a blanket of heat energy over the ground. The sun reaches its highest point around noon, after which it begins its slow journey toward the western horizon. It is around noon when Earth's surface receives the most intense solar rays. However, somewhat surprisingly, noontime is usually not the warmest part of the day. Rather, the air continues to be heated, often reaching a maximum temperature later in the afternoon. To find out why this *lag in temperature* occurs, we need to examine a shallow layer of air in contact with the ground.

DAYTIME WARMING As the sun rises in the morning, sunlight warms the ground, and the ground warms the air in contact with it by conduction. However, air is such a poor heat conductor that this process only takes place within a few centimeters of the ground. As the sun

rises higher in the sky, the air in contact with the ground becomes even warmer, and, on a windless day, a substantial temperature difference usually exists between the air at ground level and the air directly above it. This explains why runners on a clear, windless, hot summer afternoon may experience air temperatures of over 50°C (122°F) at their feet and only 35°C (95°F) at their waists (see ● Fig. 3.1).

Near the surface, convection begins, and rising air bubbles (thermals) help to redistribute heat. In calm weather, these thermals are small and do not effectively mix the air near the surface. Thus, large vertical temperature differences can exist. On windy days, however, turbulent eddies can mix hot, surface air with the cooler air above. This form of mechanical stirring, sometimes called *forced convection,* helps the thermals to transfer heat away from the surface more efficiently. Therefore, on sunny, windy days the temperature difference between the surface air and the air directly above is not as great as it is on sunny, calm days.

We can now see why the warmest part of the day is usually in the afternoon. Around noon, the sun's rays are most intense. Incoming solar radiation decreases in intensity after noon, but for a time it still exceeds outgoing heat energy from the surface. This situation yields an energy surplus for two to four hours after noon and substantially contributes to a lag between the time of maximum solar heating and the time of maximum air temperature. The warmest part of the day several feet above the surface occurs when incoming energy from the sun is balanced by outgoing energy from Earth's surface (see ● Fig. 3.2).

The exact time of the highest temperature reading varies somewhat. Where the summer sky remains cloud-free all afternoon, the maximum temperature may

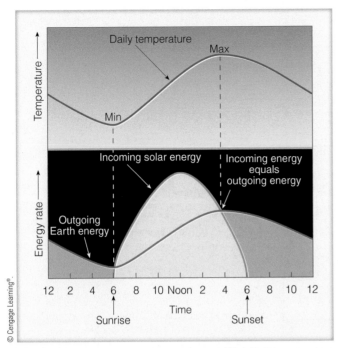

FIGURE 3.2 The daily variation in air temperature is controlled by incoming energy (primarily from the sun) and outgoing energy from Earth's surface. Where incoming energy exceeds outgoing energy (orange shade), the air temperature rises. Where outgoing energy exceeds incoming energy (gray shade), the air temperature falls.

DID YOU KNOW?

During the summer, Death Valley, California, can sizzle at any hour of the day, even at night. On July 12, 2012, the minimum temperature dipped no lower than 107°F, which tied the record for the warmest minimum temperature ever recorded on Earth.

occur sometime between 3 p.m. and 5 p.m. standard time. Where there is afternoon cloudiness or haze, the temperature maximum usually occurs an hour or two earlier. If clouds persist throughout the day, the overall daytime temperatures are usually lower, because clouds reflect a great deal of incoming sunlight.

Adjacent to large bodies of water, cool air moving inland can modify the rhythm of temperature change such that the warmest part of the day occurs at noon or before. In winter, atmospheric storms circulating warm air northward can even cause the highest temperature to occur at night.

Just how warm the air becomes depends on such factors as the type of soil, its moisture content, and vegetation cover. When the soil is a poor heat conductor (as loosely packed sand is), heat energy does not readily transfer into the ground. This allows the surface layer to reach a higher temperature, permitting more energy to warm the air above. On the other hand, if the soil is moist or covered with vegetation, much of the available energy evaporates water, leaving less to heat the air. As you might expect, the highest summer temperatures usually occur over desert regions, where clear skies coupled with low humidities and meager vegetation permit the surface and the air above to warm up rapidly.

Where the air is humid, haze and cloudiness lower the maximum temperature by preventing some of the sun's rays from reaching the ground. In humid Atlanta, Georgia, the average maximum temperature for July is 31.7°C (89°F). In contrast, Phoenix, Arizona—in the desert southwest at the same latitude as Atlanta—experiences an average July maximum of 41.1°C (106°F).

EXTREME HIGH TEMPERATURES Most people are aware of the extreme heat that exists during the summer in the Desert Southwest of the United States. But how hot does it get there? On July 10, 1913, Greenland Ranch in Death Valley, California, reported the highest temperature ever observed in the world: 57°C (134°F) (see Fig. 3.3). Here, air temperatures are persistently hot throughout the summer, with the average maximum for July being 47°C (116°F). During the summer of 1917, there was an incredible period of 43 consecutive days when the maximum temperature reached 120°F or higher.

One of the hottest urban areas in the United States is Palm Springs, California, where the average high temperature during July is 108°F. Another hot city is Yuma, Arizona. Located along the California–Arizona border, Yuma's high temperature during July also averages 108°F.

FIGURE 3.3 Death Valley, California, where the air temperature reached a world-record high of 57°C (134°F) during July 1913.

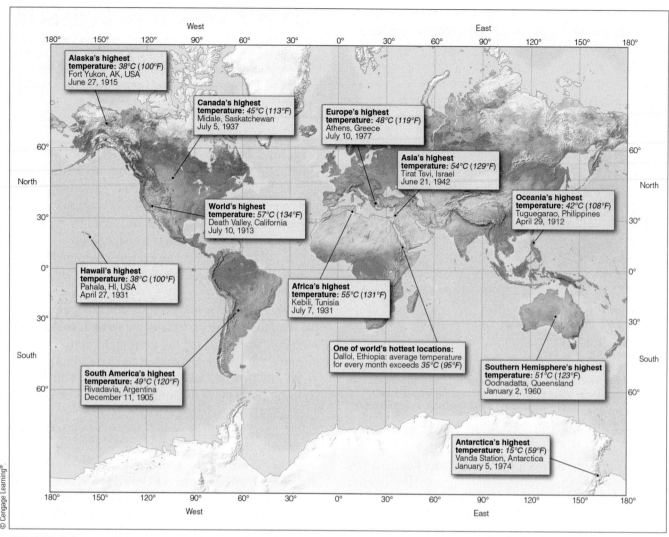

● **FIGURE 3.4** Record high temperatures throughout the world.

In 1937, the high in Yuma reached 100°F or more for 101 consecutive days.

In a more humid climate, the maximum temperature rarely climbs above 41°C (106°F). However, during the record heat wave of 1936, the air temperature reached 121°F near Alton, Kansas. And during the heat wave of 1983, which destroyed about $7 billion in crops and increased the nation's air-conditioning bill by an estimated $1 billion, Fayetteville reported North Carolina's all-time record high temperature when the mercury hit 110°F.

Although Death Valley holds the record for the highest officially measured air temperature in the world, it is not the hottest place on Earth. This distinction may belong to Dallol, Ethiopia. Dallol is located near latitude 12°N, in the hot, dry Danakil Depression (see ● Fig. 3.4). A prospecting company kept weather records at Dallol from 1960 to 1966. During this time, the average daily maximum temperature exceeded 38°C (100°F) every month of the year, except during December and January, when the average maximum lowered to 98°F and 97°F, respectively. On many days, the air temperature exceeded 120°F. The average

annual temperature for the six years at Dallol was 34°C (94°F). In comparison, the average annual temperature in Yuma is 23°C (74°F) and at Death Valley, 24°C (76°F).

On September 23, 1922, the temperature was reported to have reached a scorching 58°C (136°F) in Africa, northwest of Dallol at El Azizia, Libya (32°N). Until recently, this was considered to be the highest temperature recorded on Earth using standard measurement techniques. However, the reading was declared invalid in 2012 after an investigation by a panel of experts sponsored by the World Meteorological Organization (WMO). The panel found several major concerns with the El Azizia reading, including problematic instrumentation, an observer who was likely inexperienced, and asphalt-like material beneath the observing site that did not represent the native desert soil.

Because of these and other factors, the actual temperature in El Azizia may have been 7°C (11°F) cooler than the reported record. Consequently, the 1913 reading in Death Valley was declared to be the world's hottest officially measured temperature, returning it to the position it had held almost a century before.

NIGHTTIME COOLING We know that nights are typically much cooler than days. Nights are cooler because, as the afternoon sun lowers, its energy is spread over a larger area, which reduces the heat available to warm the ground. Look back at Fig. 3.2 on p. 57 and observe that sometime in late afternoon or early evening, Earth's surface and air above begin to lose more energy than they receive; hence, they start to cool.

Both the ground and air above cool by radiating infrared energy, a process called **radiational cooling.** The ground, being a much better radiator than air, is able to cool more quickly. Consequently, shortly after sunset, Earth's surface is slightly cooler than the air directly above it. The surface air transfers some energy to the ground by conduction, which the ground, in turn, quickly radiates away.

As the night progresses, the ground and the air in contact with it continue to cool more rapidly than the air a few meters higher. The warmer upper air does transfer *some* heat downward, a process that is slow due to the air's poor thermal conductivity. However, by late night or early morning, the coldest air is next to the ground, with slightly warmer air above (see ● Fig. 3.5).

This measured increase in air temperature just above the ground is known as a **radiation inversion** because it forms mainly through radiational cooling of the surface. Because radiation inversions occur on most clear, calm nights, they are also called *nocturnal inversions.**

COLD AIR NEAR THE SURFACE A strong radiation inversion occurs when the air near the ground is much colder than the air higher up. Ideal conditions for a strong inversion and, hence, very low nighttime temperatures exist when the air is calm, the night is long, and the air is fairly dry and cloud-free. Let's examine these conditions one by one.

A windless night is essential for a strong radiation inversion because a stiff breeze tends to mix the colder air at the surface with the warmer air above. This mixing, along with the cooling of the warmer air as it comes in contact with the cold ground, causes a vertical temperature profile that is almost isothermal (a constant temperature) in a layer several feet thick. In the absence of wind, the cooler, more-dense surface air does not readily mix with the warmer, less-dense air above, and the inversion is more strongly developed as illustrated in Fig. 3.5.

A long night also contributes to a strong inversion. Generally, the longer the night, the longer the time of radiational cooling and the better are the chances that the air near the ground will be much colder than the air above. Consequently, winter nights provide the best conditions for a strong radiation inversion, other factors being equal.

Finally, radiation inversions are more likely with a clear sky and dry air. Under these conditions, the ground

---

*Radiation (nocturnal) inversions are also called *surface inversions.*

● **FIGURE 3.5** On a clear, calm night, the air near the surface can be much colder than the air above. The increase in air temperature with increasing height above the surface is called a radiation temperature inversion.

is able to radiate its energy to outer space and thereby cool rapidly. With cloudy weather and moist air, much of the outgoing infrared energy is absorbed and radiated back to the surface, retarding the rate of cooling. Also, on humid nights, condensation in the form of fog or dew will release latent heat, which warms the air. So, radiation inversions can occur on any night. But during long winter nights, when the air is still, cloud-free, and relatively dry, these inversions can become strong and deep. On a cold, dry winter night, then, it is common to experience below-freezing temperatures near the ground, and air more than 10°F warmer at your waist. This process explains why ice or frost can appear on Earth's surface even if the official low temperature (measured a few feet above) never dips to 0°C (32°F).

It should now be apparent that how cold the night air becomes depends primarily on the length of the night, the moisture content of the air, cloudiness, and the wind. Even though wind may initially bring cold air into a region, the coldest nights usually occur when the air is clear and relatively calm.

Look back at Fig. 3.2 (p. 57) and observe that the lowest temperature on any given day is usually observed around sunrise. However, the cooling of the ground and surface air may even continue beyond sunrise for a half hour or so, as outgoing energy can exceed incoming energy, because light from the early morning sun passes through a thick section of atmosphere and strikes the ground at a low angle. Consequently, the sun's energy does not effectively warm the surface. Surface heating can be reduced further when the ground is moist and available energy is used for evaporation. Hence, the lowest temperature can occur shortly after the sun has risen.

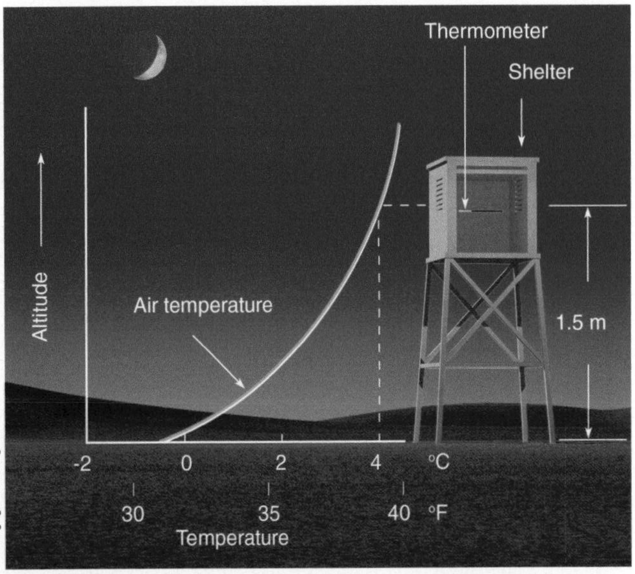

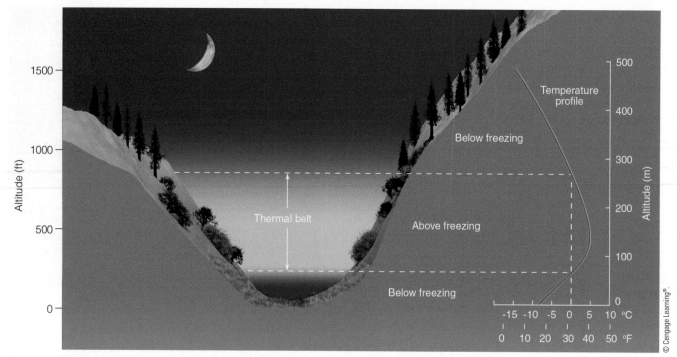

● **FIGURE 3.6** On cold, clear nights, the settling of cold air into valleys makes them colder than surrounding hillsides. The region along the side of the hill where the air temperature is above freezing is known as a *thermal belt*.

On cold nights, plants and certain crops can be damaged by the low temperatures. If the cold occurs over a widespread area for a long enough time to damage certain crops, the extreme cold is called a **freeze**.* A single freeze in California, Texas, or Florida can cause crop losses in the millions or even billions of dollars. In fact, citrus crop losses in Florida during the hard freeze of January 1977 exceeded $2 billion. In California, several freezes during the spring of 2001 caused millions of dollars in damages to California's north coast vineyards, which resulted in higher wine prices. And after extremely warm weather in March 2012 led to early blooming of fruit trees, freezing temperatures in April destroyed nearly half of the apple crop in New York and nearly 90 percent in Michigan. Another widespread freeze across the East and Midwest in April 2007 inflicted more than $2 billion in damage to fruit and field crops.

The coldest air and lowest temperatures are frequently found in low-lying areas, because cold, heavy surface air slowly drains downhill during the night and eventually settles in low-lying basins and valleys. In middle latitudes, the warmer hillsides, called **thermal belts**, are less likely to experience freezing temperatures than the valley below (see ● Fig. 3.6). This phenomenon encourages farmers to plant

on hillsides those trees and sensitive crops that are unable to survive the valley's low temperatures. Moreover, on the valley floor, the cold, dense air is unable to rise, so smoke and other pollutants trapped in this heavy air can restrict visibility. Therefore, valley bottoms are not only colder, but also are more frequently polluted than nearby hillsides.

PROTECTING CROPS FROM THE COLD NIGHT AIR On very cold nights, protect small plants or shrubs by covering them with straw, cloth, or plastic sheeting. This prevents ground heat from being radiated away to the colder surroundings. If you are a household gardener concerned about outside flowers and plants during cold weather, simply wrap them in plastic or cover each with a paper cup.

Fruit trees are particularly vulnerable to cold weather in the spring when they are blossoming. The protection of such trees presents a serious problem to the farmer. Since the lowest temperatures on a clear, still night occur near the surface, the lower branches of a tree are the most susceptible to damage. Therefore, increasing the air temperature close to the ground may prevent damage. A traditional technique for doing this is the use of **orchard heaters,** which warm the air around the trees by setting up convection currents close to the ground. In addition, heat energy radiated from oil- or gas-fired orchard heaters is intercepted by the buds of the trees, which raises their temperature. Orchard heaters that generate smoke, known as "smudge pots," were used for many decades but are now prohibited in most areas due to their effects on local air quality.

Another way to protect trees is to mix the cold air at the ground with the warmer air above, thus raising the

---

*A freeze occurs over a widespread area when the surface air temperature remains below freezing for a long enough time to damage certain agricultural crops. The terms *frost* and *freeze* are often used interchangeably by various segments of society. However, to the grower of perennial crops (such as apples and citrus) who has to protect the crop against damaging low temperatures, it makes no difference if visible "frost" is present or not. The concern is whether or not the plant tissue has been exposed to temperatures equal to or below 32°F. The actual freezing point of the plant, however, can vary because perennial plants can develop hardiness in the fall that usually lasts through the winter, then wears off gradually in the spring.

FIGURE 3.7 Wind machines mix cooler surface air with warmer air above.

● FIGURE 3.8 Ice covers citrus trees in Clermont, Florida, that were sprayed with water during the early morning to protect them from damaging low temperatures that dipped into the 20s (°F) on December 15, 2010.

temperature of the air next to the ground. Such mixing can be accomplished by using **wind machines** (see ●Fig. 3.7), which are power-driven fans that resemble airplane propellers. One significant benefit of wind machines is that they can be thermostatically controlled to turn off and on at prescribed temperatures. Farmers without their own wind machines can rent air mixers in the form of helicopters. Although helicopters are effective in mixing the air, they are expensive to operate.

If sufficient water is available, trees can be protected by irrigation. On potentially cold nights, farmers might flood the orchard. Because water has a high heat capacity, it cools more slowly than dry soil, and so the surface does not become as cold as it would if it were dry. Furthermore, wet soil has a higher thermal conductivity than dry soil. In wet soil heat is conducted upward from subsurface soil more rapidly, which helps to keep the surface warmer.

If the air temperature both at the surface and above fall below freezing, farmers are left with a difficult situation. Wind machines won't help because they would only mix cold air at the surface with the colder air above. Orchard heaters and irrigation are of little value as they would only protect the branches just above the ground. Fortunately, there is one form of protection that does work: An orchard's sprinkling system can be turned on so that it emits a fine spray of water. In the cold air, the water freezes around the branches and buds, coating them with a thin veneer of ice (see ●Fig. 3.8). As long as the spraying continues, the latent heat—given off as the water changes into ice—keeps the ice temperature at 0°C (32°F). The ice acts as a protective coating against the subfreezing air by keeping the buds (or fruit) at a temperature higher than their damaging point. Care must be taken since too much ice can cause the branches to break. The fruit can be saved from the cold air, but the tree itself may be damaged by too much protection. Sprinklers work well when the air is

fairly humid. They do not work well when the air is dry, as a good deal of the water can be lost through evaporation.

So far, we have looked at how and why the air temperature near the ground changes during the course of a 24-hour day. We saw that during the day the air near Earth's surface can become quite warm, whereas at night it can cool off dramatically. ●Figure 3.9 summarizes these observations by illustrating how the average air temperature above the ground can change over a span of 24 hours. Notice in the figure that although the air several feet above the surface both cools and warms, it does so at a slower rate than air at the surface.

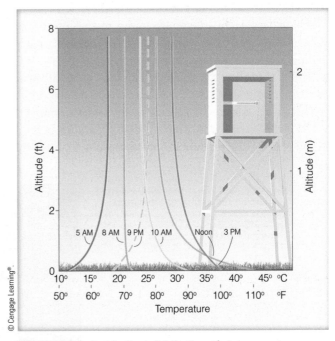

●FIGURE 3.9 An idealized distribution of air temperature above the ground during a 24-hour day. The temperature curves represent the variations in average air temperature above a grassy surface for a mid-latitude city during the summer under clear, calm conditions.

EXTREME LOW TEMPERATURES   One city in the United States that experiences very low temperatures is International Falls, Minnesota, where the average temperature for January is –15°C (4°F). Located several hundred miles to the south of International Falls, Minneapolis–St. Paul, with an average January temperature of –9°C (16°F), is the coldest major urban area in the nation. For duration of extreme cold, Minneapolis reported 186 consecutive hours of temperatures below 0°F during the winter of 1911–1912. Within the forty-eight adjacent states, however, the record for the longest duration of severe cold belongs to Langdon, North Dakota, where the thermometer remained below 0°F for 41 consecutive days (January 11 to February 20, 1936).

The most extensive cold wave in the United States occurred in February 1899. Temperatures during this cold spell fell below 0°F in every existing state, including Florida. This extreme cold event was the first and only of its kind in recorded history. Record temperatures set during this extremely cold outbreak still stand today in many cities of the United States. The official record for the lowest temperature in the forty-eight adjacent states, however, belongs to Rogers Pass, Montana, where on the morning of January 20, 1954, the mercury dropped to –57°C (–70°F). The lowest official temperature for Alaska, –62°C (–80°F), occurred at Prospect Creek on January 23, 1971.

The coldest areas in North America are found in the Yukon and Northwest Territories of Canada. Resolute, Canada (latitude 75°N), has an average temperature of –32°C (–26°F) for the month of January.

The lowest temperatures and coldest winters in the Northern Hemisphere are found in the interior of Siberia and Greenland. For example, the average January temperature in Yakutsk, Siberia (latitude 62°N), is –39°C (–38°F). There, the mean temperature for the entire year is a bitter cold –9°C (16°F). At Eismitte, Greenland, the average temperature for February (the coldest month) is –47°C (–53°F), with the mean annual temperature being a frigid –30°C (–22°F). Even though these temperatures are extremely low, they do not come close to the coldest area of the world: the Antarctic (see ●Fig. 3.10).

At the geographical South Pole, over 9000 feet above sea level, where the Amundsen–Scott scientific station has been keeping records for more than fifty years, the average temperature for the month of July (winter) is –56°C (–69°F) and the mean annual temperature is –46°C (–51°F). The lowest temperature ever recorded there (–83°C or –117°F) occurred under clear skies with a light wind on the morning of June 23, 1983. Cold as it was, it was not the record low for the world. That belongs to the Russian station at Vostok, Antarctica (latitude 78°S), where the temperature plummeted to –89°C (–129°F) on July 21, 1983. (●Figure 3.11 provides more information on record low temperatures throughout the world.)

BRIEF REVIEW

Up to this point we have examined daily temperature variations. Before going on, here is a review of some of the important concepts and facts we have covered:

● During the day, Earth's surface and the air above will continue to warm as long as incoming energy (mainly sunlight) exceeds outgoing energy from the surface.

● At night, Earth's surface cools, mainly by giving up more infrared radiation than it receives—a process called radiational cooling.

● The coldest nights of winter normally occur when the air is calm, fairly dry (a low water-vapor content), and cloud-free.

● The highest temperatures during the day and the lowest temperatures at night are normally observed at Earth's surface.

● Radiation inversions usually exist at night when the air near the ground is colder than the air above.

● The coldest air and lowest nighttime temperatures are normally found in low-lying areas. Surrounding hillsides are usually much warmer than the valley bottoms.

● Farmers use a variety of techniques to protect crops or fruit from damaging low temperatures, including heating the air, mixing the air, irrigating, and spraying water onto trees in below-freezing weather.

●FIGURE 3.10 Antarctica, the coldest continent on Earth, where air temperatures often drop below -100°F.

NASA/James Yungel

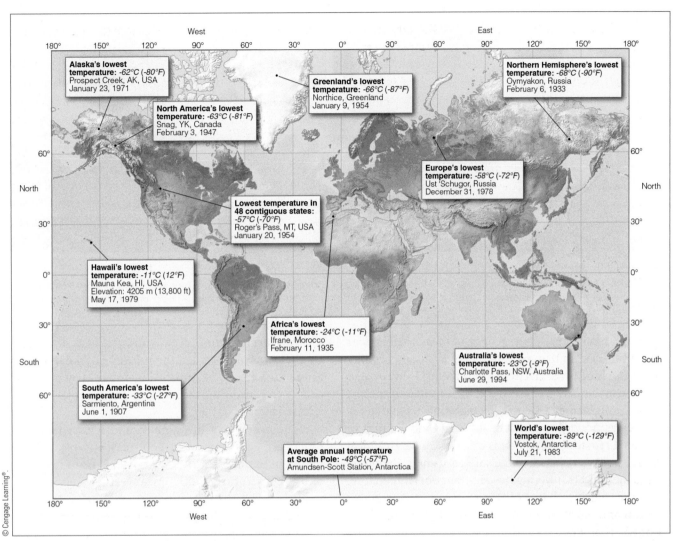

Alaska's lowest
temperature: -62°C (-80°F)
Prospect Creek, AK, USA
January 23, 1971

North America's lowest
temperature: -63°C (-81°F)
Snag, YK, Canada
February 3, 1947

Greenland's lowest
temperature: -66°C (-87°F)
Northice, Greenland
January 9, 1954

Northern Hemisphere's lowest
temperature: -68°C (-90°F)
Oymyakon, Russia
February 6, 1933

Europe's lowest
temperature: -58°C (-72°F)
Ust 'Schugor, Russia
December 31, 1978

Lowest temperature in
48 contiguous states:
-57°C (-70°F)
Roger's Pass, MT, USA
January 20, 1954

Hawaii's lowest
temperature: -11°C (12°F)
Mauna Kea, HI, USA
Elevation: 4205 m (13,800 ft)
May 17, 1979

Africa's lowest
temperature: -24°C (-11°F)
Ifrane, Morocco
February 11, 1935

Australia's lowest
temperature: -23°C (-9°F)
Charlotte Pass, NSW, Australia
June 29, 1994

South America's lowest
temperature: -33°C (-27°F)
Sarmiento, Argentina
June 1, 1907

World's lowest
temperature: -89°C (-129°F)
Vostok, Antarctica
July 21, 1983

Average annual temperature
at South Pole: -49°C (-57°F)
Amundsen-Scott Station, Antarctica

© Cengage Learning®.

● FIGURE 3.11 Record low temperatures throughout the world.

DAILY TEMPERATURE VARIATIONS The greatest variation in daily temperature occurs at Earth's surface. The difference between the daily maximum and minimum temperature—called the **daily (diurnal) range of temperature**—is greatest next to the ground and becomes progressively smaller as we move away from the surface (see ● Fig. 3.12). This daily variation in temperature is also much larger on clear days than on cloudy ones.

The largest diurnal range of temperature occurs on high deserts, where the air is fairly dry, often cloud-free, and there is little water vapor to radiate much infrared energy back to the surface. By day, clear summer skies allow the sun's energy to quickly warm the ground which, in turn, warms the air above to a temperature often exceeding 38°C (100°F). At night, the ground cools rapidly by radiating infrared energy to space, and the minimum temperature in these regions occasionally dips below 7°C (45°F), thus giving an extremely high daily temperature range of more than 31°C (55°F).

Clouds can have a large effect on the daily range in temperature. As we saw in Chapter 2, clouds (especially

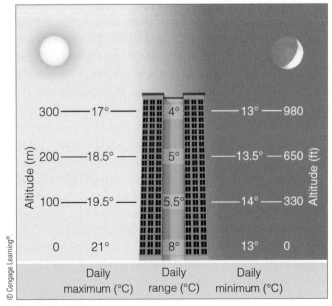

● FIGURE 3.12 The daily range of temperature decreases as we climb away from Earth's surface. Hence, there is less day-to-night variation in air temperature near the top of a high-rise apartment complex than at the ground level.

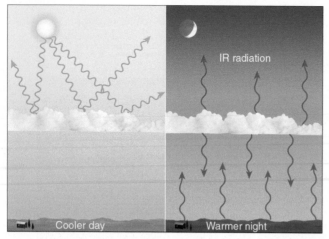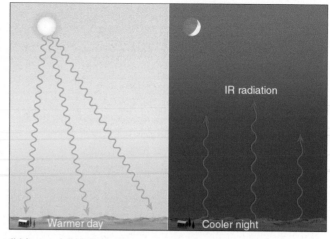

(a) Small daily temperature range

(b) Large daily temperature range

● FIGURE 3.13 (a) Clouds tend to keep daytime temperatures lower and nighttime temperatures higher, producing a small daily range in temperature. (b) In the absence of clouds, days tend to be warmer and nights cooler, producing a larger daily range in temperature.

low, thick ones) are good reflectors of incoming solar radiation, and so they prevent much of the sun's energy from reaching the surface. This effect tends to lower daytime temperatures (see ● Fig. 3.13a). If the clouds persist into the night, they tend to keep nighttime temperatures higher, as clouds are excellent absorbers and emitters of infrared radiation—the clouds actually emit a great deal of infrared energy back to the surface. Clouds, therefore, have the effect of lowering the daily range of temperature. In clear weather (Fig. 3.13b), daytime air temperatures tend to be higher as the sun's rays impinge directly upon the surface, while nighttime temperatures are usually lower due to rapid radiational cooling. Therefore, clear days and clear nights combine to promote a large daily range in temperature.

Humidity can also have an effect on diurnal temperature ranges. For example, in humid regions, the diurnal temperature range is usually small. Here, haze and clouds lower the maximum temperature by preventing some of the sun's energy from reaching the surface. At night, the moist air keeps the minimum temperature high by absorbing Earth's infrared radiation and radiating a portion of it to the ground. An example of a humid city with a small summer diurnal temperature range is Charleston, South Carolina, where the average July maximum temperature is 33°C (91°F), the average minimum is 23°C (73°F), and the diurnal range is only 10°C (18°F).

Cities near large bodies of water typically have smaller diurnal temperature ranges than cities farther inland. This phenomenon is caused in part by the additional water vapor in the air and by the fact that water warms and cools much more slowly than land.

Moreover, cities whose temperature readings are obtained at airports often have larger diurnal temperature ranges than those whose readings are obtained in downtown areas, because nighttime temperatures in cities tend to be warmer than those in outlying rural areas. This nighttime city warmth—called the *urban heat island*—forms as the sun's energy is absorbed by urban structures and concrete; then, during the night, this heat energy is slowly released into the city air.

The average of the highest and lowest temperature observed in a given 24-hour period (typically from midnight to midnight) is known as the **mean (average) daily temperature**. Sometimes weather websites, TV weathercasts, or daily newspapers will provide the mean daily temperature along with the highest and lowest temperatures for the preceding day. The average of the mean daily temperatures for a particular date across a 30-year period gives the average (or "*normal*") temperatures for that date. Currently, the National Weather Service uses the period 1981–2010 to calculate average temperatures. If the average high temperature in a particular city on a certain date is 68°F, does this mean that the high temperature on this date *should be* 68°F? If you are unsure of the answer, read Focus section 3.1.

REGIONAL TEMPERATURE VARIATIONS The main factors that cause variations in temperature from one place to another are called the **controls of temperature**. In the previous chapter, we saw that the greatest factor in determining temperature is the amount of solar radiation that reaches the surface. This amount is determined by the length of daylight hours and the intensity of incoming solar radiation. Both of these factors are a function of latitude; hence, latitude is considered an important control of temperature. The main controls are:

1. latitude
2. land and water distribution
3. ocean currents
4. elevation

When It Comes to Temperature, What's Normal?

When the weathercaster reports that "the normal high temperature for today is 68°F," does this mean that the high temperature on this day is usually 68°F? Or does it mean that we should expect a high temperature near 68°F? Actually, we should expect neither one.

Remember that the word *normal*, refers to weather data averaged over a period of 30 years. For example, ● Fig. 1 shows the high temperatures measured for 30 years (1981 to 2010) in Salt Lake City, Utah, on May 6. The average (mean) high temperature for this period is 68°F; hence, the normal high temperature for this date is 68°F (dashed line). Notice, however, that only on two days during this 30-year period did the high temperature actually measure 68°F (large red dots). In fact, the most common high temperature (called the *mode*) was 58°F, which occurred on three days (blue dots).

So what would be considered a typical high temperature for this date? Actually, any high temperature that lies between about 46°F and 90°F (two standard deviations* on either side of 68°F) would be considered

---

*A standard deviation is a statistical measure of the spread of the data. Two standard deviations for this set of data mean that 95 percent of the time the high temperature occurs between 47°F and 89°F.

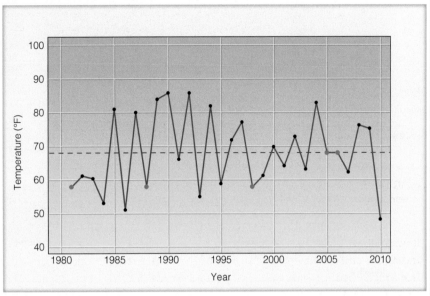

● FIGURE 1 High temperatures measured on May 6 each year from 1981 to 2010 in Salt Lake City, Utah. The dashed line represents the *normal* (average) temperature for the 30-year period.

in the range of typical high temperatures for this date. It would be truly noteworthy, and "abnormal," if the high temperature happened to be 68°F on May 6 every year, even though 68°F is the *normal* (average) high temperature for this 30-year period. While a high temperature of 86°F may be quite warm and a high temperature of 51° may be on the cool side, they are both no more uncommon (unusual) in this period than a high temperature of 68°F. This same type of reasoning applies to *normal rainfall*, as the actual amount of precipitation will likely be greater or less than the 30-year average.

We can obtain a better picture of these controls by examining ● Fig. 3.14 and ● Fig. 3.15, which show the average monthly temperatures throughout the world for January and July. (The average temperature for each month is the average of the daily mean temperatures for that month.) The lines on the map are **isotherms**—lines connecting places that have the same temperature. Because air temperature normally decreases with height, cities at very high elevations are much colder than their sea-level counterparts. Consequently, the isotherms in Fig. 3.14 and Fig. 3.15 are corrected to read at the same horizontal level (sea level) by adding to each station above sea level the equivalent average temperature change with height.*

Figures 3.14 and 3.15 show the importance of latitude on temperature. Notice that on both maps and in both

hemispheres the isotherms are oriented east-west, indicating that locations at the same latitude receive nearly the same amount of solar energy. In addition, the annual solar heat that each latitude receives decreases from low-to-high latitudes; hence, average temperatures in January and July tend to decrease from lower to higher latitudes. But there is a greater variation in solar radiation between low and high latitudes in winter than in summer. Thus, the isotherms in January (during the Northern Hemisphere winter) are closer together (a tighter gradient)* than they are in July. If you travel from New Orleans to Detroit in January, you are

---

*Gradient represents the rate of change of some quantity (in this case, temperature) over a given distance.

DID YOU KNOW?

One of the greatest daily temperature ranges ever recorded in North America (100°F), occurred at Browning, Montana, where the air temperature plummeted from a high of 44°F to a low of −56°F in less than 24 hours on January 23, 1916.

---

*The amount of change is usually less than the standard temperature lapse rate of 3.6°F per 1000 feet (6.5°C per 1000 meters). The reason is that the standard lapse rate is computed for altitudes above Earth's surface in the "free" atmosphere. In the less-dense air at high elevations, the absorption of solar radiation by the ground causes an overall slightly higher temperature than that of the free atmosphere at the same level.

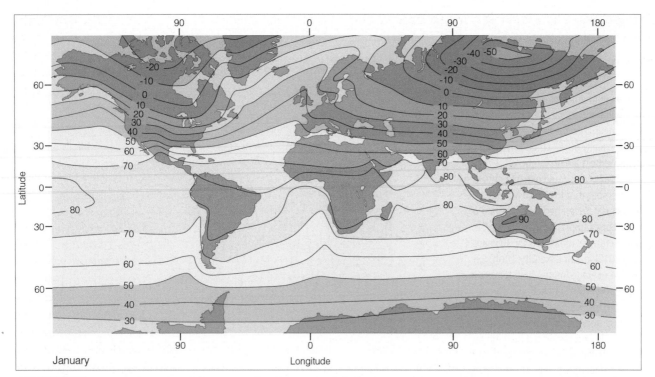

● **FIGURE 3.14** Average air temperature near sea level in January (°F). Temperatures in Central Antarctica are not visible on this map.

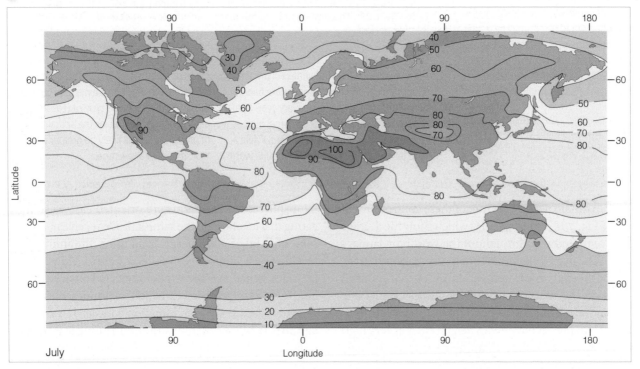

● **FIGURE 3.15** Average air temperature near sea level in July (°F). Temperatures in Central Antarctica are not visible on this map.

more likely to experience greater temperature variations than if you make the same trip in July.

Even though average temperatures tend to decrease from low latitudes toward high latitudes, notice on the July map (Fig. 3.15) that the highest average temperatures do not occur in the tropics, but rather in the subtropical deserts of the Northern Hemisphere. Here, sinking air associated with high-pressure areas generally produces clear skies and low humidity. These conditions, along with a high sun beating down upon a relatively barren landscape, produce scorching heat.

The most extreme cold over land areas in January is across the interior of Siberia (Fig. 3.14). Even colder readings, on average, occur in Antarctica during the dark winter months, as relatively dry air, high elevations, and snow-covered surfaces allow for rapid radiational cooling. Although not shown in Fig. 3.15, the average temperature for the coldest month at the South Pole is below −70°F.

And for absolute cold, the lowest average temperature for any month (−100°F) was recorded at the Plateau Station during July 1968.

So far we've seen that January temperatures in the Northern Hemisphere are much lower in the middle of continents than they are at the same latitude near the oceans. Notice on the July map that the reverse is true. One reason for these temperature differences is the unequal heating and cooling properties of land and water (as discussed in Chapter 2). For one thing, solar energy reaching land is absorbed in a thin layer of soil; reaching water, it penetrates deeply. Because water circulates, it distributes its heat through a much deeper layer. In addition, some of the solar energy striking the water evaporates it rather than heats it.

Another reason for the sharp temperature difference between oceans and interior locations is that it takes a great deal more heat to raise the temperature of a given amount of water by one degree than it does to raise the temperature of the same amount of land by one degree.* Water not only heats more slowly than land, it cools more slowly as well, allowing the oceans to act like huge heat reservoirs. Thus, mid-ocean surface temperatures change relatively little from summer to winter compared to the much larger annual temperature changes over the middle of continents.

As a result of the warming and cooling properties of water, even large lakes can modify the temperature around them. In summer, for example, the Great Lakes remain cooler than the land and refreshing breezes blow inland, bringing relief from the sometimes sweltering heat. As winter approaches, the water cools more slowly than the land. The first blast of cold air from Canada is modified as it crosses the lakes, and so the first freeze is delayed on the eastern shores of Lake Michigan.

Look closely at Figs. 3.14 and 3.15 and notice that in many places the isotherms on both maps tend to bend when they approach an ocean-continent boundary. Such bending of the isotherms along the margin of continents is due in part to the unequal heating and cooling properties of land and water, and in part to *ocean currents*. For example, along the eastern margins of continents warm ocean currents transport warm water toward the poles, whereas, along the western margins, they transport cold water toward the equator. As we will see in Chapter 7, some coastal areas also experience upwelling, which brings cold water from below to the surface.

At any location, the difference in average temperature between the warmest month (often July in the Northern Hemisphere) and coldest month (often January) is called the **annual range of temperature**. As we would expect, annual temperature ranges are largest over interior

continental landmasses and much smaller over larger bodies of water (see ● Fig. 3.16). Moreover, inland cities have larger annual temperature ranges than do coastal cities. Near the equator (because daylight length varies little and the sun is always high in the noon sky), annual temperature ranges are small, usually less than 3°C (5°F). Quito, Ecuador—on the equator at an elevation of 2850 m (9350 ft)—experiences an annual range of less than 1°C. In middle and high latitudes, annual ranges are large, especially in the middle of a continent. Yakutsk, in northeastern Siberia near the Arctic Circle, has an extremely large annual temperature range of 58°C (104°F).

The average temperature of any station for the entire year is the **mean (average) annual temperature,** which represents the average of the twelve monthly average temperatures.* When two cities have the same mean annual temperature, it might at first seem that their temperatures throughout the year are quite similar. However, often this is not the case. For example, San Francisco, California, and Richmond, Virginia, are situated at nearly the same latitude (38°N). Both have similar hours of daylight during the year; both have a mean annual temperature near 15°C (59°F). But here the similarities end. The temperature differences between the two cities are apparent to anyone who has traveled to San Francisco during the summer with a suitcase full of clothes suitable for summer weather in Richmond.

● Figure 3.17 summarizes the average temperatures for San Francisco and Richmond. Notice that the coldest month for both cities is January. Even though January in Richmond averages only about 8°C (14°F) colder than January in San Francisco, people in Richmond awaken to an average January minimum temperature of −3°C (28°F), which is the lowest temperature ever recorded in San Francisco. Trees that thrive in San Francisco's weather would find it difficult to survive a winter in Richmond. So, even though San Francisco and Richmond have the same mean annual temperature, the behavior and range of their temperatures differ greatly.

Applications of Air Temperature Data

There are a variety of applications for the mean daily temperature. An application developed by heating engineers in estimating energy needs is the **heating degree day.** The heating degree day is based on the assumption that people will begin to use their furnaces when the mean daily temperature drops below 65°F. Therefore, heating degree days are determined by subtracting the mean temperature for

---

*The amount of heat needed to raise the temperature of one gram of a substance by one degree Celsius is called *specific heat*. Water has a higher specific heat than does land.

*The mean annual temperature can be obtained by multiplying each of the 12 monthly means by the number of days in that month, adding the 12 numbers, and dividing that total by 12; or by obtaining the sum of the daily means and dividing that total by 365.

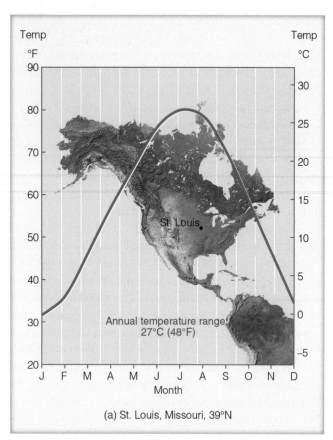

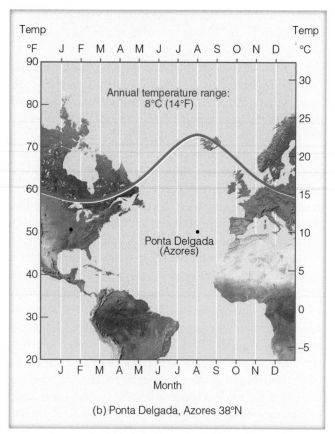

FIGURE 3.16 Monthly temperature data and annual temperature range for (a) St. Louis, Missouri, a city located near the middle of a continent and (b) Ponta Delgada, a city located in the Azores in the Atlantic Ocean. Notice that the annual temperature range is much higher in St. Louis, even though both cities are at the same latitude.

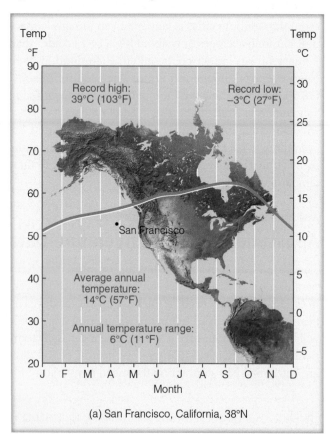

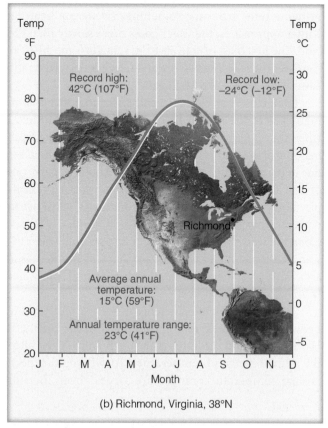

FIGURE 3.17 Temperature data for (a) San Francisco, California (38°N) and (b) Richmond, Virginia (38°N)—two cities with very similar mean temperatures.

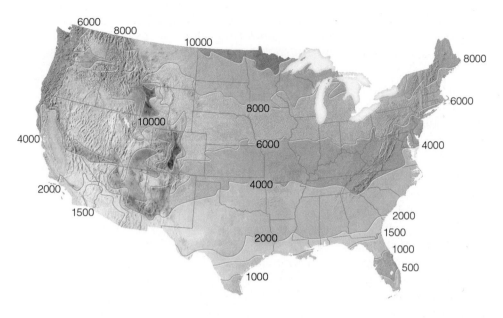

● **FIGURE 3.18** Mean annual total heating degree days across the United States (base 65°F).

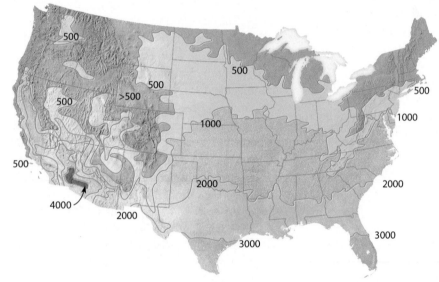

● **FIGURE 3.19** Mean annual total cooling degree days across the United States (base 65°F).

the day from 65°F. Thus, if the mean temperature for a day is 64°F, the heating degree day for this day is 1.*

On days when the mean temperature is above 65°F, there are no heating degree days. Hence, the lower the average daily temperature, the more heating degree days and the greater the predicted consumption of fuel. When the number of heating degree days for a whole year is calculated, the heating fuel requirements for any location can be estimated. ● Figure 3.18 shows the yearly average number of heating degree days in various locations throughout the United States.

As the mean daily temperature climbs above 65°F, people begin to cool their indoor environment. Consequently, an index called the **cooling degree day** is used during warm weather to estimate the energy needed to cool indoor air to a comfortable level. The forecast of mean daily temperature is converted to cooling degree

days by subtracting 65°F from the mean. The remaining value is the number of cooling degree days for that day. For example, a day with a mean temperature of 70°F would correspond to (70 − 65), or 5 cooling degree days. High values indicate warm weather and high power production for cooling (see ● Fig. 3.19).

Knowledge of the number of cooling degree days in an area allows a builder to plan the size and type of equipment that should be installed to provide adequate air conditioning. Also, the forecasting of cooling degree days during the summer gives power companies a way of predicting the energy demand during peak energy periods. A composite of heating plus cooling degree days gives a practical indication of the energy requirements over the year.

Farmers use an index called **growing degree days** as a guide to planting and for determining the approximate dates when a crop will be ready for harvesting. A growing degree day for a particular crop is defined as a day on which the mean daily temperature is one degree above the

*In the United States, the National Weather Service and the Department of Agriculture use degrees Fahrenheit in their computations.

base temperature (also known as the *zero temperature*): the minimum temperature required for growth of that crop. For sweet corn, the base temperature is 50°F and, for peas, it is 40°F.

On a summer day in Iowa, the mean temperature might be 80°F. From ▼ Table 3.1, we can see that, on this day, sweet corn would accumulate (80 – 50), or 30 growing degree days. Theoretically, sweet corn can be harvested when it accumulates a total of 2200 growing degree days. For example, if sweet corn is planted in early April and each day thereafter averages about 20 growing degree days, the corn would be ready for harvest about 110 days later, or around the middle of July.*

At one time, corn varieties were rated in terms of "days to maturity." This rating system was unsuccessful because, in actual practice, corn took considerably longer in some areas than in others. This discrepancy was the reason for defining "growing degree days." In humid Iowa, for example, where summer nighttime temperatures are high, growing degree days accumulate much faster. Consequently, the corn matures in considerably fewer days than in the drier west, where summer nighttime temperatures are lower, and each day accumulates fewer growing degree days. Although moisture and other conditions are not taken into account, growing degree days nevertheless serve as a useful guide in forecasting approximate dates of crop maturity.

Air Temperature and Human Comfort

Probably everyone realizes that the same air temperature can feel different on different occasions. For example, a temperature of 20°C (68°F) on a clear, windless March afternoon in New York City can feel almost balmy after a long, hard winter. Yet this same temperature can feel uncomfortably cool on a summer afternoon in a stiff breeze. The human body's perception of temperature— called **sensible temperature**—obviously changes with varying atmospheric conditions. The reason for these changes is related to how we exchange heat energy with our environment.

The body stabilizes its temperature primarily by converting food into heat *(metabolism)*. To maintain a constant temperature, the heat produced and absorbed by the body must be equal to the heat it loses to its surroundings. There is, therefore, a constant exchange of heat—especially at the surface of the skin—between the body and the environment.

▼ **Table 3.1** Estimated Growing Degree Days for Certain Naturally Grown Agricultural Crops to Reach Maturity

CROP (VARIETY, LOCATION)	BASE TEMPERATURE (°F)	GROWING DEGREE DAYS TO MATURITY
Beans (Snap/South Carolina)	50	1200–1300
Corn (Sweet/Indiana)	50	2200–2800
Cotton (Delta Smooth Leaf/Arkansas)	60	1900–2500
Peas (Early/Indiana)	40	1100–1200
Rice (Vegold/Arkansas)	60	1700–2100
Wheat (Indiana)	40	2100–2400

One way the body loses heat is by emitting infrared energy. But we not only emit radiant energy, we absorb it as well. Another way the body loses and gains heat is by conduction and convection, which transfer heat to and from the body by air motions. On a cold day, a thin layer of warm air molecules forms close to the skin, protecting it from the surrounding cooler air and from the rapid transfer of heat. In cold weather, then, when the air is calm, the temperature we perceive (the *sensible temperature*) is often higher than a thermometer might indicate. (Could the opposite effect occur where the air temperature is very high and a person might feel exceptionally cold? If you are not sure how to answer this question, read Focus section 3.2).

Once the wind starts to blow, the insulating layer of warm air is swept away, and heat is rapidly removed from the skin by the constant bombardment of cold air. When all other factors are the same, the faster the wind blows, the greater the heat loss, and the colder we feel. How cold the wind makes us feel is usually expressed as a **wind-chill index** (WCI).

The modern wind-chill index (see ▼ Table 3.2, p. 71) was formulated in 2001 by a joint action group of the National Weather Service and other agencies. The index takes into account the wind speed at about 1.5 m (5 ft) above the ground (close to where an adult's upper body would be) instead of the 10 m (33 ft) where official wind readings are usually taken. In addition, the index translates the capacity of the air to take heat away from a person's face (the air's cooling power) into a wind-chill equivalent temperature.* For example, notice in Table 3.2 that an air temperature of 10°F with a wind speed of 10 mi/hr produces a

---

*As a point of interest, when the air temperature climbs above 86°F in the Corn Belt of the Midwest, the hot air puts added stress on the growth of the corn. Consequently, the corn grows more slowly. Because of this fact, any maximum temperature over 86°F is reduced to 86°F when computing the mean air temperature for growing degree days.

---

*The wind-chill equivalent temperature formulas are as follows: Wind chill (°F) = 35.74 + 0.6215T − 35.75 ($V^{0.16}$) + 0.4275T ($V^{0.16}$), where T is the air temperature in °F and V is the wind speed in mi/hr. Wind chill (°C) = 13.12 + 0.6215T − 11.37 ($V^{0.16}$) + 0.3965T ($V^{0.16}$), where T is the air temperature in °C, and V is the wind speed in km/hr.

A Thousand Degrees and Freezing to Death

Is there somewhere in our atmosphere where the air temperature can be exceedingly high (say above 500°C or 900°F) yet a person might feel extremely cold? There is such a region, but it's not at Earth's surface.

You may recall from Chapter 1 (see Fig. 1.25, p. 23), that in the upper reaches of our atmosphere (in the middle and upper thermosphere), air temperatures can exceed 500°C. However, a thermometer shielded from the sun in this region of the atmosphere would indicate an extremely low temperature. This apparent discrepancy lies in the meaning of air temperature and how we measure it.

In Chapter 2, we learned that the air temperature is directly related to the average speed at which the air molecules are moving—faster speeds correspond to higher temperatures. In the middle and upper thermosphere at altitudes approaching 300 km (200 mi), air molecules are zipping about at speeds corresponding to extremely high temperatures. However, in order to transfer enough energy to heat something up by conduction (exposed skin or a thermometer bulb), an extremely large number of molecules must collide with the object. In the "thin" air of the upper atmosphere, air molecules are moving extraordinarily fast, but there are simply not enough of them bouncing against the thermometer bulb for it to register a high temperature. In fact, when properly shielded from the sun, the thermometer bulb loses far more energy than it receives and indicates a temperature near absolute zero. This explains why an astronaut, when space walking, will not only survive temperatures exceeding 500°C, but will also feel a profound coldness when shielded from the sun's radiant energy. At these high altitudes, the

NASA

● **FIGURE 2** How can an astronaut survive when the "air" temperature is 1000°C?

traditional meaning of air temperature (that is, regarding how "hot" or "cold" something feels) is no longer applicable.

wind-chill equivalent temperature of –4°F. In other words, the skin of a person's exposed face would lose as much heat in one minute in air with a temperature of 10°F and a wind speed of 10 mi/hr as it would in calm air with a temperature of –4°F. Of course, how cold we feel actually depends on a number of factors, including the fit and type of clothing we wear, the amount of sunshine striking the body, and the actual amount of exposed skin.

▼**Table 3.2** **Wind-Chill Equivalent Temperature (°F). A 20-mi/hr Wind Combined with an Air Temperature of 20°F Produces a Wind-Chill Equivalent Temperature of 4°F.***

	AIR TEMPERATURE (°F)																	
Calm	40	35	30	25	20	15	10	5	0	–5	–10	–15	–20	–25	–30	–35	–40
5	36	31	25	19	13	7	1	–5	–11	–16	–22	–28	–34	–40	–46	–52	–57
10	34	27	21	15	9	3	–4	–10	–16	–22	–28	–35	–41	–47	–53	–59	–66
15	32	25	19	13	6	0	–7	–13	–19	–26	–32	–39	–45	–51	–58	–64	–71
20	30	24	17	11	4	–2	–9	–15	–22	–29	–35	–42	–48	–55	–61	–68	–74
25	29	23	16	9	3	–4	–11	–17	–24	–31	–37	–44	–51	–58	–64	–71	–78
30	28	22	15	8	1	–5	–12	–19	–26	–33	–39	–46	–53	–60	–67	–73	–80
35	28	21	14	7	0	–7	–14	–21	–27	–34	–41	–48	–55	–62	–69	–76	–82
40	27	20	13	6	–1	–8	–15	–22	–29	–36	–43	–50	–57	–64	–71	–78	–84
45	26	19	12	5	–2	–9	–16	–23	–30	–37	–44	–51	–58	–65	–72	–79	–86
50	26	19	12	4	–3	–10	–17	–24	–31	–38	–45	–52	–60	–67	–74	–81	–88
55	25	18	11	4	–3	–11	–18	–25	–32	–39	–46	–54	–61	–68	–75	–82	–89
60	25	17	10	3	–4	–11	–19	–26	–33	–40	–48	–55	–62	–69	–76	–84	–91

(WIND SPEED (MI/HR))

*Dark shaded areas represent conditions where frostbite occurs in 30 minutes or less.

High winds in below-freezing air can remove heat from exposed skin so quickly that the skin may actually freeze and discolor. The freezing of skin, called **frostbite,** usually occurs on the body extremities first because they are the greatest distance from the source of body heat.

In cold weather, wet skin can be a factor in how cold we feel. A cold, rainy day (drizzly or even foggy) often feels colder than a "dry" one because water on exposed skin conducts heat away from the body better than air does. In fact, in cold, wet, and windy weather a person may actually lose body heat faster than the body can produce it. This can even occur in relatively mild weather with air temperatures as high as 10°C (50°F). The rapid loss of body heat can lower the body temperature below its normal level and bring on a condition known as **hypothermia**—the rapid, progressive mental and physical collapse that accompanies the lowering of human body temperature.

The first symptom of hypothermia is exhaustion. If exposure continues, judgment and reasoning power begin to disappear. Prolonged exposure, especially at temperatures near or below freezing, produces stupor, collapse, and death when the internal body temperature drops to about 26°C (79°F).

In cold weather, heat is more easily dissipated through the skin. To counteract this rapid heat loss, the peripheral blood vessels of the body constrict, cutting off the flow of blood to the outer layers of the skin. In hot weather, the blood vessels enlarge, allowing a greater loss of heat energy to the surroundings. Perspiration is also a factor. As evaporation occurs, the skin cools. When the air contains a great deal of water vapor (is very humid) and it is close to being saturated, perspiration does not readily evaporate from the skin. Less evaporational cooling causes most people to feel hotter than it "really" is, and a number of people start to complain about the "heat and humidity." A closer look at how we feel in hot, humid weather will be given in Chapter 4, after we examine the concepts of relative humidity and wet-bulb temperature.

Measuring Air Temperature

Thermometers were developed to measure air temperature. Each thermometer has a definite scale and is calibrated so that a thermometer reading 0°C will denote the same temperature whether it is in Vermont or North Dakota. If a particular temperature reading were to represent different degrees of hot or cold, depending on location, thermometers would be useless.

A very common thermometer for measuring surface air temperature is the **liquid-in-glass thermometer.** This type of thermometer has a glass bulb attached to a sealed, graduated tube about 25 cm (10 in.) long. A very small

opening, or bore, extends from the bulb to the end of the tube. A liquid in the bulb (usually mercury or red-colored alcohol) is free to move from the bulb up through the bore and into the tube. The length of the liquid in the tube represents the air temperature. When the air temperature increases, the liquid in the bulb expands, and rises up the tube. When the air temperature decreases, the liquid contracts, and moves down the tube. Because the bore is very narrow, a small temperature change shows up as a relatively large change in the length of the liquid column.

Maximum and minimum thermometers are liquid-in-glass thermometers used for determining daily maximum and minimum temperatures. The **maximum thermometer** looks like any other liquid-in-glass thermometer with one exception: It has a small constriction within the bore just above the bulb (see ● Fig. 3.20). As the air temperature increases, the mercury expands and freely moves past the constriction up the tube, until the maximum temperature occurs. However, as the air temperature begins to drop, the small constriction prevents the mercury from flowing back into the bulb. Thus, the end of the stationary mercury column indicates the maximum temperature for the day. The mercury will stay at this position until either the air warms to a higher reading or the thermometer is reset by whirling it on a special holder and pivot. Usually, the whirling is sufficient to push the mercury back into the bulb past the constriction until the end of the column indicates the present air temperature.*

A **minimum thermometer** measures the lowest temperature reached during a given period. Most minimum thermometers use alcohol as a liquid, since it freezes at a temperature of –130°C compared to –39°C

---

*Liquid-in-glass thermometers that measure body temperature are maximum thermometers, which is why they are shaken both before and after you take your temperature.

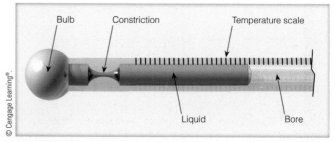

● **FIGURE 3.20** A section of a maximum thermometer.

for mercury. The minimum thermometer is similar to other liquid-in-glass thermometers except that it contains a small barbell-shaped index marker in the bore (see • Fig. 3.21). The small index marker is free to slide back and forth within the liquid. It cannot move out of the liquid because the surface tension at the end of the liquid column (the *meniscus*) holds it in.

A minimum thermometer is mounted horizontally. As the air temperature drops, the contracting liquid moves back into the bulb and brings the index marker down the bore with it. When the air temperature stops decreasing, the liquid and the index marker stop moving down the bore. As the air warms, the alcohol expands and moves freely up the tube past the stationary index marker. Because the index marker does not move as the air warms, the minimum temperature is read by observing the upper end of the marker.

To reset a minimum thermometer, simply tip it upside down. This allows the index marker to slide to the upper end of the alcohol column, which is indicating the current air temperature. The thermometer is then remounted horizontally, so that the marker will move toward the bulb as the air temperature decreases.

Highly accurate temperature measurements can be made with **electrical thermometers.** One type of electrical thermometer is the *electrical resistance thermometer.* This does not actually measure air temperature; rather, it measures the resistance of a wire, usually platinum or nickel, whose resistance increases as the temperature increases. An electrical meter measures the resistance, and is calibrated to represent air temperature.

Another type of electrical thermometer is the *thermistor.* Made of ceramic material, its electrical resistance changes as the air temperature changes. A thermistor is the temperature-measuring device used in the radiosonde, the instrument that measures air temperature from the surface up to an altitude near 30 km. (For additional information on the radiosonde, read Focus section 1.2 in Chapter 1, p. 22.)

Electrical resistance thermometers are the type used in the measurement of air temperature at the over 900 fully

• **FIGURE 3.22** The instruments that comprise the ASOS system. The max-min temperature shelter is the middle box.

automated surface weather stations (known as *ASOS* for *Automated Surface Observing System*) that exist at airports and military facilities throughout the United States. (See • Fig. 3.22.) They have replaced many of the liquid-in-glass thermometers formerly in use.

At this point it should be noted that the replacement of liquid-in-glass thermometers with electrical thermometers has raised concern among climatologists. For one thing, the response of the electrical thermometers to temperature change is faster. Thus, electrical thermometers might reach a brief extreme reading that could have been missed by the slower-responding liquid-in-glass thermometer. In addition, many temperature readings that were previously taken at airport weather offices are now taken at ASOS locations situated near or between runways at the airport. This change in instrumentation and relocation of the measurement site can sometimes introduce a small, but significant, temperature change at the reporting station. To reduce the impact of such temperature changes, the United States has created a Climate Reference Network of about 100 weather stations. These stations are carefully placed, calibrated, and maintained to produce a consistent, accurate long-term reading of air temperature.

Air temperature can also be obtained with instruments called *infrared sensors*, or **radiometers.** Radiometers do not measure temperature directly; rather, they measure emitted radiation (usually infrared). By measuring both the intensity of radiant energy and the wavelength of

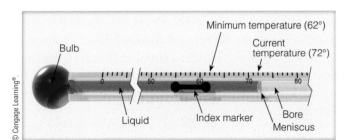

• **FIGURE 3.21** A section of a minimum thermometer showing both the current air temperature and the minimum temperature in °F.

Why Thermometers Must Be Read in the Shade

When we measure air temperature with a common liquid thermometer, an incredible number of air molecules bombard the bulb, transferring energy either to or away from it. When the air is warmer than the thermometer, the liquid gains energy, expands, and rises up the tube; the opposite will happen when the air is colder than the thermometer. The liquid stops rising (or falling) when equilibrium between incoming and outgoing energy is established. At this point, we can read the temperature by observing the height of the liquid in the tube.

It is *impossible* to measure *air temperature* accurately in direct sunlight because the thermometer absorbs radiant energy from the sun in addition to energy from the air molecules. The thermometer gains energy at a much faster rate than it can radiate it away, and the liquid keeps expanding and rising until there is equilibrium between incoming and outgoing energy. Because of the direct absorption of solar energy, the level of the liquid in the thermometer indicates a temperature *much* higher than the actual air temperature. Thus, a statement that says, "Today the air temperature measured 100 degrees in the sun," has no meaning; a thermometer must be kept in a shady place to measure the temperature of the air accurately.

© Ross DePaola

● **FIGURE 3** Instrument shelters such as the one shown here serve as a shady place for thermometers. Thermometers inside shelters measure the temperature of the air; whereas thermometers held in direct sunlight do not.

maximum emission of a particular gas (either water vapor or carbon dioxide), radiometers in orbiting satellites are now able to obtain temperature measurements at selected levels in the atmosphere.

A **bimetallic thermometer** consists of two different pieces of metal (usually brass and iron) welded together to form a single strip. As the temperature changes, the brass expands more than the iron, causing the strip to bend. The small amount of bending is amplified through a system of levers to a pointer on a calibrated scale. The bimetallic thermometer is usually the temperature-sensing part of the **thermograph,** an instrument that measures and records temperature (see ● Fig. 3.23).

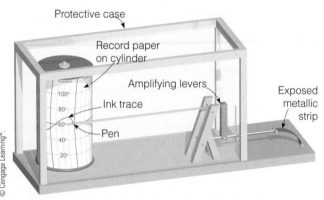
© Cengage Learning®.

● **FIGURE 3.23** The thermograph with a bimetallic thermometer.

Thermographs are gradually being replaced with *data loggers.* These small instruments have a thermistor connected to a circuit board inside the logger. A computer programs the interval at which readings are taken. The loggers are not only more responsive to air temperature than are thermographs, they also are less expensive.

Chances are, you may have heard someone exclaim something like, "Today the thermometer measured 100 degrees in the shade!" Does this mean that the air temperature is sometimes measured in the sun? If you are unsure of the answer, read Focus section 3.3 before reading the next section on instrument shelters.

Thermometers and other instruments are usually housed in an **instrument shelter.** The shelter completely encloses the instruments, protecting them from rain, snow, and the sun's direct rays. It is painted white to reflect sunlight, faces north to avoid direct exposure to sunlight, and has louvered sides, so that air is free to flow through it. This construction helps to keep the air inside the shelter at the same temperature as the air outside.

The thermometers inside a standard shelter are mounted about 1.5 to 2 m (5 to 6 ft) above the ground. As we saw in an earlier section, on a clear, calm night the air at ground level may be much colder than the air at the level of the shelter. As a result, on clear winter mornings it is possible to see ice or frost on the ground even though the minimum thermometer in the shelter did not reach the freezing point.

Many older instrument shelters (such as the one shown in Focus Fig. 3, p. 74) have been replaced by the *Max-Min Temperature* Shelter of the ASOS system (the middle white box in Fig. 3.22, p. 73). The shelter is mounted on a pipe, and wires from the electrical temperature sensor inside are run to a building. A readout inside the building displays the current air temperature and stores the maximum and minimum temperatures for later retrieval.

Because air temperatures vary considerably above different types of surfaces, shelters are placed over grass where possible, to ensure that the air temperature is measured at the same elevation over the same type of surface. Unfortunately, some shelters are placed on asphalt, others sit on concrete, while others are located on the tops of tall buildings, making it difficult to compare air temperature measurements from different locations. In fact, if either the maximum or minimum air temperature in your area seems suspiciously different from those of nearby towns, find out where the instrument shelter is situated.

SUMMARY

The daily variation in air temperature near Earth's surface is controlled mainly by the input of energy from the sun and the output of energy from the surface. On a clear, calm day, the surface air warms, as long as heat input (mainly sunlight) exceeds heat output (mainly convection and radiated infrared energy). The surface air cools at night, as long as heat output exceeds input. Because the ground at night cools more quickly than the air above, the coldest air is normally found at the surface where a radiation inversion usually forms. When the air temperature in agricultural areas drops to dangerously low readings, fruit trees and grape vineyards can be protected from the cold by a variety of means, from mixing the air to spraying the trees and vines with water.

The greatest daily variation in air temperature occurs at Earth's surface. Both the diurnal and annual range of temperature are greater in dry climates than in humid ones. Even though two cities may have similar average annual temperatures, the range and extreme of their temperatures can differ greatly. Temperature information influences our lives in many ways, from deciding what clothes to take on a trip to providing critical information for energy-use predictions and agricultural planning. We reviewed some of the many types of thermometers in use: maximum, minimum, bimetallic, electrical, radiometer. Those designed to measure air temperatures near the surface are housed in instrument shelters to protect them from direct sunlight and precipitation.

KEY TERMS

The following terms are listed (with corresponding page numbers) in the order they appear in the text. Define each. Doing so will aid you in reviewing the material covered in this chapter.

radiational cooling, 59
radiation inversion, 59
freeze, 60
thermal belt, 60
orchard heater, 60
wind machine, 61
daily (diurnal) range of temperature, 63
mean (average) daily temperature, 64
controls of temperature, 64
isotherm, 65

annual range of temperature, 67
mean (average) annual temperature, 67
heating degree day, 67
cooling degree day, 69
growing degree day, 69
sensible temperature, 70
wind-chill index, 70
frostbite, 72
hypothermia, 72

liquid-in-glass thermometer, 72
maximum thermometer, 72
minimum thermometer, 72
electrical thermometer, 73
radiometer, 73

bimetallic thermometer, 74
thermograph, 74
instrument shelter, 74

QUESTIONS FOR REVIEW

1. Explain why the warmest time of the day is usually in the afternoon, even though the sun's rays are most direct at noon.
2. On a calm, sunny day, why is the air next to the ground normally much warmer than the air several feet above?
3. Explain how incoming energy and outgoing energy regulate the daily variation in air temperature.
4. Draw a vertical profile of air temperature from the ground to an elevation of 3 m (10 ft) on (a) a clear, windless afternoon and (b) an early morning just before sunrise. Explain why the temperature curves are different.
5. Explain how radiational cooling at night produces a radiation temperature inversion.
6. What weather conditions are best suited for the formation of a cold night and a strong radiation inversion?
7. Explain why thermal belts are found along hillsides at night.
8. List four measures farmers use to protect their crops against the cold. Explain the physical principle behind each method.
9. Why are the lower branches of trees most susceptible to damage from low temperatures?
10. Describe each of the controls of temperature.
11. Look at Fig. 3.14, p. 66 (temperature map for January) and explain why the isotherms dip southward (equatorward) over the Northern Hemisphere continents.
12. During the winter, white frost can form on the ground when the minimum thermometer in an instrument shelter indicates a low temperature above freezing. Explain.
13. Why do the first freeze in autumn and the last freeze in spring occur in bottomlands?
14. Explain why the daily range of temperature is normally greater (a) in drier regions than in humid regions and (b) on clear days than on cloudy days.
15. Why are the largest annual ranges of temperatures normally observed over continents away from large bodies of water?

16. Two cities have the same mean annual temperature. Explain why this fact does not mean that their temperatures throughout the year are similar.

17. What is a heating degree day? A cooling degree day? How are these units calculated?

18. During a cold, calm, sunny day, why do we usually feel warmer than a thermometer indicates?

19. (a) Assume the wind is blowing at 30 mi/hr and the air temperature is 5°F. Determine the wind-chill equivalent temperature using Table 3.2, p. 71.
    (b) Under the conditions listed in (a) above, explain why an ordinary thermometer would measure a temperature of 5°F, and not a much lower temperature.

20. What atmospheric conditions can bring on hypothermia?

21. Someone says, "Today, the air temperature measured 99°F in the sun." Why does this statement have no meaning?

22. Explain why the minimum thermometer is the one with a small barbell-shaped index marker in the bore.

23. Briefly describe how the following thermometers measure air temperature:
    (a) liquid-in-glass
    (b) bimetallic
    (c) radiometer
    (d) electrical

QUESTIONS FOR THOUGHT AND EXPLORATION

1. How do you think a thick layer of low clouds would affect the lag in daily temperature?

2. Which location is most likely to have the greater daily temperature range: a tropical rain forest near the equator or a desert site in Nevada? Explain.

3. Explain why putting on a heavy winter jacket would be effective in keeping you warm, even if the jacket had been outside in sub-freezing temperatures for several hours.

4. Why is the air temperature displayed on a bank or building marquee usually inaccurate?

5. If you were forced to place a meteorological instrument shelter over asphalt rather than over grass, what modification(s) would you have to make so that the temperature measurements inside the shelter were more representative of the actual air temperature?

6. The average temperature in San Francisco, California, for December, January, and February is 52°F. During the same three-month period the average temperature in Richmond, Virginia, is 40°F. Yet, San Francisco and Richmond have nearly the same yearly total of heating degree days. Explain why. (Hint: See Fig. 3.17, p. 68.)

7. How would the lag in daily temperature experienced over land compare to the daily temperature lag over water?

8. In Pennsylvania and New York, wine grapes are planted on the sides of hills rather than in valleys. Explain why this practice is so common in these areas.

9. Suppose peas are planted in Indiana on May 1. If the peas need 1200 growing degree days before they can be picked, and if the average maximum temperature for May and June is 80°F and the average minimum is 60°F, on about what date will the peas be ready to pick? (Assume a base temperature of 55°F.)

ONLINE RESOURCES

Visit www.cengagebrain.com to view additional resources, including video exercises, practice quizzes, an interactive eBook, and more.

CHAPTER 4

Humidity, Condensation, and Clouds

Contents

It's 9 a.m. on April 26, 2005, in Bangkok, Thailand, one of the hottest and most humid major cities in the world. The streets are clogged with traffic and on this hot, muggy morning perspiration streams down the faces of anxious people struggling to get to work. What makes this day so eventful is that a rare weather event is occurring: Presently, the air temperature is 91°F, the relative humidity is 94 percent, and the heat index, which tells us how hot it really feels, is a staggering 130°F.

We know from Chapter 1 that, in our atmosphere, the concentration of the invisible gas water vapor is normally less than a few percent of all the atmospheric molecules. Yet water vapor is exceedingly important, for it transforms into cloud droplets and ice crystals—particles that grow in size and fall to Earth as precipitation. The term *humidity* can describe the amount of water vapor in the air. To most of us, a moist day suggests high humidity. However, there is usually more water vapor in the hot, "dry" air of the Sahara Desert than in the cold, "damp" winter air of New England, which raises an interesting question: Does the desert air have a higher humidity? As we will see later in this chapter, the answer to this question is both yes and no, depending on the type of humidity we mean.

So that we may better understand the concept of humidity, we will begin this chapter by examining the circulation of water in the atmosphere. Then we will look at different ways to express humidity. Near the end of the chapter, we will investigate various forms of condensation, including dew, fog, and clouds.

Circulation of Water in the Atmosphere

Within the atmosphere, there is an unending circulation of water. Since the oceans occupy over 70 percent of Earth's surface, we can think of this circulation as beginning over the ocean. Here, the sun's energy transforms enormous quantities of liquid water into water vapor in

a process called **evaporation.** Winds then transport the moist air to other regions, where the water vapor changes back into liquid (or ice), forming clouds, in a process called **condensation.** Under certain conditions, the liquid cloud particles (or solid ice crystals) may grow in size and fall to the surface as **precipitation**—rain, snow, or hail. If the precipitation falls into an ocean, the water begins its cycle again. If, on the other hand, the precipitation falls on a continent, a great deal of the water returns to the ocean only after a complex journey. This cycle of moving and transforming water molecules from liquid to vapor and back to liquid again is called the **hydrologic** (water) **cycle.** In the most simplistic form of this cycle, water molecules travel from ocean to atmosphere to land and then back to the ocean.

● Figure 4.1 illustrates the complexities of the hydrologic cycle. For example, before falling rain ever reaches the ground, a portion of it evaporates back into the air. Some of the precipitation may be intercepted by vegetation, where it evaporates or drips to the ground long after a storm has ended. Once on the surface, a portion of the water soaks into the ground by percolating downward through small openings in the soil and rock, forming groundwater that can be tapped by wells. What does not soak into the ground collects in puddles of standing water or runs off into streams and rivers, which find their way back to the ocean. Even the underground water moves slowly and eventually surfaces, only to evaporate or be carried seaward by rivers.

Over land, a considerable amount of water vapor is added to the atmosphere through evaporation from the

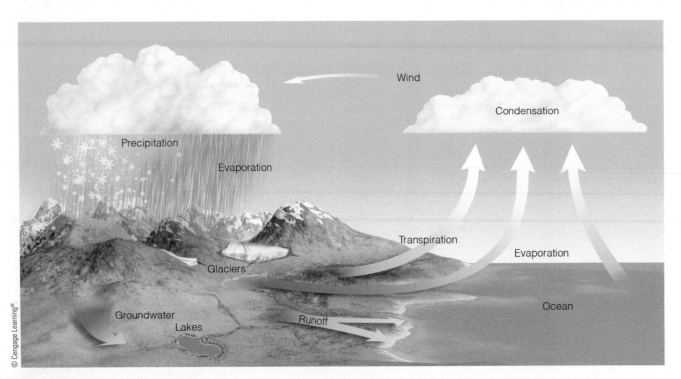

● **FIGURE 4.1** The hydrologic cycle.

soil, lakes, and streams. Even plants give up moisture by a process called *transpiration*. The water absorbed by a plant's root system moves upward through the stem and emerges from the plant through numerous small openings on the underside of the leaf. In all, evaporation and transpiration from continental areas amount to only about 15 percent of the nearly 1.5 quintillion (10^{18}) gallons of water vapor that annually evaporate into the atmosphere; the remaining 85 percent evaporates from the oceans. If all of this water vapor were to suddenly condense and fall as rain, it would be enough to cover the entire globe with about 2.5 centimeters (or 1 inch) of water. The total mass of water vapor stored in the atmosphere at any moment adds up to only a little over a week's supply of the world's precipitation. Since this amount varies only slightly from day to day, the hydrologic cycle is exceedingly efficient in circulating water in the atmosphere.

Evaporation, Condensation, and Saturation

To obtain a slightly different picture of water in the atmosphere, suppose we examine water in a beaker similar to the one shown in ● Fig. 4.2a. If we were able to magnify the surface water about a billion times, we would see water molecules fairly close together, jiggling, bouncing, and moving about. We would also see that the molecules are not all moving at the same speed—some are moving much faster than others. Recall from Chapter 2 that the *temperature* of the water is a measure of the average motion of its molecules. At the surface, molecules with enough speed (and traveling in the right direction) will occasionally break away from the liquid surface and enter into the air above. These molecules, changing *from the liquid state into the vapor state*, are *evaporating*. While some water molecules are leaving the liquid, others are returning. Those returning are *condensing*, as they are changing from a *vapor state to a liquid state*.

When a cover is placed over the dish (Fig. 4.2b), after a while the total number of molecules escaping from the liquid (evaporating) is balanced by the number returning (condensing). When this condition exists, the air is said to be **saturated** with water vapor. Under saturated conditions, for every molecule that evaporates, one must condense, and no net loss of liquid or vapor molecules results.

If we remove the cover and blow across the top of the water, some of the vapor molecules already in the air above will be blown away, creating a difference between the actual number of vapor molecules and the total number required for *saturation*. This helps prevent saturation from

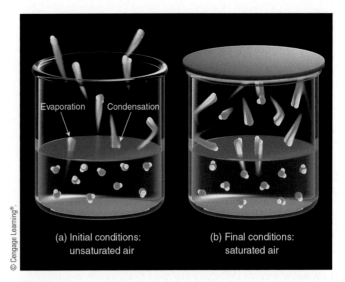

● **FIGURE 4.2** (a) Water molecules at the surface of the water are evaporating (changing from liquid into vapor) and condensing (changing from vapor into liquid). Since more molecules are evaporating than condensing, net evaporation is occurring. (b) When the number of water molecules escaping from the liquid (evaporating) balances those returning (condensing), the air above the liquid is saturated with water vapor. (For clarity, only water molecules are illustrated.)

occurring and allows for a greater amount of evaporation, just as if wind were blowing atop the water surface. Wind, therefore, enhances evaporation.

The temperature of the water also influences evaporation. All else being equal, warm water will evaporate more readily than cool water, the reason being that when the water molecules are heated, they will speed up. At higher temperatures, a greater fraction of the molecules have sufficient speed to break through the surface tension of the water and zip off into the air above. In other words, the warmer the water, the greater the rate of evaporation.

If we could examine the air above the water in Fig. 4.2b, we would observe the water vapor molecules freely darting about and bumping into each other as well as into neighboring molecules of oxygen and nitrogen. We would also observe that mixed in with all of the air molecules are microscopic bits of dust, smoke, and salt from ocean spray. Since many of these serve as surfaces on which water vapor may condense, they are called **condensation nuclei.** In the warm air above the water, fast-moving vapor molecules strike the nuclei with such impact that they simply bounce away (see ● Fig. 4.3a). However, if the air is chilled (Fig. 4.3b), the molecules move more slowly and are more apt to stick and condense to the nuclei. When many billions of these water vapor molecules condense onto the nuclei, tiny liquid cloud droplets form.

We can see, then, that condensation is more likely to happen as the air cools and the speed of the water vapor molecules decreases. As the air temperature increases, condensation is less likely because most of the water vapor molecules have sufficient speed (sufficient energy)

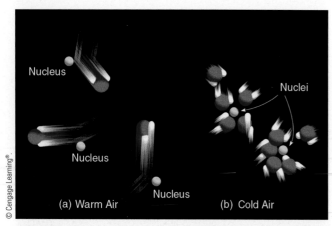

● **FIGURE 4.3** Condensation is more likely to occur as the air cools. (a) In the warm air, fast-moving H₂O vapor molecules tend to bounce away after colliding with nuclei. (b) In the cool air, slow-moving vapor molecules are more likely to join together on nuclei. The condensing of many billions of water molecules produces tiny liquid water droplets.

to remain as a vapor. As we will see in this and other chapters, *condensation occurs primarily when the air is cooled.*

Even though condensation is more likely to occur when the air cools, it is important to note that no matter how cold the air becomes, there will always be a few water vapor molecules with sufficient speed (sufficient energy) to remain as a vapor. It should be apparent, then, that with the same number of water vapor molecules in the air, saturation is more likely to occur in cool air than in warm air. This fact often leads to the statement that "warm air can hold more water vapor molecules before becoming saturated than can cold air" or, simply, "warm air has a greater

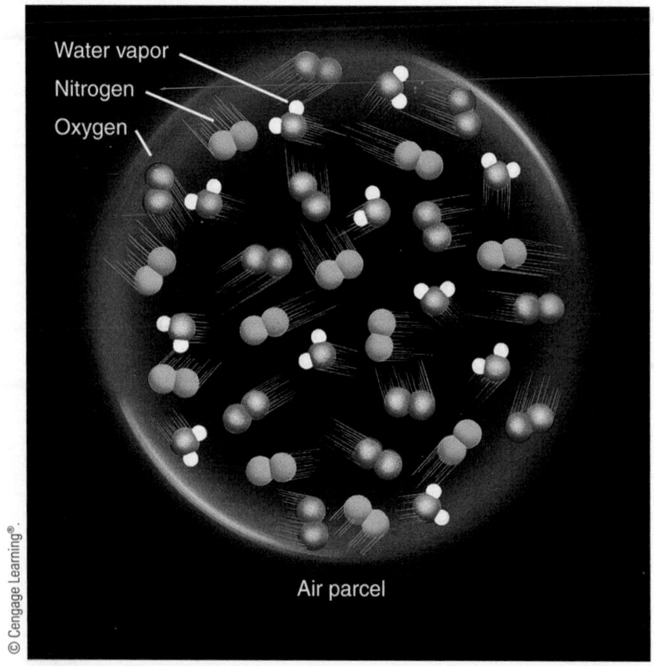

● **FIGURE 4.4** The water vapor content (humidity) inside this air parcel can be expressed in a number of ways.

capacity for water vapor than does cold air." At this point, it is important to realize that although these statements are correct, the use of such words as "hold" and "capacity" are misleading when describing water vapor content, as air does not really "hold" water vapor in the sense of making "room" for it.

Humidity

Humidity refers to any one of a number of ways of specifying the amount of water vapor in the air. Most people have heard of *relative humidity*, which we will examine later, but there are several other ways to express atmospheric water vapor content.

Imagine, for example, that we enclose a volume of air (about the size of a large balloon) in a thin elastic container—a *parcel*—as illustrated in ● Fig. 4.4. If we extract the water vapor from the parcel, we can specify the humidity in the following ways:

1. We can compare the weight (mass) of the water vapor with the volume of air in the parcel and obtain the *water vapor density,* or *absolute humidity.*
2. We can compare the weight (mass) of the water vapor in the parcel with the total weight (mass) of all the air in the parcel (including vapor) and obtain the *specific humidity.*
3. Or, we can compare the weight (mass) of the water vapor in the parcel with the weight (mass) of the remaining dry air and obtain the *mixing ratio.*

Absolute humidity is normally expressed as grams of water vapor per cubic meter of air (g/m³), whereas both specific humidity and mixing ratio are expressed as grams of water vapor per kilogram of air (g/kg).

Look at Fig. 4.4 and notice that we could also express the humidity of the air in terms of *water vapor pressure*—the push (force) that the water vapor molecules are exerting against the inside walls of the parcel.

VAPOR PRESSURE Suppose the air parcel in Fig. 4.4 is near sea level and the air pressure inside the parcel is 1000 millibars (mb).* The total air pressure inside the parcel is due to the collision of all the molecules against the walls of the parcel. In other words, the total pressure inside the parcel is equal to the sum of the pressures of the individual gases. Since the total pressure inside the parcel is 1000 millibars, and the gases inside include nitrogen (78 percent), oxygen (21 percent), and water vapor (1 percent), the partial pressure exerted by nitrogen would then be 780 mb and, by oxygen, 210 mb. The partial

---

*You may recall from Chapter 1 that the millibar is the unit of pressure most commonly found on surface weather maps, and that it expresses atmospheric pressure as a force over a given area.

pressure of water vapor, called the **actual vapor pressure,** would be only 10 mb (one percent of 1000).* It is evident, then, that because the number of water vapor molecules in any volume of air is small compared to the total number of air molecules in the volume, the actual vapor pressure is normally a small fraction of the total air pressure.

Everything else being equal, the more air molecules in a parcel, the greater the total air pressure. When you blow up a balloon, you increase its pressure by putting in more air. Similarly, an increase in the number of water vapor molecules will increase the total vapor pressure. Hence, the actual vapor pressure is a fairly good measure of the total amount of water vapor in the air: *High actual vapor pressure indicates large numbers of water vapor molecules, whereas low actual vapor pressure indicates comparatively small numbers of vapor molecules.*

Actual vapor pressure indicates the air's total water vapor content, whereas **saturation vapor pressure** describes how much water vapor is necessary to make the air saturated at any given temperature.** Put another way, *saturation vapor pressure is the pressure that the water vapor molecules would exert if the air were saturated with vapor at a given temperature.*

We can obtain a better picture of the concept of saturation vapor pressure by imagining molecules evaporating from a water surface. Look back at Fig. 4.2b and recall that when the air is saturated, the number of molecules escaping from the water's surface equals the number returning. Since the number of "fast-moving" molecules increases as the temperature increases, the number of water molecules escaping per second increases also. In order to maintain equilibrium, this situation causes an increase in the number of water vapor molecules in the air above the liquid. Consequently, *at higher air temperatures, it takes more water vapor to saturate the air.* And more vapor molecules exert a greater pressure. *Saturation vapor pressure, then, depends primarily on the air temperature.* From the graph in ●Fig. 4.5, we can see that at 10°C, the saturation vapor pressure is about 12 mb, whereas at 30°C, it is about 42 mb.

RELATIVE HUMIDITY While relative humidity is the most commonly used way of describing atmospheric moisture, it is also, unfortunately, the most misunderstood. The concept of relative humidity may at first seem confusing because it does *not* indicate the actual amount of water vapor in the air. Instead, it tells us how close the air is to being saturated. The **relative humidity** (RH) is the *ratio of the amount of water vapor actually in the air to*

---

*When we use the percentages of various gases in a volume of air, these percentages only give us an approximation of the actual vapor pressure. The point here is that, near Earth's surface, the actual vapor pressure is often close to 10 mb.

**When the air is saturated, the amount of water vapor is the maximum possible at the existing temperature and pressure.

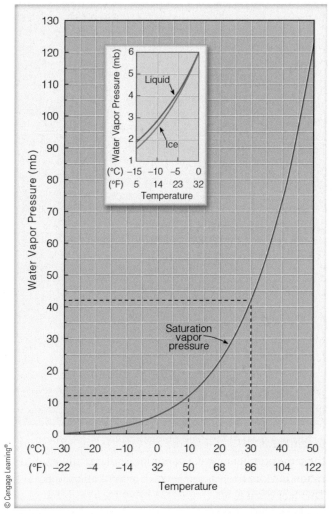

●**FIGURE 4.5** Saturation vapor pressure increases with increasing temperature. At a temperature of 10°C, the saturation vapor pressure is about 12 mb, whereas at 30°C, it is about 42 mb. The insert illustrates that the saturation vapor pressure over water is greater than the saturation vapor pressure over ice.

the maximum amount of water vapor required for saturation at that particular temperature (and pressure). It is the *ratio* of the air's water vapor *content* to its *capacity;* thus

$$RH = \frac{water\ vapor\ content}{water\ vapor\ capacity}.$$

We can think of the actual vapor pressure as a measure of the air's actual water vapor content, and the saturation vapor pressure as a measure of air's total capacity for water vapor. Hence, the relative humidity can be expressed as

$$RH = \frac{actual\ vapor\ pressure}{saturation\ vapor\ pressure} \times 100\ percent.$$

Relative humidity is given as a percent. Air with a 50 percent relative humidity contains one-half the amount of water vapor required for saturation. Air with a 100 percent relative humidity is said to be *saturated* because it is filled to capacity with water vapor. Air with a relative humidity greater than 100 percent is said to be

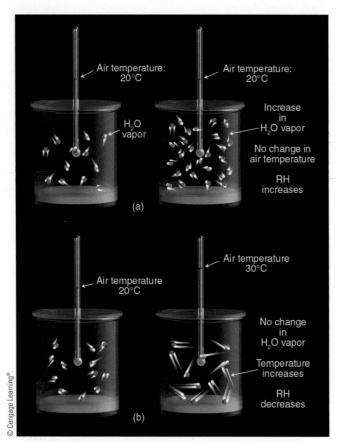

summary, with no change in air temperature, adding water vapor to the air increases the relative humidity; removing water vapor from the air lowers the relative humidity.

Figure 4.6b illustrates that as the air temperature increases (with no change in water vapor content), the relative humidity decreases. This decrease in relative humidity occurs because in the warmer air the water vapor molecules are zipping about at such high speeds they are unlikely to join together and condense. The higher the temperature, the faster the molecular speed, the less likely saturation will occur, and the lower the relative humidity.* As the air temperature lowers, the water vapor molecules move more slowly. Condensation becomes more likely as the air approaches saturation, and the relative humidity increases. In summary, with no change in water vapor content, an increase in air temperature lowers the relative humidity, while a decrease in air temperature raises the relative humidity.

In many places, the air's total vapor content varies only slightly during an entire day, and so it is the changing air temperature that primarily regulates the daily variation in relative humidity (see ● Fig. 4.7). As the air cools during the night, the relative humidity increases. Normally, the highest relative humidity occurs in the early morning, during the coolest part of the day. As the air warms during the day, the relative humidity decreases, with the lowest values usually occurring during the warmest part of the afternoon.

These changes in relative humidity are important in determining the amount of evaporation from vegetation and wet surfaces. If you water your lawn on a hot afternoon, when the relative humidity is low, much of the water will evaporate quickly from the lawn, instead of soaking into the ground. Watering the same lawn in the evening,

● **FIGURE 4.6** (a) At the same air temperature, an increase in the water vapor content of the air increases the relative humidity as the air approaches saturation. (b) With the same water vapor content, an increase in air temperature causes a decrease in relative humidity as the air moves farther away from being saturated.

supersaturated, a condition that does not tend to occur often or last long.

A change in relative humidity can be brought about in two primary ways:

1. by changing the air's water vapor content
2. by changing the air temperature

In ● Fig. 4.6a, we can see that an increase in the water vapor content of the air (with no change in air temperature) increases the air's relative humidity. As more water vapor molecules are added to the air, there is a greater likelihood that some of the vapor molecules will stick together and condense. Condensation takes place in saturated air. So, as more and more water vapor molecules are added to the air, the air gradually approaches saturation, and the relative humidity of the air increases.* Conversely, removing water vapor from the air decreases the likelihood of saturation, which lowers the air's relative humidity. In

*Another way to look at this concept is to realize that, as the air temperature increases, the air's saturation vapor pressure also increases. As the saturation vapor pressure increases, with no change in water vapor content, the air moves farther away from saturation, and the relative humidity decreases.

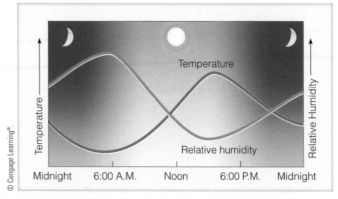

● **FIGURE 4.7** When the air is cool (morning), the relative humidity is high. When the air is warm (afternoon), the relative humidity is low. These conditions exist in clear weather when the air is calm or of constant wind speed.

*We can also see in Fig. 4.6a that as the total number of vapor molecules increases (at a constant temperature), the actual vapor pressure increases and approaches the saturation vapor pressure at 20°C. As the actual vapor pressure approaches the saturation vapor pressure, the air approaches saturation, and the relative humidity rises.

or during the early morning when the relative humidity is higher, will cut down the evaporation and increase the effectiveness of the watering.

RELATIVE HUMIDITY AND DEW POINT

Suppose it is early morning and the outside air is saturated. The air temperature is 10°C (50°F) and the relative humidity is 100 percent. We know from the previous section that relative humidity can be expressed as

$$RH = \frac{actual\ vapor\ pressure}{saturation\ vapor\ pressure} \times 100\ percent.$$

Looking back at Fig. 4.5, p. 83, we can see that air with a temperature of 10°C has a saturation vapor pressure of about 12 mb. Since the air is saturated and the relative humidity is 100 percent, the actual vapor pressure *must* be the same as the saturation vapor pressure (12 mb), since

$$RH = \frac{12\ mb}{12\ mb} \times 100\ \% = 100\ percent.$$

Suppose during the day the air warms to 30°C (86°F), with no change in water vapor content (or air pressure). Because there is no change in water vapor content, the actual vapor pressure must be the same (12 mb) as it was in the early morning when the air was saturated. The saturation vapor pressure, however, has increased because the air temperature has increased. From Fig. 4.5, note that air with a temperature of 30°C has a saturation vapor pressure of about 42 mb. The relative humidity of this unsaturated, warmer air is now much lower, as

$$RH = \frac{12\ mb}{42\ mb} \times 100\ \% = 29\ percent.$$

To what temperature must the outside air, with a temperature of 30°C, be cooled so that it is once again saturated? The answer, of course, is 10°C. For this amount of water vapor in the air, 10°C is called the **dew-point temperature** or, simply, the **dew point.** It represents *the temperature to which air would have to be cooled (with no change in air pressure or moisture content) for saturation to occur.* Since atmospheric pressure varies only slightly at Earth's surface, *the dew point is a good indicator of the air's actual water vapor content. High dew points indicate high water vapor content; low dew points, low water vapor content.* Adding water vapor to the air increases the dew point; removing water vapor lowers it.

● Figure 4.8 shows the average dew-point temperatures across the United States and southern Canada for January. Notice that the dew points are highest (the greatest amount of water vapor in the air) over the Gulf Coast states and lowest over the interior. Compare New Orleans with Fargo. Cold, dry winds from northern Canada flow relentlessly into the Central Plains during the winter, keeping

this area dry. But warm, moist air from the Gulf of Mexico helps maintain a higher dew-point temperature in the southern states.

● Figure 4.9 is a similar diagram showing the average dew-point temperatures for July. Again, the highest dew points are observed along the Gulf Coast, with some areas experiencing average dew-point temperatures near 75°F. In fact, most people consider it to be "humid" when the dew-point temperature exceeds 65°F, and "oppressive" when it equals or exceeds 75°F. Note, too, that the dew points over the eastern and central portion of the United States are much higher in July, meaning that the July air contains between 3 and 6 times more water vapor than the January air. The reason for the high dew points is that, in summertime, this region is almost constantly receiving humid air from the warm Gulf of Mexico. The lowest dew point, and hence the driest air, is found in the West, with the lowest values observed in Nevada—a region surrounded by mountains that effectively shield it from significant amounts of moisture moving in from the southwest and northwest.

The difference between air temperature and dew point can indicate whether the relative humidity is low or high. When the air temperature and dew point are far apart, the relative humidity is low; when they are close to the same value, the relative humidity is high. When the air temperature and dew point are equal, the air is saturated and the relative humidity is 100 percent.*

Under certain conditions, the air can be considered "dry" even though the relative humidity may be 100 percent. Observe, for example, in ●Fig. 4.10a that because the air temperature and dew point are the same in the polar air, the air is saturated and the relative humidity is 100 percent. On the other hand, the desert air (Fig. 4.10b), with a large separation between air temperature and dew point, has a much lower relative humidity, 21 percent.** However, since dew point is a measure of the amount of water vapor in the air, the desert air (with a higher dew

---

*As a general rule of thumb, the difference between air temperature and dew point is roughly 10°F when the relative humidity is around 70%, and roughly 20°F when the relative humidity is around 50%.

**The relative humidity can be computed from Fig. 4.5 (p. 83). The desert air with an air temperature of 35°C has a saturation vapor pressure of about 56 mb. A dew-point temperature of 10°C gives the desert air an actual vapor pressure of about 12 mb. These values produce a relative humidity of 12/56 × 100, or 21 percent.

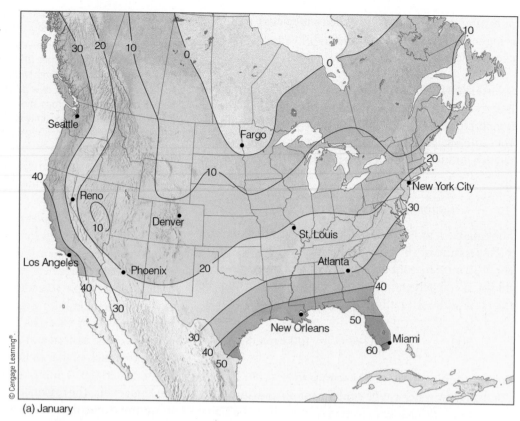

● FIGURE 4.8 Average surface dew-point temperatures (°F) across the United States and Canada for January.

(a) January

© Cengage Learning®

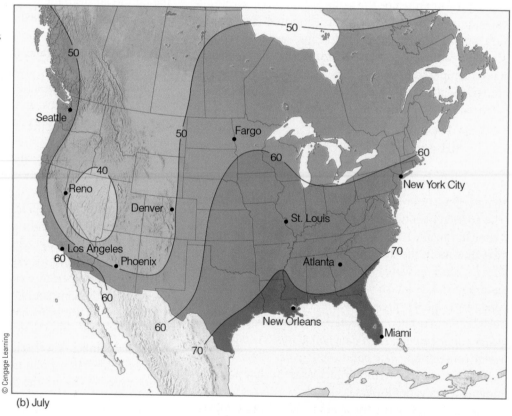

● FIGURE 4.9 Average surface dew-point temperatures across the United States and Canada (°F) for July.

(b) July

© Cengage Learning

point) must contain *more* water vapor. So even though the polar air has a higher relative humidity, the desert air that contains more water vapor has a higher water vapor density, or *absolute humidity*. (The specific humidity and mixing ratio are also higher in the desert air.)

Now we can see why polar air is often described as being "dry" when the relative humidity is high (often close to 100 percent). In cold, polar air, the dew point and air temperature are normally close together. But the low dew-point temperature means that there is little water vapor

(a) POLAR AIR: Air temperature –2°C (28°F)
Dew point –2°C (28°F)
Relative humidity 100 percent

(b) DESERT AIR: Air temperature 35°C (95°F)
Dew point 10°C (50°F)
Relative humidity 21 percent

● **FIGURE 4.10** In this example, the polar air has the higher relative humidity, whereas the desert air, with the higher dew point, contains more water vapor.

in the air. Consequently, the air is "dry" even though the relative humidity is high.

There is a misconception that if it is raining (or snowing), the outside relative humidity must be 100 percent. Look at ● Fig. 4.11 and observe that inside the cloud the relative humidity is 100 percent, but at the ground the relative humidity is much less than 100 percent. As the rain falls into the drier air near the surface, some of the drops evaporate, a process that chills the air and increases the air's water vapor content. The lowering air temperature and rising dew point cause the relative humidity to rise. If the falling rain persists, the air at the surface may become saturated and the relative humidity may reach 100 percent.

$T = 40°F$
$T_d = 40°F$
RH = 100%

$T = 70°F$
$T_d = 65°F$
RH = 84%

● **FIGURE 4.11** Inside the cloud the air temperature (T) and dew point (T_d) are the same, the air is saturated, and the relative humidity (RH) is 100 percent. However, at the surface where the air temperature and dew point are not the same, the air is not saturated (even though it is raining), and the relative humidity is considerably less than 100 percent.

BRIEF REVIEW

Up to this point we have looked at the different ways of describing humidity. Before going on, here is a review of some of the important concepts and facts we have covered:

● Relative humidity tells us how close the air is to being saturated.

● Relative humidity can change when the air's water-vapor content changes, or when the air temperature changes.

● With a constant amount of water vapor, cooling the air raises the relative humidity and warming the air lowers it.

● The dew-point temperature is a good indicator of the air's water-vapor content: High dew points indicate high water-vapor content; and low dew points, low water-vapor content.

● Where the air temperature and dew point are close together, the relative humidity is high; when they are far apart, the relative humidity is low.

● Dry air can have a high relative humidity when the air is very cold and the air temperature and dew point are close together.

RELATIVE HUMIDITY AND HUMAN DISCOMFORT On a hot, muggy day when the relative humidity is high, it is common to hear someone exclaim (often in exasperation), "It's not the heat, it's the humidity!" Actually, this statement has validity. In warm weather, the main source of body cooling is through evaporation of perspiration. Recall from Chapter 2 that evaporation is a cooling process, so when the air temperature is high and the relative humidity low, perspiration on the skin evaporates quickly, often making us feel that the air temperature is lower than it really is. However, when both the air temperature and relative humidity are high and the air is nearly saturated with water vapor, body moisture does not readily evaporate; instead, it collects on the skin as beads of perspiration. Less evaporation means less cooling, and so we usually feel warmer than we did with a similar air temperature but a lower relative humidity. A good measure of how cool the skin can become is the **wet-bulb temperature**—*the lowest temperature that can be reached by evaporating water into the air.** On a hot day when the wet-bulb temperature is low, rapid evaporation (and, hence, cooling) takes place at the skin's surface. As the wet-bulb temperature approaches the air temperature, less cooling occurs, and the skin temperature may begin to rise. When the wet-bulb temperature exceeds the skin's temperature, no net evaporation occurs, and the body temperature can rise quite rapidly. Fortunately, the wet-bulb temperature is almost always considerably below the temperature of the skin. When the weather is hot and muggy, a number of heat-related problems can occur. For example, in hot weather when the human body temperature rises, the *hypothalamus* (a gland in the brain that regulates body temperature) activates the body's heat-regulating mechanism, and more than ten million sweat glands wet the body with as much as two liters of liquid per hour. As this perspiration evaporates, rapid loss of water and salt can result in a chemical imbalance that may lead to painful *heat cramps.* Excessive water loss through perspiring coupled with increasing body temperature can result in *heat exhaustion*—fatigue, headache, nausea, and even fainting. If one's body temperature rises above about 41°C (106°F), **heat stroke** can occur, resulting in complete failure of the circulatory functions. If the body temperature continues to rise, death may result. In fact, each year across North America, hundreds of people die from heat-related maladies. Even strong, healthy individuals can succumb to heat stroke, as did the Minnesota Vikings' all-pro offensive lineman, Korey Stringer, who collapsed after practice in Mankato, Minnesota, on July 31, 2001, and died 15 hours later. Before Stringer fainted, temperatures on the practice

field were in the 90s (°F) with the relative humidity above 55 percent.

In an effort to draw attention to this serious weather-related health hazard, an index called the **heat index (HI)** is used by the National Weather Service. The index combines air temperature with relative humidity to determine an **apparent temperature**—what the air temperature "feels like" to the average person for various combinations of air temperature and relative humidity. For example, in ● Fig. 4.12, an air temperature of 102°F and a relative humidity of 60 percent produce an apparent temperature of 137°F. Heat stroke is very likely when the index reaches this level. However, as we can see from the preceding paragraph, deaths related to heat stroke can occur when the heat index value is considerably lower than 137°F. Also, the values shown in Fig. 4.12 are calculated assuming shade. Under full sunshine, the effective heat index can climb as much as 15°F higher than indicated.

Tragically, hundreds and even thousands of people can be killed by a single heat wave. One example is the great Chicago heat wave of July 1995. Dew-point temperatures were extremely high, which led to several days of unusually warm overnight readings that kept residents from gaining relief from the heat. On July 13, the afternoon air temperature reached 106°F at Midway Airport, followed by an overnight low temperature of only 84°F. Many residents either had no air conditioning or could not afford to use it. More than 700 deaths occurred over five days. Since the time of that disaster, Chicago and many other cities in the United States have added neighborhood "cooling centers" and taken other steps to address the danger of heat waves.

In a closed vehicle, temperatures can soar far above outdoor readings in a matter of minutes. Since 1998, in the United States, more than 600 children have died after being left in parked vehicles. When sunshine is strong, temperatures need not be blistering outside to cause such a tragedy. On an 80°F day in full sun, the air temperature inside a closed car can rise to 99°F in just ten minutes and to 114°F in half an hour. On a 100°F day, interior temperatures can top 140° within an hour. Unfortunately, "cracking" the windows does little to reduce the buildup of heat.

At this point it is important to dispel a common myth about hot, humid weather. Often people will recall a particularly sultry day as having been "90 degrees with 90 percent humidity" or even "95 degrees with 95 percent humidity." We see in Fig. 4.12 that a temperature of 90°F

---

*Notice that the wet-bulb temperature and the dew-point temperature are different. The wet-bulb temperature is attained by *evaporating water* into the air, whereas the dew-point temperature is reached by *cooling* the air.

Relative humidity (%)																	Possible heat-related risks
	20	25	30	35	40	45	50	55	60	65	70	75	80	85	90	95	100	
120	130	138	148															**Extreme danger** Heat stroke or sunstroke highly likely
118	126	134	142															
116	122	129	137	146														
114	119	125	132	140	148													
112	116	121	127	134	142	150												
110	112	117	122	129	136	143												
108	109	113	118	123	130	137	144											**Danger** Sunstroke muscle cramps and/or heat exhaustion likely
106	106	109	114	119	124	130	137	145										
104	103	106	110	114	119	124	131	137	145									
102	100	103	106	110	114	119	124	130	137	144								
100	97	100	102	106	109	114	118	124	129	136	143	150						**Extreme Caution** Sun stroke muscle cramps and/or heat exhaustion possible
98	95	97	99	102	105	109	113	117	123	128	134	141	148					
96	93	94	96	98	101	104	108	112	116	121	126	132	138	145				
94	90	91	93	95	97	100	103	106	110	114	119	124	129	135	141	148		
92	88	89	90	92	94	96	99	101	105	108	112	116	121	126	131	137	143	
90	86	87	88	89	91	92	95	97	100	103	106	109	113	117	122	127	132	
88	85	85	86	87	88	89	91	93	95	98	100	103	106	110	113	117	121	**Caution** Fatigue possible
86	83	83	84	85	85	87	88	89	91	93	95	97	100	102	105	109	112	
84	81	82	82	83	83	84	85	86	88	89	90	92	94	96	98	101	104	
82	80	80	80	81	81	82	83	84	84	85	86	88	89	90	92	94	96	
80	79	79	79	80	80	80	81	81	82	82	83	84	84	85	86	88	89	

Temperature (°F)

● **FIGURE 4.12** Air temperature (°F) and relative humidity are combined to determine an apparent temperature or heat index (HI). An air temperature of 96°F with a relative humidity of 55 percent produces an apparent temperature (HI) of 112°F.

with 90 percent relative humidity would produce a heat index of 122°F. Although this weather situation is remotely possible, it is *extremely unlikely*, as a temperature of 90°F and a relative humidity of 90 percent can occur only if the dew-point temperature is incredibly high (nearly 87°F), and a dew point this high rarely occurs anywhere in the United States, even on the muggiest of days.

Similarly, in hot muggy weather, there are people who will remark about how "heavy" or how dense the air feels. Is hot, humid air really more dense than hot, dry air? If you are interested in the answer, read Focus section 4.1.

Up to this point we've only looked at the discomfort brought on by high humidity. Can a very low relative humidity have an adverse effect on humans, too?

During the winter, the relative humidity inside a home can drop to an extremely low value and the inhabitants are usually unaware of it. When cold polar air is brought indoors and heated, its relative humidity decreases dramatically. Notice in ● Fig 4.13 that when outside air with a temperature and dew point of 5°F is brought indoors and heated to 68°F, the relative humidity of the heated air drops to 8 percent—a value lower than what you would normally experience in a desert during the hottest time of the day.

Very low relative humidities in a house can have an adverse effect on living things inside. For example, house plants have a difficult time surviving because the moisture from their leaves and the soil evaporates rapidly. Thus, they usually need watering more frequently in winter than

in summer. People suffer, too, when the relative humidity is quite low. The rapid evaporation of moisture from exposed flesh causes skin to crack, dry, flake, or itch. These low humidities also irritate the mucous membranes in the nose and throat, producing an "itchy" or "scratchy" throat.

INSIDE AIR
$T = 68°F$
$T_d = 5°F$
RH = 8%

OUTSIDE AIR
$T = 5°F$
$T_d = 5°F$
RH = 100%

● **FIGURE 4.13** When outside air with an air temperature and a dew point of 5°F is brought indoors and heated to a temperature of 68°F (without adding water vapor to the air), the relative humidity drops to 8 percent, placing stress on plants, animals, and humans living inside. (T represents temperature; T_d, dew point; and *RH*, relative humidity.)

Humid Air and Dry Air Do Not Weigh the Same

Does a volume of hot, humid air really weigh more than a similar-size volume of hot, dry air? The answer is no! At the same temperature and at the same level in the atmosphere, hot, humid air is lighter (less dense) than hot, dry air. This is because a molecule of water vapor (H_2O) weighs appreciably less than a molecule of either nitrogen (N_2) or oxygen (O_2). (Keep in mind that we are referring strictly to water vapor—a gas—and not suspended liquid droplets.)

Consequently, in a given volume of air, as lighter water vapor molecules replace either nitrogen or oxygen molecules one for one, the number of molecules in the volume does not change, but the total weight of the air becomes slightly less. Since air density is the mass of air in a volume, the more humid air must be lighter than the drier air. Hence, *hot, humid air at the surface is lighter (less dense) than hot, dry air.*

This fact can have an important influence in the weather. The lighter the air becomes, the more likely it is to rise. All other factors being equal, hot, humid (less-dense) air will rise more readily than hot, dry (more-dense) air (see Fig. 1). It is, of course, the water vapor in the rising air

● **FIGURE 1** On this summer afternoon in Maryland, lighter (less-dense) hot, humid air rises and condenses into towering cumulus clouds.

that changes into liquid cloud droplets and ice crystals, which, in turn, grow large enough to fall to Earth as precipitation.

Of lesser importance to weather but of greater importance to sports is the fact that a baseball will "carry" farther in less-dense air. Consequently, without the influence of wind, a ball

will travel slightly farther on a hot, humid day than it will on a hot, dry day. So when the sports announcer proclaims that "the air today is heavy because of the high humidity," remember that this statement is not true and, in fact, a 404-foot home run on this humid day might simply be a 400-foot out on a very dry day.

Similarly, dry nasal passages permit inhaled bacteria to incubate, causing persistent infections. The remedy for most of these problems is simply to increase the relative humidity. Inside the home, the relative humidity can be increased simply by heating water and allowing it to evaporate into the air. The added water vapor raises the relative humidity to a more comfortable level. In modern homes, a humidifier, installed near the furnace, adds moisture to the air at a rate of about one gallon per room per day. This air, with its increased water vapor, is circulated throughout the home by a forced-air heating system. In this way, all rooms get their fair share of moisture, not just the room where the vapor is added. So, if your throat begins to feel scratchy while you are inside during a cold winter day, boil a pot of water and see if increasing the relative humidity of the air can help soothe your irritated throat.

MEASURING HUMIDITY Humidity is most often measured today with automated instruments (though some measurements are still taken manually at some observing sites). One common instrument used to obtain dew point

and relative humidity is the **psychrometer,** which consists of two liquid-in-glass thermometers mounted side by side and attached to a piece of metal that has either a handle or chain at one end (see Fig. 4.14). The thermometers are exactly alike except that one has a piece of cloth (wick) covering the bulb. The wick-covered thermometer—called

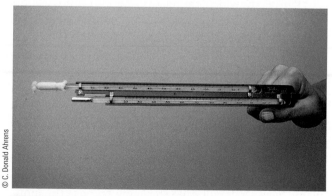

● **FIGURE 4.14** The sling psychrometer consists of two thermometers, one of which has a wick covering the bulb, the *wet bulb* thermometer.

the *wet bulb*—is dipped in clean water, whereas the other thermometer is kept dry. Both thermometers are ventilated for a few minutes, either by being whirled *(sling psychrometer)* or by having air drawn past it with an electric fan *(aspirated psychrometer)*. Water evaporates from the wick and that thermometer cools. The drier the air, the greater the amount of evaporation and cooling. After a few minutes, the wick-covered thermometer will cool to the lowest value possible. Recall from an earlier section that this is the *wet-bulb temperature*—the lowest temperature that can be attained by evaporating water into the air.

The dry thermometer (commonly called the *dry bulb*) gives the current air temperature, or *dry-bulb temperature*. The temperature difference between the dry bulb and the wet bulb is known as the *wet-bulb depression*. A large depression indicates that a great deal of water can evaporate into the air and that the relative humidity is low. A small depression indicates that little evaporation of water vapor is possible, so the air is close to saturation and the relative humidity is high. If there is no depression, the dry bulb, the wet bulb, and the dew point are the same; the air is saturated and the relative humidity is 100 percent.

Instruments that measure humidity are commonly called **hygrometers.** One type—called the *hair hygrometer*—uses human or horse hair or a synthetic fiber to measure relative humidity. It is constructed on the principle that, as the relative humidity increases, the length of hair increases and, as the relative humidity decreases, so does the hair length. A number of strands of hair (with oils removed) are attached to a system of levers. A small change in hair length is magnified by a linkage system and transmitted to a dial (Fig. 4.15) calibrated to show relative humidity, which can then be read directly or recorded on a chart. (Often, the chart is attached to a clock-driven rotating drum that gives a continuous record of relative humidity.) Because the hair hygrometer is not as accurate as the psychrometer (especially at very high and very low relative humidities), it requires frequent calibration, principally in areas that experience large daily variations in relative humidity.

An automated instrument that measures humidity is the *electrical hygrometer*. It consists of a flat plate coated with a film of carbon. An electric current is sent across the plate. As water vapor is absorbed, the electrical resistance of the carbon coating changes. These changes are translated into relative humidity. This instrument is commonly used in the radiosonde, which gathers atmospheric data at various levels above Earth. The *dew-point hygrometer* measures the dew-point temperature by cooling the surface of a mirror until condensation (dew) forms. This sensor is the type that measures dew-point temperature in the hundreds of fully automated weather stations—Automated Surface Observing System (ASOS)—that exist throughout the United States. (A picture of ASOS is shown in Fig. 3.22, p. 73.)

Over the last several sections we have seen that, as the air cools, the air temperature approaches the dew-point temperature and the relative humidity increases. When the air temperature reaches the dew point, the air is saturated with water vapor and the relative humidity is 100 percent. Continued cooling, however, causes some of the water vapor to condense into liquid water. The cooling may take place in a thick portion of the atmosphere, or it may occur near Earth's surface. In the next section, we will examine condensation that forms near the ground.

Dew and Frost

On clear, calm nights, objects near Earth's surface cool rapidly by emitting infrared radiation. The ground and objects on it often become much colder than the surrounding air. Air that comes in contact with these cold surfaces cools by conduction. Eventually, the air cools to the dew point. As surfaces (such as twigs, leaves, and blades of grass) cool below this temperature, water vapor begins to condense upon them, forming tiny visible specks of water called **dew** (see Fig. 4.16). If the air temperature should drop to freezing or below, the dew will freeze, becoming tiny beads of ice called *frozen dew*. Because the coolest air is usually at ground level, dew is more likely to form on blades of grass than on objects several feet above the surface. This thin coating of dew, of course, dampens bare feet, but, more importantly, it is a valuable source of moisture for many plants during periods of low rainfall.

Dew is more likely to form on nights that are clear and calm than on nights that are cloudy and windy. Clear nights allow objects near the ground to cool rapidly, and calm winds mean that the coldest air will be located at ground level. These atmospheric conditions are usually associated with large fair-weather, high-pressure systems.

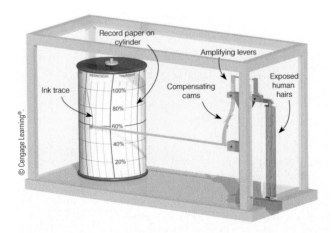

 FIGURE 4.15 The hair hygrometer measures relative humidity by amplifying and measuring changes in the length of human (or horse) hair or a synthetic fiber.

● **FIGURE 4.16** Dew forms on clear nights when objects on the surface cool to a temperature below the dew point. If these beads of water should freeze, they would become frozen dew.

● **FIGURE 4.17** These are the delicate ice-crystal patterns that frost exhibits on a window during a cold winter morning.

On the other hand, the cloudy, windy weather that inhibits rapid cooling near the ground and the forming of dew often signifies the approach of a rain-producing storm system. These observations inspired the following folk rhyme:

> When the dew is on the grass,
> rain will never come to pass.
> When grass is dry at morning light,
> look for rain before the night!

Visible white frost forms on cold, clear, calm mornings when the dew-point temperature is at or below freezing. When the air temperature cools to the dew point (now called the *frost point*) and further cooling occurs, water vapor can change directly to ice without becoming a liquid first—a process called *deposition*.* The delicate, white crystals of ice that form in this manner are called *hoarfrost, white frost,* or simply **frost**. Frost has a treelike branching

---
*When the ice changes back into vapor without melting, the process is called *sublimation*.

pattern that easily distinguishes it from the nearly spherical beads of frozen dew (see ● Fig. 4.17).

In very dry weather, the air may become quite cold and drop below freezing without ever reaching the frost point, and no visible frost forms. *Freeze* and *black frost* are words denoting this situation, one that can severely damage certain crops (see Chapter 3, p. 60).

As a deep layer of air cools during the night, its relative humidity increases. When the air's relative humidity reaches about 75 percent, some of its water vapor may begin to condense onto tiny floating particles of sea salt and other substances—*condensation nuclei*—that are *hygroscopic* ("water-seeking") in that they allow water vapor to condense onto them when the relative humidity is considerably below 100 percent. As water collects onto these nuclei, their size increases and the particles, although still small, are now large enough to scatter visible light in all directions, becoming **haze**—a layer of particles dispersed through a portion of the atmosphere (see ● Fig. 4.18).

● **FIGURE 4.18** The high relative humidity of the cold air above the lake is causing a layer of haze to form on a still winter morning.

As the relative humidity gradually approaches 100 percent, the haze particles grow larger, and condensation begins on the less-active nuclei. Now water is condensing onto a large fraction of the available nuclei, causing the droplets to grow even bigger, until eventually they become visible to the naked eye. The increasing size and concentration of droplets further restrict visibility. When the visibility lowers to less than 1 km (or 0.62 mi) and the air is wet with millions of tiny floating water droplets, the haze becomes a cloud resting near the ground, which we call **fog.***

Fog

Fog, like any cloud, usually forms in one of two ways:

1. by cooling—air is cooled below its saturation point (dew point); and
2. by evaporation and mixing—water vapor is added to the air by evaporation, and the moist air mixes with relatively dry air.

Once fog forms it is maintained by new fog droplets, which constantly form on available nuclei. In other words, the air must maintain its degree of saturation either by continual cooling or by evaporation and mixing of vapor into the air.

Fog produced by Earth's radiational cooling is called **radiation fog,** or *ground fog.* It forms best on clear nights when a shallow layer of moist air near the ground is overlain by drier air. Under these conditions, the ground cools rapidly since the shallow, moist layer does not absorb much of Earth's outgoing infrared radiation. As the ground cools,

so does the air directly above it, and a surface inversion forms, with colder air at the surface and warmer air above. The moist, lower layer (chilled rapidly by the cold ground) quickly becomes saturated, and fog forms. The longer the night, the longer the time of cooling and the greater the likelihood of fog. Hence, radiation fogs are most common over land in late fall and winter.

Another factor promoting the formation of radiation fog is a light breeze of less than five knots. Although radiation fog may form in calm air, slight air movement brings more of the moist air in direct contact with the cold ground and the transfer of heat occurs more rapidly. A strong breeze tends to prevent a radiation fog from forming by mixing the air near the surface with the drier air above. The ingredients of clear skies and light winds are associated with large high-pressure areas (anticyclones). Consequently, during the winter, when a high becomes stagnant over an area, radiation fog may form on consecutive days.

Because cold, heavy air drains downhill and collects in valley bottoms, we normally see radiation fog forming in low-lying areas. Hence, radiation fog is frequently called *valley fog.* The cold air and high moisture content in river valleys make them susceptible to radiation fog. Since radiation fog normally forms in lowlands, hills may be clear all day long, while adjacent valleys are fogged in (see ●Fig. 4.19).

Radiation fogs are normally deepest around sunrise. Usually, however, a shallow fog layer will dissipate or *burn off* by afternoon. Of course, the fog does not "burn"; rather, sunlight penetrates the fog and warms the ground, causing the temperature of the air in contact with the ground to increase. The warm air rises and mixes with the foggy air above, which increases the temperature of the foggy air. In the slightly warmer air, some of the fog droplets evaporate, allowing more sunlight to reach the ground, which produces more heating, and soon the fog completely evaporates and disappears. If the fog layer is quite thick, it may

*This is the official international definition of *fog.* The United States National Weather Service reports fog as a restriction to visibility when fog restricts the visibility to 6 miles or less and the spread between the air temperature and dew point is 5°F or less. When the visibility is less than one-quarter of a mile, the fog is considered *dense.*

●**FIGURE 4.19** Radiation fog nestled in a valley in central Oregon.

© C. Donald Ahrens

Herbert Spichtinger/Bridge/Corbis

●**FIGURE 4.20** Advection fog rolling in past the Golden Gate Bridge in San Francisco. As fog moves inland, the air warms and the fog lifts above the surface. Eventually, the air becomes warm enough to totally evaporate the fog.

not completely dissipate and a layer of low clouds (called *stratus*) covers the region. This type of fog is sometimes called *high fog*.

When warm, moist air moves over a sufficiently colder surface, the moist air may cool to its saturation point, forming **advection fog**. A good example of advection fog can be observed along the Pacific Coast during summer. The main reason fog forms in this region is that the surface water near the coast is much colder than the surface water farther offshore. Warm, moist air from the Pacific Ocean is carried (advected) by westerly winds over the cold coastal waters. Chilled from below, the air temperature drops to the dew point, and fog forms. Advection fog, unlike radiation fog, always involves the movement of air, so when there is a stiff summer breeze in San Francisco, it's common to watch advection fog roll in past the Golden Gate Bridge (see ●Fig. 4.20).

As summer winds carry the fog inland over warmer land, the fog near the ground dissipates, leaving a sheet of low-lying gray clouds that block out the sun. Farther inland, the air is sufficiently warm, so that even these low clouds evaporate and disappear.

Because they provide moisture to the coastal redwood trees, advection fogs are important to the scenic beauty of the Pacific Coast. The needles and branches of the redwoods absorb moisture from the fog, allowing the trees to grow very tall without having to draw moisture from their roots far below. Additional moisture drips to the ground (*fog drip*), where it is utilized by the tree's shallow root system. Without the summer fog, the coast's redwood trees would have trouble surviving the dry California summers. Advection fogs also prevail where two ocean currents with different temperatures flow next to one another. Such is the case in the Atlantic Ocean off the coast of Newfoundland, where the cold, southward-flowing Labrador Current

lies almost parallel to the warm northward-flowing Gulf Stream. Warm southerly air moving over the cold water produces fog in that region—so frequently that fog occurs on about two out of three days during summer.

Advection fog also forms over land. In winter, warm, moist air from the Gulf of Mexico moves northward over progressively colder and slightly elevated land. As the air cools to its saturation point, fog will form in the southern or central United States. Because the cold ground is often the result of radiational cooling, fog that forms in this manner is sometimes called *advection-radiation fog*. During this same time of year, air moving across the warm Gulf Stream encounters the colder land of the British Isles and produces the thick fogs of England. Similarly, fog forms as marine air moves over an ice or snow surface. In extremely cold arctic air, ice crystals form instead of water droplets, producing an *ice fog*.

Keep in mind that advection fog forms when wind blows moist air over a cooler surface, whereas radiation fog forms under relatively calm conditions. ●Figure 4.21 visually summarizes the formation of these two types of fog.

Fog that forms as moist air flows up along an elevated plain, hill, or mountain is called **upslope fog.** Typically, upslope fog forms during the winter and spring on the eastern side of the Rockies, where the eastward-sloping plains are nearly a kilometer higher than the land farther east. Occasionally, cold air moves from the lower eastern plains westward. The air gradually rises, expands, becomes cooler, and—if sufficiently moist—a fog forms (see ●Fig. 4.22). Upslope fogs that form over an extensive area can last for days.

So far, we have seen how the cooling of air produces fog. But remember that fog can also form from the mixing of two unsaturated masses of air. Fog that forms in this

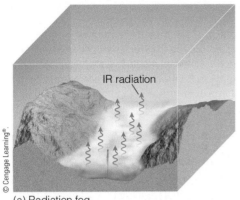

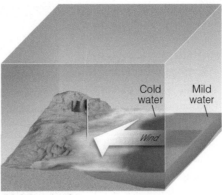

(a) Radiation fog

(b) Advection fog

● FIGURE 4.21 (a) Radiation fog tends to form on clear, relatively calm nights when cool, moist surface air is overlain by drier air and rapid radiational cooling occurs. (b) Advection fog forms when the wind moves moist air over a cold surface and the moist air cools to its dew point.

manner is usually called *evaporation fog* because evaporation initially enriches the air with water vapor. Probably a more appropriate name for the fog is **evaporation (mixing) fog.** On a cold day, you may have unknowingly produced evaporation (mixing) fog. When moist air from your mouth or nose meets the cold air and mixes with it, the air becomes saturated, and a tiny cloud forms with each exhaled breath.

A common form of evaporation-mixing fog is the *steam fog,* which forms when cold air moves over warm water. This type of fog forms above a heated outside swimming pool in winter. As long as the water is warmer than the unsaturated air above, water will evaporate from the pool into the air. The increase in water vapor raises the dew point, and, if mixing is sufficient, the air above becomes saturated. The colder air directly above the water is heated from below and becomes warmer than the air directly above it. This warmer air rises and, from a distance, the rising condensing vapor appears as "steam."

It is common to see steam fog forming over lakes on autumn mornings, as cold air settles over water still warm from the long summer. On occasion, over the Great Lakes and other warm bodies of water, columns of condensed vapor rise from the fog layer, forming whirling *steam devils,* which appear similar to the dust devils observed on land. If you travel to Yellowstone National Park, you will see steam fog forming above thermal ponds all year long (see ●Fig. 4.23). Over the ocean in polar regions, steam fog is referred to as *arctic sea smoke.*

Steam fog may form above a wet surface on a sunny day. This type of fog is commonly observed after a rain shower as sunlight shines on a wet road, heats the asphalt, and quickly evaporates the water. This added vapor mixes with the air above, producing steam fog. Fog that forms in this manner is short-lived and disappears as the road surface dries.

A warm rain falling through a layer of cold, moist air can produce fog. As a warm raindrop falls into a cold layer of air, some of the water evaporates from the raindrop into the air. This process may saturate the air,

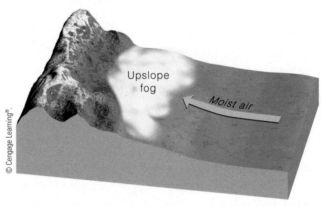

● FIGURE 4.22 Upslope fog forms as moist air slowly rises, cools, and condenses over elevated terrain.

● FIGURE 4.23 Even in summer, warm air rising above thermal pools in Yellowstone National Park condenses into a type of steam fog.

and if mixing occurs, fog forms. Fog of this type is often associated with warm air riding up and over a mass of colder surface air. The fog usually develops in the shallow layer of cold air just ahead of an approaching warm front or behind a cold front, which is why this type of evaporation fog is also known as *precipitation fog,* or *frontal fog.*

Foggy Weather

The foggiest regions in the United States are shown in ● Fig. 4.24. Notice that dense fog is more prevalent in coastal margins (especially those regions lapped by cold ocean currents and near the Great Lakes and Appalachian Mountains) than in the center of the continent. In fact, the foggiest spot near sea level in the United States is Cape Disappointment, Washington. Located at the mouth of the Columbia River, it averages 2556 hours or the equivalent of 106.5 twenty-four hour days of dense fog each year. Anyone hoping to enjoy the sun during August and September by traveling to this spot would find its name appropriate indeed.

Notice in Fig. 4.24 that the coast of Maine is also foggy. In fact, Moose Peak Lighthouse on Mistake Island averages 1580 hours (66 equivalent days) of dense fog. To the south, Nantucket Island has on average 2040 hours (85 equivalent days) of fog each year.

Extremely limited visibility exists while driving at night in thick fog with the high-beam lights on. The light scattered

back to the driver's eyes from the fog droplets makes it difficult to see very far down the road. Along a gently sloping highway, the elevated sections may have excellent visibility, while in lower regions—only a few miles away—fog can cause poor visibility. Driving from a clear area into fog on a major freeway can be extremely dangerous. In fact, every winter many people are involved in fog-related auto accidents. These usually occur when a driver enters fog and, because of the reduced visibility, puts on the brakes to slow down. The car behind then slams into the slowed vehicle, causing a chain-reaction accident with many cars involved.

Airports suspend flight operations when fog causes visibility to drop below a prescribed minimum. The

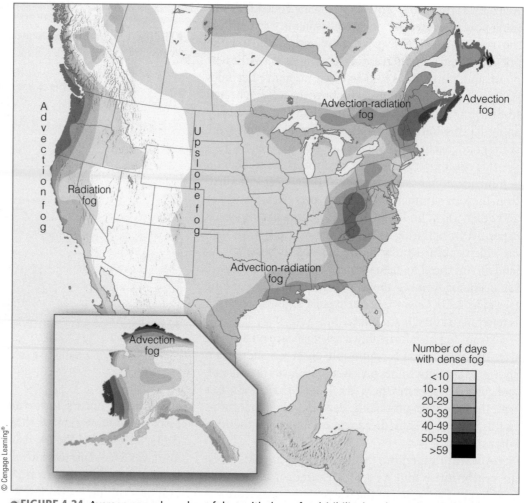

● **FIGURE 4.24** Average annual number of days with dense fog (visibility less than 0.25 miles) across North America. (Dense fog observed in small mountain valleys and on mountaintops is not shown.)

Fog Dispersal

In any airport fog-clearing operation, the problem is to improve visibility so that aircraft can take off and land. Experts have tried various methods, which can be grouped into four categories: (1) increase the size of the fog droplets, so that they become heavy and settle to the ground as a light drizzle; (2) seed cold fog with dry ice (solid carbon dioxide), so that fog droplets are converted into ice crystals; (3) heat the air, so that the fog evaporates; and (4) mix the cooler saturated air near the surface with the warmer unsaturated air above.

To date, only one of these methods has been reasonably successful—the seeding of cold fog. *Cold fog* forms when the air temperature is below freezing, and most of the fog droplets remain as liquid water. (Liquid fog in below-freezing air is also called *supercooled fog*.) The fog can be cleared by injecting several hundred pounds of dry ice into it. As the tiny pieces of cold ($-78°C$) dry ice descend, they freeze some of the super-cooled fog droplets in their path, producing ice crystals. As we will see in Chapter 5, these crystals then grow larger at the expense of the remaining liquid fog droplets. Hence, the fog droplets evaporate and the larger ice crystals fall to the ground, which leaves

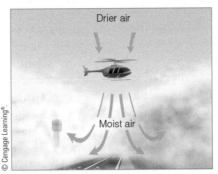

● FIGURE 2 Helicopters hovering above an area of shallow fog can produce a clear area by mixing the drier air into the foggy air below.

a "hole" in the fog for aircraft takeoffs and landings.

Unfortunately, most of the fogs that close airports in the United States are *warm fogs* that form when the air temperature is above freezing. Since dry ice seeding does not work in warm fog, other techniques must be tried.

One method involves injecting hygroscopic (water-absorbing) particles into the fog. Large salt particles and other chemicals absorb the tiny fog droplets and form into larger drops. A mix of more large drops and fewer small drops improves the visibility; plus, the larger drops are more likely to fall as a light drizzle. However,

since the chemicals are expensive and the fog clears for only a short time, this method of fog dispersal is not economically feasible.

Another technique for fog dispersal is to warm the air enough so that the fog droplets evaporate and visibility improves. Tested at Los Angeles International Airport in the early 1950s, this technique was abandoned because it was smoky, expensive, and not very effective. In fact, the burning of hundreds of dollars worth of fuel only cleared the runway for a short time. And the smoke particles, released during the burning of the fuel, provided abundant nuclei for the fog to recondense upon.

A final method of warm fog dispersal uses helicopters to mix the air. The chopper flies across the fog layer, and the turbulent downwash created by the rotor blades brings drier air above the fog into contact with the moist fog layer (see ● Fig. 2). The aim, of course, is to evaporate the fog. Experiments show that this method works well, as long as the fog is a shallow radiation fog with a relatively low liquid water content. But many fogs are thick, have a high liquid water content, and form by other means. An inexpensive and practical method of dispersing warm fog has yet to be discovered.

---

resulting delays and cancellations become costly to the airline industry and irritate passengers. With fog-caused problems such as these, it is no wonder that scientists have been seeking ways to disperse or, at least, "thin," fog. (For more information on fog-thinning techniques, read Focus section 4.2 entitled "Fog Dispersal.")

Up to this point, we have looked at the different forms of condensation that occur on or near Earth's surface. In particular, we learned that fog is simply many billions of tiny liquid droplets (or ice crystals) that form near the ground. In the following sections, we will see how these same particles, forming well above the ground, are classified and identified as clouds.

BRIEF REVIEW

Before we go on to the section on clouds, here is a brief review of some of the important concepts and facts we have covered so far:

- Dew, frost, and frozen dew generally form on clear nights when the temperature of objects on the surface cools below the air's dew-point temperature.

- Visible white frost forms in saturated air when the air temperature is at or below freezing. Under these conditions, water vapor can change directly to ice, in a process called *deposition*.

- Condensation nuclei act as surfaces on which water vapor condenses. Those nuclei that have an affinity for water vapor are called *hygroscopic*.

- Fog is a cloud resting on the ground. It can be composed of water droplets, ice crystals, or a combination of both.

- Radiation fog, advection fog, and upslope fog all form as the air cools. The cooling for radiation fog is mainly radiational cooling at Earth's surface; for advection fog, the cooling is mainly warm air moving over a colder surface; for upslope fog, the cooling occurs as moist air gradually rises and expands along sloping terrain. Evaporation (mixing) fog, such as steam fog and frontal fog, forms as water evaporates and mixes with drier air.

Clouds

Clouds are aesthetically appealing and add excitement to the atmosphere. Without them, there would be no rain or snow, thunder or lightning, rainbows or halos. How monotonous it would be if there were only a clear blue sky to look at.

A *cloud* is a visible aggregate of tiny water droplets or ice crystals suspended in the air. Some are found only at high elevations, whereas others nearly touch the ground (or are classified as fog if they do touch the ground). Clouds can be thick or thin, big or little—they exist in a seemingly endless variety of forms. To impose order on this variety, we divide clouds into ten basic types. With a careful and practiced eye, you can become reasonably proficient in correctly identifying them.

CLASSIFICATION OF CLOUDS   Although ancient astronomers named the major stellar constellations about 2000 years ago, clouds were not formally identified and classified until the early nineteenth century. The French naturalist Jean-Baptiste Lamarck (1744–1829) proposed the first system for classifying clouds in 1802; however, his work did not receive wide acclaim. One year later, Luke Howard, an English naturalist, developed a cloud classification system that found general acceptance. In essence, Howard's innovative system employed Latin words to describe clouds as they appear to a ground observer. He named a sheetlike cloud *stratus* (Latin for "layer"); a puffy cloud *cumulus* ("heap"); a wispy cloud *cirrus* ("curl of hair"); and a rain cloud *nimbus* ("violent rain"). In Howard's system, these were the four basic cloud forms. Other clouds could be described by combining the basic types. For example, nimbostratus is a rain cloud that shows layering, whereas cumulonimbus is a rain cloud having pronounced vertical development.

In 1887, Ralph Abercromby and Hugo Hildebrandsson expanded Howard's original system and published a classification system that, with only slight modification, is still in use today. Ten principal cloud forms are divided into four primary cloud groups. Each group is identified by the height of the cloud's base above the surface: high clouds, middle clouds, and low clouds. The fourth group contains clouds showing more vertical than horizontal development. Within each group, cloud types are identified by their appearance. ▼ Table 4.1 lists these four groups and their cloud types.

The approximate base height of each cloud group is given in ▼ Table 4.2. Note that the altitude separating the high and middle cloud groups overlaps and varies with latitude. Large temperature changes cause most of this latitudinal variation. For example, high cirriform clouds are composed almost entirely of ice crystals. In subtropical regions, air temperatures low enough to freeze all liquid water usually occur only above about 20,000 feet. In polar regions, however, these same temperatures may be found at altitudes as low as 10,000 feet. Although you may observe cirrus clouds at 12,000 feet over northern Alaska, you will not see them at that elevation above southern Florida.

Clouds cannot be accurately identified strictly on the basis of elevation. Other visual clues are necessary. Some of these are explained in the following section.

HIGH CLOUDS   High clouds in middle and low latitudes generally form above 16,000 ft (5000 m). Because the air at these elevations is quite cold and dry, high clouds are composed almost exclusively of ice crystals and are also rather

▼ Table 4.1   **The Four Major Cloud Groups and Their Types**

1. High clouds
   Cirrostratus (Cs)
   Cirrus (Ci)
   Cirrocumulus (Cc)

2. Middle clouds
   Altostratus (As)
   Altocumulus (Ac)

3. Low clouds
   Stratus (St)
   Stratocumulus (Sc)
   Nimbostratus (Ns)

4. Clouds with vertical development
   Cumulus (Cu)
   Cumulonimbus (Cb)

▼ Table 4.2   **Approximate Height* of Cloud Bases Above the Surface for Various Locations**

CLOUD GROUP	TROPICAL REGION	MID-LATITUDE REGION	POLAR REGION
High Ci, Cs, Cc	20,000 to 60,000 ft (6000 to 18,000 m)	16,000 to 43,000 ft (5000 to 13,000 m)	10,000 to 26,000 ft (3000 to 8000 m)
Middle As, Ac	6500 to 26,000 ft (2000 to 8000 m)	6500 to 23,000 ft (2000 to 7000 m)	6500 to 13,000 ft (2000 to 4000 m)
Low St, Sc, Ns	surface to 6500 ft (0 to 2000 m)	surface to 6500 ft (0 to 2000 m)	surface to 6500 ft (0 to 2000 m)

*Note that the height of a cloud base in each region varies by season.

© C. Donald Ahrens

● **FIGURE 4.25** Cirrus clouds. Notice the silky "mare's tail" appearance.

© C. Donald Ahrens

● **FIGURE 4.26** Cirrocumulus clouds.

thin.* High clouds usually appear white, except near sunrise and sunset, when the unscattered (red, orange, and yellow) components of sunlight are reflected from the underside of the clouds.

The most common high clouds are **cirrus** (Ci), which are thin, wispy clouds blown by high winds into long streamers called *mares' tails*. Notice in ● Fig. 4.25 that they can look like a white, feathery patch with a faint wisp of a tail at one end. Cirrus clouds usually move across the sky from west to east, indicating the prevailing winds at their elevation, and they generally occur during periods of fair, pleasant weather.

Cirrocumulus (Cc) clouds, seen less frequently than cirrus, appear as small, rounded, white puffs that may occur individually, or in long rows (see ● Fig. 4.26). When in rows, the cirrocumulus cloud has a rippling appearance that distinguishes it from the silky look of cirrus and the sheetlike cirrostratus. Cirrocumulus seldom cover more than a small portion of the sky. The dappled cloud elements that reflect the red or yellow light of a setting sun make this one of the most beautiful of all clouds. The small ripples in the cirrocumulus strongly resemble the scales of a fish; hence, the expression "*mackerel sky*" commonly describes a sky full of cirrocumulus clouds.

---

*Small quantities of liquid water in cirrus clouds at temperatures as low as −36°C (−33°F) were discovered during research conducted above Boulder, Colorado.

The thin, sheetlike, high clouds that often cover the entire sky are **cirrostratus** (Cs), which are so thin that the sun and moon can be clearly seen through them (see ● Fig. 4.27). The ice crystals in these clouds bend the light passing through them and will often produce a *halo*— a ring of light that encircles the sun or moon. In fact, the veil of cirrostratus may be so thin that a halo is the only clue to its presence. Thick cirrostratus clouds give the sky a glary white appearance and frequently form ahead of an advancing mid-latitude cyclonic storm; hence, they can be used to predict rain or snow within twelve to twenty-four hours, especially if they are followed by middle-type clouds.

MIDDLE CLOUDS   The middle clouds have bases between about 6500 and 23,000 ft (2000 and 7000 m) in the middle latitudes. These clouds are composed of water droplets and—when the temperature becomes low enough—some ice crystals. Precipitation can form in middle clouds if they become thick enough.

Altocumulus (Ac) clouds are middle clouds that appear as gray, puffy masses, sometimes rolled out in parallel waves or bands (see ● Fig. 4.28). Usually, one part of each

© C. Donald Ahrens

● **FIGURE 4.27** Cirrostratus clouds. Notice the faint halo encircling the sun. The sun is the bright white area in the center of the circle.

● **FIGURE 4.28** Altocumulus clouds. Notice the dark-to-light contrasting patterns that distinguish these clouds from cirrocumulus clouds.

● **FIGURE 4.29** Altostratus clouds. The appearance of a dimly visible "watery sun" through a deck of light gray clouds is usually a good indication that the clouds are altostratus.

cloud element is darker than another, which helps to distinguish it from the higher cirrocumulus. Also, the individual puffs of the altocumulus appear larger than those of the cirrocumulus. A layer of altocumulus can sometimes be confused with altostratus; in case of doubt, clouds are called altocumulus if there are rounded masses or rolls present. Altocumulus clouds that look like "little castles" (*castellanus*) in the sky indicate the presence of rising air at cloud level. The appearance of these clouds on a warm, humid summer morning often portends thunderstorms by late afternoon.

Altostratus (As) are gray or blue-gray clouds that often cover the entire sky over an area that extends over many hundreds of square kilometers. In the thinner section of the cloud, the sun (or moon) may be dimly

visible as a round disk, which is sometimes referred to as a "watery sun" (see ● Fig. 4.29). Thick cirrostratus clouds are occasionally confused with thin altostratus clouds. The gray color, height, and dimness of the sun are good clues to identifying an altostratus. The fact that halos only occur with cirriform clouds also helps one to distinguish them. Another way to separate the two is to look at the ground for shadows. If there are none, it is a good bet that the cloud is altostratus because cirrostratus are usually transparent enough to produce shadows. Altostratus clouds often form ahead of mid-latitude cyclonic storms having widespread and relatively continuous precipitation. If precipitation falls from altostratus, the cloud base usually lowers, and the precipitation is steady and not showery as found with cumuliform clouds. If the precipitation reaches the ground, the cloud is then classified as *nimbostratus*.

LOW CLOUDS Low clouds, with their bases lying below 6500 ft (or 2000 m) are almost always composed of water droplets, although in cold weather, they may contain ice particles.

Nimbostratus (Ns) are dark-gray, "wet"-looking cloud layers associated with more or less continuously falling rain or snow (see ● Fig. 4.30). The intensity of this precipitation is usually light or moderate; it is never of the heavy, showery variety, unless well-developed cumuliform clouds are embedded within the nimbostratus cloud. Precipitation often makes the base of the nimbostratus cloud impossible to identify clearly. The distance from the cloud's base to the top can be over 3 km (10,000 feet). Nimbostratus is easily confused with altostratus. Thin nimbostratus is usually darker gray than thick altostratus, and you normally cannot see the sun or moon through a layer of nimbostratus. Visibility below a nimbostratus cloud deck

● **FIGURE 4.30** The nimbostratus is the sheetlike cloud from which light rain is falling. The ragged-appearing clouds beneath the nimbostratus are stratus fractus, or scud.

● FIGURE 4.31 Stratocumulus clouds forming along the south coast of Florida. Notice that the rounded masses are larger than those of the altocumulus.

is usually quite poor because rain will evaporate and mix with the air in this region. If this air becomes saturated, a lower layer of clouds or fog may form beneath the original cloud base, as seen in Fig. 4.30. Since these lower clouds drift rapidly with the wind, they form irregular shreds with a ragged appearance called *stratus fractus,* or *scud.*

Stratocumulus (Sc) are low, lumpy clouds that appear in rows, in patches, or as rounded masses with blue sky visible between individual cloud elements (see ● Fig. 4.31). Often they appear near sunset as the spreading remains of a much larger cumulus cloud. Occasionally, the sun will shine through the cloud breaks, producing bands of light (called *crepuscular rays*) that appear to reach down to the ground. The color of stratocumulus ranges from light to dark gray. This cloud type differs from altocumulus in that it has a lower base and larger individual clouds. (Compare Fig. 4.28 with Fig. 4.31.) To distinguish between the two, hold your hand at arm's length and point toward one of these clouds. Altocumulus cloud elements will generally be about the size of your thumbnail, whereas stratocumulus will usually be about the size of your fist. Although precipitation rarely falls from stratocumulus, light rainshowers or winter snow flurries can occur if the cloud develops vertically into a much thicker cloud with a top colder than about –5°C (23°F).

Stratus (St) is a uniform grayish cloud that often covers the entire sky. It resembles a fog that does not reach the ground (see ● Fig. 4.32). Actually, when a thick fog "lifts," the resulting cloud is a deck of low stratus. Normally, no precipitation falls from stratus, but sometimes it is accompanied by a light mist or drizzle. This cloud commonly occurs over Pacific and Atlantic coastal waters in summer.

● FIGURE 4.32 A layer of low-lying stratus clouds hides the mountains in Iceland.

● **FIGURE 4.33** Cumulus clouds. Small cumulus clouds such as these are sometimes called *fair weather cumulus,* or *cumulus humilis.*

A thick layer of stratus might be confused with nimbostratus, but the distinction between them can be made by observing the base of the cloud. Often, stratus has a more uniform base than does nimbostratus. Also, a deck of stratus may be confused with a layer of altostratus. However, if you remember that stratus clouds are lower and darker gray, the distinction can be made.

CLOUDS WITH VERTICAL DEVELOPMENT

Familiar to almost everyone, the puffy **cumulus** (Cu) cloud takes on a variety of shapes, but most often it looks like a piece of floating cotton with sharp outlines and a flat base (see ● Fig. 4.33). The base appears white to light gray, and, on a humid day, may be only a few thousand feet above the ground and half a mile or so wide. The top of the cloud—often in the form of rounded towers—denotes the limit of rising air and is usually not very high. These clouds can be distinguished from stratocumulus by the fact that cumulus clouds are detached (usually with a great deal of blue sky between each cloud) whereas stratocumulus usually occur in groups or patches. Also, the cumulus has a dome- or tower-shaped top as opposed to the generally flatter tops of the stratocumulus. Cumulus clouds that show only slight vertical growth *(cumulus humilis)* are associated with fair weather; therefore, we call these clouds "fair-weather cumulus." If the cumulus clouds are small and appear as broken fragments of a cloud with ragged edges, they are called *cumulus fractus.*

Harmless-looking cumulus often develop on warm summer mornings and, by afternoon, become much larger and more vertically developed. When the growing cumulus resembles a head of cauliflower, it becomes a *cumulus congestus,* or *towering cumulus* (Tcu). Most often, it is a single large cloud, but, occasionally, several grow into each other, forming a line of towering clouds, as shown in ● Fig. 4.34. Precipitation that falls from a cumulus congestus is always showery with frequent changes in intensity.

● **FIGURE 4.34** Cumulus congestus clouds are frequently called towering cumulus. These clouds are taller than cumulus clouds and are more likely to produce showers. Here a line of cumulus congestus clouds is building along Maryland's eastern shore.

● FIGURE 4.35 A cumulonimbus cloud. Strong upper-level winds blowing from right to left produce a well-defined anvil shape. Sunlight scattered by falling ice crystals produces the white (bright) area beneath the anvil. Notice the heavy rainshower falling from the base of the cloud.

© T. Ansel Toney

If a cumulus congestus continues to grow vertically, it develops into a giant **cumulonimbus** (Cb)—a thunderstorm cloud (see ● Fig. 4.35). While its dark base may be no more than 2000 ft above Earth's surface, its top can extend upward to the tropopause, over 39,000 ft higher, and sometimes thousands of feet beyond that. A cumulonimbus can occur as an isolated cloud or as part of a line or "wall" of clouds.

The tremendous amounts of energy released by the condensation of water vapor within a cumulonimbus results in the development of violent updrafts and downdrafts, which can exceed 70 knots. The lower (warmer) part of the cloud is usually composed only of water droplets. Higher up in the cloud, water droplets and ice crystals both abound, while, toward the cold top, there are only ice crystals. Swift winds at these higher altitudes can reshape the top of the cloud into a huge flattened *anvil*.* These great thunderheads may contain all forms of precipitation—large raindrops, snowflakes, snow pellets, and sometimes hailstones—all of which can fall to Earth in the form of heavy showers. Lightning, thunder, and even tornadoes are associated with cumulonimbus. (More information on the violent nature of thunderstorms and tornadoes is given in Chapter 10.)

Cumulus congestus and cumulonimbus frequently look alike. However, you can usually distinguish them by looking at the top of the cloud. If the sprouting upper part of the cloud is sharply defined and not fibrous, it is usually a cumulus congestus; conversely, if the top of the cloud loses its sharpness and becomes fibrous in texture, it is usually a cumulonimbus. (Compare Fig. 4.34 with Fig. 4.35.) The weather associated with these clouds also differs: Lightning, thunder, and large hail typically occur with cumulonimbus.

So far, we have discussed the ten primary cloud forms, summarized pictorially in ● Fig. 4.36. This figure, along with the cloud photographs and descriptions, should help you learn to identify the more common cloud forms. Don't worry if you find it hard to estimate cloud heights. This is a difficult procedure, requiring much practice. You can use local objects (hills, mountains, tall buildings) of known height as references on which to base your height estimates.

To better describe a cloud's shape and form, a number of descriptive words can be used in conjunction with its name. We mentioned a few in the previous section; for example, a stratus cloud with a ragged appearance is a stratus fractus, and a cumulus cloud with marked vertical growth is a cumulus congestus. ▼Table 4.3 lists some of the more common terms that are used in cloud identification.

SOME UNUSUAL CLOUDS Although the ten basic cloud forms are the most frequently seen, there are some unusual clouds that deserve mentioning. For example, moist air crossing a mountain barrier often forms into waves. The clouds that form in the wave crest usually have a lens shape and are, therefore, called **lenticular clouds** (see ● Fig. 4.37). Frequently, they form one above the other like a stack of pancakes, and at a distance they can resemble hovering spacecraft. Hence, it is no wonder that many UFO sightings take place when lenticular clouds are present.

Similar to the lenticular cloud is the *cap cloud,* or **pileus,** which usually resembles a silken scarf capping the top of a sprouting cumulus cloud (see ● Fig. 4.38). Pileus clouds form when moist winds are deflected up and over

*An anvil is a heavy block of iron or steel with a smooth, flat top on which metals are shaped by hammering.

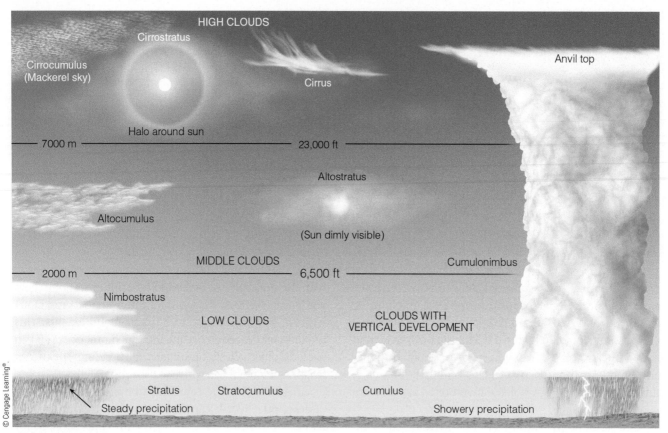

HIGH CLOUDS

Cirrostratus

Cirrocumulus
(Mackerel sky)

Cirrus

Anvil top

Halo around sun

— 7000 m —————————— 23,000 ft —

Altostratus

Altocumulus

(Sun dimly visible)

MIDDLE CLOUDS

Cumulonimbus

— 2000 m —————————— 6,500 ft —

Nimbostratus

LOW CLOUDS

CLOUDS WITH
VERTICAL DEVELOPMENT

Stratus   Stratocumulus   Cumulus

Steady precipitation   Showery precipitation

© Cengage Learning®.

● **FIGURE 4.36** A generalized illustration of basic cloud types based on height above the surface and vertical development.

▼ Table 4.3 **Common Terms Used in Identifying Clouds**

TERM	LATIN ROOT AND MEANING	DESCRIPTION
Lenticularis	(*lens, lenticula,* lentil)	Clouds having the shape of a lens; often elongated and usually with well-defined outlines, a term that applies mainly to cirrocumulus, altocumulus, and stratocumulus
Fractus	(*frangere,* to break or fracture)	Clouds that have a ragged or torn appearance; applies only to stratus and cumulus
Humilis	(*humilis,* of small size)	Cumulus clouds with generally flattened bases and slight vertical growth
Congestus	(*congerere,* to bring together; to pile up)	Cumulus clouds of great vertical extent that, from a distance, may resemble a head of cauliflower
Undulatus	(*unda,* wave; having waves)	Clouds in patches, sheets, or layers showing undulations
Translucidus	(*translucere,* to shine through; transparent)	Clouds that cover a large part of the sky and are sufficiently translucent to reveal the position of the sun or moon
Mammatus	(*mamma,* mammary)	Baglike clouds that hang like a cow's udder on the underside of a cloud; may occur with cirrus, altocumulus, altostratus, stratocumulus, and cumulonimbus
Pileus	(*pileus,* cap)	A cloud in the form of a cap or hood above or attached to the upper part of a cumuliform cloud, particularly during its developing stage
Castellanus	(*castellum,* a castle)	Clouds that show vertical development and produce towerlike extensions, often in the shape of small castles

● FIGURE 4.38 A pileus cloud forming above a developing cumulus cloud.

● FIGURE 4.39 Mammatus clouds forming beneath a thunderstorm anvil.

the top of a building cumulus congestus or cumulonimbus. If the air flowing over the top of the cloud condenses, a pileus often forms.

Most clouds form in rising air, but the mammatus forms in sinking air. **Mammatus clouds** derive their name from their appearance—baglike sacs that hang beneath the cloud and resemble mammary glands (see ● Fig. 4.39). Although mammatus most frequently form on the underside of cumulonimbus, they may develop beneath cirrus, cirrocumulus, altostratus, altocumulus, and stratocumulus.

Jet aircraft flying at high altitudes often produce a cirruslike trail of condensed vapor called a *condensation trail* or **contrail** (see ● Fig. 4.40). The condensation may come directly from the water vapor added to the air from engine exhaust. In this case, there must be sufficient mixing of the hot exhaust gases with the cold air to produce saturation. Contrails evaporate rapidly when the relative humidity of the surrounding air is low. If the relative

● **FIGURE 4.40** A contrail forming behind a jet aircraft.

humidity is high, however, contrails can persist for many hours. Contrails can also form by a cooling process as the reduced pressure produced by air flowing over the wing causes the air to cool.

Aside from the cumulonimbus cloud that sometimes penetrates into the stratosphere, all of the clouds described so far are observed in the lower atmosphere, the troposphere. Occasionally, however, clouds can be seen above the troposphere. For example, soft pearly looking clouds called **nacreous clouds,** or *mother-of-pearl clouds,* form in the stratosphere at altitudes above 30 km or 100,000 ft (see ● Fig. 4.41). They are best viewed in polar latitudes during the winter months when the sun, being just below the horizon, is able to illuminate them because of their high altitude. Their exact composition is not known, although they appear to be composed of water in either solid or liquid (supercooled) form.

Wavy bluish-white clouds, so thin that stars shine brightly through them, can sometimes develop in the upper mesosphere, at altitudes above 75 km (46 mi). These clouds are at such a high altitude that they appear bright against a dark background. For this reason, they are called **noctilucent clouds,** meaning "luminous night clouds" (see ● Fig. 4.42). They are best seen at twilight, during

the summer at latitudes poleward of 50°, although they have been observed in recent years in Utah and Colorado. Studies reveal that these clouds are composed of tiny ice crystals. The water to make the ice may originate in meteoroids that disintegrate when entering the upper atmosphere or from the chemical breakdown of methane gas at high levels in the atmosphere.

Our system of classifying clouds may never be completely finalized. One identified formation, *asperitas,* has only recently been recognized as a specific cloud type (see Did You Know on p. 103 and ● Fig. 4.43).

CLOUDS AND SATELLITE IMAGERY The weather satellite is a cloud-observing platform in Earth's orbit. Satellites provide extremely valuable cloud images of areas where there are no ground-based observations. Because water covers over 70 percent of Earth's surface, there are vast regions where few (if any) surface cloud observations are made. Before weather satellites were in use, tropical storms, such as hurricanes and typhoons, often went undetected until they moved dangerously near inhabited areas. Residents of the regions affected had little advance warning. Today, satellites spot these storms while they are still far out in the ocean and track them accurately.

Two primary types of weather satellites are used for observing clouds. The first are called **geostationary satellites** (or *geosynchronous satellites*) because they orbit the equator at the same rate Earth spins and, hence, remain at nearly 36,000 km (22,300 mi) above a fixed spot on Earth's surface (see ● Fig. 4.44). This positioning allows continuous monitoring of a specific region.

Geostationary satellites are also important because they use a "real time" data system, meaning that the satellites transmit images to the receiving system on the ground as soon as the image is taken. Successive cloud images from these satellites can be put into a time-lapse movie sequence to show the cloud movement, dissipation, or development associated with weather fronts and storms. This information is a great help in forecasting the progress

● **FIGURE 4.41** The clouds in this photograph are nacreous clouds. They form in the stratosphere and are most easily seen at high latitudes.

of large weather systems. Wind directions and speeds at various levels can also be approximated by monitoring cloud movement with the geostationary satellite.

Complementing the geostationary satellites are **polar-orbiting satellites,** which closely parallel Earth's meridian (longitude) lines. These satellites pass over the north and south polar regions on each revolution. As Earth rotates to the east beneath the satellite, each pass monitors an area to the west of the previous pass (see ● Fig. 4.45). Eventually, the satellite covers the entire Earth.

Polar-orbiting satellites have the advantage of scanning clouds directly beneath them. Thus, they provide sharp images in polar regions, where images from a geostationary satellite are distorted because of the low angle at which the satellite "sees" those regions. Polar orbiters also circle Earth at a much lower altitude (about 850 km, or 530 mi) than geostationary satellites. The lower altitude allows them to provide detailed images of phenomena such as violent storms and cloud systems.

Continuously improved detection devices make weather observation by satellites more versatile than ever. Early satellites, such as *TIROS I,* launched in 1960, used television cameras to photograph clouds. Contemporary satellites use radiometers, which can observe clouds during both day and night by detecting radiation that emanates from the top of the clouds. When the radiometer measures only visible sunlight reflected from the cloud, the image is called a *visible cloud image.* Additionally, satellites have the capacity to obtain cloud images and, at the same time, provide vertical profiles of atmospheric temperature and moisture by detecting emitted radiation from atmospheric gases, such as water vapor. In modern satellites, a special type of advanced radiometer (called an *imager*) provides satellite images with much better resolution than did previous imagers. Moreover, another type of special radiometer (called a *sounder*) gives a more accurate profile of temperature and moisture at different levels in the atmosphere than did earlier instruments. The latest

● **FIGURE 4.42** The wavy clouds in this photograph are noctilucent clouds. They are usually observed at high latitudes, at altitudes above 75 km.

● **FIGURE 4.43** This asperitas, photographed over eastern Colorado, is a cloud that looks ominous but does not produce stormy weather.

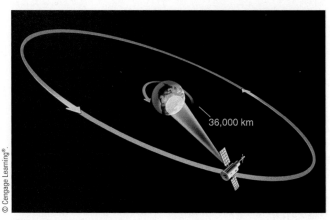

● **FIGURE 4.44** The geostationary satellite moves through space at the same rate that Earth rotates, so it remains above a fixed spot on the equator and monitors one area constantly.

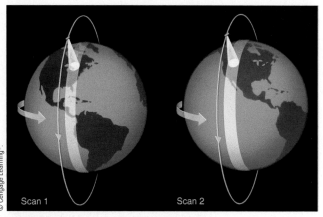

● **FIGURE 4.45** Polar-orbiting satellites scan from north to south, and on each successive orbit the satellite scans an area farther to the west.

in the Geostationary Operational Environmental Satellite (GOES) series, GOES-R, includes many additional features, including higher spatial resolution and the ability to map lightning activity from space.

Information on cloud thickness and height can be deduced from satellite images. Visible images show the sunlight reflected from a cloud's upper surface. Because thick clouds have a higher reflectivity (albedo) than thin clouds, they appear brighter on a visible satellite image. However, middle and low clouds have just about the same reflectivity, so it is difficult to distinguish among them simply by viewing them in visible light. To make this distinction, *infrared cloud images* are used. Such pictures produce a better image of the actual radiating surface because they do not show the strong visible reflected light. Since warm objects radiate more energy than cold objects, high temperature regions can be artificially made to appear darker on an infrared image. Because the tops of low clouds are warmer than those of high clouds, cloud observations made in the infrared can distinguish between warm low clouds (dark) and cold high clouds (light)—see ● Fig. 4.46. Moreover, cloud temperatures can be converted by a computer into a three-dimensional image of the cloud. These are the 3-D cloud photos presented on television by many weathercasters.

● Figure 4.47 shows a visible satellite image (from a geostationary satellite) of a mid-latitude cyclonic storm in the eastern Pacific. Notice that all of the clouds in the image appear white. However, in the infrared image (● Fig. 4.48), taken on the same day (and at just about the same time), the clouds appear to have many shades of gray. In the visible image, the clouds covering part of Oregon and northern California appear relatively thin compared to the thicker, bright clouds to the west. Furthermore, these thin clouds must be high because they also appear bright in the infrared image.

Along the elongated band of clouds associated with the occluded front and cold front, the clouds appear white and bright in both images, indicating a zone of thick, heavy clouds. Behind the front, the lumpy clouds are probably cumulus because they appear gray in the infrared image (Fig. 4.48), suggesting that their tops are low and relatively warm.

When temperature differences are small, it is difficult to directly identify significant cloud and surface features on an infrared image. Some way must be found to increase the contrast between features and their backgrounds. This can be accomplished by a process called *computer enhancement*. Certain temperature ranges in the infrared image are assigned specific shades of gray, grading from black to

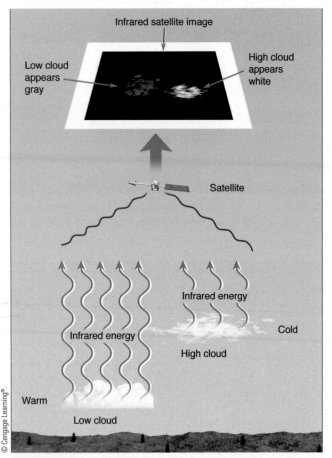

● FIGURE 4.46 Generally, the lower the cloud, the warmer its top. Warm objects emit more infrared energy than do cold objects. Thus, an infrared satellite picture can distinguish warm, low (gray) clouds from cold, high (white) clouds.

● FIGURE 4.47 A visible image of the eastern Pacific Ocean taken at just about the same time on the same day as the image in Fig. 4.48. Notice that the clouds in the visible image appear white. Superimposed on the image are the cold, warm, and occluded fronts.

white. Often, clouds with cold tops, and those with tops near freezing, are assigned the darkest gray color.

● Figure 4.49 is an *enhanced infrared image* for the same day and area as shown in Figs. 4.47 and 4.48. To make cloud features more obvious, colors such as dark blue, red, or purple are assigned to clouds with the coldest (highest) tops. Hence, the dark red areas embedded along the occluded front and cold front in Fig. 4.49 represent the region where the coldest and, therefore, highest and thickest clouds are found. It is here where the stormiest weather is probably occurring. Also notice that, near the southern tip of the picture, the dark red blotches surrounded by areas of white are thunderstorms that have developed over warm tropical waters. They show up clearly as white, thick clouds in both the visible and infrared images. By examining the movement of these clouds on successive satellite images, forecasters can predict the arrival of clouds and storms, and the passage of weather fronts.

In regions where there are no clouds, it is difficult to observe the movement of the air. To help with this situation, geostationary satellites are equipped with water-vapor sensors that can profile the distribution of atmospheric water vapor in the middle and upper troposphere (see ● Fig. 4.50). In time-lapse films, the swirling patterns

● **FIGURE 4.50** Infrared water-vapor image. The darker areas represent dry air aloft; the brighter the gray, the more moist the air in the middle or upper troposphere. Bright white areas represent dense cirrus clouds or the tops of thunderstorms. The area in color represents the coldest cloud tops. The swirl of moisture off the West Coast represents a well-developed mid-latitude cyclonic storm.

● **FIGURE 4.48** Infrared satellite image of the eastern Pacific Ocean taken at just about the same time on the same day as the image in Fig. 4.47. Notice that the low clouds in this infrared image appear in various shades of gray.

● **FIGURE 4.49** An enhanced infrared image of the eastern Pacific Ocean taken on the same day as the images shown in Figs. 4.47 and 4.48.

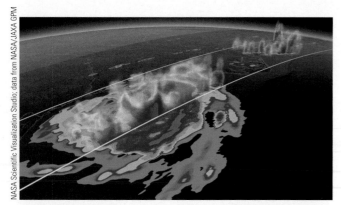

NASA Scientific Visualization Studio; data from NASA/JAXA GPM

● **FIGURE 4.51** Precipitation in Hurricane Arthur as sensed by the Global Positioning Mission (*GPM*) satellite on July 3, 2014, off the South Carolina coast. Colors on the surface ranging from light green to dark red to purple indicate areas ranging from low to high rainfall. The violet areas aloft indicate frozen precipitation.

of moisture clearly show wet regions and dry regions, as well as middle tropospheric swirling wind patterns and jet streams.

Specialized satellites have gathered data on clouds and precipitation for more than 20 years. From 1997 to 2015, the long-lived *Tropical Rainfall Measuring Mission* (TRMM) satellite provided information on clouds and precipitation from about 35°S to 35°N. A joint venture of NASA and the Japan Aerospace Exploration Agency (JAXA), this satellite orbited Earth at an altitude of about 400 km (250 mi), sensing individual cloud features as small as about 1.5 miles in diameter. TRMM gathered three-dimensional images of clouds and storms, details on the intensity and distribution of precipitation, and data on Earth's energy budget and lightning discharges within storms. TRMM has been succeeded by another NASA/JAXA project called the *Global Precipitation Mission* (GPM), whose core observatory was launched in 2014. GPM covers a much broader swath than TRMM—from about 65°S to 65°N—and it includes advanced sensors that can distinguish precipitation intensity and type as well as cloud characteristics (see ● Fig. 4.51).

Another specialized satellite has also provided enhanced detail on clouds and precipitation. Launched in 2006, the NASA *CloudSat* satellite circles Earth in an orbit

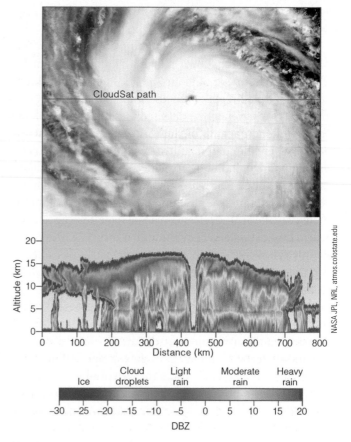

NASA JPL, NRL, atmos.colostate.edu

● **FIGURE 4.52** (*Top*) Visible satellite image of super-typhoon Choi-Wan over the tropical eastern Pacific Ocean on September 15, 2009. (*Bottom*) *CloudSat* vertical radar profile through super-typhoon Choi-Wan. The location of the profile is shown by the red line in the top view.

about 700 km (435 mi) above the surface. Onboard *CloudSat*, a very sensitive radar (called *Cloud Profiling Radar*, or *CPR*) uses microwave radiation to peer into a cloud and unveil its very fine structures, including the altitude of the cloud's top and base, its thickness, optical properties, the abundance of liquid and ice particles, along with the intensity of precipitation inside the cloud. *CloudSat* provides this information in a vertical view, as shown in ● Fig. 4.52. Such vertical profiling of a cloud's makeup will potentially provide scientists with a better understanding of precipitation processes that go on inside the cloud and the role that clouds play in Earth's global climate system.

SUMMARY

In this chapter, we examined the hydrologic cycle and saw how water is circulated within our atmosphere. We then looked at some of the ways of describing humidity and found that relative humidity does not tell us how much water vapor is in the air but, rather, how close the air is to being saturated. A good indicator of the air's actual water vapor content is the dew-point temperature. When the air temperature and dew point are close together, the relative humidity is high; when they are far apart, the relative humidity is low.

When the air temperature drops below the dew point in a shallow layer of air near the surface, dew forms. If the dew freezes, it becomes frozen dew. Visible white frost forms when the air cools to a below-freezing dew-point temperature. As the air cools in a deeper layer near the surface, the relative humidity increases and water vapor begins to condense upon "water-seeking" hygroscopic condensation nuclei, forming haze. As the relative humidity approaches 100 percent, the air can become filled with tiny liquid droplets (or ice crystals) called fog. Upon examining fog, we found that it forms in two primary ways: (1) when air cools and evaporates and (2) when water evaporates mixes into the air.

Condensation above Earth's surface produces clouds. When clouds are classified according to their height and physical appearance, they are divided into four main groups: high, middle, and low clouds, and clouds with vertical development. Since each cloud type has physical characteristics that distinguish it from others, careful observation normally leads to correct identification.

Satellites enable scientists to obtain a bird's-eye view of clouds on a global scale. Polar-orbiting satellites obtain data covering Earth from pole to pole, while geostationary satellites located above the equator continuously monitor a desired portion of Earth. Both types of satellites use radiometers (imagers) that detect emitted radiation. As a consequence, clouds can be observed both day and night.

Visible satellite images, which show sunlight reflected from a cloud's upper surface, can distinguish thick clouds from thin clouds. Infrared images show an image of the cloud's radiating top and can distinguish low clouds from high clouds. To increase the contrast between cloud features, infrared photographs are enhanced. Specialized instruments aboard satellites can be used to analyze cloud characteristics and precipitation.

KEY TERMS

The following terms are listed (with corresponding page numbers) in the order they appear in the text. Define each. Doing so will aid you in reviewing the material covered in this chapter.

evaporation, 80
condensation, 80
precipitation, 80
hydrologic cycle, 80
saturated air, 81
condensation nuclei, 81
humidity, 82
actual vapor pressure, 83
saturation vapor pressure, 83
relative humidity, 84
supersaturated air, 84
dew-point temperature
    (dew point), 85
wet-bulb temperature, 88
heat stroke, 88
heat index (HI), 88
apparent temperature, 88
psychrometer, 90
hygrometer, 91
dew, 91
frost, 92
haze, 92
fog, 93
radiation fog, 93
advection fog, 94
upslope fog, 94
evaporation (mixing) fog, 95

cirrus clouds, 99
cirrocumulus clouds, 99
cirrostratus clouds, 99
altocumulus clouds, 99
altostratus clouds, 100
nimbostratus clouds, 100
stratocumulus clouds, 101
stratus clouds, 101
cumulus clouds, 102
cumulonimbus clouds, 103
lenticular clouds, 103
pileus clouds, 103
mammatus clouds, 105
contrail, 105
nacreous clouds, 106
noctilucent clouds, 106
geostationary satellites, 106
polar-orbiting satellites, 107

QUESTIONS FOR REVIEW

1. Briefly explain the movement of water in the hydrologic cycle.
2. How does condensation differ from precipitation?
3. What are condensation nuclei and why are they important in our atmosphere?
4. In a volume of air, how does the actual vapor pressure differ from the saturation vapor pressure? When are they the same?
5. What does saturation vapor pressure primarily depend upon?

6. (a) What does the relative humidity represent?
   (b) When the relative humidity is given, why is it also important to know the air temperature?
   (c) Explain two ways the relative humidity can be changed.
   (d) During what part of the day is the relative humidity normally lowest? Normally highest?
7. Why do hot, humid summer days usually feel hotter than hot, dry summer days?
8. Why is cold polar air described as "dry" even when the relative humidity of that air is very high?
9. Why is the wet-bulb temperature a good measure of how cool human skin can become?
10. (a) What is the dew-point temperature?
    (b) How is the difference between dew point and air temperature related to the relative humidity?
11. How can you obtain both the dew point and the relative humidity using a sling psychrometer?
12. Explain how dew, frozen dew, and visible frost form.
13. List the two primary ways in which fog forms.
14. Describe the conditions that are necessary for the formation of:
    (a) radiation fog
    (b) advection fog
15. How does evaporation (mixing) fog form?
16. Clouds are most generally classified by height above Earth's surface. List the major height categories and the cloud types associated with each.
17. How can you distinguish altostratus clouds from cirrostratus clouds?
18. Which clouds are normally associated with each of the following characteristics?
    (a) mackerel sky
    (b) lightning
    (c) halos
    (d) hailstones
    (e) mares' tails
    (f) anvil top
    (g) light continuous rain or snow
    (h) heavy rain showers
19. Name the clouds that form above the troposphere.
20. How do geostationary satellites differ from polar-orbiting satellites?
21. How can you distinguish a visible satellite image from an infrared satellite image?
22. Why are infrared satellite images enhanced?

QUESTIONS FOR THOUGHT AND EXPLORATION

1. Use the concepts of condensation and saturation to explain why eyeglasses often fog up after you come indoors on a cold day.
2. After completing a grueling semester of meteorological course work, you contact a travel agent to arrange a much-needed summer vacation. When the agent suggests a trip to the desert, you decline because of a concern that the dry air will make your skin feel uncomfortable. The travel agent assures you that almost daily "desert relative humidities are above 90 percent." Could the agent be correct? Explain.
3. Can the actual vapor pressure ever be greater than the saturation vapor pressure? Explain.
4. Suppose while measuring the relative humidity using a sling psychrometer, you accidentally moisten both the dry-bulb and the wet-bulb thermometers. Will the relative humidity you determine be higher or lower than the air's true relative humidity?
5. A large family lives in northern Minnesota. This family gets together for a huge dinner three times a year: on Thanksgiving, on Christmas, and on the March equinox. The Thanksgiving and Christmas dinners consist of turkey, ham, mashed potatoes, and lots of boiled vegetables. The March dinner is pizza. The air temperature inside the home is about the same for all three meals (70°F), yet everyone remarks on how "warm, cozy, and comfortable" the air feels during the Thanksgiving and Christmas dinners, and how "cool" the inside air feels during the equinox meal. Explain to the family members why they might feel "warmer" inside the house during Thanksgiving and Christmas, and "cooler" during the March equinox. (The answer has nothing to do with the amount or type of food consumed.)
6. Why is advection fog more common along the coast of southern California than along the coast of southern Virginia?
7. With all other factors being equal, would you expect a lower minimum temperature on a night with cirrus clouds or on a night with stratocumulus clouds? Explain your answer.
8. Explain why icebergs are frequently surrounded by fog.

9. While driving from cold air (well below freezing) into much warmer air (well above freezing), frost forms on the windshield of the car. Does the frost form on the inside or outside of the windshield? How can the frost form when the air is so warm?

10. Why do relative humidities seldom reach 100 percent in polluted air?

11. If all fog droplets gradually settle earthward, explain how fog can last (without disappearing) for many days at a time.

12. The air temperature during the night cools to the dew point in a deep layer, producing fog. Before the fog formed, the air temperature cooled each hour about 3°F. After the fog formed, the air temperature cooled by only 1°F each hour. Give *two* reasons why the air cooled more slowly after the fog formed.

13. Why can you see your breath on a cold morning? Does the air temperature have to be below freezing for this to occur?

14. The sky is overcast and it is raining. Explain how you can tell if the cloud above you is a nimbostratus or a cumulonimbus.

15. You are sitting inside your house on a sunny afternoon. The shades are drawn and you look at the window and notice the sun disappears for about 10 seconds. The alternating light and dark periods last for nearly 30 minutes. Are the clouds passing in front of the sun cirrocumulus, altocumulus, stratocumulus, or cumulus? Give a reasonable explanation for your answer.

GLOBAL **GEOSCIENCE** WATCH Go the the Basic Search field and search for news items using the keywords "cloud classification." Review at least three news articles that describe techniques used to classify clouds based on data from weather satellites. What are the common elements among these techniques? How do they differ?

ONLINE RESOURCES

 Visit www.cengagebrain.com to view additional resources, including video exercises, practice quizzes, an interactive eBook, and more.

CHAPTER 5

Cloud Development and Precipitation

Contents

The young boy pushed his nose against the cold window-pane, hoping to see snowflakes glistening in the light of the streetlamp across the way. Perhaps if it snowed, he thought, it would be deep enough to cancel school, maybe for a day, possibly a week, or perhaps, forever. But clear skies and a full moon gave little hope for snow on this evening. Nor did the voice from the back room that insisted, "Don't even think about snow. You know it won't snow tonight; it's too cold to snow." With hopes dashed, the boy pondered: could it really be too cold to snow?

Clouds, spectacular features in the sky, add beauty and color to the natural landscape. Yet clouds are important for nonaesthetic reasons, too. As they form, vast quantities of heat are released into the atmosphere. Clouds help regulate Earth's energy balance by reflecting and scattering solar radiation and by absorbing infrared energy from Earth's surface. And, of course, without clouds there would be no precipitation. But clouds are also significant because they visually indicate the physical processes taking place in the atmosphere; to a trained observer, they are signposts in the sky. In the beginning of this chapter, we will look at the atmospheric processes these signposts point to, the first of which is atmospheric stability. Later, we will examine the different mechanisms responsible for the formation of most clouds. Toward the end of the chapter, we will peer into the tiny world of cloud droplets to see how rain, snow, and other types of precipitation form. And yes, we will answer the question raised in our opener, "Is it ever too cold to snow?"

● FIGURE 5.2 The dry adiabatic rate. As long as the air parcel remains unsaturated, it expands and cools by 10°C per 1000 m; the sinking parcel compresses and warms by 10°C per 1000 m.

Atmospheric Stability

We know that most clouds form as air rises, expands, and cools. But why does the air rise on some occasions and not on others? And why do the size and shape of clouds vary so much when the air does rise? To answer these questions, let's focus on the concept of atmospheric stability.

When we speak of atmospheric stability, we are referring to a condition of equilibrium. For example, rock A resting in the depression in ● Fig. 5.1 is in *stable* equilibrium. If the rock is pushed up along either side of the hill and then let go of, it will quickly return to its original position. On the other hand, rock B, resting on the top of the hill, is in a state of *unstable* equilibrium, as a slight push will set it moving away from its original position. Applying these concepts to the atmosphere, we can see that air is in stable equilibrium when, after being lifted or lowered, it tends to return to its original position—it resists upward and downward air motions. Air that is in unstable equilibrium will, when given a little push, move farther away

from its original position—it favors upward and downward motion.

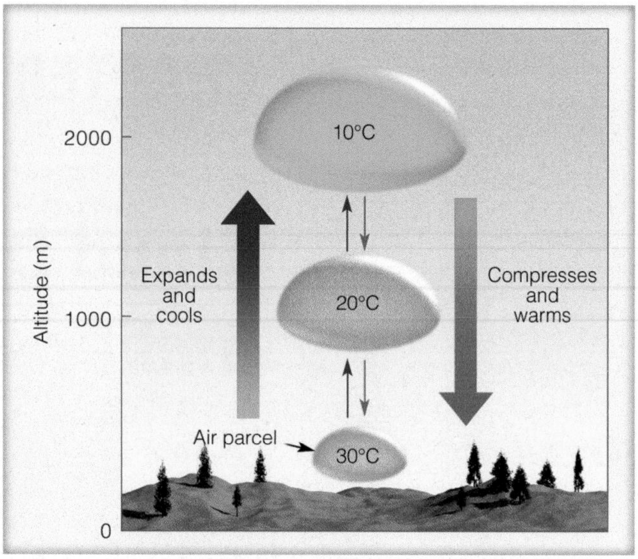

In order to explore the behavior of rising and sinking air, we must first review some concepts we learned in earlier chapters. Recall that a balloonlike blob of air is called an *air parcel*. (The concept of air parcels is illustrated in Fig. 4.4, p. 82, and in the Focus section on p. 32.) When an air parcel rises, it moves into a region where the air pressure surrounding it is lower. This situation allows the air molecules inside to push outward on the parcel walls, expanding it. As the air parcel expands, the air inside cools. If the same parcel is brought back to the surface, the increasing pressure around the parcel squeezes (compresses) it back to its original volume, and the air inside warms. Hence, *a rising parcel of air expands and cools, while a sinking parcel is compressed and warms.*

If a parcel of air expands and cools, or compresses and warms, and there is no interchange of heat with its outside surroundings, this situation is called an **adiabatic process.** As long as the air in the parcel is unsaturated (the relative humidity is less than 100 percent), the rate of adiabatic cooling or warming remains constant and is about 10°C for every 1000 meters of change in altitude, or about 5.5°F for every 1000 feet. Since this rate of cooling or warming only applies to unsaturated air, it is called the **dry adiabatic rate*** (see ● Fig. 5.2).

As the rising air cools, its relative humidity increases as the air temperature approaches the dew-point temperature. If the rising air cools to its dew-point temperature, the relative humidity becomes 100 percent. Further lifting results in condensation, a cloud forms, and latent heat is

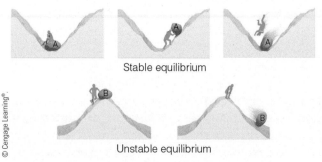

Stable equilibrium

Unstable equilibrium

● FIGURE 5.1 When rock A is disturbed, it will return to its original position; rock B, however, will accelerate away from its original position.

---

*For aviation purposes, the dry adiabatic rate is sometimes expressed as 3°C per 1000 ft.

released into the rising air. Because the heat added during condensation offsets some of the cooling due to expansion, the air no longer cools at the dry adiabatic rate but at a lesser rate called the **moist adiabatic rate.** (Because latent heat is added to the rising saturated air, the process is not really adiabatic.*) If a saturated parcel containing water droplets were to sink, it would compress and warm at the moist adiabatic rate because evaporation of the liquid droplets would offset the rate of compressional warming. Hence, the rate at which rising or sinking saturated air changes temperature—the moist adiabatic rate—is less than the dry adiabatic rate.

Unlike the dry adiabatic rate, the moist adiabatic rate is not constant, but varies greatly with temperature and, hence, with moisture content, because warm saturated air produces more liquid water than cold saturated air. The added condensation in warm, saturated air liberates more latent heat. Consequently, the moist adiabatic rate is much less than the dry adiabatic rate when the rising air is quite warm; however, the two rates are nearly the same when the rising air is very cold. Although the moist adiabatic rate does vary, to make the numbers easy to deal with we will use an average of 6°C per 1000 m (3.3°F per 1000 ft) in most of our examples and calculations.

Determining Stability

We determine the stability of the air by comparing the temperature of a rising parcel to that of its surroundings. If the rising air is colder than its environment, it will be more dense** (heavier) and tend to sink back to its original level. In this case, the air is *stable* because it resists upward movement. If the rising air is warmer and, therefore, less dense (lighter) than the surrounding air, it will continue to rise until it reaches the same temperature as its environment. This is an example of *unstable* air. To figure out the air's stability, we need to measure the temperature both of the rising air and of its environment at various levels above Earth.

A STABLE ATMOSPHERE Suppose we release a balloon-borne instrument called a radiosonde. (A photo of a radiosonde is found in Fig. 2 on p. 22.) As the balloon carries the radiosonde up into the atmosphere, it sends back temperature data, as shown in ● Fig. 5.3. Notice that the air temperature measured by the radiosonde decreases by 4°C for every 1000 meters rise in altitude. Remember

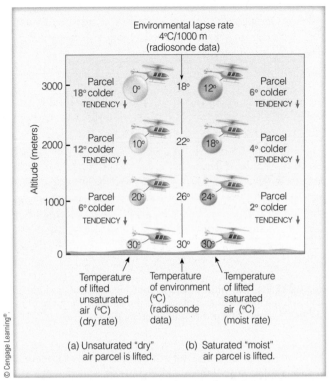

● **FIGURE 5.3** A stable atmosphere. Imagine that a helicopter could lift parcels of air, as shown above. An *absolutely stable atmosphere* exists when a rising air parcel is colder and heavier (i.e., more dense) than the air surrounding it. If given the chance (i.e., released), the air parcel in both situations would return to its original position, the surface. (In both situations, the helicopter shows that the air is being lifted. In the real world, this type of parcel lifting, of course, would be impossible.)

from Chapter 1 that the rate at which the air temperature changes with altitude is called the *lapse rate.* Because this rate is the one at which the air temperature surrounding us would be changing if we were to climb upward into the atmosphere, we refer to it as the **environmental lapse rate.**

Notice in Fig. 5.3a that (with an environmental lapse rate of 4°C per 1000 m) a rising parcel of unsaturated, "dry" air is colder and heavier than the air surrounding it at all levels. Even if the parcel is initially saturated (Fig. 5.3b), as it rises it, too, will be colder than its environment at all levels. In both cases, the atmosphere is **absolutely stable** because the lifted parcel of air is colder and heavier than the air surrounding it. If released, the parcel will have a tendency to return to its original position.

Since air in a stable atmosphere strongly resists upward vertical motion, it will, *if forced to rise,* tend to spread out horizontally. If clouds form in this rising air, they, too, will spread horizontally in relatively thin layers and usually have flat tops and bases. We might expect to see stratiform clouds—such as cirrostratus, altostratus, nimbostratus, or stratus—forming in a stable atmosphere.

The atmosphere is stable when the environmental lapse rate is small, that is, when there is a relatively small difference in temperature between the surface air and the

---

*If condensed water or ice is removed from the rising saturated parcel, the cooling process is called an *irreversible pseudoadiabatic process.*

**When, at the same level in the atmosphere, we compare parcels of air that are equal in size but vary in temperature, we find that cold air parcels are more dense than warm air parcels; that is, in the cold parcel, there are more molecules that are crowded closer together.

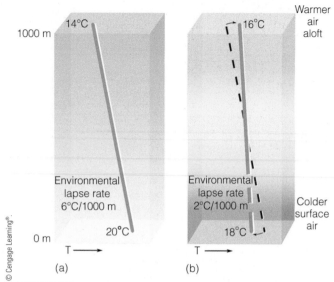

1000 m 14°C

Environmental
lapse rate
6°C/1000 m

0 m   20°C

T →

(a)

16°C   Warmer
air
aloft

Environmental
lapse rate
2°C/1000 m

                                    Colder
                  18°C   surface
                                    air

T →

(b)

● **FIGURE 5.4** The initial environmental lapse rate in diagram (a) will become more stable (stabilize) as the air aloft warms and the surface air cools, as illustrated in diagram (b).

air aloft. Consequently, the atmosphere tends to become more stable—it *stabilizes*—as the air aloft warms or the surface air cools (see ● Fig. 5.4). The *cooling* of the *surface air* can be due to:

1. nighttime radiational cooling of the surface
2. an influx of cold surface air brought in by the wind
3. air moving over a cold surface

So, on a typical day, the atmosphere is usually most stable in the early morning around sunrise, when the lowest surface air temperature is recorded. If the surface air becomes saturated in a stable atmosphere, a persistent layer of fog may form (see ● Fig. 5.5).

The air aloft may warm as winds bring in warmer air or as the air slowly sinks over a large area. Recall that sinking (subsiding) air warms as it is compressed. The warming can produce an *inversion*, where the air aloft is actually warmer than the air at the surface. (Recall from Chapter 3 that an inversion represents an atmospheric condition where the air becomes warmer with height.) An inversion that forms by slow, sinking air is termed a *subsidence inversion*. Because inversions represent a very stable atmosphere, they act as a lid on vertical air motion. When an inversion exists near the ground, stratus, fog, haze, and pollutants are all kept close to the surface. In fact, as we will see in Chapter 14, most air pollution episodes occur with subsidence inversions.

AN UNSTABLE ATMOSPHERE The atmosphere is unstable when the environmental air temperature decreases rapidly with height. For example, in ● Fig. 5.6, notice that the measured air temperature decreases by 11°C for every 1000-meter rise in altitude, which means that the environmental lapse rate is 11°C per 1000 meters. Also notice that a lifted parcel of unsaturated "dry" air in Fig. 5.6a, as well as a lifted parcel of saturated "moist" air in Fig. 5.6b, will, at each level above the surface, be warmer than the air surrounding them. Since, in both cases, the rising air parcels are warmer and less dense than the air around them, once the parcels start upward, they will continue to rise on their own, away from the surface. Thus, we have an **absolutely unstable atmosphere**. In an unstable environment, parcels of air are "buoyant" because the contrast between their warmer temperature and their cooler surroundings exerts an upward force on them. The warmer the air

● **FIGURE 5.5** On this morning near Boulder, Colorado, cold surface air has produced a stable atmosphere that inhibits vertical air motions and allows the fog to linger close to the ground.

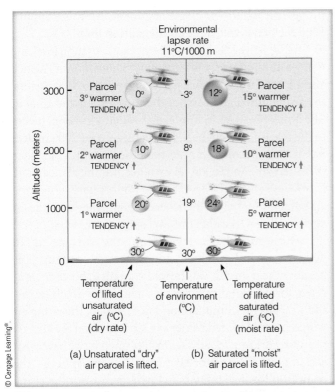

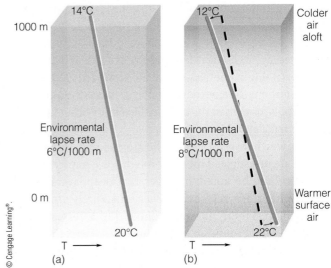

● **FIGURE 5.7** The initial environmental lapse rate in diagram (a) will become more unstable (that is, destabilize) as the air aloft cools and the surface air warms, as illustrated in diagram (b).

● **FIGURE 5.6** An unstable atmosphere. An absolutely unstable atmosphere exists when a rising air parcel is warmer and lighter (i.e., less dense) than the air surrounding it. If given the chance (i.e., released), the lifted parcel in both (a) and (b) would continue to move away (accelerate) from its original position. (As in Fig. 5.3, the lifting of air parcels by a helicopter would be impossible.)

parcels compared to their surroundings, the greater this *buoyant force* and the more rapidly they rise.*

The atmosphere becomes more unstable as the environmental lapse rate steepens; that is, as the

---

*A good example of buoyant force occurs when you submerge a balloon filled with air into a tub of water. As you push the lighter (less dense) balloon into the heavier (more dense) water, you can feel the upward-directed buoyant force acting on the balloon. Let go of the balloon and watch how rapidly it rises to the surface.

temperature of the air drops rapidly with increasing height. This circumstance can be brought on either by the air aloft becoming colder or by the surface air becoming warmer (see ● Fig. 5.7). The *warming* of the *surface air* may be due to:

1. daytime solar heating of the surface
2. an influx of warm surface air brought in by the wind
3. air moving over a warm surface

The combination of cold air aloft and warm surface air can produce a steep lapse rate and an unstable atmosphere (see ● Fig. 5.8).

Generally, then, as the surface air warms during the day, the atmosphere becomes more unstable—it *destabilizes*. The air aloft may cool as winds bring in colder air or as the air (or clouds) emit infrared radiation to space (radiational cooling). Just as sinking air produces

● **FIGURE 5.8** The warmth from this forest fire in Idaho during August 2003 heats the air, causing instability near the surface. Warm, less-dense air (and smoke) bubbles upward, expanding and cooling as it rises. Eventually the rising air cools to its dew point, condensation begins, and a cumulus cloud forms. If the rising air parcels are large and strong enough, the resulting clouds (sometimes called "pyrocumulus") may produce lightning and precipitation.

warming and a more stable atmosphere, rising air, especially an entire layer where the top is dry and the bottom is humid, produces cooling and a more unstable atmosphere. The lifted layer becomes more unstable as it rises and stretches out vertically in the less-dense air aloft. This stretching effect steepens the environmental lapse rate as the top of the layer cools more than the bottom. Instability brought on by the lifting of air is often associated with the development of severe weather, such as thunderstorms and tornadoes, which are investigated more thoroughly in Chapter 10.

It should be noted, however, that deep layers in the atmosphere are seldom, if ever, absolutely unstable. Absolute instability is usually limited to a very shallow layer near the ground on hot, sunny days. Here, the environmental lapse rate can exceed the dry adiabatic rate, and the lapse rate is called *superadiabatic.*

A CONDITIONALLY UNSTABLE ATMOSPHERE Suppose an unsaturated (but humid) air parcel is somehow forced to rise from the surface, as shown in ● Fig. 5.9. (What causes the air parcel to rise will be covered in a later section.) As the parcel rises, it expands, and cools at the *dry adiabatic rate* until its air temperature cools to its dew point. At this level, the air is saturated, the relative humidity is 100 percent, and further lifting results in condensation and the formation of a cloud. The elevation above

the surface where the cloud first forms (in this example, 1000 meters) is called the **condensation level.**

In Fig. 5.9, notice that above the condensation level, the rising saturated air cools at the *moist adiabatic rate.* Notice also that from the surface up to a level near 2000 meters, the rising, lifted air is colder than the air surrounding it. The atmosphere up to this level is *stable.* However, owing to the release of latent heat, the rising air near 2000 meters has actually become warmer than the air around it. Since the lifted air can rise on its own accord, the atmosphere is now *unstable.* The level in the atmosphere where the air parcel, after being lifted, becomes warmer than the air surrounding it, is called the *level of free convection.*

The atmospheric layer from the surface up to 4000 meters in Fig. 5.9 has gone from stable to unstable because the rising air was humid enough to become saturated, form a cloud, and release latent heat, which warms the air. Had the cumulus cloud not formed, the rising air would have remained colder at each level than the air surrounding it. From the surface to 4000 meters, we have what is said to be a **conditionally unstable atmosphere**—the condition for instability being whether or not the rising air becomes saturated. Therefore, *conditional instability* means that if unsaturated stable air is somehow lifted to a level where it becomes saturated, instability may result. In Fig. 5.9, we can see that the environmental lapse rate is 9°C

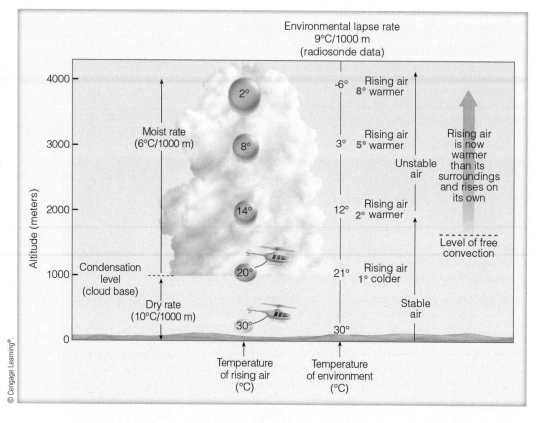

●**FIGURE 5.9** Conditionally unstable atmosphere. The atmosphere is conditionally unstable when unsaturated, stable air is lifted to a level where it becomes saturated and warmer than the air surrounding it. If the atmosphere remains unstable, vertical developing cumulus clouds can build to great heights.

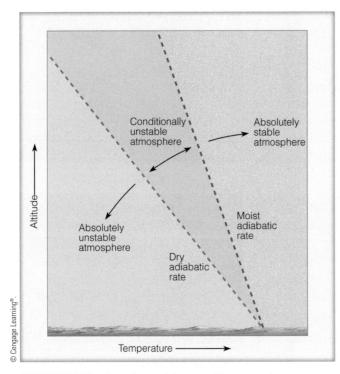

● **FIGURE 5.10** When the environmental lapse rate is greater than the dry adiabatic rate (blue region), the atmosphere is absolutely unstable. When the environmental lapse rate is less than the moist adiabatic rate (red region), the atmosphere is absolutely stable. And when the environmental lapse rate lies between the dry adiabatic rate and the moist adiabatic rate (green region), the atmosphere is conditionally unstable.

per 1000 meters. This value is between the dry adiabatic rate (10°C/1000 m) and the moist adiabatic rate (6°C/1000 m). Consequently, *conditional instability exists whenever the environmental lapse rate is between the dry and moist adiabatic rates.* Recall from Chapter 1 that the average lapse rate in the troposphere is about 6.5°C per 1000 m (3.6°F per 1000 ft). Since this value lies between the dry adiabatic rate and the average moist rate, *the atmosphere is ordinarily in a state of conditional instability.* ●Figure 5.10 summarizes how the three categories of stability (absolutely stable, conditionally unstable, and absolutely unstable) relate to the dry and moist adiabatic rates.

At this point, it should be apparent that the stability of the atmosphere changes during the course of a day. In clear, calm weather around sunrise, surface air is normally colder than the air above it, a radiation inversion exists, and the atmosphere is quite stable, as indicated by smoke or haze lingering close to the ground. As the day progresses, sunlight warms the surface and the surface warms the air above. As the air temperature near the ground increases, the lower atmosphere gradually becomes more unstable, with maximum instability usually occurring during the hottest part of the day. On a humid summer afternoon this phenomenon can be witnessed by the development of cumulus clouds.

Before going on to the next section, here is a brief review of some of the facts and concepts concerning atmospheric stability:

- The air temperature in a rising parcel of *unsaturated* air decreases at the dry adiabatic rate, whereas the air temperature in a rising parcel of *saturated* air decreases at the moist adiabatic rate.

- The dry adiabatic rate and moist adiabatic rate of cooling are different due to the fact that latent heat is released in a rising parcel of saturated air.

- In a *stable atmosphere*, a lifted parcel of air will be colder (heavier) than the air surrounding it. Because of this fact, the lifted parcel will tend to sink back to its original position.

- In an *unstable atmosphere*, a lifted parcel of air will be warmer (lighter) than the air surrounding it, and thus will continue to rise upward, away from its original position.

- The atmosphere becomes more stable (stabilizes) as the surface air cools, the air aloft warms, or a layer of air sinks (subsides) over a vast area.

- The atmosphere becomes more unstable (destabilizes) as the surface air warms, the air aloft cools, or a layer of air is lifted.

- A conditionally unstable atmosphere exists when a parcel of air can be lifted to a level where it becomes saturated, a cloud forms, and the rising parcel becomes warmer than the air surrounding it.

- The atmosphere is normally most stable in the early morning and most unstable in the afternoon.

- Layered clouds tend to form in a stable atmosphere, whereas cumuliform clouds tend to form in a conditionally unstable atmosphere.

Cloud Development and Stability

We know that most clouds form as air rises and cools and its water vapor condenses. Since air normally needs a "trigger" to start it moving upward, what is it that causes the air to rise so that clouds can form? The following mechanisms are primarily responsible for the development of the majority of clouds we observe:

1. surface heating and free convection
2. uplift along topography
3. widespread ascent due to the flowing together (convergence) of surface air
4. uplift along weather fronts (see ●Fig. 5.11).

The first mechanism that can cause the air to rise is *convection*. Although we briefly looked at convection in Chapter 2 when we examined rising thermals and how they transfer heat upward into the atmosphere, we will now look at convection from a slightly different perspective: how rising thermals can produce cumulus clouds.

CONVECTION AND CLOUDS Some areas of Earth's surface are better absorbers of sunlight than others and,

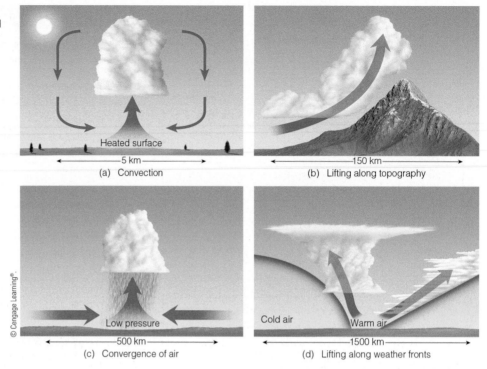

● **FIGURE 5.11** The primary ways clouds form: (a) surface heating and convection; (b) forced lifting along topographic barriers; (c) convergence of surface air; (d) forced lifting along weather fronts.

(a) Convection — 5 km

(b) Lifting along topography — 150 km

(c) Convergence of air — 500 km

Low pressure

(d) Lifting along weather fronts — 1500 km

Cold air   Warm air

therefore, heat up more quickly. The air in contact with these "hot spots" becomes warmer than its surroundings. A hot "bubble" of air—a *thermal*—breaks away from the warm surface and rises, expanding and cooling as it ascends. As the thermal rises, it mixes with the cooler, drier air around it and gradually loses its identity. Its upward movement now slows. Before the thermal is completely diluted, subsequent rising thermals often penetrate it and help the air rise a little higher. If the rising air cools to its saturation point, the moisture will condense, and the thermal becomes visible to us as a cumulus cloud.

Observe in ● Fig. 5.12 that the air motions are downward on the outside of the cumulus cloud. The downward motions are caused in part by evaporation around the outer edge of the cloud, which cools the air, making it heavy (more dense). Another reason for the downward motion is the completion of the convection current started by the thermal. Cool air slowly descends to replace the rising warm air. Therefore, we have rising air in the cloud and sinking air around it. Since subsiding air greatly inhibits the growth of thermals beneath it, small cumulus clouds usually have a great deal of blue sky between them (see ● Fig. 5.13).

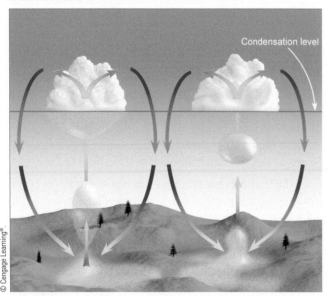

Condensation level

● **FIGURE 5.12** Cumulus clouds form as warm, invisible air bubbles detach themselves from the surface, then rise and cool to the condensation level. Below and within the cumulus clouds, the air is rising. Around the cloud, the air is sinking.

© UCAR

● **FIGURE 5.13** Cumulus clouds building on a warm summer afternoon. Each cloud represents a region where thermals are rising from the surface. The clear areas between the clouds are regions where the air is sinking.

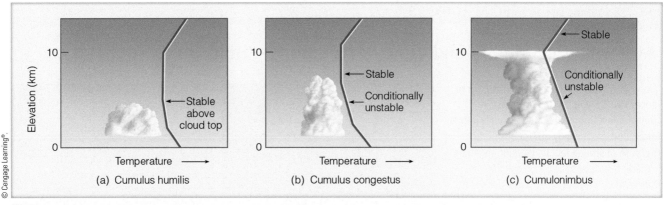

● **FIGURE 5.14** Variations in the air's stability, as indicated by the environmental lapse rate, greatly influence the growth of cumulus clouds.

As the cumulus clouds grow, they shade the ground from the sun. This, of course, cuts off surface heating and upward convection. Without the continual supply of rising air, the cloud begins to erode as its droplets evaporate. Unlike the sharp outline of a growing cumulus, the cloud now has indistinct edges, with cloud fragments extending from its sides. As the cloud dissipates (or moves along with the wind), surface heating begins again and regenerates another thermal, which becomes a new cumulus. This is why you often see cumulus clouds form, gradually disappear, then reform in the same spot.

The stability of the atmosphere plays an important part in determining the vertical growth of cumulus clouds. Notice in ● Fig. 5.14 that when a stable layer (such as an inversion) exists near the top of the cumulus cloud, the cloud would have a difficult time rising much higher, and it would remain as a "fair-weather" cumulus cloud, *cumulus humilis*. However, if a deep, conditionally unstable layer exists above the cloud, then the cloud can develop vertically into a towering cumulus congestus with a cauliflowerlike top. When the conditionally unstable air is several miles deep, the cumulus congestus can even develop into a cumulonimbus with a flat anvil-shaped top.

Notice in ● Fig. 5.15 that the distant thunderstorm has an anvil-shaped top. The cloud is shaped this way because it has reached the stable part of the atmosphere, and the rising air is unable to puncture very far into this stable layer, so the top of the cloud spreads laterally as high winds at this altitude (usually above 10 km or 33,000 ft) blow the cloud's ice crystals horizontally.

Atmospheric stability also plays a role in making many afternoons windier than mornings. This topic is discussed further in Focus section 5.1.

TOPOGRAPHY AND CLOUDS Horizontally moving air obviously cannot go through a large obstacle, such as a mountain, so the air must go over it. Forced lifting along a topographic barrier is called **orographic uplift**. Often, large masses of air rise when they approach a long chain of mountains such as the Sierra Nevada and Rockies. This lifting produces cooling, and if the air is humid, clouds form. Clouds produced in this manner are called *orographic clouds.*

An example of orographic uplift and cloud development is given in ● Fig. 5.16. Notice that, after having risen over the mountain, the air at the surface on the leeward (downwind) side is considerably warmer than it was at the surface on the windward (upwind) side. The higher air temperature on the leeward side is the result of latent heat being converted into sensible heat during condensation on the windward side. In fact, the rising air at the top of the mountain is considerably warmer than it would have been had condensation not occurred.

Notice also in Fig. 5.16 that the dew-point temperature of the air on the leeward side is lower than it was

● **FIGURE 5.15** Cumulus clouds developing into thunderstorms in a conditionally unstable atmosphere over the Great Plains. Notice that, in the distance, the cumulonimbus with the flat anvil-shaped top has reached a stable layer of the atmosphere.

Atmospheric Stability and Windy Afternoons—Hold On to Your Hat

On warm days when the weather is clear or partly cloudy, you may have noticed that the windiest time of the day is usually in the afternoon. Such windy afternoons occur because of several factors working together, including surface heating, convection, and atmospheric stability.

We know that in the early morning the atmosphere is most stable, meaning that the air resists up-and-down motions. As an example, consider the flow of air in the early morning as illustrated in ● Fig. 1a. Notice that weak winds exist near the surface with much stronger winds aloft. Because the atmosphere is stable, there is little vertical mixing between the surface air and the air higher up.

As the day progresses and the sun rises higher in the sky, the surface heats up and the lower atmosphere becomes more unstable. Over hot surfaces, the air begins to rise in the form of thermals that carry the slower-moving air with them (see Fig. 1b). At some level above the surface, the rising air links up with the faster-moving air aloft. If the air begins to sink as part of a convective

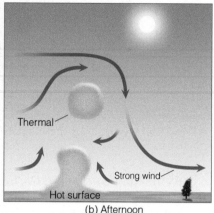

circulation, it may pull some of the stronger winds aloft downward with it. If this sinking air should reach the surface, it produces a momentary gust of strong wind. This exchange of air also increases the average wind speed at the surface. Because this type of air exchange is greatest on a clear day in

the afternoon when the atmosphere is most unstable, we tend to experience the strongest, most gusty winds in the afternoon. At night, when the atmosphere stabilizes, the interchange between the surface air and the air aloft is at a minimum, and the winds at the surface tend to die down.

● **FIGURE 1** (a) During the early morning, there is little exchange between the surface winds and the winds aloft. (b) In the afternoon, when the atmosphere is usually most unstable, convection in the form of rising thermals links surface air with the air aloft, causing strong winds from aloft to reach the ground and produce strong, gusty surface winds.

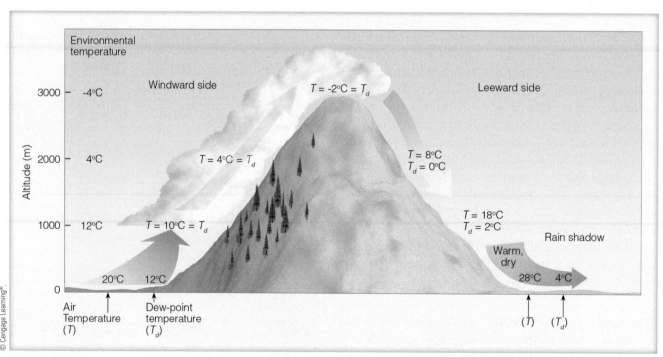

● **FIGURE 5.16** Orographic uplift, cloud development, and the formation of a rain shadow. (Note: The reason for the decrease in the dew-point temperature of the rising, unsaturated air on the windward side, and for the increase in dew-point temperature of the unsaturated sinking air on the leeward side is given in the footnote at the bottom of p. 125.)

before the air was lifted over the mountain. The lower dew point and, hence, drier air on the leeward side is the result of water vapor condensing and then remaining as liquid cloud droplets and precipitation on the windward side.* This region on the leeward side of a mountain, where precipitation is noticeably low, and the air is often drier, is called a **rain shadow.**

From Fig. 5.16 we have two important concepts to remember:

1. Air descending a mountain warms by compressional heating and, upon reaching the surface, can be much warmer than the air at the same level on the upwind side.
2. Air on the leeward side of a mountain is normally drier (has a lower dew point) than the air on the windward side. The lower dew point and higher air temperature on the leeward side produce a lower relative humidity, a greater potential for evaporation of water, and a rain shadow desert.

Although clouds are more prevalent on the windward side of mountains, they may, under certain atmospheric conditions, form on the leeward side as well. For example, stable air flowing over a mountain often moves in a series of waves that may extend for several hundred miles on the leeward side. Such waves often resemble the waves that form in a river downstream from a large boulder. Recall

---

*You may have noticed in Fig. 5.16 that the dew-point temperature of the rising unsaturated air on the windward side of the mountain beneath the cloud decreases by 2°C per 1000 m, and increases by 2°C per 1000 m in the descending unsaturated air on the leeward side. The decrease in dew-point temperature on the windward side is caused by the rapid decrease in air pressure of the rising air. Since the dew point is directly related to the actual vapor pressure, a decrease in total air pressure of the rising air causes a corresponding decrease in vapor pressure and, hence, a lowering of the dew-point temperature. Likewise, the rapid increase in air pressure of the sinking air on the leeward side causes an increase in the dew-point temperature.

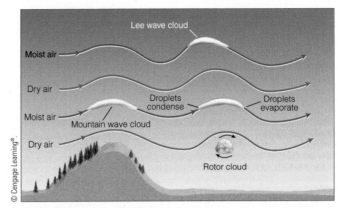

● **FIGURE 5.17** Lenticular clouds that form in the wave directly over the mountain are called mountain wave clouds, whereas those that form downwind of the mountain are called lee wave clouds. On the underside of the lee wave's crest a turbulent rotor may form.

from Chapter 4 that wave clouds often have a characteristic lens shape and are called *lenticular clouds.*

The formation of lenticular clouds is shown in ● Fig. 5.17. As moist air rises on the upwind side of the wave, it cools and condenses, producing a cloud. On the downwind side, the air sinks and warms, and the cloud evaporates. Viewed from the ground, the clouds appear motionless as the air rushes through them. When the air between the cloud-forming layers is too dry to produce clouds, lenticular clouds will form one above the other, sometimes extending into the stratosphere and appearing as a fleet of hovering spacecraft (see ● Fig. 5.18).

Notice in Fig. 5.17 that beneath the lenticular cloud downwind of the mountain range, a large swirling eddy forms. The rising part of the swirling air may cool enough to produce a visible cloud called a *rotor cloud*. The air in the rotor is extremely turbulent and presents a major hazard

● **FIGURE 5.18** Lenticular clouds tend to form over and downwind of mountains. They also tend to remain in one place as air rushes through them. Here, lenticular clouds are forming over mountainous terrain in Argentina's Los Glaciares National Park.

to aircraft in the vicinity. Dangerous flying conditions also exist near the lee side of the mountain, where strong downward air motions are present.

Now, having examined the concept of stability and the formation of clouds, we are ready to see how minute cloud particles are transformed into rain and snow. The next section, therefore, takes a look at the processes that produce precipitation.

Precipitation Processes

As we all know, cloudy weather does not necessarily mean that it will rain or snow. In fact, clouds may form, linger for many days, and never produce **precipitation.*** In Eureka, California, the August daytime sky is overcast more than 50 percent of the time, yet the average precipitation there for August is merely one-tenth of an inch. How, then, do cloud droplets grow large enough to produce rain? And why do some clouds produce rain, but not others?

In ● Fig. 5.19, we can see that an ordinary cloud droplet is extremely small, having an average diameter of 0.02 millimeters (mm), which is less than one-thousandth of an inch. Also, notice in Fig. 5.19 that a typical cloud droplet is 100 times smaller in diameter than a typical raindrop. Clouds, then, are composed of many droplets too small to fall as rain. These minute droplets require only slight upward air currents to keep them suspended. Those

---

*Recall from Chapter 4 that precipitation is any form of water (liquid or solid) that falls from a cloud and reaches the ground.

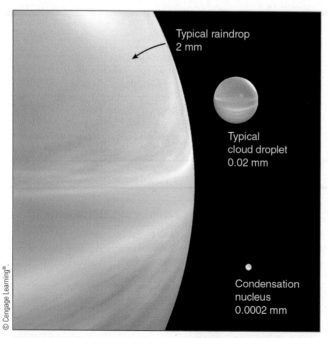

● **FIGURE 5.19** Relative sizes of raindrops, cloud droplets, and condensation nuclei, with diameters shown in millimeters (mm).

droplets that do fall, descend slowly and evaporate in the drier air beneath the cloud.

In Chapter 4, we learned that condensation begins on tiny particles called *condensation nuclei.* The growth of cloud droplets by condensation is slow and, even under ideal conditions, it would take several days for this process alone to create a raindrop. It is evident, then, that the condensation process by itself is entirely too slow to produce rain. Yet, observations show that clouds can develop and begin to produce rain in less than an hour. Since it takes about 1 million average-size cloud droplets to make an average-size raindrop, there must be some other process by which cloud droplets grow large and heavy enough to fall as precipitation.

Even though all the intricacies of how rain is produced are not yet fully understood, two important processes stand out: (1) the collision-coalescence process and (2) the ice-crystal (or Bergeron) process.

COLLISION AND COALESCENCE PROCESS   In clouds with tops warmer than −15°C (5°F), the **collision-coalescence process** can play a significant role in producing precipitation. To produce the many collisions necessary to form a raindrop, some cloud droplets must be larger than others. Larger drops can form on large condensation nuclei, such as salt particles, or through random collisions of droplets. Studies also suggest that turbulent mixing between the cloud and its drier environment can play a role in producing larger droplets.

As cloud droplets fall, air slows them down. The amount of air resistance depends on the size of the drop and on its rate of fall: The greater its speed, the more air molecules the drop encounters each second. The speed of the falling drop increases until the air resistance equals the pull of gravity. At this point, the drop continues to fall, but at a constant speed, which is called its *terminal velocity.* Because larger drops have a smaller surface area–to-weight ratio, they must fall faster before reaching their terminal velocity. Thus, *larger drops fall faster than smaller drops.*

Eventually, large droplets overtake and collide with smaller drops in their path. This merging of cloud droplets by collision is called **coalescence.** Within a cloud, there can be both updrafts and downdrafts. If an updraft is especially strong, then droplets of various sizes may all be pushed upward. Coalescence may now occur as smaller droplets are pushed upward more quickly, colliding with larger droplets in their path. Laboratory studies show that collision does not always guarantee coalescence; sometimes the droplets actually bounce apart during collision. For example, the forces that hold a tiny droplet together (*surface tension*) are so strong that if a droplet were to collide with another tiny droplet, chances are the two would not stick together (coalesce) (see ● Fig. 5.20). Coalescence

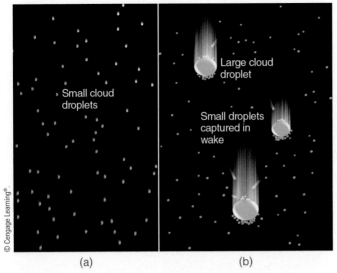

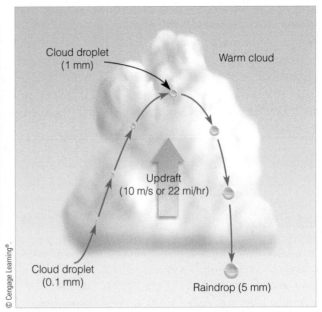

● **FIGURE 5.20** Collision and coalescence. (a) In a warm cloud composed only of small cloud droplets of uniform size, the droplets are less likely to collide as they all fall very slowly at about the same speed. Those droplets that do collide, frequently do not coalesce because of the strong surface tension that holds together each tiny droplet. (b) In a cloud composed of different-size droplets, larger droplets fall faster than smaller droplets. Although some tiny droplets are swept aside, some collect on the larger droplet's forward edge, while others (captured in the wake of the larger droplet) coalesce on the droplet's backside.

● **FIGURE 5.21** A cloud droplet rising, then falling through a warm cumulus cloud can grow by collision and coalescence and emerge from the cloud as a large raindrop.

appears to be enhanced if colliding droplets have opposite (hence, attractive) electrical charges.*

An important factor influencing cloud droplet growth by the collision process is the amount of time the droplet spends in the cloud. Since rising air currents slow the rate at which droplets fall, a thick cloud with strong updrafts will maximize the time cloud droplets spend in a cloud and, hence, the size to which they grow.

Clouds that have above-freezing temperatures at all levels are called *warm clouds*. In tropical regions, where warm cumulus clouds build to great heights, strong convective updrafts frequently occur. In ● Fig. 5.21, suppose a cloud droplet is caught in a strong updraft. As the droplet rises, smaller droplets rise more quickly and collide with it, allowing the droplet to reach a size of about 1 mm. At this point, the updraft in the cloud is just able to balance the pull of gravity on the drop, so the drop remains suspended until it grows just a little bigger. Once the fall velocity of the drop is greater than the updraft velocity in the cloud, the drop slowly descends. As the drop falls, some of the smaller droplets get caught in the airstream around it, and are swept aside. Larger cloud droplets are captured by the

falling drop, which then grows larger. By the time this drop reaches the bottom of the cloud, it will be a large raindrop with a diameter of over 5 mm. Because raindrops of this size fall faster and reach the ground first, they typically occur at the beginning of a rainshower originating in these warm, convective cumulus clouds.

So far, we have examined the way cloud droplets in warm clouds (that is, those clouds with temperatures above freezing) grow large enough by the collision-coalescence process to fall as raindrops. The most important factor in the production of raindrops is the cloud's liquid water content. In a cloud with sufficient water, other significant factors are:

1. the range of droplet sizes
2. the cloud thickness
3. the updrafts of the cloud
4. the electric charge of the droplets and the electric field in the cloud

Relatively thin stratus clouds with slow, upward air currents are, at best, only able to produce drizzle (the lightest form of rain), whereas the towering cumulus clouds associated with rapidly rising air can cause heavy showers. Now, let's turn our attention to how clouds with temperatures below freezing are able to produce precipitation.

ICE-CRYSTAL PROCESS The **ice-crystal** (or **Bergeron**) **process*** of rain formation proposes that both ice crystals and liquid cloud droplets are present in clouds at

---

*It was once thought that atmospheric electricity played a significant role in the production of rain. Today, evidence suggests that the difference in electrical charge that exists between cloud droplets results from the bouncing collisions between them. It is felt that the weak separation of charge and the weak electrical fields in developing relatively warm clouds are not significant in initiating precipitation. However, studies show that coalescence is often enhanced in thunderstorms where strongly charged droplets exist in a strong electrical field.

---

*The ice-crystal process is also known as the *Bergeron process* after the Swedish meteorologist Tor Bergeron, who proposed that essentially all raindrops begin as ice crystals.

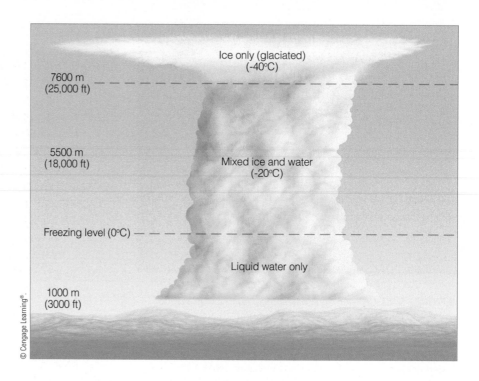

● **FIGURE 5.22** The distribution of ice and water in a typical cumulonimbus cloud.

7600 m (25,000 ft)

Ice only (glaciated) (-40°C)

5500 m (18,000 ft)

Mixed ice and water (-20°C)

Freezing level (0°C)

Liquid water only

1000 m (3000 ft)

© Cengage Learning®.

temperatures below freezing. This process of rain formation is extremely important in middle and high latitudes, where clouds are able to extend upwards into regions where air temperatures are below freezing. Such clouds are called *cold clouds.* ● Figure 5.22 illustrates a typical cumulonimbus cloud that has formed over the Great Plains of North America.

In the warm region of the cloud (below the freezing level) where only water droplets exist, we might expect to observe cloud droplets growing larger by the collision and coalescence process described in the previous section. Surprisingly, in the cold air just above the freezing level, almost all of the cloud droplets are still composed of liquid water. Water droplets existing at temperatures below freezing are referred to as **supercooled droplets.** At higher levels, ice crystals become more numerous, but are still outnumbered by water droplets. Ice crystals exist overwhelmingly in the upper part of the cloud, where air temperatures drop to well below freezing. Why are there so few ice crystals in the middle of the cloud, even though temperatures there, too, are below freezing? Laboratory studies reveal that the smaller the amount of pure water, the lower the temperature at which water freezes. Since cloud droplets are extremely small, it takes very low temperatures to turn them into ice.

Just as liquid cloud droplets form on condensation nuclei, ice crystals can form in subfreezing air if there are ice-forming particles present called **ice nuclei.** The number of ice-forming nuclei available in the atmosphere is small, especially at temperatures above −10°C (14°F). Although some uncertainty exists regarding the principal source of ice nuclei, it is known that certain clay minerals are excellent ice nuclei, as are some types of bacteria in decaying plant leaf material, along with ice crystals themselves and other particles whose geometry resembles that of an ice crystal.

We can now understand why there are so few ice crystals in the subfreezing region of some clouds. Liquid cloud droplets can freeze, but only at very low temperatures. Ice nuclei can initiate the growth of ice crystals, but they do not abound in nature. Therefore, we are left with a cold cloud that contains many more liquid droplets than ice particles, even at low temperatures. Neither the tiny liquid nor solid particles are large enough to fall as precipitation. How, then, does the ice-crystal process produce rain and snow?

In the subfreezing air of a cloud, many supercooled liquid droplets will surround each ice crystal. Suppose that the ice crystal and liquid droplet in ● Fig. 5.23 are part of a cold (−15°C), supercooled, saturated cloud. Since the air is saturated, both the liquid droplet and the ice crystal are in equilibrium, meaning that the number of molecules leaving the surface of both the droplet and the ice crystal must equal the number of molecules returning. Observe, however, that there are more vapor molecules above the liquid, because molecules escape the surface of water much more easily than they escape the surface of ice. Consequently, more molecules escape the water surface at a given temperature, requiring more in the vapor phase to maintain saturation. Therefore, it takes more vapor molecules to saturate the air directly above the water droplet than it does to saturate the air directly above the ice crystal. Put another way, at the same subfreezing temperature, *the saturation vapor pressure just above the water surface is greater than the saturation vapor pressure above the ice surface.*[*]

---

[*]This concept is illustrated in the insert in Fig. 4.5, p. 83.

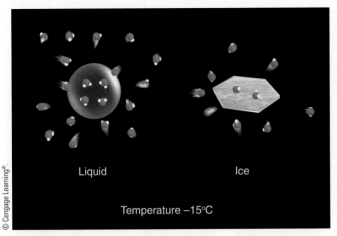

© Cengage Learning®.

● **FIGURE 5.23** In a saturated environment, the water droplet and the ice crystal are in equilibrium, as the number of water molecules leaving the surface of each droplet and ice crystal equals the number returning. There are more water vapor molecules above the droplet than above the ice, which produces a greater vapor pressure above the droplet. At saturation, then, the pressure exerted by the water molecules is greater over the water droplet than above the ice crystal.

This difference in saturation vapor pressure causes water vapor molecules to move (diffuse) from the droplet toward the ice crystal. The removal of vapor molecules reduces the vapor pressure above the droplet. Since the droplet is now out of equilibrium with its surroundings, it evaporates to replenish the diminished supply of water vapor above it. This process provides a continuous source of moisture for the ice crystal, which absorbs the water vapor and grows rapidly (see ● Fig. 5.24). Hence, *during the ice-crystal (Bergeron) process, ice crystals grow larger at the expense of the surrounding water droplets.*

The ice crystals may now grow even larger. For example, in some clouds, ice crystals might collide with supercooled liquid droplets. Upon contact, the liquid droplets freeze into ice and stick together. This process of ice crystals growing larger as they collide with super-cooled cloud droplets is called **accretion**. The icy matter that forms is called *graupel* (or *snow pellets*). As the graupel falls, it may fracture or splinter into tiny ice particles when it collides with cloud droplets. These splinters may then go on themselves to become new graupel, which, in turn, may produce more splinters. In colder clouds, the delicate ice crystals may collide with other crystals and fracture into smaller ice particles, or tiny seeds, which freeze hundreds of supercooled droplets on contact. In both cases a chain reaction can develop, producing many ice crystals (see ● Fig. 5.25). As they fall, they may collide and stick to one another, forming an aggregate of ice crystals called a *snowflake*. If the snowflake melts before reaching the ground, it continues its fall as a raindrop. Much of the rain falling in middle and northern latitudes—even in summer—actually begins as snowflakes.

CLOUD SEEDING AND PRECIPITATION The primary goal in many experiments concerning **cloud seeding** is to inject (or seed) a cloud with small particles that will act as nuclei, so that the cloud particles will grow large enough to fall to the surface as precipitation. The first ingredient in any seeding project is, of course, the presence of clouds, as seeding does not generate clouds. In clouds where the air temperature is below freezing, at least a portion of the cloud (preferably the upper part) must be supercooled, because in this situation cloud seeding uses the ice-crystal process to cause the cloud particles to grow. The idea is to find clouds that have too low a ratio of ice crystals to droplets and then to add enough artificial ice nuclei so that the ratio of crystals to droplets is optimal (about 1:100,000) for producing precipitation.

Some of the first experiments in cloud seeding were conducted by Vincent Schaefer and Irving Langmuir during the late 1940s. To seed a cloud, they dropped crushed pellets of *dry ice* (solid carbon dioxide) from a plane. Because dry ice has a temperature of −78°C (−108°F), it acts as a cooling agent. As the extremely cold pellets of dry ice fall through the cloud, they quickly cool the air around them. This cooling causes the air around the pellet to become supersaturated. In this supersaturated air, water

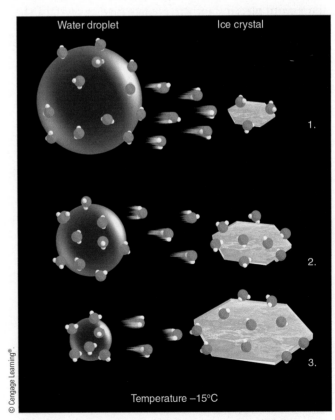

© Cengage Learning®.

● **FIGURE 5.24** The ice-crystal (Bergeron) process. (1) The greater number of water vapor molecules around the liquid droplet causes water molecules to diffuse from the liquid droplet toward the ice crystal. (2) The ice crystal absorbs the water vapor and grows larger, while (3) the water droplet grows smaller.

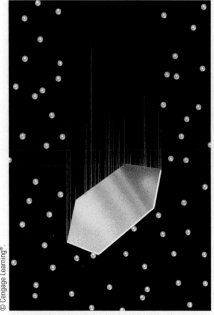

(a) Falling ice crystals may freeze supercooled droplets on contact (accretion), producing larger ice particles.

(b) Falling ice particles may collide and fracture into many tiny (secondary) ice particles.

(c) Falling ice crystals may collide and stick to other ice crystals (aggregation), producing snowflakes.

● FIGURE 5.25 Ice particles in clouds.

vapor directly forms into many tiny cloud droplets. In the very cold air created by the falling pellets (below −40°C), the tiny droplets instantly freeze into tiny ice crystals. The newly formed ice crystals then grow larger by deposition as the water vapor molecules attach themselves to the ice crystals at the expense of the nearby liquid droplets and, upon reaching a sufficiently large size, fall as precipitation.

In 1947, Bernard Vonnegut demonstrated that silver iodide (AgI) could be used as a cloud-seeding agent. Because silver iodide has a crystalline structure similar to an ice crystal, it acts as an effective ice nucleus at

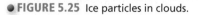

● FIGURE 5.26 Ice crystals falling from a dense cirriform cloud into a lower nimbostratus cloud. This photo was taken at an altitude near 6 km (19,700 ft) above western Pennsylvania. At the surface, moderate rain was falling over the region.

Thick cirrus

Regions of falling ice crystals

Nimbostratus

temperatures of −4°C (25°F) and lower. Silver iodide causes ice crystals to form in two primary ways:

1. Ice crystals form when silver iodide crystals come in contact with supercooled liquid droplets.
2. Ice crystals grow in size as water vapor is deposited onto the silver iodide crystal.

Silver iodide is much easier to handle than dry ice, since it can be supplied to the cloud from burners located either on the ground or on the wing of a small aircraft. Although other substances, such as lead iodide and cupric sulfide, are also effective ice nuclei, silver iodide still remains the most commonly used substance in cloud-seeding projects. (Additional information on the controversial topic of cloud seeding's effectiveness is given in Focus section 5.2.)

Under certain conditions, clouds can be seeded naturally. For example, when cirriform clouds lie directly above a lower cloud deck, ice crystals may descend from the higher cloud and seed the cloud below (see ● Fig. 5.26). As the ice crystals mix into the lower cloud, supercooled droplets are converted to ice crystals, and the precipitation process is enhanced. Sometimes the ice crystals in the lower cloud may settle out, leaving a clear area or "hole" in the cloud (see Fig. 2 in Focus section 5.2). When the cirrus clouds form waves downwind from a mountain chain, bands of precipitation often form—producing heavy precipitation in some areas and practically no precipitation in others (see ● Fig. 5.27).

Does Cloud Seeding Enhance Precipitation?

Just how effective is artificial seeding with silver iodide or other substances in increasing precipitation? In any given year, dozens of cloud-seeding projects are taking place around the world. However, the likelihood that cloud seeding will reliably increase precipitation is a much-debated question among meteorologists. First of all, it is difficult to evaluate the results of any cloud-seeding experiment. When a seeded cloud produces precipitation, the question always remains as to how much precipitation would have fallen had the cloud not been seeded. Other factors must be considered when evaluating cloud-seeding experiments: the type of cloud, its temperature and moisture content, the droplet-size distribution, and the updraft velocities in the cloud.

Some experiments have suggested that cloud seeding in some areas, *under the right conditions,* may enhance precipitation by anywhere from 5 percent to 20 percent or more. However, the results vary from project to project, and it can be difficult to confirm or disprove claims of success. And so the controversy continues.

Some cumulus clouds show an "explosive" growth after being seeded. The latent heat given off when the droplets freeze acts to warm the cloud, causing it to become more buoyant. It grows rapidly and becomes a longer-lasting cloud, which may produce more precipitation.

The business of cloud seeding can be a bit tricky, though, since overseeding can produce too many ice crystals. When this happens, the cloud becomes glaciated (all liquid droplets become ice) and

● **FIGURE 2** When an aircraft flies through a layer of altocumulus clouds composed of supercooled droplets, a hole in the cloud layer may form. The cirrus-type cloud in the center is probably the result of inadvertent cloud seeding by the aircraft.

the ice particles, being very small, do not fall as precipitation. Since few liquid droplets exist, the ice crystals cannot grow by the ice-crystal (Bergeron) process; rather, they disappear by changing from ice into water vapor (sublimating) and leaving a clear area in a thin, stratified cloud (see ● Fig. 2). Because dry ice can produce the most ice crystals in a supercooled cloud, it is the substance most suitable for deliberate overseeding. Hence, it is the substance most commonly used to dissipate cold fog at airports (see Chapter 4, p. 97).

Warm clouds with temperatures above freezing have also been seeded in an attempt to produce rain. Tiny water drops and particles of hygroscopic salt are injected into the base (or top) of the cloud. These particles (called *seed drops*), when carried into the cloud by updrafts, create large cloud droplets, which grow even larger by the collision-coalescence process. Apparently, the seed-drop size plays a major role in determining the effectiveness of seeding with hygroscopic particles. To date, however, the results obtained using this method are inconclusive.

In summary, cloud seeding in certain instances may lead to more precipitation; in others, to less precipitation; and, in still others, to no change in precipitation amounts. Many of the questions about cloud seeding have yet to be resolved.

PRECIPITATION IN CLOUDS In cold, strongly convective clouds, precipitation may begin only minutes after the cloud forms, and can be initiated by either the collision-coalescence or the ice-crystal (Bergeron) process. Once either process begins, most precipitation growth is by accretion, as supercooled liquid droplets freeze on impact with snowflakes and ice crystals. Although precipitation is commonly absent in warm layered clouds, such as stratus, it is often associated with such cold-layered clouds as nimbostratus and altostratus. This precipitation is thought to form principally by the ice-crystal (Bergeron) process because the liquid water content of these clouds is generally lower than that in convective clouds, thus making the collision-coalescence process much less effective. Nimbostratus clouds are normally thick enough to extend to levels where air temperatures are quite low, and they usually last long enough for the ice-crystal process to initiate precipitation.

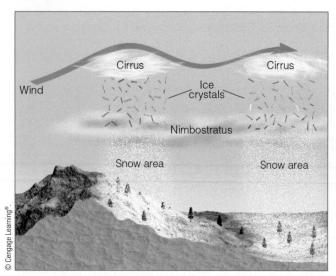

● **FIGURE 5.27** Natural seeding by cirrus clouds may form bands of precipitation downwind of a mountain chain. Notice that heavy snow is falling only in the seeded areas.

BRIEF REVIEW

In the last few sections we encountered a number of important concepts and ideas about how cloud droplets can grow large enough to fall as precipitation. Before examining the various types of precipitation, here is a summary of some of the important ideas presented so far:

- Cloud droplets are very small, much too small to fall as rain.

- Cloud droplets form on cloud condensation nuclei. Hygroscopic nuclei, such as salt, allow condensation to begin when the relative humidity is less than 100 percent.

- Cloud droplets, in above-freezing air, can grow larger as faster-falling, bigger droplets collide and coalesce with smaller droplets in their path.

- In the ice-crystal (Bergeron) process of rain formation, both ice crystals and liquid cloud droplets must coexist at below-freezing temperatures. The difference in saturation vapor pressure between liquid droplets and ice crystals causes water vapor to diffuse from the liquid droplets (which shrink) toward the ice crystals (which grow).

- Most of the rain that falls over middle latitudes results from melted snow that formed from the ice-crystal (Bergeron) process.

- Cloud seeding with silver iodide can only be effective in coaxing precipitation from a cloud if the cloud is supercooled and the correct ratio of cloud droplets to ice crystals exists.

Precipitation Types

Up to now, we have seen how cloud droplets can grow large enough to fall to the ground as rain or snow. While falling, raindrops and snowflakes can be altered by atmospheric conditions encountered beneath the cloud and transformed into other forms of precipitation that can profoundly influence our environment.

RAIN   Most people consider **rain** to be any falling drop of liquid water. To the meteorologist, however, that falling drop must have a diameter equal to, or greater than, 0.5 mm (0.02 in.) to be considered rain. Fine, uniform drops of water with diameters smaller than 0.5 mm (about one-half the width of the letter "o" in the print version of this page) are called **drizzle**. Most drizzle falls from stratus clouds; however, small raindrops may fall through air that is unsaturated, partially evaporate, and reach the ground as drizzle. Surfaces can be moistened by fog or mist, especially in windy conditions, even though the droplets in fog and mist are too tiny to fall to the ground as precipitation.

Occasionally, the rain falling from a cloud never reaches the surface because the low humidity causes rapid evaporation. As the drops become smaller, their rate of fall decreases, and they appear to hang in the air as a rain streamer. These evaporating streaks of precipitation are called **virga*** (see ● Fig. 5.28).

Raindrops may also fall from a cloud and not reach the ground if they encounter the rapidly rising air of an updraft. If the updraft weakens or changes direction and becomes a downdraft, the suspended drops will fall to the ground as a sudden **rainshower.** The showers falling from cumuliform clouds, such as cumulonimbus or cumulus congestus, are usually brief and sporadic as the cloud moves overhead and then drifts on by. If the shower is excessively heavy, it may be informally called a *cloudburst.* Beneath a cumulonimbus cloud, which normally contains strong, deep convection currents of rising and descending air, it is entirely possible for one side of a street to be dry (updraft side), while a heavy shower is occurring across the street (downdraft side) (see ● Fig. 5.29). Continuous rain, on the other hand, usually falls from a layered cloud that covers a large area and has weaker, more shallow vertical air currents. These are the conditions normally associated with nimbostratus clouds.

Raindrops that reach Earth's surface are seldom much larger than about 5 mm (0.2 in.), the reason being that the collisions (whether glancing or head-on) between raindrops tend to break them up into many smaller drops. Additionally, when raindrops grow too large they become unstable and break apart. What is the shape of the falling raindrop? Is it tear-shaped or is it round? You may be surprised at the answer, which is given in Focus section 5.3.

After a rainstorm, visibility usually improves, primarily because precipitation removes (scavenges) many of the suspended particles. When rain combines with

---

*Studies suggest that the "rain streamer" is actually caused by ice (which is more reflective) changing to water (which is less reflective). Apparently, most evaporation occurs below the virga line.

● FIGURE 5.28 The streaks of falling precipitation that evaporate before reaching the ground are called *virga*.

gaseous pollutants, such as oxides of sulfur and nitrogen, it becomes acidic. *Acid rain,* which has an adverse effect on plants and water resources, has become a major problem in many industrialized regions of the world over the past few decades. We will examine the acid rain problem more thoroughly in Chapter 14, which emphasizes air pollution.

SNOW   We know that much of the precipitation reaching the ground actually begins as **snow**. In summer, the freezing level is usually high and the snowflakes falling from a cloud melt before reaching the surface. In winter, however, the freezing level is much lower, and falling snowflakes have a better chance of survival. In fact, snowflakes can generally fall about 300 m (or 1000 ft) below the freezing level before completely melting. Occasionally, you can

spot the melting level when you look in the direction of the sun, if it is near the horizon. Because snow scatters incoming sunlight better than rain, the darker region beneath the cloud contains falling snow, while the lighter region is falling rain. The melting zone, then, is the transition between the light and dark areas (see ● Fig. 5.30).

When the warmer air beneath the cloud is relatively dry, the snowflakes partially melt. As the liquid water evaporates, it chills the snowflake, which retards its rate of melting. Consequently, in air that is relatively dry, snowflakes can reach the ground even when the air temperature is considerably above freezing, even above 40°F.

Is it ever "too cold to snow"? Although many believe this expression, the fact remains that it is *never* too cold to snow. For one thing, snow may actually fall from cold

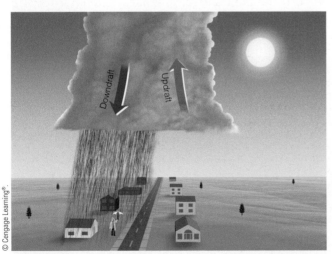

● **FIGURE 5.29** Strong updrafts and downdrafts of a cumulonimbus cloud can cause rain to fall on one side of a street but not on the other.

Snow

Melting level - - - - - - - - -

Rain

● **FIGURE 5.30** Snow scatters sunlight more effectively than rain. Consequently, when you look toward the sun, the region of falling precipitation looks darker above the melting level than below it.

Are Raindrops Tear-Shaped?

As rain falls, the drops take on a characteristic shape. Choose the shape in Fig. 3 that you think most accurately describes that of a falling raindrop. Did you pick number 1? The tear-shaped drop has been depicted by artists for many years. Unfortunately, *raindrops are not tear-shaped*. Actually, the shape depends on the drop size. Raindrops less than 2 mm (0.08 in.) in diameter are nearly spherical and look like raindrop number 2. The attraction among the molecules of the liquid (surface tension) tends to squeeze the drop into a shape that has the smallest surface area for its total volume—a sphere.

● FIGURE 3 Which of the three drops shown here represents the real shape of a falling raindrop?

Large raindrops, with diameters exceeding 2 mm, take on a different shape as they fall. Believe it or not, they look like number 3, slightly elongated, flattened on the bottom, and rounded on top. As the larger drop falls, the air pressure against the drop is greatest on the bottom and least on the sides. The pressure of the air on the bottom flattens the drop, while the lower pressure on its sides allows it to expand a little. This mushroom shape has been described as resembling everything from a falling parachute to a loaf of bread, or even a hamburger bun. You may call it what you wish, but remember: It is not tear-shaped.

air into extremely cold surface air. True, colder air cannot "hold" as much water vapor as warmer air, but no matter how cold the air becomes, it always contains some water vapor that could produce snow. In fact, tiny ice crystals have been observed falling at temperatures as low as −47°C (−53°F). We usually associate extremely cold air with "no snow" because the coldest winter weather occurs on clear, calm nights—conditions that normally prevail with strong high-pressure areas that have few if any clouds.

When ice crystals and snowflakes fall from high cirrus clouds they are called **fallstreaks.** Fallstreaks behave in much the same way as virga: As the ice particles fall into drier air, they usually disappear as they change from ice into vapor (called *sublimation*). Because the wind at higher levels moves the cloud and ice particles horizontally more quickly than do the slower winds at lower levels, fallstreaks often appear as dangling white streamers (see ● Fig. 5.31). Moreover, fallstreaks descending into lower, supercooled clouds may actually seed them.

Snowflakes falling through moist air that is slightly above freezing slowly melt as they descend. A thin film of water forms on the edge of the flake, acting like glue when other snowflakes come in contact with it. In this way, several flakes can join to produce giant snowflakes that often measure an inch or more in diameter. These large, soggy snowflakes are associated with moist air

● FIGURE 5.31 The dangling white streamers of ice crystals beneath these cirrus clouds are known as *fallstreaks*. The bending of the streaks is due to the changing wind speed with height.

● **FIGURE 5.32** Computer color-enhanced image of dendrite snowflakes.

● **FIGURE 5.33** High winds, blowing and falling snow, along with low temperatures, produced this blizzard over the Great Plains.

and temperatures near freezing. However, when snowflakes fall through extremely cold air with a low moisture content, they do not readily stick together and small, powdery flakes of "dry" snow accumulate on the ground.

If you catch falling snowflakes on a dark object and examine them closely, you will see that the most common snowflake form is a fernlike branching shape called *dendrite* (see ● Fig. 5.32). As ice crystals fall through a cloud, they are constantly exposed to changing temperatures and moisture conditions. Since many ice crystals can join together *(aggregate)* to form a much larger snowflake, ice crystals can assume many complex patterns.

Snow falling from developing cumulus clouds is often in the form of **flurries.** These are usually light showers that fall intermittently for short durations and produce only light accumulations. A more intense snow shower is called a **snow squall.** These brief but heavy falls of snow are comparable to summer rainshowers and, like snow flurries, usually fall from cumuliform clouds. If the snow falls from intense cumuliform clouds producing thunder and lightning, the snow is often referred to as **thundersnow.** A more continuous snowfall (sometimes persisting for several

hours) accompanies nimbostratus and altostratus clouds. The intensity of snow is based on its reduction of horizontal visibility at the time of observation (see ▼Table 5.1). However, the intensity of snow as measured by visibility does not tell us how much water the snow is bringing to the surface. A moderate snow made up of dense, small flakes can deposit more frozen liquid than a heavy snow consisting of large, fluffy flakes.

When a strong wind is blowing at the surface, snow can be picked up and deposited into huge drifts. Drifting snow is usually accompanied by *blowing snow;* that is, snow lifted from the surface by the wind and blown about in such quantities that horizontal visibility is greatly restricted. The combination of drifting and blowing snow, after falling snow has ended, is called a *ground blizzard.* A true **blizzard** is a weather condition characterized by low temperatures and strong winds (greater than 30 knots) bearing large amounts of fine, dry, powdery particles of snow that reduces visibility to less than one-quarter mile (and sometimes to as little as a few feet) for at least three hours (see ● Fig. 5.33).

● Figure 5.34 shows the annual average snowfall across the United States and Canada. As you would expect, annual snowfall totals tend to be low in the southern United States and higher as you move north. Notice that in areas of the northeast United States, eastern

▼ **Table 5.1   Snowfall Intensity**

SNOWFALL DESCRIPTION	VISIBILITY
Light	Greater than ½ mile*
Moderate	Greater than ¼ mile, less than or equal to ½ mile
Heavy	Less than or equal to ¼ mile

*In the United States, the National Weather Service determines visibility (the greatest distance you can see) in miles.

DID YOU KNOW?

On a sunny day, when the atmosphere is conditionally unstable, thermals break away from the surface and rapidly rise into the atmosphere carrying surface air with them. If the thermals form over a field of alfalfa, a feed lot, or a garbage dump, the smell of the surface air can be carried upward for thousands of feet.

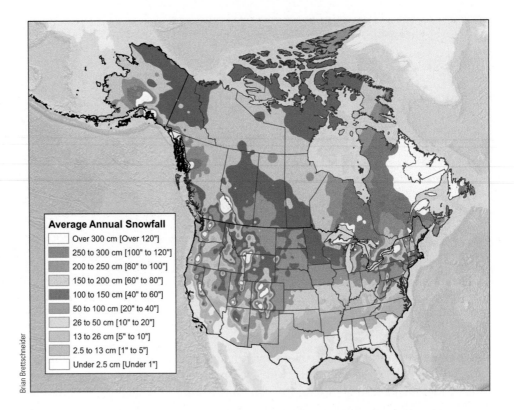

Brian Brettschneider

Average Annual Snowfall
- Over 300 cm [Over 120"]
- 250 to 300 cm [100" to 120"]
- 200 to 250 cm [80" to 100"]
- 150 to 200 cm [60" to 80"]
- 100 to 150 cm [40" to 60"]
- 50 to 100 cm [20" to 40"]
- 26 to 50 cm [10" to 20"]
- 13 to 26 cm [5" to 10"]
- 2.5 to 13 cm [1" to 5"]
- Under 2.5 cm [Under 1"]

Canada, and the mountainous west, annual snowfall totals exceed 183 cm (72 in.). In fact, Paradise Ranger Station on Mount Rainier, Washington, receives an annual average of 1758 cm (692 in.) of snow, making it one of the snowiest places in the world. Extremely heavy snowfalls also occur downwind from large bodies of water, such as the Great Lakes of North America. These so-called *lake effect snows* will be covered more completely in Chapter 8 (p. 215).

SLEET AND FREEZING RAIN Consider the falling snowflake in ● Fig. 5.35. As it falls into warmer air, it begins to melt. When it falls through the deep subfreezing surface

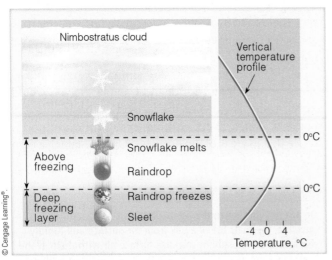

© Cengage Learning®.

Nimbostratus cloud

Vertical temperature profile

Snowflake

0°C

Above freezing

Snowflake melts

Raindrop

Deep freezing layer

0°C

Raindrop freezes

Sleet

-4 0 4
Temperature, °C

● FIGURE 5.35 Sleet forms when a partially melted snowflake or a cold raindrop freezes into a pellet of ice before reaching the ground.

layer of air, the partially melted snowflake or cold raindrop turns back into ice, not as a snowflake, but as a tiny transparent (or translucent) *ice pellet* called **sleet**.* Generally, these ice pellets bounce when striking the ground and produce a tapping sound when they hit a window or piece of metal.

The cold surface layer beneath a cloud may be too shallow to freeze raindrops as they fall. In this case, they reach the surface as supercooled liquid drops. Upon striking a cold object, the drops spread out and almost immediately freeze, forming a thin veneer of ice. This form of precipitation is called **freezing rain,** or *glaze.* If the drops are quite small, the precipitation is called *freezing drizzle.* When small, supercooled cloud or fog droplets strike an object whose temperature is below freezing, the tiny droplets freeze, forming an accumulation of white or milky granular ice called **rime** (see ● Fig. 5.36).

Occasionally, light rain, drizzle, or supercooled fog droplets come into contact with such surfaces as bridges and overpasses that have cooled to a temperature below freezing. The tiny liquid droplets freeze on contact to road surfaces or pavements, producing a sheet of ice that often appears relatively dark. Such ice, usually called **black ice**, can produce extremely hazardous driving conditions.

Freezing rain can create a beautiful winter wonderland by coating everything with silvery, glistening ice. At the same time, highways turn into skating rinks for automobiles, and the destructive weight of the ice—which

*Occasionally, news media in the United States will use the term "sleet" to describe a mixture of rain and snow. While this is not the American definition, it is correct in some other nations, including the United Kingdom.

Aircraft Icing

The formation of ice on an aircraft—called *aircraft icing*—can be extremely dangerous, sometimes leading to tragic accidents. Icing is believed to be the probable cause for the crash-landing of a passenger plane as it approached the Detroit airport on January 9, 1997. All 29 people aboard the flight were killed. Fortunately, the number of icing-related aircraft fatalities in the United States has decreased substantially in recent years with increased safety measures. Just how does aircraft icing develop?

Consider an aircraft flying through an area of freezing rain or through a region of large supercooled droplets in a cumuliform cloud. As the large, supercooled drops strike the leading edge of the wing, they break apart and form a film of water, which quickly freezes into a solid sheet of ice. This smooth, transparent ice—called *clear ice*—is similar to the freezing rain or glaze that coats trees during ice storms. Clear ice can build up quickly; it is heavy and difficult to remove, even with modern de-icers.

When an aircraft flies through a cloud composed of tiny, supercooled liquid droplets, *rime ice* may form. Rime ice forms when some of the cloud droplets strike the wing and freeze before they have time to spread, thus leaving a rough and brittle coating of ice on the wing. Because the small, frozen droplets trap air between them, rime ice usually appears white

● **FIGURE 4** An aircraft undergoing de-icing during inclement winter weather.

(see Fig. 5.35). Even though rime ice redistributes the flow of air over the wing more than clear ice does, it is lighter in weight and is more easily removed with de-icers.

Because the raindrops and cloud droplets in most clouds vary in size, a mixture of clear and rime ice usually forms on aircraft. Also, because concentrations of liquid water tend to be greatest in warm air, icing is usually heaviest and most severe when the air temperature is between 0°C and –10°C (32°F and 14°F).

A major hazard to aviation, icing reduces aircraft efficiency by increasing weight. Icing has other adverse effects,

depending on where it forms. On a wing or fuselage, ice can disrupt the airflow and decrease the plane's flying capability. When ice forms in the air intake of the engine, it robs the engine of air, causing a reduction in power. Icing may also affect the operation of brakes, landing gear, and instruments. Because of the hazards of ice on an aircraft, its wings are usually sprayed with a type of antifreeze before taking off during cold, inclement weather (see ● Fig. 4). In a snowstorm, the amount of liquid in the snow is the primary factor in determining how much de-icing fluid is needed for aircraft.

●**FIGURE 5.36** An accumulation of rime forms on tree branches as supercooled fog droplets freeze on contact in the below-freezing air.

can be many tons on a single tree—breaks tree branches, power lines, and telephone cables. When there is a substantial accumulation of freezing rain, the result is an **ice storm** (see ●Fig. 5.37). A case in point is the catastrophic ice storm that struck an area from Oklahoma to West Virginia in January 2009. More than 2 million people lost power, and 65 people died. Many of the deaths were attributed to the use of emergency heaters without adequate ventilation, which led to carbon monoxide poisoning. The area most frequently hit by these storms extends over a broad region from Texas into Minnesota and eastward into the middle Atlantic states and New England. Such storms are extremely rare in most of California and Florida. (For information on freezing rain and its effect on aircraft, read Focus section 5.4.)

● **FIGURE 5.37** A heavy coating of freezing rain (glaze) covers Syracuse, New York, causing tree limbs to break and power lines to sag. This massive ice storm in January 1998 left millions without power and caused over $1 billion in damage in northern New England and southeast Canada.

In summary, ● Fig. 5.38 shows various winter temperature profiles and the type of precipitation associated with each. In profile (a), the air temperature is below freezing at all levels, and snowflakes reach the surface. In (b), a zone of above-freezing air causes snowflakes to partially melt; then, in the deep, subfreezing air at the surface, the liquid freezes into sleet. In the shallow sub-freezing surface air in (c), the melted snowflakes, now supercooled liquid drops, freeze on contact, producing freezing rain. In (d), the air temperature is above freezing in a sufficiently deep layer so that precipitation reaches the surface as rain.

SNOW GRAINS AND SNOW PELLETS **Snow grains** are small, opaque grains of ice, the solid equivalent of drizzle. They fall in small quantities from stratus clouds, and never in the form of a shower. Upon striking a hard surface, they neither bounce nor shatter. **Snow pellets**, on the other hand, are white, opaque grains of ice about the size of an average raindrop. They are sometimes confused with snow grains. The distinction is easily made, however, by remembering that, unlike snow grains, snow pellets are brittle, crunchy, and bounce (or break apart) upon hitting a hard surface. They usually fall as showers, especially from cumulus congestus clouds.

Snow pellets form as ice crystals collide with supercooled water droplets that freeze into a spherical aggregate of icy matter *(rime)* containing many air spaces. When the ice particle accumulates a heavy coating of rime, it is called *graupel*. During the winter, when the freezing level is at a low elevation, the graupel reaches the surface as a *snow pellet*, a light, round clump of snowlike ice (see ● Fig. 5.39).

On the surface, the accumulation of snow pellets sometimes gives the appearance of tapioca pudding; hence, it can be referred to as *tapioca snow*. In a thunderstorm, when the freezing level is well above the surface, graupel that reaches the ground is sometimes called *soft hail*. During summer, graupel may melt and reach the surface as large raindrops. In vigorously convective clouds, however, graupel may develop into full-fledged hailstones.

HAIL **Hailstones** are pieces of ice, either transparent or partially opaque, ranging in size from that of small peas to that of golf balls or larger (see ● Fig. 5.40). Some are round; others take on irregular shapes. In the United States, the hailstone with the greatest measured circumference (18.7 inches) fell on Aurora, Nebraska, on June 22, 2003. This giant hailstone, almost as large as a soccer ball, had a measured diameter of 7 inches and probably weighed over 1.75 pounds. The hailstone in the United States with the

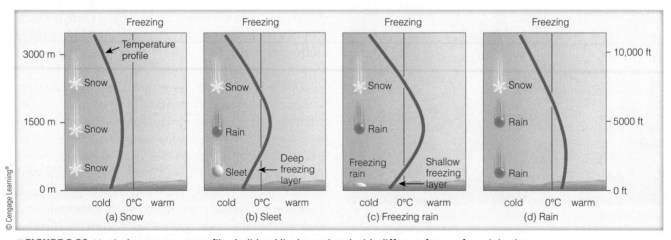

● **FIGURE 5.38** Vertical temperature profiles (solid red line) associated with different forms of precipitation.

● **FIGURE 5.39** A snowflake becoming a rimed snowflake, then finally graupel (a snow pellet).

largest diameter ever measured (8 inches) and the greatest weight (1.94 lb.) fell on Vivian, South Dakota, on July 23, 2010 (see ● Fig. 5.41). The world's heaviest confirmed hailstone, at 2.25 pounds, fell in a Bangladesh thunderstorm on April 14, 1986. Needless to say, large hailstones are quite destructive. They can break windows, dent cars, batter roofs of homes. Even small hail can injure livestock and cause extensive damage to crops if it falls heavily, especially in a high wind. A single hailstorm can destroy a farmer's crop in a matter of minutes, which is why farmers sometimes call it "*the white plague.*"

Estimates are that, in the United States alone, hail damage amounts to hundreds of millions of dollars annually. Although hailstones are potentially lethal, only a few fatalities from falling hail have been documented in the United States since 1900. Deaths due to hail are somewhat more frequent in parts of Asia, where 92 people were killed in the Bangladesh thunderstorm of April 1986.

Hail forms in a cumulonimbus cloud—usually an intense thunderstorm—when graupel, or large frozen raindrops, or just about any particles (even insects) act as *embryos* that grow by *accretion*, the accumulation of supercooled liquid droplets. For a hailstone to grow to golf-ball size, it must remain in the cloud between five and ten minutes. Violent, upsurging air currents within the storm carry small ice particles high above the freezing level where the ice particles grow by colliding with supercooled liquid cloud droplets. Violent rotating updrafts in severe thunderstorms (especially the long-lived storms called *supercells*) are even capable of sweeping the growing ice particles laterally through the cloud. In fact, it appears that the best trajectory for hailstone growth is one that is nearly horizontal through the storm (see ● Fig. 5.42).

As growing ice particles pass through regions of varying liquid water content, a coating of ice forms around them, causing them to grow larger and larger. In a strong updraft, the larger hailstones ascend very slowly, and may appear to "float" in the updraft, where they continue to grow rapidly by colliding with numerous supercooled liquid droplets. When winds aloft carry the large hailstones away from the updraft or when the hailstones reach appreciable size, they become too heavy to be supported by the rising air, and they begin to fall.

● **FIGURE 5.40** Hailstones of varying sizes fell over west Texas during an intense thunderstorm.

● **FIGURE 5.41** This whopping hailstone fell on Vivian, South Dakota, on July 23, 2010. It had a record diameter of 8 inches, weighed a record 1.94 pounds, and had a circumference of 18.6 inches.

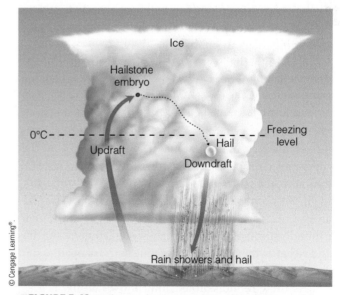

● **FIGURE 5.42** Hailstones begin as embryos (usually ice particles called *graupel*) that remain suspended in the cloud by violent updrafts. The updrafts (or a single broad updraft as illustrated here) can sweep the ice particles horizontally through the cloud, producing the optimal trajectory for hailstone growth. Along their path, the ice particles collide with supercooled liquid droplets, which freeze on contact. The ice particles eventually grow large enough and heavy enough to fall toward the ground as hailstones.

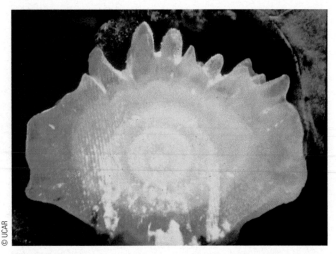

● **FIGURE 5.43** A large hailstone cut and then photographed under regular light. The layered structure of the hailstone reveals that it traveled through a cloud of varying water content and temperature.

© UCAR

In the warmer air below the cloud, the hailstones begin to melt. Small hail often completely melts before reaching the ground, but in the violent thunderstorms of late spring and summer, hailstones often grow large enough to reach the surface before completely melting. Strangely, then, we find the largest form of frozen precipitation occurring during the warmest time of the year.

● Figure 5.43 shows a cut section of a very large hailstone. Notice that it has distinct concentric layers of milky white and clear ice. We know that a hailstone grows by accumulating supercooled water droplets. If the growing hailstone enters a region inside the storm where the liquid-water content is relatively low (called the *dry growth regime*), supercooled droplets will freeze immediately on the stone, producing a coating of white or opaque rime ice containing many air bubbles.

Should the hailstone get swept into a region of the storm where the liquid-water content is higher (called the *wet growth regime*), supercooled water droplets will collect so rapidly on the stone that, due to the release of latent heat, the stone's surface temperature remains at freezing, even though the surrounding air may be much colder. Now the supercooled droplets no longer freeze on impact; instead, they spread a coating of water around the hailstone, filling in the porous regions, which leaves a layer of clear ice around the stone. Therefore, as a hailstone passes through a thunderstorm of changing liquid-water content (the dry and wet growth regimes), alternating layers of opaque and clear ice form, as illustrated in Fig. 5.43.

As the cumulonimbus cloud moves along, it may deposit its hail in a long, narrow band known as a *hail-streak*. If the cloud should remain almost stationary for a period of time, substantial accumulation of hail is possible. A rare British hailstorm in the town of Ottery St. Mary dropped 9 to 10 inches of hail in two hours in October 2008, and drifts of up to 6 feet of hail occurred during an intense thunderstorm in Cheyenne, Wyoming, in August 1985. During November 2003, an unusual hailstorm dumped more than 12.5 cm (5 in.) of hail over sections of Los Angeles, California, causing gutters to clog and floods to occur. In addition to its destructive effect, accumulation of hail on a roadway is a hazard to traffic as when, for example, four people lost their lives near Soda Springs, California, in a 15-vehicle pileup on a hail-covered freeway in September 1989.

Because hailstones are so damaging, various methods have been tried to prevent them from forming in thunderstorms. One method employs the seeding of clouds with large quantities of silver iodide. These nuclei freeze supercooled water droplets and convert them into ice crystals. The ice crystals grow larger as they come in contact with additional supercooled cloud droplets. In time, the ice crystals grow large enough to be called *graupel*, which then becomes a hailstone embryo. Large numbers of embryos are produced by seeding in hopes that competition for the remaining supercooled droplets will be so great that none of the embryos can grow into large and destructive hailstones. In the United States, the results of most hail-suppression experiments are still inconclusive, although such cloud seeding has been carried out in many areas for a number of years.

Measuring Precipitation

INSTRUMENTS Any instrument that can collect and measure rainfall is called a *rain gauge*. A **standard rain gauge** consists of a funnel-shaped collector attached to a long measuring tube (see ● Fig. 5.44). The cross-sectional area of the collector is 10 times that of the tube. Hence, rain falling into the collector is amplified tenfold in the tube, permitting measurements as precise as one-hundredth (0.01) of an inch. An amount of rainfall less than one-hundredth of an inch is called a **trace**.

Another instrument that measures rainfall is the *tipping bucket rain gauge*. In ● Fig. 5.45, notice that this gauge has a receiving funnel leading to two small metal collectors (buckets). The bucket beneath the funnel collects the rainwater. When it accumulates the equivalent of one-hundredth of an inch of rain, the weight of the water causes it to tip and empty itself. The second bucket immediately moves under the funnel to catch the water. When it fills, it also tips and empties itself, while the original bucket moves back beneath the funnel. Each time a bucket tips, an electric contact is made, causing a pen to register a mark on a remote recording chart. Adding up the total number of marks gives the rainfall for a certain time period. A problem with the tipping bucket rain gauge is that during each "tip" it loses some rainfall and, therefore, under-measures rainfall amounts, especially during heavy downpours. The tipping bucket is the rain gauge used in the automated (ASOS) weather stations.

Remote recording of precipitation can also be made with a *weighing-type rain gauge.* With this gauge, precipitation is caught in a cylinder and accumulates in a bucket. The bucket sits on a sensitive weighing platform. Special gears translate the accumulated weight of rain or snow into millimeters or inches of precipitation. The precipitation totals are recorded by a pen on chart paper, which covers a clock-driven drum. By using special electronic equipment, this information can be transmitted from various weighing types of rain gauges in remote areas to satellites or land-based stations, thus providing precipitation totals from previously inaccessible regions.

Snow is challenging to measure, since accumulations can vary greatly from one spot to the next, especially when winds are strong. Traditionally, the depth of snow in a region is determined by physically measuring its depth at three or more representative areas. The amount of snowfall is defined as the average of these measurements. Snow accumulation rates can be measured at a single site by using a *snow measuring board,* a small wooden platform resting near the ground that is cleared off after each measurement (normally every six hours). Snow depth may also be measured by removing the collector and inner cylinder of a standard rain gauge and allowing snow to accumulate in the outer tube. Automated gauges are now used in many areas to collect snowfall and measure the amount of liquid water it contains. Typically, these gauges are surrounded by one or more octagonal fences that help to block wind and produce more accurate readings.

Remote sensing techniques are becoming a popular way to measure snow depth, especially where harsh winter conditions make it difficult to reach observing stations. Laser beams and pulses of ultrasonic energy can be sent from a transmitter to a snowpack where the energy bounces off the snowpack back to the transmitter. This procedure allows the height of the snow accumulation to be measured in much the same way that radar measures the distance falling rain is from a transmitter. (The topic of radar and precipitation is discussed in the following section.) Snow height can also be obtained by measuring how long it takes

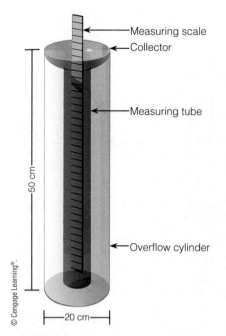

● **FIGURE 5.44** Components of the standard rain gauge.

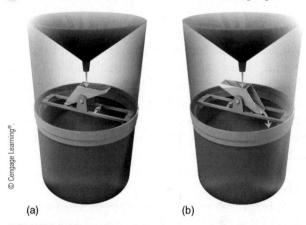

(a)   (b)

● **FIGURE 5.45** The tipping bucket rain gauge. Each time the bucket fills with one-hundredth of an inch of rain, it tips, sending an electric signal to the remote recorder.

a GPS signal to travel from a satellite to a snow field and then be reflected upward to a receiver located in the snow field.

On average, about 10 inches of snow will melt down to about 1 inch of water, giving a typical fresh snowpack a **water equivalent*** of 10:1. This ratio, however, will vary greatly, depending on the texture and packing of the snow. Knowing the water equivalent of snow can provide valuable information about spring runoff and the potential for flooding, especially in mountain areas.

DOPPLER RADAR AND PRECIPITATION **Radar** (*ra*dio *d*etection *a*nd *r*anging) is an essential tool of the atmospheric scientist, because it gathers information about storms and precipitation in otherwise inaccessible regions. Atmospheric scientists use radar to examine the inside of

---

*Water equivalent is the depth of water that would result from the melting of a snow sample.

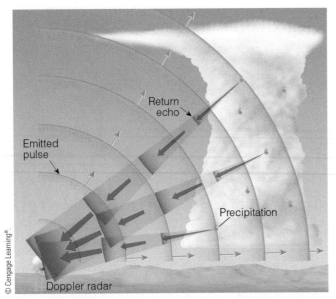

● **FIGURE 5.46** A microwave pulse is sent out from the radar transmitter. The pulse strikes raindrops and a fraction of its energy is reflected back to the radar unit, where it is detected and displayed, as shown in Fig. 5.47.

a cloud in much the same way that physicians use X rays to examine the inside of a human body. Essentially, the radar unit consists of a transmitter that sends out short microwave pulses. When this energy encounters a foreign object—called a *target*—a fraction of the energy is scattered back toward the transmitter and is detected by a receiver (see ● Fig. 5.46). The returning signal is amplified and displayed on a screen, producing an image or "echo" from the target. The elapsed time between transmission and reception indicates the target's distance.

The brightness of the echo is directly related to the amount (intensity) of rain, snow, or both falling through the cloud. So, the radar screen shows not only where precipitation is most likely occurring, but also how intense it is. Typically the radar image is displayed using various colors,

usually ranging from green or blue to dark red, to denote the intensity of precipitation within the range of the radar unit.

During the 1990s, **Doppler radar** replaced the conventional radar units that were put into service shortly after World War II. Doppler radar is like conventional radar in that it can detect areas of precipitation and measure rainfall intensity (see ● Fig. 5.47a). Using special computer programs called *algorithms,* the rainfall intensity, over a given area for a given time, can be computed and displayed as an estimate of total rainfall over that particular area (see Fig. 5.47b). But the Doppler radar can do more than conventional radar.

Because the Doppler radar uses the principle called *Doppler shift,** it has the capacity to measure the speed at which falling rain is moving horizontally toward or away from the radar antenna. Falling rain moves with the wind. Consequently, Doppler radar allows scientists to peer into a tornado-generating thunderstorm and observe its wind. We will investigate these ideas further in Chapter 10, when we consider the formation of severe thunderstorms and tornadoes.

In some instances, radar displays indicate precipitation where there is none reaching the surface. This situation happens because the radar beam travels in a nearly straight line while Earth's surface curves away from it. Hence, the return echo is not necessarily that of precipitation reaching the ground, but is that of raindrops in the cloud. So, if Doppler radar indicates that it's raining in your area, and outside you observe that it is not, remember that it is most likely raining, but the raindrops are probably evaporating before reaching the ground.

---

*The *Doppler shift* (or effect) is the change in the frequency of waves that occurs when the observer or the target is moving toward or away from the other. As an example, suppose a high-speed train is approaching you. The higher-pitched (higher frequency) whistle you hear as the train approaches will shift to a lower pitch (lower frequency) after the train passes.

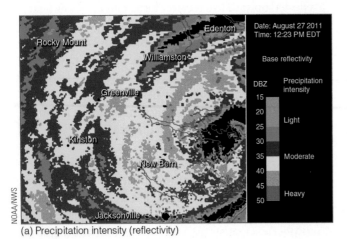

(a) Precipitation intensity (reflectivity)

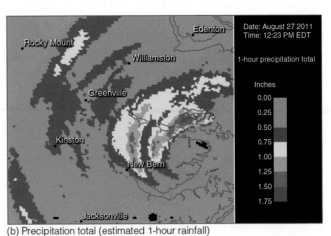

(b) Precipitation total (estimated 1-hour rainfall)

● **FIGURE 5.47** (a) Doppler radar display showing precipitation intensity over North Carolina for August 27, 2011, as Hurricane Irene moves onshore. (b) Doppler radar display showing one-hour rainfall estimates over North Carolina for August 27, 2011. Notice that in some places Doppler radar estimated that more than 1.50 inches of rain had fallen in one hour.

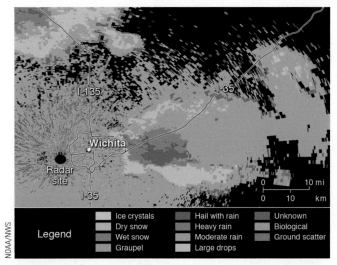

NOAA/NWS

Legend

Ice crystals	Hail with rain	Unknown
Dry snow	Heavy rain	Biological
Wet snow	Moderate rain	Ground scatter
Graupel	Large drops	

● **FIGURE 5.48** Doppler radar with dual-polarization technology shows precipitation and cloud particles inside a thunderstorm near Wichita, Kansas, on May 30, 2012. This technology was added to the NWS radars in order to more clearly identify different types of precipitation and to more precisely identify tornadic circulations.

The most recent improvement for Doppler radar is *polarimetric radar*, also referred to as *dual-polarization radar*. This form of Doppler radar, which was installed around the United States in the early 2010s, transmits both a vertical and horizontal pulse that makes it easier to determine whether falling precipitation is in the form of rain or snow. ● Fig. 5.48 shows an example of a dual-polarization radar display. Another technology being explored is *phased-array radar*, which can gather a much greater amount of data using a grid of small transmitters instead of a single large one.

As we discussed in Chapter 4 (p. 110), specialized satellites can also be used to observe precipitation from space. Up through 2015, the *Tropical Rainfall Measuring Mission* (TRMM) satellite gathered a massive amount of three-dimensional data on precipitation. The newer *Global*

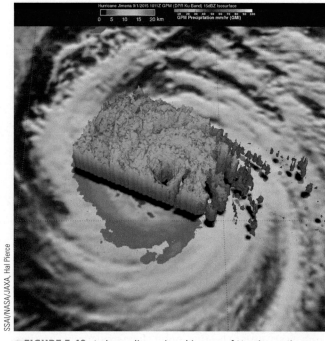

SSAI/NASA/JAXA, Hal Pierce

● **FIGURE 5.49** A three-dimensional image of Hurricane Jimena over the Central Pacific Ocean, collected on September 1, 2015, by the Global Precipitation Mission satellite. Reds and oranges indicate the heaviest rainfall at the surface. The eye of Jimena is visible as the hollow area in the middle of intense showers and thunderstorms.

Precipitation Mission (GPM) satellite has taken the place of TRMM, gathering precipitation data over a much larger area. GPM is able to gather more detail on lighter precipitation than TRMM, and it includes a dual-polarization radar that can deduce the sizes of raindrops, hailstones, and other precipitation elements (see ● Fig. 5.49). Another satellite that gathers three-dimensional data on precipitation intensity, NASA's *CloudSat* satellite, was launched in 2006. An image from CloudSat is presented in Fig. 4.52 on p. 110.

SUMMARY

In this chapter, we tied together the concepts of stability, cloud formation, and precipitation. We learned that because stable air tends to resist upward vertical motions, clouds forming in a stable atmosphere often spread horizontally and have a stratified appearance. A stable atmosphere can be caused either by the surface air being cooled or by the air aloft being warmed.

An unstable atmosphere tends to favor vertical air currents and produce cumuliform clouds. Instability can be produced either by the surface air being warmed or by the air aloft being cooled. In a conditionally unstable atmosphere, rising unsaturated air can be lifted to a level where condensation begins, latent heat is released, and instability results.

We learned that the development of most clouds results from either surface heating, uplift along topographic barriers (orographic uplift), convergence of surface air, or lifting along weather fronts.

We looked at cloud droplets and found that, individually, they are too small and light to reach the ground as rain. They can grow in size as large cloud droplets falling through a cloud collide and merge with smaller droplets in their path. In clouds where the air temperature is below freezing, ice crystals can grow larger at the expense of the surrounding liquid cloud droplets. As an ice crystal begins to fall, it can grow larger by colliding with supercooled liquid droplets, which freeze on contact. In an attempt to coax more precipitation from them, some clouds are seeded with silver iodide.

We examined the various forms of precipitation, from raindrops that freeze on impact (producing freezing rain) to raindrops that freeze into tiny ice pellets called sleet. We learned that strong updrafts in a cumulonimbus cloud can keep ice particles suspended above the freezing level, where they acquire an additional coating of ice and form destructive hailstones. We looked at instruments and found that although the rain gauge is still the most commonly used method of measuring precipitation, Doppler radar has become an important tool for determining precipitation intensity and estimating rainfall amount. Rainfall estimates can also be obtained from radar and microwave scanners onboard satellites.

KEY TERMS

The following terms are listed (with corresponding page numbers) in the order they appear in the text. Define each. Doing so will aid you in reviewing the material covered in this chapter.

adiabatic process, 116
dry adiabatic rate, 116
moist adiabatic rate, 117
environmental lapse rate, 117
absolutely stable atmosphere, 117
absolutely unstable atmosphere, 118
condensation level, 120
conditionally unstable atmosphere, 120
orographic uplift, 123
rain shadow, 125
precipitation, 126
collision-coalescence process, 126
coalescence, 126
ice-crystal (Bergeron) process, 127
supercooled droplet, 128
ice nuclei, 128
accretion, 129
cloud seeding, 129
rain, 132
drizzle, 132
virga, 132
shower (rain), 132
snow, 133
fallstreaks, 134
flurries (of snow), 135
snow squall, 135
thundersnow, 135
blizzard, 135
sleet, 136
freezing rain (glaze), 136
rime, 136
black ice, 136
ice storm, 137
snow grains, 138
snow pellets, 138
hailstones, 138
standard rain gauge, 140
trace (of precipitation), 140
water equivalent, 141
radar, 141
Doppler radar, 142

QUESTIONS FOR REVIEW

1. What is an adiabatic process?
2. How would one normally obtain the environmental lapse rate?
3. Why are the moist and dry adiabatic rates of cooling different?
4. How can the atmosphere be made more stable? More unstable?
5. If the atmosphere is conditionally unstable, what does this mean? What condition is necessary to bring on instability?
6. Explain why an inversion represents an extremely stable atmosphere.
7. What type of clouds would you most likely expect to see in a stable atmosphere? In a conditionally unstable atmosphere?
8. Why are cumulus clouds more frequently observed during the afternoon?
9. There are usually large spaces of blue sky between cumulus clouds. Explain why this is so.
10. Why do most thunderstorms have flat tops?
11. List four primary ways in which clouds form.
12. Explain why rain shadows form on the downwind (leeward) side of mountains.
13. On which side of a mountain (windward or leeward) would lenticular clouds most likely form?
14. What is the primary difference between a cloud droplet and a raindrop?
15. Why do typical cloud droplets seldom reach the ground as rain?

16. Describe how the process of collision and coalescence produces rain.
17. How does the ice-crystal (Bergeron) process produce precipitation? What is the *main* premise behind this process?
18. Explain the main principle behind cloud seeding.
19. Explain how clouds can be seeded naturally.
20. How does rain differ from drizzle?
21. Why do heavy showers usually fall from cumuliform clouds? Why does steady precipitation normally fall from stratiform clouds?
22. Why is it *never* too cold to snow?
23. How would you be able to distinguish between virga and fallstreaks?
24. What is the difference between freezing rain and sleet?
25. How do the atmospheric conditions that produce sleet differ from those that produce hail?
26. Describe how a standard rain gauge measures precipitation.
27. How can the depth of snow be measured?
28. (a) What is Doppler radar? (b) How does Doppler radar measure the intensity of precipitation?
29. How are satellites able to measure precipitation intensity inside clouds?

QUESTIONS FOR THOUGHT AND EXPLORATION

1. Suppose a mountain climber is scaling the outside of a tall skyscraper. Two thermometers (shielded from the sun) hang from the climber's belt. One thermometer hangs freely, while the other is enclosed in a partially inflated balloon. As the climber scales the building, describe the change in temperature measured by each thermometer.
2. Where would you expect the moist adiabatic rate to be *greater*: in the tropics or near the North Pole? Explain why.

3. What changes in weather conditions near Earth's surface are needed to transform an absolutely stable atmosphere into an absolutely unstable atmosphere?
4. In the middle latitudes, under what circumstances can a rain shadow be formed on the *western side* of a mountain range?
5. A major snowstorm occurred in northern New Jersey. Three volunteer weather observers measured the snowfall. Observer #1 measured the depth of newly fallen snow every hour. At the end of the storm, Observer #1 added up the measurements and came up with a total of 12 inches of new snow. Observer #2 measured the depth of new snow twice: once in the middle of the storm and once at the end, and came up with a total snowfall of 10 inches. Observer #3 measured the new snowfall only once, after the storm had stopped, and reported 8.4 inches. Which of the three observers do you feel has the correct snowfall total? List *at least five* possible reasons why the snowfall totals were different.
6. Why is a warm, tropical cumulus cloud more likely to produce precipitation than a cold, stratus cloud?
7. Suppose a thick nimbostratus cloud contains ice crystals and supercooled cloud droplets all about the same size. Which precipitation process will be most important in producing rain from this cloud? Why?
8. Clouds that form over water are usually more efficient in producing precipitation than clouds that form over land. Why do you think this is so?
9. Everyday in summer a blizzard occurs over the Great Plains. Explain where and why.
10. It is −12°C (10°F) in Albany, New York, and freezing rain is falling. Can you explain why? Draw a vertical profile of the air temperature (a sounding) that illustrates why freezing rain is occurring at the surface.
11. When falling snowflakes become mixed with sleet, why is this condition often followed by the snowflakes changing into rain?
12. Why are ice storms *not* associated with cumuliform clouds, such as cumulus congestus and cumulonimbus?

GLOBAL **GEOSCIENCE** WATCH   Go to the Academic Journals section of the Meteorology portal. Use the search box at the left to bring up journal articles related to "cloud seeding." Find an article that describes a cloud-seeding experiment. Do the authors claim that the seeding produced a measurable result? Is the result statistically significant? What other factors might have produced the result aside from the cloud seeding?

ONLINE RESOURCES

 Visit www.cengagebrain.com to view additional resources, including video exercises, practice quizzes, an interactive eBook, and more.

CHAPTER 6

Air Pressure and Winds

Contents

December 19, 1980, was a cool day in Lynn, Massachusetts, but not cool enough to dampen the spirits of more than 2000 people who gathered in Central Square—all hoping to catch at least one of the 1500 dollar bills that would be dropped from a small airplane at noon. Right on schedule, the aircraft circled the city and dumped the money onto the people below. However, to the dismay of the onlookers, a westerly wind caught the currency before it reached the ground and carried it out over the cold Atlantic Ocean. Had the pilot or the sponsoring leather manufacturer examined the weather charts beforehand, they might have been able to predict that the wind would ruin their advertising scheme.

The scenario on the previous page raises two questions: (1) Why does the wind blow? and (2) How can one tell its direction by looking at weather charts? Chapter 1 has already answered the first question: Air moves in response to horizontal differences in pressure. This is what happens when we open a vacuum-packed can: air rushes from the higher pressure region outside the can toward the region of lower pressure inside. In the atmosphere, the wind blows in an attempt to equalize imbalances in air pressure. Does this mean that the wind always blows directly from high to low pressure? Not really, because the movement of air is controlled not only by pressure differences but by other forces as well.

In this chapter, we will first consider how and why atmospheric pressure varies. Then we will look at the forces that influence atmospheric motions aloft and at the surface. Through studying these forces, we will be able to tell how the wind should blow in a particular region by examining surface and upper-air charts.

Atmospheric Pressure

In Chapter 1, we learned several important concepts about atmospheric pressure. One is that **air pressure** is simply the mass of air above a given level. As we climb in altitude above Earth's surface, there are fewer air molecules above us; hence, atmospheric pressure always decreases with increasing height. Another concept we learned is that our atmosphere is highly compressible. This means that most of our atmosphere is squeezed (compressed) close to Earth's surface, a situation that causes air pressure to decrease with height, rapidly at first, then more slowly at higher altitudes.

So one way to change air pressure is to simply move up or down in the atmosphere. But what causes the air pressure to change in the horizontal? And why does the air pressure change at the surface?

HORIZONTAL PRESSURE VARIATIONS—A TALE OF TWO CITIES
To answer these questions, we eliminate some of the complexities of the atmosphere by constructing *models*. ● Figure 6.1 shows a simple atmospheric model, a column of air extending well up into the atmosphere. In the column, the dots represent air molecules. Our model assumes: (1) that the air molecules are not crowded close to the surface and, unlike in the real atmosphere, the air density remains constant from the surface up to the top of the column, (2) that the width of the column does not change with height, and (3) that the air cannot freely move into or out of the column.

Suppose we somehow force more air into the column in Fig. 6.1. What would happen? If the air temperature in the column does not change, the added air would make

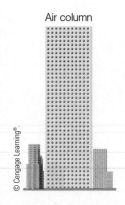

Air column

● **FIGURE 6.1** A model of the atmosphere where air density remains constant with height. The air pressure at the surface is related to the number of molecules above. When air of the same temperature is stuffed into the column, the surface air pressure rises. When air is removed from the column, the surface pressure falls. (In the actual atmosphere, unlike in this model, density decreases with height.)

© Cengage Learning®.

the column more dense, and the added weight of the air in the column would increase the surface air pressure. Likewise, if a great deal of air were removed from the column, the surface air pressure would decrease. Consequently, to change the surface air pressure, we need to change the mass of air in the column above the surface. But how can this feat be accomplished?

Look at the air columns in ● Fig 6.2a.* Suppose both columns are located at the same elevation, both have the same air temperature, and both have the same surface air pressure. There must therefore be the same number of molecules (same mass of air) in each column above both cities. Further suppose that the surface air pressure for both cities remains the same, while the air above city 1 cools and the air above city 2 warms (see Fig. 6.2b).

As the air in column 1 cools, the molecules move more slowly and crowd closer together, so the air becomes more dense. In the warm air above city 2, the molecules move faster and spread farther apart, and the air becomes less dense. Since the width of the columns does not change (and if we assume an invisible barrier exists between the columns), the total number of molecules above each city remains the same, and the surface air pressure does not change. Therefore, in the more-dense cold air above city 1, the height of the column decreases, while in the less-dense warm air above city 2, the height of the column increases.

We now have a cold, shorter, more-dense column of air above city 1 and a warm, taller, less-dense air column above city 2. From this, we can conclude that *it takes a shorter column of cold, more-dense air to exert the same surface pressure as a taller column of warm, less-dense air.* This concept has a great deal of meteorological significance.

Atmospheric pressure decreases more rapidly with height in the cold column of air. In the cold air above city 1 (Fig. 6.2b), move up the column and observe how quickly you pass through the densely packed molecules. This activity indicates a rapid change in pressure. In the warmer, less-dense air, the pressure does not decrease as

---

*We will keep our same assumption as in Fig. 6.1; that is, (1) the air molecules are not crowded close to the surface, (2) the width of each column does not change, and (3) air is unable to move into or out of each column.

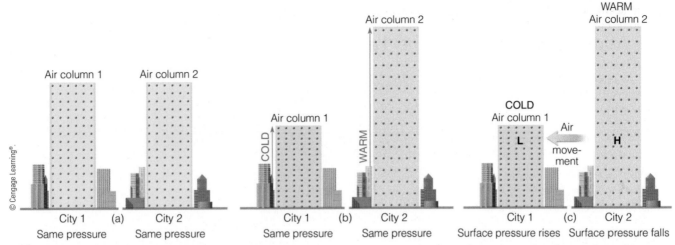

● **FIGURE 6.2** Illustration of how variations in temperature can produce horizontal pressure forces. (Note that, for simplicity, this model assumes that the air density is constant with height, whereas in the actual atmosphere, density decreases with height.) (a) Two air columns, each with identical mass, have the same surface air pressure. (b) Because it takes a shorter column of cold air to exert the same surface pressure as a taller column of warm air, as column 1 cools, it must shrink, and as column 2 warms, it must rise. (c) Because at the same level in the atmosphere there is more air above the H in the warm column than above the L in the cold column, warm air aloft is associated with high pressure and cold air aloft with low pressure. The pressure differences aloft create a force that causes the air to move from a region of higher pressure toward a region of lower pressure. The removal of air from column 2 causes its surface pressure to drop, whereas the addition of air into column 1 causes its surface pressure to rise. (The difference in height between the two columns is greatly exaggerated.)

rapidly with height, simply because you climb above fewer molecules in the same vertical distance.

In Fig. 6.2c, move up the warm, red column until you come to the letter *H*. Now move up the cold, blue column the same distance until you reach the letter *L*. Notice that there are more molecules above the letter *H* in the warm column than above the letter *L* in the cold column. The fact that the number of molecules above any level is a measure of the atmospheric pressure leads to an important concept: *Warm air aloft is normally associated with high atmospheric pressure, and cold air aloft is associated with low atmospheric pressure.*

In Fig. 6.2c, the horizontal difference in temperature creates a horizontal difference in pressure. The pressure difference establishes a force (called the *pressure gradient force*) that causes the air to move from higher pressure toward lower pressure. Consequently, if we remove the invisible barrier between the two columns near the top of Column 1 and allow the air aloft to move horizontally, the air will move from column 2 toward column 1. As the air aloft leaves column 2, the mass of the air in the column decreases, and so does the surface air pressure. Meanwhile, the accumulation of air in column 1 causes the surface air pressure to increase.

Higher air pressure at the surface in column 1 and lower air pressure at the surface in column 2 causes the surface air to move from city 1 toward city 2 (see ● Fig. 6.3). As the surface air moves out away from city 1, the air aloft slowly sinks to replace this outwardly spreading surface air. As the surface air flows into city 2, it slowly rises to replace the depleted air aloft. In this manner, a complete

circulation of air is established due to the heating and cooling of air columns. As we will see in Chapter 7, this type of thermal circulation is the basis for a wide range of wind systems throughout the world.

In summary, we can see how heating and cooling columns of air can establish horizontal variations in air pressure both aloft and at the surface. It is these horizontal differences in air pressure that cause the wind to blow.

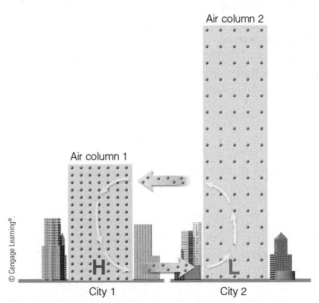

● **FIGURE 6.3** The heating and cooling of air columns causes horizontal pressure variations aloft and at the surface. These pressure variations force the air to move from areas of higher pressure toward areas of lower pressure. In conjunction with these horizontal air motions, the air slowly sinks above the surface high and rises above the surface low.

The Atmosphere Obeys the Gas Law

The relationship among the pressure, temperature, and density of air can be expressed by

Pressure = temperature × density × constant.

This simple relationship, often referred to as the *gas law* (or *equation of state*), tells us that the pressure of a gas is equal to its temperature times its density times a constant. When we ignore the constant and look at the gas law in symbolic form, it becomes

$$p \sim T \times \rho$$

where, of course, p is pressure, T is temperature, and ρ (the Greek letter rho, pronounced "row") represents air density. The line $\sim$ is a symbol meaning "is proportional to." A change in one variable causes a corresponding change in the other two variables. Thus, it will be easier to understand the behavior of a gas if we keep one variable from changing and observe the behavior of the other two.

Suppose, for example, we hold the temperature constant. The relationship then becomes

$$p \sim \rho \text{ (temperature constant).}$$

This expression says that the pressure of the gas is proportional to its density, as long as its temperature does not change. Consequently, if the temperature of a gas (such as air) is held constant: As the pressure increases the density increases, and

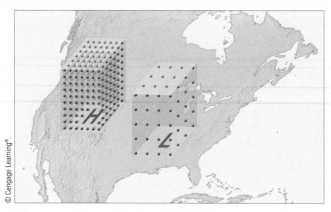

● FIGURE 1 Air above a region of surface high pressure is more dense than air above a region of surface low pressure (at the same temperature). (The dots in each column represent air molecules.)

as the pressure decreases, the density decreases. In other words, *at the same temperature, air at a higher pressure is more dense than air at a lower pressure.* If we apply this concept to the atmosphere, then with nearly the same temperature and elevation, air above a region of surface high pressure is more dense than air above a region of surface low pressure (see ● Fig. 1).

We can see, then, that for surface high-pressure areas (anticyclones) and surface low-pressure areas (mid-latitude cyclones) to form, the air density (mass of air) above these systems must change.

We just considered how pressure and density are related when the temperature is not changing. What happens to the gas law when the pressure of a gas remains constant? In shorthand notation, the relationship becomes

$$\text{(Constant pressure)} \times \text{constant} = T \times \rho.$$

This relationship tells us that when the pressure of a gas is held constant, the gas becomes less dense as the temperature goes up, and more dense as the temperature goes down. Therefore, *at a given atmospheric pressure, air that is cold is more dense than air that is warm.* Keep in mind that the idea that cold air is more dense than warm air applies only when we compare volumes of air at the same level, where pressure changes are relatively small in any horizontal direction.

Before we examine how air pressure is measured, you may wish to look at Focus section 6.1, which describes how air pressure, air density, and air temperature are interrelated.

MEASURING AIR PRESSURE   Up to this point, we have described air pressure as the mass of the atmosphere above any level. We can also define air pressure as the *force exerted by the air molecules over a given area.* Billions of air molecules constantly push on the human body. This force is exerted equally in all directions. We are not crushed by the force because billions of molecules inside the body push outward just as hard.

Even though we do not actually feel the constant bombardment of air, we can detect quick changes in it.

For example, if we climb rapidly in elevation our ears may "pop." This experience happens because air collisions outside the eardrum lessen as the air pressure decreases. The popping comes about as air collisions between the inside and outside of the ear equalize.

Instruments that detect and measure pressure changes are called *barometers,* which literally means an instrument that measures bars. In meteorology, the *bar* is a unit of pressure that describes a force over a given area.* Because the bar is a relatively large unit, and because surface

---

*By definition, a bar is a force of 100,000 newtons acting on a surface area of 1 square meter. A *newton* is the amount of force required to move an object with a mass of 1 kilogram so that it increases its speed at a rate of 1 meter per second each second.

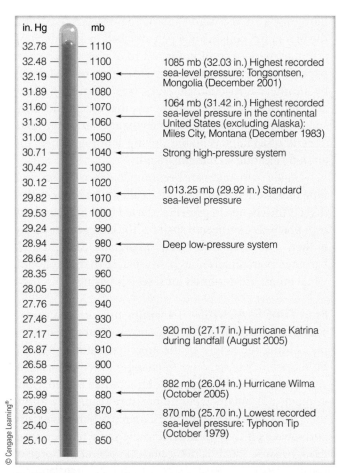

● **FIGURE 6.4** Atmospheric pressure in inches of mercury and in millibars.

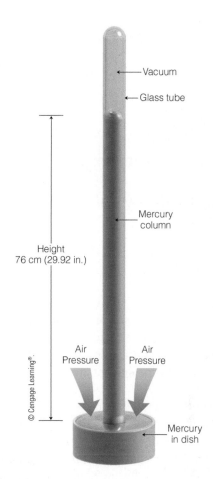

● **FIGURE 6.5** The mercury barometer. The height of the mercury column is a measure of atmospheric pressure.

pressure changes are normally small, the unit of pressure most commonly found on surface weather maps is the **millibar** (mb), where one millibar is equal to one-thousandth of a bar. The *hectopascal* (hPa)* is gradually replacing the millibar as the preferred unit of pressure on surface and upper-air maps. (No conversion is needed in moving between millibars and hectopascals.) A common pressure unit used in aviation and on television and radio weather broadcasts is *inches of mercury* (Hg). At sea level, **standard atmospheric pressure**** is

1013.25 mb = 1013.25 hPa = 29.92 in. Hg.

Because we measure atmospheric pressure with an instrument called a **barometer,** atmospheric pressure is also referred to as *barometric pressure.* ● Figure 6.4 compares pressure readings in millibars and in inches of mercury. Why do we refer to "inches of mercury"? Evangelista Torricelli, a student of Galileo's, invented the **mercury**

barometer in 1643. His barometer, similar to those used today, consisted of a long glass tube open at one end and closed at the other (see ● Fig. 6.5). Removing air from the tube and covering the open end, Torricelli immersed the lower portion into a dish of mercury. He removed the cover, and the mercury rose up the tube to nearly 30 inches above the level in the dish. Torricelli correctly concluded that the column of mercury in the tube was balancing the weight of the air above the dish, and, hence, its height was a measure of atmospheric pressure. Mercury barometers are still used today in many settings, although in some areas, including Europe, they are no longer manufactured or sold because mercury use can pose a risk to human health.

The most common type of home barometer—the **aneroid barometer**—contains no fluid. Inside this instrument is a small, flexible metal box called an *aneroid cell.* Before the cell is tightly sealed, air is partially removed, so

---

*The unit of pressure designed by the International System (SI) of measurement is the *pascal* (Pa), where 1 pascal is the force of 1 newton acting on a surface of 1 square meter. A more common unit is the *hectopascal* (hPa), as 1 hectopascal equals 1 millibar. (Additional pressure units and conversions are given in Appendix A.)

**Standard atmospheric pressure at sea level is the pressure extended by a column of mercury 29.92 in. (760 mm) high, having a density of 1.36×10^4 kg/m^3, and subject to an acceleration of gravity of 9.80 m/sec^2.

DID YOU KNOW?

Although 1013.25 mb (29.92 in.) is the *standard atmospheric* pressure at sea level, it is not the average sea-level pressure. Earth's average sea-level pressure is 1011.0 mb (29.85 in.). Because much of Earth's surface is above sea level, Earth's annual average *surface pressure* is estimated to be 984.43 mb (29.07 in.).

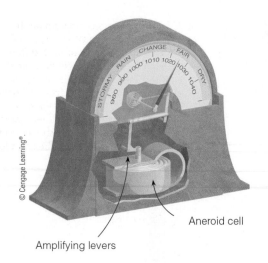

FIGURE 6.6 The aneroid barometer.

that small changes in external air pressure cause the cell to expand or contract. The size of the cell is calibrated to represent different pressures, and any change in its size is amplified by levers and transmitted to an indicating arm, which points to the current atmospheric pressure (see ● Fig. 6.6).

Notice that the aneroid barometer often has descriptive weather-related words printed above specific pressure values. These descriptions indicate the most likely weather conditions when the needle is pointing to that particular pressure reading. Generally, the higher the reading, the more likely clear weather will occur, and the lower the reading, the better are the chances for inclement weather. This situation occurs because surface high-pressure areas are associated with sinking air and normally fair weather, whereas surface low-pressure areas are associated with rising air and usually cloudy, wet weather. A steady rise in atmospheric pressure (a rising barometer reading) usually indicates clearing weather or fair weather, whereas a steady drop in atmospheric pressure (a falling barometer reading) often signals the approach of a cyclonic storm with inclement weather.

The *altimeter* and *barograph* are two types of aneroid barometers. Altimeters are aneroid barometers that measure pressure, but are calibrated to indicate altitude.

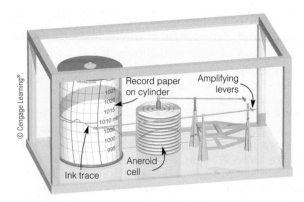

FIGURE 6.7 A recording barograph.

Barographs are recording aneroid barometers. Basically, the barograph consists of a pen attached to an indicating arm that marks a continuous record of pressure on chart paper. The chart paper is attached to a drum rotated slowly by an internal mechanical clock (see ● Fig. 6.7). Many aircraft use radar-based altimeters that take rapid measurements of the distance between the aircraft and ground. This is especially valuable during landing, when the aircraft's altitude is changing rapidly.

Digital barometers are becoming more common. This type of barometer uses a device called a transducer that detects the change in pressure exerted by the atmosphere on a precisely engineered surface. The change in pressure is then converted into electrical signals. Some digital barometers are small enough to be placed in smartphones, while others are designed for research settings.

PRESSURE READINGS Obtaining the correct air pressure from a mercury barometer involves more than simply reading the height of the mercury column. Being a fluid, mercury is sensitive to changes in temperature; it will expand when heated and contract when cooled. Consequently, to obtain accurate pressure readings without the influence of temperature, all mercury barometers are corrected as if they were read at the same temperature. Also, because Earth is not a perfect sphere, the force of gravity is not a constant. Since small gravity differences influence the height of the mercury column, they must be taken into account when reading the barometer. Finally, each mercury barometer has its own "built-in" error, called *instrument error,* which is caused, in part, by the surface tension of the mercury against the glass tube. After being corrected for temperature, gravity, and instrument error, the barometer reading at a particular location and elevation is termed **station pressure**.

● Figure 6.8a gives the station pressure measured at four locations only a few hundred kilometers apart. The different station pressures of the four cities are due primarily to the cities being at different elevations above sea level. This fact becomes even clearer when we realize that atmospheric pressure changes much more quickly when we move upward than it does when we move sideways. A small vertical difference between two observation sites can yield a large difference in station pressure. Thus, to properly monitor horizontal changes in pressure, barometer readings must be corrected for altitude.

Altitude corrections are made so that a barometer reading taken at one elevation can be compared with a barometer reading taken at another. Station pressure observations are normally adjusted to a level of mean sea level, the level representing the average surface height of the ocean. The adjusted reading is called **sea-level pressure.** The size of the correction depends primarily on how high the station is above sea level.

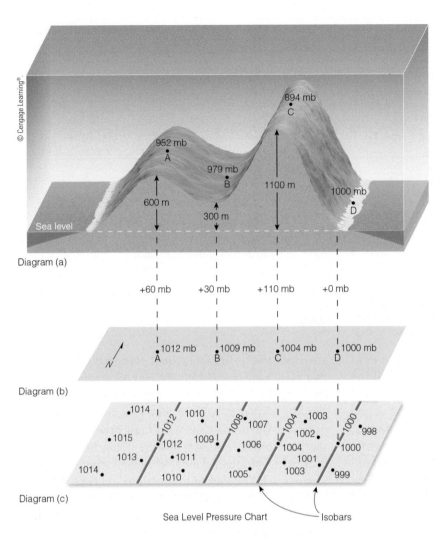

© Cengage Learning®.

Diagram (a)

+60 mb   +30 mb   +110 mb   +0 mb

N

•1012 mb   •1009 mb   •1004 mb   •1000 mb
A   B   C   D

Diagram (b)

•1014   1010 •   1007 •   1003 •   998 •
                  1002 •
•1015   •1012   1009 •   1006 •   1000 •
        1012   1008
1013 •   •1011   1005 •   1004 •   1000
        1010   1001
1014 •   1003
                 1005 •   999 •

Diagram (c)

Sea Level Pressure Chart   Isobars

● **FIGURE 6.8** The top diagram (a) shows four cities (A, B, C, and D) at varying elevations above sea level, all with different station pressures. The middle diagram (b) represents sea-level pressures of the four cities plotted on a sea-level chart. The bottom diagram (c) shows sea-level pressure readings of the four cities plus other sea-level pressure readings at locations not shown in (a) and (b). Isobars are drawn on the chart (gray lines) at intervals of 4 millibars.

Near Earth's surface, atmospheric pressure decreases on the average by about 10 millibars (mb) for every 100 meters of increase in elevation (about 1 in. of mercury for each 1000-ft rise).* Notice in Fig. 6.8a that city A has a station pressure of 952 mb. Notice also that city A is 600 meters above sea level. Adding 10 mb per 100 m to its station pressure yields a sea-level pressure of 1012 mb (Fig. 6.8b). After all the station pressures are adjusted to sea level (Fig. 6.8b), we are able to see the horizontal variations in sea-level pressure—something we were not able to see from the station pressures alone in Fig. 6.8a.

When more pressure data are added (Fig. 6.8c), the chart can be analyzed and the pressure pattern visualized. **Isobars** (lines connecting points of equal pressure) are drawn as solid dark lines at intervals of 4 mb, with 1000 mb being the base value. Note that the isobars do not pass through each point, but, rather, between many of them, with the exact values being interpolated from the data given on the chart. For example, follow the 1008-mb

line from the top of the chart southward and observe that there is no plotted pressure of 1008 mb. The 1008-mb isobar, however, comes closer to the station with a sea-level pressure of 1007 mb than it does to the station with a pressure of 1010 mb. With its isobars, the bottom chart (Fig. 6.8c) is now called a *sea-level pressure chart*, or, simply, a **surface map**. When weather data are plotted on the map, it becomes a *surface weather map*.

Surface and Upper-Air Charts

● Figure 6.9a is a simplified surface map that shows areas of high and low pressure and arrows that indicate *wind*

---

*This decrease in atmospheric pressure with height (10 mb/100 m) occurs when the air temperature decreases at the standard lapse rate of 6.5°C/1000 m. Because atmospheric pressure decreases more rapidly with height in cold (more dense) air than it does in warm (less dense) air, the vertical rate of pressure change is typically greater than 10 mb per 100 m in cold air and less than that in warm air.

DID YOU KNOW?

On October 12, 1962, one of the most powerful mid-latitude cyclonic storms ever to develop along the Pacific Northwest coast slammed into Oregon and Washington. With a central pressure of 960 mb (28.35 in.), the storm had such a strong pressure gradient that it produced winds on its eastern side in excess of 75 mi/hr along the Pacific Coast from California to British Columbia with a reported peak wind gust of 179 mi/hr at Cape Blanco, Oregon.

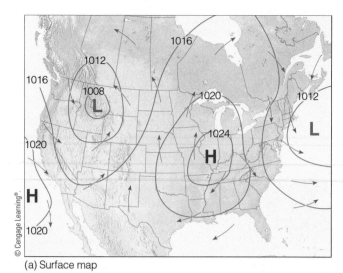

(a) Surface map

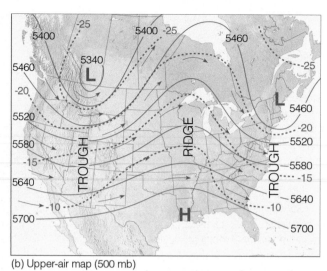

(b) Upper-air map (500 mb)

● **FIGURE 6.9** (a) Surface map showing areas of high and low pressure. The solid lines are isobars drawn at 4-mb intervals. The arrows represent wind direction—the direction from which the wind is blowing. Notice that the wind blows *across* the isobars. (b) The upper-level (500-mb) map for the same day as the surface map. Solid lines on the map are contour lines in meters above sea level. Dashed red lines are isotherms in °C. Arrows show wind direction. Notice that, on this upper-air map, the wind blows *parallel* to the contour lines.

direction, which is the direction *from which the wind is blowing.* The large blue *H*'s on the map indicate the centers of high pressure, which are also called **anticyclones.** The large red *L*'s represent centers of low pressure, also known as **mid-latitude cyclonic storms,** or *extratropical cyclones,* because they form in the middle latitudes, outside of the tropics. The solid dark lines are isobars with units in millibars. Notice that the surface winds tend to blow across the isobars toward regions of lower pressure. In fact, as we briefly observed in Chapter 1, in the Northern Hemisphere the winds blow counterclockwise and inward toward the center of the lows and clockwise and outward from the center of the highs.

Figure 6.9b shows an upper-air chart for the same day as the surface map in Fig. 6.9a. The upper-air map is a *constant pressure chart* because it is constructed to show height variations along a constant pressure *(isobaric)* surface, which is why these maps are also known as **isobaric maps.** This particular isobaric map shows height variations at a pressure level of 500 millibars and so is called a *500-millibar map.* In Fig. 6.9b, the height of the 500-mb level varies from 5340 m to more than 5700 m. A typical height in mid-latitudes is about 5600 m, or 18,000 ft, above sea level. The solid dark lines on the map are **contour lines**, lines that connect points of equal altitude above sea level. Although the contour lines are height lines, they illustrate pressure much like isobars do. Consequently, *contour lines of low height represent a region of lower pressure, and contour lines of high height represent a region of higher pressure.* (Additional information on isobaric maps is given in Focus section 6.2.)

Notice on the 500-mb map (Fig. 6.9b) that the contour lines typically decrease in value from south to north. This decrease in value is due to a decrease in temperature as illustrated by the dashed red lines, which

are *isotherms*—lines of equal temperature. Observe that colder air is generally to the north and warmer air to the south, and recall from our earlier discussion (see pp. 148-150) that cold air aloft is associated with low pressure, warm air aloft with high pressure. The contour lines are not straight, however. They bend and turn, indicating **ridges** *(elongated highs)* where the air is warmer and indicating depressions, or **troughs** *(elongated lows)* where the air is colder. The arrows on the 500-mb map show the wind direction. Notice that, unlike the surface winds that cross the isobars in Fig. 6.9a, the winds on the 500-mb chart tend to flow *parallel* to the contour lines in a wavy west-to-east direction.

Surface and upper-air charts are valuable tools for the meteorologist. Surface maps describe where the centers of high and low pressure are found, as well as the winds and weather associated with these systems, whereas upper-air charts are extremely important in forecasting the weather. The upper-level winds not only determine the movement of surface pressure systems but, as we will see in Chapter 8, they determine whether these surface systems will intensify or weaken.

At this point, however, our interest lies mainly in the movement of air. Now that we have looked at surface and upper-air maps, we will use them to study why the wind blows the way it does, at both the surface and aloft.

Why the Wind Blows

Our understanding of why the wind blows stretches back through several centuries, with many scientists contributing to our knowledge. When we think of the movement of air, however, one great scholar stands out: Isaac Newton (1642–1727), who formulated several fundamental laws of motion.

Isobaric Maps

● Figure 2 shows a column of air where warm, less-dense air lies to the south and cold, more-dense air lies to the north. The area shaded gray at the top of the column represents a constant pressure (isobaric) surface, where the atmospheric pressure at all points along the surface is 500 millibars.

Notice that the height of the pressure surface varies. In the warmer air, a pressure reading of 500 mb is found at a higher level, while in the colder air, 500 mb is observed at a much lower level. The variations in height of the 500-mb constant pressure surface are shown as contour lines on the constant pressure (500-mb) map, situated at the bottom of the column. Each contour line tells us the altitude above sea level at which we would obtain a pressure reading of 500 mb. As we would expect from our earlier discussion of pressure in Figs. 6.2 and 6.3, p. 149, the contour lines on the chart are higher in the warm air and lower in the cold air.

Although contour lines are height lines, keep in mind that they illustrate pressure in the same manner as do isobars, as contour lines of high height (warm air aloft) represent regions of higher pressure, and contour lines of low height (cold air aloft) represent regions of low pressure.

In many instances, the contour lines on an isobaric map are not straight, but rather appear as wavy lines. ● Figure 3 illustrates how these wavy contours on the map relate to the change in altitude of the isobaric surface.

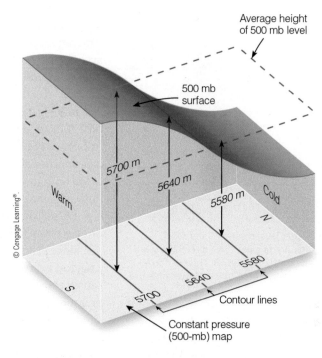

● **FIGURE 2** The area shaded gray in the diagram represents a surface of constant pressure. Because of the changes in air density, a surface of constant pressure rises in warm, less-dense air and lowers in cold, more-dense air. These changes in height of a constant pressure (500-mb) surface show up as contour lines on a constant pressure (isobaric) 500-mb map.

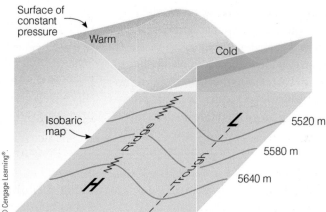

● **FIGURE 3** The wavelike patterns of an isobaric surface reflect the changes in air temperature. An elongated region of warm air aloft shows up on an isobaric map as higher heights and a ridge; the colder air shows as lower heights and a trough.

NEWTON'S LAWS OF MOTION Newton's first law of motion states that *an object at rest will remain at rest and an object in motion will remain in motion (and travel at a constant velocity along a straight line) as long as no force is exerted on the object.* For example, a baseball in a pitcher's hand will remain there until a force (a push) acts upon the ball. Once the ball is pushed (thrown), it would continue to move in that direction forever if it were not for the force of air friction (which slows it down), the force of gravity (which pulls it toward the ground), and the catcher's mitt (which exerts an equal but opposite force to bring it to a halt). Similarly, to start air moving, to speed it up, to slow it down, or even to change its direction requires the action of an external force. This brings us to Newton's second law.

Newton's second law states that *the force exerted on an object equals its mass times the acceleration produced.** In symbolic form, this law is written as

$$F = ma.$$

*Newton's second law can also be stated in this way: The acceleration of an object (times its mass) is caused by all of the forces acting on it.

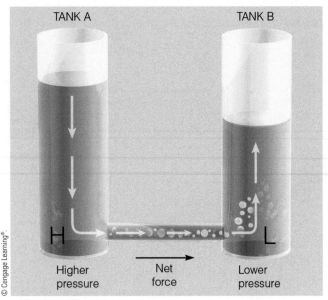

TANK A   TANK B

H   L

Higher   Net   Lower
pressure   force   pressure

● **FIGURE 6.10** The higher water level creates higher fluid pressure at the bottom of tank A and a net force directed toward the lower fluid pressure at the bottom of tank B. This net force causes water to move from higher pressure toward lower pressure.

From this relationship we can see that, when the mass of an object is constant, the force acting on the object is directly related to the acceleration that is produced. A force in its simplest form is a push or a pull. *Acceleration is the speeding up, the slowing down, and/or the changing of direction of an object.* (More precisely, acceleration is the change of velocity* over a period of time.)

Because more than one force may act upon an object, Newton's second law always refers to the *net,* or total, force that results. An object will always accelerate in the direction of the total force acting on it. Therefore, to determine in which direction the wind will blow, we must identify and examine forces that affect the horizontal movement of air. These forces include:

1. pressure gradient force
2. Coriolis force
3. friction

---

*Velocity specifies both the speed of an object and its direction of motion.

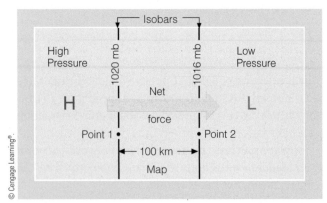

● **FIGURE 6.11** The pressure gradient between point 1 and point 2 is 4 mb per 100 km. The net force directed from higher toward lower pressure is the *pressure gradient force.*

We will first study the forces that influence the flow of air aloft. Then we will see which forces modify winds near the ground.

FORCES THAT INFLUENCE THE WIND We have already learned that horizontal differences in atmospheric pressure cause air to move and, hence, the wind to blow. Since air is an invisible gas, it may be easier to see how pressure differences cause motion if we examine a visible fluid, such as water.

In ● Fig. 6.10, the two large tanks are connected by a pipe. Tank A is two-thirds full and tank B is only one-half full. Since the water pressure at the bottom of each tank is proportional to the weight of water above, the pressure at the bottom of tank A is greater than the pressure at the bottom of tank B. Moreover, since fluid pressure is exerted equally in all directions, there is a greater pressure in the pipe directed from tank A toward tank B than from B toward A.

Since pressure is force per unit area, there must also be a net force directed from tank A toward tank B. This force causes the water to flow from left to right, from higher pressure toward lower pressure. The greater the pressure difference, the stronger the force, and the faster the water moves. In a similar way, horizontal differences in atmospheric pressure cause air to move.

Pressure Gradient Force ● Figure 6.11 shows a region of higher pressure on the map's left side, with lower pressure on the right. The isobars show how the horizontal pressure is changing. If we compute the amount of pressure change that occurs over a given distance, we have the **pressure gradient**; thus

$$\text{Pressure gradient} = \frac{\text{difference in pressure}}{\text{distance}}.$$

In Fig. 6.11, the pressure gradient between points 1 and 2 is 4 millibars per 100 kilometers.

Suppose the pressure in Fig. 6.11 were to change, and the isobars become closer together. This condition would produce a rapid change in pressure over a relatively short distance, or what is called a *steep* (or *strong*) *pressure gradient.* However, if the pressure were to change such that the isobars spread farther apart, then the difference in pressure would be small over a relatively large distance. This condition is called a *gentle* (or *weak*) *pressure gradient.*

Notice in Fig. 6.11 that when differences in horizontal air pressure exist there is a net force acting on the air. This force, called the **pressure gradient force** *(PGF), is directed from higher toward lower pressure at right angles to the isobars.* The magnitude of the force is directly related to the pressure gradient. Steep pressure gradients correspond to strong pressure gradient forces and vice versa. ● Figure 6.12 shows the relationship between pressure gradient and pressure gradient force.

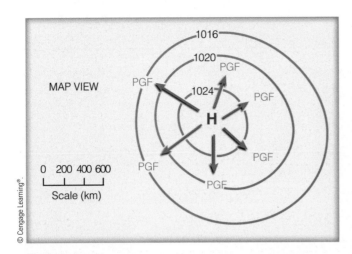

● **FIGURE 6.12** The closer the spacing of the isobars, the greater the pressure gradient. The greater the pressure gradient, the stronger the pressure gradient force *(PGF)*. The stronger the *PGF*, the greater the wind speed. The red arrows represent the relative magnitude of the force, which is always directed from higher toward lower pressure.

The *pressure gradient force is the force that causes the wind to blow.* Closely spaced isobars on a weather chart indicate steep pressure gradients, strong forces, and high winds. In contrast, widely spaced isobars indicate gentle pressure gradients, weak forces, and light winds. An example of a steep pressure gradient and strong winds is illustrated on the surface weather map in ● Fig. 6.13. Notice that the tightly packed isobars around Hurricane Sandy are producing a steep pressure gradient and strong surface winds, gusting to more than 80 knots over portions of Long Island.

If the pressure gradient force were the only force acting upon air, we would always find winds blowing directly from higher toward lower pressure. However, the moment air starts to move, it is deflected in its path by the *Coriolis force.*

Coriolis Force The **Coriolis force** describes an apparent force that is due to the rotation of Earth. To understand

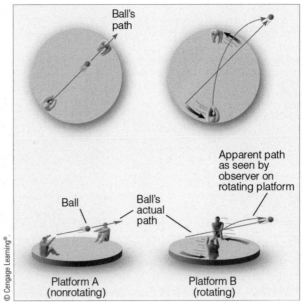

● **FIGURE 6.14** On nonrotating platform A, the thrown ball moves in a straight line. On platform B, which rotates counter-clockwise, the ball continues to move in a straight line. However, platform B is rotating while the ball is in flight; thus, to the person throwing the ball on platform B, the ball appears to deflect to the right of its intended path.

how it works, consider two people playing catch as they sit opposite one another on the rim of a merry-go-round (see ● Fig. 6.14, platform A). If the merry-go-round is not moving, each time the ball is thrown, it moves in a straight line to the other person.

Suppose the merry-go-round starts turning counter-clockwise, the same direction Earth spins as viewed from above the North Pole. If we watch the game of catch from above, we see that the ball moves in a straight-line path just as before. However, to the people playing catch on the merry-go-round, the ball seems to veer to its right each time it is thrown, always landing to the right of the point intended by the thrower (see Fig. 6.14, platform B).

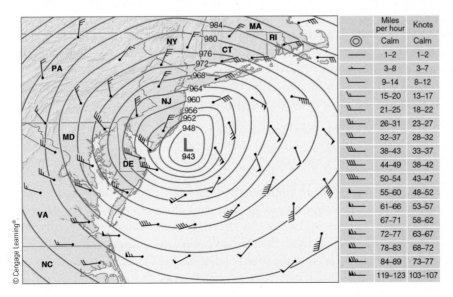

	Miles per hour	Knots
◎	Calm	Calm
—	1–2	1–2
⌐	3–8	3–7
⌐	9–14	8–12
⌐	15–20	13–17
⌐	21–25	18–22
⌐	26–31	23–27
⌐	32–37	28–32
⌐	38–43	33–37
⌐	44–49	38–42
⌐	50–54	43–47
⌐	55–60	48–52
⌐	61–66	53–57
⌐	67–71	58–62
⌐	72–77	63–67
⌐	78–83	68–72
⌐	84–89	73–77
⌐	119–123	103–107

● **FIGURE 6.13** Surface map for 4 p.m. (EST) Monday, October 29, 2012, as Hurricane (Superstorm) Sandy approaches the New Jersey shore from the east. Isobars are dark gray lines with units in millibars. The interval between isobars is 4 mb. Sandy's central pressure is 943 mb or 27.85 in. The tightly packed isobars are associated with a strong pressure gradient, a strong pressure gradient force, and high winds, gusting to over 80 knots over portions of Long Island, New York. Wind directions at each station are shown by lines that parallel the wind. Wind speeds are indicated by barbs and flags, where a wind indicated by the symbol ⌐ would be a wind from the northwest at between 23 and 27 knots. (See blue insert.) Wind speeds over the ocean, where no observations are available, are based on computer model calculations.

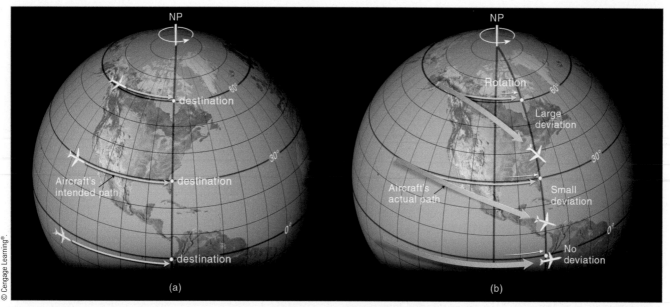

● **FIGURE 6.15** Except at the equator, a free-moving object heading either east or west (or any other direction) will appear from Earth to deviate from its path as Earth rotates beneath it. The deviation (Coriolis force) is greatest at the poles and decreases to zero at the equator. Notice that the aircraft's deviation from its intended destination is greatest at high latitudes and nonexistent at the equator.

This perspective is accounted for by the fact that, while the ball moves in a straight-line path, the merry-go-round rotates beneath it, and by the time the ball reaches the opposite side, the catcher has moved. To anyone on the merry-go-round, it seems as if there is some force causing the ball to deflect to the right of its intended path. This apparent force is called the *Coriolis force* after Gaspard Coriolis, a nineteenth-century French scientist who worked it out mathematically. (Because it is an *apparent force* due to the rotation of Earth, it is also called the *Coriolis effect*.) This effect occurs not only on merry-go-rounds but on rotating Earth, too. All free-moving objects, such as ocean currents, aircraft, artillery projectiles, and air molecules seem to deflect from a straight-line path because Earth rotates under them.

The Coriolis force *causes the wind to deflect to the right of its intended path in the Northern Hemisphere and to the left of its intended path in the Southern Hemisphere.* To illustrate, consider a satellite in polar circular orbit. If Earth were not rotating, the path of the satellite would be observed to move directly from north to south, parallel to Earth's meridian lines. However, Earth *does* rotate, carrying us and meridians eastward with it. Because of this rotation, in the Northern Hemisphere we see the satellite moving southwest instead of due south; it seems to veer off its path and move toward *its right*. In the Southern Hemisphere, Earth's direction of rotation is clockwise as viewed from above the South Pole. Consequently, a satellite moving northward from the South Pole would appear to move northwest and, hence, would veer to the *left* of its path.

As the wind speed increases, the Coriolis force increases; hence, *the stronger the wind, the greater the deflection*. Additionally, the Coriolis force increases for all wind

speeds *from a value of zero at the equator to a maximum at the poles*. This phenomenon is illustrated in ● Fig. 6.15 where three aircraft, each at a different latitude, are flying along a straight-line path, with no external forces acting on them. The destination of each aircraft is due east and is marked on the illustration in Fig. 6.15a. Each plane travels in a straight path relative to an observer positioned at a fixed spot in space. Earth rotates beneath the moving planes, causing the destination points at latitudes 30° and 60° to change direction slightly when seen by an observer in space (see Fig. 6.15b). To an observer standing on Earth, however, it is the plane that appears to deviate. The amount of deviation is greatest toward the pole and nonexistent at the equator. Therefore, the Coriolis force has a far greater effect on the plane at high latitudes (large deviation) than on the plane at low latitudes (small deviation). On the equator, it has no effect at all. The same is true of its effect on winds.

In summary, to an observer on Earth, objects moving in *any direction* (north, south, east, or west) are deflected to the *right* of their intended path in the Northern Hemisphere and to the *left* of their intended path in the Southern Hemisphere. The amount of deflection depends upon:

1. the rotation of Earth
2. the latitude
3. the object's speed

In addition, *the Coriolis force acts at right angles to the wind, only influencing wind direction and never wind speed.*

The Coriolis force is present in all motions relative to Earth's surface. However, because its strength is directly related to the speed of motion, the Coriolis force tends to be weak over relatively small areas. In most of our everyday experiences, the force is so weak (compared to other forces involved in those experiences) that it is negligible. Contrary to popular belief, the Coriolis force does not cause water to turn clockwise or counterclockwise when draining from a sink.

The Coriolis force is also minimal on small-scale winds, such as those that blow inland along coasts in summer. Here, the Coriolis force itself might be strong because of high winds, but the force cannot produce much deflection over the relatively short distances. Only where winds blow over long distances is the effect significant. With this information in mind, we will first examine how the pressure gradient force and the Coriolis force produce straight-line winds aloft, above the frictional influence of Earth's surface. We will then look at what other forces come into play as winds blow along a curved path.

BRIEF REVIEW

In summary, we know that:

- Atmospheric (air) pressure is the pressure exerted by the mass of air above a region.

- A change in surface air pressure can be brought about by changing the mass (amount of air) above the surface.

- Heating and cooling columns of air can establish horizontal variations in atmospheric pressure aloft and at the surface.

- A difference in horizontal air pressure produces a horizontal pressure gradient force.

- The pressure gradient force is always directed from higher pressure toward lower pressure and it is the pressure gradient force that causes the air to move and the wind to blow.

- Steep pressure gradients (tightly packed isobars on a weather map) indicate strong pressure gradient forces and high winds; gentle pressure gradients (widely spaced isobars) indicate weak pressure gradient forces and light winds.

- Once the wind starts to blow, the Coriolis force causes it to bend to the right of its intended path in the Northern Hemisphere and to the left of its intended path in the Southern Hemisphere.

STRAIGHT-LINE FLOW ALOFT   Earlier in this chapter, we saw that the winds aloft on an upper-level chart blow more or less parallel to the isobars or contour lines. We can see why this phenomenon happens by carefully looking at ● Fig. 6.16, which shows a map in the Northern Hemisphere, above Earth's frictional influence,* at an altitude of about 1 kilometer above Earth's surface. Horizontal pressure changes are shown by isobars. The evenly spaced isobars indicate a constant pressure gradient force *(PGF)* directed from south toward north as indicated by the red arrow at the left. Why, then, does the map show a wind blowing from the west? We can answer this question

*The friction layer (the layer where the wind is influenced by frictional interaction with objects on Earth's surface) usually extends from the surface up to about 1000 m (3300 ft) above the ground.

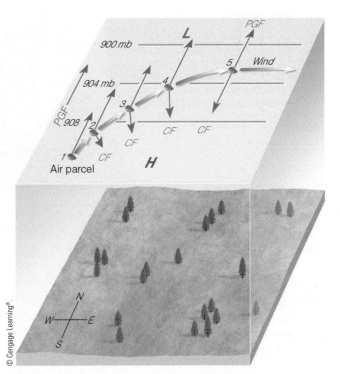

● **FIGURE 6.16** Above the level of friction, air initially at rest will accelerate until it flows parallel to the isobars at a steady speed with the horizontal pressure gradient force *(PGF)* balanced by the Coriolis force *(CF)*. Wind blowing under these conditions is called *geostrophic*.

by placing a parcel of air at position 1 in the diagram and watching its behavior.

At position 1, the *PGF* acts immediately upon the air parcel, accelerating it northward toward lower pressure. However, the instant the air begins to move, the Coriolis force deflects the air toward its right, curving its path. As the parcel of air increases in speed (positions 2, 3, and 4), the magnitude of the Coriolis force increases (as shown by the blue arrows), bending the wind more and more to its right. Eventually, the wind speed increases to a point where the Coriolis force just balances the *PGF*. At this point (position 5), the wind no longer accelerates because the net force is zero. Here the wind flows in a straight path, parallel to the isobars at a constant speed.* This flow of air is called a **geostrophic** (*geo*: "earth;" *strophic*: "turning") **wind.** Notice that the geostrophic wind blows in the Northern Hemisphere with lower pressure to its left and higher pressure to its right.

When the flow of air is purely geostrophic, the isobars (or contour lines) are straight and evenly spaced, and the wind speed is constant. In the real world, isobars are rarely straight or evenly spaced, and the wind normally changes speed as it flows along. So, the geostrophic wind is usually only an approximation of the actual wind. However, the

*At first, it may seem odd that the wind blows at a constant speed with no net force acting on it. But when we remember that the net force is necessary only to accelerate ($F = ma$) the wind, it makes more sense. For example, it takes a considerable net force to push a car and get it rolling from rest. But once the car is moving, it only takes a force large enough to counterbalance friction to keep it going. There is no net force acting on the car, yet it rolls along at a constant speed.

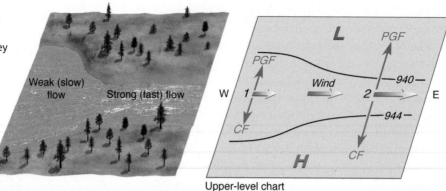

● **FIGURE 6.17** The isobars and contours on an upper-level chart are like the banks along a flowing stream. When they are widely spaced, the flow is weak; when they are narrowly spaced, the flow is stronger. The increase in winds on the chart results in a stronger Coriolis force *(CF)*, which balances a larger pressure gradient force *(PGF)*.

approximation is generally close enough to help us more clearly understand the behavior of the winds aloft.

As we would expect from our previous discussion of winds, the speed of the geostrophic wind is directly related to the pressure gradient. In ● Fig. 6.17, we can see that a geostrophic wind flowing parallel to the isobars is similar to water in a stream flowing parallel to its banks.

At position 1, the wind is blowing at a low speed; at position 2, the pressure gradient increases and the wind speed picks up. Notice also that at position 2, where the wind speed is greater, the Coriolis force is greater and balances the stronger pressure gradient force.

Because the geostrophic wind blows parallel to isobars (or contour lines), we can estimate the geostrophic wind directly by observing the orientation of the isobars on an upper-level chart. Notice in Fig. 6.17 that the west-to-east orientation of the isobars produces a westerly wind. In a similar way, it is possible to estimate the wind flow and pressure patterns aloft by watching the movement of clouds. This concept is explained further in Focus section 6.3 (p. 161).

We know that the winds aloft do not always blow in a straight line; frequently, they curve and bend into meandering loops. In the Northern Hemisphere, winds blow counterclockwise around areas of low pressure and clockwise around areas of high pressure. The next section explains why.

CURVED WINDS AROUND LOWS AND HIGHS ALOFT Because lows are also known as cyclones, the flow of air around them (which is counterclockwise in the Northern Hemisphere) is often called *cyclonic flow.* Likewise, the flow of air around a high (which is clockwise in the Northern Hemisphere) is called *anticyclonic flow.* Look at the wind flow around the upper-level low (Northern Hemisphere) in ● Fig. 6.18a. At first, it appears as though the wind is defying the Coriolis force by bending to the left as it moves counterclockwise around the system. Let's see why the wind blows this way.

Suppose we consider a parcel of air initially at rest at position 1 in Fig. 6.18a. The pressure gradient force accelerates the air inward toward the center of the low and the Coriolis force deflects the moving air to its right, until the air is moving parallel to the isobars at position 2. If the isobars were straight lines ahead of this point, the wind would move northward at a constant speed, bringing it to position 3. In reality, the isobars are curved, and the wind follows that curvature (position 4). A wind that blows at a constant speed parallel to *curved isobars* above the level of frictional influence is termed a gradient wind. Why does the wind follow this curved path? Look closely at position 2 (Fig. 6.18a), where a parcel is moving northward, and

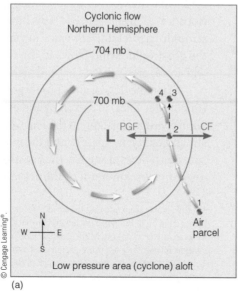

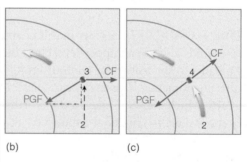

● **FIGURE 6.18** Winds and related forces around an area of low pressure above the friction level in the Northern Hemisphere. Notice that the pressure gradient force *(PGF)* is in red, while the Coriolis force *(CF)* is in blue

Estimating Wind Direction and Pressure Patterns Aloft by Watching Clouds

Both the wind direction and the orientation of the isobars aloft can be estimated by observing middle- and high-level clouds from Earth's surface. Suppose, for example, we are in the Northern Hemisphere watching clouds directly above us move from southwest to northeast at an elevation of about 3000 m or 10,000 ft (see ● Fig. 4a). This indicates that the geostrophic wind at this level is southwesterly. Looking downwind, the geostrophic wind blows parallel to the isobars with lower pressure on the left and higher pressure on the right. Thus, if *we stand with our backs to the direction from which the clouds are moving, lower pressure aloft will always be to our left and higher pressure to our right*. From this observation, we can draw a rough upper-level chart (Fig. 4b), which shows isobars and wind direction for an elevation of approximately 10,000 ft.

The isobars aloft will not continue in a southwest-northeast direction indefinitely;

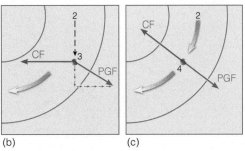

rather, they will often bend into wavy patterns. We may carry our observation one step further, then, by assuming a bending of the lines (Fig. 4c). Thus, with a southwesterly wind aloft, a trough of low pressure will be found to our west and a ridge

of high pressure to our east. What would be the pressure pattern if the winds aloft were blowing *from* the northwest? Answer: A trough would be to the east and a ridge to the west.

● **FIGURE 4** The movement of clouds can help determine the geostrophic wind direction and orientation of the isobars. Upper-level clouds moving from the southwest (a) indicate isobars and winds aloft (b). When extended horizontally, the upper-level chart appears as in (c), where lower pressure is to the west and higher pressure is to the east.

---

observe that if the wind were in geostrophic balance, the inward-directed pressure gradient force (*PGF*) would be in balance with the outward-directed Coriolis force (*CF*) and the wind should continue blowing in a straight line toward position 3. Suppose the parcel reaches position 3. Notice in Fig. 6.18b that here the *PGF* would now be angled toward the southwest, because the *PGF* is always directed toward lower pressure at right angles to the isobars. This means that part of the *PGF* will be directed southward, against the northward motion of the parcel. This situation slows

the wind down just a bit. Recall that the Coriolis force is directly related to wind speed, so the slower wind weakens the Coriolis force. As a result of the weaker Coriolis force, the *PGF* causes the wind to bend the parcel toward the left and thus move in a circular, counterclockwise path, parallel to curved isobars around the low, as illustrated in Fig. 6.18c.

Look at ● Fig. 6.19a and notice that above the level of frictional influence, the winds blow clockwise around an area of high pressure. The same spacing of the isobars tells us that the magnitude of the *PGF* is the same as in

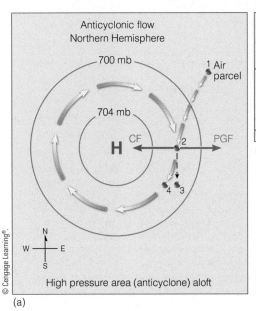

● **FIGURE 6.19** Winds and related forces around an area of high pressure above the friction level in the Northern Hemisphere. Notice that the pressure gradient force (*PGF*) is in red, while the Coriolis force (*CF*) is in blue.

Fig. 6.18a. Suppose at position 2 in Fig. 6.19a the wind is in geostrophic balance with the *PGF* directed outward and the Coriolis force directed inward. If the wind were in geostrophic balance, the air would move in a straight line parallel to straight-line isobars. But the isobars are curved. Let's further suppose that the air parcel moves from position 2 directly southward to position 3. We can see in Fig. 6.19b that at position 3 the *PGF* crosses the isobars toward the southeast, producing a southerly component of the *PGF*, which increases the wind speed just a bit. The increase in wind speed increases the magnitude of the Coriolis force, thereby bending the parcel to the right, which causes the wind to blow clockwise, parallel to the isobars in a circular path around the high.*

The greater Coriolis force around the high as compared to around the low results in an interesting relationship. Because the Coriolis force (at any given latitude) can increase only when the wind speed increases, we can see that for the same pressure gradient (the same spacing

of the isobars), the winds around a high-pressure area (or a ridge) must be greater than the winds around a low-pressure area (or a trough). Typically, however, this difference in speed is overshadowed by the stronger winds that occur around low-pressure areas (cyclonic storms) because of the closer spacing of the isobars and stronger pressure gradients associated with them.

In the Southern Hemisphere, the pressure gradient force starts the air moving and the Coriolis force deflects the moving air to the *left,* thereby causing the wind *to blow clockwise around lows* and *counterclockwise around highs.* So far we have seen how winds blow in theory, but how do they appear on an actual map?

WINDS ON UPPER-LEVEL CHARTS On the upper-level 500-mb map (● Fig. 6.20), notice that, as we would expect, the winds tend to parallel the contour lines in a wavy west-to-east direction. Notice also that the contour lines tend to decrease in height from south to north. This situation occurs because the air at this level is warmer to the south and colder to the north. On the map, where horizontal temperature contrasts are large, there is also a large height gradient—the contour lines are close together and the winds are strong. Where the horizontal temperature contrasts are small, there is a small height gradient—the contour lines are spaced farther apart and the winds are weaker. In general, on maps such as this we find stronger north-to-south temperature contrasts in winter than in summer, which is why the winds aloft are usually stronger in winter.

In Fig. 6.20, the wind is geostrophic where it blows in a straight path, parallel to evenly spaced contour lines; it

---

*Earlier in this chapter we learned that an object accelerates when there is a change in its speed or direction (or both). Therefore, the gradient wind blowing *around* the low-pressure center is constantly accelerating because it is constantly changing direction. This acceleration, called the *centripetal acceleration,* is directed at right angles to the wind, inward toward the low center. Remember from Newton's second law that, if an object is accelerating, there must be a *net force* acting on it. In this case, the net force acting on the wind must be directed toward the center of the low, so that the air will keep moving in a circular path. This inward-directed force is called the *centripetal force* (*centri:* "center;" *petal:* "to push toward"). In some cases, it is more convenient to express the centripetal force (and the centripetal acceleration) as the *centrifugal force,* an apparent force that is equal in magnitude to the centripetal force, but directed outward from the center of rotation.

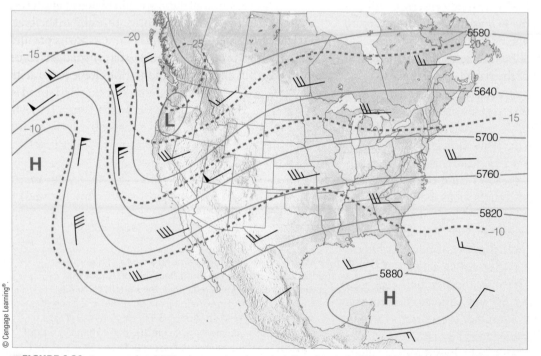

● **FIGURE 6.20** An upper-level 500-mb map showing wind direction, as indicated by lines that parallel the wind. Wind speeds are indicated by barbs and flags. (See the blue insert.) Solid gray lines are contours in meters above sea level. Dashed red lines are isotherms in °C.

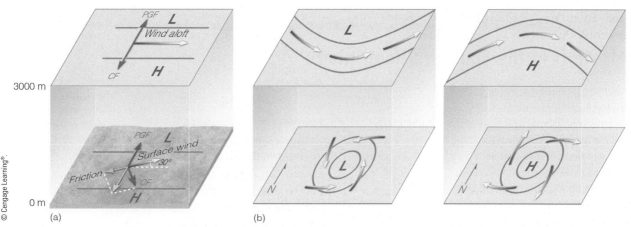

● **FIGURE 6.21** (a) The effect of surface friction is to slow down the wind so that, near the ground, the wind crosses the isobars and blows toward lower pressure. (b) This phenomenon at the surface produces an inflow of air around a low and an outflow of air around a high. Aloft, the winds blow parallel to the lines, usually in a wavy west-to-east pattern. Both diagram (a) and (b) are in the Northern Hemisphere.

is gradient where it blows parallel to curved contour lines. Where the wind flows in large, looping meanders, following a more or less north-south trajectory (as it does in Fig. 6.20 off the west coast of North America), the wind-flow pattern is called **meridional.** Where the winds are blowing in a west-to-east direction (as seen in Fig. 6.20 over the eastern third of the United States), the flow is termed **zonal.**

Because the winds aloft in middle and high latitudes generally blow from west to east, planes flying in this direction have a beneficial tail wind, which explains why a flight from San Francisco to New York City takes, on average, about 30 to 45 minutes less than the return flight. If the flow aloft is zonal, clouds, storms, and surface anticyclones tend to move more rapidly from west to east. However, where the flow aloft is meridional, as we will see in Chapter 8, surface storms tend to move more slowly, often intensifying into major storm systems.

We know that because of the pressure gradient force and the Coriolis force (which, as we have seen, bends moving air to the right in the Northern Hemisphere), the winds aloft in the middle latitudes of the Northern Hemisphere tend to blow in a west-to-east pattern. Because the Coriolis force bends moving air to the left in the Southern Hemisphere, does this situation mean that the winds aloft in the Southern Hemisphere blow from east to west? The answer to this question is given in Focus section 6.4.

Take a minute and look back at Fig. 6.13 on p. 157 and observe that the winds on the surface map tend to cross the isobars, blowing from higher pressure toward lower pressure. Observe also that the tightly packed isobars are indicating strong winds. Yet this same distribution of pressure (same pressure gradient) and same temperature on an upper-level chart would produce even stronger winds. Why, then, do surface winds normally cross the isobars and why do they blow more slowly than the winds aloft? The answer to both of these questions is *friction*.

SURFACE WINDS   Because of surface friction, winds on a surface weather map do not blow exactly parallel to the isobars; instead, they cross the isobars, moving from higher to lower pressure. The angle at which the wind crosses the isobars varies, but averages about 30°.

The frictional drag of the ground slows the wind down. Because the effect of friction decreases as we move away from Earth's surface, wind speeds tend to increase with height above the ground. The atmospheric layer that is influenced by friction, called the **friction layer** (or *planetary boundary layer*), usually extends upward to an altitude near 1000 m or 3000 ft above the surface, but this altitude can vary somewhat since both strong winds and rough terrain can extend the region of frictional influence.

In ● Fig. 6.21a, the wind aloft is blowing at a level above the frictional influence of the ground. At this level, the wind is approximately geostrophic and blows parallel to the isobars with the horizontal pressure gradient force *(PGF)* on its left balanced by the Coriolis force *(CF)* on its right. Notice, however, that at the surface the wind speed is slower. Apparently, the same pressure gradient force aloft will not produce the same wind speed at the surface, and the wind at the surface will not blow in the same direction as it does aloft.

Near the surface, *friction reduces the wind speed, which in turn reduces the Coriolis force.* Consequently, the weaker Coriolis force no longer balances the pressure gradient force, and the wind blows across the isobars toward lower

DID YOU KNOW?

The difference in air pressure from the base to the top of a New York skyscraper about 0.5 km (1600 ft) tall is typically near 55 millibars, a value much greater than the typical horizontal difference in air pressure between New York and Miami, Florida—a distance of more than 1800 km (1100 mi).

Winds Aloft in the Southern Hemisphere

In the Southern Hemisphere, just as in the Northern Hemisphere, the winds aloft blow because of horizontal differences in pressure. The pressure differences, in turn, are due to variations in temperature. Recall from an earlier discussion of pressure that warm air aloft is associated with high pressure and cold air aloft with low pressure. Look at ● Fig. 5. It shows an upper-level chart that extends from the Northern Hemisphere into the Southern Hemisphere. Over the equator where the air is warmer, the pressure aloft is high. North and south of the equator, where the air is colder, the pressure aloft is lower.

Let's assume, to begin with, that there is no wind on the chart. In the Northern Hemisphere, the pressure gradient force directed northward starts the air moving toward lower pressure. Once the air is set in motion, the Coriolis force bends it to the *right* until it is a *west wind*, blowing parallel to the isobars.

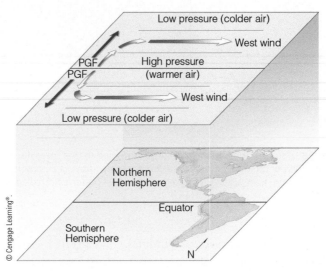

● FIGURE 5 Upper-level chart that extends over the Northern and Southern Hemispheres. Solid gray lines on the chart are isobars.

In the Southern Hemisphere, the pressure gradient force directed southward starts the air moving south. But notice that the Coriolis force in the Southern Hemisphere bends the moving air to its *left*, until the wind is blowing parallel to the isobars *from the west*. Hence, in the middle and high latitudes of both hemispheres, we generally find westerly winds aloft.

pressure. The pressure gradient force is now balanced by the sum of the frictional force and the Coriolis force. Therefore, in the Northern Hemisphere, we find surface winds blowing counterclockwise and *into* a low; they flow clockwise and *out* of a high (see Fig. 6.21b).

In the Southern Hemisphere, winds blow clockwise and inward around surface lows, counterclockwise and outward around surface highs. See ● Fig. 6.22, which shows a surface weather map and the general wind-flow pattern on a day in December in South America.

Winds and Vertical Air Motions

Up to this point, we have seen that surface winds blow in toward the center of low pressure and outward away from the center of high pressure. As air moves inward toward the center of a low-pressure area (see ● Fig. 6.23), it must go somewhere. This converging air cannot go into the ground, so it slowly rises. Above the surface low (at about 6 km or so), the air begins to spread apart (diverge) to compensate for the converging surface air.

As long as the upper-level diverging air balances the converging surface air, the central pressure in the low does not change. However, the surface pressure *will change* if upper-level divergence and surface convergence

are not in balance. For example, as we saw earlier in this chapter, when we examined the air pressure above two cities, the surface air pressure will change if the mass of air above the surface changes. Consequently, if upper-level divergence exceeds surface convergence (that is, more air is removed at the top than is taken in at the surface), the air pressure at the center of the low will decrease, and isobars around the low will become more tightly packed. This situation increases the pressure gradient (and, hence, the pressure gradient force), which, in turn, increases the surface winds.

Surface winds move outward (diverge), away from the center of a high-pressure area. Observe in Fig. 6.23 that to replace this laterally spreading surface air the air aloft converges and slowly descends. Again, as long as upper-level converging air balances surface diverging air, the air pressure in the center of the high will not change. Should surface diverging air exceed upper-level converging air (more air removed at the surface than is brought in at the top), the air pressure at the center of the high will decrease, the pressure gradient force will weaken, and the surface winds will blow more slowly. (Convergence and divergence of air are so important to the development or weakening of surface pressure systems that we will examine this topic again in Chapter 8, when we look more closely at the vertical structure of pressure systems.)

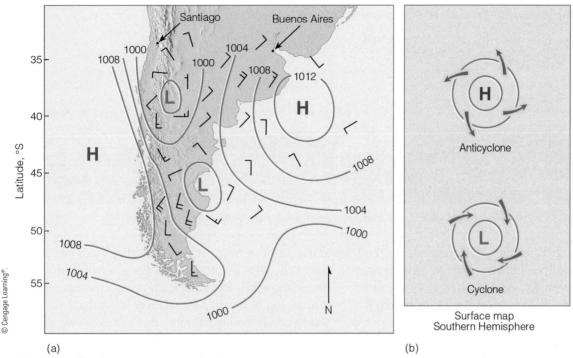

© Cengage Learning®.

(a)   (b)

● **FIGURE 6.22** (a) Surface weather map showing isobars and winds on a day in December in South America. (b) This boxed area shows the idealized flow around surface pressure systems in the Southern Hemisphere.

The rate at which air rises above a low or descends above a high is small compared to the horizontal winds that spiral about these systems. Generally, the vertical motions are usually only about several centimeters per second, or about 1.5 km (or 1 mi) per day.

Earlier in this chapter we learned that air moves in response to pressure differences. Because air pressure decreases rapidly with increasing height above the surface, there is always a strong pressure gradient force directed upward, much stronger than in the horizontal. Why, then, doesn't the air rush off into space?

Air does not rush off into space because the upward-directed pressure gradient force is nearly always exactly balanced by the downward force of gravity. When these two forces are in exact balance, the air is said to be in **hydrostatic equilibrium.** When air is in hydrostatic equilibrium, there is no net vertical force acting on it,

and so there is no net vertical acceleration. (Remember the example of the moving car in the footnote on p. 159.) Most of the time, the atmosphere approximates hydrostatic balance, even when air slowly rises or descends at a constant speed. However, this balance does not exist in violent thunderstorms and tornadoes, where the air shows appreciable vertical acceleration. Such storms occur over relatively small vertical distances, considering the total vertical extent of the atmosphere.

Determining Wind Direction and Speed

Wind is characterized by its direction, speed, and gustiness. Because air is invisible, we cannot really see it. Rather, we see things being moved by it. Thus, we can determine wind direction by watching the movement of objects as air passes them. For example, the rustling of small leaves, smoke drifting near the ground, and flags waving on a pole all indicate wind direction. In a light breeze, a tried and true method for an observer to determine wind direction is to raise a wet finger into the air. The dampness quickly evaporates on the windward side, cooling the skin. Traffic sounds carried from nearby railroads or airports can also be used to help figure out the direction of the wind. Even your nose can alert you to the wind direction as the smell of fried chicken or broiled hamburgers drifts with the wind from a local restaurant.

We already know that *wind direction* is given as the direction from which it is blowing—a north wind blows

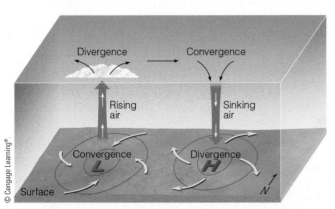

© Cengage Learning®.

● **FIGURE 6.23** Winds and air motions associated with surface highs and lows in the Northern Hemisphere.

● **FIGURE 6.24** An onshore wind blows from water to land, whereas an offshore wind blows from land to water.

from the north toward the south. However, near large bodies of water and in hilly regions, wind direction is expressed differently. For example, wind blowing from the water onto the land is referred to as an **onshore wind** (or *onshore breeze*), whereas wind blowing from land to water is called an **offshore wind** (or *offshore breeze*). (See ● Fig. 6.24.) Air moving uphill is an *upslope wind;* air moving downhill is a *downslope wind.* The wind direction can also be given as degrees about a 360° circle. These directions are expressed by the numbers shown in ● Fig. 6.25. For example: A wind direction of 360° is a north wind; an east wind is 90°; a south wind is 180°; and calm is expressed as zero. It is also common practice to express the wind direction in terms of compass points, such as N, NW, NE, and so on.

THE INFLUENCE OF PREVAILING WINDS At many locations, the wind blows more frequently from one direction than from any other. The **prevailing wind** is the name given to the wind direction most often observed

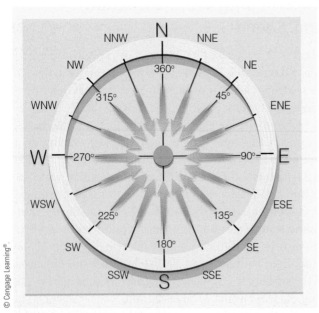

● **FIGURE 6.25** Wind direction can be expressed in degrees about a circle or as compass points.

during a given time period. Prevailing winds can greatly affect the climate of a region. For example, where the prevailing winds are upslope, the rising, cooling air makes clouds, fog, and precipitation more likely than where the winds are downslope. Prevailing onshore winds in summer carry moisture, cool air, and fog into coastal regions, whereas prevailing offshore breezes carry warmer and drier air into the same locations.

In city planning, the prevailing wind is typically considered when planning where industrial centers, factories, and city dumps should be built. All of these, of course, must be located so that the wind will not carry pollutants into populated areas. Sewage disposal plants must be situated downwind from large housing developments, and major runways at airports must be aligned with the prevailing wind to assist aircraft in taking off or landing. In the high country, strong prevailing winds can bend and twist tree branches toward the downwind side, producing sculpted *"flag" trees* (see ● Fig. 6.26).

The prevailing wind can even be a significant factor in building an individual home. In the northeastern half of the United States, the prevailing wind in winter is northwest and in summer it is southwest. Thus, to maximize comfort and energy efficiency, houses built in the northeastern United States should have windows facing southwest to provide summertime ventilation and few, if any, windows facing the cold winter winds from the northwest. The northwest side of the house should be thoroughly insulated and, perhaps, even protected by a windbreak.

The prevailing wind can be represented by a **wind rose** (see ● Fig. 6.27), which indicates the percentage of time the wind blows from different directions. Extensions from the center of a circle point to the wind direction, and the length of each extension indicates the percentage of time the wind blew from that direction. Wind speed does not affect the length of each extension. However, the strength of the wind blowing from different directions is very helpful to know, so wind roses may include this information, as shown by the wind rose illustrated in Fig. 6.27. In this case, the prevailing wind is from the south, but winds from the west and

● FIGURE 6.26 In the high country of Colorado, these trees standing unprotected from the wind are often sculpted into "flag" trees.

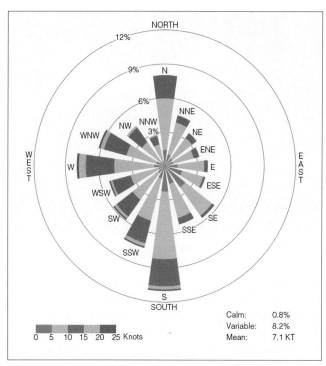

● FIGURE 6.27 This wind rose represents the percentage of time the wind blew from various directions at the International Airport in Louisville, Kentucky, between 1977 and 2006. The most common directions are south (S) and north (N); the least common is north-northwest (NNW). Each bar includes color categories showing how often the wind blew at a particular speed from a given direction. The wind was variable or calm about 9 percent of the time (not shown in the wind rose). (Midwest Regional Climate Center)

north are almost as common. While this wind rose covers all four seasons and all times of day, a wind rose can be made for any particular time of the day, and it can represent the wind direction for any month or season of the year.

WIND INSTRUMENTS A very old, yet reliable, weather instrument for determining wind direction is the **wind vane**. Most wind vanes consist of a long arrow with a tail, which is allowed to move freely about a vertical post (see ● Fig. 6.28). The arrow always points into the wind and, hence, always gives the wind direction. Wind vanes can be made of almost any material. At airports, a cone-shaped bag, opened at both ends so that it extends horizontally as the wind blows through it, is situated near the runway. This form of wind vane, called a *wind sock,* enables pilots to tell the surface wind direction when landing.

The instrument that measures wind speed is the **anemometer.** *Cup anemometers* consist of three (or more) hemispheric cups mounted on a vertical shaft as shown in Fig. 6.28. The difference in wind pressure from one side of a cup to the other causes the cups to spin about the shaft. The rate at which they rotate is directly proportional to the speed of the wind. The spinning of the cups is usually translated through a system of gears into wind speed, which can be read from a dial or transmitted to a recorder.

The **aerovane** *(skyvane)* is an instrument that indicates both wind speed and direction. It consists of a bladed propeller that rotates at a rate proportional to the wind speed. Its streamlined shape and a vertical fin keep the blades facing into the wind (see ● Fig. 6.29). When attached to a recorder, a continuous record of both wind speed and direction can be obtained.

Over the last few years, the national ASOS network of automated weather stations has been replacing the original cup anemometers with *sonic anemometers*

(see ● Fig. 6.30). These anemometers measure changes in ultrasonic signals that are produced when the wind blows across three sets of transmitters and receivers. The changes in signal speed are then converted into wind speeds in each of the three directions.

The wind-measuring instruments described thus far are "ground-based" and only give wind speed or direction

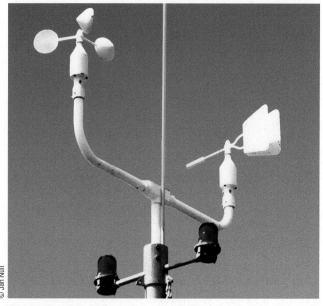

● FIGURE 6.28 A wind vane and a cup anemometer.

● **FIGURE 6.29** The aerovane (skyvane).

at a particular fixed location. But the wind is influenced by local conditions, such as buildings, trees, and so on. Also, wind speed normally increases rapidly with height above the ground. Thus, wind instruments should be exposed to freely flowing air well above the roofs of buildings. In practice, unfortunately, anemometers are placed at various levels often resulting in erratic wind observations. Thus, placement of anemometers is critical in order to avoid erratic and unrepresentative wind readings.

Wind information can also be obtained by using instruments that either ascend or descend through the atmosphere. In the first example, a balloon rises from the surface carrying a *radiosonde* (an instrument package designed to measure the vertical profile of temperature, pressure, and humidity—see Chapter 1, p. 22). Equipment located on the ground constantly tracks the balloon, measuring its vertical and horizontal angles as well as its height above the ground. From this information, a computer determines and prints

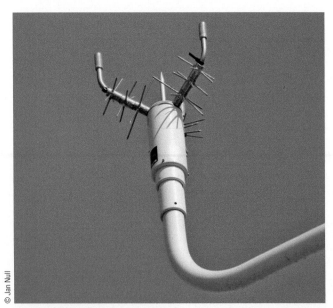

● **FIGURE 6.30** The sonic anemometer shown here measures wind speed as part of the ASOS system.

the vertical profile of wind from the surface up to where the balloon normally pops, typically in the stratosphere near 30 km or 100,000 ft. The observation of winds using a radiosonde balloon is called a *rawinsonde observation.*

Rawindsonde data, taken from balloons launched around the world every 12 hours has been an important tool in tracking upper-level winds for more than 70 years. On special occasions, such as during field studies or when hurricanes are being monitored, an airplane is used to deploy an instrument package called a *dropsonde*, which falls to the ground via parachute. The resulting data are *dropwindsonde observations.*

With the aid of an upward-pointing Doppler radar, a vertical profile of wind speed and direction up to an altitude of 16 km or so above the ground can be obtained. Such a profile is called a *wind sounding,* and the radar, a **wind profiler** (or simply a *profiler*). Doppler radar, like conventional radar, emits pulses of microwave radiation that are returned (backscattered) from various targets, in this case the irregularities in moisture and temperature created by turbulent, twisting eddies that move with the wind. Doppler radar works on the principle that, as these eddies move toward or away from the receiving antenna, the returning radar pulse will change in frequency. The Doppler radar wind profilers are so sensitive that they can translate the backscattered energy from these eddies into a vertical picture of wind speed and direction in a column of air 16 km (10 mi) thick. Wind profilers have been used at NASA's Kennedy Space Center in Florida and across the Great Plains.

Observations of upper-level winds are also made by satellites. As we saw in Chapter 4, geostationary satellites positioned above a particular location can show the movement of clouds, which is translated into wind direction and speed. Satellites now measure surface winds above the ocean by observing the roughness of the sea. An instrument called a *scatterometer* sends out a microwave pulse of energy that travels through the clouds, down to the sea surface. A portion of this energy is bounced back to the satellite. The amount of energy returning depends on the roughness of the sea: rougher seas return more energy. Since the sea's roughness depends upon the strength of the wind blowing over it, the intensity of the returning energy can be translated into the surface wind speed and direction, as illustrated in ●Fig. 6.31. Scatterometers operate aboard two European satellites called MetOp-A and MetOp-B, and a NASA scatterometer called *RapidScat* was placed aboard the International Space Station in 2014.

The wind is an essential weather element that affects our environment in many ways. It can shape the landscape, transport material from one area to another, and generate ocean waves. It can also turn the blades of a windmill and blow down a row of trees. The reason the wind is capable of such feats is that, as the wind blows against an object, it exerts a force upon it. The amount of force exerted by

Wind Energy

For centuries, thousands of small windmills—their arms spinning in a stiff breeze—pumped water, sawed wood, and even supplemented the electrical needs of homes and farms around the world. It was not until the energy crisis of the early 1970s, however, that we seriously considered wind-driven turbines, called *wind turbines*, to run generators that produce electricity. Recently, the amount of wind energy in the United States has doubled every few years. As of 2015, only China had more capacity in its wind energy installations.

Wind seems an attractive way of producing energy. It is nonpolluting and, unlike the sun, is not restricted to daytime use. A single commercial wind turbine, which may cost several million dollars to build and install, can generate power for hundreds of homes or businesses at a cost that is comparable in some areas to electricity generated by fossil fuels. However, the sight of large wind turbines is not to everyone's liking. (Probably, though, it is no more of an eyesore than other types of energy-related infrastructure, such as oil pumps and electrical transmission towers.) Unfortunately, each year the blades of spinning turbines kill countless birds. To help remedy this problem, many wind turbine companies hire avian specialists to study bird behavior,

● FIGURE 6 A wind farm near Palm Springs, California, generates electricity that is sold to Southern California.

and some turbines are actually shut down during nesting time. And the blades of modern high-capacity turbines turn more slowly, thereby helping birds to avoid them.

If the wind turbine is to produce electricity, there must be wind—and not just any wind, but a flow of air neither too weak nor too strong. A slight breeze will not turn the blades, and a powerful wind gust could severely damage the machine. Thus, regions with the greatest potential for wind-generated power have moderate, steady winds. So much of the Great Plains of the United States is well-suited to wind power, so much so that the region has been dubbed "the Saudi Arabia of wind energy."

As of 2015, more than 48,000 wind turbines had the capacity of generating more than 74,000 megawatts of electricity in the United States, which is enough energy to supply the annual needs of more than 15 million homes. In California alone, there are thousands of wind turbines, many of which are on *wind farms*, clusters of 50 or more wind turbines (see ● Fig. 6). Wind provided more than 4 percent of the electricity needs in the United States in 2015, roughly double the percentage of 2010. The United States Department of Energy has estimated that wind energy using current technology could provide 20 percent of the nation's electricity by 2030.

the wind over an area increases as the square of the wind velocity. So when the wind velocity doubles, the force it exerts on an object goes up by a factor of four. To harness some of the wind's energy and turn it into electricity, large

wind turbines and *wind farms* are becoming increasingly common around the world. More information on this topic is given in Focus section 6.5.

● FIGURE 6.31 A satellite image of wind speed and cloud-top temperature associated with Cyclone Ita on April 10, 2014. This intense tropical cyclone eventually struck far northeast Australia (left side of image). The colored background shows the temperature of the cyclone's cloud tops, which is correlated with the intensity of convection. Each wind barb represents 10 m/s (19 knots), with the satellite-observed winds reaching 50 m/s (95 knots) near the center of Ita. Wind speeds were gathered by the European Advanced Scatterometer (*ASCAT*) aboard the MetOp-A satellite.

SUMMARY

This chapter gives us a broad view of how and why the wind blows. Aloft where horizontal variations in temperature exist, there is a corresponding horizontal change in pressure. The difference in pressure establishes a force, the pressure gradient force *(PGF)*, which starts the air moving from higher toward lower pressure.

Once the air is set in motion, the Coriolis force bends the moving air to the right of its intended path in the Northern Hemisphere and to the left in the Southern Hemisphere. Above the level of surface friction, the wind is bent enough so that it blows nearly parallel to the isobars, or contours. Where the wind blows in a straight-line path, and a balance exists between the pressure gradient force and the Coriolis force, the wind is termed geostrophic. Where the wind blows parallel to curved isobars (or contours), the wind is called gradient wind. When the wind-flow pattern aloft is more west-to-east, the flow is called *zonal*; when the wind-flow pattern aloft is more north-to-south or south-to-north, the flow is called *meridional*.

The interaction of the forces causes the wind in the Northern Hemisphere to blow clockwise around regions of high pressure and counterclockwise around areas of low pressure. In the Southern Hemisphere, the wind blows counterclockwise around highs and clockwise around lows. The effect of surface friction is to slow down the wind. This causes the surface air to blow across the isobars from higher pressure toward lower pressure. Consequently, in both hemispheres, surface winds blow outward, away from the center of a high, and inward, toward the center of a low.

We also looked at various methods and instruments used to determine and measure wind speed and direction.

KEY TERMS

The following terms are listed (with corresponding page numbers) in the order they appear in the text. Define each. Doing so will aid you in reviewing the material covered in this chapter.

air pressure, 148
millibar, 151
standard atmospheric
 pressure, 151
barometer, 151
mercury barometer, 151
aneroid barometer, 151
station pressure, 152
sea-level pressure, 152

isobar, 153
surface map, 153
anticyclone, 154
mid-latitude cyclonic
 storm, 154
isobaric map, 154
contour line, 154
ridge, 154
trough, 154

pressure gradient, 156
pressure gradient force, 156
Coriolis force, 157
geostrophic wind, 159
gradient wind, 161
meridional (wind flow), 163
zonal (wind flow), 163
friction layer, 163
hydrostatic equilibrium, 165

onshore wind, 166
offshore wind, 166
prevailing wind, 166
wind rose, 166
wind vane, 167
anemometer, 167
aerovane, 167
wind profiler, 168

QUESTIONS FOR REVIEW

1. Explain why atmospheric pressure always decreases with increasing altitude.
2. What might cause the air pressure to change at the bottom of an air column?
3. Why is the decrease of air pressure with increasing altitude more rapid when the air is cold?
4. What is considered standard sea-level atmospheric pressure in millibars? In inches of mercury? In hectopascals?
5. With the aid of a diagram, describe how a mercury barometer works.
6. How does an aneroid barometer measure atmospheric pressure?
7. How does sea-level pressure differ from station pressure? Can the two ever be the same? Explain.
8. Why will Denver, Colorado, *always* have a lower *station pressure* than Chicago, Illinois?
9. What are isobars? In what increment are they usually drawn on a surface weather map?
10. On an upper-level map, is cold air aloft generally associated with low or high pressure? What about warm air aloft?
11. What do Newton's first and second laws of motion tell us?
12. What does a steep (or strong) pressure gradient mean? How would it appear on a surface map?
13. What is the name of the force that initially sets the air in motion and, hence, causes the wind to blow?
14. Explain why, on a map, closely spaced isobars (or contours) indicate strong winds, and widely spaced isobars (or contours) indicate weak winds.
15. What does the Coriolis force do to moving air (a) in the Northern Hemisphere? (b) in the Southern Hemisphere?
16. Explain how each of the following influences the Coriolis force: (a) wind speed (b) latitude.

17. Why do upper-level winds in the middle latitudes of both hemispheres generally blow from west to east?

18. What is a geostrophic wind? On an upper-level chart, how does it blow?

19. What are the forces that affect the horizontal movement of air?

20. Describe how the wind blows around high-pressure areas and low-pressure areas aloft and near the surface
    (a) in the Northern Hemisphere; and
    (b) in the Southern Hemisphere.

21. How does zonal flow differ from meridional flow?

22. If the clouds overhead are moving from north to south, would the upper-level center of low pressure be to the east or west of you?

23. On a surface map, why do surface winds tend to cross the isobars and flow from higher pressure toward lower pressure?

24. Since there is always an upward-directed pressure gradient force, why doesn't air rush off into space?

25. List as many ways as you can of determining wind direction and wind speed.

26. Below is a list of instruments. Describe how each one measures wind speed, wind direction, or both.
    (a) wind vane (b) cup anemometer (c) aerovane (skyvane) (d) radiosonde (e) satellite (f) wind profiler

27. An upper wind direction is reported as 225°. From what compass direction is the wind blowing?

QUESTIONS FOR THOUGHT AND EXPLORATION

1. The gas law states that pressure is proportional to temperature times density. Use the gas law to explain why a basketball seems to deflate when placed in a refrigerator.

2. Can the station pressure ever *exceed* the sea-level pressure? Explain.

3. The pressure gradient force causes air to move from higher pressures toward lower pressures (perpendicular to the isobars), yet actual winds rarely blow in this fashion. Explain why they don't.

4. The Coriolis force causes winds to deflect to the right of their intended path in the Northern Hemisphere, yet around a surface low-pressure area, winds blow counterclockwise, appearing to bend to their left. Explain why.

5. Explain why, on a sunny day, an aneroid barometer would indicate "stormy" weather when carried to the top of a hill or mountain.

6. Pilots often use the expression "high to low, look out below." In terms of upper-level temperature and pressure, explain what this can mean.

7. If Earth were not rotating, how would the wind blow with respect to centers of high and low pressure?

8. Why are surface winds that blow over the ocean closer to being geostrophic than those that blow over the land?

9. In the Northern Hemisphere, you observe surface winds shift from N to NE to E, then to SE. From this observation, you determine that a west-to-east moving high-pressure area (anticyclone) has passed north of your location. Describe with the aid of a diagram how you were able to come to this conclusion.

10. As a cruise ship crosses the equator, the entertainment director exclaims that water in a tub will drain in the opposite direction now that the ship is in the Southern Hemisphere. Give *two* reasons to the entertainment director why this assertion is not so.

GLOBAL **GEOSCIENCE** WATCH   Go to the Wind Energy portal. Under the Primary Sources section, open the most recent "Year-End Market Report" from the American Wind Energy Association. Examine the map of wind power capacity by state. Which states show the most and least capacity? Do you think the differences are mainly due to the availability of strong, reliable winds? What other factors might be playing an important role?

ONLINE RESOURCES

 Visit www.cengagebrain.com to view additional resources, including video exercises, practice quizzes, an interactive eBook, and more.

CHAPTER 7

Atmospheric Circulations

Contents

On December 28, 1997, a United Airlines' Boeing 747 carrying 374 passengers was over the Pacific Ocean en route to Hawaii from Japan. About two hours into the flight, the aircraft was at a cruising altitude of 31,000 feet when suddenly, east of Tokyo, this routine, uneventful flight turned harrowing. Seat-belt signs were turned on because of reports of severe air turbulence nearby. The plane hurtled upward, then quickly dropped by about 30 meters (100 feet) before stabilizing. Screaming, terrified passengers not fastened to their seats were flung against the walls of the aircraft, then dropped. Bags, serving trays, and luggage that slipped out from under the seats were flying about inside the plane. Within seconds, the entire ordeal was over. A total of 160 people were injured. Tragically, there was one fatality: A 32-year-old woman who had been hurled against the ceiling of the plane died of severe head injuries. What sort of atmospheric phenomenon could cause such turbulence?

The aircraft in our opener on the previous page encountered a turbulent eddy—an "air pocket"—in perfectly clear weather. Such eddies are not uncommon, especially in the vicinity of jet streams. In this chapter, we will examine a variety of eddy circulations. First, we will look at the formation of small-scale winds. Then we will examine slightly larger circulations—local wind—such as the sea breeze and the chinook, describing how they form and the type of weather they generally bring. Finally, we will look at the general wind-flow pattern around the world.

Scales of Atmospheric Motion

The air in motion—what we commonly call wind—is a powerful phenomenon. It is invisible, yet we see evidence of it nearly everywhere we look. It sculpts rocks, moves leaves, blows smoke, and lifts water vapor upward to where it can condense into clouds. The wind is with us wherever we go. On a hot day, it can cool us off; on a cold day, it can make us shiver. A breeze can sharpen our appetite when it blows the aroma from the local bakery in our direction. The wind is a powerful element. The workhorse of weather, it moves storms and large fair-weather systems around the globe. It transports heat, moisture, dust, insects, bacteria, and pollens from one area to another.

Circulations of all sizes exist within the atmosphere. Little whirls form inside bigger whirls, which encompass even larger whirls—one huge mass of turbulent, twisting *eddies*.* For clarity, meteorologists arrange circulations according to their size. This hierarchy of motion from tiny gusts to giant storms is called the **scales of motion.**

Consider smoke rising into otherwise clean air from a chimney in an industrial section of a large city

*Eddies are spinning globs of air that have a life history of their own.

(see ● Fig. 7.1a). Within the smoke, small chaotic motions—tiny eddies—cause it to tumble and turn. These eddies constitute the smallest scale of atmospheric motion—the **microscale.** At the microscale level, eddies with diameters of a few meters or less not only disperse smoke, but they also sway branches and swirl dust and papers into the air. They form by convection or by the wind blowing past obstructions and are usually short-lived, lasting only a few minutes at best.

In Fig. 7.1b, observe that, as the smoke rises, it drifts many kilometers downwind. This circulation of city air constitutes the next larger scale—the **mesoscale** ("middle scale"). Typical mesoscale circulations range from a few kilometers to about a hundred kilometers in diameter. Generally, they last longer than microscale motions, often many minutes, hours, or in some cases as long as a day. Mesoscale circulations include local winds (which form along shorelines and mountains), as well as thunderstorms, tornadoes, and small tropical cyclones.

When we look for the smokestack on a surface weather map (Fig. 7.1c), neither the smokestack nor the circulation of city air shows up. All that we see are the circulations around high- and low-pressure areas—the cyclones and anticyclones of the middle latitudes—as well as the large tropical cyclones of lower latitudes. We are now looking at the **synoptic scale,** or weather-map scale. Circulations of this magnitude dominate regions of hundreds to even thousands of square kilometers and, although the life spans of these features vary, they typically last for days and sometimes weeks. When we look at wind patterns over the entire Earth, we are looking at the **global scale,** or *planetary scale.* Together, the synoptic and global scales are referred to as the **macroscale**—the largest scale of atmospheric motion. ●Figure.7.2 summarizes the various scales of motion and their average life span.

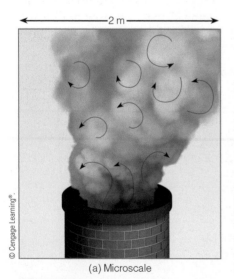
(a) Microscale

(b) Mesoscale

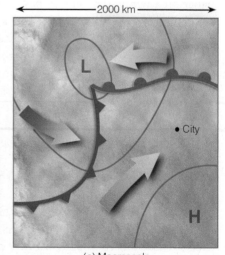
(c) Macroscale

● **FIGURE 7.1** Scales of atmospheric motion. The tiny microscale motions constitute a part of the larger mesoscale motions, which, in turn, are part of the much larger macroscale. Notice that as the scale becomes larger, motions observed at the smaller scale are no longer visible.

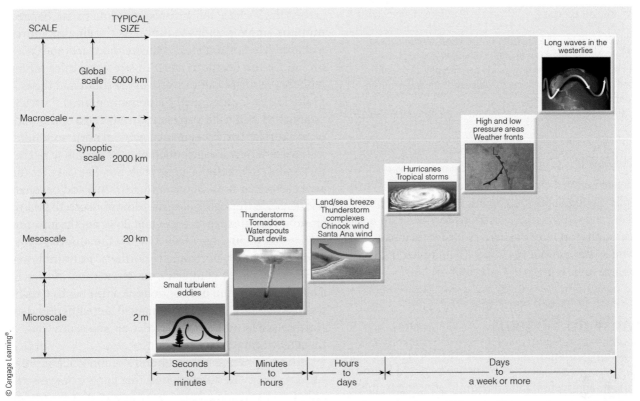

SCALE   TYPICAL SIZE

Global scale   5000 km

Macroscale - - - - - - - - - - - -

Synoptic scale   2000 km

Mesoscale   20 km

Microscale   2 m

Long waves in the westerlies

High and low pressure areas
Weather fronts

Hurricanes
Tropical storms

Land/sea breeze
Thunderstorm complexes
Chinook wind
Santa Ana wind

Thunderstorms
Tornadoes
Waterspouts
Dust devils

Small turbulent eddies

Seconds to minutes | Minutes to hours | Hours to days | Days to a week or more

© Cengage Learning®.

● **FIGURE 7.2** The scales of atmospheric motion with the phenomenon's average size and life span. (Because the actual size of certain features can vary, some of the features fall into more than one category.)

Eddies—Big and Small

When the wind encounters a solid object, a whirl of air, or *eddy*, forms on the object's downwind side.* The size and shape of the eddy often depend upon the size and shape of the obstacle and on the speed of the wind. Light winds produce small stationary eddies. Wind moving past trees, shrubs, and even your body produces small eddies. (You may have had the experience of dropping a piece of paper on a windy day only to have it carried away by a swirling eddy as you bend down to pick it up.) Air flowing over a building produces larger eddies that will, at best, be about the size of the building. Strong winds blowing past an open sports stadium can produce eddies that may rotate in such a way as to create surface winds on the playing field that move in a direction opposite to the wind flow above the stadium. Wind blowing over a fairly smooth surface produces few eddies, but when the surface is rough, many eddies form.

The eddies that form downwind from obstacles can produce a variety of interesting effects. For instance, wind moving over a mountain range in stable air with a speed greater than 40 knots usually produces waves and eddies, such as those shown in ● Fig. 7.3. We can see that eddies form both close to the mountain and beneath each wave

crest. These so-called **rotors** have violent vertical motions that produce extreme turbulence and hazardous flying conditions. Strong winds blowing over a mountain in stable air can produce a *mountain wave eddy* on the downwind side, with a reverse flow near the ground. On a much smaller scale, the howling of wind on a blustery night is believed to be caused by eddies that are constantly being shed around obstructions, such as chimneys and roof corners.

Turbulent eddies form aloft as well as near the surface. Turbulence aloft can occur suddenly and unexpectedly, especially where the wind changes its speed or direction (or both) abruptly. Such a change is called **wind shear.**

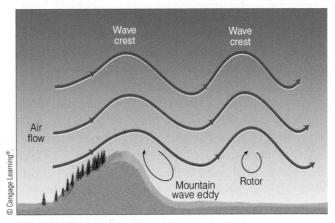

© Cengage Learning®.

● **FIGURE 7.3** Under stable conditions, air flowing past a mountain range can create eddies many kilometers downwind of the mountain itself.

---

*The irregular, disturbed flow of gusty winds and eddies is called *turbulence.*

The shearing creates forces that produce eddies along a mixing zone. If the eddies form in clear air, this form of turbulence is called **clear air turbulence,** or **CAT.** When an airplane is flying through such turbulence, the bumpiness can range from small vibrations to violent up-and-down motions that force passengers against their seats and toss objects throughout the cabin. (Additional information on this topic is given in Focus section 7.1.)

Local Wind Systems

Every summer, millions of people flock to the New Jersey shore, hoping to escape the oppressive heat and humidity of the inland region. On hot, humid afternoons, these travelers often encounter thunderstorms about twenty miles or so from the ocean, thunderstorms that invariably last for only a few minutes. In fact, by the time the vacationers arrive at the beach, skies are generally clear and air temperatures are much lower, as cool ocean breezes greet them. If the travelers return home in the afternoon a few days later, these "mysterious" showers often occur at just about the same location as before.

The showers are not really mysterious, of course. They are caused by a local wind system, the *sea breeze.* As cooler ocean air pours inland, it forces the warmer (less dense), unstable humid air to rise and condense, producing majestic clouds and rainshowers along a line that separates the contrasting temperatures.

The sea breeze forms as part of a thermally driven circulation, so we will begin our study of local winds by examining the formation of thermal circulations.

THERMAL CIRCULATIONS   Consider the vertical distribution of pressure shown in ● Fig. 7.4a. The isobars* all lie parallel to Earth's surface; thus, there is no horizontal variation in pressure (or temperature), and there is no pressure gradient and no wind. Suppose in Fig. 7.4b the atmosphere is cooled to the north and warmed to the south. In the cold, dense air above the surface, the isobars bunch closer together, while in the warm, less-dense air, they spread farther apart. This dipping of the isobars produces a horizontal pressure gradient force *(PGF)* aloft that causes the air to move from higher pressure toward lower pressure.

*The isobars depicted here actually represent a surface of constant pressure (an *isobaric surface*) rather than a line, or isobar. Information on isobaric surfaces is given in Focus section 6.2 on p. 155.

At the surface, the air pressure remains unchanged until the air aloft begins to move. As the air aloft moves from south to north, air leaves the southern area and "piles up" above the northern area. This redistribution of air reduces the surface air pressure to the south and raises it to the north. Consequently, a pressure gradient force is established at Earth's surface from north to south and, hence, surface winds begin to blow from north to south.

We now have a distribution of pressure and temperature and a circulation of air, as shown in Fig. 7.4c. As the cool surface air flows southward, it warms and becomes less dense. In the region of surface low pressure, the warm air slowly rises, expands, cools, and flows out the top at an elevation of about 1 km (3300 ft) above the surface. At this level, the air flows horizontally northward toward lower pressure, where it completes the circulation by slowly sinking and flowing out the bottom of the surface high. Circulations brought on by changes in air temperature, in which warmer air rises and colder air sinks, are termed **thermal circulations.**

The regions of surface high and low atmospheric pressure created as the atmosphere either cools or warms are called *thermal* (cold core) *highs* and *thermal* (warm core) *lows*. In general, they are shallow systems, usually extending no more than a few kilometers above the ground.

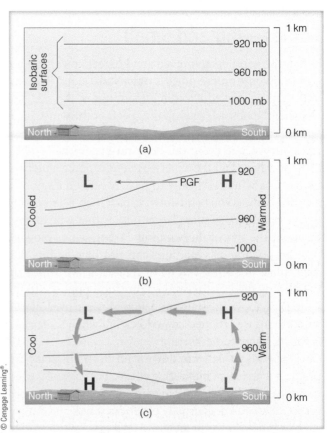

● **FIGURE 7.4**   A thermal circulation produced by the heating and cooling of the atmosphere near the ground. The H's and L's refer to atmospheric pressure. The lines represent surfaces of constant pressure (isobaric surfaces), where 1000 is 1000 millibars.

Eddies and "Air Pockets"

To better understand how eddies form along a zone of wind shear, imagine that, high in the atmosphere, there is a stable layer of air having vertical wind speed shear (changing wind speed with height) as depicted in ● Fig. 1a. The top half of the layer slowly slides over the bottom half, and the relative speed of both halves is low. As long as the wind shear between the top and bottom of the layer is small, few if any eddies form. However, if the shear and the corresponding relative speed of these layers increases (Figs. 1b and 1c), wavelike undulations may form. When the shearing exceeds a certain value, the waves break into large swirls, with significant vertical movement (Fig. 1d). Eddies such as these often form in the upper troposphere near jet streams, where large wind speed shears exist. If wavelike clouds form in the region of wind shear, they are often called *billow clouds* (see ● Fig. 2).

Turbulent eddies also occur in conjunction with mountain waves, which may extend upward into the stratosphere. As we learned earlier, when these huge eddies develop in clear air, this form of turbulence is referred to as *clear air turbulence*, or *CAT*.

The eddies that form in clear air can have diameters ranging from a couple of meters to several hundred meters. An unsuspecting aircraft entering such a region can be in for more than just a bumpy ride. If the aircraft flies into a zone of descending air, it can drop suddenly, producing the sensation that there is no air to support the wings. Consequently, these regions have come to be known as *air pockets*.

Commercial aircraft entering air pockets have dropped hundreds of meters, injuring passengers and flight attendants not strapped into their seats. Five passengers and crew members had to be hospitalized in February 2014 after severe turbulence struck a Boeing 737 jetliner as it approached Billings, Montana. In April 1981, a DC-10 jetliner flying at 11,300 m (37,000 ft) over central Illinois encountered a region of severe clear air turbulence and reportedly plunged about 600 m (2000 ft) toward Earth before stabilizing. Twenty-one of the 154 people aboard were injured; one person sustained a fractured hip and another person, after hitting the ceiling, jabbed himself in the nose with a fork, then landed in the seat in front of him.* Clear air turbulence has occasionally caused structural damage to aircraft by breaking off vertical stabilizers and tail structures. Fortunately, the effects are usually not this dramatic.

The potential adverse effects of clear air turbulence is one important reason why passengers are frequently told to "fasten your seat belts" while flying, even when there are no thunderstorms or obvious hazards in sight.

---

*Another example of an aircraft that experienced severe turbulence as it flew into an air pocket is given in the opening vignette on p. 173.

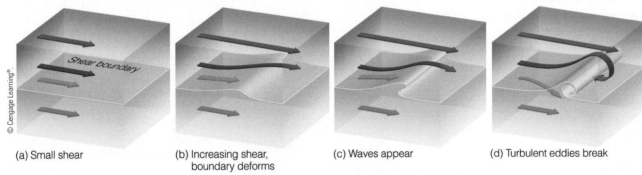

© Cengage Learning®.

(a) Small shear

(b) Increasing shear, boundary deforms

(c) Waves appear

(d) Turbulent eddies break

● **FIGURE 1** The formation of clear air turbulence (CAT) along a boundary of increasing wind speed shear. The wind in the top layer increases in speed from (a) through (d) as it flows over the bottom layer.

© C. Donald Ahrens

● **FIGURE 2** Billow clouds forming in a region of rapidly changing wind speed, called *wind shear*.

SEA AND LAND BREEZES The sea breeze is a type of thermal circulation. The uneven heating rates of land and water (described in Chapter 3) cause these mesoscale coastal winds. During the day, the land heats more quickly than the adjacent water, and the intensive heating of the air above produces a shallow thermal low. The air over the water remains cooler than the air over the land; hence, a shallow thermal high exists above the water. The overall effect of this pressure distribution is a **sea breeze** that blows at the surface from the sea toward the land as illustrated in ● Fig. 7.5. Since the strongest gradients of temperature and pressure occur near the land-water boundary, the strongest winds typically occur right near the beach and diminish inland. Further, since the greatest contrast in temperature between land and water usually occurs in the afternoon, sea breezes are strongest at this time. (The same type of breeze that develops along the shore of a large lake is called a *lake breeze*.)

At night, the land cools more quickly than the water, and the air above the land becomes cooler than the air over the water, producing a distribution of pressure such as the one shown in Fig. 7.5b. With higher surface pressure now over the land, the wind reverses itself and becomes a **land breeze**, a breeze at the surface that flows from the land toward the water. Temperature contrasts between land and water are generally much smaller at night; hence, land breezes are usually weaker than their daytime counterpart, the sea breeze. In regions where greater nighttime temperature contrasts exist, stronger land breezes occur over the water, off the coast. They are not usually noticed much on-shore, but are frequently observed by ships in coastal waters.

Look at Fig. 7.5 again and observe that the rising air is over the land during the day and over the water during the night. Therefore, along the humid east coast of the United States, daytime clouds tend to form over land and night-time clouds over water. This explains why, at night, distant lightning flashes are sometimes seen over the ocean.

The leading edge of the sea breeze is called the *sea breeze front*. As the front moves inland, a rapid drop in temperature usually occurs just behind it. In some locations, this temperature change can be 5°C (9°F) or more during the first hours—a refreshing experience on a hot, sultry day. In regions where the water temperature is warm, the cooling effect of the sea breeze is hardly evident. Since cities near

● **FIGURE 7.5** Development of a sea breeze and a land breeze. (a) At the surface, a sea breeze blows from the water onto the land, whereas (b) the land breeze blows from the land out over the water. Notice that the pressure at the surface changes more rapidly with the sea breeze. This situation indicates a stronger pressure gradient force and higher winds with a sea breeze.

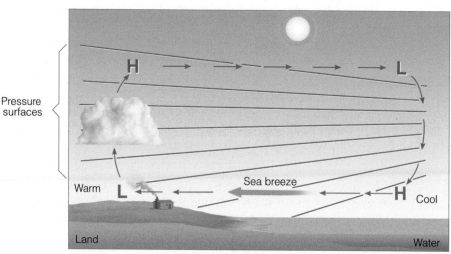

(a) Sea breeze

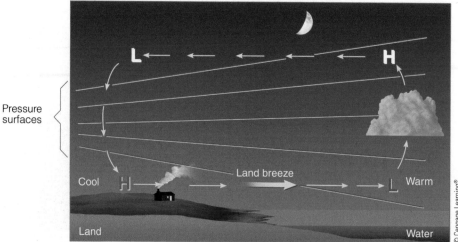

(b) Land breeze

© Cengage Learning®.

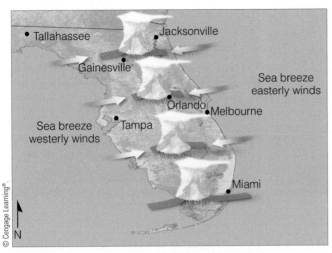

● **FIGURE 7.6** Typically, during the summer over Florida, converging sea breezes in the afternoon produce uplift that enhances thunderstorm development and rainfall. However, when westerly surface winds dominate and a ridge of high pressure forms over the area, thunderstorm activity diminishes, and dry conditions prevail.

the ocean usually experience the sea breeze by noon, their highest temperature usually occurs much earlier than in inland cities. Along the east coast of North America, the passage of the sea breeze front is marked by a wind shift, usually from west to east. In the cool ocean air, the relative humidity rises as the temperature drops. If the relative humidity increases beyond about 70 percent, water vapor begins to condense upon particles of sea salt or industrial smoke, producing haze. When the ocean air is highly concentrated with pollutants, the sea breeze front may meet relatively clear air and thus appear as a *smoke front,* or a *smog front.* If the ocean air becomes saturated, a mass of low clouds and fog will mark the leading edge of the marine air.

When there is a sharp contrast in air temperature across the frontal boundary, the warmer, lighter air will converge and rise. In many regions, this makes for good sea breeze glider soaring. If this rising air is sufficiently moist, a line of cumulus clouds will form along the sea breeze front, and, if the air is also conditionally unstable, thunderstorms may form. As previously mentioned, on a hot, humid day one can drive toward the shore, encounter heavy showers several miles from the ocean, and arrive at the beach to find it sunny with a steady onshore breeze.

When cool, dense, stable marine air encounters an obstacle, such as a row of hills, the heavy air tends to flow around them rather than over them. When the opposing breezes meet on the opposite side of the obstruction, they form what is called a *sea breeze convergence zone.* Such conditions are common along the mountainous Pacific coast of North America.

Sea breezes in Florida help produce that state's abundant summertime rainfall. On the Atlantic side of the state, the sea breeze blows in from the east; on the Gulf shore, it moves in from the west (see ● Fig. 7.6). The convergence of these two moist wind systems, coupled with daytime convection, produces cloudy conditions and showery weather over the land (see ● Fig. 7.7). Over the water (where cooler, more stable air lies close to the surface), skies often remain cloud-free. On many days during June and July 1998, however, Florida's converging wind system did not materialize. The lack of converging surface air and its accompanying showers left much of the state parched. Huge fires broke out over northern and central Florida, which left hundreds of people homeless and burned many thousands of acres of grass and woodlands. A weakened sea breeze and dry conditions have produced wildfires on numerous other occasions, including during the spring of 2006.

Convergence of coastal breezes is not restricted to ocean areas. Both Lake Michigan and Lake Superior are capable of producing well-defined lake breezes. In upper Michigan, where these large bodies of water are separated by a narrow strip of land, the two breezes push inland and converge near the center of the peninsula, creating afternoon clouds and showers, while the lakeshore areas remains sunny, pleasantly cool, and dry.

● **FIGURE 7.7** Surface heating and lifting of air along a converging sea breeze combine to form thunderstorms almost daily during the summer in southern Florida.

● **FIGURE 7.8** Valley breezes blow uphill during the day; mountain breezes blow downhill at night. (The L's and H's represent pressure, whereas the purple lines represent surfaces of constant pressure.)

Valley breeze

Mountain breeze

MOUNTAIN AND VALLEY BREEZES   Mountain and valley breezes develop along mountain slopes. Observe in ● Fig. 7.8 that, during the day, sunlight warms the valley walls, which in turn warm the air in contact with them. The heated air, being less dense than the air of the same altitude above the valley, rises as a gentle upslope wind known as a **valley breeze.** At night, the flow reverses. The mountain slopes cool quickly, chilling the air in contact with them. The cooler, more-dense air glides downslope into the valley, providing a **mountain breeze.** (Because gravity is the force that directs these winds downhill, they are also referred to as *gravity winds,* or *nocturnal drainage winds.*) This daily cycle of wind flow is best developed in clear, summer weather when prevailing winds are light.

When the upslope valley winds are well developed and have sufficient moisture, they can reveal themselves through cumulus clouds that build above mountain summits (see ● Fig. 7.9). Since valley breezes usually reach their maximum strength in the early afternoon, cloudiness, showers, and even thunderstorms are common over mountains during the warmest part of the day—a fact well known to seasoned hikers and climbers.

KATABATIC WINDS   Although any downslope wind is technically a **katabatic wind,** the name is usually reserved for downslope winds that are much stronger than mountain breezes. Katabatic (or *fall*) winds can rush down elevated slopes at hurricane speeds, but most are not that intense and many are on the order of 10 knots or less.

The ideal setting for a katabatic wind is an elevated plateau surrounded by mountains, with an opening that slopes rapidly downhill (see ● Fig. 7.10). When winter snows accumulate on the plateau, the overlying air grows extremely cold. Along the edge of the plateau the cold, dense air begins to descend through gaps and saddles in the hills, usually as a gentle or moderate cold breeze. If

● **FIGURE 7.9** As mountain slopes warm during the day, air rises and often condenses into cumuliform clouds, such as the ones shown here.

the breeze, however, is confined to a narrow canyon or channel, the flow of air can increase, often destructively, as cold air rushes downslope like water flowing over a fall.

Katabatic winds are observed in various regions of the world. For example, along the northern Adriatic coast in the former Yugoslavia, a polar invasion of cold air from Russia descends the slopes from a high plateau and reaches the lowlands as the *bora*, a cold, gusty, northeasterly wind with speeds sometimes in excess of 100 knots. A similar, but often less violent, cold wind known as the *mistral* descends the western mountains into the Rhone Valley of France, and then out over the Mediterranean Sea. It frequently causes frost damage to exposed vineyards and makes people bundle up in the otherwise mild climate along the Riviera. Strong, cold katabatic winds also blow downslope off the ice sheets in Greenland and Antarctica, occasionally with speeds greater than 100 knots.

In North America, when cold air accumulates over the Columbia Plateau of Idaho, Oregon, and Washington,* it may flow westward through the Columbia River Gorge as a strong, gusty, and sometimes violent wind. Even though the sinking air warms by compression, it is so cold to begin with that it reaches the ocean side of the Cascade Mountains much colder than the marine air it replaces. The *Columbia Gorge wind* (called the *coho*) is often the harbinger of a prolonged cold spell.

Strong downslope katabatic-type winds funneled through a mountain canyon can do extensive damage. In January 1984, a ferocious downslope wind blew through Yosemite National Park in California at speeds estimated at 100 knots. The wind toppled trees and, unfortunately, caused a fatality when a tree fell on a park employee sleeping in a tent.

CHINOOK (FOEHN) WINDS
The **chinook wind** is a warm, dry, downslope wind that descends the eastern slope of the Rocky Mountains. The region of the chinook is rather narrow and extends from northeastern New Mexico northward into Canada. Similar winds occur along the leeward slopes of mountains in other regions of the world. The general term for chinook wind is the *foehn* (a name that originated in the European Alps), but there are many local names, such as the *zonda* in Argentina. When foehn winds move through an area, the temperature rises sharply, sometimes 20°C (36°F) or more in less than an hour, and a corresponding sharp drop in the relative humidity occurs, occasionally to less than 5 percent.

In North America, chinooks occur when strong westerly winds aloft flow over a north-south–trending mountain range, such as the Rockies and Cascades. Such conditions can produce a trough of low pressure on the mountain's eastern side, a trough that tends to force the

*Information on geographic features and their location in North America is provided at the back of the book.

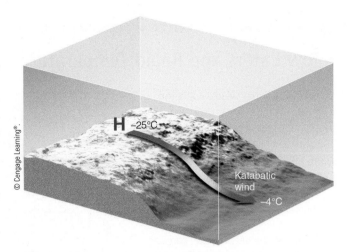

● **FIGURE 7.10** Strong katabatic winds can form where cold winds rush downhill from an elevated plateau covered with snow.

air downslope. As the air descends, it is compressed and warms. So the main source of warmth for a chinook is *compressional heating,* as potentially warmer (and drier) air is brought down from aloft.

Clouds and precipitation on the mountain's windward side can enhance the chinook. For example, as the cloud forms on the upwind side of the mountain in ● Fig. 7.11, the release of latent heat inside the cloud supplements the compressional heating on the downwind side. This phenomenon makes the descending air at the base of the mountain on the downwind side warmer than it was before it started its upward journey on the windward side. The air is also drier, since much of its moisture was removed as precipitation on the windward side. (More information on temperature changes associated with chinooks is given in Focus section 7.2.)

Along the Front Range of the Rockies, a bank of clouds forming on the windward side of the mountains is a telltale sign of an impending chinook. This cloud feature, called a *foehn wall,* usually remains stationary as air rises, condenses, and then rapidly descends the leeward slopes, often causing strong winds in foothill communities. These

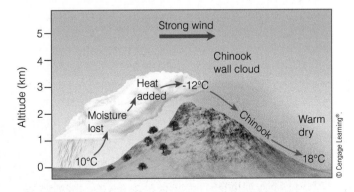

● **FIGURE 7.11** A chinook wind can be enhanced when clouds form on the mountain's windward side. Heat added and moisture lost on the upwind side produce warmer and drier air on the downwind side.

Snow Eaters and Rapid Temperature Changes

Chinooks are thirsty winds. As they move over a heavy snow cover, they can melt and evaporate a foot of snow in less than a day. This situation has led to some tall tales about these so-called "snow eaters." Canadian folklore has it that a sled-driving traveler once tried to outrun a chinook. During the entire ordeal, as the story has it, his front runners were supposedly in snow while his back runners were on bare soil.

Actually, the chinook is important economically. It not only brings relief from the winter cold, but it also uncovers prairie grass, so that livestock can graze on the open range. Also, these warm winds can help keep railroad tracks clear of snow, so that trains can keep running. On the other hand, the drying effect of a chinook can create an extreme fire hazard. And when a chinook follows spring planting, the seeds can die in the parched soil. Along with the dry air comes a buildup of static electricity, making a simple handshake a shocking experience. These warm, dry winds have sometimes adversely affected human behavior. During periods of chinook winds some people feel irritable and depressed and others become ill. The exact reason for this phenomenon is not clearly understood.

Chinook winds have been associated with very rapid temperature changes. On

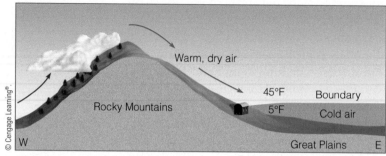

● FIGURE 3 Cities near the warm air–cold air boundary can experience sharp temperature changes if cold air should slosh back and forth like water in a bowl.

January 11, 1980, the air temperature in Great Falls, Montana, rose from −32°F to 17°F (a 49°F rise in temperature) in just seven minutes. The nation's largest temperature swing within a 24-hour period occurred in Loma, Montana, on January 15, 1972, when the temperature soared from −54°F to 48°F. How such rapid changes in temperature can occur is illustrated in ● Fig. 3. Notice that a shallow layer of extremely cold air has moved out of Canada and is now resting against the Rocky Mountains.

The cold air behaves just as any fluid, and, in some cases, atmospheric conditions can cause the air to slosh up and down much like water does when a bowl is rocked back and forth. This rocking motion

can cause extreme temperature variations for cities located at the base of the hills along the periphery of the cold air–warm air boundary, as they are alternately in and then out of the cold air.

Such a situation is probably responsible for the extremely rapid two-minute temperature change of 49°F recorded at Spearfish, South Dakota, during the morning of January 22, 1943. On the same morning, in nearby Rapid City, the temperature fluctuated from −4°F at 5:30 A.M. to 54°F at 9:40 A.M., then down to 11°F at 10:30 A.M. and up to 55°F just 15 minutes later. At nearby cities, the undulating cold air produced similar temperature variations that recurred over several hours.

---

strong winds are especially notorious in winter in Boulder, Colorado, where windstorms have caused millions of dollars in damage. ● Figure 7.12 shows how a foehn wall appears as one looks west toward the Rockies from the Colorado plains. The photograph was taken on a winter afternoon with the air temperature about −7°C (20°F). That evening, the chinook moved downslope at high speeds through foothill valleys, picking up sand and pebbles (which dented cars and cracked windshields). The chinook spread out over the plains like a warm blanket, raising the air temperature the following day to a mild 15°C (59°F). The chinook and its wall of clouds remained for several days, bringing with it a welcomed break from the cold grasp of winter.

SANTA ANA WINDS A warm, dry wind that blows downhill from the east or northeast into southern California is the **Santa Ana wind.** As the air descends from the elevated desert plateau, it funnels through

mountain canyons in the San Gabriel and San Bernardino Mountains, finally spreading over the Los Angeles Basin and San Fernando Valley and out over the Pacific Ocean (see ● Fig. 7.13). The wind often blows with exceptional speed—occasionally over 90 knots—in the Santa Ana Canyon (the canyon from which it derives its name).

These warm, dry winds develop as a region of high pressure builds over the Great Basin. The clockwise circulation around the anticyclone forces air downslope from the high plateau. Thus, *compressional heating* provides the primary source of warming. The air is dry, since it originated in the desert, and it dries out even more as it is heated. ● Figure 7.14 shows a typical wintertime Santa Ana situation.

As the wind rushes through canyon passes, it lifts dust and sand and dries out vegetation, which sets the stage for serious brush fires, especially in autumn, when

● FIGURE 7.12 A foehn wall forming over the Colorado Rockies (viewed from the plains).

H

Los Angeles

Pacific Ocean

N

● FIGURE 7.13 Warm, dry Santa Ana winds sweep downhill through mountain canyons into Southern California. The large H represents higher air pressure over the elevated desert.

chaparral-covered hills are already parched from the dry summer.* One such fire in November 1961—the infamous Bel Air fire—burned for three days, destroying 484 homes and causing over $25 million in damage in 1961 dollars (close to $200 million in today's dollars). During October 2003, massive wildfires driven by strong Santa Ana winds swept through Southern California. The fires charred more than 750,000 acres, destroyed over 2800 homes, took 20 lives, and caused over $2 billion in property damage. Only four years later (after one of the driest years on record) in October 2007, wildfires broke out again in

*Chaparral denotes a shrubby environment, in which many of the plant species contain highly flammable oils.

● FIGURE 7.14 Surface weather map showing Santa Ana conditions in January. Maximum temperatures for this particular day are given in °F. Observe that the downslope winds blowing into Southern California raised temperatures into the upper 80s, while elsewhere temperature readings were much lower.

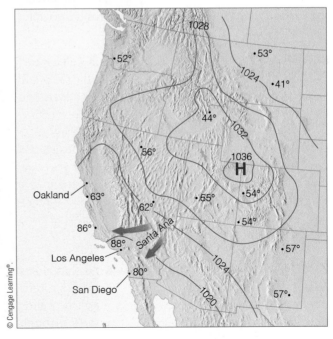

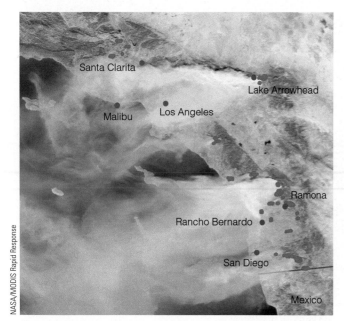

● **FIGURE 7.15** Satellite view showing strong northeasterly Santa Ana winds on October 23, 2007, blowing smoke from massive wild fires (red dots) across Southern California and out over the Pacific Ocean.

rains wash away topsoil and, in some areas, create serious mudslides. The adverse effects of a wind-driven Santa Ana fire may be felt long after the fire itself has been put out.

DESERT WINDS   Winds of all sizes develop over the desert. Huge *dust storms* form in dry regions, where strong winds are able to lift and fill the air with particles of fine dust. In February 2001, an exceptionally large dust storm (about the size of Spain) formed over the African Sahara and then swept westward off the African coast, then northeastward for thousands of miles. During the drought years of the 1930s, large dust storms formed over the Great Plains of the United States. Some individual storms lasted for three days and spread dust for hundreds of miles to the east, to the Atlantic coast and beyond. In desert areas where loose sand is more prevalent, *sandstorms* develop, as high winds enhanced by surface heating rapidly carry sand particles close to the ground.

A spectacular example of a storm composed of dust or sand is the **haboob** (from Arabic *habb:* "to blow"). The haboob forms as cold downdrafts along the leading edge of a thunderstorm lift dust or sand into a huge, tumbling dark cloud that can extend horizontally for more than 100 kilometers and rise vertically to the base of the thunderstorm. Haboobs are most common in the African Sudan (where about twenty-four occur each year) and in the Desert Southwest of the United States, especially in southern Arizona. A particularly strong haboob swept into Phoenix, Arizona, in July 2011 (see ● Fig. 7.17), moving vast amounts of dust from a landscape parched by drought.

On a smaller scale, in dry areas, the wind can also produce rising, spinning columns of air that pick up dust or sand from the ground. Called **dust devils** or *whirlwinds,** these rotating vortices generally form on clear, hot days over a dry surface where most of the sunlight goes into

Southern California. Pushed on by hellacious Santa Ana winds that gusted to over 80 knots, the fires raced through dry vegetation, scorching virtually everything in their paths. The fires, which extended from north of Los Angeles to the Mexican border (see ● Fig. 7.15), burned over 500,000 acres, destroyed more than 1800 homes, and took 9 lives. The total costs of the fires exceeded $1.5 billion.

Four hundred miles to the north in Oakland, California, a ferocious Santa Ana–type wind was responsible for the disastrous *Oakland hills fire* during October 1991, which damaged or destroyed over 3000 dwellings, caused over $1.5 billion in damage, and took 25 lives (see ● Fig. 7.16). When a Santa Ana fire removes the protective cover of vegetation, the land is ripe for erosion, as winter

---

*In Australia, the Aboriginal word *willy-willy* refers to a dust devil.

● **FIGURE 7.16** From atop his roof, a resident of Oakland's Rockridge district looks on in disbelief as his neighbors' homes are consumed in a raging firestorm on October 20, 1991.

AP Photo/Amanda Lee Myers

● **FIGURE 7.17** A large haboob (dust storm) moves through Phoenix, Arizona, on July 5, 2011.

heating the surface, rather than into evaporating water from vegetation. The atmosphere directly above the hot surface becomes unstable, convection sets in, and the heated air rises, often lifting dust, sand, and dirt high into the air. Wind, often deflected by small topographic barriers, flows into this region, rotating the rising air as depicted in ● Fig. 7.18. Depending on the nature of the topographic feature, the spin of a dust devil around its central core may be cyclonic or anticyclonic, and both directions occur with about equal frequency. (Dust devils are too small and fleeting to be substantially influenced by the Coriolis force.)

Having diameters of only a few meters and heights usually of less than 100 m (330 ft), most dust devils last only a short time (see ● Fig. 7.19). However, some dust devils reach sizable dimension, extending upward from the surface for several hundred meters. Such whirlwinds are capable of considerable damage; winds exceeding 75 knots can overturn mobile homes and tear the roofs off buildings. Fortunately, the majority of dust devils are small. Also keep in mind that dust devils are *not* tornadoes. The circulations of most tornadoes (as we will see in Chapter 10) develop near the base of a thunderstorm,

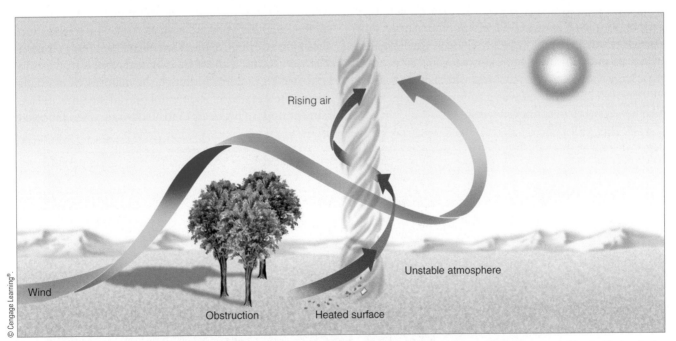

© Cengage Learning®.

● **FIGURE 7.18** The formation of a dust devil. On a hot, dry day, the atmosphere next to the ground becomes unstable. As the heated air rises, wind blowing past an obstruction twists the rising air, forming a rotating column, or *dust devil*. Air from the sides rushes into the rising column, lifting sand, dust, leaves, or any other loose material from the surface.

● FIGURE 7.19 A well-developed dust devil moves over a hot desert landscape on a clear summer day.

SEASONALLY CHANGING WINDS—THE MONSOON

On our planet there are thermal circulations that are much larger than those of the more local sea and land breezes described earlier. A good example of such a circulation is the **monsoon,** which derives from the Arabic *mausim* ("seasons"). A **monsoon wind system** is one that *changes direction seasonally,* blowing from one direction in summer and from the opposite direction in winter. This seasonal reversal of winds is especially well developed in eastern and southern Asia.

In some ways, the monsoon is similar to a large-scale sea breeze. During the winter, the air over northern Asia becomes much colder than the air over the ocean. A large, shallow high-pressure area develops over continental Siberia, producing a *clockwise* circulation of air that flows out over the Indian Ocean and South China Sea (see ● Fig. 7.20a). Subsiding air of the anticyclone and the downslope movement of northeasterly winds from the inland plateau provide eastern and southern Asia with generally fair weather and the dry season. Hence, the *winter monsoon* means clear skies and generally dry weather, with surface winds that blow from land to sea.

In summer, the wind-flow pattern reverses itself as air over the continents becomes warmer than air above the water. A shallow thermal low develops over the continental interior. The heated air within the low rises, and the surrounding air responds by flowing *counterclockwise* into the low center. This condition brings moisture-bearing winds that sweep into the continent from the ocean. The humid air converges with drier continental air, causing the humid air to rise; further lifting is provided by hills and mountains. Lifting cools the air to its saturation point, resulting in heavy showers and thunderstorms. Thus, the *summer monsoon* of southeastern Asia, which lasts from about June through September, means wet, rainy weather (the wet season) with surface winds that blow from sea to land (see Fig. 7.20b). Although the majority of rain falls during the wet season, it does not rain all the time. In fact, rainy periods of between 15 to 40 days are often followed

whereas the circulation of a dust devil begins at the surface, normally in sunny weather (although some dust devils form beneath convective-type clouds).

Desert winds are not confined to planet Earth; they form on the planet Mars as well. Most of the Martian dust storms are small, and only cover a relatively small portion of that planet. However, some dust storms can actually grow large enough to encircle Mars with a dusty haze, as happened in 2001. Dust devils also form on Mars when high winds sweep over uneven terrain.

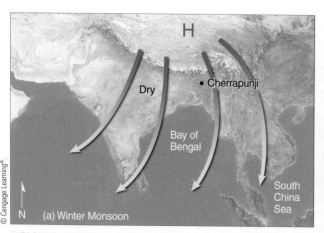
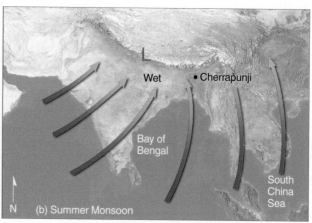

● FIGURE 7.20 Changing annual surface wind-flow patterns associated with the winter and summer Asian monsoons.

by 3 to 15 days of hot, sunny weather known as *monsoon breaks*.

The strength of the Indian monsoon appears to be related to the reversal of surface air pressure that occurs at irregular intervals about every two to seven years at opposite ends of the tropical South Pacific Ocean. As we will see later in this chapter, this reversal of pressure (which is known as the *Southern Oscillation*) is linked to the ocean warming phenomenon known as *El Niño*. During an El Niño event, surface water near the equator becomes warmer over the central and eastern Pacific. Over the region of warm water we find rising air, huge convective clouds, and heavy rain. Meanwhile, to the west of the warm water (over the region influenced by the summer monsoon), sinking air inhibits cloud formation and convection. Hence, during El Niño years, monsoon rainfall is more likely to be deficient.

Summer monsoon rains over southern Asia can be truly extreme. The town of Cherrapunji (also known as Sohra), located about 300 km inland on the southern slopes of the Khasi Hills in northeastern India, receives an average of 1176 cm (463 in.) of rainfall each year, most of it between April and October (see ● Fig. 7.21). The town also holds world records for the heaviest rainfall measured anywhere in a 12-month period—2,647 cm (1042 in.) from August 1860 to July 1861—and the heaviest 48-hour rainfall, 249 cm (98 in.) on June 15–16, 1995.

The summer monsoon rains are essential to the agriculture of southern and eastern Asia. More than two billion people rely on the summer rains so that crops will grow. The people also depend on the rains for drinking water. Unfortunately, the monsoon can be unreliable in both duration and intensity, and these are difficult to predict. Since the monsoon is vital to the survival of so many people, it is no wonder that meteorologists have investigated it extensively. They have tried to develop methods of accurately forecasting the intensity and duration of the monsoon. With the aid of current research projects and the latest climate models (which tie in the interaction of ocean and atmosphere), there is hope that monsoon forecasts will improve in accuracy.

Monsoon wind systems exist in other regions of the world, such as in Australia, Africa, and North and South America, where large contrasts in temperature develop between oceans and continents. Usually, however, these systems are not as pronounced as in southeast Asia. For example, the North American monsoon affects northwest Mexico and the southwestern United States, especially Arizona, New Mexico, Nevada, and the southern part of California. In this region, spring and early summer are normally dry, as warm westerly winds predominate. By mid-July, however, southerly or southeasterly winds are more common, and so are afternoon showers and thunderstorms (see ● Fig. 7.22 and ● Fig. 7.23).

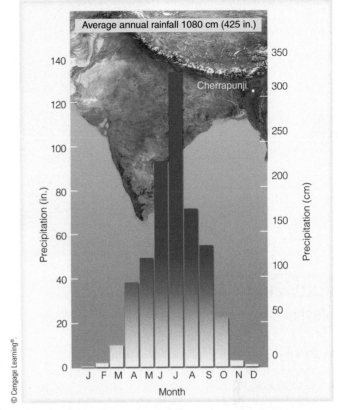

● **FIGURE 7.21** Average annual precipitation for Cherrapunji, India. Note the abundant rainfall during the summer monsoon (April through October) and the lack of rainfall during the winter monsoon (November through March).

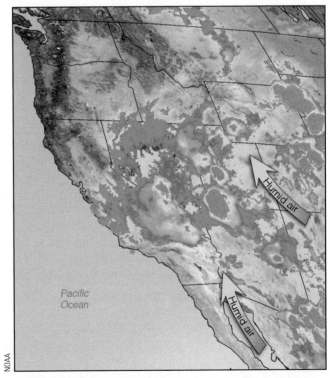

● **FIGURE 7.22** Enhanced infrared satellite image with heavy arrows showing strong monsoonal circulation. Moist, southerly winds from the *Gulf of California* and southeasterly winds (far right arrow) from the *Gulf of Mexico* are causing showers and thunderstorms (yellow and red areas) to form over the southwestern section of the United States during July 2001.

● **FIGURE 7.23** Clouds and thunderstorms forming over Arizona, as humid monsoonal air flows northward over the region during July 2007.

© C. Donald Ahrens

BRIEF REVIEW

Before moving on to the next section here is a brief review of some of the main points so far:

- The size of atmospheric circulations range from the smallest *microscale* to the larger *mesoscale*, to the largest *macroscale*.

- Thermal pressure systems are shallow pressure systems that are driven by the unequal heating and cooling of Earth's surface.

- The sea breeze and the land breeze are types of thermal circulations that are due to uneven heating and cooling rates of land and water.

- At the surface, a sea breeze blows from water to land; whereas a land breeze blows from land to water.

- A valley breeze blows uphill during the day and a mountain breeze blows downhill at night.

- Chinook (foehn) winds are warm, dry winds that blow downhill along the eastern side of the Rocky Mountains.

- The main source of warmth for the chinook is compressional heating.

- Santa Ana winds are warm, dry downslope winds that warm by compressional heating and blow from the east or northeast into Southern California.

- Dust devils tend to form over dry terrain on clear, hot days. They are not tornadoes, although the winds of a large dust devil may cause minor damage to structures.

- Monsoon winds are winds that change direction seasonally. In southern Asia, the winter monsoon, which blows from land to water, is dry; the summer monsoon, which blows from water to land, is wet.

Global Winds

Up to now, we have seen that local winds vary considerably from day to day and from season to season. As you may suspect, these winds are part of a much larger circulation, the little whirls within larger whirls that we spoke of earlier in this chapter. Indeed, if the rotating high- and low-pressure areas in our atmosphere are like spinning eddies in a huge river, then the flow of air around the globe is like the meandering river itself. When winds throughout the world are averaged over a long period of time, the local wind patterns vanish, and what we see is a picture of the winds on a global scale—what is commonly called the **general circulation of the atmosphere.**

GENERAL CIRCULATION OF THE ATMOSPHERE Before we study the general circulation, we must remember that it only represents the *average* air flow around the world. Actual winds at any one place and at any given time may vary considerably from this average. Nevertheless, the average can answer why and how the winds blow around the world the way they do—why, for example, prevailing surface winds are northeasterly in Honolulu, Hawaii, and westerly in New York City. The average can also give a picture of the mechanism driving these winds, as well as a model of how heat is transported from equatorial regions poleward, keeping the climate in middle latitudes tolerable.

The underlying cause of the general circulation is the unequal heating of Earth's surface. We learned in Chapter 2 that, averaged over the entire earth, incoming solar radiation is roughly equal to outgoing earth radiation. However, we also know that this energy balance is not maintained for each latitude, since the tropics experience a net gain in energy, while polar regions suffer a net loss. To balance these inequities, the atmosphere transports warm air poleward and cool air equatorward. Although seemingly simple, the actual flow of air is complex; certainly not everything is known about it. In order to better understand it, we will first look at some models (that is, artificially constructed analogies) that eliminate some of the complexities of the general circulation.

SINGLE-CELL MODEL The first model is the single-cell model, in which we assume that:

1. Earth's surface is uniformly covered with water (so that differential heating between land and water does not come into play).

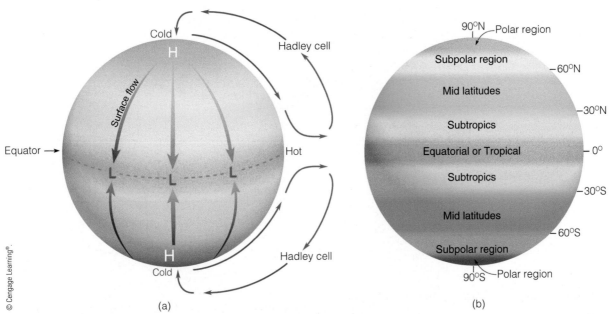

● **FIGURE 7.24** Diagram (a) shows the general circulation of air on the side of Earth facing the sun on a nonrotating Earth uniformly covered with water and with the sun directly above the equator. (Vertical air motions are highly exaggerated in the vertical.) Diagram (b) shows the names that apply to the different regions of the world and their approximate latitudes.

2. The sun is always directly over the equator (so that the winds will not shift seasonally).

3. Earth does not rotate (so that the only force we need deal with is the pressure gradient force).

With these assumptions, the general circulation of the atmosphere on the side of Earth facing the sun would look much like the representation in ● Fig. 7.24a— a huge thermally driven convection cell in each hemisphere. (For reference, the names of the different regions of the world and their approximate latitudes are given in Figure 7.24b.)

The circulation of air described in Fig. 7.24a is the **Hadley cell** (named after the eighteenth-century English meteorologist George Hadley, who first proposed the idea). It is driven by energy from the sun. Excessive heating of the equatorial area produces a broad region of surface low pressure, while at the poles excessive cooling creates a region of surface high pressure. In response to the horizontal pressure gradient, cold surface polar air flows equatorward, while at higher levels air flows toward the poles. The entire circulation consists of a closed loop with rising air near the equator, sinking air over the poles, an equatorward flow of air near the surface, and a return flow aloft. In this manner, some of the excess energy of the tropics is transported as sensible and latent heat to the regions of energy deficit at the poles.

Such a simple cellular circulation does not actually exist on Earth. For one thing, Earth rotates, so the Coriolis force would deflect the southwardmoving surface air in the Northern Hemisphere to the right, producing easterly surface winds at practically all latitudes north of the equator. We know that this does not happen and that

prevailing winds in middle latitudes actually blow from the west. Therefore, observations alone tell us that a closed circulation of air between the equator and the poles is not an accurate model for a rotating earth. But this model does show us how a non-rotating planet would balance an excess of energy at the equator and a deficit at the poles. How, then, does the wind blow on a rotating planet? To answer, we will keep our model simple by retaining our first two assumptions—that is, that Earth is covered with water and that the sun is always directly above the equator.

THREE-CELL MODEL   If we allow Earth to spin, the simple convection system breaks into a series of cells as shown in ● Fig. 7.25. Although this model is considerably more complex than the single-cell model, there are some similarities. The tropical regions still receive an excess of heat and the poles a deficit. In each hemisphere, three cells instead of one have the task of energy redistribution. A surface high-pressure area is located at the poles, and a broad trough of surface low pressure still exists at the equator. From the equator to latitude 30°, the circulation is the Hadley cell. Let's look at this model more closely by examining what happens to the air above the equator. (Refer to Fig. 7.25 as you read the following section.)

Over equatorial waters, the air is warm, horizontal pressure gradients are weak, and winds are light. This region is referred to as the **doldrums.** (The monotony of the weather in this area has given rise to the expression "down in the doldrums.") Here, warm, humid air rises, often condensing into huge cumulus clouds and thunderstorms

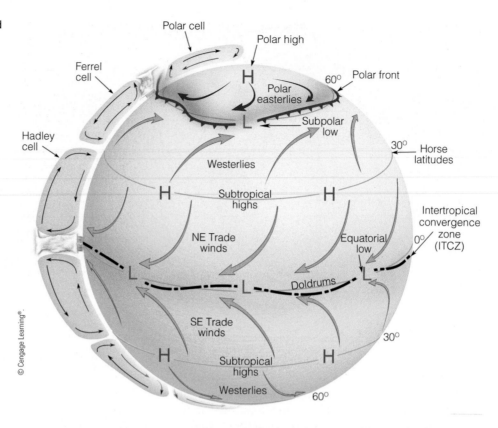

● **FIGURE 7.25** The idealized wind and surface-pressure distribution over a uniformly water-covered rotating earth.

© Cengage Learning®.

that liberate an enormous amount of latent heat. This heat makes the air more buoyant and provides energy to drive the Hadley cell. The rising air reaches the tropopause, which acts like a barrier, causing the air to move laterally toward the poles. The Coriolis force deflects this poleward flow toward the right in the Northern Hemisphere and to the left in the Southern Hemisphere, providing westerly winds aloft in both hemispheres. (We will see later that these westerly winds reach maximum velocity and produce jet streams near 30° latitude and 60° latitude.)

Air aloft moving poleward from the tropics constantly cools by giving up infrared radiation, and at the same time it also begins to converge, especially as it approaches the middle latitudes.* This convergence (piling up) of air aloft increases the mass of air above the surface, which in turn causes the air pressure at the surface to increase. Hence, at latitudes near 30°, the convergence of air aloft produces belts of high pressure called **subtropical highs** (or anticyclones). As the converging, relatively dry air above the highs slowly descends, it warms by compression. This subsiding air produces generally clear skies and warm surface temperatures; hence, on earth it is here that we find the major deserts of the world, such as the Sahara of Africa and the Sonoran of North America (see ● Fig. 7.26).

Over the ocean, the weak pressure gradients in the center of the high produce only weak winds. According

to legend, sailing ships traveling to the New World were frequently becalmed in this region; and, as food and supplies dwindled, horses were either thrown overboard or eaten. Because of this, the region is sometimes called the *horse latitudes.*

From the horse latitudes near latitude 30°, some of the surface air moves back toward the equator. It does not flow straight back, however, because the Coriolis force deflects the air, causing it to blow from the northeast in the Northern Hemisphere and from the southeast in the Southern Hemisphere. These steady winds provided sailing ships with an ocean route to the New World; hence, these winds are called the **trade winds.** Near the equator, the *northeast trades* converge with the *southeast trades* along a boundary called the **intertropical convergence zone (ITCZ).** In this region of surface convergence, air rises and continues its cellular journey. Along the ITCZ, it is usually very wet as the rising air develops into huge thunderstorms that drop copious amounts of rain in the form of heavy showers (see ● Fig. 7.27).

Meanwhile, at latitude 30°, not all of the surface air moves equatorward. Some air moves toward the poles

DID YOU KNOW?

Christopher Columbus was a lucky man. The year he set sail for the New World, the trade winds had edged unusually far north, and a steady northeast wind eased his ships along. Only for about ten days did he encounter the light and variable wind more typical of this notorious region (30°N)— the horse latitudes.

*You can see why the air converges if you have a globe of the world. Put your fingers on meridian lines at the equator and then follow the meridians poleward. Notice how the lines and your fingers bunch together in the middle latitudes.

● **FIGURE 7.26** Subtropical deserts, such as the Sonoran Desert shown here, are mainly the result of sinking air associated with subtropical high-pressure areas.

and deflects toward the east, resulting in a more or less westerly air flow—called the *prevailing westerlies* or, simply, **westerlies**—in both hemispheres. Consequently, from Texas northward into Canada, it is much more common to experience winds blowing out of the west than from the east. The westerly flow in the real world is not constant as migrating areas of high and low pressure break up the surface flow pattern from time to time. In the middle latitudes of the Southern Hemisphere, where the surface is mostly water, winds blow more steadily from the west.

As this mild surface air travels poleward from latitude 30°, it encounters cold air moving down from the poles. These two air masses of contrasting temperature do not readily mix. They are separated by a boundary called the **polar front,** a zone of low pressure—the **subpolar low**—where surface air converges and rises, and storms and clouds develop. In our model in Fig. 7.25, some of the rising air returns at high levels to the horse latitudes, where it sinks back to the surface in the vicinity of the subtropical high. This middle cell (called the *Ferrel cell,* after the American meteorologist William Ferrel) is completed when surface air from the horse latitudes flows poleward toward the polar front.

Notice in Fig. 7.25, p. 190, that in the Northern Hemisphere, behind the polar front the cold air from the poles is deflected by the Coriolis force, so that the general flow of air is from the northeast. Hence, this is the region of the **polar easterlies.** In winter, the polar front, with its cold air, can move into middle and subtropical latitudes, producing a cold polar outbreak. Along the front, a portion of the

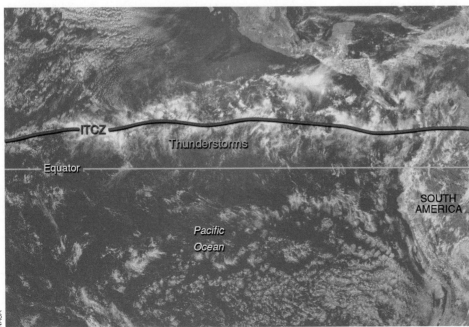

● **FIGURE 7.27** The solid red line in this visible satellite image marks the position of the ITCZ in the eastern Pacific. The bright white clouds are huge thunderstorms forming along the ITCZ.

ITCZ
Thunderstorms
Equator
Pacific Ocean
SOUTH AMERICA

ATMOSPHERIC CIRCULATIONS 191

rising air moves poleward, and the Coriolis force deflects the air into a westerly wind at high levels. Air aloft eventually reaches the poles, slowly sinks to the surface, and flows back toward the polar front, completing the weak *polar cell.*

We can summarize all of this by referring back to Fig. 7.25 on p. 190 and noting that, at the surface, there are two major areas of high pressure and two major areas of low pressure. Areas of high pressure exist near latitude 30° and the poles; areas of low pressure exist over the equator and near 60° latitude in the vicinity of the polar front. Knowing the way the surface winds blow around these pressure systems on the three-cell model gives us a generalized picture of how surface winds blow throughout the world. The trade winds extend from the subtropical high to the equator, the westerlies from the subtropical high to the polar front, and the polar easterlies from the poles to the polar front.

How does this three-cell model compare with actual observations of winds and pressure in the real world? We know, for example, that upper-level winds at middle latitudes generally blow from the west. The middle cell in our model, however, suggests an east wind aloft as air flows equatorward. Discrepancies do exist between this model and atmospheric observations, but the model does agree closely with the winds and pressure distribution at the *surface,* so we will examine this next.

AVERAGE SURFACE WINDS AND PRESSURE: THE REAL WORLD When we examine the real world with its continents and oceans, mountains and ice fields, we obtain an average distribution of sea-level pressure and winds for January and July, as shown in ● Figs. 7.28a and 7.28b. Look closely at both maps and observe that there are regions where pressure systems appear to persist throughout the year. These systems are referred to as *semipermanent highs and lows* because they move only slightly during the course of a year.

In Fig. 7.28a, we can see that there are four semipermanent pressure systems in the Northern Hemisphere during January. In the eastern Atlantic, between latitudes 25° and 35°N is the *Bermuda-Azores high,* often called the **Bermuda high,** and, in the Pacific Ocean, its counterpart, the **Pacific high.** These are the subtropical anticyclones that develop in response to the convergence of air aloft. Since surface winds blow clockwise around these systems, we find the trade winds to the south and the prevailing westerlies to the north. In the Southern Hemisphere, where there is relatively less land area, there is less contrast between land and water, and the subtropical highs show up as well-developed systems with a clearly defined circulation.

Where we would expect to observe the polar front (between latitudes 40° and 65°), there are two semipermanent subpolar lows. In the North Atlantic, there is the *Greenland-Icelandic low* or simply **Icelandic low,**

which covers Iceland and southern Greenland, while the **Aleutian low** sits over the Gulf of Alaska and the Bering Sea near the Aleutian Islands in the North Pacific. These zones of cyclonic activity actually represent regions where numerous storms, having traveled eastward, tend to converge, especially in winter. In the Southern Hemisphere, where there is very little land to disrupt the flow, the subpolar low forms a continuous trough that completely encircles the globe.

The January map (Fig. 7.28a) shows other pressure systems that are not semipermanent in nature but are still observed often. Over Asia, for example, there is a huge (but shallow) thermal anticyclone called the **Siberian high,** which forms because of the intense cooling of the land. South of this system, the winter monsoon shows up clearly, as air flows away from the high across Asia and out over the ocean. A similar (but less intense) anticyclone (called the *Canadian high*) is evident over North America.

As summer approaches, the land warms and the cold, shallow highs disappear. In some regions, areas of surface low pressure replace areas of high pressure. The lows that form over the warm land are shallow *thermal lows.* On the July map (Fig. 7.28b), warm thermal lows are found over the Desert Southwest of the United States, over the plateau of Iran, and north of India. As the thermal low over India intensifies, warm, moist air from the ocean is drawn into it, producing the wet summer monsoon so characteristic of India and Southeast Asia.

When we compare the January and July maps, we can see several changes in the semipermanent pressure systems. The strong subpolar lows so well developed in January over the Northern Hemisphere are hardly discernible on the July map. The subtropical highs, however, remain dominant in both seasons. Because the sun is overhead in the Northern Hemisphere in July and overhead in the Southern Hemisphere in January, the zone of maximum surface heating shifts seasonally. In response to this shift, the major pressure systems, wind belts, and ITCZ (heavy red line in Fig. 7.28) *shift toward the north in July and toward the south in January.**

THE GENERAL CIRCULATION AND PRECIPITATION PATTERNS The position of the major features of the general circulation and their latitudinal displacement (which annually averages about 10° to 15°) strongly influence the climate of many areas. For example, on the global scale, we would expect abundant rainfall where the air rises and very little where the air sinks. Thus, areas of high rainfall exist in the tropics, where humid air rises in conjunction with the ITCZ, and between about 40° and 55° latitude, where middle-latitude cyclonic storms and the polar front

---

*An easy way to remember the seasonal shift of surface pressure systems is to think of birds. In the Northern Hemisphere, they migrate south in the winter and north in the summer.

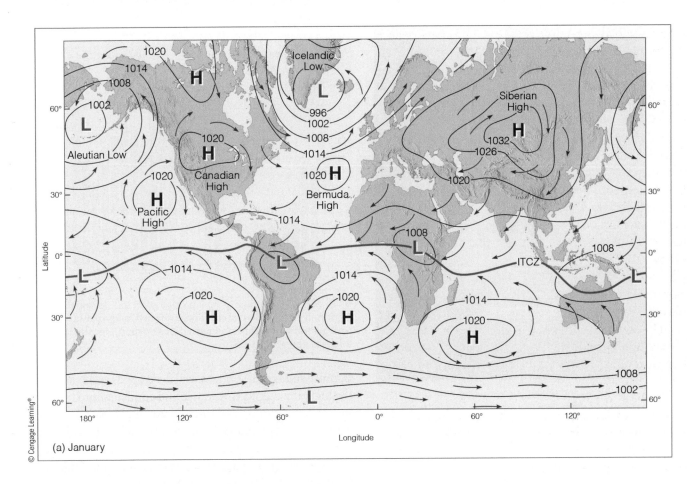

(a) January

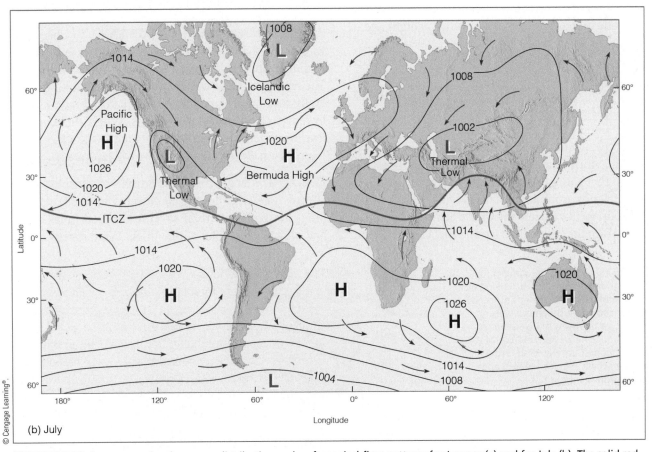

(b) July

● **FIGURE 7.28** Average sea-level pressure distribution and surface wind-flow patterns for January (a) and for July (b). The solid red line represents the position of the ITCZ.

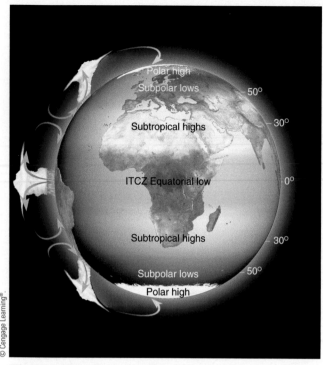

© Cengage Learning®.

● **FIGURE 7.29** Rising and sinking air associated with the major pressure systems of Earth's general circulation. Where the air rises, precipitation tends to be abundant (blue shade); where the air sinks, drier regions prevail (tan shade). Note that the sinking air of the subtropical highs produces the major desert regions of the world.

force air upward. Areas of low precipitation are found near 30° latitude in the vicinity of the subtropical highs and in polar regions where the air is cold and dry (see ● Fig. 7.29).

During the summer, the Pacific high drifts northward to a position off the California coast (see ● Fig. 7.30). Sinking air on its eastern side produces a strong upper-level subsidence inversion, which tends to keep summer weather along the West Coast relatively dry. The rainy season typically occurs in winter when the high moves south and storms can penetrate the region. Observe in Fig. 7.30 that along the East Coast, the clockwise circulation of winds around the Bermuda high brings warm, tropical air northward into the United States and southern Canada from the Gulf of Mexico and the Atlantic Ocean.

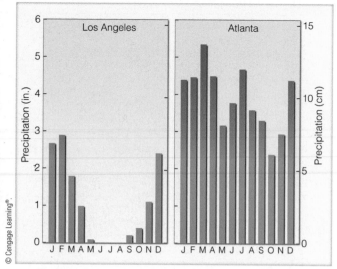

© Cengage Learning®.

● **FIGURE 7.31** Average annual precipitation for Los Angeles, California (Los Angeles International Airport), and Atlanta, Georgia (Hartsfield-Jackson International Airport), for the climatological period 1981–2010.

Because sinking air is not as well developed on this side of the high, the humid air can rise and condense into towering cumulus clouds and thunderstorms. In part, then, it is the air motions associated with the subtropical highs that keep summer weather dry in California and moist in Georgia. (Compare the patterns for Los Angeles, California, and Atlanta, Georgia, in ● Fig. 7.31.)

WESTERLY WINDS AND THE JET STREAM   In Chapter 6, we learned that the winds above the middle latitudes in both hemispheres blow in a wavy west-to-east direction. The reason for these westerly winds is that, aloft, we generally find higher pressure over equatorial regions and lower pressures over polar regions. Where these upper-level winds tend to concentrate into narrow bands, we find rivers of fast-flowing air—what we call **jet streams.**

Characteristics of Jet Streams   Atmospheric jet streams are swiftly flowing air currents hundreds of miles long, normally less than several hundred miles wide, and typically less than a mile thick. Wind speeds in the

● **FIGURE 7.30** During the summer, the Pacific high moves northward. Sinking air along its eastern margin (over California) produces a strong subsidence inversion, which causes relatively dry weather to prevail. Along the western margin of the Bermuda high, southerly winds bring in humid air, which rises, condenses, and produces abundant rainfall.

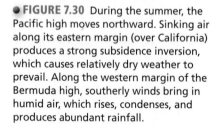

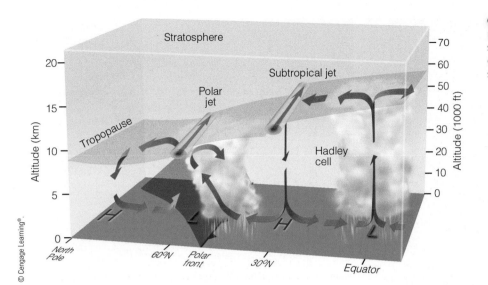

Stratosphere

Subtropical jet

Polar jet

Tropopause

Hadley cell

Altitude (km)

Altitude (1000 ft)

North Pole   60°N   Polar front   30°N   Equator

H   L   H   L

© Cengage Learning®

FIGURE 7.32 Average position of the polar jet stream and the subtropical jet stream, with respect to a model of the general circulation in winter. Both jet streams are flowing from west to east.

central core of a jet stream often exceed 100 knots and occasionally 200 knots. Jet streams are usually found at the tropopause at elevations between 10 and 14 km (33,000 and 46,000 ft) although they can occur at both higher and lower altitudes.

Jet streams were first encountered by high-flying military aircraft during World War II, but their existence was suspected before that time. Ground-based observations of fast-moving cirrus clouds had revealed that westerly winds aloft must be moving rapidly indeed.

Figure 7.32 illustrates the average position of two jet streams, the tropopause, and the general circulation of air for the Northern Hemisphere in winter. Both jet streams are located at tropopause gaps, where mixing between tropospheric and stratospheric air takes place. The jet stream situated near 30° latitude at about 13 km (43,000 ft) above the subtropical high is the **subtropical jet stream.*** To the north, the jet stream situated at a lower altitude of about 10 km (33,000 ft) near the polar front is known as the **polar front jet stream** or, simply, the *polar jet stream.*

In Fig. 7.32, the wind in the center of the jet stream would be flowing as a westerly wind away from the viewer. This direction, of course, is only an average, as jet streams often flow in a wavy west-to-east pattern. When the polar jet stream flows in broad loops that sweep north and south, it may even merge with the subtropical jet. Occasionally, the polar jet splits into two jet streams. The jet stream to the north is often called the *northern branch* of the polar jet, whereas the one to the south is called the *southern branch.* Figure 7.33 illustrates how the polar jet stream and the subtropical jet stream might appear as they sweep around Earth in winter.

We can better see the looping pattern of the jet by studying Fig. 7.34a, which shows the position of the polar jet stream and the subtropical jet stream at the 300-mb level (near 9 km or 30,000 ft) on March 9, 2005. The fastest

flowing air, or *jet core,* is represented by the heavy dark arrows. The map shows a strong polar jet stream sweeping south over the Great Plains with an equally strong subtropical jet over the Gulf states. Notice that the polar jet has a number of loops, with one off the west coast of North America and another over eastern Canada. Observe in the satellite image (Fig. 7.34b) that the polar jet stream (blue arrow) is directing cold, polar air into the Plains states, while the subtropical jet stream (orange arrow) is sweeping subtropical moisture, in the form of a dense cloud cover, over the southeastern states.

The looping pattern of the polar jet stream has an important function. In the Northern Hemisphere, where

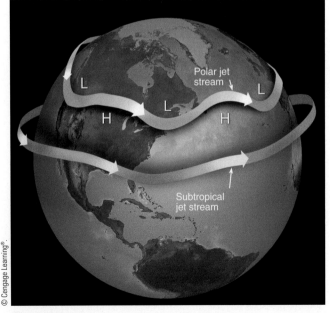

Polar jet stream

Subtropical jet stream

L   H   L   H

© Cengage Learning®

FIGURE 7.33 Jet streams are swiftly flowing currents of air that move in a wavy west-to-east direction. The figure shows the position of the polar jet stream and subtropical jet stream in winter. Although jet streams are shown as one continuous river of air, in reality they are discontinuous, with their position varying from one day to the next.

---

*The subtropical jet stream is normally found between 20° and 30° latitude.

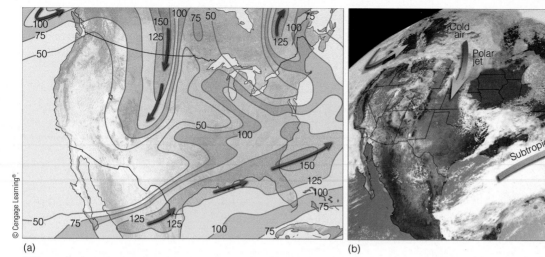

(a)

(b)

● FIGURE 7.34 (a) Position of the polar jet stream and the subtropical jet stream at the 300-mb level (about 9 km or 30,000 ft above sea level) on March 9, 2005. Solid lines are lines of equal wind speed (isotachs) in knots. (b) Satellite image showing clouds and positions of the jet streams for the same day.

the air flows southward, swiftly moving air directs cold air equatorward; where the air flows northward, warm air is carried toward the poles. Jet streams, therefore, play a major role in the global transfer of heat. Moreover, since jet streams tend to meander around the world, we can easily understand how pollutants or volcanic ash injected into the atmosphere in one part of the globe could eventually settle to the ground many thousands of kilometers downwind. And, as we will see in Chapter 8, the looping nature of the polar jet stream has an important role in the development of mid-latitude cyclonic storms.

The Formation of Jet Streams Since jet streams are bands of strong winds, they form in the same manner as all winds do—from horizontal differences in air pressure. In ● Fig. 7.35a, notice that the polar jet stream forms along the polar front where sharp contrasts in temperature produce rapid horizontal pressure changes and strong winds. Notice also in Fig. 7.35a that as the 20°C isotherm crosses the frontal boundary, it dips sharply. This rapid change in temperature causes the constant pressure (isobaric) 500-mb surface to bend sharply as it passes through the front.

The bending of the 500-mb surface in Fig. 7.35a shows up as tightly packed contour lines and strong winds along the front on the 500-mb chart (Fig. 7.35b). Because north-to-south temperature contrasts along the front are greater in winter than they are in summer, the polar jet stream shows seasonal variations. In winter, the polar jet stream winds are stronger and the jet moves farther south, sometimes as far south as Florida and Mexico. In summer, the polar jet stream is weaker and forms over higher latitudes.

Look back at Fig. 7.32 on p. 195 and see that the subtropical jet stream forms on the poleward (north) side of the Hadley cell, at a higher altitude than the polar jet stream. Here, warm air aloft carried poleward by the Hadley cell produces sharp temperature differences, strong pressure gradients, and high winds.

Although the polar and subtropical jets are the two most frequently in the news, there are other jet streams that deserve mentioning. For example, there is a *low-level jet stream* that forms just above the Central Plains of the

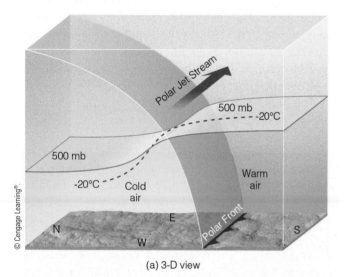

(a) 3-D view

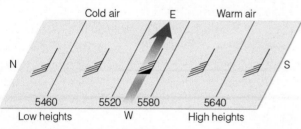

(b) 500-mb chart

● FIGURE 7.35 Diagram (a) is a model that shows a vertical 3-D view of the polar front in association with a sharply dipping 500-mb pressure surface, an isotherm (dashed line), and the position of the polar front jet stream in winter. The diagram is highly exaggerated in the vertical. Diagram (b) represents a 500-mb chart that cuts through the polar front as illustrated by the dipping 500-mb surface in (a). Sharp temperature contrasts along the front produce tightly packed contour lines and strong winds (contour lines are in meters above sea level).

United States. During the summer, this jet (which can attain wind speeds of more than 60 knots) often contributes to the formation of nighttime thunderstorms by transporting moisture and warm air northward. Higher up in the atmosphere, over the subtropics, a summertime easterly jet called the *tropical easterly jet* forms at the base of the tropopause. And during the dark polar winter, a *stratospheric polar jet* forms near the top of the stratosphere.

BRIEF REVIEW

Before going on to the next section, which describes the many interactions between the atmosphere and the ocean, here is a review of some of the important concepts presented so far:

- The two major semipermanent subtropical highs that influence the weather of North America are the Pacific high situated off the west coast and the Bermuda high situated off the southeast coast.

- The polar front is a zone of low pressure where cyclonic storms often form. It separates the mild westerlies of the middle latitudes from the cold, polar easterlies of the high latitudes.

- In equatorial regions, the intertropical convergence zone (ITCZ) is a boundary where air rises in response to the convergence of the northeast trades and the southeast trades. Along the ITCZ huge thunderstorms produce heavy rain showers.

- In the Northern Hemisphere, the major global pressure systems and wind belts shift northward in summer and southward in winter.

- The northward movement of the Pacific high in summer tends to keep summer weather along the west coast of North America relatively dry.

- Jet streams exist where strong winds become concentrated in narrow bands. The polar-front jet stream is associated with the polar front. The polar jet meanders in a wavy west-to-east pattern, becoming strongest in winter when the contrast in temperature along the front is greatest.

- The subtropical jet stream is found on the poleward side of the Hadley cell, between 20° and 30° latitude. It is normally observed at a higher altitude than the polar jet stream.

- The general flow of air around the globe in both the Northern and Southern Hemispheres finds surface winds blowing from the east in the tropics, from the west in the middle latitudes, and from the east in polar regions.

- At the surface, in both the Northern and Southern Hemispheres, areas of high pressure tend to persist at the poles and along a belt near 30° latitude. Areas of surface low pressure are generally found near the equator and between about 40° and 55° latitude.

Atmosphere-Ocean Interactions

The atmosphere and oceans are both dynamic fluid systems that interact with one another in many complex ways. For example, evaporation of ocean water provides the atmosphere with surplus water that falls as precipitation. The latent heat that is taken up by the water vapor during evaporation goes into the atmosphere during condensation to fuel storms. The storms, in turn, produce winds that blow over the ocean, which causes waves and currents. The currents, in turn, can modify the weather and climate of a region by bringing in vast quantities of warm or cold water.

The complexity of the interaction between the atmosphere and ocean makes our scientific understanding of how one influences the other on a global scale far from complete. What we will focus on in the remainder of this chapter is what we do know, beginning with ocean currents. Later, we will concentrate on some of the most important weather and climate oscillations that result from atmosphere-ocean interactions.

GLOBAL WIND PATTERNS AND SURFACE OCEAN CURRENTS As the wind blows over the oceans, it causes the surface water to drift along with it. The moving water gradually piles up, creating pressure differences within the water itself. This leads to further motion several hundreds of meters down into the water. In this manner, the general wind flow around the globe starts the major surface ocean currents moving. The relationship between the general wind flow and ocean currents can be seen by comparing ● Fig. 7.36 with ● Fig. 7.37.

Because of the larger frictional drag in water, ocean currents move more slowly than the prevailing winds above. Typically, they range in speed from several kilometers per day to several kilometers per hour. However, comparing Fig. 7.36 with Fig. 7.37, we can see that ocean currents do not follow the wind pattern exactly; rather, they spiral in semiclosed circular whirls. On the eastern edge of continents there is usually a warm current that flows from the equator to the pole. For example, in the North Atlantic, flowing northward along the east coast of the United States, is a tremendous warm water current called the **Gulf Stream**, which carries vast quantities of tropical water into higher latitudes. Off the coast of North Carolina, the Gulf Stream provides warmth and moisture for developing mid-latitude cyclonic storms.

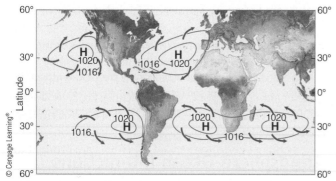

● **FIGURE 7.36** Annual average global wind patterns and surface high-pressure areas over the oceans.

Notice in Fig. 7.37 that as the Gulf Stream moves northward, the prevailing westerlies steer it away from the coast of North America and eastward toward Europe. Generally, it widens and slows as it merges into the broader *North Atlantic Drift*. As this current approaches Europe, part of it flows northward along the coasts of Great Britain and Norway, bringing with it warm water (which helps keep winter temperatures much warmer than one would expect this far north). The other part flows southward as the *Canary Current*, which transports cool, northern water equatorward. In the Pacific Ocean, the counterpart

to the *Canary Current* is the *California Current* that carries cool water southward along the coastline of the western United States. Hence, on the western edge of the major continents, there is usually a cool current that flows from the pole toward the equator.

Up to now, we have seen that atmospheric circulations and ocean circulations are closely linked; wind blowing over the oceans produces surface ocean currents. The currents, along with the wind, transfer heat from tropical areas, where there is a surplus of energy, to polar regions, where there is a deficit. This helps to equalize the latitudinal energy imbalance with about 40 percent of the total heat transport in the Northern Hemisphere coming from surface ocean currents. The environmental implications of this heat transfer are tremendous. If the energy imbalance were to go unchecked, yearly temperature differences between low and high latitudes would increase greatly, and the climate would gradually change.

WINDS AND UPWELLING Earlier, we saw that the cool California Current flows roughly parallel to the west coast of North America. From this, we might conclude that summer surface water temperatures will be cool along the coast of Washington and gradually warmer as

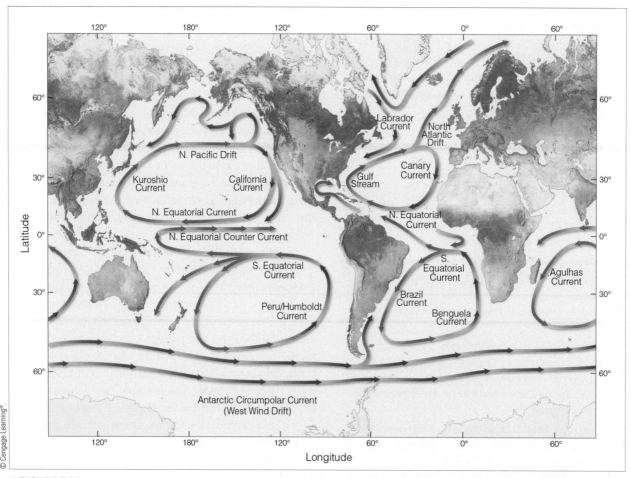

● **FIGURE 7.37** Average position and extent of the major surface ocean currents. Cold currents are shown in blue; warm currents are shown in red.

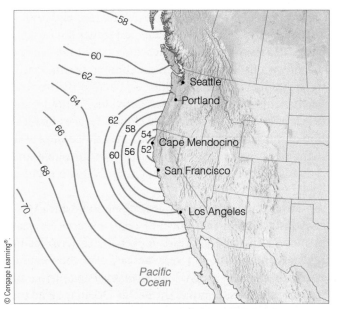

● **FIGURE 7.38** Average sea surface temperatures (°F) along the west coast of North America during August.

we move south. A quick glance at the water temperatures along the west coast of the United States during August (see ● Fig. 7.38) quickly alters that notion. The coldest water is observed along the northern California coast near Cape Mendocino. The reason for the cold, coastal water is **upwelling**—the rising of cold water from below.

For upwelling to occur, the wind must flow more or less parallel to the coastline. Notice in ● Fig. 7.39 that summer winds tend to parallel the coastline of California. As the wind blows over the ocean, the surface water beneath it is set in motion. As the surface water moves, it bends slightly to its right due to the Coriolis effect. (Remember, it would bend to the left in the Southern Hemisphere.) The water beneath the surface also moves, and it too bends slightly to its right. The net effect of this phenomenon is that a rather shallow layer of surface water moves at right angles to the surface wind and heads seaward. As the surface water drifts away from the coast, cold, nutrient-rich

water from below rises (upwells) to replace it. Upwelling is strongest and surface water is coolest where the wind parallels the coast, such as it does in summer along the coast of northern California.

Because of the cold coastal water, summertime weather along the West Coast often consists of low clouds and fog, as the air over the water is chilled to its saturation point. Upwelling produces good fishing conditions, however, as higher concentrations of nutrients are brought to the surface. But swimming is only for the hardiest of souls, as the average surface water temperature along the coast of northern California in summer is nearly 10°C (18°F) colder than the average coastal water temperature found at the same latitude along the Atlantic coast.

Between the ocean surface and the atmosphere, there is an exchange of heat and moisture that depends, in part, on temperature differences between water and air. In winter, when air-water temperature contrasts are greatest, there is a substantial transfer of sensible and latent heat from the ocean surface into the atmosphere. This energy helps to maintain the global air flow. Consequently, even a relatively small change in surface ocean temperatures can modify atmospheric circulations and have far-reaching effects on global weather and climate patterns. The next section describes how weather events can be linked to surface ocean temperature changes in the tropical Pacific.

EL NIÑO, LA NIÑA, AND THE SOUTHERN OSCILLATION

Along the west coast of South America, where the cool Peru Current sweeps northward, southerly winds promote upwelling of cold, nutrient-rich water that gives rise to large fish populations, especially anchovies. The abundance of fish supports a large population of sea birds whose droppings (called *guano*) produce huge phosphate-rich deposits, a valuable source of fertilizer. Every two to five years or so, a warm current of nutrient-poor tropical water moves southward, replacing the cold, nutrient-rich surface water. Because this condition frequently occurs

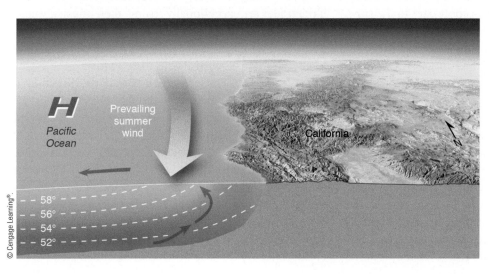

● **FIGURE 7.39** As winds blow parallel to the west coast of North America, surface water is transported to the right (out to sea). Cold water moves up from below (upwells) to replace the surface water. The large H represents the position of the Pacific high in summer. Blue arrows show the movement of water.

around Christmas, local fishermen (more than a century ago) called this warm current *Corriente del Niño,* which translated means "current of the Christ Child;" hence, the warm current's name—*El Niño.*

It was once thought that El Niño was a local event that occurred only along the west coast of Peru and Ecuador. It is now known that the ocean-warming can cover an area of the tropical Pacific much larger than the continental United States. Although in recent decades the term **El Niño** has gained global prominence, the large, prolonged warming that develops at irregular intervals every three to seven years is often referred to as an *El Niño event.* During such an event, the surface water temperature across much of the tropical Pacific rises by 0.5°C (0.9°F) or more for periods of a few months to a year or more.

During a El Niño event, because of the unusually warm water, large numbers of fish and marine plants may die and other marine species may travel far from the tropics. A very strong El Niño event in 2015–2016 resulted in whale and hammerhead sharks moving up the California coast, and California squid were reported in southeast Alaska. The El Niño of 1972–1973 was one of the first to be studied in depth. It reduced the annual Peruvian anchovy catch by more than 50 percent in 1972. Since much of the harvest of this fish is converted into fishmeal and exported for use in feeding livestock and poultry, the world's fish meal production in 1972 was greatly reduced. Countries such as the United States that rely on fish meal for animal feed had to use soybeans as an alternative. This raised poultry prices in the United States by more than 40 percent. Peru's fishing industry is now more carefully managed, so that the decline of anchovy during El Niño is not causing such severe repercussions.

Why does the ocean become so warm over such a large area during an El Niño event? Normally in the tropical Pacific Ocean, trade winds are persistently blowing westward from a region of higher pressure over the eastern Pacific toward a region of lower pressure over the western Pacific, centered near Indonesia (see ● Fig. 7.40a). The trades create upwelling that brings cool water to the surface over the eastern tropical Pacific and along the west coast of South America. As this cool water moves westward, it is heated by sunlight and the atmosphere. Consequently, surface water along the equator usually is cooler in the eastern Pacific and warmer in the western Pacific. In addition, the dragging of surface water by the trades raises sea level by a few inches in the western Pacific and lowers it in the eastern Pacific, which produces a thick layer of warm water over the tropical western Pacific Ocean. The higher level of water in the western Pacific causes warm water to flow slowly eastward toward South America as a weak, narrow ocean current called the *countercurrent.*

Every few years, the surface atmospheric pressure patterns break down. Air pressure rises over the western

Pacific and falls over the eastern Pacific (see Fig. 7.40b). This change in pressure weakens the trades, and over a period of a few months, easterly winds may be replaced by westerly winds that strengthen the countercurrent. Warm water gradually extends farther east across the tropical Pacific. The warm surface water provides energy for the development of huge convective clouds. Hence, the location of showers and thunderstorms may shift east as well. If the ocean warming reaches a certain threshold and remains at that level long enough, an El Niño event is declared to be in place.

Toward the end of the warming period, which typically lasts about a year but may recur for another year or more, atmospheric pressure over the eastern Pacific reverses and begins to rise, whereas, over the western Pacific, it falls. This seesaw pattern of reversing surface air pressure at opposite ends of the Pacific Ocean is called the **Southern Oscillation.** Because the reversals in air pressure and the ocean warming are more or less simultaneous, scientists call this phenomenon the *El Niño–Southern Oscillation* or **ENSO** for short. Although most El Niño episodes follow a similar evolution, each event has its own personality, differing in both strength and behavior.

During especially strong El Niño events (such as in 1982–1983, 1997–1998, and 2015–2016) the easterly trades may actually become westerly winds, as illustrated in Fig. 7.40b. As these winds push eastward, they drag surface water with them. This dragging raises sea level in the eastern Pacific and lowers sea level in the western Pacific. The eastward-moving water gradually warms under the tropical sun, becoming as much as 6°C (11°F) warmer than normal in the eastern equatorial Pacific. Gradually, a thick layer of warm water pushes into coastal areas of Ecuador and Peru, choking off the upwelling that supplies cold, nutrient-rich water to South America's coastal region. The unusually warm water can extend from South America's coastal region for many thousands of kilometers westward along the equator (see ● Fig. 7.41a). Following an El Niño event, the trade winds usually return to normal. However, if the trades are exceptionally strong, then unusually cold surface water develops over the central and eastern Pacific, and warm water and rainy weather are confined mainly to the western tropical Pacific (see Figure 7.41b). This cold-water episode is termed **La Niña** ("the girl child"). In some ways, La Niña can be thought of as the opposite of El Niño. Water temperatures in the eastern tropical Pacific are colder than average during La Niña, while they are warmer than average during El Niño. Keep in mind, however, that while an El Niño event features a *reversal* of the typical east-to-west wind flow and ocean currents across the tropical Pacific, La Niña features a *strengthening* of the same east-to-west wind flow and ocean currents. La Niña events sometimes occur immediately after El Niño events, but they can also develop independently.

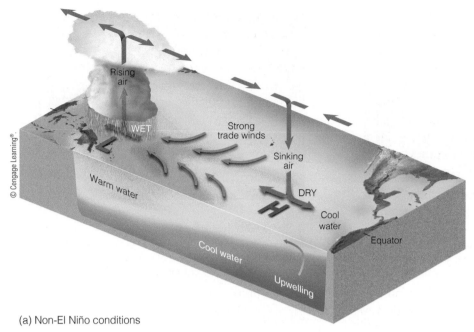

(a) Non-El Niño conditions

© Cengage Learning®

● **FIGURE 7.40** In diagram (a), under non-El Niño conditions higher pressure over the southeastern Pacific and lower pressure near Indonesia produce easterly trade winds along the equator. These winds promote upwelling and cooler ocean water in the eastern Pacific, while warmer water prevails in the western Pacific. The trades are part of a circulation that typically finds rising air and heavy rain over the western Pacific and sinking air and generally dry weather over the eastern Pacific. When the trades are exceptionally strong, water along the equator in the eastern Pacific becomes quite cool. This cool event is called La Niña. During El Niño conditions—diagram (b)—atmospheric pressure decreases over the eastern Pacific and rises over the western Pacific. This change in pressure causes the trades to weaken or reverse direction, which enhances the countercurrent that carries warm water from the west over a vast region of the eastern tropical Pacific.

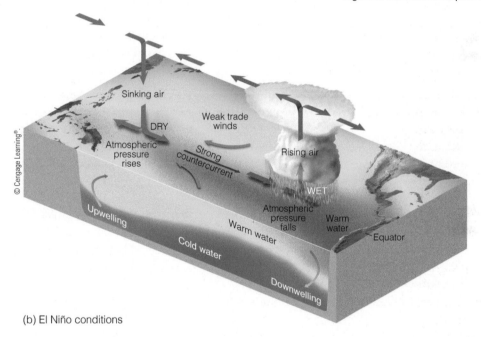

(b) El Niño conditions

The large areas of abnormally warm water associated with El Niño and abnormally cold water associated with La Niña can have effects far beyond South America. During El Niño, the usual region of warm water in the western tropical Pacific is moved thousands of kilometers to the east. There, the water fuels the atmosphere with additional warmth and moisture, which feeds into storminess and rainfall. Along with the added warmth from the oceans, latent heat is released during condensation. All of these factors help change the regional atmospheric circulation, effects that can propagate for thousands of miles, rearranging the locations where rising and sinking air tend to predominate. As a result, some regions of the world experience too much rainfall during El Niño, whereas others have too little. Over the warm tropical central Pacific, the

frequency of typhoons usually increases. However, over the tropical Atlantic, between Africa and Central America, the winds aloft tend to disrupt the organization of thunderstorms that is necessary for hurricane development; hence, there are usually fewer hurricanes in this region during strong El Niño events. And, as we saw earlier in this chapter, during a strong El Niño, summer monsoon conditions tend to weaken over India.

Although the actual mechanism by which changes in surface ocean temperatures influence global wind patterns is not fully understood, the by-products are plain to see. For example, during exceptionally warm El Niños, drought is normally felt in Indonesia, southern Africa, and Australia, while heavy rains and flooding often occur in Ecuador and Peru. In the Northern Hemisphere, a strong

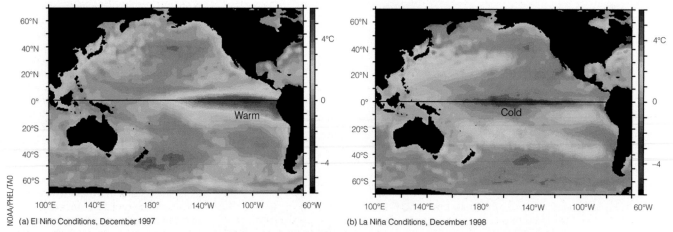

NOAA/PHEL/TAO

(a) El Niño Conditions, December 1997

(b) La Niña Conditions, December 1998

● **FIGURE 7.41** (a) Average sea surface temperature departures from normal as measured by satellite. During El Niño conditions, upwelling is greatly diminished and warmer than normal water (deep red color) extends from the coast of South America westward, across the Pacific. (b) During La Niña conditions, strong trade winds promote upwelling, and cooler than normal water (dark blue color) extends over the eastern and central Pacific.

subtropical westerly jet stream often directs mid-latitude cyclonic storms into California and heavy rain into the Gulf Coast states. The total global impact of an El Niño event due to flooding, winds, and drought may exceed many billions of dollars. A major El Niño event can also pump so much heat into the atmosphere that global temperatures rise by

several tenths of a degree Fahrenheit for a few months. The very strong El Niño of 2015–2016 helped produce record-warm global surface temperatures in 2015 and 2016.

● Figure 7.42 shows warm events (El Niño years) in red and cold events (La Niña years) in blue. Notice in Fig. 7.42 that it is possible to have two or more El Niño or

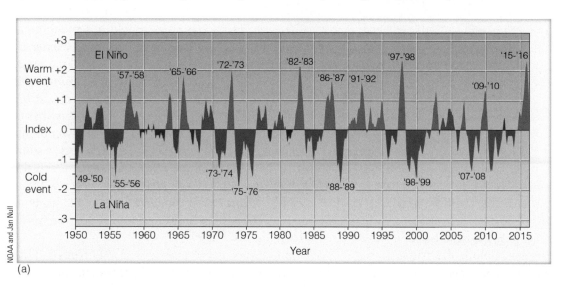

NOAA and Jan Null

(a)

● **FIGURE 7.42** The Oceanic Niño Index (ONI). The numbers on the left side of the diagram represent a running 3-month mean for sea surface temperature variations (from normal, in degrees Celsius) over the tropical Pacific Ocean from latitude 5°N to 5°S and from longitude 120°W to 170°W (area outlined in adjacent map). Warm El Niño episodes are in red; cold La Niña episodes are in blue. Warm and cold events occur when the deviation from the normal is 0.5 or greater (faint red and blue shading in graph). An index value between 0.5 and 0.9 is considered weak; an index value between 1.0 and 1.4 is considered moderate; an index value between 1.5 and 2.0 is considered strong; and an index value greater than 2.0 is considered very strong.

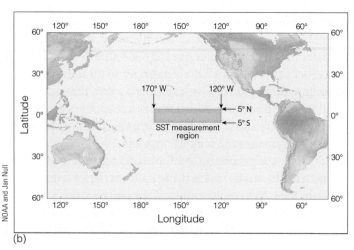

NOAA and Jan Null

(b)

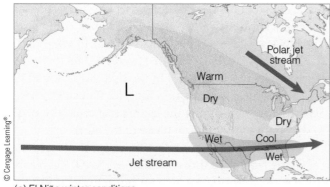
(a) El Niño winter conditions

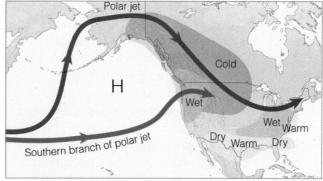
(b) La Niña winter conditions

● **FIGURE 7.43** Typical winter weather patterns across North America during an El Niño warm event (a) and during a La Niña cold event (b). During El Niño conditions, a persistent trough of low pressure forms over the north Pacific and, to the south of the low, the jet stream (from off the Pacific) steers wet weather and cyclonic storms into California and the southern part of the United States. During La Niña conditions, a persistent high-pressure area forms south of Alaska forcing the polar jet stream and accompanying cold air over much of western North America. The southern branch of the polar jet stream directs moist air from the ocean into the Pacific Northwest, producing a wet winter for that region.

La Niña periods in a row, separated by neutral conditions. El Niño events typically develop in the autumn (Northern Hemisphere), peak in the winter, and weaken in spring and summer. Although El Niño events rarely last more than a year, La Niña conditions are more likely to recur or persist for two or three years in a row. Over a long time frame (several decades), the tropical Pacific is divided roughly equally among El Niño, La Niña, and neutral conditions.

As we have seen, El Niño and the Southern Oscillation are part of a large-scale ocean-atmosphere interaction that can take several years to run its course. During this time, certain regions in the world can expect significant climatic responses to an ENSO event. ● Figure 7.43 shows how typical winter weather patterns over North America will change between El Niño conditions and La Niña conditions. Such ocean-atmosphere interactions, where a warmer or colder ocean surface can influence weather patterns in distant parts of the world, are called **teleconnections.** As an example of teleconnections, note in Fig. 7.43 that in the Pacific Northwest, La Niña winters tend to be wetter than usual, while drier-than-average winters are more likely during El Niño.

Notice also in Fig. 7.43 that in some areas, the teleconnections related to La Niña are opposite to those of El Niño. However, this is not the case in every location, because the mechanisms behind El Niño and La Niña are not exact opposites. Also, it is important to keep in mind that these preferred patterns are not *guaranteed* to occur with every El Niño or La Niña event, since each one is different. For example, Los Angeles, California, had a drier-than-average winter during the strong El Niño event of 2015–2016, even though strong El Niño winters tend to be wetter than average in Los Angeles. Over the long term, though, the effects shown in Fig. 7.43 are the most *likely* ones to expect. So while we cannot say with certainty that an El Niño winter will bring more rainfall than average to

California, we can say that the odds are higher during a strong El Niño than they might otherwise be. For several decades, atmospheric scientists have developed and improved techniques for predicting the future state of the El Niño–Southern Oscillation and the onset of El Niño and La Niña events. This is an especially difficult challenge because ENSO events are not like typical day-to-day weather features. Instead, they unfold in a much more gradual way through a complex sequence of events involving both the ocean and atmosphere. Using long-range climate models, the National Oceanic and Atmospheric Administration and several other agencies around the world now issue outlooks giving the probability that El Niño or La Niña will develop over the next few months. When it becomes clear in the Northern Hemisphere summer or autumn that an El Niño or La Niña event is actually developing, then forecasters can provide several months' notice of the type of conditions that are most likely to occur in the winter, the time of year when the events usually reach their full strength.

Up to this point, we have looked at El Niño and the Southern Oscillation, and at how the reversal of surface ocean temperatures and atmospheric pressure combine to influence regional and global weather and climate patterns. There are other atmosphere-ocean interactions that can have an effect on large-scale weather patterns. Some of these are described in the following section.

OTHER ATMOSPHERE-OCEAN INTERACTIONS Over the North Pacific Ocean, periodic changes in surface ocean temperatures can influence weather along the west coast of North America for much longer periods of time than El Niño and La Niña. The **Pacific Decadal Oscillation (PDO)** is like ENSO in that it has a warm phase and a cool phase, and the temperature patterns produced by the PDO are similar in some locations to those produced by ENSO. However, the PDO has more of an influence in the mid-latitudes of the North Pacific than in the tropical Pacific, and it operates on a much longer time scale than ENSO. Each PDO phase tends to predominate for 20 to 30 years before switching to the other phase.

During the warm (or positive) phase of the PDO, unusually warm surface water exists along the west coast of North America, while over the central North Pacific, cooler than normal surface water prevails (see ● Fig. 7.44a). At the same time, the Aleutian low in the Gulf of Alaska strengthens, which causes more Pacific storms to move into Alaska and California. This situation causes winters, as a whole, to be warmer and drier over northwestern North America. Elsewhere, winters tend to be drier over the Great Lakes, and cooler and wetter in the southern United States.

Cool (or negative) PDO phases have cooler-than-average surface water along the west coast of North America and an area of warmer-than-normal surface water extending from Japan into the central North Pacific (see Fig. 7.44b). Winters in the cool phase tend to be cooler and wetter than average over northwestern North America, wetter over the Great Lakes, and warmer and drier in the southern United States.

These climate patterns only represent average conditions, as individual years within either PDO phase may vary considerably, sometimes reverting to the opposite phase for short periods. These variations, which can last several months to a year or more, make it more difficult to decipher exactly when the PDO has changed from one long-term phase to the other. Knowing the long-term phase of the PDO can help improve seasonal climate prediction, because the warm phase of PDO tends to reinforce the climate patterns of El Niño, which helps strengthen the

impact of an El Niño event. Likewise, La Niña events are generally reinforced when the PDO is in its cool phase. The PDO was generally positive from 1922 to 1947, negative from 1947 to 1977, positive from 1977 to 1998, and negative from 1998 to 2013. In the mid-2010s, the PDO appeared to be entering a new long-term positive phase. Scientists are researching the factors that cause the PDO to shift from one long-term phase to the other.

Over the Atlantic, a periodic reversal of pressure called the **North Atlantic Oscillation** (**NAO**) has a substantial effect on the weather in Europe and along the east coast of North America, especially during the winter. The NAO is measured by the difference in atmospheric pressure between the vicinity of the Icelandic low and the region of the Bermuda-Azores high. (Note that the NAO is defined by atmospheric change, whereas ENSO and the PDO are defined by oceanic change.) When the air pressure in the North Atlantic is unusually high near the Azores and unusually low near Iceland, the increased pressure gradient leads to stronger westerlies. These winds, in turn, drive frequent, powerful cyclonic storms into northern Europe, which tends to make winters wet and mild. During this *positive phase* of the NAO, winters in the eastern United States also tend to be wet and relatively mild, while northern Canada and eastern Europe are typically cold and dry (see ● Fig. 7.45a).

The *negative phase* of the NAO occurs when the atmospheric pressure in the vicinity of the Icelandic low rises, while the pressure drops in the region of the Bermuda high (see Fig. 7.45b). This results in a reduced pressure gradient and weaker westerlies, which steer fewer and weaker winter storms across the Atlantic. The weaker jet stream also allows for storms to move or develop unusually far south, bringing wet weather to southern Europe and other areas

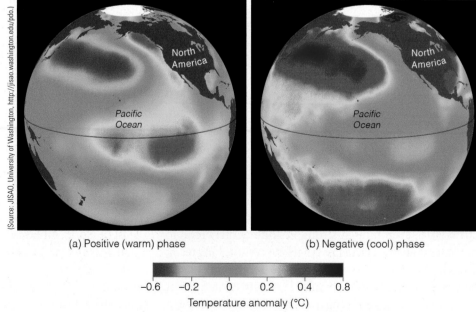

● **FIGURE 7.44** Typical winter sea surface temperature departure from normal in °C during the Pacific Decadal Oscillation's warm phase (a) and cool phase (b).

(Source: JISAO, University of Washington, http://jisao.washington.edu/pdo.)

(a) Positive (warm) phase

(b) Negative (cool) phase

−0.6   −0.2   0   0.2   0.4   0.8

Temperature anomaly (°C)

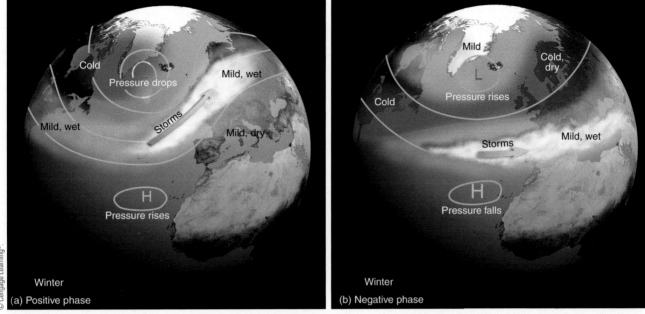

● **FIGURE 7.45** Change in surface atmospheric pressure and typical winter weather patterns associated with the (a) positive phase and (b) negative phase of the North Atlantic Oscillation.

near the Mediterranean Sea. The weaker, more variable jet stream during a negative NAO phase also allows cold air masses to move southward more readily into northern Europe and the eastern United States, which tends to make these regions generally colder and drier than usual (although the presence of cold air can sometimes lead to intense winter storms in the eastern United States). Far eastern Canada and eastern Europe often experience milder-than-average winters during a negative NAO. Although the NAO often varies from month to month, it can also tend to favor one phase or the other for several years.

Closely related to the North Atlantic Oscillation, but analyzed at a more northerly latitude, is the **Arctic Oscillation (AO).** The AO is defined by variations in atmospheric pressure between the Arctic and the North Pacific and Atlantic. During the *positive (warm) phase* of the *AO,* higher pressures to the south and lower pressures across the Arctic produce strong westerly winds aloft. These winds wrap around the semipermanent zone of upper-level low pressure over the North Pole that is sometimes referred to as the **polar vortex.** When a positive AO is in place, the polar vortex is typically stronger than usual, and cold arctic air generally remains bottled up in and near the polar regions. Thus, winters in the United States tend to be warmer than normal, while winters over Newfoundland and Greenland tend to be very cold. Meanwhile, strong winds over the

Atlantic direct storms into northern Europe, bringing with them wet, mild weather.

During the *negative (cold) phase* of the *AO,* pressure differences are smaller between the Arctic and regions to the south, leading to weaker and more variable westerly winds aloft. The weaker westerlies mean that the polar vortex can more easily shift to lower latitudes, and cold arctic air is now able to penetrate farther south, often producing colder-than-normal winters over much of the United States. Cold air also invades northern Europe and Asia, while Newfoundland and Greenland normally experience warmer than normal winters. Much like the NAO, the AO switches from one phase to another on an irregular basis, and one phase may predominate for several years in a row, bringing with it a succession of either cold or mild winters.

Unlike El Niño and La Niña, the NAO and AO can switch modes over just a few weeks' time, and these changes cannot be predicted more than a couple of weeks in advance. This influence makes it more challenging to anticipate winter conditions over eastern North America and Europe than over western North America, where the more slowly varying ENSO and the PDO play a larger role. As our knowledge of the interactions between the ocean and atmosphere improves, we can expect scientists to gain skill at predicting shifts in all of these phenomena, as well as the resulting influences they have on regional weather and climate.

SUMMARY

In this chapter, we examined a variety of atmospheric circulations. We looked at small-scale winds and found that eddies can form in a region of strong wind shear, especially in the vicinity of a jet stream. On a slightly larger scale, land and sea breezes blow in response to local pressure differences created by the uneven heating and cooling rates of land and water. Monsoon winds change direction seasonally, while mountain and valley winds change direction daily.

A warm, dry wind that descends the eastern side of the Rocky Mountains is the chinook. The same type of wind in the Alps is the foehn. A warm, dry downslope wind that blows into southern California is the Santa Ana wind. Local intense heating of the surface can produce small rotating winds, such as the dust devil, while downdrafts in a thunderstorm are responsible for the desert haboob.

The largest pattern of winds that persists around the globe is called the general circulation. At the surface in both hemispheres, winds tend to blow from the east in the tropics, from the west in the middle latitudes, and from the east in polar regions. Where upper-level westerly winds tend to concentrate into narrow bands, we find jet streams. The annual shifting of the major pressure systems and wind belts—northward in July and southward in January—strongly influences the annual precipitation of many regions.

Toward the end of the chapter we examined the interaction between the atmosphere and oceans. Here we found the interaction to be an ongoing process where everything, in one way or another, seems to influence everything else. On a large scale, winds blowing over the surface of the water drive the major ocean currents; the oceans, in turn, release energy to the atmosphere, which helps to maintain the general circulation of winds.

When atmospheric circulation patterns change over the tropical Pacific, and the trade winds weaken or reverse direction, warm tropical water is able to flow eastward toward South America where it chokes off upwelling. When the warm water extends over a vast area of the tropical Pacific, and persists for several months to a year or more, the warming is called an El Niño event, and the associated reversal of pressure over the Pacific Ocean is called the Southern Oscillation. The large-scale interaction between the atmosphere and the ocean during El Niño and the Southern Oscillation (ENSO) affects global atmospheric circulation patterns, resulting in too much rain in some areas and not enough in others.

Over the northern central Pacific and along the west coast of North America the surface water temperature reverses every 20 to 30 years, a phenomenon called the Pacific Decadal Oscillation (PDO). Over the Atlantic Ocean there is a periodic reversal of air pressure called the North Atlantic Oscillation that influences weather in various regions of the world. Atmospheric pressure changes over the Arctic produce a related phenomenon, the Arctic Oscillation which causes winter weather patterns to change across the United States, Greenland, and Europe. Researchers are studying how the interchange between atmosphere and ocean can produce such events.

KEY TERMS

The following terms are listed (with corresponding page numbers) in the order they appear in the text. Define each. Doing so will aid you in reviewing the material covered in this chapter.

scales of motion, 174
microscale, 174
mesoscale, 174
synoptic scale, 174
global scale, 174
macroscale, 174
rotor, 175
wind shear, 175
clear air turbulence (CAT), 176
thermal circulation, 176
sea breeze, 178
land breeze, 178
valley breeze, 180
mountain breeze, 180
katabatic wind, 180
chinook wind, 181
Santa Ana wind, 182
haboob, 184
dust devils (whirlwinds), 184
monsoon, 186
monsoon wind system, 184
general circulation of the atmosphere, 188
Hadley cell, 189
doldrums, 189
subtropical highs, 190

trade winds, 190
intertropical convergence zone (ITCZ), 190
westerlies, 191
polar front, 191
subpolar low, 191
polar easterlies, 191
Bermuda high, 192
Pacific high, 192
Icelandic low, 192
Aleutian low, 192
Siberian high, 192
jet stream, 195
subtropical jet stream, 195
polar front jet stream, 195
Gulf Stream, 197
upwelling, 199
El Niño, 200
Southern Oscillation, 200
ENSO, 200
La Niña, 200
teleconnections, 203
Pacific Decadal Oscillation (PDO), 203
North Atlantic Oscillation (NAO), 204
Arctic Oscillation (AO), 205
polar vortex, 205

QUESTIONS FOR REVIEW

1. Describe the various scales of motion and give an example of each.
2. What is wind shear and how does it relate to clear air turbulence?
3. Using a diagram, explain how a thermal circulation develops.
4. Why does a sea breeze at the surface blow from sea to land and a land breeze from land to sea?
5. Which wind will produce clouds: a valley breeze or a mountain breeze? Why?
6. What are katabatic winds? How do they form?
7. Explain why chinook winds are warm and dry.
8. Describe how dust devils usually form.

9. (a) Briefly explain how the monsoon wind system develops over eastern and southern Asia.
   (b) Why in India is the summer monsoon wet and the winter monsoon dry?
10. Draw a large circle. Now, place the major surface semipermanent pressure systems and the wind belts of the world at their appropriate latitudes.
11. According to Fig. 7.25 (p. 190), most of the United States is located in what wind belt?
12. Explain how and why the average surface pressure features shift from summer to winter.
13. Explain why summers along the West Coast of the United States tend to be dry, whereas along the East Coast summers tend to be wet.
14. How does the polar front influence the development of the polar front jet stream?
15. Why is the polar jet stream typically stronger in winter?
16. Explain the relationship between the general circulation of air and the circulation of surface ocean currents.
17. Describe how the winds along the west coast of North America produce upwelling.
18. What is an El Niño event? What happens to the surface pressure at opposite ends of the Pacific Ocean during the Southern Oscillation?
19. Describe how an ENSO event may influence the weather in different parts of the world.
20. What are the conditions over the tropical eastern and central Pacific Ocean during La Niña?
21. Describe the ocean surface temperatures associated with the Pacific Decadal Oscillation.
22. How does the positive (warm) phase of the Northern Atlantic Oscillation differ from the negative (cold) phase?
23. During the negative (cold) phase of the Arctic Oscillation when Greenland is experiencing mild winters, what type of winters (cold or mild) is Northern Europe usually experiencing?

QUESTIONS FOR THOUGHT AND EXPLORATION

1. Suppose you are fishing in a mountain stream during the early morning. Is the wind more likely to be blowing upstream or downstream? Explain why.
2. Why, in Antarctica, are winds on the high plateaus usually lighter than winds in steep, coastal valleys?
3. What atmospheric conditions must change so that the westerly flowing polar front jet stream reverses direction and becomes an easterly flowing jet stream?
4. After a snowstorm, Cheyenne, Wyoming, reports a total snow accumulation of 48 cm (19 in.), while the maximum depth in the surrounding countryside is only 28 cm (11 in.). If the storm's intensity and duration were practically the same for a radius within 50 km of Cheyenne, explain why Cheyenne received so much more snow.
5. The prevailing winds in southern Florida are northeasterly. Knowing this, would you expect the strongest sea breezes to be along the east or west coast of southern Florida? What about the strongest land breezes?
6. Explain why icebergs tend to move at right angles to the direction of the wind.
7. Give *two* reasons why pilots would prefer to fly in the core of a jet stream rather than just above or below it.
8. Why do the major ocean currents in the North Indian Ocean reverse direction between summer and winter?
9. Why is the surface water along the northern California coast warmer in winter than in summer?
10. The Coriolis force deflects moving water to the right of its intended path in the Northern Hemisphere and to the left of its intended path in the Southern Hemisphere. Why, then, does upwelling tend to occur along the western margin of continents in both hemispheres?

GLOBAL **GEOSCIENCE** WATCH   Go to the Reference section of the Global Environment Watch: Meteorology portal. Search under the term "El Niño" and consult the document "El Niño and La Niña," from *Environmental Science: In Context* (or other reference works in this portal) to answer these questions: What is the earliest recorded use of the term El Niño? Have scientists found evidence for El Niño and La Niña in prehistoric times? Are El Niño and La Niña expected to change as Earth's climate warms?

ONLINE RESOURCES

 Visit www.cengagebrain.com to view additional resources, including video exercises, practice quizzes, interactive eBooks, and more.

CHAPTER 8

Air Masses, Fronts, and Middle-Latitude Cyclones

Contents

About two o'clock in the afternoon it began to grow dark from a heavy, black cloud which was seen in the northwest. Almost instantly the strong wind, traveling at the rate of 70 miles an hour, accompanied by a deep bellowing sound, with its icy blast, swept over the land, and everything was frozen hard. The water in the little ponds in the roads froze in waves, sharp-edged and pointed, as the gale had blown it. The chickens, pigs, and other small animals were frozen in their tracks. Wagon wheels ceased to roll, frozen to the ground. Men, going from their barns or fields a short distance from their homes, in slush and water, returned a few minutes later walking on the ice. Those caught out on horseback were frozen to their saddles and had to be lifted off and carried to the fire to be thawed apart. Two young men were frozen to death near Rushville. One of them was found with his back against a tree, with his horse's bridle over his arm and his horse frozen in front of him. The other was partly in a kneeling position, with a tinder box in one hand and a flint in the other, with both eyes wide open, as if intent on trying to strike a light. Many other casualties were reported. As to the exact temperature, however, no instrument has left any record; but the ice was frozen in the stream, as variously reported, from six inches to a foot in thickness in a few hours.

John Moses, *Illinois: Historical and Statistical*

The opening quotation of this chapter details the passage of a spectacular cold front as it moved through Illinois on December 21, 1836. While some of the incidents described are likely exaggerated, at least one observing station, in Augusta, Illinois, reported air temperatures dropping from 40°F at dawn to 0°F by 2 p.m., and many people and animals are believed to have perished in the sudden cold. Fortunately, temperature changes of this magnitude with cold fronts are uncommon.

In this chapter, we will examine the more typical weather associated with cold fronts and warm fronts. We will address questions such as: Why are cold fronts usually associated with showery weather? How can warm fronts during the winter cause freezing rain and sleet to form over a vast area? And how can one read the story of an approaching warm front by observing its clouds? We will also see how weather fronts are an integral part of a mid-latitude cyclonic storm. But, first, so that we may better understand fronts and storms, we will examine air masses. We will look at where and how they form and the type of weather usually associated with them.

Air Masses

An **air mass** is an extremely large body of air whose properties of temperature and humidity are fairly similar in any horizontal direction at any given altitude. A single air mass may cover more than a million square kilometers. In ● Fig. 8.1, a large winter air mass, associated with a high-pressure area, covers over half of the United States. Note that although the surface air temperature and dew point vary somewhat, everywhere the air is cold and dry, with the exception of the zone of snow showers on the eastern shores of the Great Lakes. This cold, shallow anticyclone will drift eastward, carrying with it the temperature and moisture characteristic of the region where the air mass formed; hence, in a day or two, cold air will be located over the central Atlantic Ocean. Part of weather forecasting is, then, a matter of determining air mass characteristics, predicting how and why they change, and in what direction the systems will move.

SOURCE REGIONS   Regions where air masses originate are known as **source regions**. In order for a huge mass of air to develop uniform characteristics, its source region should be generally flat and of uniform composition, with light surface winds. The longer the air remains stagnant over its source region, or the longer the path over which the air moves, the more likely it will acquire properties of the surface below. Ideal source regions are usually those areas dominated by surface high pressure, which include the ice- and snow-covered arctic plains in winter and subtropical oceans in summer. The middle latitudes, where surface temperatures and moisture characteristics vary considerably, are not good source regions. Instead, the middle latitudes are a transition zone where air masses with different physical properties move in, clash, and produce an exciting array of weather activity.

● **FIGURE 8.1** Here, a large, extremely cold winter air mass is dominating the weather over much of the United States. At almost all cities, the air is cold and dry. Upper number is air temperature (°F); bottom number is dew point (°F).

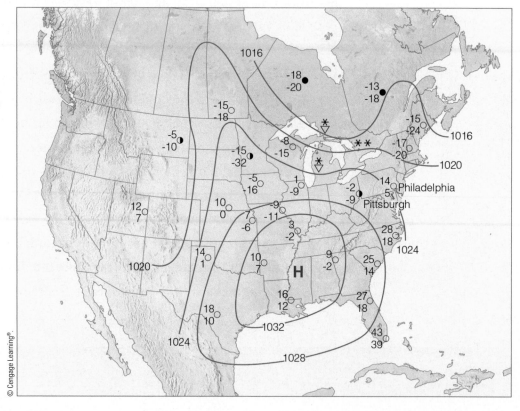

Table 8.1 Air Mass Classification and Characteristics

SOURCE REGION	ARCTIC REGION (A)	POLAR (P)	TROPICAL (T)
Land	cA	cP	cT
Continental (c)	extremely cold, dry, stable; ice- and snow-covered surface	cold, dry, stable	hot, dry, stable air aloft; unstable surface air
Water		mP	mT
Maritime (m)		cool, moist, unstable	warm, moist; usually unstable

CLASSIFICATION Air masses are usually classified according to their temperature and humidity, both of which usually remain fairly uniform in any horizontal direction. There are cold and warm air masses, humid and dry air masses. Air masses are grouped into five general categories according to their source region (see ▼ Table 8.1). Air masses that originate in polar latitudes are designated by the capital letter "P" (for *polar*); those that form in warm tropical regions are designated by the capital letter "T" (for *tropical*). If the source region is land, the air mass will be dry and the lowercase letter "c" (for *continental*) precedes the P or T. If the air mass originates over water, it will be moist—at least in the lower layers—and the lowercase letter "m" (for *maritime*) precedes the P or T. We can now see that polar air originating over land will be classified cP on a surface weather map, whereas tropical air originating over water will be marked as mT. In winter, an extremely cold air mass that forms over the Arctic is designated as cA, *continental arctic*. Sometimes, however, it is difficult to distinguish between arctic and polar air masses, especially when the arctic air mass has traveled over warmer terrain. Table 8.1 lists the five basic air masses.

After the air mass spends some time over its source region, it may begin to move in response to strengthening winds aloft. As it moves away from its source region, the air mass encounters surfaces that may be warmer or colder than itself. When the air mass is colder than the underlying surface, it is warmed from below, which produces instability at low levels. In this case, increased convection and turbulent mixing near the surface usually result in good visibility, cumuliform clouds, and showers of rain or snow. On the other hand, when the air mass is warmer than the surface below, the lower layers are chilled by contact with the cold Earth. Warm air above cooler air produces stable air with little vertical mixing. This situation causes the accumulation of dust, smoke, and pollutants, which restricts surface visibilities. In moist air, stratiform clouds accompanied by drizzle or fog may form.

AIR MASSES OF NORTH AMERICA The principal air masses (with their source regions) that enter the United States are shown in ● Fig. 8.2. We are now in a position to study the formation and modification of each of these air masses and the variety of weather that accompanies them.

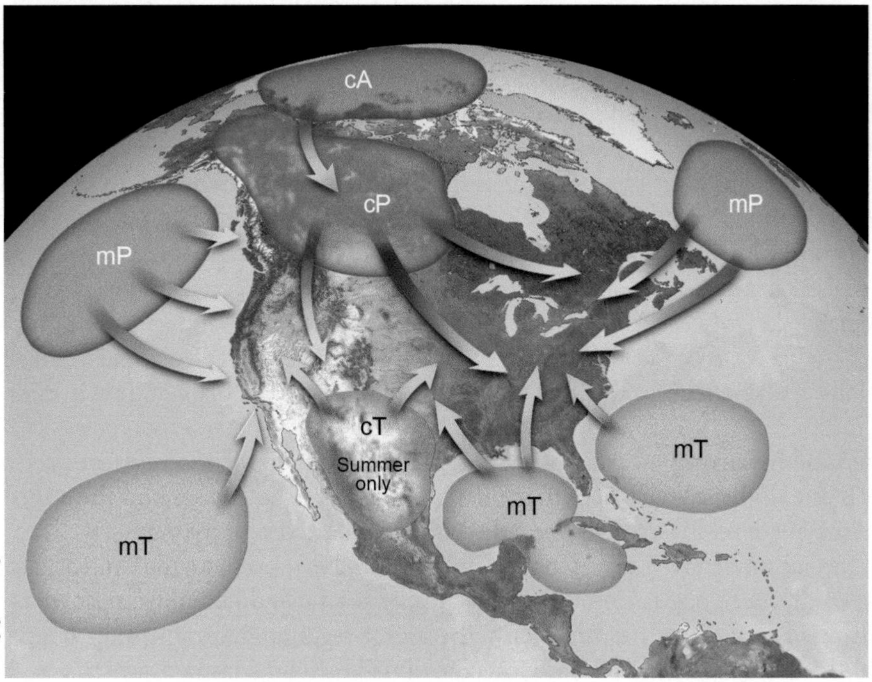

●FIGURE 8.2 Air mass source regions and their paths.

© Cengage Learning®.

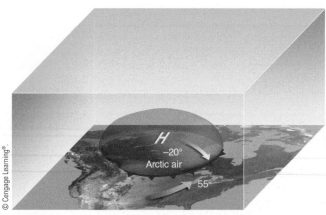

● **FIGURE 8.3** A shallow but large dome of extremely cold air—a continental arctic air mass—moves slowly southeastward across the upper plains. The leading edge of the air mass is marked by a cold front. (Numbers represent air temperature, °F.)

Continental Polar (cP) and Continental Arctic (cA) Air Masses

The bitterly cold weather that invades southern Canada and the United States in winter is associated with **continental polar** and **continental arctic** air masses. These air masses originate over the ice- and snow-covered regions of the Arctic, northern Canada, and Alaska where long, clear nights allow for strong radiational cooling of the surface. Air in contact with the surface becomes quite cold and stable. Since little moisture is added to the air, it is also quite dry. Eventually a portion of this cold air breaks away and, under the influence of the airflow aloft, moves southward as an enormous shallow high-pressure area, as illustrated in ● Fig. 8.3. (As we will see later in this chapter, air masses can be more than 100 times wider than they are tall.)

As the cold air moves into the interior plains, there are no topographic barriers to restrain it, so it continues southward, bringing with it frigid temperatures. As the air mass moves over warmer land to the south, the air temperature moderates slightly. Even during the afternoon, when the surface air is most unstable, cumulus clouds are rare because of the extreme dryness of the air. At night, when the winds die down, rapid radiational surface cooling and clear skies combine to produce low minimum temperatures. If the cold air moves as far south as central or southern Florida, or south Texas, the winter vegetable crop may be severely damaged. When the cold, dry air mass moves over a relatively warm body of water, such as the Great Lakes, heavy snow showers—called **lake-effect snows**—often form on the downwind shores. (More information on lake-effect snows is provided in Focus section 8.1.)

In winter, the generally fair weather accompanying polar continental and arctic air masses is due to the stable nature of the atmosphere aloft. Sinking air develops above the large dome of high pressure. The subsiding air warms by compression and creates warmer air, which lies above colder surface air, often causing a strong upper-level temperature inversion to form. Should the anticyclone stagnate over a region for several days, visibility gradually drops as pollutants become trapped in the cold air near the ground. Usually, however, winds aloft move the cold air mass either eastward or southeastward.

The Rockies, Sierra Nevada, and Cascades normally protect the Pacific Northwest from the onslaught of arctic air, but, occasionally, very cold air masses do invade these regions. When the upper-level winds over Washington and Oregon blow from the north or northeast on a trajectory beginning over northern Canada or Alaska, cold air can slip over the mountains and extend its icy fingers all the way to the Pacific Ocean. As the air moves off the high plateau, over the mountains, and on into the lower valleys, compressional heating of the sinking air causes its temperature to rise, so that by the time it reaches the lowlands, it is considerably warmer than it was originally. However, in no way would this air be considered warm. In some cases, the subfreezing temperatures slip over the Cascades and extend southward into the coastal areas of southern California.

A similar but less dramatic warming of continental polar and arctic air masses occurs along the east coast of the United States. Air rides up and over the lower Appalachian Mountains. Turbulent mixing and compressional heating increase the air temperatures on the downwind side, with the result that cities located to the east of the Appalachian Mountains usually do not experience temperatures as low as those on the west side. In Fig. 8.1 on p. 210, notice that for the same time of day—in this case 7 a.m. (EST)—Philadelphia, on the eastern side of the mountains, with an air temperature of 14°F, is 16°F warmer than Pittsburgh, at −2°F, on the western side of the mountain.

● Figure 8.4 shows two upper-air patterns that led to extremely cold outbreaks of arctic air during December 1989 and 1990. Upper-level winds typically blow from west to east, but, in both of these cases, the flow, as indicated by the heavy, dark arrows, had a strong north-south (meridional) trajectory. The H represents the positions of the cold surface anticyclones. Numbers on the map represent minimum temperatures (°F) recorded during the cold spells. East of the Rocky Mountains, over 350 record low temperatures were set between December 21 and 24, 1989, with the arctic outbreak causing an estimated $480 million in damage to the fruit and vegetable crops in Texas and Florida. Along the West Coast, the frigid air during December 1990 caused over $300 million in damage to the vegetable and citrus crops, as temperatures over parts of California plummeted to their lowest readings in more than fifty years. Notice in both cases how the upper-level wind directs the paths of the air masses.

Continental polar air that moves into the United States in summer has properties much different from its winter counterpart. The source region remains the same but the air is now accompanied by long summer days that melt

Lake-Effect (Enhanced) Snows

During the winter, when the weather in the Midwest is dominated by clear and cold polar or arctic air, people living on the eastern shores of the Great Lakes brace themselves for heavy snow showers. Snowstorms that form on the downwind side of one of these lakes are known as *lake-effect snows*. (Since the lakes are responsible for enhancing the amount of snow that falls, these snowstorms are also called *lake-enhanced snows*, especially when the snow is associated with a cold front or mid-latitude cyclone.) Such storms are highly localized, extending from just a few kilometers to more than 100 km inland. The snow usually falls as a heavy shower or squall in a concentrated zone. So centralized is the region of snowfall that one part of a city may accumulate many inches of snow, while, in another part, the ground is bare. The amount of snow that falls can be enormous: for example, 65 inches fell near Buffalo, New York (on the eastern side of Lake Ontario), in less than 48 hours during November 2014.

Lake-effect snows are most numerous from November to January. During these months, cold air moves over the lakes when they are relatively warm and often not frozen. The contrast in temperature between water and air can be as much as 25°C (45°F). Studies show that the greater the contrast in temperature, the greater the potential for snow showers. In ● Fig. 1, we can see that as the cold air moves over the warmer water, the air mass is quickly warmed from below, making it more

buoyant and less stable. Rapidly, the air sweeps up moisture, soon becoming saturated. Out over the water, the vapor condenses into steam fog. As the air continues to warm, it rises and forms billowing cumuliform clouds, which continue to grow as the air becomes more unstable. Eventually, these clouds produce heavy showers of snow, which make the lake seem like a snow factory. Once the air and clouds reach the downwind side of the lake, additional lifting is provided by low hills and the convergence of air as it slows down over the rougher terrain. In late winter, the frequency and intensity of lake-effect snows often taper off as the temperature contrast between water and air diminishes and larger portions of the lakes freeze.

Generally, the longer the stretch of water over which the air mass travels (the longer the fetch), the greater the amount of warmth and moisture derived from the lake, and the greater the potential for heavy snow showers. Consequently, forecasting lake-effect snowfalls depends to a large degree on determining the trajectory of the air as it flows over the lake. Regions that experience heavy lake-effect snowfalls are shown in ● Fig. 2.*

As the cold air moves farther east, the heavy snow showers usually taper off;

● **FIGURE 2** Areas shaded white show regions that experience heavy lake-effect snows.

however, the western slope of the Appalachian Mountains produces further lifting, enhancing the possibility of more and heavier showers. The heat given off during condensation warms the air and, as the air descends the eastern slope, compressional heating warms it even more. Snowfall ceases, and by the time the air arrives in Philadelphia, New York, or Boston, the only remaining trace of the snow showers occurring on the other side of the mountains is the puffy cumulus clouds drifting overhead.

Lake-effect (or enhanced) snows are not confined to the Great Lakes. In fact, any large unfrozen lake (such as the Great Salt Lake) can enhance snowfall when cold, relatively dry air sweeps over it. Moreover, a type of lake-effect snow occurs when cold air moves over a relatively warm ocean, then lifts slightly as it moves over a landmass. Such *ocean-effect snows* are common over Cape Cod, Massachusetts, in winter.

---

*Buffalo, New York, is a city that experiences heavy lake-effect snows. Visit the National Weather Service website in Buffalo at http://www.weather.gov/buf/lakeeffect and read about lake-effect snowstorms measured in feet, and other interesting weather stories.

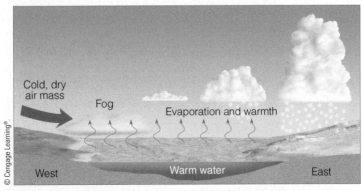

● **FIGURE 1** The formation of lake-effect snows. Cold, dry air crossing the lake gains moisture and warmth from the water. The more buoyant air now rises, forming clouds that deposit large quantities of snow on the lake's leeward (downwind) shores.

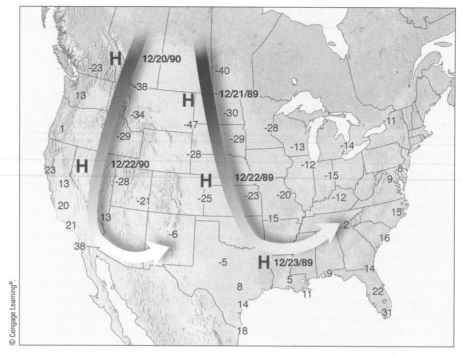

● **FIGURE 8.4** Average upper-level wind flow (heavy arrows) and surface position of anticyclones (H) associated with two extremely cold outbreaks of arctic air during late December 1989 and 1990. Numbers on the map represent some minimum temperatures (°F) measured during each cold snap.

snow and warm the land. The air is only moderately cool, and surface evaporation adds water vapor to the air. A summertime continental polar air mass usually brings relief from the oppressive heat often occurring in the central and eastern states, as cooler air lowers the air temperature to more comfortable levels. Daytime heating warms the lower layers, producing surface instability. With its added moisture, the water vapor in the rising air may condense and create a sky dotted with fair-weather cumulus clouds.

When an air mass moves over a large body of water, its original properties can change considerably. For instance, cold, dry continental polar air moving over the Gulf of Mexico warms rapidly and gains moisture. The air quickly assumes the qualities of a maritime air mass. Notice in ● Fig. 8.5 that rows of cumulus clouds are forming over the Gulf of Mexico parallel to northerly surface winds as continental polar air is being warmed by the water beneath

it, causing the air mass to destabilize. As the air continues its journey southward into Mexico and Central America, strong, moist northerly winds build into heavy clouds (bright white area) and showers along the northern coast. In this way, a once cold, dry, and stable air mass can be modified to such an extent that its original characteristics are no longer discernible. When this happens, the air mass is often given a new designation.

In summary, polar and arctic air masses are responsible for the bitterly cold winter weather that can cover wide sections of North America. When the air mass originates over the Canadian Northwest Territories, frigid air can bring record-breaking low temperatures. Such was the case on Christmas Eve 1983, when arctic air covered most of North America. (A detailed look at this air mass and its accompanying record-setting low temperatures is given in Focus section 8.2.)

● **FIGURE 8.5** Visible satellite image showing the modification of cold continental polar air as it moves over the warmer Gulf of Mexico and the Atlantic Ocean.

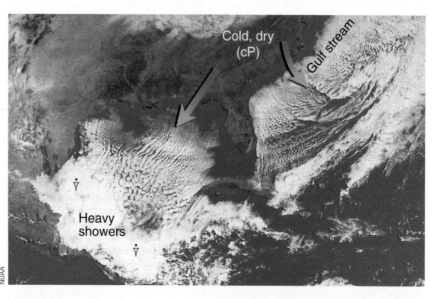

The Return of the Siberian Express

The winter of 1983–1984 was one of the coldest on record across North America. Unusually frigid weather arrived in December, which ended up as the coldest December for the contiguous 48 states since records began in 1895 (a title it still holds). During the first part of the month, continental polar air covered most of the northern and central plains. As the cold air moderated slightly, far to the north a huge mass of bitterly cold arctic air was forming over the frozen reaches of the Canadian Northwest Territories.

By mid-month, the frigid air, associated with a massive high-pressure area, covered all of northwest Canada. Meanwhile, aloft, strong northerly winds directed the leading edge of the frigid air southward over the prairie provinces of Canada and southward into the United States. The extraordinarily cold air was accompanied in some regions by winds gusting to 45 knots. At least one news reporter labeled the onslaught, "The Siberian Express." (The term has since become common in news coverage of cold waves in the United States, although it does not necessarily mean the air mass originates in Siberia.)

In many locations across the United States, the Siberian Express dropped temperatures to the lowest readings ever recorded during the month of December. On December 22, Elk Park, Montana, recorded an unofficial low of –64°F, only 6°F higher than the all-time low of –70°F for the United States (excluding Alaska), which was recorded at Rogers Pass, Montana, on January 20, 1954.

The center of the massive anticyclone gradually pushed southward out of Canada. By December 24, its center was over eastern Montana (see ●Fig. 3), where the sea-level pressure at Miles City reached an incredible 1064 mb (31.42 in.), a record for the 48 contiguous states. An enormous ridge of high pressure stretched from the Canadian arctic coast to the Gulf of Mexico. On the east side of the ridge, cold westerly winds brought lake-effect snows to the eastern shores of the Great Lakes. To the south of the high-pressure center, cold easterly winds, rising along the elevated plains, brought light amounts of *upslope*

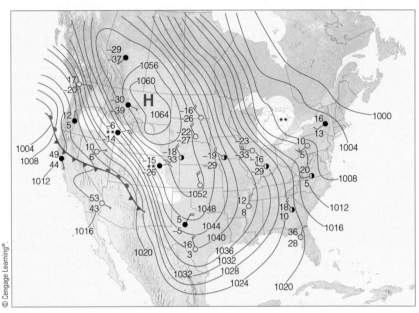

●FIGURE 3 Surface weather map for 7 a.m., EST, December 24, 1983. Solid lines are isobars. Areas shaded white represent snow. An extremely cold arctic air mass covers nearly 90 percent of the United States. (Weather symbols for the surface map are given in Appendix C.)

*snow** to sections of the Rocky Mountain states. Notice in Fig. 3 that, on Christmas Eve, arctic air covered almost 90 percent of the United States. As the cold air swept eastward and southward, a hard freeze caused hundreds of millions of dollars in damage to the fruit and vegetable crops in Texas, Louisiana, and Florida. On Christmas Day, 125 record low temperature readings were set in twenty-four states. That afternoon, at 1:00 p.m., it was actually colder in Atlanta, Georgia, at 9°F, than it was in Fairbanks, Alaska (10°F). One of the worst cold waves to occur in December during the twentieth century continued through the week, as many new record lows were established in the Deep South from Texas to Louisiana.

By January 1, the extreme cold had moderated, as the upper-level winds became more westerly. These winds brought milder Pacific air eastward into the Great Plains. The warmer pattern continued until about January 10, when the Siberian Express decided to make a return visit. Driven by strong upper-level northerly winds, impulse after impulse of arctic air from Canada swept across the United

*Upslope snow forms as cold air moving from east to west gradually rises (and cools even more) as it approaches the Rocky Mountains.

States. On January 18, a low of –65°F was recorded at Middle Sinks, Utah. On January 19, temperatures plummeted to a new low of –7°F for the airports in Philadelphia and Baltimore, tying the coldest readings observed on any date at these locations. Toward the end of the month, the upper-level winds once again became more westerly. Over much of the nation, the cold air moderated. But the Express was to return at least one more time.

The beginning of February saw relatively warm air covering much of the United States from California to the Atlantic coast. However, on February 4, an arctic outbreak spread southward and eastward across the United States. Although freezing air extended southward into central Florida, the Express ran out of steam, and a February heat wave soon engulfed most of the United States east of the Rocky Mountains as warm, humid air from the Gulf of Mexico spread northward.

Even though February 1984 was a warmer-than-normal month over much of the United States, the winter of 1983–1984 (December, January, and February) will go down in the record books as one of the coldest winters for the United States as a whole since reliable record keeping began.

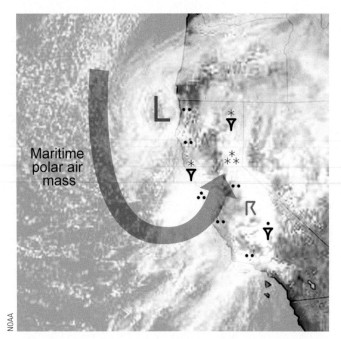

● **FIGURE 8.6** Clouds and airflow aloft (large blue arrow) associated with maritime polar air moving into California. The large L shows the position of an upper-level low. Regions experiencing precipitation are also shown. The small, white clouds over the open ocean are cumulus clouds forming in the conditionally unstable air mass. (Precipitation symbols are given in Appendix C at the back of the book.)

Maritime Polar (mP) Air Masses During the winter, polar and arctic air originating over Asia and frozen polar regions is carried eastward and southward over the Pacific Ocean by the circulation around the prevailing Aleutian low. The ocean water modifies these cold air masses by adding warmth and moisture to them. Since this air travels over water many hundreds or even thousands of kilometers, it gradually changes into a **maritime polar** air mass.

By the time this air mass reaches the Pacific coast, it is cool, moist, and conditionally unstable. The ocean's effect is to keep air near the surface warmer than the air aloft. Temperature readings in the 40s and 50s (°F) are common near the surface, while air at an altitude of about a kilometer or so may be at the freezing point. Within this colder air, characteristics of the original cold, dry air mass may still prevail. As the air moves inland, coastal mountains force it to rise,

and much of its water vapor condenses into rain-producing clouds. In the colder air aloft, the rain changes to snow, with heavy amounts accumulating in mountain regions. Over the relatively warm open ocean, the cool moist air mass produces cumulus clouds that show up as tiny white splotches on a visible satellite image (see ● Fig. 8.6).

When maritime polar air from the Pacific moves into North America, it loses much of its moisture as it crosses a series of mountain ranges. Beyond these mountains, it travels over a cold, elevated plateau that chills the surface air and slowly transforms the lower level into dry, stable continental polar air. East of the Rockies this air mass is referred to as **Pacific air** (see ● Fig. 8.7). Here, it often brings fair weather and temperatures that are cool but not nearly as cold as the continental polar and arctic air that invades this region from northern Canada. In fact, when Pacific air from the west replaces retreating cold air from the north, chinook winds often develop (see p. 181). Furthermore, when the modified maritime polar air replaces moist subtropical air, thunderstorms can form along the boundary separating the two air masses.

Along the east coast of North America, maritime polar air originates in the North Atlantic when modified continental polar air moves over a large body of water. Because the water of the North Atlantic is very cold and the air mass travels only a short distance over water, wintertime Atlantic maritime polar air masses are usually much colder than their Pacific counterparts. Because the prevailing winds aloft are westerly, Atlantic maritime polar air masses are also much less likely to move into the United States, although they often affect Europe. ● Figure 8.8 illustrates a typical late winter or early spring surface weather pattern that carries maritime polar air from the Atlantic into the New England and middle Atlantic states. A slow-moving, cold anticyclone drifting to the east (north of New England) causes a northeasterly onshore flow of cold, moist air to the south. The boundary separating this invading colder air from warmer air even farther south is marked by a stationary front. North of this front, northeasterly winds provide generally undesirable weather, consisting of damp air and low, thick clouds from which light precipitation falls in the form of rain, drizzle, or snow. As we will see

● **FIGURE 8.7** After crossing several mountain ranges, cool, moist, maritime polar air from off the Pacific Ocean descends the eastern side of the Rockies as modified, relatively dry Pacific air.

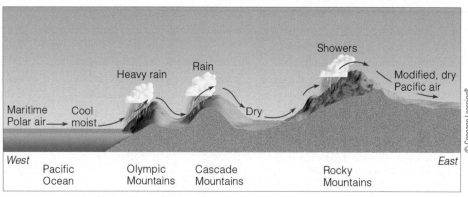

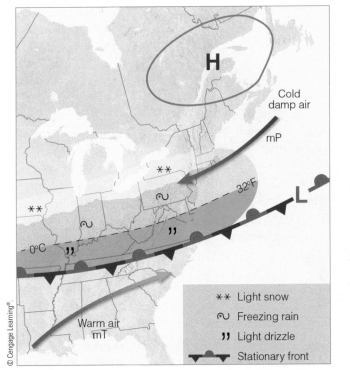

● **FIGURE 8.8** Winter and early spring surface weather pattern that usually prevails during the invasion of cold, moist maritime polar (mP) air into the mid-Atlantic and New England states. (Green-shaded area represents light rain and drizzle; pink-shaded region represents freezing rain and sleet; white-shaded area is experiencing snow.)

later in this chapter, when upper atmospheric conditions are right, mid-latitude cyclonic storms may develop along the stationary front, move eastward, and intensify near the shores of Cape Hatteras. Such storms, called *Hatteras*

lows (see Fig. 8.28a, p. 232), sometimes swing northeastward along the coast, where they become *northeasters* (or *nor'easters*) bringing with them strong northeasterly winds, heavy rain or snow, and coastal flooding. (We will examine northeasters later in this chapter when we examine mid-latitude cyclonic storms.)

Maritime Tropical (mT) Air Masses The wintertime source region for Pacific **maritime tropical** air masses is the subtropical central and eastern Pacific Ocean. (Look back at Fig. 8.2, p. 211.) Air from this region must travel over many kilometers of water before it reaches the California coast. Consequently, these air masses are often very warm and moist by the time they arrive along the West Coast. In winter, the warm air produces heavy precipitation usually in the form of rain, even at high elevations. Melting snow and rain quickly fill rivers, which overflow into the low-lying valleys. The rapid snowmelt leaves local ski slopes barren, and the heavy rain can cause disastrous mud slides in the steep canyons.

● Figure 8.9 shows maritime tropical air (usually referred to as *subtropical air*) streaming into northern California on January 1, 1997. The flow of humid, subtropical air, which originated near the Hawaiian Islands, was termed by at least one forecaster the "Pineapple Express," a term that has since become common in media coverage. After battering the Pacific Northwest with heavy rain, the Pineapple Express roared into northern and central California, causing catastrophic floods that sent over 100,000 people fleeing from their homes, landslides that closed roads, property damage (including crop losses)

● **FIGURE 8.9** An infrared satellite image that shows maritime tropical air (heavy yellow arrow) moving into northern California on January 1, 1997. The warm, humid airflow (sometimes called the "Pineapple Express") produced heavy rain and extensive flooding in northern and central California.

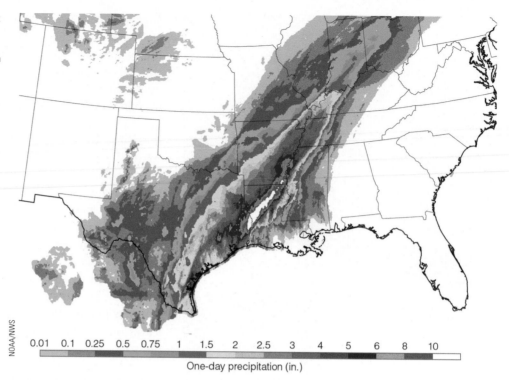

● FIGURE 8.10 An atmospheric river streaming north from the Gulf of Mexico brought torrential rain to parts of Louisiana and nearby states on March 10, 2016. Rainfall totals for the month of March (not shown) exceeded 20 inches in parts of Louisiana.

NOAA/NWS

| 0.01 | 0.1 | 0.25 | 0.5 | 0.75 | 1 | 1.5 | 2 | 2.5 | 3 | 4 | 5 | 6 | 8 | 10 |

One-day precipitation (in.)

that amounted to more than $1.5 billion, and eight fatalities. Yosemite National Park, which sustained over $170 million in damages due mainly to flooding, was forced to close for more than two months.

The Pineapple Express is one example of the feature known as an **atmospheric river**. Like a river of moisture far above ground, an atmospheric river transports huge amounts of water vapor from one place to another, typically from the warm, humid tropics and subtropics into the mid-latitudes. A typical atmospheric river is 400 to 600 km wide and a few kilometers deep. The strongest atmospheric rivers can transport up to 15 times more moisture in the form of water vapor than the amount of water flowing into the Gulf of Mexico at the mouth of the Mississippi River. Along the west coast of the United States, a few atmospheric river events each year can provide as much as half of all annual precipitation. A powerful atmospheric river in early December 2012 brought more than 20 in. of rain to parts of northern California, with wind gusts as high as 150 mi/hr. Atmospheric rivers also flow north from lower latitudes to strike the central and eastern United States, such as the one that brought 10 to 20 in. of rain and devastating floods to parts of Texas and Louisiana in March 2016 (see ● Fig. 8.10).

The humid subtropical air that influences much of the weather east of the Rockies originates over the Gulf of Mexico and Caribbean Sea. In winter, cold polar and arctic air tends to dominate the continental weather scene, so maritime tropical air is usually confined to the Gulf and extreme southern states. Occasionally, a slow-moving cyclonic storm system over the Central Plains draws warm, humid air northward. Gentle south or southwesterly

winds carry this air into the central and eastern parts of the United States in advance of the system. Since the land is still extremely cold, air near the surface is chilled to its dew point. Fog and low clouds form in the early morning, dissipate by midday, and re-form in the evening. This mild winter weather in the Mississippi and Ohio valleys lasts, at best, only a few days. Soon cold air will move down from the north behind the eastward-moving storm system. Along the boundary between the two air masses, the warm, humid air is lifted above the more-dense cold, polar air, which often leads to heavy and widespread precipitation and storminess.

When a large, mid-latitude cyclonic storm system stalls over the Central Plains, a constant supply of warm, humid air from the Gulf of Mexico can bring record-breaking maximum temperatures to the eastern half of the country. Sometimes the air temperatures are higher in the mid-Atlantic states than they are in the Deep South, as compressional heating warms the air even more as it moves downslope after crossing the Appalachian Mountains.

● Figure 8.11 shows a surface weather map and the associated upper-level airflow (heavy arrow) that brought unseasonably warm maritime tropical air into the central and eastern states during March 2012. A large surface high-pressure area centered off the east coast coupled with a strong southwesterly flow aloft carried warm, moist air into the Midwest and East, causing a record-breaking March heat wave. The flow aloft prevented the surface low and the cooler air behind it from making much eastward progress, so that the warm spell lasted a week or more in some locations. Michigan saw its first ever 90°F reading in March, and Burlington, Vermont, reached 80°F. More

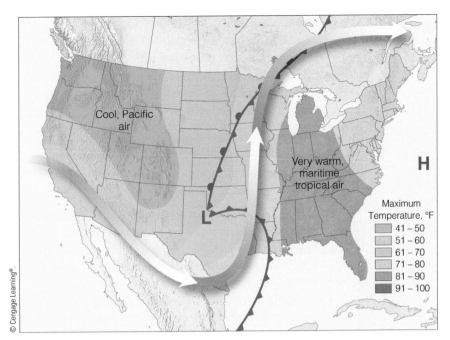

● FIGURE 8.11 Weather conditions during an unseasonably warm spell in the eastern portion of the United States that occurred in the second half of March 2012. The surface low-pressure area, fronts, and maximum temperatures are shown for March 21. The heavy arrow is the average upper-level flow during the warm period.

than 7000 daily record highs were set or tied in the United States during March. Note that, on the west side of the surface low, the winds aloft funneled cool Pacific air into the western states, bringing chilly conditions from California to the Rockies. Hence, while people in the Northwest were huddled around heaters, others several thousand miles away in the Midwest and Northeast were soaking up sunshine and warmth. We can see that it is the upper-level flow, directing cooler air southward and warm subtropical air northward, that makes these contrasts in temperature possible.

In summer, the circulation of air around the Bermuda High (which sits off the southeast coast of North America—see Fig. 7.30, p. 194) pumps warm, humid maritime tropical air northward from off the Gulf of Mexico and from off the Atlantic Ocean into the eastern half of the United States. As this humid air moves inland, it warms even more, rises, and frequently condenses into cumuliform clouds, which produce afternoon showers and thunderstorms. You can count on thunderstorms developing along the Gulf Coast almost every summer afternoon. As evening approaches, thunderstorm activity typically dies off. Nighttime cooling lowers the temperature of this hot, muggy air only slightly. Should the air become saturated, fog or low clouds usually form, and these normally dissipate by late morning as surface heating warms the air again.

A weak, but often persistent, flow around an upper-level anticyclone in summer will spread warm, humid tropical air from the Gulf of Mexico or from the Gulf of California into the southern and central Rockies, where it causes afternoon thunderstorms. Occasionally, this easterly flow can work its way even farther west, producing shower activity in the otherwise dry southwestern desert.

Humid subtropical air that originates over the southeastern Pacific and Gulf of California remains south of the United States during most of the year. However, during the summer monsoon period, a weak upper-level southerly flow can spread this humid air northward into the southwestern United States, most often in Arizona, Nevada, and southern California. In many places, the moist, conditionally unstable air aloft only shows up as middle and high cloudiness. However, where the moist flow meets a mountain barrier, it tends to rise and condense into towering shower-producing clouds. If the air is sufficiently moist, it can cause heavy showers over a broad area. (For an illustration of exceptionally strong flow of subtropical air into this region, see Fig. 7.22 on p. 187.)

Continental Tropical (cT) Air Masses   The only real source region for hot, dry **continental tropical** air masses in North America is found during the summer in northern Mexico and the adjacent arid southwestern United States (see Fig. 8.2, p. 211). Here, the air mass is hot, dry, and conditionally unstable at low levels, with frequent dust devils forming during the day. Because of the low relative humidity (typically less than 10 percent during

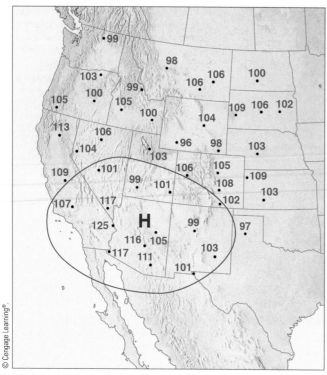

● **FIGURE 8.12** From July 14 through July 22, 2005, continental tropical air covered a large area of the southwestern United States. Numbers on the map represent maximum temperatures (°F) during this period. The large H with the isobar shows the upper-level position of the subtropical high. Sinking air associated with the high contributed to the hot, dry weather. Winds aloft were weak, with the main flow over central Canada.

the afternoon), air must rise more than 3000 m (10,000 ft) before condensation begins. Furthermore, an upper-level ridge usually produces sinking air over the region, tending to make the air aloft rather stable and the surface air even warmer. Consequently, skies are generally clear, the weather is hot, and rainfall is practically nonexistent where continental tropical air masses prevail. If this air mass moves outside its source region and into the Great Plains and stagnates over that region for any length of time, a severe drought may result. ● Fig. 8.12 shows an instance where continental tropical air covered a large portion of the southwestern United States and produced hot, dry weather during July 2005.

So far, we have examined the various air masses that enter North America annually. The characteristics of each depend upon the air mass source region and the type of surface over which the air mass moves. The winds aloft determine the trajectories of these air masses. Occasionally, an air mass will control the weather in a region for some time. These persistent weather conditions are sometimes referred to as *airmass weather*.

Airmass weather is especially common in the southeastern United States during summer as, day after day, humid subtropical air from the Gulf brings sultry conditions and afternoon thunderstorms. It is also common in the Pacific Northwest in winter when conditionally

unstable, cool maritime air accompanied by widely scattered showers dominates the weather for several days or more. The real weather action, however, usually occurs not within air masses but at their margins, where air masses with sharply contrasting properties meet—in the zone marked by weather fronts.*

BRIEF REVIEW

Before we examine fronts, here is a review of some of the important facts about air masses:

- An air mass is a large body of air whose properties of temperature and humidity are fairly similar in any horizontal direction.
- Source regions for air masses tend to be generally flat, of uniform composition, and in an area of light winds dominated by surface high pressure.
- Continental air masses form over land. Maritime air masses form over water. Polar air masses originate in cold, polar regions, and extremely cold arctic air masses form over arctic regions. Tropical air masses originate in warm, tropical regions.
- Continental polar (cP) air masses are cold and dry; continental arctic (cA) air masses are extremely cold and dry. Continental arctic air masses produce the extreme cold of winter as they move across North America.
- Continental tropical (cT) air masses are hot and dry, and are responsible for the heat waves of summer in the western half of the United States.
- Maritime polar (mP) air masses are cold and moist; these air masses are responsible for the cold, damp, and often wet weather along the northeast coast of North America, as well as for the cold, rainy winter weather along the west coast of North America.
- Maritime tropical (mT) air masses are warm and humid, and are responsible for the hot, muggy weather that frequently plagues the eastern half of the United States in summer.

Fronts

Although we briefly looked at fronts in Chapter 1, we are now in a position to study them in depth, which will aid us in forecasting the weather. We will now learn about the general nature of fronts—how they move and what weather patterns are associated with them.

A **front** is the transition zone between two air masses of different densities. Since density differences are most often caused by temperature differences, fronts usually separate air masses with contrasting temperatures. Often, they separate air masses with different humidities as well. Remember that air masses have both horizontal and vertical extent; consequently, the upward extension of a front is referred to as a *frontal surface*, or a *frontal zone*.

---

*The word *front* is used to denote the clashing or meeting of two air masses, probably because it resembles the fighting in Western Europe during World War I, when the term originated.

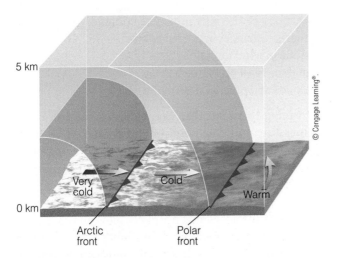

● FIGURE 8.13 The polar front represents a cold frontal boundary that separates colder air from warmer air at the surface and aloft. The more shallow arctic front separates cold air from extremely cold air.

● Figure 8.13 illustrates the vertical extent of two frontal zones—the polar front and the arctic front. The polar front boundary, which extends upward more than 5 km (3 mi), separates warm, humid air to the south from cold polar air to the north. The arctic front, which separates cold air from extremely cold arctic air, is much shallower than the polar front and only extends upward to an

altitude of about one or two kilometers. In the next several sections, as we examine individual fronts on a flat surface weather map, keep in mind that all fronts have horizontal and vertical extent.

● Figure 8.14 shows a surface weather map illustrating four different fronts. Notice that the fronts are associated with lower pressure and that the fronts separate differing air masses. As we move from west to east across the map, the fronts appear in the following order: a stationary front between points A and B; a cold front between points B and C; a warm front between points C and D; and an occluded front between points C and L. Let's examine the properties of each of these fronts.

STATIONARY FRONTS A **stationary front** has essentially no movement.* On a colored weather map, it is drawn as an alternating red and blue line. Semicircles face toward colder air on the red line and triangles point toward warmer air on the blue line. The stationary front between points A and B in Fig. 8.14 marks the boundary where cold, dense continental polar (cP) air from Canada butts up against the north-south trending Rocky Mountains. Unable to cross the barrier, the cold air shows little

---

*They are usually called *quasi-stationary fronts* because they can show some movement.

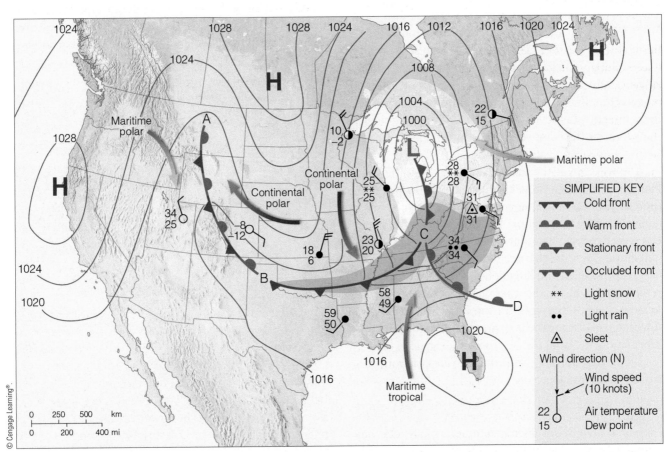

● FIGURE 8.14 A weather map showing surface pressure systems, air masses, fronts, and isobars (in millibars) as solid gray lines. Large arrows in color show airflow. (Green-shaded area represents rain; pink-shaded area represents freezing rain and sleet; white-shaded area represents snow.)

or no westward movement. The stationary front is drawn along a line separating the continental polar air from the milder, more humid maritime polar air to the west. Notice that the surface winds tend to blow parallel to the front, but in opposite directions on either side of it. Moreover, upper-level winds often blow parallel to a stationary front.

The weather along the front is clear to partly cloudy, with much colder air lying on its eastern side. Because both air masses are relatively dry, there is no precipitation. This is not, however, always the case. When warm, moist air rides up and over the cold air, widespread cloudiness with light precipitation can cover a vast area. These are the conditions that prevail north of the east-west running stationary front depicted in Fig. 8.8, p. 217. In some cases when a stationary front butts up against a mountain range, as shown in Fig. 8.14, winds blowing upslope can generate light rain or snow (called *upslope precipitation*) if there is enough moisture in the air.

If the colder air to the east begins to retreat and is replaced by the warmer air to the west, the front in Fig. 8.14 will no longer remain stationary; it will become a warm front. If, on the other hand, the colder air slides up over the mountain and replaces the warmer air on the other side, the front will become a cold front. If either a cold front or a warm front stops moving, it becomes a stationary front.

COLD FRONTS The **cold front** between points B and C on the surface weather map in Fig. 8.14 represents a zone where cold, dry, stable polar air is replacing warm, moist, conditionally unstable subtropical air. The front is drawn as a solid blue line with the triangles along the front showing its direction of movement. How did the meteorologist know to draw the front at that location? A closer look at the situation will give us the answer.

The weather in the immediate vicinity of this cold front in the southern United States is shown in ● Fig. 8.15. The data plotted on the map represent the current weather at selected cities. The station model used to represent the data at each reporting station is a simplified one that shows temperature, dew point, present weather, cloud cover, sea-level pressure, and wind direction and speed. The little line in the lower right-hand corner of each station shows the pressure change—the *pressure tendency*, whether rising (/) or falling (\)—during the last three hours. With all of this information, the front can be properly located.*

We can use the following criteria to locate a front on a surface weather map:

1. sharp temperature changes over a relatively short distance

2. changes in the air's moisture content (as shown by marked changes in the dew point)
3. shifts in wind direction
4. pressure and pressure changes
5. clouds and precipitation patterns

In Fig. 8.15, we can see a large contrast in air temperature and dew point on either side of the front. There is also a wind shift from southwesterly ahead of the front, to northwesterly behind it. Notice that each isobar kinks as it crosses the front, forming an elongated area of low pressure—a *trough*—which accounts for the wind shift. Since surface winds normally blow across the isobars toward lower pressure, we find winds with a southerly component ahead of the front and winds with a northerly component behind it.

Since the cold front is a trough of low pressure, sharp changes in pressure can be significant in locating the front's position. One important fact to remember is that the lowest pressure usually occurs just as the front passes a station. Notice that, as you move toward the front, the pressure drops, and, as you move away from it, the pressure rises.

The precipitation pattern along the cold front in Fig. 8.15 might appear similar to the Doppler radar image shown in ● Fig. 8.16. The region in color extending from northeast to southwest represents precipitation along a cold front. Notice that light-to-moderate rain (colored in green) occurs over a wide area along the front, while the heavier precipitation (yellow) tends to occur in a narrow

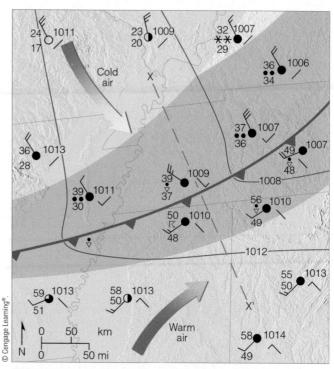

● **FIGURE 8.15** A closer look at the surface weather associated with the cold front situated in the southern United States in Fig. 8.14. (Gray lines are isobars. Green-shaded area represents rain; white-shaded area represents snow.)

---

*Locating any front on a weather map is not always a clear-cut process. Even meteorologists can disagree on an exact position.

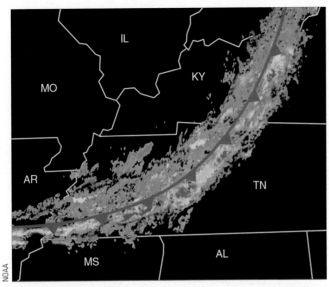

NOAA

● **FIGURE 8.16** A Doppler radar image showing precipitation patterns along a cold front similar to the cold front in Fig. 8.15. Green represents light-to-moderate precipitation; yellow represents heavier precipitation; and red the most likely areas for thunderstorms. (The cold front is superimposed on the radar image.)

freezing level dips as it crosses the front.) The winds shift from southwesterly to northwesterly, pressure rises, and precipitation ends. As the air dries out, the skies clear, except for a few lingering cumulus clouds.

Observe that the leading edge of the front is steep. The steepness is due to friction, which slows the airflow near the ground. The air aloft pushes forward, blunting the frontal surface. Even so, the leading edge of the front does not extend very high into the atmosphere. If we could walk from where the front touches the surface back into the cold air, a distance of 50 km, the front would be about 1 km above us. Thus, the slope of the front—the ratio of vertical rise to horizontal distance—is 1:50. This is typical for a fast-moving cold front, a front that moves at about 25 knots or more. In a slower-moving cold front—one that moves at about 15 knots—the slope is much more gentle. With slow-moving cold fronts, clouds and precipitation usually cover a broad area behind the front. When the ascending warm air is stable, stratiform clouds, such as nimbostratus, become the predominant cloud type and fog may even develop in the rainy area. Occasionally, along a fast-moving front, a line of active showers and thunderstorms, called a *squall line*, develops parallel to and often ahead of the advancing front. Scattered thunderstorms may also occur well ahead of a cold front.

So far, we have considered the general weather patterns of "typical" cold fronts. There are, of course, many exceptions. In fact, no two fronts are exactly alike. In some,

band along the front itself. Strong thunderstorms (red) are found only in certain areas along the front.

The cloud and precipitation patterns in Fig. 8.15 along the line X–X' are shown in a side view of the front in ● Fig. 8.17. We can see in Fig. 8.17 that, at the front, the cold, dense air wedges under the warm air, forcing the warm air upward, much like a snow shovel forces snow upward as it glides through the snow. As the moist, conditionally unstable air rises, it condenses into a series of cumuliform clouds. Strong, upper-level westerly winds blow the delicate ice crystals (which form near the top of the cumulonimbus) into cirrostratus (Cs) and cirrus (Ci). These clouds usually appear far in advance of the approaching front. At the front itself, a relatively narrow band of thunderstorms (Cb) produces heavy showers with gusty winds. Behind the front, the air cools quickly. (Notice how the

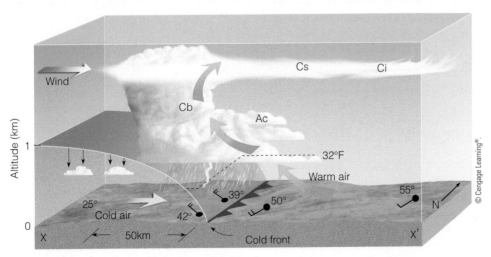

© Cengage Learning®.

● **FIGURE 8.17** A model representing vertical view of a model representing weather across the cold front in Fig. 8.15 along the line X–X′.

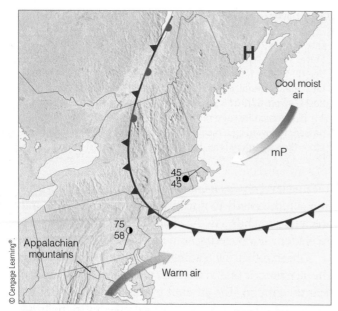

● **FIGURE 8.18** A "back door" cold front moving into New England during the spring. Notice that, behind the front, the weather is cold and damp with drizzle, while to the south, ahead of the front, the weather is partly cloudy and warm.

described in the previous section, are rarely if ever seen. In fact, as a cold front moves inland from the Pacific Ocean, the surface temperature contrast across the front may be quite small. Topographic features usually distort the wind pattern so much that locating the position of the front and the time of its passage are exceedingly difficult. In this case, the pressure tendency is the most reliable indication of a frontal passage.

Most cold fronts move toward the south, southeast, or east. But sometimes they will move southwestward out of Canada into the northeastern United States. Cold fronts that move in from the east, or northeast, are called **"back door" cold fronts**. Typically, as the front passes, westerly surface winds shift to easterly or northeasterly, and temperatures drop (see ● Fig. 8.18).

Even though cold-front weather patterns have many exceptions, learning these patterns can be to your advantage if you live in an area that experiences well-defined cold fronts. Knowing them improves your own ability to make short-range weather forecasts. For your reference, ▼ Table 8.2 summarizes idealized cold-front weather in winter in the Northern Hemisphere.

the cold air is shallow; in others, it is much deeper. If the rising warm air is dry and stable, scattered clouds are all that form, and there is no precipitation. In extremely dry weather, a marked change in the dew point, accompanied by a slight wind shift, may be the only clue to a passing cold front.

During the winter, a series of cold polar or arctic outbreaks may travel across the United States so quickly that warm air is unable to develop ahead of the front. In this case, frigid arctic air associated with the arctic front usually replaces cold polar air, and a drop in temperature is the only indication that a cold front has moved through your area. Along the West Coast, the Pacific Ocean modifies the air so much that typical cold fronts, such as those

WARM FRONTS   In Fig. 8.14, p. 221, a **warm front** is drawn along the solid red line running from points C to D. Here, the leading edge of advancing warm, moist subtropical air from the Gulf of Mexico replaces the retreating cold maritime polar air from the North Atlantic. The direction of frontal movement is given by the half circles, which point into the cold air; this front is heading toward the northeast. As the cold air recedes, the warm front slowly advances. The average speed of a warm front is about 10 knots, or about half that of an average cold front. During the day, as mixing occurs on both sides of the front, its movement may be much faster. Warm fronts often move in a series of rapid jumps, which show up on

▼ Table 8.2   **Typical Weather Conditions Associated with a Cold Front in Winter in the Northern Hemisphere**

WEATHER ELEMENT	BEFORE PASSING	WHILE PASSING	AFTER PASSING
Winds	South or southwest	Gusty, shifting	West or northwest
Temperature	Warm	Sudden drop	Steadily dropping
Pressure	Falling steadily	Minimum, then sharp rise	Rising steadily
Clouds	Increasing Ci, Cs, then either Tcu* or Cb*	Tcu or Cb	Often Cu, Sc* when ground is warm
Precipitation	Short period of showers	Heavy showers of rain or snow, sometimes with hail, thunder, and lightning	Decreasing intensity of showers, then clearing
Visibility	Fair to poor in haze	Poor, followed by improving	Good, except in showers
Dew point	High; remains steady	Sharp drop	Lowering

*Tcu stands for towering cumulus, such as cumulus congestus; whereas Cb stands for cumulonimbus. Sc stands for stratocumulus.

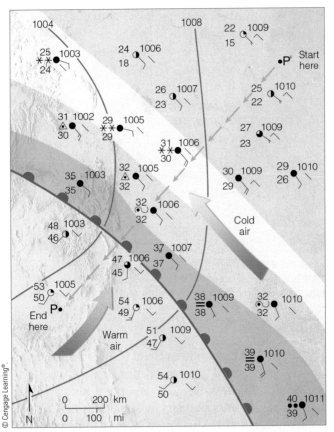

● **FIGURE 8.19** Surface weather associated with a typical warm front in winter. A vertical view along the dashed line P–P′ is shown in Fig. 8.20. (Green-shaded area represents rain; pink-shaded area represents freezing rain and sleet; white-shaded area represents snow.)

successive weather maps. At night, however, radiational cooling creates cool, dense surface air behind the front. This inhibits both lifting and the front's forward progress. When the forward surface edge of the warm front passes a station, the wind shifts, the temperature rises, and the overall weather conditions improve. To see why, we will examine the weather commonly associated with the warm front both at the surface and aloft.

● Figure 8.19 is a surface weather map showing the position of a warm front in winter and its associated weather. ● Figure 8.20 is a vertical view of a model of the warm front in Fig. 8.19. Look at these two figures and observe that, in the model, the warmer, less-dense air rides

up and over the colder, more-dense surface air. This rising of warm air over cold (called **overrunning** in this model) produces clouds and precipitation well in advance of the front's surface boundary. The warm front that separates the two air masses has an average slope of about 1:300—a much more gentle or inclined shape than that of a typical cold front.*

Suppose we are standing at the position marked P′ in Fig. 8.19 and Fig. 8.20. Note that we are over 1200 km (750 mi) ahead of where the warm front is touching the surface. Here, the surface winds are light and variable, the air is cold, and about the only indication of an approaching warm front is the high cirrus clouds overhead. We know the front is moving slowly toward us and that within a day or so it will pass our area. Suppose that, instead of waiting for the front to pass us, we drive toward it, observing the weather as we go.

Heading toward the warm front, we notice that the cirrus (Ci) clouds gradually thicken into a thin, white veil of cirrostratus (Cs) whose ice crystals cast a halo around the sun.** Almost imperceptibly, the clouds thicken and lower, becoming altocumulus (Ac) and altostratus (As) through which the sun shows only as a faint spot against an overcast gray sky. Snowflakes begin to fall, and we are still over 600 km (370 mi) from the surface front. The snow increases, and the clouds thicken into a sheetlike covering of nimbostratus (Ns). The winds become brisk out of the southeast, while the atmospheric pressure slowly falls. Within 400 km (250 mi) of the front, the cold surface air mass is now quite shallow. The surface air temperature moderates and, as we approach the front, the light snow changes first into sleet. It then becomes freezing rain and finally rain and drizzle as the air temperature climbs above freezing. Overall, the precipitation remains light or moderate but covers a broad area. Moving still closer to the front, warm, moist air mixes with cold, moist air producing ragged windblown stratus (St) and fog. (As you might deduce, flying in the vicinity of a warm front can be quite hazardous.)

---

*This slope of 1:300 is a more gentle slope than that of most warm fronts. Typically, the slope of a warm front is on the order of 1:150 to 1:200.

**If the warm air is relatively unstable, ripples or waves of cirrocumulus clouds will appear as a "mackerel sky."

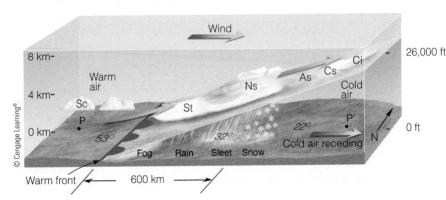

● **FIGURE 8.20** A model illustrating a vertical view of clouds, precipitation, and winds across the warm front in Fig. 8.19 along the line P–P.

Finally, after a trip of over 1200 km, we reach the warm front's surface boundary. As we cross the front, the weather changes are noticeable, but much less pronounced than those experienced with the cold front; they show up more as a gradual transition rather than a sharp change. On the warm side of the front, the air temperature and dew point rise, the wind shifts from southeast to south or southwest, and the atmospheric pressure stops falling. The light rain ends and, except for a few stratocumulus, the fog and low clouds vanish.

This scenario of an approaching warm front represents average, if not idealized, warm-front weather in winter. In some instances, the weather can differ from this example dramatically. For example, if the overrunning warm air is relatively dry and stable, only high and middle clouds will form, and no precipitation will occur. On the other hand, if the warm air is relatively moist and conditionally unstable (as is often the case during the summer), heavy showers can develop as thunderstorms become embedded in the cloud mass. Some of these thunderstorms may have bases at a relatively high level above the surface; these are called *elevated storms*.

Along the West Coast, the Pacific Ocean significantly modifies the surface air so that warm fronts are difficult to locate on a surface weather map. Also, not all warm fronts move northward or northeastward. On rare occasions, a warm front will move into the eastern seaboard from the Atlantic Ocean as it spins all the way around a deep cyclonic storm positioned off the coast. Cold northeasterly winds ahead of the front usually become warm northeasterly winds behind it. Even with these exceptions, knowing the normal sequence of warm-front weather can be useful, especially if you live where warm fronts become well developed. You can look for certain cloud and weather patterns and make reasonably accurate short-range forecasts of your own. ▼ Table 8.3 summarizes typical winter warm-front weather.

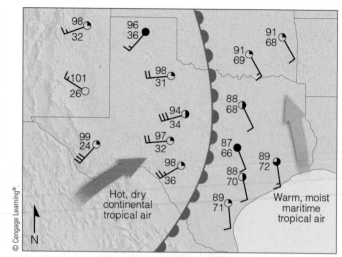

● **FIGURE 8.21** A dryline represents a narrow boundary where there is a steep horizontal change in moisture as indicated by a rapid change in dew-point temperature. Here, a dryline moving across Texas and Oklahoma separates warm, moist air from warm, dry air during an afternoon in May.

DRYLINES   In the southern Great Plains, warm, humid air may be separated from warm, dry air along a boundary called a **dryline**. Because dew-point temperatures may drop along this boundary by as much as 10°C (18°F) per kilometer, drylines have been referred to as *dew-point fronts*.* Although drylines can occur in the United States as far north as the Dakotas, and as far east as the Texas-Louisiana border, they are most frequently observed in the western half of Texas, Oklahoma, and Kansas, especially during spring and early summer. In these locations, drylines tend to move eastward during the day, then westward toward evening. ● Figure 8.21 shows a well-developed dryline moving across Texas and Oklahoma during May 2001.

Cumulus clouds and thunderstorms often form along or to the east of the dryline. This cloud development is

---

*Recall from Chapter 4 that the dew-point temperature is a measure of the amount of water vapor in the air.

▼ **Table 8.3**   **Typical Weather Conditions Associated with a Warm Front in the Northern Hemisphere**

WEATHER ELEMENT	BEFORE PASSING	WHILE PASSING	AFTER PASSING
Winds	South or southeast	Variable	South or southwest
Temperature	Cool to cold, slow warming	Steady rise	Warmer, then steady
Pressure	Usually falling	Leveling off	Slight rise, followed by fall
Clouds	In this order: Ci, Cs, As, Ns, St, and fog; occasionally Cb in summer	Stratus-type	Clearing with scattered Sc, especially in summer; occasionally Cb in summer
Precipitation	Light-to-moderate rain, snow, sleet, or drizzle; showers in summer	Drizzle or none	Usually none; sometimes light rain or showers
Visibility	Poor	Poor, but improving	Fair in haze
Dew point	Steady rise	Steady	Rise, then steady

caused in part by daytime convection and a sloping terrain. The Central Plains area of North America is higher to the west and lower to the east. Convection over the elevated western plains carries dry air high above the surface. Westerly winds sweep this dry air eastward over the lower plains where it overrides the slightly cooler but more humid air at the surface. This situation sets up a potentially unstable atmosphere that finds warm, dry air above warm, moist air. In regions where the air rises, cumulus clouds and organized bands of thunderstorms can form. We will look more closely at drylines and their effect on developing thunderstorms in Chapter 10.

OCCLUDED FRONTS If a cold front catches up to and overtakes a warm front, the frontal boundary created between the two air masses is called an **occluded front,** or, simply, an **occlusion** (meaning "closed off"). There are two main types of occlusions, *cold occlusions* and *warm occlusions*. On a surface weather map, an occluded front is represented as a purple line with alternating cold-front triangles and warm-front half circles; both symbols point in the direction toward which the front is moving. Look back at Fig. 8.14, p. 221, and notice that the air behind the occluded front is colder than the air ahead of it. This is known as a *cold-type occluded front*, or *cold occlusion*. Let's see how this front develops.

The classic model of a developing cold-type occluded front is shown in ● Fig. 8.22. Along line *A–A'*, the cold front is rapidly approaching the slower-moving warm front. Along *B–B'*, the cold front overtakes the warm front, and as we can see in the vertical view across *C–C'*, it underrides and lifts both the warm front and the warm air mass off the ground. This situation allows colder air to replace cool air at the surface. As a cold occluded front approaches, the weather sequence is similar to that of a warm front, with high clouds lowering and thickening into middle and low clouds and precipitation forming well in advance of the surface front. Since the front represents a trough of low pressure, southeasterly winds and falling atmospheric pressure occur ahead of it. The frontal passage, however, can bring weather similar to that of a cold front: heavy, showery precipitation, with winds shifting to west or northwest. After a period of wet weather, the sky begins to clear, atmospheric pressure rises, and the air turns colder. The most active weather usually occurs where the cold front is just overtaking the warm front, at the point of occlusion, where the greatest contrast in temperature occurs.

The classic model of a developing warm-type occluded front is shown in ● Fig. 8.23. Notice that continental polar air over eastern Washington and Oregon is much colder than milder maritime polar air moving inland from the Pacific Ocean. Observe also that the surface air ahead of the warm front is colder than the air behind the cold

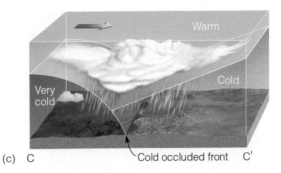
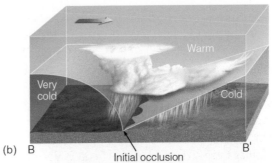
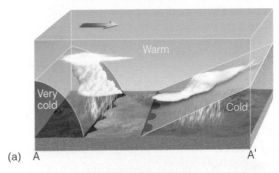

(c) C   Cold occluded front   C'

(b) B   Initial occlusion   B'

(a) A   A'

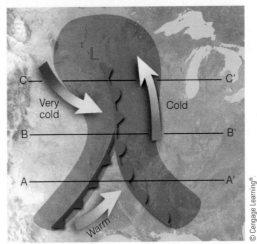

● **FIGURE 8.22** A model that shows the formation of a cold-occluded front. The faster-moving cold front in (a) catches up to the slower-moving warm front in (b) and forces it to rise off the ground in (c). (Green-shaded area on the map represents precipitation.)

front. Consequently, when the cold front catches up to and overtakes the warm front, the milder, lighter air behind the cold front is unable to lift the colder, heavier air off the ground. As a result, the cold front rides "piggyback" along the sloping warm front, producing a warm-type occluded front. The surface weather associated with the warm

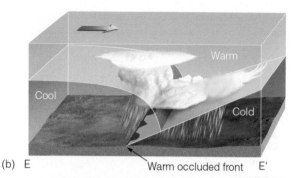

(b) E   Warm occluded front   E'

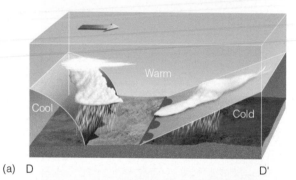

(a) D   D'

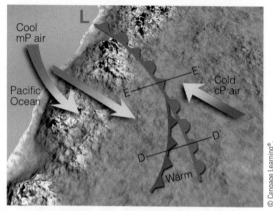

© Cengage Learning®.

● FIGURE 8.23 A model illustrating the formation of a warm-type occluded front. The faster-moving cold front in (a) over-takes the slower-moving warm front in (b). The lighter air behind the cold front rises up and over the denser air ahead of the warm front. The bottom illustration represents a surface map of a warm occlusion.

occlusion is similar to that of a warm front. In addition, the relatively mild winter air that moves into Europe from the North Atlantic causes many of the occlusions that move into this region in winter to be of the warm variety.

Contrast Fig. 8.22 and Fig. 8.23. Note that the primary difference between the cold- and warm-type occluded front is the location of the upper-level front. In a cold occlusion, there is an upper-level warm front that *follows* the surface occluded front, whereas in a warm occlusion, there is an upper-level cold front that *precedes* the surface occluded front.

So far, the ideas we have presented on the formation of occluded fronts are based on the surface air temperature on either side of the front. In recent years, the classic model of cold and warm occluded fronts forming due to temperature

differences ahead of the advancing warm front and behind the cold front has been challenged. Studies show that the *stability* of the air (rather than the surface air temperature) is the most important factor in determining the type of occluded front that forms. The air within the warm frontal zone is typically more stable than the air within the cold frontal zone, so the air behind the cold front is more likely to ride up and over the warm front than to flow beneath it. For this reason, warm-type occluded fronts are far more common than cold-type occluded fronts. In the world of weather fronts, occluded fronts are the mavericks. In fact, they may show up on a surface chart as a trough of low pressure separating two cold air masses. Because of this, locating and defining occluded fronts at the surface is often difficult for the meteorologist.* Similarly, you too may find it hard to recognize an occlusion. In spite of this, we will assume that the weather associated with occluded fronts in North America behaves in a similar way to that shown in ▼ Table 8.4.

The frontal systems described so far are actually part of a much larger storm system: the middle-latitude cyclone. ● Figure 8.24 shows the cold front, warm front,

---

*In some countries, such as Canada, the surface occlusion is seldom analyzed on a surface weather map. Instead, the location of the occluded front aloft—where the cold air lifts the warm air above the surface—is marked by a *trowal* (which stands for "trough of warm air aloft"). So the position of the *trowal* on the weather map (as indicated by a "hook") marks the location where the cold and warm fronts intersect aloft.

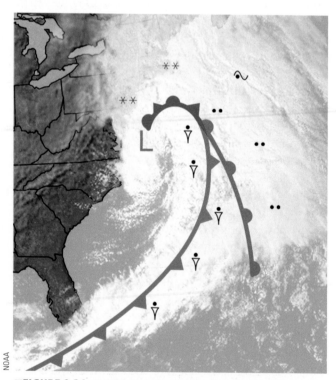

NOAA

● FIGURE 8.24 A visible satellite image showing a mid-latitude cyclonic storm with its weather fronts over the Atlantic Ocean during March 2005. Superimposed on the image is the position of the surface cold front, warm front, and occluded front. Precipitation symbols indicate where precipitation is reaching the surface.

▼ Table 8.4 Typical Winter Weather Most Often Associated with Occluded Fronts in North America

WEATHER ELEMENT	BEFORE PASSING	WHILE PASSING	AFTER PASSING
Winds	East, southeast, or south	Variable	West or northwest
Temperature			
(a) Cold-type occluded	Cold or cool	Dropping	Colder
(b) Warm-type occluded	Cold	Rising	Milder
Pressure	Usually falling	Low point	Usually rising
Clouds	In this order: Ci, Cs, As, Ns	Ns, sometimes Tcu and Cb	Ns, As, or scattered Cu
Precipitation	Light, moderate, or heavy precipitation	Light, moderate, or heavy continuous precipitation or showers	Light-to-moderate precipitation followed by general clearing
Visibility	Poor in precipitation	Poor in precipitation	Improving
Dew point	Steady	Usually slight drop, especially if cold-occluded	Slight drop, although may rise a bit if warm-occluded

and occluded front with such a storm. Notice, as we would expect, clouds and precipitation form in a rather narrow band along the cold front and in a much wider band along the warm front and the occluded front.

The next section explains where, why, and how mid-latitude cyclones form.

Mid-Latitude Cyclonic Storms

Early weather forecasters were aware that precipitation generally accompanied falling barometers and areas of low pressure. However, it was not until the early part of the twentieth century that scientists began to piece together the information that yielded the ideas of modern meteorology and cyclonic storm development.

Working largely from surface observations, a group of scientists in Bergen, Norway, developed a model explaining the life cycle of an *extratropical*, or *middle-latitude cyclonic storm;* that is, a storm that forms at middle and high latitudes outside of the tropics. This extraordinary group of meteorologists included Vilhelm Bjerknes, his son Jakob, Halvor Solberg, and Tor Bergeron. They published their *Norwegian Cyclone Model* shortly after World War I. It was widely acclaimed and became known as the "polar front theory of a developing wave cyclone," or, simply, the **polar front theory.** What these meteorologists gave to the world was a working model of how a mid-latitude cyclone progresses through the stages of birth, growth, and decay. An important part of the model involved the development of weather along the polar front. As new information became available, the original work was modified, so that, today, it serves as a convenient way to describe the structure and weather associated with a migratory middle-latitude cyclonic storm system, such as the one shown in Fig. 8.24.

POLAR FRONT THEORY The development of a mid-latitude cyclone, according to the Norwegian model, begins along the polar front. Remember (from our discussion of the general circulation in Chapter 7) that the polar front is a semicontinuous global boundary separating cold polar air from warm subtropical air. Because the mid-latitude cyclone forms and moves along the polar front in a wavelike manner, the developing storm is referred to as a **wave cyclone.** The stages of a developing wave cyclone from a surface perspective are illustrated in the sequence of surface weather maps shown in ● Fig. 8.25.

Figure 8.25a shows a segment of the polar front as a stationary front. It represents a trough of lower pressure with higher pressure on both sides. Cold air to the north and warm air to the south flow parallel to the front, but in opposite directions. This type of flow sets up a cyclonic wind shear. You can conceptualize the shear more clearly if you place a pen between the palms of your hands and move your left hand toward your body; the pen turns counterclockwise, cyclonically.

Under the right conditions, a wavelike kink forms on the front, as shown in Fig. 8.25b. The wave that forms is known as a **frontal wave.** Watching the formation of a frontal wave on a weather map is like watching a water wave from its side as it approaches a beach: It first builds, then breaks, and finally dissipates. This pattern is the reason a mid-latitude cyclonic storm system is known as a *wave cyclone.*

Figure 8.25b shows the newly formed wave with a cold front pushing southward and a warm front moving

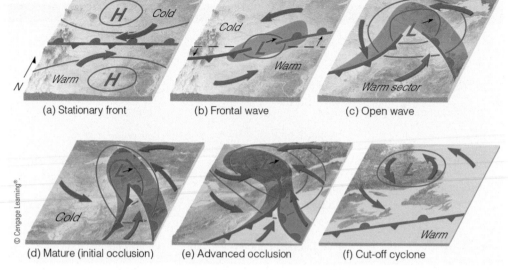

● **FIGURE 8.25** The idealized life cycle of a mid-latitude cyclonic storm (a through f) in the Northern Hemisphere based on the polar front theory. As the life cycle progresses, the system moves northeastward in a dynamic fashion. The small arrow next to each L shows the direction of storm movement.

(a) Stationary front

(b) Frontal wave

(c) Open wave

(d) Mature (initial occlusion)

(e) Advanced occlusion

(f) Cut-off cyclone

© Cengage Learning®.

northward. The region of lowest pressure is at the junction of the two fronts. As the cold air displaces the warm air upward along the cold front, and as warm air rises ahead of the warm front, a narrow band of precipitation forms (shaded green area). Steered by the winds aloft, the system typically moves east or northeastward and gradually becomes a fully developed **open wave** in 12 to 24 hours (Fig. 8.25c). The central pressure of the wave cyclone is now much lower, and several isobars encircle the wave's apex. These more tightly packed isobars create a stronger cyclonic flow, as the winds swirl counterclockwise and inward toward the low's center. Precipitation forms in a wide band *ahead* of the warm front and along a narrow band of the cold front. The region of warm air between the cold and warm fronts is known as the *warm sector*. Here, the weather tends to be partly cloudy, although scattered showers and thunderstorms may develop if the air is conditionally unstable.

Energy for the storm is derived from several sources. As the air masses try to attain equilibrium, warm air rises and cold air sinks, transforming potential energy into kinetic energy (that is, energy of motion). Condensation supplies energy to the system in the form of latent heat. And, as the surface air converges toward the low center, wind speeds may increase, producing an increase in kinetic energy.

As the open wave moves eastward, central pressures continue to decrease, and the winds blow more vigorously. The faster-moving cold front constantly inches closer to the warm front, squeezing the warm sector into a smaller area (as shown in Fig. 8.25d), and the wave quickly develops into a *mature cyclone*. In this model, the cold front eventually overtakes the warm front and the system becomes occluded. At this point, the storm is usually most intense, with clouds and precipitation covering a large area. The area of most intense weather is normally found to the northwest of the storm's center. Here, strong winds and blowing and drifting snow can create blizzard conditions in winter.

The intense storm system shown in Fig. 8.25e gradually dissipates, because cold air now lies on both sides of the occluded front. Without the supply of energy provided by the rising warm, moist air, the old storm system dies out and gradually disappears (Fig. 8.25f). Occasionally, however, a new wave will form on the westward end of the trailing cold front. We can think of the sequence of a developing wave cyclone as a whirling eddy in a stream of water that forms behind an obstacle, moves with the flow, and gradually vanishes downstream. The entire life cycle of a wave cyclone can last from a few days to more than a week.

Take a second and look back at the mid-latitude cyclonic storm depicted in the satellite image in Fig. 8.24. According to what you have just read, what is the stage of development of this storm? (Answer given in footnote below.*)

● Figure. 8.26 shows a series of wave cyclones at various stages of development along the polar front in winter. Such a succession of storms is known as a *"family" of cyclones*. Observe that to the north of the front are cold anticyclones; to the south over the Atlantic Ocean is the warm, semipermanent Bermuda high. The polar front itself has developed into a series of loops, and at the apex of each loop is a cyclone. The cyclone over the northern plains (Low 1) is just forming; the one along the east coast (Low 2) is an open wave; and the system near Iceland (Low 3) is dying out. If the average rate of movement of a wave cyclone from birth to decay is 25 knots, then it is entirely possible for a storm to develop over the central part of the United States, intensify into a large storm over New England, become occluded over the ocean, and reach the coast of England in its dissipating stage less than a week after it forms.

*The storm shown in Fig. 8.24 is in its occluded stage, and would be classified as a mature cyclone.

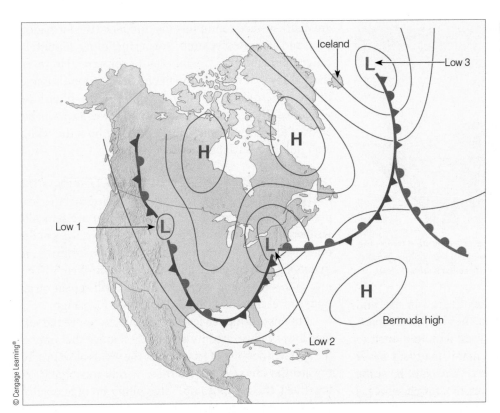

© Cengage Learning®.

●FIGURE 8.26 A series of wave cyclones (a "family" of cyclones) forming along the polar front.

Up to now, we have considered the polar front model of a developing wave cyclone, which represents a rather simplified version of the stages that a mid-latitude cyclonic storm system must go through. In fact, even though few (if any) storms adhere to the model exactly, it still can serve as a good foundation for understanding the structure of cyclonic storms. So keep the model in mind as you read the following sections.

WHERE DO MID-LATITUDE CYCLONES TEND TO FORM?

Any development or strengthening of a mid-latitude cyclone is called **cyclogenesis**. Certain regions of North America show a propensity for cyclogenesis, including the eastern slopes of the Rockies. As westerly winds blow over a mountain range, the air expands vertically on the downwind (lee) side, which can help intensify any pre-existing areas of low pressure. Troughs and developing cyclonic storms that form in this manner are called **lee-side lows** (see ● Fig. 8.27). Additional areas that exhibit cyclogenesis are the Great Basin, the Gulf of Mexico, and the Atlantic Ocean east of the Carolinas. Near Cape

DID YOU KNOW?

In mid-January 1888, a ferocious mid-latitude cyclonic storm swept across the Great Plains from Texas to the Dakotas and into Wisconsin. Strong winds, extremely low temperatures, and heavy snow on the storm's western side wiped out the Plains' free-range livestock and took 237 lives. This infamous storm has come to be known as the "Children's Blizzard" because of the many dozens of schoolchildren frozen to death on their way to school.

Hatteras, North Carolina, for example, warm Gulf Stream water can supply moisture and warmth to the region south of a stationary front, thus increasing the contrast between air masses to a point where storms can suddenly spring up along the front. As noted earlier, these cyclones normally move northeastward along the Atlantic coast, bringing high winds and heavy snow or rain to coastal areas. Before the age of modern satellite imagery and weather prediction such coastal storms would often go undetected during their formative stages; and sometimes an evening weather forecast of "fair and colder" along the eastern seaboard would have to be changed to "heavy snowfall" by morning. Fortunately, with today's weather information gathering and forecasting techniques, these storms rarely strike by surprise. (Storms that form along the eastern seaboard of the United States and then move northeastward are called **nor'easters** or *northeasters*. Additional information on nor'easters is given in Focus section 8.3)

● Figure 8.28 shows the typical paths taken in winter by mid-latitude cyclones and anticyclones (high-pressure areas). Notice in Fig. 8.28a that some of the lows are named after the region where they form, such as the *Hatteras Low*, which develops off the coast near Cape Hatteras, North Carolina. The *Alberta Clipper* forms (or redevelops) on the eastern side of the Canadian Rockies in Alberta, then rapidly skirts across the northern tier states. Similarly, the *Colorado Low* forms (or redevelops) on the eastern side of the United States Central Rockies. Notice that the lows generally move eastward or northeastward, whereas the highs (Fig. 8.28b) typically move southeastward, then eastward.

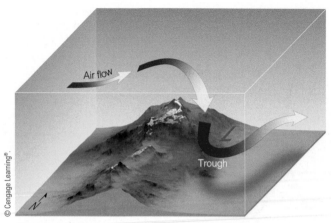

● **FIGURE 8.27** As westerly winds blow over a mountain range, the air can expand vertically on the downwind (lee) side, enhancing the development of a trough or cyclonic storm, called a *lee-side low*.

Some frontal waves form suddenly, grow in size, and develop into huge cyclonic storms, then slowly dissipate, with the entire process taking several days to a week to complete. Other frontal waves remain small and never grow into giant weather-producers. Why is it that some frontal waves develop into huge cyclonic storms, whereas others simply dissipate in a day or so?

This question poses one of the real challenges in weather forecasting. The answer is complex. Indeed, there are many surface conditions that influence the formation of a mid-latitude cyclonic storm, including mountain ranges and land-ocean temperature contrasts. However, the real key to the development of a wave cyclone is found in the *upper-wind flow,* in the region of the high-level westerlies. Therefore, before we can arrive at a reasonable answer to our question, we need to see how the winds aloft influence surface pressure systems.

DEVELOPING MID-LATITUDE CYCLONES AND ANTICYCLONES In Chapter 7, we learned that thermal pressure systems are shallow systems that are typically weaker with increasing height above the surface. On the other hand, developing mid-latitude cyclonic storms are *dynamic lows* that are usually stronger with height. This means that a surface low-pressure area will appear on an upper-level chart as either a closed low or a trough.

Suppose the upper-level low is directly above the surface low, as illustrated in ● Fig. 8.29. Notice that only at the surface (because of friction) do the winds blow inward toward the low's center. As these winds converge (flow together), the air "piles up." This piling up of air, called **convergence,** causes air density to increase directly above the surface low. This increase in mass causes surface pressures to rise; gradually, the low fills and the surface low dissipates. The same reasoning can be applied to surface anticyclones. Winds blow outward away from the center of a surface high. If a closed high or ridge lies directly over the surface anticyclone, **divergence** (the spreading out of air) at the surface will remove air from the column directly above the high. The decrease in mass causes the surface pressure to fall and the surface high-pressure area to weaken. Consequently, it appears that, if upper-level pressure systems were always located directly above those at the surface (such as shown in Fig. 8.29), cyclones and anticyclones would die out soon after they form (if they could form at all). What, then, is it that allows these systems to develop and intensify? (Before reading on, you may wish to review the additional information on convergence and divergence given in Focus section 8.4.).

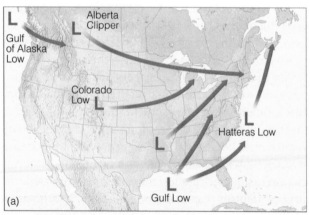

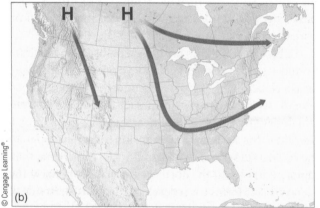

● **FIGURE 8.28** (a) Typical paths of winter mid-latitude cyclones. The lows are named after the region where they form. (b) Typical paths of winter anticyclones.

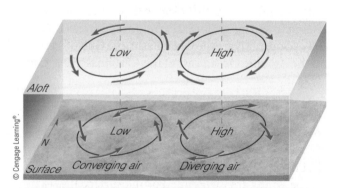

● **FIGURE 8.29** If lows and highs aloft were always directly above lows and highs at the surface, the surface systems would quickly dissipate.

Nor'easters

Northeasters, commonly called *nor'easters*, are east coast mid-latitude cyclonic storms that develop or intensify off the eastern seaboard of North America during the fall, winter, and spring. They usually move northeastward along the coast, often bringing strong northeasterly winds to coastal areas, hence the name, nor'easter. In addition to strong winds, these storms can bring heavy rain, snow, and sleet. Most often they deepen and become most intense off the coast of New England. Nor'easters are fueled by the large temperature gradient between the warm ocean and the cold continental landmass. They gain additional energy from the moisture over the ocean, especially the warm Gulf Stream that flows northeastward parallel to the east coast of the United States.

The ferocious northeaster of January 22–24, 2016 (shown in ● Fig. 4), produced the heaviest snow totals on record at several locations, including New York's Kennedy International Airport (30.6 in.) and Baltimore-Washington International Airport (29.2 in.). Much of the snow fell in less than 24 hours. High winds and 20-foot waves pounded the coastlines of Delaware and New Jersey, and high storm tides put many coastal areas and highways under water.

Studies suggest that some of the nor'easters that batter the coastline in winter may actually possess some of the characteristics of a tropical hurricane. For example, a strong nor'easter that dumped between 10 and 40 inches of snow over parts of the Northeast and New England in February 2013 developed an eye-like feature. One of the most dramatic combinations of tropical and extratropical cyclone processes ever observed was Hurricane

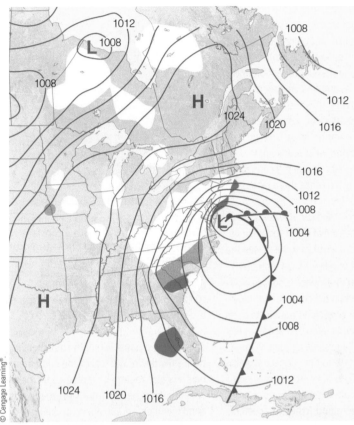

● **FIGURE 4** The surface weather map for 7 a. m. (EST), January 23, 2016, shows an intense low-pressure area (central pressure 987 mb, or 29.15 in.), which is generating strong northeasterly winds and heavy precipitation (areas shaded green for rain, white for snow, and salmon for sleet or freezing rain) from the mid-Atlantic states into New England. More than 30 million people were placed under blizzard warnings as a result of this storm, which dropped more than two feet of snow from Washington, D.C., to New York. Damage estimates were as high as $3 billion.

Sandy in late October 2012. Sandy briefly attained Category 3 strength before striking Cuba, weakened, then restrengthened as it approached the northeast United States. The storm grew to an immense size, with some circulation features comparable to a strong nor'easter, while it also maintained hurricane characteristics until just several hours before it struck New Jersey, when it was reclassified as a post-tropical cyclone. (We will examine hurricanes and their characteristics, including Sandy, in more detail in Chapter 11.)

The Role of Converging and Diverging Air   For mid-latitude cyclones and anticyclones to maintain themselves or intensify, the winds aloft must blow in such a way that zones of converging and diverging air form. For example, notice in ● Fig. 8.30 that the surface winds are converging about the center of the low; while aloft, directly above the low, the winds are diverging. For the surface low to develop into a major storm system, *upper-level divergence of air must be greater than surface convergence of air*; that is,

more air must be removed above the storm than is brought in at the surface. When this phenomenon happens, surface air pressure decreases, and we say that the storm system is *intensifying*, or *deepening*. If the reverse should occur (more air flowing in at the surface than is removed at the top), surface pressure will rise, and the storm system will weaken and gradually dissipate in a process called *filling*.

Notice also in Fig. 8.30 that surface winds are diverging about the center of the high, while aloft, directly above

A Closer Look at Convergence and Divergence

We know that *convergence* is the piling up of air above a region, while *divergence* is the spreading out of air above some region. Convergence and divergence of air may result from changes in wind direction and wind speed. For example, convergence occurs when moving air is funneled into an area, much in the way cars converge when they enter a crowded freeway. Divergence occurs when moving air spreads apart, much as cars spread out when a congested two-lane freeway becomes three lanes.

On an upper-level chart, this type of convergence (also called *confluence*) occurs when contour lines move closer together, as a steady wind flows parallel to them (see the upper-level chart in • Fig. 5). On the same chart, this type of divergence (also called *diffluence*) occurs when the contour lines move apart as a steady wind flows parallel to them. Notice that below the area of divergence lies the surface middle-latitude cyclonic storm.

Convergence and divergence may also result from changes in wind speed. *Speed convergence* occurs when the wind slows down as it moves along, whereas *speed divergence* occurs when the wind speeds up.

We can grasp these relationships more clearly if we imagine air molecules to be

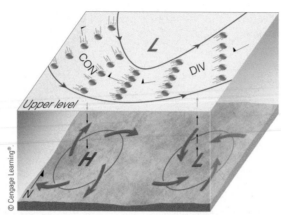

• **FIGURE 5** The formation of convergence (CON) and divergence (DIV) of air with a constant wind speed (indicated by flags) in the upper troposphere. Circles represent air parcels that are moving parallel to the contour lines on a constant pressure chart. Below the area of convergence the air is sinking, and we find the surface high (H). Below the area of divergence the air is rising, and we find the surface low (L).

marching in a band. When the marchers in front slow down, the rest of the band members squeeze together, causing convergence; when the marchers in front start to run, the band members spread apart, or diverge.

In summary, *speed convergence* takes place when the wind speed decreases downwind, and *speed divergence* takes

place when the wind speed increases downwind.

As you continue to read about mid-latitude cyclonic storms, keep in mind from Fig. 5 that converging air aloft is most likely found to the left (west) of the upper trough, whereas diverging air is most likely found to the right (east) of the trough.

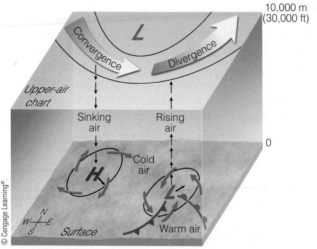

• **FIGURE 8.30** Convergence, divergence, and vertical motions associated with surface pressure systems. Notice that for the surface storm to intensify, the upper trough of low pressure must be located to the left (or west) of the surface low. Gray lines on the upper-air chart are contour lines.

the anticyclone, they are converging. In order for the surface high to strengthen, *upper-level convergence of air must exceed low-level divergence of air* (more air must be brought in above the anticyclone than is removed at the surface). When this occurs, surface air pressure increases, and we say that the high-pressure area is *building*.

In Fig. 8.29, the convergence of air aloft causes an accumulation of air above the surface high, which allows the air to sink slowly and replace the diverging surface air. Above the surface low, divergence allows the converging surface air to rise and flow out the top of the column.

We can see from Fig. 8.30 that when an upper-level trough is as sufficiently deep as is illustrated here, a region of converging air usually forms on the west side of the trough and a region of diverging air forms on the east side. (For reference, compare Fig. 8.30 with Fig. 5 above.) Aloft, the area of diverging air is directly above the surface low, and the area of convergence is directly

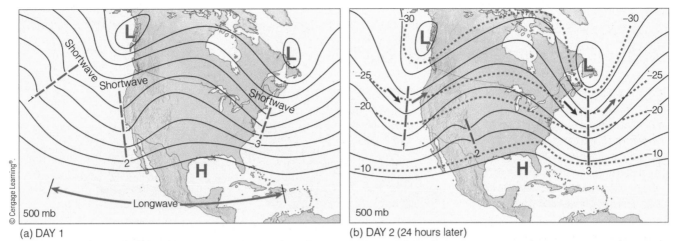

● **FIGURE 8.31** (a) Upper-air chart showing a longwave with three shortwaves (heavy dashed lines) embedded in the flow. (b) Twenty-four hours later the shortwaves have moved rapidly around the longwave. Notice that the shortwaves labeled 1 and 3 tend to deepen the longwave troughs, while shortwave 2 has weakened as it moves into a ridge. Dashed red lines are isotherms (lines of constant temperature) in °C. Solid gray lines are contours. Blue arrows indicate cold advection and red arrows, warm advection.

above the surface high. This configuration means that, for a surface mid-latitude cyclone to intensify, the upper-level trough of low pressure must be located behind (or to the *west* of) the surface low. When the upper-level trough is in this position, the atmosphere is able to redistribute its mass, as regions of low-level convergence are compensated for by regions of upper-level divergence, and vice versa.

Winds aloft steer the movement of the surface pressure systems. Since the winds above the surface low in Fig. 8.30 are blowing from the southwest, the surface low should move northeastward. The northwesterly winds above the surface high should direct it toward the southeast. These paths are typical of the average movement of surface pressure systems in the eastern two-thirds of the United States, as shown in Fig. 8.28 on p. 232.

Waves in the Westerlies   Waves that form in the flow aloft can play a major role in the development of mid-latitude cyclones. Recall from Chapter 6 that the flow above the middle latitudes usually consists of a series of waves in the form of troughs and ridges. The distance from trough to trough (or ridge to ridge) is known as the *wavelength*. When the wavelength is on the order of many thousands of kilometers, the wave is called a **longwave**. Observe in ● Fig. 8.31a that the length of the longwave is greater than the width of North America. Typically, at any given time, there are between three and six longwaves looping around Earth. These longwaves are also known as *Rossby waves*, after C. G. Rossby, a famous meteorologist who carefully studied their motion. In Fig. 8.31a, we can see that embedded in longwaves are **shortwaves**, which are small disturbances or ripples that move with the wind flow.

By comparing Fig. 8.31a with Fig. 8.31b, we can see that while the longwaves move eastward very slowly, the shortwaves move fairly quickly around the longwaves. Generally, shortwaves deepen (that is, increase in size) when they approach a longwave trough and weaken (become smaller) when they approach a ridge. Also notice in Fig. 8.31b that when a shortwave moves into a longwave trough, the trough tends to deepen. The upper flow is now capable of providing the necessary ingredients for the development or intensification of surface low- and high-pressure areas as illustrated in Fig. 8.30.

Notice in Fig. 8.31b that there are regions where the winds (blue and red arrows) cross the isotherms (dashed red lines). In these regions, there is an important process taking place called *temperature advection*.* Where the wind crosses the isotherms in such a way that colder air is replacing warmer air, the transport of colder air into a region is called **cold advection**. In Fig. 8.31b, cold advection is represented by blue arrows. Where the wind crosses the isotherms in such a way that warmer air is replacing colder air, the transport of warmer air into a region is called **warm advection**. On the map, warm advection is represented by red arrows. Because temperature advection plays a major part in the development of a mid-latitude cyclonic storm, we will examine its important role in the following section.

Upper-Air Support for the Developing Storm   To better understand how a wave cyclone can develop and intensify into a huge mid-latitude cyclonic storm, we need to examine atmospheric conditions at the surface and aloft. Suppose that a portion of a longwave

---

*Recall from Chapter 2 that the term *advection* refers to the horizontal transfer of any atmospheric property by the wind.

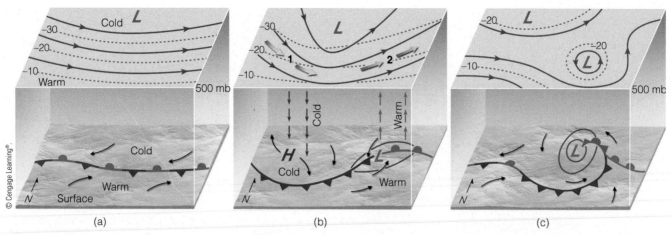

●**FIGURE 8.32** An idealized 3-D view of the formation of a mid-latitude cyclone. (a) A longwave trough at 500 mb lies parallel to and directly above the surface stationary front. (b) A shortwave (not shown) disturbs the flow aloft, initiating temperature advection (blue arrow, cold advection; red arrow, warm advection). The upper trough intensifies and provides the necessary vertical motions (as shown by vertical arrows) for the development of the surface wave cyclone. (c) As the surface storm moves northeastward, it occludes, and without upper-level diverging air to compensate for surface converging air, the cyclonic storm system dissipates.

trough at the 500-mb level lies directly above a surface stationary front, as illustrated in ● Fig. 8.32a. On the 500-mb chart, contour lines (solid lines) and isotherms (dashed lines) parallel each other and are crowded close together. Colder air is located in the northern half of the map, while warmer air is located to the south. Winds at this level are blowing at fairly high velocities. Suppose a shortwave moves through this region, disturbing the flow as shown in Fig. 8.32b.

As the flow aloft becomes disturbed, it begins to lend support for the intensification of surface pressure systems, as a region of converging air forms above position 1 in Fig. 8.32b and a region of diverging air forms above position 2. The converging air aloft causes the surface air pressure to rise in the region marked *H* in Fig. 8.32b. Surface winds begin to blow out away from the region of higher pressure, and the air aloft gradually sinks to replace it. Meanwhile, diverging air aloft causes the surface air pressure to decrease beneath position 2, in the region marked *L* on the surface map. This initiates rising air, as the surface winds blow in

toward the region of lower pressure. As the converging surface air develops cyclonic spin, cold air flows southward and warm air northward. We can see in Fig. 8.32b that the western half of the stationary front is now a cold front and the eastern half a warm front. Cold air moves in behind the cold front, while warm air slides up along the warm front. These regions of cold and warm advection occur all the way up to the 500-mb level, about 18,000 feet above sea level.

On the 500-mb chart in Fig. 8.32b, cold advection is occurring at position 1 (blue arrow) as the wind crosses the isotherms, bringing cold air into the trough. The cold advection makes the air more dense, which has the effect of deepening the trough. The deepening of the upper trough causes the contour lines to crowd closer together and the winds aloft to increase. Meanwhile, at position 2 warm advection is taking place (red arrows), which has the effect of strengthening the ridge. Therefore, *the overall effect of differential temperature advection is to amplify the upper-level wave.* As the trough aloft deepens, its curvature increases, which in turn increases the region of divergence above the developing surface storm. At this point, the surface mid-latitude cyclone rapidly develops as surface pressures fall.

In regions where there is cold advection, some of the cold, heavy air sinks; where there is warm advection, some of the warm, light air rises. The sinking of cold air and the rising of warm air provide energy for a developing cyclone, as potential energy is transformed into kinetic energy. Further, if clouds form, condensation in the ascending air releases latent heat, which warms the air. The warmer air lowers the surface pressure, which strengthens the surface low even more. So, we now have a full-fledged

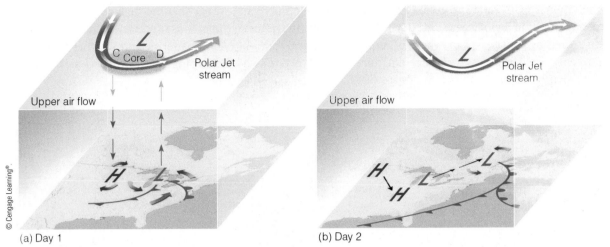

● **FIGURE 8.33** (a) As the polar jet stream and its area of maximum winds (the jet streak, or core) swings over a developing mid-latitude cyclone, an area of divergence (D) draws warm surface air upward, and an area of convergence (C) allows cold air to sink. The jet stream removes air above the surface storm, which causes surface pressures to drop and the storm to intensify. (b) When the surface storm moves northeastward and occludes, it no longer has the upper-level support of diverging air, and the surface storm gradually dies out.

middle-latitude cyclonic storm with all of the necessary ingredients for its development.

Eventually, the warm air curls around the north side of the low, and the storm system occludes (see Fig. 8.32c). Some storms may continue to deepen, but most do not as they move out from under the region of upper-level divergence. Additionally, at the surface the storm weakens as the supply of warm air is cut off and cold, dry air behind the cold front (called a *dry slot*) is drawn in toward the surface low.

The Role of the Jet Stream   Streams can play an additional part in the formation of surface mid-latitude cyclones. When the polar jet stream flows in a wavy west-to-east pattern, deep troughs and ridges exist in the flow aloft. Notice in ● Fig. 8.33a that, in the trough, the area shaded orange represents a strong core of winds called the *jet stream core,* or **jet streak**. The curving of the jet stream coupled with the changing wind speeds around the jet streak produce regions of strong convergence and divergence along the flanks of the jet. The region of diverging air above the surface low (marked D in Fig. 8.33a) draws warm surface air upward to the jet stream, which quickly sweeps the air downstream. Since the air above the mid-latitude cyclone is being removed more quickly than converging surface winds can supply air to the storm's center, the storm rapidly intensifies and surface winds increase. Above the high-pressure area, a region of converging air (marked C in Fig. 8.33a) feeds cold air downward into the anticyclone to replace the diverging surface air. Hence, we find the polar jet stream

removing air above the surface cyclonic storm and supplying air to the surface anticyclone.

As the jet stream steers the cyclonic storm along (toward the northeast, in this case), the surface storm occludes, and cold air surrounds the surface low (see Fig. 8.33b). Since the surface low has moved out from under the pocket of diverging air aloft, the occluded storm gradually fills as the surface air flows into the system.

Since the Northern Hemisphere's polar jet stream is strongest and moves farther south in winter, we can see why mid-latitude cyclonic storms are better developed and move more quickly during the coldest months. During the summer when the polar jet stream shifts northward, developing mid-latitude cyclonic storm activity shifts northward as well, occurring principally in Canada over the province of Alberta and the Northwest Territories.

In general, we now have a fairly good picture as to why some surface lows intensify into huge mid-latitude cyclones while others do not. For a surface cyclonic storm to intensify, there must be an upper-level counterpart—a trough of low pressure—that lies to the *west* of the surface low. As shortwaves disturb the flow aloft, they cause regions of differential temperature advection to appear, leading to an intensification of the upper-level trough. At the same time, the polar jet stream forms into waves and swings slightly south of the developing storm. When these conditions exist, zones of converging and diverging air, along with rising and sinking air, provide energy conversions for the storm's growth. With this atmospheric situation, storms can form even where there are no pre-existing fronts. In regions where the upper-level flow is not disturbed by shortwaves or where no upper trough or jet stream exists, the necessary vertical and horizontal motions are insufficient to enhance cyclonic storm development and we say that the surface storm does not have the necessary *upper-air support*. The horizontal and vertical motions, cloud patterns, and weather that typically occur with a developing open-wave cyclone are summarized in ● Fig. 8.34.

● **FIGURE 8.34** Summary of clouds, weather, vertical motions, and upper-air support associated with a developing mid-latitude cyclone. Dark green area represents precipitation.

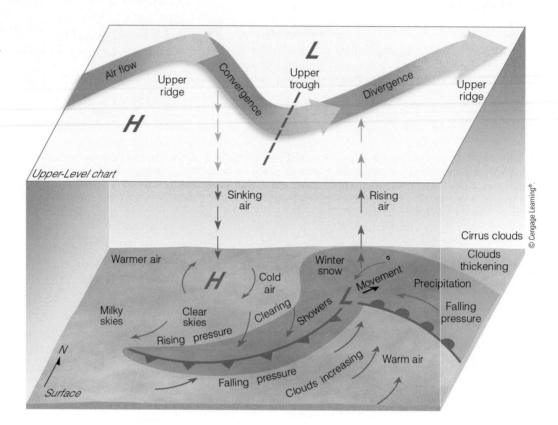

SUMMARY

In this chapter, we considered the different types of air masses and the various weather each brings to a particular region. Continental arctic air masses are responsible for the extremely cold (arctic) outbreaks of winter, whereas continental polar air masses are responsible for cold, dry weather in winter and cool, pleasant weather in summer. Maritime polar air, having traveled over an ocean for a considerable distance, brings to a region cool, moist weather. The hot, dry weather of summer is associated with continental tropical air masses, whereas warm, humid conditions are due to maritime tropical air masses. Where air masses with sharply contrasting properties meet, we find weather fronts.

Along the leading edge of a cold front, where colder air replaces warmer air, showers are prevalent, especially if the warmer air is moist and conditionally unstable. Along a warm front, warmer air rides up and over colder surface air, producing widespread cloudiness and light-to-moderate precipitation that can cover thousands of square kilometers. When the rising air is conditionally unstable (such as it often is in summer), showers and thunderstorms may form ahead of the advancing warm front. Occluded fronts, which are often difficult to locate and define on a surface weather map, may have characteristics of both cold and warm fronts.

We learned that fronts are actually part of the mid-latitude cyclone. We examined where, why, and how these storms form and found that a mid-latitude cyclone goes through a series of stages from birth, to maturity, to death as an occluded storm. An important influence on the development of a mid-latitude cyclonic storm is the upper-air flow, including the jet stream. We learned that when an upper-level low lies to the west of the surface low, and the polar jet stream bends and then dips south of the surface storm, an area of divergence above the surface low provides the necessary ingredients for the surface mid-latitude cyclone to develop into a deep low-pressure area.

KEY TERMS

The following terms are listed (with corresponding page numbers) in the order they appear in the text. Define each. Doing so will aid you in reviewing the material covered in this chapter.

air mass, 210
source regions (for air masses), 210
continental polar (air mass), 212
continental arctic (air mass), 212
continental tropical, 218
lake-effect snows, 212
maritime polar (air mass), 000
Pacific air, 216
maritime tropical (air mass), 217
continental tropical (air mass), 219
front, 220
stationary front, 221
cold front, 222
"back door" cold fronts, 224
warm front, 224
dryline, 226
overrunning, 227
occluded front, 227
occlusion, 227
polar front theory, 229
wave cyclone, 229
frontal wave, 229
open wave, 230
cyclogenesis, 231
lee-side low, 231
nor'easter, 231
convergence, 232
divergence, 232
longwave (in westerly flow aloft), 235
shortwave (in westerly flow aloft), 235
cold advection, 235
warm advection, 235
jet streak, 237

QUESTIONS FOR REVIEW

1. (a) What is an air mass?
   (b) If an area is described as a "good air-mass source region," what information can you give about it?
2. How does a continental arctic air mass differ from a continental polar air mass?
3. Why is continental polar air not welcome to the Central Plains in winter yet very welcome in summer?
4. What are lake-effect snows and how do they form? On which side of a lake do they typically occur?
5. Explain why the central United States is not a good air-mass source region.
6. List the temperature and moisture characteristics of each of the major air mass types.

7. Which air mass only forms in summer over the southwestern United States?

8. Why are maritime polar air masses along the east coast of the United States usually colder than those along the nation's west coast? Why are they also less prevalent?

9. Explain how the airflow aloft regulates the movement of air masses.

10. The boundaries between neighboring air masses tend to be more distinct during the winter than during the summer. Explain why.

11. What type of air mass would be responsible for the weather conditions listed below? (a) hot, muggy summer weather in the Midwest and the East; (b) refreshing, cool, dry breezes after a long summer hot spell on the Central Plains; (c) persistent cold, damp weather with drizzle along the East Coast; (d) drought with high temperatures over the Great Plains; (e) record-breaking low temperatures over a large portion of North America; (f) cool weather with showers over the Pacific Northwest; (g) daily afternoon thunderstorms along the Gulf Coast

12. Describe the typical characteristics of: (a) a warm front (b) a cold front (c) an occluded front

13. Sketch side views of a model showing a typical cold front, warm front, and cold-occluded front. Include in each diagram cloud types and patterns, areas of precipitation, surface winds, and relative temperature on each side of the front.

14. Describe the stages of a developing mid-latitude cyclonic storm using the polar front theory.

15. Why do mid-latitude cyclones tend to develop along the polar front?

16. List four regions in North America where mid-latitude cyclones tend to develop.

17. Why is it important that for a surface low to develop or intensify, its upper-level counterpart must be to the left (or west) of the surface storm?

18. If upper-level diverging air above a surface area of low pressure exceeds converging air around the surface low, will the surface low weaken or intensify? Explain.

19. Describe some of the necessary ingredients (upper-air support) for a wave cyclone to develop into a huge mid-latitude cyclonic storm system.

20. Explain the role that upper-level diverging air plays in the development of a mid-latitude cyclone.

21. How does the polar jet stream influence the formation of a mid-latitude cyclone?

22. Explain why, in the eastern half of the United States, a mid-latitude cyclonic storm often moves eastward or northeastward.

QUESTIONS FOR THOUGHT AND EXPLORATION

1. If Lake Erie freezes over in January, is it still possible to have lake-effect snow on its eastern shores in February? Explain your answer.

2. Explain how an autumn anticyclone can bring record low temperatures and continental polar air to the southeastern United States and, only a day or so later, bring record high temperatures and maritime tropical air to the same region.

3. During the winter, cold-front weather is typically more violent than warm-front weather. Why is this so? Explain why this is not necessarily true during the summer.

4. You are in upstate New York and observe the wind shifting from the east to the south. This wind shift is accompanied by a sudden rise in both air temperature and dew-point temperature. What type of front is passing?

5. Why does the same cold front typically produce more rain over Kentucky than over western Kansas?

6. Explain why the boundaries between neighboring air masses tend to be more distinct during the winter than during the summer?

7. Sketch a Southern Hemisphere mid-latitude cyclonic storm, complete with isobars and at least two types of fronts. Compare and contrast this Southern Hemisphere cyclone with its Northern Hemisphere counterpart.

8. Why are mid-latitude cyclones described as waves?

9. Explain how this can happen: At the same time a mid-latitude cyclonic storm over the eastern United States is moving northeastward, a large surface high-pressure area over the northern plains is moving southeastward.

10. Would a wave cyclone intensify or dissipate if the upper trough were located to the *east* of the surface low-pressure area? Explain your answer with the aid of a diagram.

GLOBAL **GEOSCIENCE** WATCH   Under the Websites and Blogs section of Global Environment Watch: Meteorology, go to the site "Arctic Climatology and Meteorology PRIMER for Newcomers to the North" (National Snow and Ice Data Center). Within the Basics section of this site, consult the "Arctic Climate" page. What are the factors that can influence local weather within a particular arctic air mass? Consult the "Optical and Acoustic Phenomena" page. Why is it that conversations can sometimes be heard more than a mile away within a continental arctic air mass?

ONLINE RESOURCES

Visit www.cengagebrain.com to view additional resources, including video exercises, practice quizzes, an interactive eBook, and more.

CHAPTER 9

Weather Forecasting

Contents

Sometimes there is no job security in weather forecasting. In fact, at least one weather forecaster actually lost his job for not altering his prediction. On April 15, 2001, a function honoring a well-known radio talk show host was scheduled for outdoors at the Madera, California, fairgrounds. The story goes that a local forecaster at the radio station that sponsored the event had called for a "chance of rain" on April 15. Upset that such a forecast might discourage people from attending the function, the station manager told the forecaster to alter his forecast and predict a greater possibility of sunshine. The forecaster refused and was promptly fired. Apparently, retribution reigned supreme—it poured on the event.

Weather forecasts are issued to save lives, to save property and crops, and to tell us what to expect in our atmospheric environment. In addition, knowing what the weather will be like in the future is vital to many human activities. For example, a summer forecast of extended heavy rain and cool weather would have construction supervisors planning work under protective cover, department stores advertising umbrellas instead of bathing suits, and ice cream vendors vacationing as their business declines. The forecast would alert farmers to harvest their crops before their fields became too soggy to support the heavy machinery needed for the job. And the commuter knows that prolonged rain could mean clogged gutters, flooded highways, stalled traffic, blocked railway lines, and late dinners.

Put yourself in the shoes of a weather forecaster: It is your responsibility to predict the weather accurately so that thousands (possibly millions) of people in your area will know whether to carry an umbrella, wear an overcoat, or prepare for a winter storm. Since weather forecasting is not an exact science, your predictions will occasionally be incorrect. If your erroneous forecast misleads many people, you may become the target of jokes, insults, and even anger. There are even people who expect you to be able to predict the unpredictable. For example, on Monday you may be asked whether two Mondays from now will be a nice day for a picnic. And, of course, what about next winter? Will it be bitterly cold?

Unfortunately, accurate answers to such questions are beyond meteorology's present technical capabilities. Will forecasters ever be able to answer such questions confidently? If so, what steps are being taken to improve the forecasting art? How are forecasts made, and why do they sometimes go awry? These are just a few of the questions we will address in this chapter.

Weather Observations

Weather forecasting basically entails predicting how the present state of the atmosphere will change. Consequently, if we wish to make a weather forecast, we must know the present weather conditions over a large area. A network of observing stations located across the world provides the forecaster with this information. Forecasters have access to many maps and charts showing the present conditions at various atmospheric heights, as well as vertical profiles (called *soundings*) of temperature, dew point, and winds. Also available are visible and infrared satellite images, as well as Doppler radar information that can detect and monitor the severity of precipitation and thunderstorms. All of these sources are used by forecasters to monitor current weather and anticipate future conditions. Many of these observations are also brought into computer-based

atmospheric models that project the weather forward, as we will see later in this chapter.

More than 10,000 land-based stations and hundreds of ships and buoys provide surface weather information at least four times a day. Most airports observe conditions hourly, and hundreds of automated stations send reports even more often. To sample the atmosphere above ground level, radiosondes are launched at more than 800 locations around the world, including almost 100 U.S sites. Special launches may occur in association with research projects or to help provide more data when a major weather threat is looming, such as a high-impact winter storm or an outbreak of tornadic thunderstorms. Data on upper-air conditions may also be gathered and provided by some aircraft as they travel their usual routes. In addition, many types of observations collected by satellites are available to forecasters, providing a clearer representation of the atmosphere (see ● Fig. 9.1).

A network of more than 100 Doppler radar units covers nearly all of the 48 contiguous United States. These radars provide round-the-clock information on the evolution of rain, snow, sleet, and hail. Because Doppler radar can track winds as well as precipitation, the network is also a valuable tool in providing warnings of destructive windstorms and tornadoes, as we will see in the following chapters. Some of the most advanced computer forecast models are now being designed to incorporate information from radars, which could help improve forecasts significantly.

Acquisition of Weather Information

Collecting weather data is only the beginning of the process that leads to a forecast. Meteorologists at government weather services and private firms across the world rely on an accurate supply of weather data in order to make predictions for their areas. A United Nations agency—the World Meteorological Organization (WMO), which includes more than 175 nations—is responsible for the international exchange of weather data. The WMO certifies that the observation procedures do not vary among nations, an extremely important task since the observations must be comparable.

Weather information from all over the world is transmitted electronically to government meteorological centers worldwide. This includes the *National Centers for Environmental Prediction* (NCEP), a branch of the *National Weather Service* (NWS) located near the University of Maryland in College Park, just outside Washington, D.C. Here, the massive job of analyzing the data, running models, preparing weather maps and charts, and predicting the weather on a global and national basis begins.* From NCEP,

*By international agreement, data are plotted using symbols illustrated in Appendix C.

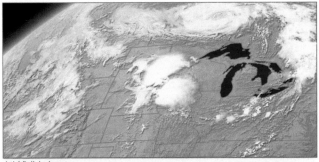

(a) Visible image

(b) Enhanced infrared image

(c) Water vapor image

NASA/MSFC Earth Science Office

● **FIGURE 9.1** These three satellite images were each collected by the GOES-East satellite at 11:15 a.m. on June 16, 2014. (a) Visible imagery, which detects solar radiation being reflected, is useful for detecting areas of snow cover as well as low-level cloud features, such as the cumulus developing in Nebraska that became thunderstorms later in the day. (b) Infrared imagery analyzes the temperature at the top of clouds, which is related to the amount of infrared energy being emitted; the bright colors over eastern South Dakota and Iowa indicate the high, cold tops of severe thunderstorms. (c) Water-vapor imagery detects the amount of energy absorbed at a particular wavelength that reveals moisture present in the middle and upper troposphere. Lighter colors indicate more moisture, while darker colors (such as in western Kansas and eastern Colorado) show where the upper troposphere is drier.

observations and computer model output are transmitted to U.S. private forecasting firms and public agencies. Many of NCEP's products are also posted on the web. Across the nation, dozens of NWS *Weather Forecast Offices* (WFOs) use the information to issue local and regional weather forecasts. Standard forecasts are prepared every 12 hours and updated as needed in between these intervals.

The public gets weather forecasts through a variety of channels, including radio, television, computers, and smartphones. Many broadcasting stations hire private meteorological companies or professional meteorologists

to make their own forecasts assisted by data and products from NCEP, or to modify an NWS forecast. Other stations hire meteorologically untrained announcers who paraphrase the NWS forecast, or read them word for word. On the web and on smartphones, the public can access local forecasts from the NWS and from private firms, often presented with eye-catching graphics. (Before going on, you may wish to read Focus section 9.1, which describes how TV weather forecasters present weather visuals.)

Weather Forecasting Tools

In the course of a single workday, a typical forecaster may examine and compare dozens or even hundreds of individual weather maps. To help forecasters handle all the available charts and maps, the NWS employs the high-speed *Advanced Weather Interactive Processing System* (**AWIPS**). A second-generation version, called *AWIPS II,* was adopted by the NWS starting in 2013 (see ● Fig. 9.2).

The AWIPS II system has data communications, storage, processing, and display capabilities (including graphical overlays) to better help the individual forecaster extract and assimilate information from the mass of available data. In addition, AWIPS is able to process information received from satellites and surface stations as well as from the Doppler radar system (the WSR-88D), which now includes dual-polarization technology. Much of the information from ASOS* and Doppler radar is processed by software according to predetermined formulas, or *algorithms,* before it goes to the forecaster. Certain criteria or combinations of measurements can alert the forecaster to an impending weather situation.

A software component of AWIPS II called the *Graphical Forecast Editor* allows forecasters to look at the daily prediction of weather elements, such as temperature and

---

*Additional information on ASOS is found in Chapter 3, on p. 73.

NOAA

● **FIGURE 9.2** Two NWS forecasters testing the AWIPS II system, adopted by the NWS in 2013.

TV WEATHERCASTERS—HOW DO THEY DO IT?

As you watch the TV weathercaster, you typically see a person describing and pointing to specific weather information, such as satellite and radar images, and weather maps, as shown in ● Fig. 1. What you may not know is that in many instances the weathercaster is actually pointing to a blank board (usually green or blue) on which there is nothing (● Fig. 2).* This

*On some stations, forecasters point to weather information that appears on a very large TV screen.

process of electronically superimposing weather information in the TV camera against a blank wall is called color-separation overlay, or *chroma key*.

The chroma key process works because the studio camera is constructed to pick up all colors except (in this case) green. The various maps, charts, satellite photos, and other graphics are electronically inserted from a computer into this green area of the color spectrum. The person in the TV studio

should not wear green clothes because such clothing would not be picked up by the camera; what you would see on your home screen would be a head and hands moving about the weather graphics!

How, then, does a TV weathercaster know where to point on the blank wall? Positioned on each side of the green wall are TV monitors (look carefully at Fig. 2) that weathercasters watch so that they know where to point.

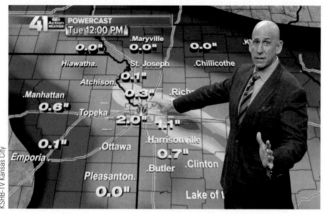

● **FIGURE 1** On your home television, this weathercaster appears to be pointing to weather information directly behind him.

● **FIGURE 2** In the studio, however, the weathercaster is actually standing in front of a blank green board.

dew point, in a gridded format with spacing as small as 2.5 km (1.6 mi). Presenting the data in this format allows the forecaster to predict the weather more precisely over a relatively small area.

With so much information at the forecaster's disposal, it is essential that the data be easily accessible and in a format that allows several weather variables to be viewed at one time. The **meteogram** is a chart that shows how one or more weather variables has changed at a station over a given period of time. As an example, the chart may represent how air temperature, dew point, and sea-level pressure have changed over the past five days, or it may illustrate how these same variables are projected to change over the next five days (see ● Fig. 9.3).

Another aid in weather forecasting is the use of **soundings.** A sounding is a two-dimensional vertical profile of temperature, dew point, and winds (see ● Fig. 9.4).* The analysis of a sounding can be especially helpful when making a short-range forecast that

*A sounding is obtained from a radiosonde or from satellite data. For additional information on the radiosonde, see the Focus section in Chapter 1 on p. 22.

covers a relatively small area, such as the mesoscale. The forecaster examines the sounding from the immediate area (or closest proximity), as well as the soundings from sites upwind, to see how the atmosphere might be changing. Computer programs automatically calculate from the sounding a number of meteorological *indexes* that can aid the forecaster in determining the likelihood of smaller-scale weather phenomena, such as thunderstorms, tornadoes, and hail. Soundings also provide information that can aid in the prediction of fog, air pollution alerts, and the downwind mixing of strong winds.

Satellite information is also a valuable tool for the forecaster. Visible, enhanced infrared, and water vapor images provide a wealth of information, some of which comes from inaccessible regions, that can be examined to analyze fast-changing conditions.* Special instruments aboard satellites can detect a wide variety of important phenomena, including lightning, sea surface temperature, and smoke from forest fires. Many kinds of satellite observations are incorporated into forecast models.

*Information provided by satellites is discussed in various sections of this book. For example, see Chapter 4, p. 106.

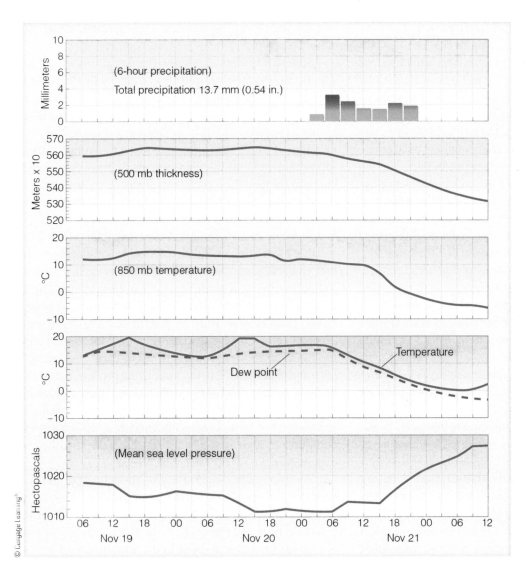

● FIGURE 9.3 Meteogram illustrating predicted weather at the surface and aloft at St. Louis, Missouri, from 6 a.m., November 19, 2007, to noon on November 21, 2007. The forecast is derived from the Global Forecast System (GFS) model. (NOAA)

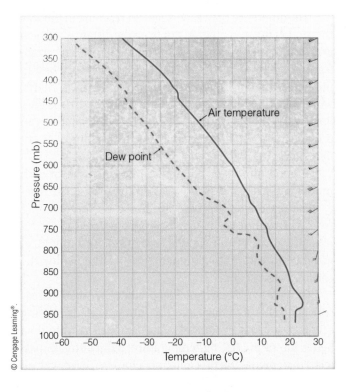

Weather Forecasting Methods

Up to this point, we have examined some of the weather data and tools a forecaster might use in making a weather prediction. With all of this information available to the forecaster, including countless charts and maps, just *how* does a meteorologist make a weather forecast?

THE COMPUTER AND WEATHER FORECASTING: NUMERICAL WEATHER PREDICTION Each day the many thousands of observations transmitted to NCEP are depicted in surface and upper-air charts. Meteorologists interpret the weather patterns and then correct any errors that may be present. The final chart is referred to as an **analysis**.

● FIGURE 9.4 A sounding of air temperature, dew point, and winds near Oklahoma City, Oklahoma, during the evening of May 20, 2013, on the same day a violent tornado (pictured in Fig. 9.12) tore through the town of Moore, Oklahoma, which lies just south of Oklahoma City.

Computers not only plot and analyze data, but they also predict the weather—a much more challenging task. Today's supercomputers can analyze large quantities of data extremely quickly, carrying out trillions of calculations per second. Because the atmosphere is so complex, some of the most powerful supercomputers on Earth are devoted to weather and climate prediction. The routine daily forecasting of weather by the computer using mathematical equations is known as **numerical weather prediction.**

The many weather variables are constantly changing, so meteorologists have devised **atmospheric models** that describe the present state of the atmosphere. These are not physical models that paint a picture of a developing storm; they are, rather, mathematical models consisting of many equations that describe how atmospheric temperature, pressure, winds, and moisture will change with time. The models do not fully represent the actual atmosphere, since the processes occurring around the world at every instant are too complex to represent completely. Instead, the models are very useful approximations, formulated to retain the most important aspects of the atmosphere's behavior.

How do these models actually work? The equations are translated into complex software, and surface and upper-air observations of temperature, pressure, moisture, winds, and air density are fed into the equations at regular intervals. The process of integrating these data into numerical models is called **data assimilation**. As more and more types of data are assimilated into the models, the quality of the model forecasts often improves.

To determine how each of these key meteorological variables will change, each equation is solved for a small increment of future time—say, five minutes—for a large number of locations called *grid points,* each situated a given distance from the next.* In addition, each equation is solved for as many as 50 levels in the atmosphere. The results of these computations are then fed back into the original equations. The computer again solves the equations with the new "data," thus predicting weather over the following five minutes. This procedure is done repeatedly until it reaches some desired time in the future. For example, one standard NWS model produces weather depictions every hour out to 18 hours. Another covers a longer period, but with larger steps in between; it produces snapshots of weather conditions every 3 hours out to 84 hours (3.5 days). And one model even forecasts the state of the atmosphere 384 hours (16 days) into the future. Once the calculations of future weather are completed, the computer then analyzes the data and draws the projected positions of pressure systems with their isobars or contour lines. The final forecast chart representing the atmosphere

at a specified future time is called a **prognostic chart,** or, simply, a **prog.**

Supercomputers can solve the equations of atmospheric motion far more quickly and efficiently than could possibly be done by hand. For example, just to produce a 24-hour forecast chart for the Northern Hemisphere requires many hundreds of millions of mathematical calculations. It would take a group of meteorologists working full time with hand calculators years to produce a single chart; by the time the forecast was available, the weather for that day would already be ancient history. Today, computer models are taken for granted in weather forecasting. They have become so important that our current level of skill in weather forecasting would be impossible without them. In fact, in some cases a computer model may do just as well as a human being in predicting high and low temperatures up to several days in advance during tranquil weather. However, forecasters must be careful not to take the prediction of any model as the gospel truth. If forecasters rely too much on models, without bringing their own knowledge and experience into the forecasting process, they may become victims of what has been termed "meteorological cancer." During the most unusual and threatening weather situations, the best forecasts occur when the output of computer models is carefully adjusted as needed by an experienced, knowledgeable meteorologist.

An ever-increasing variety of models (and, hence, progs) is now available to forecasters, each producing a slightly different interpretation of the weather for the same projected time and atmospheric level. ● Figure 9.5 shows four progs for different levels in the atmosphere 24 hours into the future. How the forecaster might use each prog in making a prediction is given in ▼ Table 9.1.

The difference between progs can result from the way the models use the equations, or from the distance between grid points, called *resolution.* Some models predict certain features better than others: One model may work best in predicting the position of troughs on upper-level charts, whereas another may forecast the position of surface lows quite well. Although a model with a higher resolution can provide more detail, such enhanced detail alone does not necessarily mean the model is more accurate.

A good forecaster knows the idiosyncrasies of each model and carefully scrutinizes all the progs. The forecaster then makes a prediction based on the *guidance* from the computer, his or her personalized practical interpretation of the weather situation, and any local geographic features that influence the weather within the specific forecast area.

Currently, forecast models predict the weather reasonably well 4 to 6 days into the future, with decreasing accuracy for longer intervals. These models tend to do a better job of predicting temperature and jet-stream patterns than predicting precipitation. However, even with all

---

*Some models have a grid spacing smaller than 0.5 km, whereas the spacing in others exceeds 100 km. There are models that actually describe the atmosphere using a set of mathematical equations with wavelike characteristics rather than a set of discrete numbers associated with grid points.

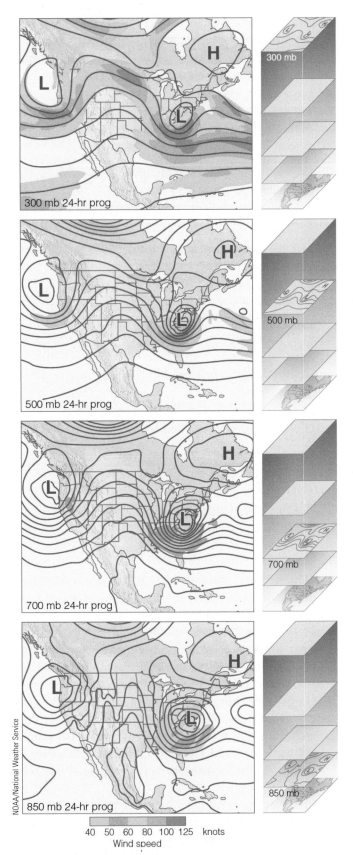

● **FIGURE 9.5** Computer-drawn 24-hour progs using the GFS (Global Forecast Systems) model for 850 mb, 700 mb, 500 mb, and 300 mb. The progs were drawn on March 5, 2013, and became valid on March 6 at 7 a.m. Solid lines represent height contours. Orange shade shows the wind speed in knots. Green shade shows where the predicted relative humidity is 70 percent or greater.

of the modern advances in weather forecasting provided by ever more powerful computers, forecasts by the National Weather Service and by private firms are sometimes wrong. (An example of making a 24-hour forecast using different progs is given toward the end of this chapter on p. 264.)

WHY COMPUTER-BASED FORECASTS CAN GO AWRY AND STEPS TO IMPROVE THEM Why do forecasts sometimes go wrong? There are a number of reasons for this unfortunate situation. For one, computer models have inherent flaws that limit the accuracy of weather forecasts. For example, computer forecast models idealize the real atmosphere, meaning that each model makes certain assumptions about the atmosphere. These assumptions may be on target for some weather situations and be way off for others. Consequently, the computer may produce a prog that comes quite close to describing the actual state of the atmosphere on one day and not so close on another. A forecaster who bases a prediction on an "off day" computer prog may find a forecast of "rain and windy" turning out to be a day of "clear and colder."

Another forecasting problem arises because the majority of models are not global in their coverage, and errors can creep in along the model's boundaries. For example, a model that predicts the weather for North America may not accurately treat weather systems that move in along its boundary from the western Pacific. Obviously, a global model would always be preferred. But a global model of similar sophistication with a high resolution requires an incredible number of computations.

Even though many thousands of weather observations are taken worldwide each day, there are still regions where observations are sparse, particularly over the oceans and at higher latitudes. To help alleviate this problem, newer satellites are providing a more accurate profile of temperature and humidity for the computer models. Wind information now comes from a variety of sources, such as Doppler radar, commercial aircraft, buoys, and satellites that translate ocean surface roughness into surface wind speed. (See Chapter 6, p. 167.)

Grid Spacing Earlier, we saw that the computer solves the equations that represent the atmosphere at many locations called *grid points*, each spaced from 100 km to as little as 0.5 km apart. On computer models with large spacing between grid points (say 60 km), larger weather systems, such as extensive mid-latitude cyclones and anticyclones, show up on computer progs, whereas much smaller systems, such as thunderstorms, do not. Because the model is too coarse to generate showers and thunderstorms directly, these features are instead *parameterized*, meaning that they are approximated for broad areas instead of being predicted for specific points. The computer models that forecast for a large area such as North America, are,

WEATHER FORECASTING 249

FORECAST CHART	APPROXIMATE ALTITUDE ABOVE SEA LEVEL	ELEMENTS THAT MAY BE SPOTTED AND TRACKED
Surface map		• Location and motion of frontal systems, centers of high and low pressure • Areas of cloudiness, precipitation, high wind, and fog • Cross-isobar winds that indicate strengthening or weakening low pressure • Warm, moist air that can foster shower and thunderstorm development if conditional instability is present
850 mb	1500 m (4900 ft)	• High moisture values that can contribute to heavy precipitation • Convergent winds associated with strengthening low pressure areas • A low-level jet stream that can help intensify thunderstorm development • Temperatures that determine whether precipitation will fall as snow, rain, sleet, or a mixture
700 mb	3000 m (9800 ft)	• Moisture to feed precipitation and mid-latitude storm systems • Temperature advection that could strengthen or weaken fronts • A dry, warm layer that can inhibit thunderstorm development • Temperatures that help determine ice crystal and snowfall type
500 mb	5600 m (18,400 ft)	• General steering flow for mid-latitude storm systems, hurricanes, and tropical cyclones • Location and motion of ridges, troughs, and short waves that generate and strengthen surface features • Areas of cold advection that can help increase conditional instability and support thunderstorm development • Large areas of high or low heights that correspond to unusually warm or cold conditions at the surface, depending on region and time of year
300 mb	9180 m (30,100 ft)	• Location of core of jet stream • Jet streaks and areas of divergence within jet stream that may correspond to intensifying low pressure at the surface • Areas of high pressure and light, divergent wind in the tropics and subtropics that can support hurricane development

therefore, better at predicting the widespread precipitation associated with a large cyclonic storm than predicting where localized showers and thunderstorms will occur. In summer, when much of the precipitation falls as local showers, a computer prog may have indicated fair weather while outside it is pouring rain.

To capture the smaller-scale weather features as well as the terrain of the region, the distance between grid points on some models is being reduced. For example, the forecast model known as *High-Resolution Rapid Refresh (HRRR)* has a grid spacing as low as 3 km. Instead of parameterizing showers and thunderstorms, a model like HRRR with its small grid spacing (high resolution) can actually incorporate radar information and simulate how showers and thunderstorms might evolve. The problem with high resolution models is that, as the horizontal spacing between grid points decreases, the number of computations increases. When the distance is halved, there are 8 times as many computations to perform, and the time (and computational expense) required to run the model goes up by a factor of 16.

Another forecasting problem is that many computer models cannot adequately interpret many of the factors that influence surface weather, such as the interactions of water, ice, surface friction, and local terrain on weather systems. Many large-scale models now take mountain regions and oceans into account. Some models (such as HRRR) take even smaller factors into account, features that large-scale computers miss due to their larger grid spacing. Given the effect of local terrain, as well as the impact of some of the other problems previously mentioned, computer models that forecast the weather over a vast area do an inadequate job of predicting local weather

DID YOU KNOW?

When a weather forecast calls for "fair weather," does the "fair" mean that the weather is better than "poor" but not up to being "good"? According to the National Weather Service, the subjective term "fair" implies a rather pleasant weather situation where there is no precipitation, temperatures are seasonable, winds are light, and visibility is good, with less than 40 percent of the sky covered by opaque clouds, such as stratus.

conditions, such as surface temperatures, winds, and precipitation.

Even with better observing techniques and high-quality computer models, countless small, unpredictable atmospheric fluctuations, referred to as **chaos**, limit model accuracy. For example, tiny eddies, much smaller than the grid spacing on the computer model, go unaccounted for in the model. These small disturbances, as well as small errors (uncertainties) in the data, generally amplify with time as the computer tries to project the weather further and further into the future. After a number of days, these initial imperfections tend to dominate, and the forecast shows little or no accuracy in predicting the behavior of the real atmosphere. In essence, what happens is that the small uncertainty in the initial atmospheric conditions eventually leads to a huge uncertainty in the model's forecast. There is, therefore, a limit to how far into the future we will ever be able to accurately forecast the weather at a specific place and time. However, it is still possible to make climatological projections that give the *likelihood* of particular types of weather far into the future.

Ensemble Forecasts Because of the atmosphere's chaotic nature, meteorologists have turned to a technique called **ensemble forecasting** to improve short- and medium-range forecasts. The ensemble approach is based on running several forecast models—or different versions (simulations) of a single model—each beginning with slightly different weather information to reflect the errors inherent in the measurements. Suppose, for example, a forecast model predicts the state of the atmosphere 24 hours into the future. For the ensemble forecast, the entire model simulation is repeated, but only after the initial conditions are "tweaked" just a little. The "tweaking," of course, represents the degree of uncertainty in the observations. Repeating this process several times creates an ensemble of forecasts for a range of small initial changes.

● Figure 9.6 shows an ensemble 500-mb forecast for March 1, 2013 (96 hours or 4 days into the future). The chart is constructed by running the model 17 different times, each time with a slightly different initial condition. Notice that the ensemble numbers are in strong agreement in some locations, such as the eastern Pacific, while there is large uncertainty in the other locations, such as

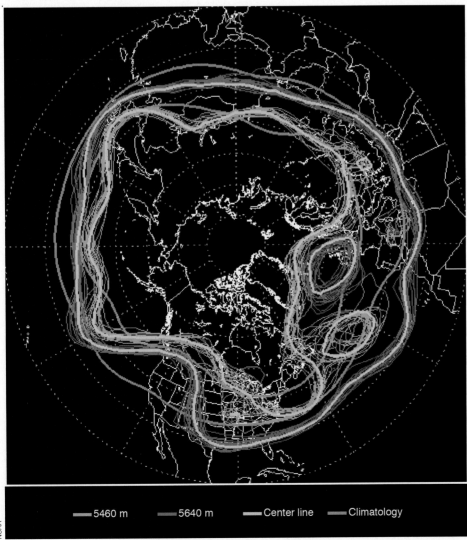

● **FIGURE 9.6** Ensemble 500-mb forecast chart for March 1, 2013, issued on the morning of February 25. The blue lines represent the 5460-meter contour; the red lines, the 5640-meter contour; and the green lines, a 16-year average, represents the average locations of the two contours. The yellow lines show where the main or "operational" member of the ensemble placed the contours.

5460 m   5640 m   Center line   Climatology

NOAA

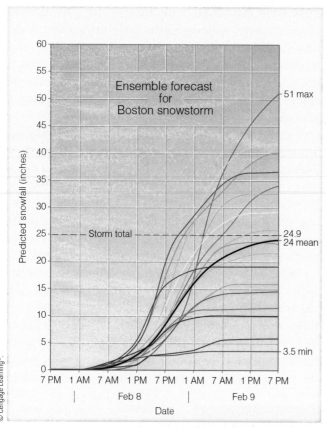

● **FIGURE 9.7** An ensemble of forecasts of total snowfall in Boston, Massachusetts, for February 8–9, 2013, issued early on February 6. Each trace represents the forecast from a different model run, all carried out at the same time but with slightly different initial conditions.

the midwestern United States. As the forecast goes further and further into the future, the lines usually look more and more like scrambled spaghetti, which is why an ensemble forecast chart is often referred to as a *spaghetti plot*.

If, at the end of a specific time, the progs, or model runs, match each other fairly well, the forecast is considered *robust*. This situation allows the forecaster to issue a prediction with a high degree of confidence. If the progs disagree, the forecaster has less faith in the computer model prediction and will issue a forecast with more limited confidence. In essence, *the less agreement among the progs, or model runs, the less predictable the weather.* Consequently, it would not be wise to make outdoor plans for Saturday when on Monday the weekend forecast calls for "sunny and warm" with a low degree of confidence. Most everyday weather forecasts do not indicate the degree of forecaster confidence, but it is sometimes mentioned in TV weather reports or special NWS advisories, especially when threatening weather is a possibility.

The example in ● Figure 9.7 shows how widely the members of an ensemble can diverge for a critical weather situation, as with the blizzard that struck New England in early February 2013. The ensemble includes three different models, each run six times with slightly different initial conditions, plus two other runs. The result is 20 predictions

for total snowfall in Boston, Massachusetts. Notice that the forecasts vary tremendously, from only 3.5 inches to an eye-opening 51 inches. The solid black line indicates the ensemble mean, which is 24 inches. Since the actual snow total in Boston in this storm was 24.9 inches, a forecast that relied on the ensemble mean would have been an excellent one. The ensemble mean is not a perfect predictor in every case, but on average it tends to perform as well or better than any single model.

In summary, imperfect numerical weather predictions may result from flaws in the computer models, from errors that creep in along the models' boundaries, from the sparseness of data, and/or from inadequate representation of many pertinent processes, interactions, and inherently chaotic behavior that occurs within the atmosphere. However, by watching carefully for potential model errors and by using ensembles made up of a number of different model runs, a forecaster can increase the likelihood of making an accurate prediction.

OTHER FORECASTING TECHNIQUES Because the weather affects every aspect of our daily lives, attempts to predict it accurately have been made for centuries. One of the earliest attempts was undertaken by Theophrastus, a pupil of Aristotle, who in 300 BC compiled all sorts of weather indicators in his *Book of Signs*. A dominant influence in the field of weather forecasting for 2000 years, this work consists of ways to foretell the weather by examining natural signs, such as the color and shape of clouds, and the intensity at which a fly bites. Some of these signs have validity and are a part of our own weather folklore—"A halo around the moon portends rain" is one of these. Today, we realize that the halo is caused by the bending of light as it passes through ice crystals and that ice crystal-type clouds (cirrostratus) are often the forerunners of an approaching cyclonic storm (see ● Fig. 9.8). If you keep

● **FIGURE 9.8** A halo around the sun (or moon) means that rain is on the way, a weather forecast made by simply observing the sky.

your eyes open and your senses keenly tuned to your environment, you should, with a little practice, be able to make fairly good short-range local weather forecasts by interpreting the messages written in the weather elements.

Official weather forecasting activities were launched by the governments of many nations in the late 1800s and expanded as observational techniques improved in the 1900s. During the years before the advent of computer-based weather models, many forecasting methods were based largely on the experience of the forecaster. Many of these techniques were of value, but typically they gave a more general overview of what the weather should be like rather than a specific forecast. As late as the mid-1950s, all weather maps and charts, including those depicting current as well as future conditions, were plotted by hand and analyzed by individuals. Meteorologists predicted the weather using certain rules that related to the particular weather system in question. For short-range forecasts of six hours or less, surface weather systems were moved along at a steady rate. Upper-air charts were used to predict where surface storms would develop and where pressure systems aloft would intensify or weaken. The predicted positions of these systems were extrapolated into the future using linear graphical techniques and current maps. Experience played a major role in making the forecast. In many cases, these forecasts turned out to be amazingly accurate. However, with the advent of modern supercomputers, along with our present observing techniques, today's forecasts are considerably better.

Probably the easiest weather forecast to make is a **persistence forecast**, which is simply a prediction that future weather will be the same as present weather. If it is snowing today, a persistence forecast would call for snow through tomorrow. Such forecasts are most accurate for time periods of several hours and become less and less accurate after that. Persistence forecasts are also more useful at those times and places where the weather tends to change less dramatically.

Another method of forecasting is the **steady-state, or trend, forecast.** The principle involved here is that surface weather systems tend to move in the same direction and at approximately the same speed as they have been moving, providing no evidence exists to indicate otherwise. Suppose, for example, that a cold front is moving eastward at an average speed of 30 mi/hr and it is 90 mi west of your home. Using the steady-state method, we might extrapolate and predict that the front should pass through your area in three hours.

The **analog method** is yet another form of weather forecasting. Basically, this method relies on the fact that existing features on a weather chart (or a series of charts) may strongly resemble features that produced certain weather conditions sometime in the past. To the forecaster, the weather map "looks familiar," and for this reason the analogue method is often referred to as *pattern recognition*. A forecaster might look at a prog and say "I've seen this weather situation before, and this happened." Previous weather events can then be utilized as a guide to the future. The problem here is that, even though weather situations may appear similar, they are never *exactly* the same. There are always sufficient differences in the variables to make applying this method a challenge.

Even so, the analog method can be used to predict a number of weather elements, such as maximum temperature. Suppose that in New York City the average maximum temperature on a particular date for the past 30 years is 10°C (50°F). By statistically relating the maximum temperatures on this date to other weather elements—such as the wind, cloud cover, and humidity—a relationship between these variables and maximum temperature can be drawn. By comparing these relationships with current weather information, the forecaster can predict the maximum temperature for the day.

Statistical forecasts of weather elements are based on the past performance of computer models. Known as *Model Output Statistics,* or MOS, these predictions, in effect, are statistically weighted analogue forecast corrections incorporated into the computer model output. For example, a forecast of tomorrow's maximum temperature for a city might be derived from a statistical equation that uses a numerical model's forecast of relative humidity, cloud cover, wind direction, and air temperature.

When the Weather Service issues a forecast calling for rain, it is usually followed by a probability. For example: "The chance of rain is 60 percent." Does this mean (a) that it will rain on 60 percent of the forecast area or (b) that there is a 60 percent chance that it will rain within the forecast area? Neither one! The expression means that there is a 60 percent chance that any random place in the forecast area, such as your home, will receive measurable rainfall. Looking at the forecast in another way, if the forecast for 10 days calls for a 60 percent chance of rain, it should rain where you live on 6 of those days. The verification of the forecast (whether it actually rained or not) is usually made at the Weather Service office, but remember that the computer models forecast for a given region, not for an individual location. When the National Weather Service issues a forecast calling for a "slight chance of rain," what is the probability (percentage) that it will rain? ▼ Table 9.2 provides this information.

An example of a **probability forecast** using climatological data is given in ● Fig. 9.9. The map shows the probability of a "White Christmas"—1 inch or more of snow on the ground—across the United States. The map is based on the average of 30 years of data and gives the likelihood of snow in terms of a probability. For instance, the chances are greater than 90 percent (9 Christmases out of 10) that portions of northern Minnesota, Michigan, and Maine will experience a White Christmas. In Chicago, it is close to 50 percent; and in Washington, D.C., about

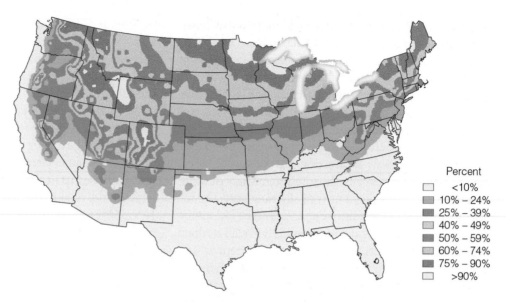

● FIGURE 9.9 Probability of a "White Christmas"—one inch or more of snow on the ground—based on a 30-year average, from 1981 to 2010, inclusive. The probabilities do not include all of the mountainous areas in the western United States. (NOAA)

Percent
- [ ] <10%
- [ ] 10% – 24%
- [ ] 25% – 39%
- [ ] 40% – 49%
- [ ] 50% – 59%
- [ ] 60% – 74%
- [ ] 75% – 90%
- [ ] >90%

20 percent. Many places in the far west and south have probabilities less than 5 percent, but nowhere is the probability exactly 0, for there is always some chance (no matter how small) that a mantle of white will cover the ground on Christmas Day. For example, on December 24–25, 2004 (see ● Fig. 9.10), Corpus Christi, Texas, reported 4.4 inches of snowfall, and Brownsville, Texas, at the very southern part of the state, had 1.5 inches of snow, making it the first snowfall in Brownsville since 1899. For both cities, it was the heaviest snowfall on record for any date.

Predicting the weather by **weather types** employs the analog method. In general, weather patterns are categorized into similar groups or "types," using such criteria as the position of the subtropical highs, the upper-level flow, and the prevailing storm track. As an example, when the Pacific high is weak or depressed southward and the flow aloft is zonal (west-to-east), surface storms tend to travel rapidly eastward across the Pacific Ocean and into the United States without developing into deep systems. But when the Pacific high is to the north of its normal position and the upper

▼ Table 9.2 **Forecast wording used by the National Weather Service to describe the percentage probability of measurable precipitation (0.01 inch or greater) for steady precipitation and for convective, showery precipitation.**

PERCENT PROBABILITY OF PRECIPITATION	FORECAST WORDING FOR STEADY PRECIPITATION	FORECAST WORDING FOR SHOWERY PRECIPITATION
10 to 20 percent	*Slight chance* of precipitation	*Isolated* showers
30 to 50 percent	*Chance* of precipitation	*Scattered* showers
60 to 70 percent	Precipitation *likely*	*Numerous* showers
≥ 80 percent	Precipitation,* rain, snow	*Showers***

*A forecast that calls for an 80 percent chance of rain in the afternoon might read like this: ". . . cloudy today with rain this afternoon. . . ." For an 80 percent chance of rain showers, the forecast might read ". . . cloudy today with rain showers this afternoon. . . ."

**The 60 percent chance of rain does not apply to a situation that involves rain showers. In the case of showers, the percentage refers to the expected area over which the showers will fall.

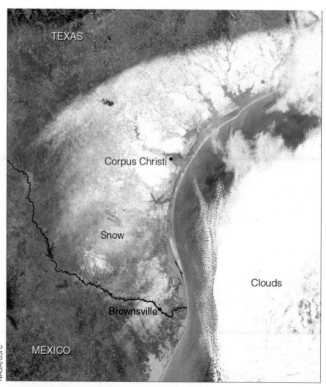

● FIGURE 9.10 Satellite view of South Texas along the Gulf Coast on Christmas Day, 2004. The white area covering Corpus Christi and Brownsville is snow. The probability of measurable snow on the ground in either of these two cities on Christmas Day is less than one percent (see Fig. 9.9). Yet, Corpus Christi received over 4 inches of snow and Brownsville about 1.5 inches. Just days later, the temperature climbed into the 80s (°F).

airflow is meridional (north-south), looping waves form in the flow with surface lows usually developing into huge storms. Since upper-level longwaves move slowly, usually remaining almost stationary for perhaps a few days to a week or more, the particular surface weather at different positions around the wave is likely to persist for some time. ● Figure 9.11 presents an example of weather conditions most likely to prevail with a winter meridional weather type.

A forecast based on the climate* of a particular region is known as a **climatological forecast.** Anyone who has lived in Los Angeles for a while knows that July and August are practically rain-free. In fact, rainfall data for the summer months taken over many years reveal that rainfall amounts of more than a trace occur in Los Angeles about 1 day in every 90, or only about 1 percent of the time. Therefore, if we predict that it will not rain on a particular date next year during July or August in Los Angeles, our chances are nearly 99 percent that the forecast will be correct based on past records. Since it is unlikely that this pattern will significantly change in the near future, we can confidently make the same forecast for the year 2025.

Time Range of Forecasts

Weather forecasts are normally grouped according to how far into the future the forecast extends. For example, a weather forecast for up to a few hours (usually not more than 6 hours) is called a **nowcast** (a very short range forecast). The techniques used in making such a forecast normally involve subjective interpretations of surface observations, satellite imagery, and Doppler radar information. Often the forecaster moves weather systems along by the steady state or trend method of forecasting, with human experience and pattern recognition coming into play.

When severe or hazardous weather is likely or is occurring, the National Weather Service issues short-range alerts in the form of weather watches, warnings, and advisories. A **watch** indicates that atmospheric conditions favor hazardous weather occurring over a particular region during a specified time period. These hazards may or may not actually develop, and their timing and location are uncertain, so a watch simply means to be on "watch" for that threat and to be prepared to act if necessary. When hazardous weather has developed, or is about to develop, two types of alerts may be issued. A **warning** indicates that the hazard now occurring or imminent is considered to be a threat to life and/or property (such as tornadoes, flash floods, severe thunderstorms, and winter storms). An *advisory* is similar to a warning, except that

*The climate of a region represents the total accumulation of daily and seasonal weather events for a specific interval of time, most often 30 years.

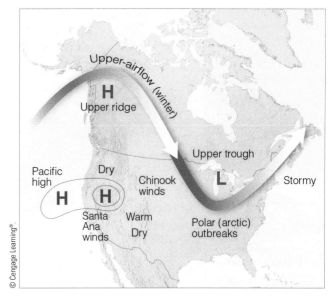

● **FIGURE 9.11** Winter weather type showing upper airflow (heavy arrow), surface position of Pacific high, and general weather conditions that should prevail.

it is used to indicate hazards that are usually less severe (such as light snow, light freezing rain, or dense fog). Note that sometimes even an advisory-level hazard can produce major problems, such as when light snow falls, melts, and quickly freezes on road surfaces.

Weather forecasts that range from about 12 hours to a few days (generally up to 3 days or 72 hours) are called short-range forecasts. The forecaster may incorporate a variety of techniques in making a **short-range forecast**, such as satellite imagery, Doppler radar, surface weather maps, upper-air wind data, and pattern recognition. As the forecast period extends beyond about 12 hours, the forecaster tends to weight the forecast heavily on computer-drawn progs and statistical information, such as Model Output Statistics (MOS).

A **medium-range forecast** is one that extends from about 3 to 8 days (192 hours) into the future. Medium-range forecasts are almost entirely based on computer-derived products, such as forecast progs and statistical forecasts (MOS). A forecast that extends beyond 3 days is often called an *extended forecast*.

DID YOU KNOW?

Weather presentations have come a long way since the early days of television. From the 1950s through the 1970s, before computer graphics were available, fads and gimmicks were common. Both men and women delivered forecasts in attire that could range from bathing suits to clown costumes. Puppets and cartoon characters were often used to deliver the forecast or to react to it. At a number of stations, the weatherperson actually wrote temperatures and drew maps while standing behind a clear Plexiglas screen. Because of the workings of the TV camera, these weathercasters had to write backwards.

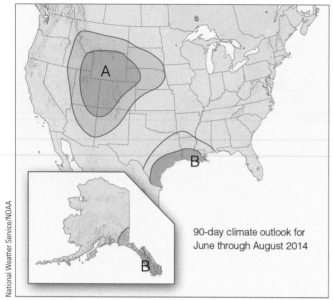

(a) Precipitation

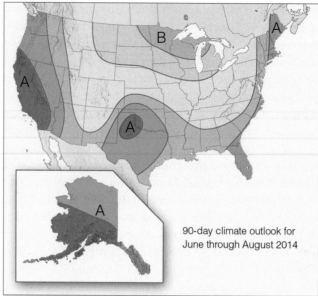

(b) Temperature

● **FIGURE 9.12** The 90-day outlook issued by NOAA in mid-May 2014 for (a) precipitation and (b) temperature for the period June through August 2014. For precipitation (a), the darker the green color, the greater the probability of precipitation being above normal, whereas the deeper the brown color, the greater the probability of precipitation being below normal. For temperature (b), the darker the orange/red colors, the greater the probability of temperatures being above normal, whereas the darker the blue color, the greater the probability of temperatures being below normal. On both maps, the letter A stands for "above normal" and the letter B for "below normal."

A forecast that extends beyond about 8 days (192 hours) is called a **long-range forecast**. Although computer progs are available for up to 16 days into the future, they are not accurate in predicting local temperature and precipitation, and at best only show the broad-scale weather features. The NOAA Climate Prediction Center summarizes these general trends in products called *outlooks* that cover 6- to 10-day and 8- to 14-day periods. These are not forecasts in the strict sense, but rather give an overview of how the expected precipitation and temperature patterns may compare with average conditions. ● Figure 9.12 gives a typical 90-day outlook.

NOAA also issues **seasonal outlooks** every month. These cover three-month periods that overlap and extend out to roughly a year. Again, rather than depicting specific weather features, these outlooks show the odds that a given area might experience temperatures or precipitation that are above or below average. Initially, these outlooks were based mainly on the relationship between the projected average upper-air flow and the surface weather conditions that the type of flow would create. Today, long-range forecasts call on models that link the atmosphere with sea-surface temperature, such as the Climate Forecast System version 2 (CFSv2). Many of the outlooks also take into account persistence statistics that carry over the general weather pattern from immediately preceding months, seasons, and years.

In Chapter 7, we saw how a vast warming (El Niño) or cooling (La Niña) of the equatorial tropical Pacific can affect the weather in different regions of the world. These interactions, where a warmer or cooler tropical Pacific can influence rainfall in California, are called *teleconnections*.* These types of interactions between widely separated regions are identified through statistical correlations. For example, over regions of North America, where temperature and precipitation patterns tend to depart from normal during El Niño and La Niña events, the Climate Prediction Center can issue—months in advance—a seasonal outlook of an impending wetter or drier winter. Seasonal outlooks using teleconnections have become increasingly useful.

Up to now we have looked at how weather forecasts are made and how forecasts can influence our daily lives. For a look at how weather forecasts can influence the marketplace, read Focus section 9.2.

In most locations throughout North America, the weather is fair more often than rainy. Consequently, there is a forecasting bias toward fair weather, which means that if you make a forecast of "no rain" where you live for each day of the year, your forecast will be correct more than 50 percent of the time. But did you show any *skill* in making your correct forecast? What constitutes skill anyway? And how accurate are the forecasts issued by the National Weather Service?

---

*Teleconnections include not only El Niño and La Niña but other indices, such as the Pacific Decadal Oscillation, the North Atlantic Oscillation, and the Arctic Oscillation. For more information on these indices, see Chapter 7, pp. 203-205.

WEATHER PREDICTION AND THE MARKETPLACE

A good forecast can not only make or break plans for a picnic, but it can also spell the difference between profit and loss for an entire business. Weather predictions are a critical tool for many parts of the economy. Short-term forecasts can help an orange grower deal with the threat of a hard freeze or tip off a construction company to the risk of work delays. On a broader scale, the prices of stocks and commodities* can swing up or down based on the approach of a major storm, the forecasts of its behavior, and the damage left behind. For example, the price of frozen concentrated orange juice rose more than 40 percent in the month after Hurricane Charley struck many of Florida's citrus groves in August 2004.

For many companies, seasonal outlooks are even more important than day-to-day forecasts. A corporation that makes bread or pasta might pay close attention to long-term outlooks for temperature and precipitation across the wheat-growing areas of North America in order to anticipate potential drops in supply. For energy companies, even a small seasonal shift can play a huge role in the demand for summer cooling or winter heating. An unusually mild winter might provide a boost to airlines and trucking companies, which would suffer fewer delays from snow and ice, but it could also cut into the sales of cold-weather clothing. Long-term outlooks for El Niño and La Niña can provide months of

valuable lead time on where winter temperatures in the United States are likely to run warmer or cooler than average.

The most direct protection against the risk of weather-related financial downturns comes from *insurance* for hail, flooding, drought, and the like. Weather insurance typically covers only the most dire meteorological threats, much like a catastrophic health-care plan that covers heart attacks but not chronic illness.

Several other tools can help a company use weather predictions to smooth out the potential ups and downs in profit linked to the atmosphere. Many commodities can be traded through contracts called *futures* (a type of *derivative**). Futures contracts are agreements to buy or sell a commodity at a fixed price at some later date. For example, a bread-baking company might buy wheat futures based on a projected precipitation outlook. This forecast would help the company plan with more confidence, knowing that the cost it will pay for wheat won't change even if a drought should strike and the price of wheat goes up dramatically.

It is also possible to trade futures contracts based on indices of the weather itself. Rather than specifying the future cost of a commodity, a weather derivative contract puts a price tag on a particular weather outcome, such as a record-hot summer that boosts demand for air conditioning. Many

● **FIGURE 3** A baking company might arrange to buy this wheat in advance at a guaranteed price if long-range weather forecasts point toward a poor crop.

such contracts are based on heating- or cooling-degree days, described in Chapter 3.

Many investors and speculators try to make a profit on the twists and turns of the atmosphere, often by looking at weather predictions and by buying and selling futures or weather derivatives. Traders keep a close eye on both seasonal weather projections and short-term forecasts, such as the track of a hurricane that could knock out oil and gas production. For example, as tropical storm Rita gathered strength on September 20, 2005, and forecasts called for Rita to approach the Gulf Coast as a major hurricane, the price of oil rose by the largest single-day amount on record—$4.39, or about 7 percent.

*Commodities represent a vast array of goods bought and sold in large quantities, from oranges to oil.

*Derivatives are contracts that derive their value from some other quantity, such as the price of a commodity.

Accuracy and Skill in Weather Forecasting

In spite of the complexity and ever-changing nature of the atmosphere, forecasts made by the National Weather Service out to between 12 and 24 hours are usually quite accurate. Those made for between 2 and 5 days are fairly good. Beyond about 7 days, computer prog forecast accuracy falls off rapidly because of the chaotic nature of the atmosphere. Although weather predictions made for up to 3 days are by no means perfect, they are far better than simply flipping a coin. But just how accurate are they?

One problem with determining forecast accuracy is deciding what constitutes a right or wrong forecast. Suppose tomorrow's forecast calls for a minimum temperature of 35°F. If the official minimum turns out to be 37°F, is the forecast incorrect? Is it as incorrect as one 10 degrees off? By the same token, what about a forecast for snow over a large city when the snow line cuts the city in half, with the southern portion receiving heavy amounts and the northern portion none? Is the forecast right or wrong? Or what if the snow started just 3 hours earlier than predicted, falling during the morning rush hour instead of at mid-morning? At present, there is no clear-cut answer to the question of determining forecast accuracy, so

● **FIGURE 9.13** This violent tornado ripped through the town of Moore, Oklahoma, on March 20, 2013. Even with a lead time of more than 15 minutes, the tornado took 24 lives, including 10 children, and injured 377 others. Even though predicting the precise location where a tornado will form is impossible at this time, it is possible to predict up to several days into the future where and when violent tornado outbreaks are likely.

VINCENT DELIGNY/AFP/Getty Images

meteorologists use a variety of mathematical techniques to measure the quality of their predictions. These might take into account how much the weather naturally varies at a given location and time of year, or how much data is actually available to determine whether a forecast was correct.

How does forecast accuracy compare with forecast skill? Suppose you are forecasting the daily summertime weather in Los Angeles. It is not raining today and your forecast for tomorrow calls for "no rain." Suppose that tomorrow it doesn't rain. You made an accurate forecast, but did you show any skill in so doing? Earlier, we saw that the chance of measurable rain in Los Angeles on any summer day is very small indeed; chances are good that day after day it will not rain. For a forecast to show skill, it should be better than one based solely on the current weather (*persistence*) or on the "normal" weather (*climatology*) for a given region. Therefore, during the summer in Los Angeles, a forecaster will have many accurate forecasts calling for "no measurable rain," but will need skill to predict correctly on which summer days it will rain. If on a sunny July day in Los Angeles you happen to forecast rain for tomorrow and it rains, you have not only made an accurate forecast, but you have also shown skill in making your forecast because your forecast was better than both persistence and climatology.

A meteorological forecast, then, shows skill when it is more accurate than a forecast utilizing only persistence or climatology. Persistence forecasts are usually difficult to improve upon for a period of time of several hours or less. Weather forecasts ranging from 12 hours to a few days generally show much more skill than persistence forecasts. However, as the range of the forecast period increases, the skill level drops quickly because of the effects of chaos discussed earlier in this chapter. The 6- to 14-day mean outlooks both show some skill (which has been increasing over the last several decades) in predicting temperature

and precipitation, although the accuracy of precipitation forecasts is less than that for temperature. Today, 7-day forecasts of major weather features are roughly as skillful as 3- to 4-day forecasts were in the 1990s. Beyond 15 days, specific forecasts are only slightly better than climatology. However, the level of skill in making forecasts of average monthly temperature and precipitation approximately doubled from 1995 to 2006.

Forecasting large-scale weather events several days in advance (such as the disruptive Groundhog Day blizzard of 2011 across the Midwest) is far more accurate than forecasting the precise evolution and movement of small-scale, short-lived weather systems, such as tornadoes and severe thunderstorms. In fact, 3-day forecasts of the development and movement of a major low-pressure system show more skill today than 36-hour forecasts did in the 1990s. Even though determining the *precise* location where a tornado will form is presently beyond modern forecasting techniques, the regions where tornadic storms are *likely* to form can often be predicted several days in advance.

With improved observing systems, such as Doppler radar and advanced satellite imagery, the lead time of watches and warnings for severe storms has increased. In fact, the lead time* for tornado warnings has more than doubled since the 1980s, with the average lead time now being close to 15 minutes, and the lead time for the most deadly and destructive tornadoes often more than 30 minutes (see ● Fig. 9.13).

Although scientists may never be able to skillfully predict the weather beyond about 15 days using available observations, the prediction of *climatic trends* is more promising. Whereas individual weather systems vary

---

Lead time is the interval of time between the issue of the warning and actual observance of the event, in this case, the tornado.

greatly and are difficult to forecast very far in advance, global-scale patterns of winds and pressure frequently show a high degree of persistence and predictable change over periods of a few weeks to a month or more. With the latest generation of high-speed supercomputers, general circulation models (GCMs) are doing a far better job of predicting large-scale atmospheric behavior than did the earlier models. (In Chapter 13, we will examine in more detail the climatic predictions based on numerical models.)

BRIEF REVIEW

Up to this point, we have looked at the various methods of weather forecasting. Before going on, here is a review of some of the important ideas presented so far:

- Available to the forecaster are a number of tools that can be used when making a forecast, including surface and upper-air maps, computer progs, meteograms, soundings, Doppler radar, and satellite information.
- The forecasting of weather by high-speed computers is known as *numerical weather prediction*. Mathematical models that describe how atmospheric temperature, pressure, winds, and moisture will change with time are programmed into the computer. The computer then draws surface and upper-air charts, and produces a variety of forecast charts called *progs*.
- Imperfections of the computer models—atmospheric chaos and small errors in the data—greatly limit the accuracy of weather forecasts for periods beyond a few days.
- Ensemble forecasting is a technique based on running several forecast models (or different versions of a single model), each beginning with slightly different weather information to approximate errors in the measurements.
- A *persistence forecast* is a prediction that future weather will be the same as the present weather, whereas a *climatological forecast* is based on the climatology of a particular region.
- For a forecast to show skill, it must be better than a persistence forecast or a climatological forecast.
- Weather forecasts that range from about 12 hours to about 3 days are called *short-range forecasts*. Those that extend from about 3 days to 8 days are called *medium-range forecasts*, and forecasts that extend beyond about 8 days are called *long-range forecasts*.
- Seasonal outlooks provide an overview of how temperature and precipitation patterns may compare with normal conditions.

Weather Forecasting Using Surface Charts

The best forecasts incorporate information about multiple layers of the atmosphere using numerical modeling, as we saw earlier in this chapter. However, even when computer models are skilled at advancing large-scale weather features forward in time, a capable forecaster also needs a strong sense of how surface features typically evolve, which can help him or her when the progs disagree. Suppose that we wish to make a short-range weather forecast and the only information available is a surface weather map. Can we make a forecast from such a chart? Most definitely. And our chances of that forecast being correct improve markedly if we have maps available from several days back. We can use these past maps to locate the previous position of surface features and predict their movement.

A simplified surface weather map is shown in Fig. 9.14. The map portrays early winter weather conditions on Tuesday morning at 6 a.m. (CST). A single isobar is drawn around the pressure centers to show their positions without cluttering the map. Note that an open wave cyclone is developing over the Central Plains with showers forming along a cold front and light rain, snow, and sleet ahead of a warm front. The dashed lines on the map represent the position of the weather systems 6 hours ago. Our first question is: How will these systems move?

DETERMINING THE MOVEMENT OF WEATHER SYSTEMS There are several rules of thumb we can use in forecasting the movement of surface pressure systems and fronts:

1. For short time intervals, mid-latitude cyclonic storms and fronts tend to move in the same direction and at approximately the same speed as they did during the previous 6 hours (providing, of course, there is no evidence to indicate otherwise).
2. Low-pressure areas tend to move in a direction that parallels the isobars in the warm air (the warm sector) ahead of the cold front.
3. Lows tend to move toward the region of the greatest drop in surface pressure, whereas highs tend to move toward the region of the greatest rise in surface pressure.
4. Surface pressure systems tend to move in the same direction as the wind at 5500 m (18,000 ft)—the 500-mb level. The speed at which surface systems move is about half the speed of the winds at this level.

When the surface map (Fig. 9.14) is examined carefully and when rules of thumb 1 and 2 are applied, it appears that—based on present trends—the low-pressure area over the Central Plains should move northeast. When we observe the 500-mb upper-air chart (Fig. 9.15), it too suggests that the surface low should move northeast at a speed of about 25 knots.

A FORECAST FOR SIX CITIES We are now in a position to make a weather forecast for six cities. To do so, we will project the surface pressure systems, fronts, and current weather into the future by assuming steady-state

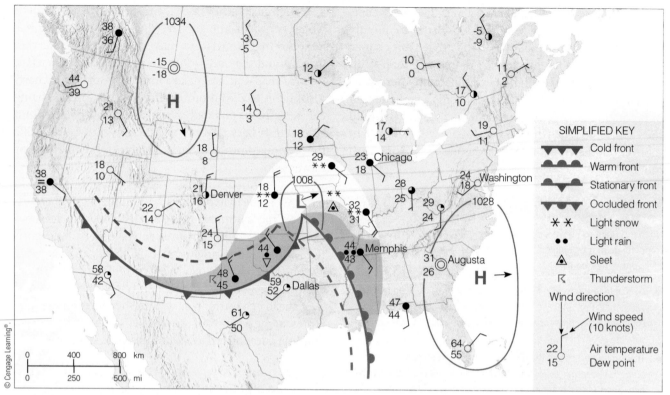

● FIGURE 9.14 Surface weather map for 6 a.m. (CST). Tuesday. Dashed lines indicate positions of weather features six hours ago. Areas shaded green are receiving rain, while areas shaded white are receiving snow, and those shaded pink, freezing rain or sleet.

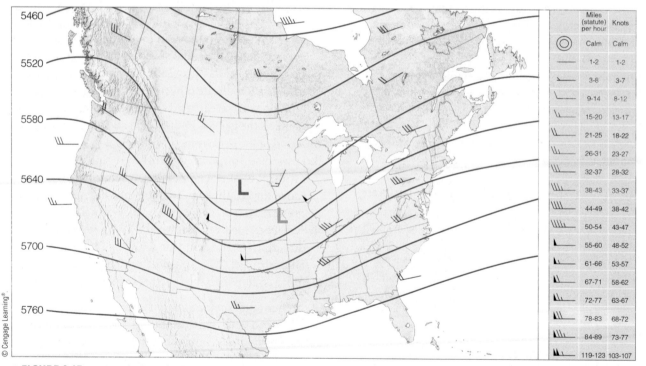

● FIGURE 9.15 A 500-mb chart for 6 a.m. (CST) Tuesday, showing wind flow. The light orange L represents the position of the surface low. The winds aloft tend to steer surface pressure systems along and, therefore, indicate that the surface low should move northeastward at about half the speed of the winds at this level, or 25 knots. Solid lines are contours in meters above sea level.

conditions. ● Fig. 9.16 gives the 12- and 24-hour projected positions of these features.

A word of caution before we make our forecasts. We are assuming that the pressure systems and fronts are moving at a constant rate, which may or may not occur.

Low-pressure areas, for example, tend to accelerate until they occlude, after which their rate of movement slows. Furthermore, the direction of moving systems can change due to "blocking" highs and lows that exist in their path or because of shifting upper-level wind patterns. We

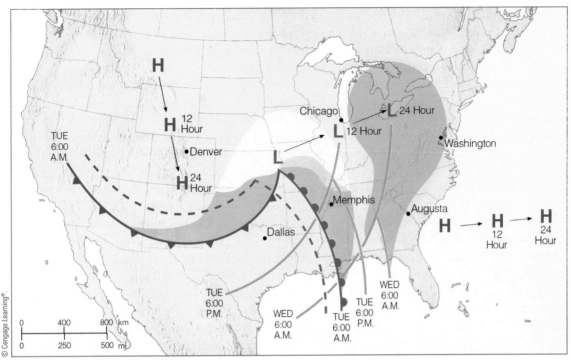

● **FIGURE 9.16** Projected 12- and 24-hour movement of fronts, pressure systems, and precipitation from 6 a.m. (CST) Tuesday until 6 a.m. (CST) Wednesday. (The dashed lines represent frontal positions 6 hours ago.)

will assume a constant rate of movement and forecast accordingly, always keeping in mind that the longer our forecasts extend into the future, the more susceptible they are to error.

If we move the low- and high-pressure areas eastward, as illustrated in Fig. 9.16, we can make a basic weather forecast for various cities. For example, the cold front moving into north Texas on Tuesday morning is projected to pass Dallas by that evening, so a forecast for the Dallas area would be "warm with showers, then turning colder." But we can do much better than this. Knowing the weather conditions that accompany advancing pressure areas and fronts, we can make more detailed weather forecasts that will take into account changes in temperature, pressure, humidity, cloud cover, precipitation, and winds. Our forecast will include the 24-hour period from Tuesday morning to Wednesday morning for the cities of Augusta, Georgia; Washington, D.C.; Chicago, Illinois; Memphis, Tennessee; Dallas, Texas; and Denver, Colorado. We will begin with Augusta.

Weather Forecast for Augusta, Georgia On Tuesday morning, cold, dry polar air associated with a high-pressure area brought freezing temperatures and fair weather to the Augusta area (see Fig. 9.14). Clear skies, light winds, and low humidities allowed rapid nighttime cooling so that, by morning, temperatures were in the low 30s (°F). Now look closely at Fig. 9.16 and observe that the high-pressure area is moving slowly eastward, away from Augusta. Southerly winds on the western side of this system will bring warmer and more humid air to the region. Therefore, afternoon temperatures will be warmer than those of the day before. As the warm front approaches from the west, clouds will increase, appearing first as cirrus, then thickening and lowering into the normal sequence of warm-front clouds. Barometric pressure should fall. Clouds and high humidity should keep minimum temperatures well above freezing on Tuesday night. Note in Fig. 9.16 that the projected area of precipitation (green-shaded region) does not quite reach Augusta. With all of this in mind, our forecast might read something like this:

> Clear and cold this morning with moderating temperatures by afternoon. Increasing high clouds with skies becoming overcast by evening. Cloudy and not nearly as cold tonight and tomorrow morning. Winds will be light and out of the south or southeast. Barometric pressure will fall slowly.

Wednesday morning we discover that the weather in Augusta is foggy with temperatures in the upper 40s (°F). But fog was not in the forecast. What went wrong? We

DID YOU KNOW?

Groundhog Day (February 2) is the day that is supposed to represent the midpoint of winter—halfway between the winter solstice and the vernal equinox. Years ago, in an attempt to forecast what the remaining half of winter would be like, people placed the burden of weather prognostication on various animals, such as the groundhog, which is actually a woodchuck. Folklore says that if the groundhog emerges from his burrow and sees (or casts) his shadow on the ground and then returns to his burrow, there will be six more weeks of winter weather. One can only wonder whether it is really the groundhog's shadow that drives him back into his burrow or the people standing around gawking at him.

forgot to consider that the ground was still cold from a recent cold snap. The warm, moist air moving over the cold surface was chilled below its dew point, resulting in fog. Above the fog were the low clouds we predicted. The minimum temperatures remained higher than anticipated because of the release of latent heat during fog formation and the absorption of infrared energy by the fog droplets. Not bad for a start. Now we will forecast the weather for Washington, D.C.

Rain or Snow for Washington, D.C.? Look at Fig. 9.16 and observe that the low-pressure area over the Central Plains is slowly approaching Washington, D.C., from the west. Hence, the clear weather, light southwesterly winds, and low temperatures on Tuesday morning (Fig. 9.14) will gradually give way to increasing cloudiness, winds becoming southeasterly, and slightly higher temperatures. By Wednesday morning, the projected band of precipitation will be over the city. Will it be in the form of rain or snow? Without a sounding (a vertical profile of temperature), this question is difficult to answer. We can see in Fig. 9.16, however, that on Tuesday morning cities south of Washington, D.C.'s latitude are receiving snow. So a reasonable forecast would call for snow, possibly changing to rain as warm air moves in aloft in advance of the approaching fronts. A 24-hour forecast for Washington, D.C., might sound like this:

> Increasing clouds today and continued cold. Snow beginning by early Wednesday morning, possibly changing to rain. Winds will be out of the southeast. Atmospheric pressure will fall.

Wednesday morning a friend in Washington, D.C., calls to tell us that the sleet began to fall but has since changed to rain. Sleet? Another fractured forecast! Well, almost. What we forgot to account for this time was the intensification of the storm. As the low-pressure area moved eastward, it deepened; central pressure lowered, pressure gradients tightened, and southeasterly winds blew stronger than anticipated. As air moved inland off the warmer Atlantic, it rode up and over the colder surface air. Snow falling into this warm layer at least partially melted; it then refroze as it entered the colder air near ground level. The influx of warmer air from the ocean slowly raised the surface temperatures, and the sleet soon became rain. Although we did not see this possibility when we made our forecast, a forecaster more familiar with local surroundings would have. Let's move on to Chicago.

Big Snowstorm for Chicago From Figs. 9.14 and 9.16, it appears that Chicago is in for a major snowstorm. Overrunning of warm air has produced a wide area of snow which, from all indications, is heading directly for the Chicago area. Since cold air north of the low's center will be over Chicago, precipitation reaching the ground should be frozen. On Tuesday morning (Fig. 9.16) the

leading edge of precipitation is less than 6 hours away from Chicago. Based on the projected path of the low-pressure area (Fig. 9.16) light snow should begin to fall around noon on Tuesday.

By evening, as the storm intensifies, snowfall should become heavy. It should taper off and finally end around midnight as the center of the low moves on east. If it snows for a total of 12 hours—6 hours as light snow (around 1 inch every 3 hours) and 6 hours as heavy snow (around 1 inch per hour)—then the total expected accumulation will be between 6 and 10 inches. As the low moves eastward, passing south of Chicago, winds on Tuesday will gradually shift from southeasterly to easterly, then northeasterly by evening. Since the storm system is intensifying, it should produce strong winds that will swirl the snow into huge drifts, which may bring traffic to a crawl.

The winds will continue to shift as they become northerly and finally northwesterly by Wednesday morning. By then the storm center will probably be far enough east so that skies should begin to clear. Cold air moving in from the northwest behind the storm will cause temperatures to drop further. Barometer readings during the storm will fall as the low's center approaches and reach a low value sometime Tuesday night, after which they will begin to rise. A weather forecast for Chicago might be:

> Cloudy and cold with light snow beginning by noon, becoming heavy by evening and ending by Wednesday morning. Total accumulations will range between 6 and 10 inches. Winds will be strong and gusty out of the east or northeast today, becoming northerly tonight and northwesterly by Wednesday morning. Barometric pressure will fall sharply today and rise tomorrow.

A text message Wednesday morning from a friend in Chicago reveals that our forecast was correct except that the total snow accumulation so far is 13 inches. We were off in our forecast because the storm system slowed as it became occluded. We did not consider this because we moved the system by the steady-state forecast method. At this time of year (early winter), Lake Michigan is not quite frozen over, and the added moisture picked up from the lake by the strong easterly and northeasterly winds also helped to enhance the snowfall. Again, a knowledge of the local surroundings would have helped make a more accurate forecast. The weather about 500 miles south of Chicago should be much different from this.

Mixed Bag of Weather for Memphis Observe in Fig. 9.16 that, within 24 hours, both a warm and a cold front should move past Memphis, Tennessee. The light rain that began Tuesday morning should saturate the cool air, creating a blanket of low clouds and fog by midday. The warm front, as it moves through sometime Tuesday afternoon, should cause temperatures to rise slightly as winds shift to the south or southwest. At night, clear to partly cloudy skies should allow the ground and air above to cool, offsetting any tendency for a rapid rise in temperature. Falling pressures should level off in the warm air, then fall once again as the cold front approaches. According to the projection in Fig. 9.16, the cold front should arrive sometime before midnight on Tuesday, bringing with it gusty northwesterly winds, showers, the possibility of thunderstorms, rising pressures, and colder air. Taking all of this into account, our weather forecast for Memphis will be:

> Cloudy and cool with light rain, low clouds, and fog early today, becoming partly cloudy and warmer by late this afternoon. Clouds increasing with possible showers and thunderstorms later tonight and turning colder. Winds southeasterly this morning, becoming southerly or southwesterly this evening and shifting to northwesterly tonight. Pressures falling this morning, leveling off this afternoon, then falling again, but rising after midnight.

A friend who lives near Memphis e-mailed us on Wednesday to inform us that our forecast was correct except that the thunderstorms did not materialize and that Tuesday night dense fog formed in low-lying valleys, but by Wednesday morning it had dissipated. Apparently, in the warm air, winds were not strong enough to mix the cold, moist air that had settled in the valleys with the warm air above. It's on to Dallas.

Cold Wave for Dallas From Fig. 9.16, it appears that our weather forecast for Dallas should be straightforward, since a cold front is expected to pass the area around noon on Tuesday. Weather along the front (Fig. 9.14) is showery with a few thunderstorms developing; behind the front the air is clear but cold. By Wednesday morning it looks as if the cold front will be far to the east and south of Dallas and an area of high pressure will be centered over southern Colorado. North or northwesterly winds on the east side of the high will bring cold arctic air into Texas, dropping temperatures as much as 40°F within a 24-hour period. With minimum temperatures well below freezing, Dallas will be in the grip of a cold wave. Our weather forecast should therefore read something like this:

> Increasing cloudiness and mild this morning with the possibility of showers and thunderstorms this afternoon. Clearing and turning much colder tonight and tomorrow. Winds will be southwesterly today, becoming gusty north or northwesterly this afternoon and tonight. Pressures falling this morning, then rising later today.

How did our forecast turn out? A quick check of Dallas weather on your smartphone on Wednesday morning reveals that the weather there is cold but not as cold as expected, and the sky is overcast. Cloudy weather? How can this be?

The cold front moved through on schedule Tuesday afternoon, bringing showers, gusty winds, and cold weather with it. Moving southward, the front gradually slowed and became stationary along a line stretching from the Gulf of Mexico westward through southern Texas and northern Mexico. (From the surface map alone, we had no way of knowing this would happen.) Along the stationary front a wave of low pressure formed. This wave caused warm, moist Gulf air to slide northward up and over the cold surface air. Clouds formed, minimum temperatures did not go as low as expected, and we are left with a fractured forecast. Let's try Denver.

Clear but Cold for Denver In Fig. 9.14, we can see that, based on our projections, the cold high-pressure area will be centered slightly to the south of Denver by Wednesday morning. Sinking air aloft associated with this high-pressure area should keep the sky relatively free of clouds. Weak pressure gradients will produce only weak winds and this, coupled with dry air, will allow for intense radiational cooling. Minimum temperatures will probably drop to well below 0°F. Our forecast should therefore read:

> Clear and cold through tomorrow. Northerly winds today becoming light and variable by tonight. Temperatures tomorrow morning will be below zero. Barometric pressure will continue to rise.

Almost reluctantly Wednesday morning, we look up the weather conditions at Denver and find that it is clear and very cold. A successful forecast at last! We find out, however, that the minimum temperature did not go below zero; in fact, 13°F was as cold as it got. A downslope wind coming off the mountains to the west of Denver kept the air mixed and the minimum temperature higher than expected. Again, a forecaster familiar with the local topography of the Denver area would have foreseen the conditions that lead to such downslope winds and would have taken this into account when making the forecast.

A complete picture of the surface weather systems for 6 a.m. Wednesday morning is given in ● Fig. 9.17. By comparing this chart to Fig. 9.16, we can summarize why our forecasts did not turn out exactly as we had predicted. For one thing, the center of the low-pressure area over the Central Plains moved slower than expected. This slow movement allowed a southeasterly flow of mild Atlantic air to overrun cooler surface air ahead of the storm while, behind the low, cities remained in the snow area for a longer time. The weak wave that developed along the trailing cold front over South Texas brought cloudiness and precipitation to Texas and prevented the really cold air from penetrating deep into the south. Farther west, the high-pressure area originally over Montana moved more southerly than southeasterly, which set up a pressure gradient that brought westerly downslope winds to eastern Colorado.

The subjective forecasting techniques demonstrated for these six cities are those one might use when making

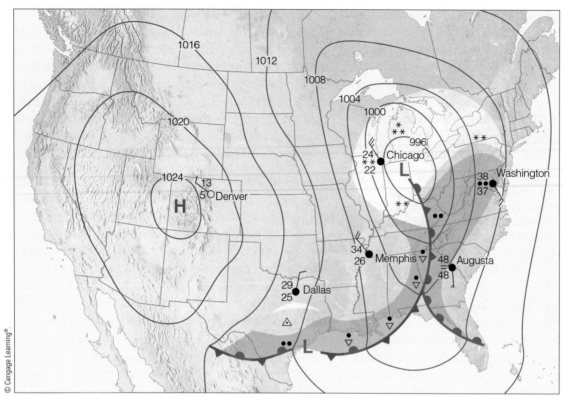

● **FIGURE 9.17** Surface weather map for 6 a.m. CDT Wednesday.

a short-range weather prediction with limited resources. The following section describes how a weather forecaster might predict the weather in a region where, to the west, surface weather features are greatly modified by a vast body of water, and only scanty surface and upper-air data are available. Here the forecaster must rely heavily on experience as well as more sophisticated tools, which include satellite data, upper-air charts, Doppler radar, and computer progs.

Using Forecasting Tools to Predict the Weather

It is late afternoon, and outside the weather forecast office near San Francisco the meteorologist mulls over what is going on in the sky. Overhead is a thin covering of cirrostratus; to the west, draped over the foothills, is the ever-present stratus and fog. The air is cool and the winds are westerly. It is Sunday, March 25, and the forecaster's task is to make a 24-hour weather forecast for the coastal area of central California.

What will tomorrow's weather be like? Will it be similar to today's or will it change markedly? A slowly falling barometer of 1016 mb (30 in.) and the high clouds moving in from the west point to an approaching storm system. A persistence forecast might be good for the next several hours, but what about tomorrow morning or tomorrow

afternoon? One of the biggest challenges for modern forecasters is to zero in on the elements that are most important on any given day.

The **forecast funnel** (illustrated in ● Fig. 9.18) outlines the steps used by forecasters to steer their attention from large scales to smaller scales and from short time frames to longer periods. The forecast funnel allows forecasters to produce the best possible predictions in a limited amount of time, by starting out with an examination of large-scale features (top of the funnel) and ending with local-scale forecasts (bottom of the funnel). Before looking at the large scale, however, let's take a quick look at current surface conditions.

The surface map for 4 p.m. (PST) Sunday, ● Fig. 9.19, shows there are no weather fronts approaching the West Coast. In fact, the nearest front is a stationary one that has stalled over the Rockies. There is, however, a region of low pressure centered about 1100 km (700 mi) west of San Francisco, which (according to previous maps) has been there for several days. With a central pressure of only about 1012 mb (29.88 in.), the system is fairly weak. Could this weak storm system be causing the increase in high cloudiness and the falling barometer? And will this pattern lead to rain tomorrow? A look at the 500-mb chart may help with these questions.

HELP FROM THE 500-MB CHART ● Figure 9.20 shows the 500-mb analysis for 4 p.m. Sunday afternoon. While examining the chart, the meteorologist recognizes certain clues that will aid in making the forecast. For one thing, the 5640-m height contour is over northern California. The forecaster knows that when this contour line is situated here or farther south, the statistical probability of receiving measurable rainfall over central California increases greatly.

West of San Francisco the flow is meridional with a warm, upper high situated just south of Alaska. To the south both east and west of the high are troughs. Because the shape of this flow around the high resembles the Greek letter omega (Ω), the high and its accompanying ridge is known as an *omega high*. The forecaster recognizes the omega high as a *blocking high,* one that tends to persist in the same geographic location for many days. This blocking pattern also tends to keep the troughs in their respective positions, which has been the case for several days now. But the chart indicates that the cold upper trough located west of San Francisco may be changing somewhat.

Observe the spacing of the contour lines around this trough. Even with a limited number of actual wind observations, the close spacing of the contours to the west and northwest of the trough, and the more widely spaced contours to the east of the trough, hint that stronger winds

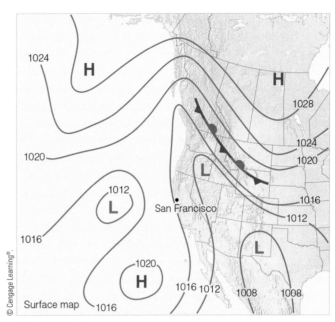

● **FIGURE 9.19** Surface weather map for 4 p.m. (PST) Sunday, March 25.

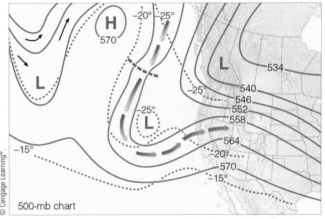

● **FIGURE 9.20** The 500-mb chart for 4 p.m. (PST) Sunday, March 25. Arrows indicate wind flow. Red arrows indicate warm advection and blue arrows, cold advection. Solid lines are height contours where 564 equals 5640 meters above sea level. Dashed lines are isotherms (lines of equal temperature) in °C. The heavy dashed purple line shows position of a shortwave trough.

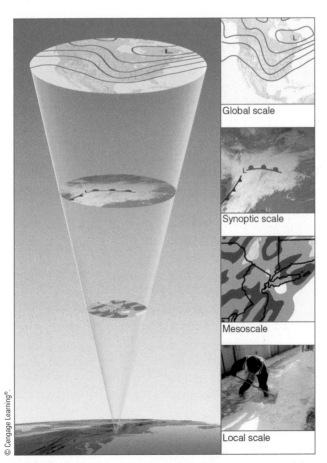

● **FIGURE 9.18** The forecast funnel, applied to a region of the northeast coast of the United States.

exist to the west of the trough. The forecaster knows from past experience that this usually means the trough will deepen. Also note that on the west side of the trough, cold air is flowing southward (blue arrows), indicating that cold advection is occurring here. The heavy dashed purple line on the west side of the trough represents the position of a shortwave trough, which is moving rapidly southward. The injection of cold air and the shortwave into the main trough should cause it to intensify. To the east of the main trough, warm moist air is moving northeastward (red arrows), indicating that warm advection is occurring. It is the lifting and condensing of this moist air that is producing the high clouds over San Francisco. All of these conditions—high wind speeds, cold air moving southward, and warm air moving northward, and a shortwave moving into a long-wave trough—manifest themselves as a deepening of the longwave trough. As the upper trough deepens, it should be capable of providing the necessary conditions favorable for the development of the surface low into a major mid-latitude cyclonic storm. The forecaster will likely consult maps for other levels of the atmosphere as well, such as 700 and 850 mb (see p. 250). These maps can reveal, for example, where moisture and warm air is flowing into the developing low-pressure center at lower levels of the atmosphere. Such knowledge gives the forecaster information on how quickly and strongly the surface storm will develop.

As we saw in Chapter 8, one of the main ingredients necessary for the development and intensification of a surface low is divergence of the airflow aloft. The forecaster knows that divergence aloft is associated with a decrease in surface pressure. This decrease, in turn, causes surface air to converge and rise, and its moisture to potentially condense into widespread cloudiness. But where will regions of divergence, convergence, and rising air be found on tomorrow's map? And how will tomorrow's map be different from today's? This is where the computer and the forecaster work together to come up with a prediction.

THE MODELS PROVIDE ASSISTANCE

The computer progs predict the future positions of weather systems. Some of the progs also predict where shortwave troughs will be located. It is important to know where the short-waves will be found, because to the east of them there is usually upper-level divergence, lower-level convergence, rising air, clouds, and precipitation. Hence, predicting the position of a shortwave means predicting regions of inclement weather.

Three 24-hour forecast models that predict the positions of the shortwaves, upper-level pressure systems, and the flow aloft at the 500-mb level for Monday, March 26, at 4 p.m. (PST) are shown in ● Fig. 9.21.* (Each prediction is made on Sunday afternoon.) Observe that there is good agreement among the models in that each model moves the upper trough slowly eastward and positions it off the coast. However, the actual positioning of the trough and the shortwaves (heavy dashed lines) differ only slightly for each model. For example, model A and model C move the upper trough eastward more quickly than does model B.

After examining each prog carefully, the forecaster must decide which model most accurately describes the future state of the atmosphere. Over the years, the forecaster knows that model A has performed well in predicting the positions of upper troughs that develop off the coast. Likewise, model C—because it uses more closely spaced grid points and a greater number of data points, and thus has better resolution—has done an admirable job of forecasting the positions of both upper troughs and shortwaves. On the other hand, although model B has its own strengths, it tends to move

---

*Explaining the differences among the three models is beyond the scope of this book. Each model treats the atmosphere in a slightly different way. Some models have closer grid points. Some models have better resolution in the lower part of the atmosphere, whereas others have better resolution in the higher regions of the troposphere. The idea in this forecasting example is *not* to illustrate the different models in use, but rather to show how a forecaster might use *any* numerical computer model as a forecasting tool.

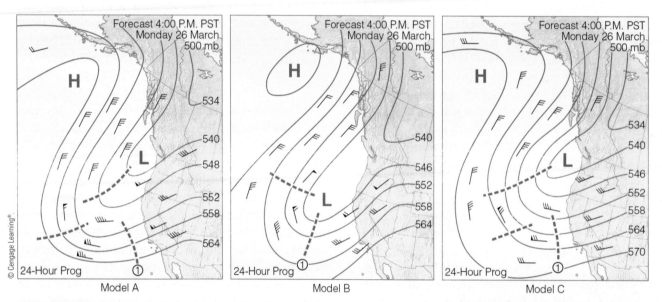

● **FIGURE 9.21** Three computer-drawn progs (model A, model B, and model C) that show the 24-hour projected 500-mb chart. Solid lines are contours, where 564 represents a height of 564 decameters (5640 meters) above sea level. Dashed lines represent projected positions of shortwaves. (Predictions were made on Sunday, March 25, at 4 p.m., PST.)

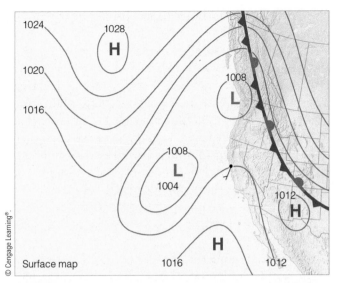

● **FIGURE 9.22** Surface weather map for 4 a.m. (PST) Monday, March 26.

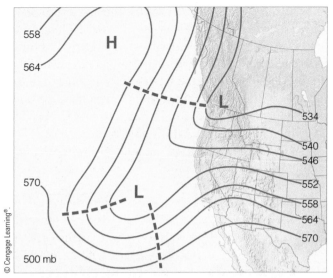

● **FIGURE 9.23** The 500-mb analysis for 4 a.m. (PST) Monday, March 26. Heavy dashed lines show position of shortwaves. Solid lines are height contours where 564 equals 5640 meters above sea level. (Compare with Fig. 9.21, the 24-hour progs for model A and model C.)

the shortwaves along too slowly. Consequently, the forecaster puts more confidence into model A and model C for this particular situation.

Using experience and the progs, the meteorologist sets out to predict the weather for the next 24 hours. The 24-hour progs for both model A and model C show a shortwave (labeled 1) approaching the California coast. As the shortwave approaches the coastline, clouds will increase and thicken and the likelihood of rain will increase. Therefore, a forecast for the next 24 hours might read like this:

> Increasing cloudiness Sunday night with rain beginning Monday morning and continuing through Monday afternoon.

A VALID FORECAST By early Monday morning, the maps begin to show the changes that the computer progs predicted. The surface map for 4 a.m. (PST) Monday morning (● Fig. 9.22) shows that the surface low in the Pacific has moved eastward and developed into a broad trough west of California. (Compare its position with Fig. 9.19.) The surface low has deepened considerably, as indicated by its central pressure of 1004 mb (29.65 in.). The approach of the storm is evidenced in San Francisco by thick middle clouds, southerly winds, and a falling barometer, nearly 4 mb lower than 12 hours ago. All these signs suggest that rain is on the way.

On the 500-mb chart for 4 a.m. Monday morning (● Fig. 9.23), we can see that the movement of cold air southward around the upper trough, along with the swift movement of the shortwave, has caused the upper trough to deepen. Note that the height contours are now displaced farther south and that the contour in the middle of the trough is lower than on the previous 500-mb map (Fig. 9.20). Compare Fig. 9.20 with the 24-hour progs in

Fig. 9.21 and notice that both model A and model C are projecting the movement of the shortwave (number 1) quite well. The forecaster made a wise choice in showing confidence in these two computer models as they did a good job predicting the position of the upper-level low and shortwave. Since the shortwave is moving with the flow toward San Francisco, it should rain today. But at what time will the rain begin? The forecaster must now go deeper into the forecast funnel and examine local conditions in greater detail. Here is where satellite and radar information come in.

SATELLITE AND UPPER-AIR ASSISTANCE The infrared satellite image taken at 6:45 a.m. Monday (see ● Fig. 9.24) shows that the middle clouds presently over California will soon give way to an organized band of cumuliform clouds in the shape of a comma. Such comma clouds tell the forecaster that the surface low-pressure area off the coast is developing into a mature mid-latitude cyclone. Observe that this comma-shaped cloud band is associated with the surface low shown in Fig. 9.22.

A quick glance at the 300-mb chart for 4 a.m. Monday morning (● Fig. 9.25) shows strong jet stream winds over Northern California and off the coast, suggesting that southwesterly winds aloft will carry the large comma-shaped cloud and its weather directly into California. And an area of strong divergence aloft (pink and red color on the map) associated with a jet streak will aid in deepening the cyclonic storm.

By examining the movement of the cloud mass on successive satellite images, the forecaster can predict its arrival time and, hence, when rainfall will begin. According to satellite images, the leading edge of the comma

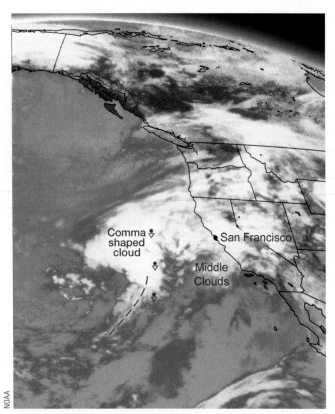

● FIGURE 9.24 Infrared satellite image taken at 6:45 a.m. (PST) Monday, March 26. The cloud in the shape of a comma indicates that the mid-latitude cyclonic storm is deepening. (The heavy dashed line shows the tail of the comma cloud.)

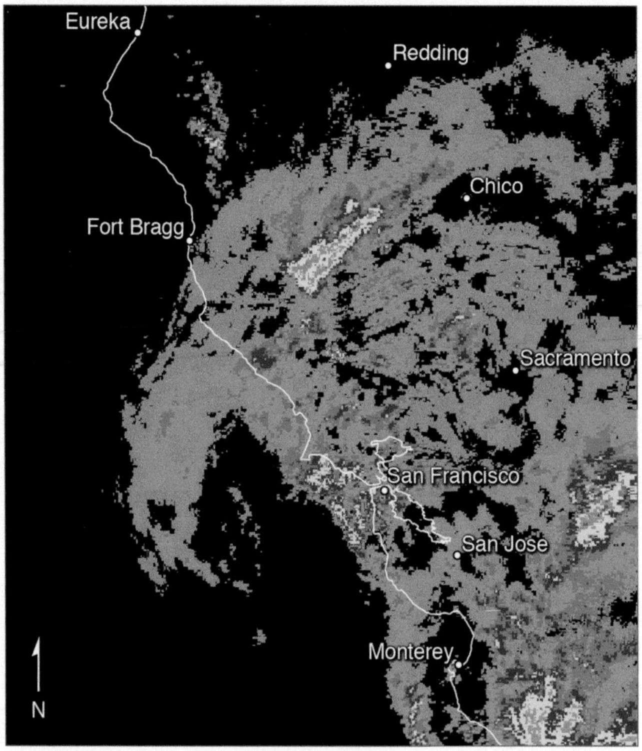

● FIGURE 9.26 Doppler radar image showing precipitation falling across northern and central California at 5 p.m. (PST) on Monday, March 26. Areas in blue and green indicate light to moderate precipitation; yellow indicates heavier precipitation; orange and red, heaviest precipitation.

cloud should be just offshore by Monday afternoon. Also, Doppler radar indicates that, just off the coast, light rain is now falling from the middle cloud layer. The position of the upper shortwave trough and the jet stream above the area of surface low pressure should provide enough support for the low to continue to intensify. Pressure gradients around the low will likely increase, creating strong, gusty winds from the south as the storm approaches. An amended forecast for San Francisco might read:

> Rain beginning this morning, becoming heavy by this afternoon. Strong and gusty southerly winds.

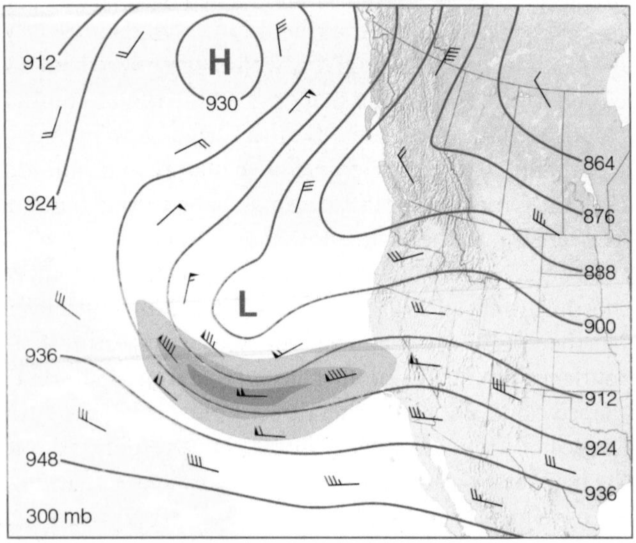

A DAY OF RAIN AND WIND The first raindrops falling from altostratus clouds dampen city streets near the end of the morning rush hour. Quickly, the rain spreads inland, and by late Monday afternoon, Doppler radar shows that precipitation is falling throughout northern and central California (see ● Fig. 9.26), as gusty southerly winds and moderate rain greet commuters on their way home.

The barometer has fallen sharply all day at San Francisco and by 4 p.m. the barometer reading is 1004 mb, a drop of 7 mb in just 6 hours. We can see the reason for this on the surface map for 4 p.m. Monday afternoon (● Fig. 9.27). The surface low has not only moved closer to the coast, it has also intensified considerably, as indicated by the drop of 11 mb in central pressure in just 12 hours. Spiraling around the low, a cold front marks the position of the comma-shaped cloud. At first, this may seem surprising, since no fronts were drawn on the previous map in Fig. 9.22. However, strong diverging air aloft caused

● FIGURE 9.25 The 300-mb chart for 4 a.m. (PST) Monday, March 26. Solid lines are height contours, where 900 equals 9000 meters above sea level. The darkest color on the map indicates the jet stream core, or jet streak.

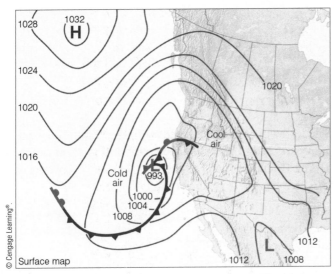

● **FIGURE 9.27** Surface weather map for 4 p.m. (PST) Monday, March 26.

surface air with contrasting temperatures and moisture properties to spiral counterclockwise into the surface low. With the coldest air on the western side of the surface low, the meteorologist saw fit to draw a cold front on the weather map.

As the surface low moves onshore, will it intensify or weaken? Will the unstable cold air behind the front develop into towering clouds and thunderstorms? It's back to the drawing board—to the computer progs, the forecast charts, Doppler radar, and the satellite images. The challenge and anticipation of making another forecast are at hand.

SUMMARY

Forecasting tomorrow's weather entails a variety of techniques and methods. Persistence, surface maps, satellite imagery, and Doppler radar are all useful when making a very short range (0–6 hour) prediction. For short- and medium-range forecasts, the current analysis, satellite data, pattern recognition, meteorologist intuition, and experience, along with statistical information and guidance from the many computer progs supplied by the National Weather Service, all go into making a prediction. For monthly and seasonal long-range forecasts, meteorologists incorporate changes in sea-surface temperature in the Pacific and Atlantic Oceans into seasonal outlooks of temperature and precipitation in North America.

Different computer progs are based upon different atmospheric models that describe the state of the atmosphere and how it will change with time. The atmosphere's chaotic behavior, along with flaws in the models and tiny errors (uncertainties) in the data, generally amplify as the computer tries to project weather further and further into the future. At present, computer progs that predict the weather over a vast region are better at forecasting the position of mid-latitude highs and lows and their development than at forecasting local showers and thunderstorms. To skillfully forecast smaller features, the grid spacing on some models is being reduced.

As new information from atmospheric research programs is fed into the latest generation of computers, it is hoped that the progs will be able to show more skill in predicting the weather up to 10 days into the future. More promising at this time is the simulation of large-scale climatic trends by the most recent general circulation models.

In the latter part of this chapter, we learned how one can make short-range weather predictions even with limited information, such as several days' worth of surface weather maps. We also saw some of the problems facing anyone who attempts to predict the behavior of the churning mass of air we call our atmosphere.

Most of the forecasting methods in this chapter apply mainly to skill in predicting events associated with large-scale weather systems, such as fronts and mid-latitude cyclones. The next chapter on severe weather deals with the formation and forecasting of smaller-scale (mesoscale) systems, such as thunderstorms, squall lines, and tornadoes.

KEY TERMS

The following terms are listed (with corresponding page numbers) in the order they appear in the text. Define each. Doing so will aid you in reviewing the material covered in this chapter.

AWIPS, 245
meteogram, 246
soundings, 246
analysis, 247
numerical weather
    prediction, 248
atmospheric models, 248
data assimilation, 248
prognostic chart (prog), 248
chaos, 251
ensemble forecasting, 251
persistence forecast, 253
steady-state (trend)
    forecast, 253
analog method, 253

statistical forecasts, 253
probability forecast, 253
weather types, 253
climatological
    forecast, 255
nowcast, 255
watch (weather), 255
warning (weather), 255
short-range
    forecasts, 255
medium-range
    forecast, 255
long-range forecast, 255
seasonal outlooks, 256
forecast funnel, 264

QUESTIONS FOR REVIEW

1. What is the function of the National Center for Environmental Prediction?
2. List at least four tools a weather forecaster might use when making a short-range forecast.
3. How does a prog differ from an analysis?
4. In what ways have high-speed computers assisted the meteorologist in making weather forecasts?
5. How are computer-generated weather forecasts prepared?
6. What are some of the problems associated with computer-model forecasts?
7. Describe four methods of forecasting the weather and give an example for each one.
8. How does pattern recognition aid a forecaster in making a prediction?
9. Suppose that where you live, the middle of January is typically several degrees warmer than the rest of the month. If you forecast this "January thaw" for the middle of next January, what type of weather forecast will you have made?
10. (a) Look out the window and make a persistence forecast for tomorrow at this time.
    (b) Did you use any skill in making this prediction?

11. How can ensemble forecasts improve medium-range forecasts?

12. Do monthly and seasonal forecasts make specific predictions of rain or snow? Explain.

13. Explain how teleconnections are used in making a long-range seasonal outlook.

14. If today's weather forecast calls for a "chance of snow," what is the percentage probability that it will snow today? (Hint: See Table 9.2, p. 254).

15. Do all accurate forecasts show skill on the part of the forecaster? Explain.

16. How does a *weather watch* differ from a *weather warning*?

17. How does chaos present a problem in forecasting the weather 15 days into the future?

18. List three methods that you would use to predict the movement of a surface mid-latitude cyclonic storm.

QUESTIONS FOR THOUGHT AND EXPLORATION

1. What types of watches and warnings are most commonly issued for your area?

2. Since computer models have difficulty in adequately considering the effects of small-scale geographic features on a weather map, why don't numerical weather forecasters simply reduce the grid spacing to, say, 1 kilometer on all models?

3. Suppose it's warm and raining outside. A cold front will pass your area in 3 hours. Behind the front, it is cold and snowing. Make a persistence forecast for your area 6 hours from now. Would you expect this forecast to be correct? Explain. Now, make a forecast for your area using the steady-state or trend forecasting method.

4. Why isn't the steady-state method very accurate when forecasting the weather more than a few hours into the future? What considerations can be taken into account to improve a steady-state forecast?

5. Go outside and observe the weather. Make a weather forecast using the weather signs you observe. Explain the rationale for your forecast.

6. Explain how the phrase "sensitive dependence on initial conditions" relates to the final outcome of a computer-based weather forecast.

7. Suppose the chance for a "White Christmas" at your home is 10 percent. Last Christmas was a white one. If for next year you forecast a "nonwhite" Christmas, will you have shown any skill if your forecast turns out to be correct? Explain.

GLOBAL **GEOSCIENCE** WATCH   Go to the Basic Search field and search for the phrase "weather forecasting." Find three articles from academic journals that evaluate a method of weather prediction. What are the phenomena being studied and the time periods of interest? What would be the practical benefits of improving each type of forecast?

ONLINE RESOURCES

 Visit www.cengagebrain.com to view additional resources, including video exercises, practice quizzes, an interactive eBook, and more.

CHAPTER 10

Thunderstorms and Tornadoes

Contents

Wednesday, March 18, 1925, was a day that began uneventfully, but within hours it turned into a day that changed the lives of thousands of people and made meteorological history. Shortly after 1:00 P.M., the sky turned a dark greenish-black and the wind began whipping around the small town of Murphysboro, Illinois. Arthur and Ella Flatt lived on the outskirts of town with their only son, Art, who would be four years old in two weeks. Arthur was working in the garage when he heard the roar of the wind and saw the threatening dark clouds whirling overhead.

Instantly concerned for the safety of his family, he ran toward the house as the tornado began its deadly pass over the area. With debris from the house flying in his path and the deafening sound of destruction all around him, Arthur reached the front door. As he struggled in vain to get to his family, whose screams he could hear inside, the porch and its massive support pillars caved in on him. Inside the house, Ella had scooped up young Art in her arms and was making a panicked dash down the front hallway towards the door when the walls collapsed, knocking her to the floor, with Art cradled beneath her. Within seconds, the rest of the house fell down upon them. Both Arthur and Ella were killed instantly, but Art was spared, nestled safely under his mother's body.

As the dead and survivors were pulled from the devastation that remained, the death toll mounted. Few families escaped the grief of lost loved ones. The infamous Tri-State Tornado killed 234 people in Murphysboro and leveled 40 percent of the town.

The devastating tornado described in our opening cut a mile-wide path for a distance of more than 150 miles through the states of Missouri, Illinois, and Indiana. The tornado (most likely a series of tornadoes) obliterated 4 towns, killed an estimated 695 persons, and left over 2000 injured. Tornadoes such as these, as well as much smaller ones, are associated with severe thunderstorms. Consequently, we will first examine the different types of thunderstorms, then focus on tornadoes, examining how and where they form and why they are so destructive.

Thunderstorms

It probably comes as no surprise that a *thunderstorm* is merely a storm containing lightning and thunder. Sometimes a thunderstorm produces gusty surface winds with heavy rain and hail. The storm itself may be a single cumulonimbus cloud, or several thunderstorms may form into a cluster. In some cases, a line of thunderstorms can extend for hundreds of miles.

Thunderstorms are *convective storms* that form with rising air. So the birth of a thunderstorm typically involves warm, moist air rising in a conditionally unstable environment.* The rising air may be a parcel of air ranging in size from a large balloon to a city block, or an entire layer, or slab of air, may be lifted. As long as the rising air is warmer (less dense) than the air surrounding it, there is an upward-directed *buoyant force* acting on it. The warmer the parcel compared to its surroundings, the greater the buoyant force and the stronger the convection. The trigger (or "forcing mechanism") needed to start air moving upward can be:

1. random, turbulent eddies that lift small bubbles of air
2. unequal heating at the surface

---

*As described in Chapter 5, a conditionally unstable atmosphere exists when cold, dry air aloft overlies warm, moist surface air. Additional information on atmospheric instability is given in Chapter 5, beginning on p. 117.

3. the effect of terrain (such as small hills) or the lifting of air along shallow boundaries of converging surface winds
4. diverging upper-level winds, coupled with converging surface winds and rising air
5. large-scale uplift along mountain barriers or gently rising terrain
6. warm air rising along a frontal zone

Usually, several of these mechanisms work together with vertical wind shear* to generate severe thunderstorms. The vertical wind shear may be produced by winds increasing with height (*speed shear*), changing direction with height (*directional shear*), or both.

Although we often see thunderstorms forming where the surface air is quite warm and humid, they can also form when the surface air temperature is no more than 10°C (50°F) or even lower. This latter situation often occurs in winter along the west coast of North America, when cold air aloft moves over the region. The cold air aloft destabilizes the atmosphere to the point where air parcels, given an initial push upward, can continue their upward journey because they remain warmer (less dense) than the colder air surrounding them. The cold air aloft may even cause sufficient instability to generate *thundersnow*—thunder and lightning observed in wintertime snowstorms.

Most thunderstorms that form over North America are short-lived, producing rainshowers, gusty surface winds, thunder and lightning, and sometimes small hail. Many have an appearance similar to the mature thunderstorm shown in ● Fig. 10.1. The majority of these storms do not reach severe status. *Severe thunderstorms* are defined by the National Weather Service as storms that produce at least one of the following: large hail with a diameter of at least one inch, surface wind gusts of at least 50 knots (58 mi/hr), or a tornado.

---

*Recall from Chapter 7 that wind shear represents the rate of change in wind speed, wind direction, or both across a given distance.

● FIGURE 10.1 An ordinary thunderstorm in its mature stage. Note the distinctive anvil top.

C. Donald Ahrens

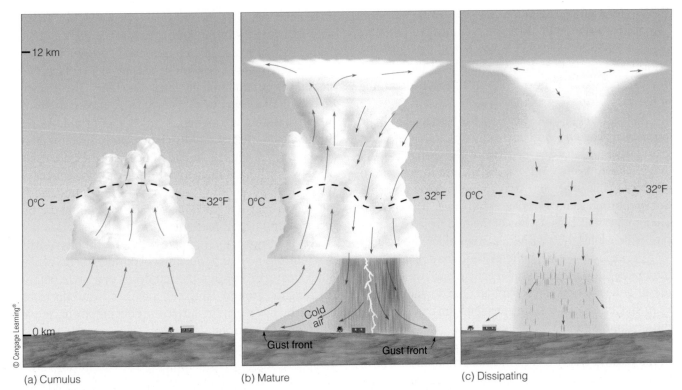

12 km

0°C — — — — 32°F

0 km

(a) Cumulus

0°C — — — — 32°F

Cold air

Gust front   Gust front

(b) Mature

0°C — — — — 32°F

(c) Dissipating

● FIGURE 10.2 Simplified model depicting the life cycle of an ordinary thunderstorm that is nearly stationary. (Arrows show vertical air currents. Dashed line represents freezing level, 0°C isotherm.)

Scattered thunderstorms (sometimes called "pop-up" or "popcorn" storms) that typically form on warm, humid days are often referred to as *ordinary cell thunderstorms** or *air-mass thunderstorms* because they tend to form in warm, humid air masses away from significant weather fronts. These thunderstorms can be considered "simple storms" because they rarely become severe, typically are less than a kilometer wide, and they go through a rather predictable life cycle from birth to maturity to decay that usually takes less than an hour to complete. However, under certain atmospheric conditions (described later in this chapter), more intense "complex thunderstorms" can form, such as the *multicell thunderstorm* and the *supercell thunderstorm*. A supercell is an intense, rotating storm that can last for hours and produce severe weather such as strong surface winds, large damaging hail, flash floods, and violent tornadoes.

ORDINARY CELL THUNDERSTORMS The **ordinary cell (air mass) thunderstorm** or, simply, *ordinary thunderstorm,* tends to form in a region where there is limited vertical wind shear, that is, where the wind speed and wind direction do not abruptly change with increasing height above the surface. Many ordinary thunderstorms appear to form as parcels of air are lifted from the surface by turbulent overturning in the presence of wind. Moreover, ordinary storms often form along shallow zones

where surface winds converge. Such zones can be due to any number of things, such as topographic irregularities, sea-breeze fronts, or the cold outflow of air from inside a thunderstorm that reaches the ground and spreads horizontally. These converging wind boundaries are normally zones of contrasting air temperature and humidity and, hence, air density.

Ordinary thunderstorms go through a fairly predictable cycle of development from birth to maturity to decay. The first stage is known as the **cumulus stage,** or *growth stage.* As a parcel of warm, humid air rises, it cools and condenses into a single cumulus cloud or a cluster of clouds (see ● Fig. 10.2a). If you have ever watched a thunderstorm develop, you may have noticed that at first the cumulus cloud grows upward only a short distance before the top of the cloud dissipates. The dissipation is because the cloud droplets evaporate as the drier air surrounding the cloud mixes with it. However, after the water drops evaporate, the air is more moist than before. So, the rising air is now able to condense at successively higher levels, and the cumulus cloud grows taller, often appearing as a rising dome or tower.

As the cloud builds, the transformation of water vapor into liquid or solid cloud particles releases large quantities of latent heat, a process that keeps the rising air inside the cloud warmer (less dense) than the air surrounding it. The cloud continues to grow in the unstable atmosphere as long as it is constantly fed by rising air from below. In this manner, a cumulus cloud can show extensive vertical

---

*In convection, the cell may be a single updraft or a single downdraft, or a combination of the two.

development and grow into a towering cumulus cloud (cumulus congestus) in just a few minutes. The cumulus stage normally does not last long enough for precipitation to form, and the updrafts keep water droplets and ice crystals suspended within the cloud. Also, there is no lightning or thunder during this stage.

As the cloud builds well above the freezing level, the cloud particles collide and join with one another and thus grow larger and heavier. Eventually, the rising air is no longer able to keep them suspended, and they begin to fall. Meanwhile, drier air from around the cloud is being drawn into it in a process called *entrainment*. The entrainment of drier air causes some of the raindrops to evaporate, which chills the air. This air, now colder and heavier than the air around it, begins to descend as a *downdraft*. As the air descends, the ice crystals begin to melt, which chills the air and enhances the downdraft. The downdraft may be further enhanced as falling precipitation drags some of the air along with it.

The appearance of the downdraft marks the beginning of the **mature stage.** The downdraft and updraft within the mature thunderstorm now constitute the cell. In some storms, there are several cells, each of which may last for less than 30 minutes.

During its mature stage, the thunderstorm is most intense. The top of the cloud, having reached a stable region of the atmosphere (which may be as high as the stratosphere), begins to take on the familiar anvil shape, as upper-level winds spread the cloud's ice crystals horizontally (see Fig. 10.2b). The cloud itself may extend upward to an altitude more than 12 km (40,000 ft) and be several kilometers in diameter near its base. Updrafts and downdrafts are strongest in the middle of the cloud, creating severe turbulence. Lightning and thunder are also present in the mature stage. Heavy rain (and occasionally small hail) often falls from the cloud. And, at the surface, there is often a downrush of cooler air with the onset of precipitation.

Where the cold downdraft reaches the surface, the air spreads out horizontally in all directions. The surface boundary that separates the advancing cooler air from the surrounding warmer air is called a *gust front*. Along the gust front, winds rapidly change both direction and speed. Look at Fig. 10.2b and notice that the gust front forces warm, humid air up into the storm, which enhances the cloud's updraft. In the region of the downdraft, rainfall may or may not reach the surface, depending on the relative humidity beneath the storm. In the dry air of the Desert Southwest, for example, a mature thunderstorm may look ominous and contain all of the ingredients of any other storm, except that the raindrops evaporate before reaching the ground. However, intense downdrafts from the storm may reach the surface, producing strong, gusty winds and a gust front.

After the storm enters the mature stage, it begins to dissipate in about 15 to 30 minutes. The **dissipating stage** occurs when the updrafts weaken as the gust front moves away from the storm and no longer enhances the updrafts. At this stage, as illustrated in Fig. 10.2c, downdrafts tend to dominate throughout much of the cloud. The reason an ordinary cell thunderstorm does not normally last very long is that the downdrafts inside the cloud tend to cut off the storm's fuel supply by destroying the humid updrafts. Deprived of a rich supply of warm, humid air, cloud droplets no longer form. Light precipitation now falls from the cloud, accompanied by only weak downdrafts. As the storm dies, the lower-level cloud particles evaporate rapidly, sometimes leaving only the cirrus anvil as the reminder of the once mighty presence (see ● Fig. 10.3). A

● **FIGURE 10.3** A dissipating thunderstorm near Naples, Florida. Most of the cloud particles in the lower half of the storm have evaporated.

Howard B. Bluestein

© C. Donald Ahrens

● **FIGURE 10.4** This multicell storm complex is composed of a series of cells in successive stages of growth. The thunderstorm in the middle is in its mature stage, with a well-defined anvil. Heavy rain is falling from its base. To the right of this cell, a thunderstorm is in its cumulus stage. To the left, a well-developed cumulus congestus cloud is about ready to become a mature thunderstorm. With new cells constantly forming, the multicell storm complex can exist for hours.

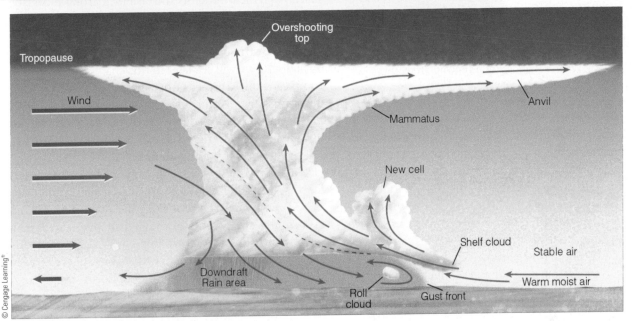

© Cengage Learning®

● **FIGURE 10.5** A simplified model describing air motions and other features associated with an intense multicell thunderstorm that has a tilted updraft. The severity depends on the intensity of the storm's circulation pattern.

single ordinary thunderstorm may go through its three stages in one hour or less.

Not only do thunderstorms produce summer rainfall for a large portion of the United States, but they can also bring with them momentary cooling after an oppressively hot day. The cooling comes during the mature stage, as the downdraft reaches the surface in the form of a blast of welcome relief. Sometimes, the air temperature can drop by more than 10°C (18°F) in just a few minutes. Unfortunately, the cooling effect often is short-lived, as the downdraft diminishes or the thunderstorm moves on. In fact, after the storm has ended, the air temperature usually rises, and as the moisture from the rainfall evaporates into the air, the humidity increases—sometimes to a level that actually feels more oppressive after the storm than it did before.

Up to this point, we've looked at ordinary cell thunderstorms that are short-lived, rarely become severe, and form in a region with weak vertical wind shear. As these storms develop, the updraft eventually gives way to the downdraft, and the storm ultimately collapses on itself. However, in a region where strong vertical wind shear exists, thunderstorms often take on a more complex structure. Strong vertical wind shear can cause the storm to tilt in such a way that it becomes a *multicell thunderstorm*—a thunderstorm with more than one cell.

MULTICELL THUNDERSTORMS   Thunderstorms that contain a number of cells, each in a different stage of development, are called **multicell thunderstorms** (see ● Fig. 10.4). Such storms tend to form in a region of moderate-to-strong vertical wind speed shear. Look at ● Fig. 10.5 and notice that on the left side of the illustration the wind speed increases rapidly with height, producing strong wind speed shear. This type of shearing causes the

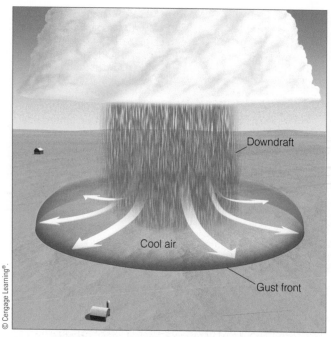

● FIGURE 10.6 When a thunderstorm's downdraft reaches the ground, the air spreads out, forming a gust front.

clouds. At the surface, below the thunderstorm's cold downdraft, the cold, dense air can cause the surface air pressure to rise, sometimes by several millibars. The relatively small, shallow area of high pressure is called a *mesohigh* (meaning "mesoscale high"). The mesohigh increases the pressure gradient between the storm-cooled air and the warmer, unstable air that lies beyond the storm, a situation that raises the risk of high winds.

The Gust Front When the cold downdraft reaches Earth's surface, it pushes outward in all directions, producing a strong **gust front** that represents the leading edge of the cold outflowing air (see ● Fig. 10.6). To an observer on the ground, the passage of the gust front resembles that of a cold front. During its passage, the temperature drops sharply and the wind shifts and becomes strong and gusty, with speeds occasionally exceeding 60 mi/hr. These high winds behind a strong gust front are called **straight-line winds** to distinguish them from the rotating winds of a tornado. As we will see later in this chapter, straight-line winds can inflict a great deal of damage, such as by blowing down trees and overturning mobile homes.

Along the leading edge of the gust front, the air is quite turbulent. Here, strong winds can pick up loose dust and soil and lift them into a huge tumbling cloud.* The cold surface air behind the gust front may linger close to the ground for hours, well after thunderstorm activity has ceased.

As warm, moist air rises along the forward edge of the gust front, a **shelf cloud** (also called an *arcus cloud*) may form, such as the one shown in ● Fig. 10.7. These clouds are especially prevalent when the atmosphere is very stable near the base of the thunderstorm. Look again at Figs. 10.5 and 10.7 and notice that the shelf cloud is

cell inside the storm to tilt in such a way that the updraft actually rides up and over the downdraft. Note that the rising updraft is capable of generating new cells that go on to become mature thunderstorms. Notice also that precipitation inside the storm does not fall into the updraft (as it does in the ordinary cell thunderstorm), so the storm's fuel supply is maintained and the storm complex can survive for a long time. Long-lasting multicell storms can become intense and produce severe weather for brief periods.

When convection is strong and the updraft intense (as it is in Fig. 10.5), the rising air may actually intrude well into the stable stratosphere, producing an **overshooting top**. As the air spreads laterally into the anvil, sinking air in this region of the storm can produce beautiful mammatus

---

*In dry, dusty areas or desert regions, the leading edge of the gust front is the haboob, described in Chapter 7, p. 184.

● FIGURE 10.7 A dramatic example of a shelf cloud (or arcus cloud) associated with an intense thunderstorm. The photograph was taken in central Oklahoma as the thunderstorm approached from the northwest.

● FIGURE 10.8 A roll cloud moves over Calgary, Alberta, on the morning of June 18, 2013.

attached to the base of the thunderstorm. Occasionally, an elongated, ominous-looking cloud forms just behind the gust front. These clouds, which appear to slowly spin about a horizontal axis, are called **roll clouds** (see ● Fig. 10.8).

When the atmosphere is conditionally unstable, the leading edge of the gust front may force the warm, moist air upward, producing a complex of multicell storms, each with new gust fronts. These gust fronts may then merge into a huge gust front called an **outflow boundary**. Along the outflow boundary, air is forced upward, often generating new thunderstorms (see ● Fig. 10.9).

Microbursts Beneath an intense thunderstorm, the downdraft can become localized so that it hits the ground and spreads horizontally in a radial burst of wind, much in the manner of water pouring from a tap and striking the

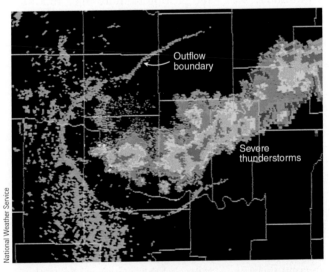

● FIGURE 10.9 Radar image of an outflow boundary. As cool (more-dense) air from inside the severe thunderstorms (red and orange colors) spreads outward, away from the storms, it comes in contact with the surrounding warm, humid (less-dense) air, forming a density boundary (blue line) called an *outflow boundary* between cool air and warm air. Along the outflow boundary, new thunderstorms often form.

sink below. (Look at the downdraft in Fig. 10.6.) Such downdrafts are called **downbursts**. A downburst with winds extending only 4 km or less is termed a **microburst**. In spite of its small size, an intense microburst can induce damaging straight-line winds well over 100 knots (115 mi/hr). (A larger downburst with winds extending more than 4 kilometers is termed a *macroburst*.) ● Figure 10.10 shows the dust clouds generated from a microburst north of Denver, Colorado. Since a microburst is an intense downdraft, its leading edge can evolve into a gust front. The leading edge of a microburst can contain an intense, horizontally rotating vortex that, in a relatively dry region, is often filled with dust.

Microbursts can blow down trees and inflict heavy damage upon poorly built structures as well as upon sailing vessels that encounter microbursts over open water. In fact, microbursts may be responsible for some damage once attributed to tornadoes. They pose an especially serious hazard to aircraft, largely because of the accompanying horizontal wind shear. Recall that wind shear is caused by rapid changes in wind speed and/or wind direction across a short horizontal distance. When an aircraft flies through a microburst at a relatively low altitude, say 300 m (1000 ft) above the ground, it first encounters a headwind that generates extra lift. This is position (a) in ● Fig. 10.11. At this point, the aircraft tends to climb (it gains lift), and if the pilot noses the aircraft downward, there could be grave consequences: In a matter of seconds the aircraft encounters the powerful downdraft (position b), and

DID YOU KNOW?

On August 13, 2011, as hundreds of fans waited for the country band *Sugarland* to perform at the Indiana State Fair in Indianapolis, strong gust front winds exceeding 60 miles per hour blew over the fairgrounds. The winds were so strong they toppled the stage, sending metal scaffolding, lights, and stage equipment into the crowd, where five people died and dozens were seriously injured.

● **FIGURE 10.10** Dust clouds rising in response to the outburst winds of a microburst north of Denver, Colorado.

© C. Donald Ahrens

the headwind is replaced by a tailwind (position c). This situation causes a sudden loss of lift and a subsequent decrease in the performance of the aircraft, which is now accelerating toward the ground.

In the United States, hundreds of air passengers died in microburst-related accidents in the 1970s and 1980s. Recognizing the danger that microbursts posed to aviation, scientists carried out intensive research that led to a warning system installed at airports in the 1990s throughout the United States. This system includes automated weather stations, Doppler radars, and computer algorithms designed to detect microbursts and low-level wind shear. The system has virtually eliminated microburst-related accidents in United States commercial flights.

Microbursts can be associated with severe thunderstorms, producing strong, damaging winds. But studies show that they can also occur with ordinary cell thunderstorms and with clouds that produce only isolated showers—clouds that may or may not contain thunder

and lightning. Microbursts can also be classified as *wet microbursts* or *dry microbursts*, based on whether they are accompanied by heavy rain or by little or no rain. In the western United States, many microbursts emanate from virga—rain falling from a cloud but evaporating before reaching the ground. In these "dry" microbursts, evaporating rain cools the air. The cooler heavy air then plunges downward through the warmer lighter air below. In humid regions, many microbursts are "wet" in that they are accompanied by blinding rain.

Up to this point, you might think that thunderstorm-related downdrafts are always cool. Most are cool, but occasionally they can be extremely hot. For example, just after midnight on June 9, 2011, a blast of hot, dry air from a dissipating thunderstorm raised the surface air temperature in Wichita, Kansas, from 85°F to 102°F in only 20 minutes, with wind gusting to more than 40 knots. Such sudden warm downbursts are called **heat bursts**. Studies indicate that the heat burst originates high up in the thunderstorm and warms by compressional heating as it plunges toward the surface. Some research suggests that the source of heat bursts may be air from outside a thunderstorm that is forced downward as it encounters the storm and light precipitation falls into it. However, the exact cause is not yet known.

SQUALL-LINE THUNDERSTORMS   Multicell thunderstorms may form as a line of thunderstorms, called a **squall line**. The line of storms may form directly along a cold front and extend for hundreds of kilometers, or the storms may form in the warm air 100 to 300 km out ahead of the cold front. These *pre-frontal squall-line thunderstorms* of the middle latitudes represent the largest and most severe type of squall line, with huge thunderstorms causing severe weather over much of its length (see ● Fig. 10.12).*

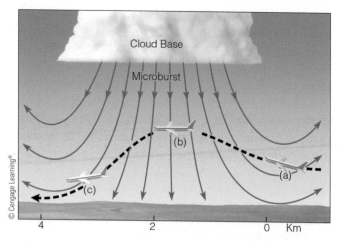

Cloud Base

Microburst

(b)

(a)

(c)

4   2   0   Km

© Cengage Learning®.

● **FIGURE 10.11** Flying into a microburst. At position (a), the pilot encounters a headwind; at position (b), a strong downdraft; and at position (c), a tailwind that reduces lift and causes the aircraft to lose altitude. (This horizontal wind shear is different from the vertical wind shear that acts to increase storm severity.)

---

*Within a squall line there may be multicell thunderstorms, as well as supercell storms—violent thunderstorms that contain a single rapidly rotating updraft. We will look more closely at supercells in the next section.

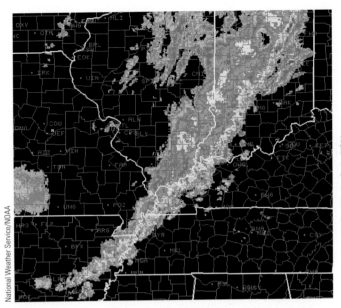

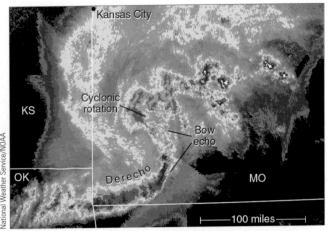

● **FIGURE 10.12** A Doppler radar composite showing a pre-frontal squall line extending from Indiana southwestward into Arkansas. Severe thunderstorms (red and orange colors) associated with the squall line produced large hail and high winds during October 2001.

● **FIGURE 10.14** A Doppler radar image showing an intense squall line in the shape of a bow—called a bow echo—moving eastward across Missouri on the morning of May 8, 2009. The strong thunderstorms (red and orange in the image) are producing damaging straight-line winds over a wide area. Damaging straight-line wind that extends for a good distance along a squall line is called a *derecho*.

There is still debate about exactly how pre-frontal squall lines form. Models that simulate their formation suggest that, initially, convection begins along the cold front, then re-forms farther away. Moreover, the surging nature of the main cold front itself, or developing cumulus clouds along the front, may cause the air aloft to develop into waves (called *gravity waves*), much like the waves that form downwind of a mountain chain. Out ahead of the cold front, the rising motion of the wave may be the trigger that initiates the development of cumulus clouds and a pre-frontal squall line.

Rising air along the frontal boundary (and along the gust front), coupled with the tilted nature of the updraft, promotes the development of new cells as the storm moves along. Hence, as old cells decay and die out, new ones constantly form, and the squall line can maintain itself for hours on end. Occasionally, a new squall line will actually form out ahead of the front as the gust front pushes forward, beyond the main line of storms.

Strong downdrafts often form to the rear of the squall line, as some of the falling precipitation evaporates and chills the air. The heavy, cooler air then descends, dragging some of the surrounding air with it. If the cool air rapidly descends, it may concentrate into a rather narrow band of fast-flowing air called the *rear-flank inflow jet*, because it enters the storm from the west, as shown in ● Fig. 10.13. Sometimes the rear-inflow jet will bring with it the strong upper-level winds from aloft. Should these winds reach the surface, they rush outward producing damaging *straight-line winds* that may exceed 90 knots.

As the strong winds rush forward along the ground, they sometimes push the squall line outward so that it appears as a *bow* (or a series of bows) on a radar screen. Such a bow-shaped squall line is called a **bow echo** (see ● Fig. 10.14). The strongest winds tend to form near the center of the bow, where the sharpest bending occurs. Tornadoes can form, especially near the left (northern) end of the bow, but they are usually small and short-lived.

If straight-line winds, gusting to more than 50 knots (58 mi/hr) persist along a path at least 400 km (250 mi)

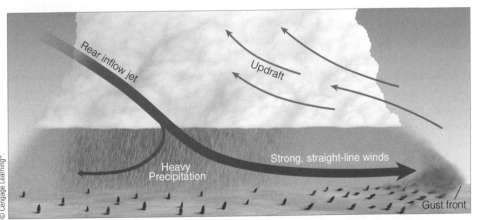

● **FIGURE 10.13** A side view of the lower half of a squall-line thunderstorm with the rear-inflow jet carrying strong winds from high altitudes down to the surface. These strong winds push forward along the surface, causing damaging straight-line winds that may reach 100 knots or more.

● **FIGURE 10.15** An enhanced infrared satellite image showing the cold cloud tops (dark red and orange colors) of a mesoscale convective complex extending from central Kansas across western Missouri. This organized mass of multicell thunderstorms brought hail, heavy rain, and flooding to this area.

long, the windstorm is called a **derecho** (day-ray-sho), after the Spanish word for "straight ahead." In an average year about 20 derechos occur in the United States. Typically, derechos form in the early evening and last throughout the night. An especially powerful derecho roared through New York State during the early morning of July 15, 1995, where it blew down millions of trees in Adirondack State Park. Damage from Ontario to New England totaled more than $500 million. Another extremely strong derecho swept from the Midwest into the Washington, D.C., area on June 29, 2012. With winds gusting to more than 70 knots (80 mi/hr), the derecho killed 22 people and left more than 4 million without power in some places for days. It is common for the damaging effects of a derecho to be attributed to a tornado. However, with a derecho, debris is blown in one direction, generally over a wide area, whereas debris with a tornado is usually thrown in many directions along a narrow, circular swath.

Squall lines are one type of convective phenomenon called a *mesoscale convective system* (MCS). Squall lines come under this heading because they are driven by convective processes and because they are mesoscale (middle scale) in size. Mesoscale convective systems are organized thunderstorms that can take on a variety of configurations, from the elongated squall line to the more circular *mesoscale convective complex* described in the next section.

Mesoscale Convective Complexes Where conditions are favorable for convection, a number of individual multicell thunderstorms may occasionally grow in size

and organize into a large circular convective weather system. These convectively driven systems, called **mesoscale convective complexes (MCCs)**, are quite large, sometimes as much as 1000 times larger in area than an individual ordinary cell thunderstorm. In fact, they often span an area in excess of 100,000 square kilometers, large enough to cover some entire states (see ● Fig. 10.15).

Within the MCCs, individual thunderstorms work together to generate a large weather system that appears on satellite to be relatively circular. MCCs typically last at least 6 hours, and sometimes more than 12 hours. Thunderstorms within MCCs support the growth of new thunderstorms as well as a region of widespread precipitation. These systems are beneficial in that they provide a significant portion of the growing season rainfall over much of the corn and wheat belts of the United States. However, MCCs can also produce a wide variety of severe weather, including hail, high winds, destructive flash floods, and tornadoes.

Mesoscale convective complexes tend to form during the summer in regions where the upper-level winds are weak, which is often beneath a ridge of high pressure. If a weak cold front should stall beneath the ridge, surface heating and moisture may be sufficient to generate thunderstorms on the cool side of the front. Moisture from the south often is brought into the system by a low-level jet stream frequently found within 5000 ft of the surface. Within the multicell storm complex, new thunderstorms form as older ones dissipate. With only weak upper-level winds, most MCCs move very slowly toward the east or southeast while dumping torrential rain.

SUPERCELL THUNDERSTORMS In a region where there is strong vertical wind shear (speed shear, directional shear, or both), the thunderstorm may form in such a way that the outflow of cold air from the downdraft never undercuts the updraft. In such a storm, the wind shear may be so strong as to create horizontal spin, which, when tilted into the updraft, causes it to rotate. An intense, long-lasting thunderstorm with a single violently rotating updraft is called a **supercell**.* As we will see later in this chapter, it is the rotating aspect of the supercell that can lead to the formation of tornadoes.

The internal structure of a supercell is organized in such a way that the storm can maintain itself as a single entity for hours. Storms of this type are capable of producing an updraft that may exceed 90 knots (104 mi/hr) and can produce damaging surface winds and large tornadoes. In some cases, the top of the storm can extend to as high as 18 km (60,000 ft) above the surface, and its width can exceed 40 km (25 mi).

A model of a classic supercell with many of its features is given in ● Fig. 10.16. In the diagram, we are viewing the storm from the southeast, and the storm is moving from

---

*Smaller thunderstorms that occur with rotating updrafts are referred to as *mini supercells*.

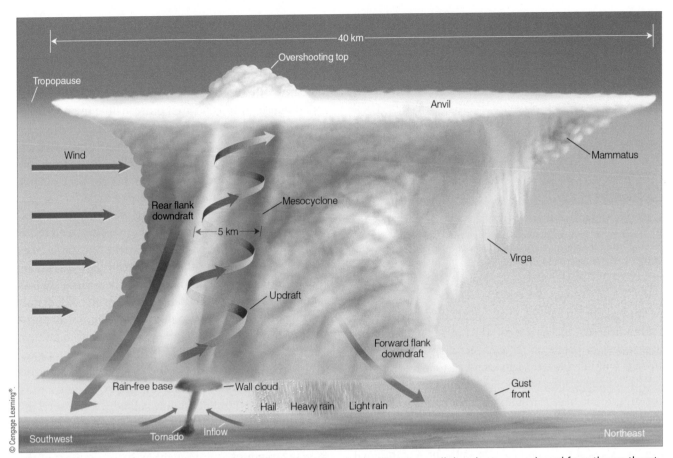

© Cengage Learning®

● **FIGURE 10.16** Some of the features associated with a classic tornado-breeding supercell thunderstorm as viewed from the southeast. The storm is moving to the northeast.

southwest to northeast. The rotating air column on the south side of the storm, usually 5 to 10 kilometers across, is called a **mesocyclone** (meaning "mesoscale cyclone"). The rotating updraft associated with the mesocyclone is so strong that precipitation cannot fall through it, thus producing a rain-free area (called a *rain-free base*) beneath the updraft. Strong southwesterly winds aloft typically blow the precipitation northeastward. Notice that large hail, having remained in the cloud for some time, usually falls just north

of the updraft. The heaviest rain occurs just north of the falling hail, with the lighter rain falling in the northeast quadrant of the storm. If humid low-level air is drawn into the updraft, a rotating cloud, called a **wall cloud**, may descend from the base of the storm (see ● Fig. 10.17).

The largest hail observed on Earth forms within supercells because of updrafts that are both wide and intense. In rare cases, such hailstones can grow to the size of grapefruits or even larger. One hailstone that fell in Vivian,

● **FIGURE 10.17** A wall cloud develops above the Kansas prairie on April 29, 2010.

Melanie Metz/Shutterstock.com

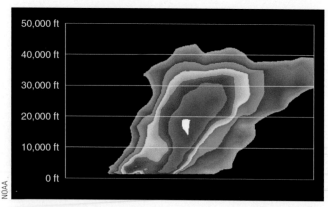

● **FIGURE 10.18** A vertical cross section of reflectivities observed by Doppler radar in a supercell thunderstorm that produced hail as large as golf balls in the El Paso, Texas, area on September 16, 2009. With damage of $150 million, it was the most destructive hailstorm in the city's history.

South Dakota, on July 23, 2010, weighed nearly 2 pounds and measured 8 inches across, which made it the heaviest and widest hailstone on record. The broad and powerful updraft within a supercell can keep hailstones airborne for relatively long periods, which allows many water droplets to accumulate and freeze on them. Once the hailstones are large enough, they can fall out the bottom of the cloud with the downdraft, or the violent spinning updraft may whirl them out the side of the cloud or even from the base of the anvil. Aircraft have actually encountered hail in clear air several kilometers from a storm. ● Fig. 10.18 shows a cross section through a supercell thunderstorm producing large hail.

Although no two supercells are exactly alike, for convenience they are often divided into three types. *Classic supercells* (as illustrated in Fig. 10.16) produce heavy rain, large hail, high surface winds, and the majority of tornadoes. When a supercell becomes dominated by heavy precipitation, strong downdrafts (downbursts), and large hail, it is referred to as an *HP (high precipitation) supercell* (see ● Fig. 10.19). Tornadoes in an HP supercell may be wrapped in heavy rain and difficult to see. On the other hand, tornadoes and cloud features are often quite visible in an *LP (low precipitation) supercell* (see ● Fig. 10.20). Although LP supercells tend to produce little rain, they are still capable of producing large hail as well as tornadoes. The rotation of an LP supercell is often visible as a bell-shaped tower, with a corkscrew-like pattern along its sides.

We can obtain a better picture of how vertical wind shear plays a role in the development of supercell thunderstorms by observing ● Fig. 10.21. The illustration represents atmospheric conditions often observed during the spring over the Central Plains. At the surface, we find an open-wave middle-latitude cyclone with cold, dry air moving in behind a cold front, and warm humid air pushing northward from the Gulf of Mexico behind a warm front. Above the warm surface air, a wedge, or "tongue," of warm, moist air is streaming northward. It is in this region we find a relatively narrow band of strong winds, sometimes exceeding 50 knots, called the *low-level jet.* Directly above the moist layer is a wedge of cooler, drier air moving in from the southwest. Higher up, at the 500-mb level, a trough of low pressure exists to the west of the surface low. At the 300-mb level, the polar front jet stream swings over the region, often with an area of maximum wind (a jet streak) above the surface low. At this level, the jet stream provides an area of divergence that enhances surface convergence and rising air. The stage is now set for the development of supercell thunderstorms.

The yellow area on the surface map (Fig. 10.21) shows where supercells are likely to form. They tend to form in

● **FIGURE 10.19** A high-precipitation (HP) supercell thunderstorm bearing heavy rain, large hail, strong winds, and lightning plows across eastern Nebraska on June 14, 2013.

● FIGURE 10.20 A tornado descends from beneath a low-precipitation (LP) supercell thunderstorm in eastern Colorado on June 10, 2010.

this region because (1) the position of cold air above warm air produces a conditionally unstable atmosphere and because (2) strong vertical wind shear induces rotation.

Rapidly increasing wind speed from the surface up to the low-level jet provides strong wind speed shear. Within this region, wind shear causes the air to spin about a horizontal axis. You can obtain a better idea of this spinning by placing a pen (or pencil) in your left hand, parallel to the edge of the table in front of you. Now take your right hand and push it over the pen away from you. The pen rotates much like the air rotates. If you tilt the spinning pen into the vertical by lifting its left side, then the pen rotates counterclockwise from the perspective of looking down on it. A similar situation occurs with the rotating air. As the spinning air rotates counterclockwise about a horizontal axis, an updraft from a developing thunderstorm can draw the spinning air into the cloud, causing the updraft to rotate. It is this rotating updraft that is characteristic of all supercells. The increasing wind speed with height up to the 300-mb level, coupled with the changing wind direction with height from more southerly at low levels to more westerly at high levels, further induces storm rotation.

● FIGURE 10.21 Conditions leading to the formation of severe thunderstorms, and especially supercells. The area in yellow shows where supercell thunderstorms are likely to form.

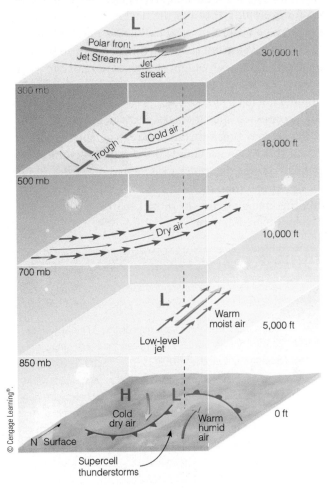

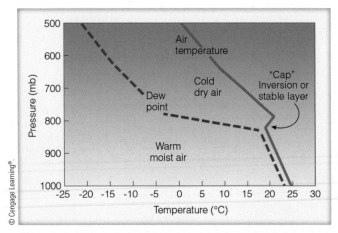

● **FIGURE 10.22** A typical sounding of air temperature and dew point that frequently precedes the development of supercell thunderstorms. The thickness of the warm, moist air from the surface up to the cap at 800 mb is usually on the order of about 2000 m (about 6000 ft).

In the warm air out ahead of the advancing cold front, we might expect to observe many supercells forming as warm, conditionally unstable air rises from the surface. However, there may be a persistent inversion above the layer of surface mixing. This layer, called a **cap**, serves as a lid on rising thermals, often allowing only small cumulus clouds to form. ● Fig. 10.22 shows the atmospheric structure of the cap. Sometimes the cap is strong enough to prevent any thunderstorms from developing through the day. At other times, the cap may be absent or weak, allowing many thunderstorms to develop but reducing the chance of an isolated supercell. The most violent storms tend to occur when a cap is just strong enough to prevent a large number of thunderstorms from developing early in the day. As the day progresses (and the surface air heats even more), rising air may break through the cap at isolated places. When this happens, clouds build rapidly—sometimes explosively—as the moist air is vented upward through the opening. When the surface air is finally able to puncture the inversion, one or more supercells may quickly develop to great height.

BRIEF REVIEW

In the last several sections, we examined different types of thunderstorms. Listed below for your review are important concepts we considered:

● All thunderstorms need three basic ingredients: (1) moist surface air; (2) a conditionally unstable atmosphere; and (3) a "trigger" that forces the air to rise.

● Ordinary cell (air-mass) thunderstorms tend to form where warm, humid air rises in a conditionally unstable atmosphere and where vertical wind shear is weak. They are usually short-lived and go through their life cycle of growth (cumulus stage), maturity (mature stage), and decay (dissipating stage) in less than an hour. They rarely produce severe weather.

● As wind shear increases (and the winds aloft become stronger), multicell thunderstorms are more likely to form as the storm's updraft rides up and over the downdraft. The tilted nature of the storm allows new cells to form as old ones die out.

● Multicell storms often form as a complex of storms, such as the squall line (a long line of thunderstorms that form along or out ahead of a frontal boundary) and the mesoscale convective complex (a large circular cluster of thunderstorms).

● Supercell thunderstorms are long-lasting violent thunderstorms, with a single rotating updraft that forms in a region of strong vertical wind shear. A rotating supercell is more likely to develop when (a) the winds aloft are strong and change direction from southerly at the surface to more westerly aloft and (b) a low-level jet exists just above Earth's surface.

● A gust front, or outflow boundary, represents the leading edge of cool air that originates inside a thunderstorm, reaches the surface as a downdraft, and moves outward away from the thunderstorm.

● Strong downdrafts from a thunderstorm, called downbursts (or microbursts if the downdrafts are smaller than 4 km), have been responsible for several airline crashes, because upon striking the surface, these winds produce extreme horizontal wind shear.

● A derecho is a long-lived straight-line wind produced by strong downbursts from intense thunderstorms that often appear as a bow (bow echo) on a radar screen.

THUNDERSTORMS AND FLOODING   Intense thunderstorms can be associated with **flash floods**—floods that develop rapidly with little or no advance warning. Such flooding often results when thunderstorms stall or move very slowly, causing heavy rainfall over a relatively small area. Flooding may also occur when thunderstorms move quickly, but keep passing over the same area, a phenomenon called *training,* much like the way that railroad cars, one after another, pass over the same tracks. In recent years, floods and flash floods in the United States have claimed an average of close to 100 lives a year and have accounted for untold property and crop damage. (An example of a terrible flash flood that took the lives of more than 135 people is given in Focus section 10.1.)

If thunderstorms bring repeated heavy rain to a region for days or weeks, the result can be one or more *river floods.* A river flood occurs when a major river system rises slowly but floods a large area, whereas a flash flood may devastate a smaller area in minutes to hours. During the summer of 1993, dozens of mesoscale convective complexes rumbled across the upper Midwest, causing the worst flood ever recorded in that part of the United States (see ● Fig. 10.23). Estimates are that $6.5 billion in crops were lost as millions of acres of valuable farmland were inundated by floodwaters. The flooding, much of it along the Mississippi River, took 45 lives, damaged or destroyed 45,000 homes, and forced the evacuation of 74,000 people.

Sometimes a region may experience both flash flooding and river flooding. This was the case across much of

The Terrifying Flash Flood in the Big Thompson Canyon

Saturday, July 31, 1976, was a banner day in Colorado, as residents and visitors were celebrating the weekend of the state's centennial. Meteorologically speaking, though, July 31 was like any other summer day in the Colorado Rockies, as small cumulus clouds with flat bases and domeshaped tops began to develop over the eastern slopes near the Big Thompson and Cache La Poudre rivers. At first glance, there was nothing unusual about these clouds, as almost every summer afternoon they form along the warm mountain slopes. Normally, strong upper-level winds push them over the plains, causing rainshowers of short duration. But the cumulus clouds on this day were different. For one thing, their bases were much lower than usual, indicating that the southeasterly surface winds were bringing in a great deal of moisture. Also, their tops were somewhat flattened, suggesting that an inversion (or "cap") aloft was stunting their growth. But these harmless-looking clouds gave no clue that later that evening in the Big Thompson Canyon more than 135 people would lose their lives in a terrible flash flood.

By late afternoon, a few of the cumulus clouds were able to break through the cap. Fed by moist southeasterly winds, these clouds soon developed into gigantic multicell thunderstorms with tops exceeding 18 km (60,000 ft). By early evening, these same clouds were producing incredible downpours in the mountains.

In the narrow canyon of the Big Thompson River, some places received as much as 30.5 cm (12 in.) of rain in the four hours between 6:30 p.m. and 10:30 p.m. local time. This is truly an impressive amount of precipitation, considering that the area normally receives about 40.5 cm (16 in.) for an entire year. The heavy downpours turned small creeks into raging torrents, and the Big Thompson River was quickly filled to capacity. Where the canyon narrowed, the river overflowed its banks and water covered the road. The relentless pounding of water caused the road to give way.

Soon cars, tents, mobile homes, resort homes, and campgrounds were being claimed by the river (see Fig. 1), affecting many hundreds of people who had planned

● FIGURE 1 This car is one of more than 400 destroyed by floodwaters in the Big Thompson Canyon on July 31, 1976.

Robert J. Jarrett/USGS

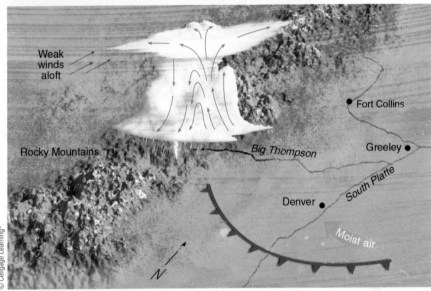

© Cengage Learning®

● FIGURE 2 Weather conditions that led to the development of intense multicell thunderstorms that remained nearly stationary over the Big Thompson Canyon in the Colorado Rockies. The arrows within the thunderstorm represent air motions.

to spend the weekend celebrating the state centennial. Where the debris entered a narrow constriction, it became a dam. Water backed up behind it, then broke through, causing a wall of water to rush downstream.

● Figure 2 shows the weather conditions during the evening of July 31, 1976. A cool front moved through earlier in the day and is now south of Denver. The weak inversion layer associated with the front kept the cumulus clouds from building to great heights earlier in the afternoon. However, the strong southeasterly flow behind the cool front pushed unusually moist air upslope along the mountain range. Heated from below, the conditionally unstable air eventually punctured the inversion and

developed into a huge multicell thunderstorm complex that remained nearly stationary for several hours due to the weak southerly winds aloft. The deluge may have deposited 19 cm (7.5 in.) of rain on the main fork of the Big Thompson River in about one hour. Of the approximately 2000 people in the canyon that evening, more than 140 lost their lives, and property damage exceeded $35 million. Nearly all of the deaths were from blunt trauma due to the raging waters and debris, rather than from drowning. As a result of this disaster, signs have been placed in major river canyons encouraging people to leave their homes or vehicles and "climb to safety" if they encounter floodwaters.

● **FIGURE 10.23** Flooding during the summer of 1993 covered a vast area of the upper Midwest. Here, floodwaters near downtown Des Moines, Iowa, during July 1993, inundate buildings of the Des Moines waterworks facility. Flood-contaminated water left 250,000 people without drinking water.

© AP Associated Press

Texas and Oklahoma in May 2015. Many rivers overflowed for days, as persistent heavy rain led to the wettest month on record in both states. On the night of May 24, thunderstorms dumped more than 30 cm (12 in.) of rain in the Texas Hill Country, turning the normally calm Blanco River into a raging torrent that destroyed hundreds of homes. At least 31 people were killed in Texas and Oklahoma. Catastrophic flooding also struck northeast Colorado in September 2013. Several days of rainfall included periods when showers and thunderstorms passed over the same areas repeatedly for hours, producing flash flooding and huge rainfall totals. Boulder, Colorado, reported 23 cm (9.08 in.), almost doubling its previous 24-hour record. The waters eventually merged to cause record flooding along the South Platte River from northeast Colorado into Nebraska. In all, the multiday event produced an estimated $2 billion in damage, caused at least 10 deaths, and destroyed more than 1800 homes and 750 businesses, as well as some 200 miles of state highway.

DISTRIBUTION OF THUNDERSTORMS It is estimated that more than 50,000 thunderstorms occur each day throughout the world. Hence, over 18 million occur annually. The combination of warmth and moisture make equatorial landmasses especially conducive to thunderstorm formation. Thunderstorms occur here on about one out of every three days. Thunderstorms are also prevalent over

DID YOU KNOW?

The Great Flood of 1993 in the Mississippi and Missouri river basins had an impact on the living and the dead as more than 700 graves opened in the Hardin Cemetery in Missouri. Some caskets were swept away by raging floodwaters and deposited many miles downstream, and some were never found.

water along the intertropical convergence zone, where the low-level convergence of air helps to initiate uplift. The heat energy liberated in these storms helps Earth maintain its heat balance by distributing heat poleward (see Chapter 7). Thunderstorms are much less prevalent in dry climates, such as the polar regions and the desert areas dominated by subtropical highs.

● Figure 10.24 shows the average annual number of days having thunderstorms in the United States and southern Canada. Notice that they occur most frequently in the southeastern United States along the Gulf Coast with a maximum in Florida. The region with the fewest thunderstorms is the Pacific coast and interior valleys.

In many areas, thunderstorms form primarily in summer during the warmest part of the day when the surface air is most unstable. There are some exceptions, however. During the summer in the valleys of central and southern California, dry, sinking air produces an inversion that inhibits the development of towering cumulus clouds. In these regions, thunderstorms are most frequent in winter and spring, particularly when cold, moist, conditionally unstable air aloft moves over moist, mild surface air. The surface air remains relatively warm because of its proximity to the ocean.

On many summer days, thunderstorms develop near the Rocky Mountains in the afternoon, then intensify in the evening as they move eastward across the Central Plains. Often these thunderstorms congeal into large mesoscale convective systems and continue through the night. One of the most common times to experience thunder and lightning in parts of Iowa and Missouri is between midnight and dawn. Such storms can be fueled by a southerly low-level jet that often strengthens after sunset, bringing humid air northward and helping trigger convergence

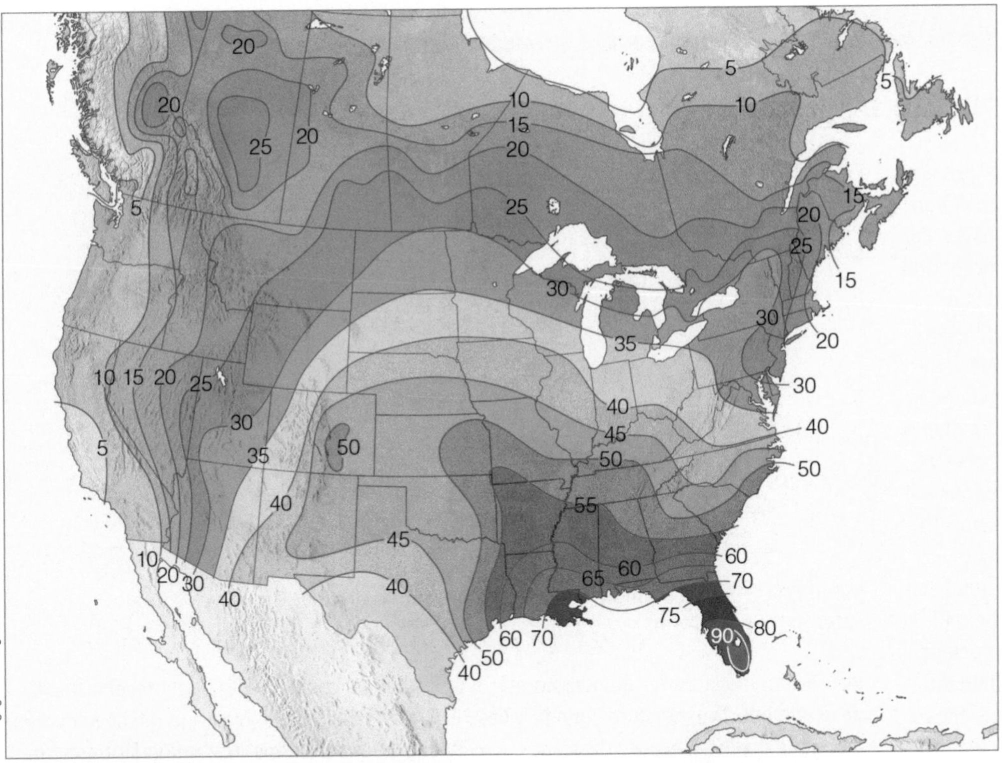

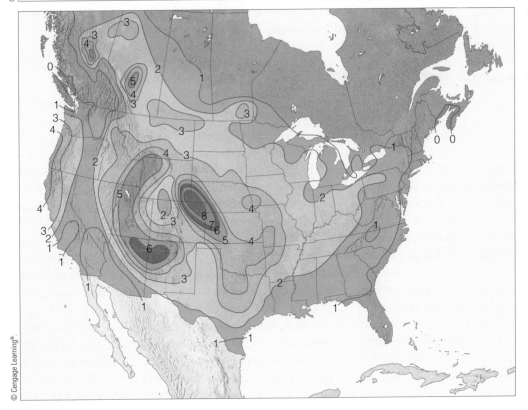

● **FIGURE 10.24** The average number of days each year on which thunderstorms are observed throughout the United States and southern Canada. (Due to the scarcity of data, the number of thunderstorms is underestimated in the mountainous far west.)

● **FIGURE 10.25** The average number of days each year on which hail is observed throughout the United States and southern Canada.

and uplift of surface air. As the thunderstorms build, their tops cool by radiating infrared energy to space. This cooling process tends to destabilize the atmosphere around the storms, making it even more suitable for nighttime thunderstorm development.

At this point, it is interesting to compare Fig. 10.24 and ● Fig. 10.25. Notice that even though the greatest frequency of thunderstorms is near the Gulf Coast, the greatest frequency of hailstorms is over the western Great Plains. One reason for this situation is that conditions over the Great Plains are more favorable for the development of severe thunderstorms and especially supercells that have strong updrafts capable of keeping hailstones suspended within the cloud for a long time so that they can grow to an appreciable size before plunging to the ground. We also find that, in summer along the Gulf Coast, a thick layer of

Thunderstorms and the Dryline

Thunderstorms may form along or just east of a boundary called a *dryline*. Recall from Chapter 8 that the dryline represents a narrow zone where there is a sharp horizontal change in moisture. In the United States, drylines are most frequently observed in the western half of Texas, Oklahoma, and Kansas. In this region, drylines occur most frequently during spring and early summer.

● Figure 3 shows springtime weather conditions that can lead to the development of a dryline and intense thunderstorms. The map shows a developing mid-latitude cyclone with a cold front, a warm front, and three distinct air masses. Behind the cold front, cold dry continental polar air or modified cool dry Pacific air pushes in from the northwest. In the warm air, ahead of the cold front, hot, dry continental tropical air moves in from the southwest. Farther east, warm but very humid air sweeps northward from the Gulf of Mexico. The dryline is the north–south oriented boundary that separates the hot, dry air and the warm, humid air.

Along the cold front—where cold, dry air replaces warm, dry air—there is

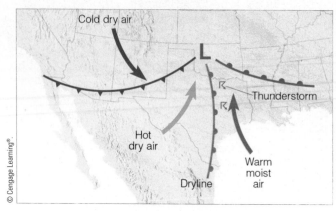

● **FIGURE 3** Surface conditions that can produce a dryline with intense thunderstorms.

insufficient moisture for thunderstorm development. The moisture boundary lies along the dryline. Because the Central Plains of North America are elevated to the west, some of the hot, dry air from the southwest is able to ride over the slightly cooler, more humid air from the Gulf. This condition sets up a potentially unstable atmosphere just east of the dryline. Converging surface winds in the vicinity of the

dryline, coupled with upper-level outflow, may result in rising air and the development of thunderstorms. As thunderstorms form, the cold downdraft from inside the storm may produce a blast of cool air that moves along the ground as a gust front and initiates the uplift necessary for generating new (possibly more severe) thunderstorms.

---

warm, moist air extends upward from the surface. Most hailstones falling into this warm layer will melt before reaching the ground. In contrast, when a thunderstorm is located on the high plains east of the Rocky Mountains, a much larger portion of the storm will typically be located above the freezing level, so hail has a much better chance of occurring.*

Now that we have looked at the development and distribution of thunderstorms, we are ready to examine an interesting, though not yet fully understood, aspect of all thunderstorms—lightning. (However, before studying lightning and thunder, you may want to read Focus section 10.2, which discusses thunderstorm development along the dryline.)

LIGHTNING AND THUNDER **Lightning** is simply a discharge of electricity, a giant spark, which usually occurs in mature thunderstorms.** Lightning may travel within a cloud, from one cloud to another, from a cloud to the surrounding air, or from a cloud to the ground (see

● Fig. 10.26). (The majority of lightning strikes occur within the cloud, while only about 20 percent or so occur between cloud and ground.) The lightning stroke can heat the air through which it travels to an incredible 30,000°C (54,000°F), which is 5 times hotter than the surface of the sun. This extreme heating causes the air to expand explosively, thus initiating a shock wave that becomes a booming sound wave—called **thunder**—that travels outward in all directions from the flash.

A sound occasionally mistaken for thunder is the **sonic boom.** Sonic booms are produced when an aircraft exceeds the speed of sound at the altitude at which it is flying. (The speed of sound at sea level is 661 knots, or 761 mi/hr, but it decreases gradually with height.) The aircraft compresses the air, forming a shock wave that trails out as a cone behind the aircraft. Along the shock wave, the air pressure changes rapidly over a short distance. The rapid pressure change causes the distinct boom. (Exploding fireworks generate a similar shock wave and a loud bang.)

How Far Away Is the Lightning?—Start Counting
When you see a flash of lightning, how can you tell how far away (or how close) it is? Light travels so fast that we

---

*The formation of hail is described in Chapter 5 on p. 138.

**Lightning may also occur in snowstorms, in duststorms, on rare occasions in nimbostratus clouds, and in the gas cloud of an erupting volcano.

● **FIGURE 10.26** The lightning stroke can travel in a number of directions. It can occur within a cloud, from one cloud to another cloud, from a cloud to the air, or from a cloud to the ground. Notice that the cloud-to-ground lightning can travel out away from the cloud, then turn downward, striking the ground many miles from the thunderstorm. When lightning behaves in this manner, it is often described as a *"bolt from the blue."*

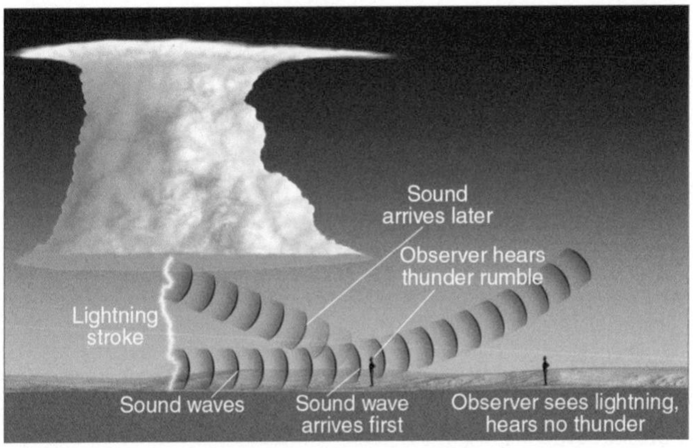

● **FIGURE 10.27** Thunder travels outward from the lightning stroke in the form of waves. If the sound waves from the lower part of the stroke reach an observer before the waves from the upper part of the stroke, the thunder appears to rumble. If the sound waves bend upward away from an observer, the lightning stroke may be seen, but the thunder will not be heard.

see light instantly after a lightning flash. But the sound of thunder, traveling at only about 1100 ft/sec, takes much longer to reach the ear. If we start counting seconds from the moment you see the lightning until you hear the thunder, you can determine how far away the stroke is. Because it takes sound about 5 seconds to travel 1 mile, if you see lightning and hear the thunder 15 seconds later, the lightning stroke (and the thunderstorm) is about 3 miles away.

When the lightning stroke is very close (several hundred feet or less) thunder sounds like a clap or a crack followed immediately by a loud bang. When it is farther away, it often rumbles. The rumbling can be due to the sound emanating from different areas of the stroke (see ● Fig. 10.27). Moreover, the rumbling is accentuated when the sound wave reaches an observer after having bounced off obstructions, such as hills and buildings.

In some instances, lightning is seen but no thunder is heard. Does this mean that thunder was not produced by the lightning? Actually, there is thunder, but the atmosphere has refracted (bent) and attenuated the sound waves, making the thunder inaudible. Sound travels faster in warm air than in cold air. Because thunderstorms form in a conditionally unstable atmosphere, where the temperature normally drops rapidly with height, a sound wave moving outward away from a lightning stroke will often bend upward, away from an observer at the surface. Consequently, an observer closer than about 5 km (3 mi) to a lightning stroke usually will hear thunder, while an observer 15 km (about 9 mi) away usually will not.

As for lightning, what causes it? The normal fair weather electric field of the atmosphere is characterized by a negatively charged surface and a positively charged upper atmosphere. For lightning to occur, separate regions containing opposite electrical charges must exist within a cumulonimbus cloud. Exactly how this charge separation comes about is not fully understood; however, many theories have tried to account for it.

Electrification of Clouds   One theory proposes that clouds become electrified when graupel (small ice particles also called *soft hail*) and hailstones fall through a region of supercooled liquid droplets and ice crystals. As liquid droplets collide with a hailstone, they freeze on contact and release latent heat. This process keeps the surface of the hailstone warmer than that of the surrounding ice crystals. When the warmer hailstone comes in contact with a colder ice crystal, an important phenomenon occurs: *There is a net transfer of positive ions (charged molecules) from the warmer object to the colder object.* Hence, the hailstone (larger, warmer particle) becomes negatively charged and the ice crystal (smaller, cooler particle) positively charged, as the positive ions are incorporated into the ice crystal (see ● Fig. 10.28).

The same effect occurs when colder, supercooled liquid droplets freeze on contact with a warmer hailstone and tiny splinters of positively charged ice break off. These lighter, positively charged particles are then carried to the upper part of the cloud by updrafts. The larger hailstones (or graupel), left with a negative charge, either remain suspended in an updraft or fall toward the bottom of the cloud. By this mechanism, the cold upper part of the cloud becomes positively charged, while the middle of the cloud becomes negatively charged. The lower part of the cloud is generally of negative and mixed charge except for an occasional positive region located in the falling precipitation near the melting level (see ● Fig. 10.29).

Another school of thought proposes that during the formation of precipitation, regions of separate charge exist within tiny cloud droplets and larger precipitation

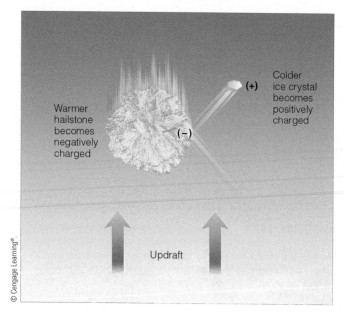

● FIGURE 10.28 When the tiny colder ice crystals come in contact with the much larger and warmer hailstone (or graupel), the ice crystal becomes positively charged and the hailstone negatively charged. Updrafts carry the tiny positively charged ice crystal into the upper reaches of the cloud, while the heavier hailstone falls through the updraft toward the lower region of the cloud.

● FIGURE 10.29 The generalized charge distribution in a mature thunderstorm.

particles. In the upper part of these particles we find negative charge, while in the lower part we find positive charge. When falling precipitation collides with smaller particles, the larger precipitation particles become negatively

A Canadian couple got quite the "shock" in May 2014 when their truck was hit by lightning as they drove along a highway in central Alberta. Video taken from outside the truck shows a quick, fiery blast, after which smoke filled the vehicle. Unable to open windows or doors because of malfunctioning electronic systems, the couple was finally rescued by a passing police officer.

charged and the smaller particles, positively charged. Updrafts within the cloud then sweep the smaller positively charged particles into the upper reaches of the cloud, while the larger negatively charged particles either settle toward the lower part of the cloud or updrafts keep them suspended near the middle of the cloud. These two theories of cloud electrification do not rule each other out. It is possible that both processes are at work during the evolution of a thunderstorm.

The Lightning Stroke   Because unlike charges attract one another, the negative charge at the bottom of the cloud causes a region of the ground beneath it to become positively charged. As the thunderstorm moves along, this region of positive charge follows the cloud like a shadow. The positive charge is most dense on protruding objects, such as trees, poles, and buildings. The difference in charges causes an electric potential between the cloud and ground. In dry air, however, a flow of current does not occur because the air is a good electrical insulator. Gradually, the electrical potential gradient builds, and when it becomes sufficiently large (on the order of one million volts per meter), the insulating properties of the air break down, a current flows, and lightning occurs.

Cloud-to-ground lightning begins within the cloud when the localized electric potential gradient exceeds roughly 3 million volts per meter along a path perhaps 50 meters long. This situation causes a discharge of electrons to rush toward the cloud base and then toward the ground in a series of steps (see ● Fig. 10.30a). Each discharge covers about 50 to 100 meters, then stops for about 50-millionths of a second, then occurs again over another 50 meters or so. This **stepped leader** is very faint and is usually invisible to the human eye. As the tip of the stepped leader approaches the ground, the potential gradient (the voltage per meter) increases, and a current of positive charge starts upward from the ground (usually along elevated objects) to meet it (see Fig. 10.30b). After they meet, large numbers of electrons flow to the ground and a much larger, more luminous **return stroke** several centimeters in diameter surges upward to the cloud along the path followed by the stepped leader (Fig. 10.30c). Hence, the downward flow of electrons establishes the bright channel of upward propagating current. Even though the bright return stroke travels from the ground up to the cloud, it happens so

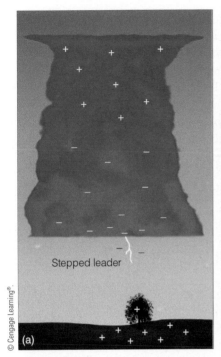
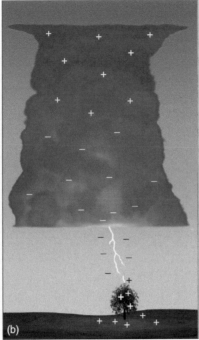
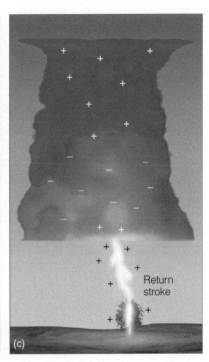

Stepped leader

Return stroke

(a) (b) (c)

© Cengage Learning®.

● **FIGURE 10.30** The development of a lightning stroke. (a) When the negative charge near the bottom of the cloud becomes large enough to overcome the air's resistance, a flow of electrons—the stepped leader—rushes toward Earth. (b) As the electrons approach the ground, a region of positive charge moves up into the air through any conducting object, such as trees, buildings, and even humans. (c) When the downward flow of electrons meets the upward surge of positive charge, a strong electric current—a bright return stroke—carries positive charge upward into the cloud.

quickly—in one ten-thousandth of a second—that our eyes cannot resolve the motion, and we see what appears to be a continuous bright flash of light (see ● Fig. 10.31).

Sometimes there is only one lightning stroke, but more often the leader-and-stroke process is repeated in the same ionized channel at intervals of about four-hundredths of a second. The subsequent leader, called a **dart leader,**

proceeds from the cloud along the same channel as the original stepped leader; however, it proceeds downward more quickly because the electrical resistance of the path is now lower. As the leader approaches the ground, normally a less energetic return stroke than the first one travels from the ground to the cloud. Typically, a lightning flash will have three or four leaders, each followed by a return

© Richard Lee Kaylin

● **FIGURE 10.31** Time exposure of an evening thunderstorm with an intense lightning display near Denver, Colorado. The bright flashes are return strokes. The lighter forked flashes are probably stepped leaders that did not make it to the ground.

stroke. A lightning flash consisting of many strokes (one photographed flash had 26 strokes) usually lasts less than a second. During this short period of time, our eyes may barely be able to perceive the individual strokes, and the flash appears to flicker.

The lightning described so far (where the base of the cloud is negatively charged and the ground positively charged) is called *negative cloud-to-ground-lightning*, because the stroke carries negative charges from the cloud to the ground. About 90 percent of all cloud-to-ground lightning is negative. However, when the base of the cloud is positively charged and the ground negatively charged, a *positive cloud-to-ground lightning* flash may result. Positive lightning, most common with supercell thunderstorms, has the potential to cause more damage because it generates a much higher current level and its flash lasts longer than negative lightning.

Types of Lightning   Notice in Fig. 10.31 that lightning can take on a variety of shapes and forms. When a dart leader moving toward the ground deviates from the original path taken by the stepped leader, the lightning appears crooked or forked as shown in ● Fig. 10.32a. Lightning that takes on this shape is called *forked lightning*. An interesting type of lightning is *ribbon lightning*, which forms when wind moves the ionized channel between each return stroke, causing the lightning to appear as a ribbon hanging from the cloud (see Fig. 10.32b). If the lightning channel breaks up, or appears to break up, the

lightning, called *bead lightning*, looks like a series of beads tied to a string (see Fig. 10.32c). *Ball lightning* looks like a luminous sphere, often about the size of a football, that appears to float in the air or slowly dart about for several seconds as illustrated in 10.32d). For centuries, scientists were unable to confirm the existence of ball lightning, despite many reports from observers of luminous spheres striking or entering buildings. Finally, the first detailed observation of a ball lightning event, including high-speed video, was made over the Qinghai Plateau in western China in 2012. Although many theories have been proposed, the actual cause of ball lightning remains an enigma. *Sheet lightning* forms when either the lightning flash occurs inside a cloud or intervening clouds obscure the flash, such that a portion of the cloud (or clouds) appears as a luminous white sheet (see Fig. 10.32e).

Distant lightning from thunderstorms that is seen but not heard is commonly called **heat lightning** because it frequently occurs on hot summer nights when the overhead sky is clear. As the light from distant electrical storms is refracted through the atmosphere, air molecules and fine dust scatter the shorter wavelengths of visible light, often causing heat lightning to appear orange to a distant observer (see Fig. 10.32f). When cloud-to-ground lightning occurs with thunderstorms that do not produce rain, the lightning is often called **dry lightning**. Such lightning often starts forest fires in regions of dry timber.

Lightning may also shoot upward from the tops of thunderstorms into the upper atmosphere as a dim red

(a) Forked lightning

(b) Ribbon lightning

(c) Bead lightning

(d) Ball lightning

(e) Sheet lightning

(f) Heat lightning

● **FIGURE 10.32** Different forms of lightning.

flash called a *red sprite,* or as a narrow blue cone called a *blue jet.* These phenomena were seen for years by pilots, but they were not widely studied until they were photographed by sensitive low-light cameras, beginning in 1989.

As the electric potential near the ground increases during a thunderstorm, a current of positive charge moves up pointed objects, such as antennas and masts of ships. However, instead of a lightning stroke, a luminous greenish or bluish halo may appear above them, as a continuous supply of sparks—a *corona discharge*—is sent into the air. This electric discharge, which can cause the top of a ship's mast to glow, is known as **St. Elmo's fire,** named after the patron saint of sailors (see ● Fig. 10.33). St. Elmo's fire is also seen around power lines and the wings of aircraft. When St. Elmo's fire is visible and a thunderstorm is nearby, a lightning flash may occur in the near future, especially if the electric field of the atmosphere is increasing.

Lightning rods are placed on buildings to protect them from lightning damage. The rod is made of metal and has a pointed tip, which extends well above the structure. The positive charge concentration will be maximum on the tip of the rod, thus increasing the probability that the lightning will strike the tip and follow the metal rod harmlessly down into the ground, where the other end is deeply buried.

When lightning strikes an object such as a car, the bolt typically leaves the passengers unharmed because it usually takes the quickest path to the ground along the outside metal casing of the vehicle. The lightning then jumps to the road through the air, or it enters the roadway through the tires. The same type of protection is provided by the metal skin of a jet airliner, as hundreds of aircraft are struck by lightning each year with no harm to passengers.

If you should be caught in the open in a thunderstorm, what should you do? Of course, seek shelter immediately, but under a tree? If you are not sure, please read Focus section 10.3.

Lightning Detection and Suppression

For many years, lightning strokes were detected primarily by visual observation. In recent decades, cloud-to-ground lightning has been located by means of an instrument called a *lightning direction-finder,* which works by detecting the radio waves produced by lightning. Networks of these magnetic devices can be used to pinpoint the location of cloud-to-ground flashes throughout the United States and Canada. Lightning detection devices allow scientists to examine in detail the lightning activity inside a storm as it intensifies and moves. Such investigation gives forecasters a better idea where intense lightning strokes might be expected.*

---

*In fact, with the aid of these instruments, together with satellite images, radar displays, and computer models of the atmosphere, the National Weather Service currently issues experimental lightning probability forecasts for the contiguous United States.

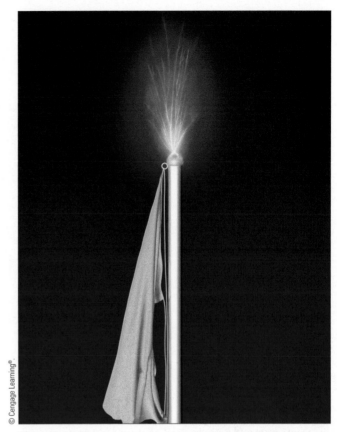

© Cengage Learning®.

● **FIGURE 10.33** St. Elmo's fire tends to form above objects, such as aircraft wings, ships' masts, and flagpoles.

Satellites now have the capability of providing even more lightning information than ground-based sensors, because satellites can continuously detect all forms of lightning over land and over water (see ● Fig. 10.34). Lightning information correlated with satellite images provides a more complete and precise structure of a thunderstorm. The new GOES-R series of satellites includes a new Geostationary Lightning Mapper that will provide continuous monitoring of lightning activity across the Americas.

Each year, fire departments respond to more than 20,000 fires started by lightning in the United States alone, with damages in both timber and property value averaging more than $400 million annually. For this reason, tests have been conducted to see whether the number of cloud-to-ground lightning discharges can be reduced. One technique that has shown some success in suppressing lightning involves seeding a cumulonimbus cloud with hair-thin pieces of aluminum about 10 cm long. The idea is that these pieces of metal will produce many tiny sparks, or corona discharges, and prevent the electrical potential in the cloud from building to a point where lightning occurs. While the results of this experiment are inconclusive, many forestry specialists point out that nature itself may use a similar mechanism to prevent excessive lightning damage. The long, pointed needles of a conifer tree may act as tiny lightning rods, diffusing the concentration of electric charges and preventing massive lightning strokes.

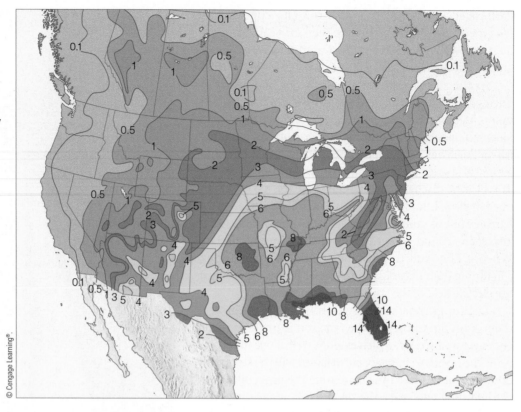

● **FIGURE 10.34** Average lightning flash density per square kilometer per year from 1997 to 2010. Notice that in the United States, Florida is the most lightning-prone state. (Data from the North American Lightning Detection Network. Courtesy of Vaisala.)

© Cengage Learning®.

Now that we have looked at thunderstorms, we are ready to explore a product of a thunderstorm that is one of nature's most awesome phenomena: the tornado, a rapidly spiraling column of air that usually extends down from the base of a cumulonimbus cloud and can strike sporadically and violently.

Tornadoes

A **tornado** is a rapidly rotating column of air, extending down from a cumuliform cloud, that blows around a small area of intense low pressure with a circulation that reaches the ground. A tornado's circulation may be visible as a funnel-shaped cloud extending from a cumulonimbus cloud to the surface, or as a swirling cloud of dust and debris on the ground. Sometimes called *twisters* or *cyclones,* tornadoes can assume a variety of shapes and forms that range from twisting rope-like funnels, to cylinder-shaped funnels, to massive black wedge-shaped funnels, to funnels that resemble an elephant's trunk hanging from a large cumulonimbus cloud (see ● Fig. 10.35).

Sometimes the term **funnel cloud** is used to refer to a visible funnel that does not touch the ground. However, a tornado may still be present even if the visible cloud does not extend to the surface. Perhaps only about 30 percent of funnel clouds become active tornadoes. When viewed from above, most North American tornadoes rotate counterclockwise about their central core of low pressure. Some have been seen rotating clockwise, but those are infrequent, accounting for as few as two percent of all tornadoes.

Tornadoes can vary greatly in their size, strength, speed, and duration. About 75 percent of all reported tornadoes in the United States have wind speeds of less than 110 knots, although violent tornadoes may have winds exceeding 220 knots. The diameter of most tornadoes is between 100 and 600 m (about 300 to 2000 ft), although some are just a few meters wide and others have diameters exceeding 1600 m (1 mi). The largest tornado on record, spanning 4.2 km (2.6 mi), took the lives of three tornado researchers as it plowed across western Oklahoma near El Reno on May 31, 2013.

Tornadic thunderstorms that form ahead of an advancing cold front are often steered by southwesterly winds and, therefore, these thunderstorms and their associated tornadoes tend to move from the southwest toward the northeast at speeds usually between 20 and 40 knots. However, some tornadoes have been clocked at speeds greater than 70 knots. Most tornadoes last only a few minutes and have an average path length of about 7 km (4 mi), yet there are cases where they have reportedly traveled for hundreds of kilometers over several hours. The 1925 Tri-State Tornado mentioned at the start of this chapter featured the longest official track on record for a single tornado: 352 km (219 mi).*

* A recent analysis found that the damage path of the 1925 Tri-State Tornado was bracketed by destruction that may have been produced by two other tornadoes, which brings the total damage swath to as much as 278 km (235 mi).

Don't Sit Under the Apple Tree

Because a single lightning stroke may involve a current as great as 100,000 amperes, animals and humans can be electrocuted when struck by lightning. Although the per-capita rate of lightning deaths in the United States has decreased by more than 90 percent in the last century, several dozen Americans are killed by lightning each year, with Florida accounting for the most fatalities. Many victims are struck in open places, riding on farm equipment, playing golf, attending sports events, or sailing in a small boat. Some live to tell about it, as did the retired champion golfer Lee Trevino. Others are less fortunate, as about 10 percent of people struck by lightning are killed. Most die from cardiac arrest. Consequently, when you see someone struck by lightning, immediately give CPR (cardiopulmonary resuscitation), as lightning normally leaves its victims unconscious without heartbeat and without respiration. Those who do survive often suffer from long-term psychological disorders, such as personality changes, depression, and chronic fatigue.

Many lightning fatalities occur in the vicinity of relatively isolated trees (see Fig. 4). As a tragic example, during June 2004, three people were killed near Atlanta, Georgia, while seeking shelter under a tree. Because a positive charge tends to concentrate in upward projecting objects, the upward return stroke that meets the stepped leader is most likely to originate from such objects. Clearly, sitting under a tree during an electrical storm is not wise. What *should* you do?

When you are caught outside in a thunderstorm, the best protection, of course, is to get inside a building. But stay away from electrical appliances and corded phones, and avoid taking a shower. Automobiles with metal frames and trucks (but not golf carts) may also provide protection. If no such shelter exists, be sure to avoid elevated places and isolated trees. If you are on level ground, try to keep your head as low as possible, but do not lie down. Because lightning channels usually emanate outward through the ground at the point of a lightning strike, a surface current may travel through your body and injure or kill you. Therefore, crouch down as low as possible and minimize the contact area you have with the ground by touching it with only your toes or your heels.

● **FIGURE 5** Lightning can be both hair-raising and deadly. This photograph, taken by Mary McQuilken, shows her younger brother, Sean (on the left), and older brother, Michael (on the right), standing beneath a thunderstorm atop Moro Rock in California's Sequoia National Park. Shortly after this photo was taken, Sean was struck by lightning and seriously injured. A nearby hiker was killed by the same lightning strike.

There are some warning signs to alert you to a strike. If your hair begins to stand on end or your skin begins to tingle and you hear clicking sounds, beware: Lightning may be about to strike. And if you are standing upright, you may be acting as a lightning rod (see ● Fig. 5).

What happens when lightning strikes a vehicle? As long as the windows are rolled up, the occupants may be unharmed by the flash, because the lightning will typically travel across the car's outer surface or through wiring. However, the complex electronic systems of a modern vehicle can easily be damaged by a lightning strike, and tires can be blown out. During an electrical storm, occupants should avoid touching any metallic interior object that may be connected to the exterior, such as a door handle or a gear shifter. If it is possible to do so safely, pull over and wait until the storm has passed.

● **FIGURE 4** A cloud-to-ground lightning flash hitting a 65-foot sycamore tree. It should be apparent why one should not seek shelter under a tree during a thunderstorm.

TORNADO LIFE CYCLE   As we will see later in this chapter, a variety of factors must come together in order to produce a tornado. Major tornadoes usually evolve through a series of stages. The first stage is the *dust-whirl stage*, where dust swirling upward from the surface marks the tornado's circulation on the ground and a short funnel often extends downward from the thunderstorm's base. Damage during this stage is normally light. As the tornado increases in strength, it enters its *mature stage*. During this stage, damage normally is most severe as the funnel reaches its greatest width and is almost vertical (see ● Fig. 10.36). The final stage, called the *decay stage*, usually finds the tornado stretched into the shape of a rope. Normally, the tornado becomes greatly contorted before it finally dissipates. Although these are the typical stages of a major tornado, minor tornadoes may skip the mature stage and go directly

● **FIGURE 10.35** A large wedge-shaped violent tornado moves northwestward directly for Windsor, Colorado, on May 22, 2008. The photo (taken by a webcam) shows hail the size of golf balls falling from the thunderstorm and covering the ground. It also illustrates how an approaching tornado can appear as a massive dark cloud.

into the decay stage. However, when a tornado reaches its mature stage, its circulation usually stays in contact with the ground until it dissipates. Sometimes a new tornado will emerge from a supercell thunderstorm just before or shortly after an existing tornado dissipates.

TORNADO OCCURRENCE AND DISTRIBUTION Tornadoes occur in many parts of the world, but no country experiences more tornadoes than the United States, which in recent years has averaged more than 1000 annually. In 2004, a record was set with 1,819 tornadoes observed. The number of total tornadoes reported each year has more than doubled since the 1950s (see ● Fig. 10.37), even though the number of strong tornadoes (those with winds exceeding 117 knots, or 135 mi/hr) has shown no significant trend. The difference in tornado numbers is most likely because many

more weaker, short-lived tornadoes are being reported (and photographed) than was the case decades ago. Although tornadoes have occurred in every state, including Alaska and Hawaii, the greatest number occur in the tornado belt or **Tornado Alley** of the Central Plains, which stretches from central Texas to Nebraska (see ● Fig. 10.38). The belt of tornadoes centered across Mississippi and Alabama is sometimes called *Dixie Alley*.*

The Central Plains region is most susceptible to tornadoes because it often provides the proper atmospheric setting for the development of the severe thunderstorms that spawn tornadoes. You may recall from Fig. 10.21 on p. 285, that over the Central Plains (especially in spring)

---

*Many of the tornadoes that form along the Gulf Coast are generated by thunderstorms embedded within the circulation of hurricanes.

● **FIGURE 10.36** A narrow tornado descends from a rain-free thunderstorm base in eastern Colorado on May 9, 2015.

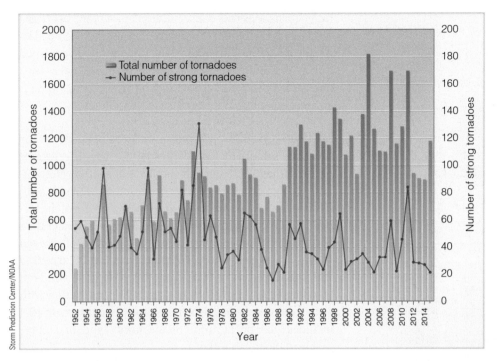

Storm Prediction Center/NOAA

● **FIGURE 10.37** Total number of tornadoes reported in the United States for each year from 1952 to 2015 (blue bar); and the total number of strong tornadoes with winds exceeding 117 knots, or 135 mi/hr, reported during the same period (red line). (Note: Tornadoes that occurred before the introduction of the Enhanced Fujita (EF) Scale in 2007 have been converted to the new scale.)

warm, humid surface air is overlain by cooler, drier air aloft, producing a conditionally unstable atmosphere. When a strong vertical wind shear exists (usually provided by a low-level jet and the polar jet stream) and the surface air is forced upward, large supercell thunderstorms capable of spawning tornadoes may form. Therefore, tornado frequency is highest during the spring and lowest during the winter when the warm surface air is normally absent.

In ● Fig. 10.39, we can see that about 70 percent of all tornadoes in the United States develop from March to July. The month of May normally has the greatest number of tornadoes* (the average is about 9 per day), while many of the most violent tornadoes occur in April, when

*During May 2003, 516 tornadoes touched down in the United States, an average of more than 16 per day. However, during April 2011, a record of 748 tornadoes was reported, the most in any month since records began in 1950.

vertical wind shear tends to be stronger and horizontal and vertical temperature and moisture contrasts are often high. Since 1950, the month holding the record for the greatest number of tornadoes is April 2011, when 748 were reported. Although tornadoes have occurred at all times of the day and night, they are most frequent in the late afternoon (between 3 p.m. and 7 p.m.), when the surface air is most unstable; they are least frequent in the early morning before sunrise, when the atmosphere is most stable.

Although large, destructive tornadoes are most common in the Central Plains, they can develop anywhere in the United States (or the world, for that matter) if conditions are right. For example, a series of at least 36 tornadoes, more typical of those that form over the plains, marched through North and South Carolina on March 28, 1984, claiming 59 lives and causing hundreds of millions

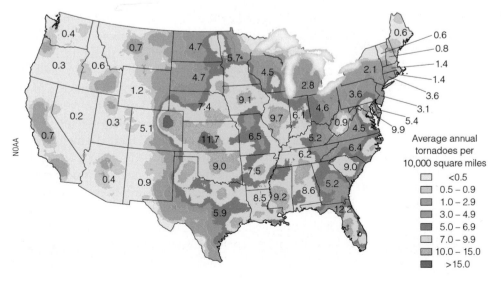

● **FIGURE 10.38** The average annual number of observed tornadoes per 10,000 square miles in each state from 1991 to 2012.

Average annual tornadoes per 10,000 square miles

▢	<0.5
▢	0.5 – 0.9
▢	1.0 – 2.9
▢	3.0 – 4.9
▢	5.0 – 6.9
▢	7.0 – 9.9
▢	10.0 – 15.0
■	>15.0

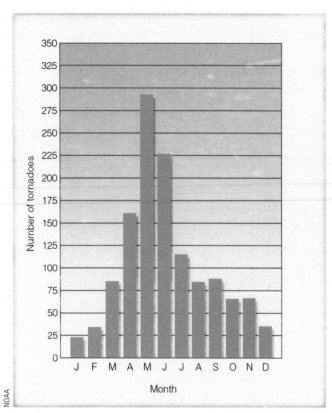

● FIGURE 10.39 Average number of tornadoes during each month in the United States from 2000 to 2010.

NOAA

of dollars in damage. One tornado was enormous, with a diameter over 2 miles and winds that exceeded 200 knots. No place is totally immune to a tornado's destructive force. On March 1, 1983, a rare tornado cut a 5-km swath of destruction through downtown Los Angeles, California, damaging more than 100 homes and businesses and injuring 33 people. And during the summer of 2010, two tornadoes actually touched down in New York City. One tornado (on July 25) caused only minimal damage but injured seven people.

Even in the central part of the United States, the statistical chance that a tornado will strike a particular place this year is quite small. However, tornadoes can provide many exceptions to statistics. Oklahoma City, for example, has been struck by tornadoes at least 35 times in the past 100 years. The adjacent suburb of Moore experienced destructive tornadoes on May 3, 1999; May 8, 2003; and May 20, 2013. And the little town of Codell, Kansas, was

DID YOU KNOW?

Although the United States and Canada rank one and two in the world in annual number of tornadoes, Bangladesh has experienced the deadliest tornadoes. About 1300 people died when a violent tornado struck north of Dacca on April 26, 1989, and on May 13, 1996, over 700 lives were lost when a violent tornado touched down in Tangail.

hit by tornadoes in 3 consecutive years—1916, 1917, and 1918—and each time on the same date: May 20! Considering the many millions of tornadoes that must have formed during the geological past, it is very probable that at least one actually moved across the land where your home is located, especially if it is in the Central Plains.

TORNADO WINDS   At one point, our knowledge of the furious winds of a tornado came mainly from observations of the damage done and the analysis of motion pictures. Today, more accurate wind measurements are made with Doppler radar. Because of the destructive nature of the tornado, it was once thought that it packed winds greater than 500 knots. However, radar measurements reveal that even the most powerful twisters seldom have winds exceeding 220 knots, and most tornadoes probably have winds of less than 125 knots. Nevertheless, being confronted with even a weak tornado can be terrifying. (Focus section 10.4 includes more background on the quirky types of damage that tornadoes can inflict.)

When a tornado is approaching from the southwest, its strongest winds are on its southeast side. We can see why in ● Fig. 10.40. The tornado is heading northeast at 50 knots. If its rotational speed is 100 knots, then its forward speed will add 50 knots to its southwestern side (position D) and subtract 50 knots from its northwestern side (position A). Hence, the most destructive and extreme winds will be on the tornado's southeastern side.

Many violent tornadoes (with winds exceeding 180 knots) contain smaller whirls that rotate within them. Such tornadoes are called *multi-vortex tornadoes* and the smaller whirls are called **suction vortices** (see ● Fig. 10.41). Suction vortices are only about 10 m (330 ft) in diameter, but they rotate very fast and apparently do a great deal of damage.

Seeking Shelter   The high winds of a tornado cause the most damage, as walls of buildings may buckle and collapse when blasted by the extreme wind force and by debris carried by the wind. Also, as high winds blow over a roof, lower air pressure forms above the roof. The greater air pressure inside the building then lifts the roof just high enough for the strong winds to carry it away. A similar effect occurs when the tornado's intense low-pressure center passes overhead. Because the pressure in the center of a tornado may be more than 100 mb (3 in.) lower than that of its surroundings, there is a momentary drop in outside pressure when the tornado is above the structure. It was once thought that opening windows and allowing inside and outside pressures to equalize would minimize the chances of the building exploding. However, it is now known that opening windows during a tornado actually increases the pressure on the opposite wall and

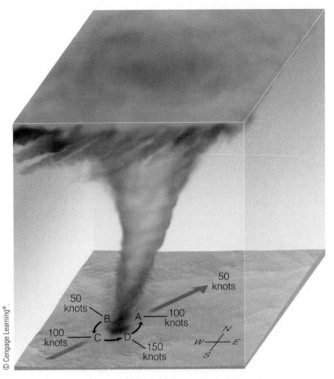

● **FIGURE 10.40** The total wind speed of a tornado is greater on one side than on the other. When facing an onrushing tornado, the strongest winds will be on your left side.

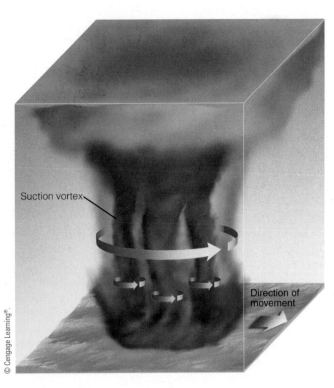

● **FIGURE 10.41** A powerful multi-vortex tornado with three suction vortices.

increases the chances that the building will collapse. (The windows are usually shattered by flying debris anyway.) Damage from tornadoes may also be inflicted on people and structures by flying debris. Hence, the wisest course to take when confronted with an approaching tornado is to *seek shelter immediately.*

At home, take shelter in a basement, storm shelter, or a dedicated "safe room" (a hardened structure built to FEMA standards to survive high winds with as little damage as possible). In a large building without a basement, the safest place is usually in a small room, such as a bathroom, closet, or interior hallway, preferably on the lowest floor and near the middle of the edifice. Pull a mattress or sleeping bag around you. Wear a bike or football helmet, if one is available, to protect your head from flying debris, and stay away from windows. At school, move to the hallway and lie flat with your head covered. In a mobile home, leave immediately and seek substantial shelter. If none exists, lie flat on the ground in a depression or ravine.

Don't try to outrun an oncoming tornado in a car or truck, as tornadoes often cover erratic paths with speeds sometimes exceeding 45 knots (50 mi/hr). Instead, if the tornado is still at some distance, drive at a right angle to the tornado's path. (If the tornado appears to be stationary, but getting bigger, it is most likely moving directly toward you.) If there is any question about how close the tornado is, abandon your vehicle and seek shelter immediately—but

not under a freeway overpass, as the tornado's winds may be funneled (strengthened) by the overpass structure. If you are caught outdoors in an open field, look for a ditch, streambed, or ravine, and lie flat with your head covered.

The National Weather Service (NWS) issues several types of bulletins related to tornadoes. A **tornado watch** is issued by the *Storm Prediction Center* in Norman, Oklahoma (in collaboration with local NWS offices) to alert the public that tornadoes may develop within a specific area during a certain time period, usually a few hours long. Many communities have trained volunteer spotters, who look for tornadoes after the watch is issued. Once a tornado is spotted visually or its telltale signs are evident on radar, a **tornado warning** is issued by the local NWS office, typically covering parts of one or several counties and lasting from 30 to 45 minutes. A **tornado emergency** may also be issued on rare occasions when an especially strong tornado threatens a populated area. In some communities, sirens are sounded to alert people to an approaching storm. Radio and television stations (including NOAA Weather Radio) interrupt regular programming to broadcast the warning, and most newer cellphones will display the warning automatically through the Wireless Emergency Alert system (assuming the phone's owner has not blocked this feature). Although not completely effective, the current warning system is saving many lives. Despite the large

The Weird World of Tornado Damage

The strong winds of a tornado can destroy buildings, uproot trees, and hurl all sorts of lethal missiles into the air. People, animals, and home appliances all have been picked up, carried several kilometers, then deposited. Even motor vehicles have been tossed a kilometer or more, and lighter objects such as checkbooks have been blown 100 km (62 mi) or more, sometimes from one state to another. One tornado lifted a railroad coach with its 117 passengers and dumped it in a ditch about 25 m (82 ft) away. In one instance, a schoolhouse was demolished, and the 85 students inside were carried more than 90 m (295 ft) without any of them being killed. In another freakish example, a house in Michigan survived a tornado but was turned onto its side so that the front door couldn't be reached without a ladder. Showers of toads and frogs have poured out of a cloud after tornadic winds sucked them up from a nearby pond. Other oddities include pieces of straw being driven into metal pipes, frozen hot dogs being driven into concrete walls, and chickens losing all of their feathers. (Actually, the most likely explanation for the de-feathering is a process called flight molt, in which chickens involuntarily release feathers when threatened.)

● FIGURE 6 bicycle tire was wrapped around a utility pole by the fierce winds of an EF4 tornado that struck Picher, Oklahoma, on May 10, 2008.

The power delivered by the winds of a strong tornado is far beyond anything in our everyday experience. The pressure of the wind (the force per unit area) increases with the square of the wind speed, which means that if you double the wind speed, you get four times as much destructive potential. Imagine a very windy day, with gusts of 40 knots (45 mi/hr) flinging dust into your eyes and tossing loose objects down the street. In the most violent tornadoes on Earth, winds can exceed 174 knots (200 mi/hr), which means they carry more than 16 times the power of the gusts making your day miserable. When a long object such as a straw or a two-by-four board (or a frozen hot dog) is tossed lengthwise in a violent tornado, its force is packed into the tiny area at its tip. This concentration gives the wind-blown projectile a surprising ability to penetrate solid objects.

Another factor that makes tornado damage so quirky is the huge variability of tornado behavior. Tornadoes can strengthen or weaken dramatically, and shrink or enlarge, in just a few seconds. The suction vortices inside a tornado illustrated in Fig. 10.39 can be as small as 10 m (30 ft) in diameter, so it is quite possible for a suction vortex to hit one house and miss the next-door neighbor's house entirely. Even as winds destroy a building, they may leave a few objects untouched due to the complex flow around the disintegrating structure. (Such cases often make the news when the intact item is a cherished keepsake.) In a large tornado, pieces of debris may be spun a mile or more away from the main tornadic circulation. These objects may cause localized pockets of damage in areas that are otherwise unscathed.

increase in population in the tornado belt during the past 30 years, tornado-related deaths have generally held steady during this period, with 2011 being a major exception (see ▼ Table 10.1).* (For additional information on tornado watches and warnings, read Focus section 10.5.)

The Enhanced Fujita Scale   In the 1960s, the late Dr. T. Theodore Fujita, a noted authority on tornadoes, at the University of Chicago, proposed a scale, called the **Fujita scale**, for classifying tornadoes according to their rotational wind speed. The tornado winds are estimated based on the damage caused by the storm. The original Fujita scale, implemented in 1971, was based mainly on the extent of tornado damage to frame houses. Because many types of structures are susceptible to tornado damage, and not all tornadoes strike frame homes, a new scale

▼ Table 10.1   Average Annual Number of Tornadoes and Tornado Deaths by Decade

DECADE	TORNADOES/ YEAR	DEATHS/YEAR
1950–59	480	148
1960–69	681	94
1970–79	858	100
1980–89	819	52
1990–99	1220†	56
2000–09	1277	56
2010–15	1149†	136*

†More tornadoes are being reported in recent decades as populations increase, more citizens watch for tornadoes, and tornado-spotting technology improves.

*This six-year average includes the especially deadly year of 2011, when 553 deaths occurred. Otherwise, the average for 2010-15 was around 53 deaths per year.

*The year 2011 was an exceptionally deadly year for United States tornadoes, as 553 people perished in these storms.

▼ Table 10.2 **Enhanced Fujita (EF) Scale for Damaging Tornado Winds**

EF SCALE	CATEGORY	MI/HR*	KNOTS*
EF0	Weak	65–85	56–74
EF1		86–110	75–95
EF2	Strong	111–135	96–117
EF3		136–165	118–143
EF4	Violent	166–200	144–174
EF5		>200	>174

*The wind speed is a 3-second gust estimated at the point of damage, based on a judgment of damage indicators.

● **FIGURE 10.42** A house situated on the Great Plains. Observe in Fig. 10.43 how tornadoes of varying EF intensity can damage this house and its surroundings.

came into effect in February 2007. Called the **Enhanced Fujita Scale**, or simply the **EF Scale**, the new scale attempts to provide a wide range of criteria in estimating a tornado's winds by using a set of 28 damage indicators, including small barns, mobile homes, schools, and trees. The quality of building construction is also taken into account. Each structure or object is then examined for the degree of damage it sustained. The combination of the damage indicators along with the degree of damage provides a range of probable wind speeds and an EF rating for the tornado. The wind estimates for the Enhanced Fujita Scale are given in ▼ Table 10.2.

● Figure 10.42 shows a house situated somewhere on the Great Plains of the United States or Canada, and ● Fig. 10.43 shows the damaging effect that tornadoes ranging in intensity from EF0 to EF5 can have on this structure and its surroundings. Notice that an EF0

tornado causes only minimal damage, whereas an EF5 completely demolishes the house and sweeps it off its foundation. In this example, we assume that the home is well constructed; otherwise, the damage for a given strength of tornado could be even greater than shown. A devastating tornado that struck central Arkansas on April 27, 2014, was rated EF4 rather than EF5, even though some homes were swept off their foundations, because damage surveyors found that these homes were generally fastened to their foundations with nails rather than anchor bolts. Manufactured or mobile homes are especially vulnerable in tornadoes: During recent years, roughly 45 percent of all U.S. tornado fatalities have occurred in mobile homes. Note that damage can vary greatly along a tornado's path; for example, one section may exhibit EF2 or EF3 damage

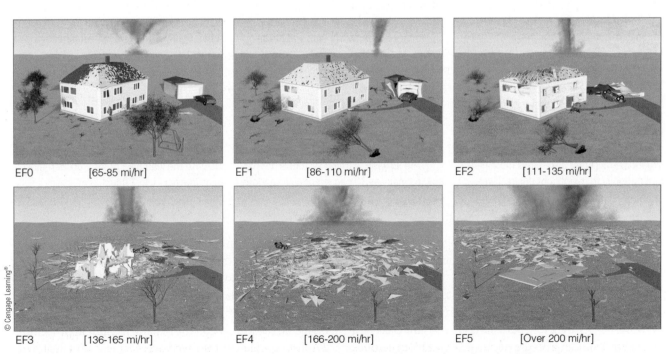

| EF0 | [65-85 mi/hr] | EF1 | [86-110 mi/hr] | EF2 | [111-135 mi/hr] |

| EF3 | [136-165 mi/hr] | EF4 | [166-200 mi/hr] | EF5 | [Over 200 mi/hr] |

● **FIGURE 10.43** Damage to the house in Fig. 10.42 and its surroundings caused by tornadoes of varying EF intensity.

The Evolution of Tornado Watches and Warnings

Few weather situations are as terrifying as the approach of a churning tornado. Fortunately, the United States has a well-established system for letting people know when twisters are possible and when one is imminent. The system of tornado watches and warnings in the United States was created in the 1950s, shortly after a remarkable coincidence. On March 20, 1947, a destructive tornado struck Tinker Air Force Base, just southeast of Oklahoma City. A total of 50 aircraft were destroyed, at a cost of $10 million (likely more than $100 million in today's dollars). The next day, two Air Force meteorologists, E. J. Fawbush and Robert Miller, were asked to develop a technique for predicting when tornadoes were likely. Fawbush and Miller quickly created a scheme based on such factors as instability, wind shear, and the approach of a front. Incredibly, another tornado struck the base only five days after the first one. This time the brand-new Fawbush-Miller technique provided notice that tornadoes were very likely, so aircraft were safely sheltered and the damage toll was far less.

As we previously learned, a tornado watch is issued when conditions favor the formation of severe thunderstorms and tornadoes. The average tornado watch covers a period of 6 to 8 hours and an area of roughly 65,000 square km, about the size of South Carolina. Tornado watches are issued by the National Weather Service's Storm Prediction Center in Norman, Oklahoma, in coordination with local NWS offices. When conditions are most threatening, a watch will receive the "PDS" tag, which denotes a *particularly dangerous situation*. The center also issues more general outlooks that highlight where severe

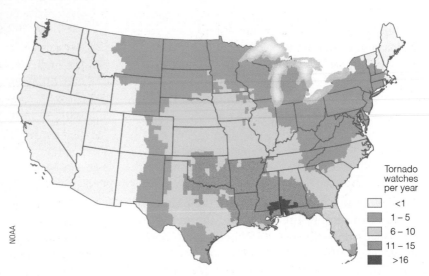

● **FIGURE 7** Average annual number of tornado watches per county issued by the Storm Prediction Center from 1993 to 2012.

weather is possible up to eight days in advance. If you are in a region under a tornado watch, stay aware of changing weather conditions and watch or listen for possible warnings.

When a tornado is actually sighted, or when a tornadic circulation is evident on radar, the local NWS office will issue a *tornado warning*. Warnings typically cover parts of one to several counties and are in effect for 30 to 60 minutes. If your smartphone is equipped to receive Wireless Emergency Alerts (WEA), you will normally receive a text message if a tornado warning is issued in your area. Should you find yourself in a tornado warning area, you need to take cover immediately—ideally in an interior room on the lowest floor of a well-built structure, but not in a vehicle or a mobile home.

The tornado warning system is not perfect. As many as 30 percent of all tornadoes

come and go before a warning can be issued, and more than half of all warnings are false alarms, typically because a tornadic signature on radar has failed to produce a twister. Even when a tornado is accurately warned, it will usually strike only part of a warned area. Still, the risk of injury or death is very real in these situations, and many lives are saved by quick response to a tornado warning. The advent of the United States Next Generation Doppler Radar (NEXRAD) network helped improve the warning system substantially. The average lead time for a tornado warning rose from 3 minutes in the late 1970s to 13 minutes by the late 1990s, and the lead time is now often 20 minutes or more for the longest-lived and most violent tornadoes. A new NWS initiative is exploring whether high-resolution computer models that incorporate radar data can help increase the lead time of tornado warnings to as much as 1 to 2 hours.

while another area may show EF4 damage. Such a tornado would be given an EF4 rating, corresponding to the most severe damage observed.

Statistics reveal that the majority of tornadoes are relatively weak, with wind speeds less than about 110 knots (115 mi/hr). Only a few percent each year are classified as violent, with an average of one or two EF5 tornadoes reported annually (although several years may pass

without an EF5 in the United States). However, it is the violent tornadoes that account for the majority of tornado-related deaths. For example, a powerful EF5 tornado roared through the town of Greensburg, Kansas, on the evening of May 4, 2007. The tornado, with winds estimated at 180 knots (205 mi/hr) and a width approaching 2 miles, completely destroyed over 95 percent of the town. The tornado took 11 lives, and probably more would have perished had it not been for the tornado warning issued by the National Weather Service and the sirens in the town, signaling "take cover," that sounded about 20 minutes before the tornado struck.

On May 22, 2011, a violent EF5 multi-vortex tornado struck Joplin, Missouri, completely demolishing part of the city.* The tornado injured almost 1000 people and took 159 lives, the greatest death toll attributed to a single tornado in the United States since the Woodward, Oklahoma, tornado on April 9, 1947. However, several hundred more people were killed just weeks before the Joplin tornado by a swarm of twisters—an event referred to as a *tornado outbreak*.

TORNADO OUTBREAKS   As we have seen, tornadoes take the lives of many people each year. The annual average is less than 100, although more than 100 people can die in a single tornado, as tragically occurred in Joplin, Missouri. The deadliest tornadoes are often those that occur in *families*, a group or series of tornadoes spawned by the same thunderstorm. (Some thunderstorms produce a sequence of several tornadoes over 2 or more hours and over distances of 100 km or more.) Tornado families are typically the result of a single, long-lived supercell thunderstorm. Sometimes a large, well-organized frontal system will spawn multiple supercells, each producing its own family of tornadoes. When a large number of tornadoes develops in association with a particular weather system (typically six tornadoes or more, although there is no strict definition), it is referred to as a **tornado outbreak**. An outbreak can extend over multiple days. The most severe tornado outbreaks are typically very well predicted because they involve large areas of conditionally unstable air and extremely strong vertical wind shear, along with an upper-level trough that helps trigger widespread supercell formation. These features can often be spotted by computer forecast models several days in advance. The NOAA Storm Prediction Center in Norman, Oklahoma, issues convective outlooks each day that show where outbreaks of severe weather and tornadoes are possible over the next eight days. Researchers are now exploring techniques that could provide several weeks' notice of when there may be an elevated risk of tornado outbreaks.

A particularly devastating outbreak occurred on May 3, 1999, when 78 tornadoes marched across parts of Texas, Kansas, and Oklahoma. One tornado, whose width at times reached one mile and whose wind speed was measured by a portable Doppler radar at 262 knots (301 mi/hr), moved through parts of Oklahoma City and the suburb of Moore. Within its 38-mile path, it damaged or destroyed thousands of homes, injured nearly 600 people, claimed 36 lives, and caused over $1 billion in property damage. Ironically, on May 20, 2013, an EF5

tornado with maximum winds estimated at 183 knots (210 mi/hr) took a path similar to that of the deadly 1999 tornado, with the paths actually crossing at one point. The 2013 tornado was on the ground for 27 km (17 mi) and cut a destructive swath through a highly populated section of Moore, killing 24 people, including 10 children, and causing an estimated $2 billion in damage.

One of the most violent outbreaks ever recorded occurred on April 3 and 4, 1974. During a 16-hour period, 148 tornadoes cut through parts of 13 states and one Canadian province, killing 319 people, injuring more than 5000, and causing an estimated $600 million in damage in the United States. Some of these tornadoes were among the most powerful ever recorded, as 7 tornadoes were rated at F5 intensity and 23 tornadoes at F4 intensity. The combined paths of all the tornadoes during this *"Super Outbreak"* amounted to 4181 km (2598 mi), well over half of the total path for an average year. The greatest one-day loss of life attributed to tornadoes in the United States occurred during the 1925 Tri-State Outbreak mentioned previously, when an estimated 695 people died as at least 7 tornadoes traveled across portions of Missouri, Illinois, and Indiana.

The only recorded outbreak on par with the 1974 Super Outbreak occurred on April 25–28, 2011, when 357 tornadoes (4 of which reached EF5 intensity) moved across portions of the eastern United States, plus one in southeastern Canada. The tornadoes claimed 316 lives, injured thousands of people, and caused more than $10 billion in damages. One particularly strong EF4 tornado, with winds estimated at 190 mi/hr, moved through Tuscaloosa, Alabama, on April 27 (see ● Fig. 10.44). The tornadoes, which had a damage path width of about 1.5 miles, left 43 dead in Tuscaloosa and injured more than 1000 (see ● Fig. 10.45).

Tornado Formation

Although not everything is known about the formation of a tornado, we do know that tornadoes develop only within thunderstorms and that a conditionally unstable atmosphere is essential for their development. Most often, tornadoes form in supercell thunderstorms in an environment with strong vertical wind shear.* The rotating air of the tornado may begin within a thunderstorm and work its way downward, or it may begin at the surface and work its way upward. First, we will examine supercell tornadoes; then we will examine nonsupercell tornadoes.

SUPERCELL TORNADOES   Tornadoes that form with supercell thunderstorms are called **supercell tornadoes**. Earlier we learned that a supercell is a thunderstorm

*A photo of the Joplin tornado is shown in Fig. 1.10, on p. 12. The destruction caused by this tornado is shown in Fig. 1.11.

*A description of atmospheric conditions favorable for the formation of supercell thunderstorms is presented beginning on p. 282.

● FIGURE 10.44 This huge EF4 multi-vortex tornado devastated sections of Tuscaloosa, Alabama, on April 27, 2011. (See also Fig. 10.45.)

AP Photo/The Tuscaloosa News, Dusty Compton, File

NOAA

● FIGURE 10.45 Damage in Tuscaloosa, Alabama, after a massive EF4 tornado (shown in Fig. 10.44) plowed through the city on April 27, 2011.

that has a single rotating updraft that can exist for hours. ● Figure 10.46 illustrates this updraft and the pattern of precipitation associated with the storm. Notice that as warm, humid air is drawn into the supercell, it spins counterclockwise as it rises. Near the top of the storm, strong winds push the rising air to the northeast. Heavy precipitation falling northeast of the updraft mixes with drier air. Evaporative cooling chills the air. The heavy rain-chilled air then descends as a strong downdraft called the *forward-flank downdraft*. The separation of the updraft from the downdraft means that the downdraft is unable to fall into the updraft and suppress it. This is why the storm is able to maintain itself as a single entity for hours.

Tornadoes are rapidly rotating columns of air, so what is it that starts the air rotating? We can see how rotation can develop by looking at ● Fig. 10.47a. Notice that there is vertical directional wind shear, because the surface winds are southerly and a kilometer or so above the surface they are westerly. There is also vertical wind speed shear, because the wind speed increases rapidly with height. This

vertical wind shear causes the air near the surface to rotate about a horizontal axis much like a pencil would rotate around its long axis. Such horizontal tubes of spinning air are called *vortex tubes*. (These spinning vortex tubes also form when a southerly low-level jet exists just above weaker surface winds.) If the strong updraft of a developing thunderstorm should *tilt* the rotating tube upward and draw it into the storm, as illustrated in Fig. 10.47b, the tilted rotating tube then becomes a rotating air column inside the storm. The rising, spinning air is now part of the storm's structure called the *mesocyclone*—an area of lower pressure (a small cyclone) perhaps 5 to 10 kilometers across. The rotation of the updraft lowers the pressure in the mid-levels of the thunderstorm, which acts to increase the strength of the updraft.*

As we learned earlier in the chapter, the updraft is so strong in a supercell (sometimes exceeding 90 knots, or

*You can obtain an idea of what might be taking place in the supercell by stirring a cup of coffee or tea with a spoon and watching the low pressure form in the middle of the beverage.

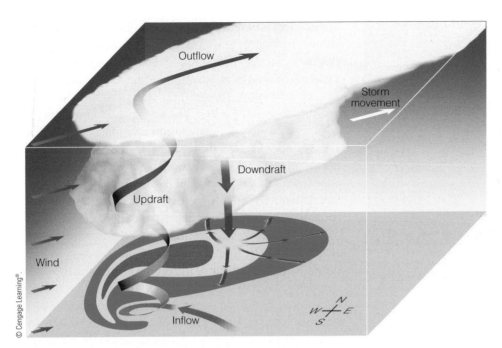

● **FIGURE 10.46** A simplified view of a supercell thunderstorm with a strong updraft and downdraft forming in a region of strong vertical wind speed shear. Regions beneath the supercell receiving precipitation are shown in color: green for light rain, yellow for heavier rain, and red for very heavy rain and hail.

104 mi/hr) that precipitation cannot fall through it. South-westerly winds aloft usually blow the precipitation north-eastward. If the mesocyclone persists, it can circulate some of the precipitation counterclockwise around the updraft. This swirling precipitation shows up on the radar screen, whereas the area inside the mesocyclone (nearly void of precipitation at lower levels) does not. The region inside the supercell where radar is unable to detect precipitation is known as the *bounded weak echo region (BWER)*. Mean-while, as the precipitation is drawn into a cyclonic spiral around the mesocyclone, the rotating precipitation may, on the Doppler radar screen, unveil itself in the shape of a hook, called a **hook echo**, as shown in ● Fig. 10.48.

At this point in the storm's development, the updraft, the counterclockwise swirling precipitation, and the surrounding air may all interact to produce the *rear-flank downdraft* (to the south of the updraft), as shown in ● Fig. 10.49. The strength of the downdraft is driven in part by the amount of precipitation-induced cooling in the upper levels of the storm. Downdrafts appear to be essential to the process of tornado formation. However, if a downdraft is too cold and strong, it may inhibit the lift-ing needed for tornado formation. In fact, relatively warm rear-flank downdrafts appear to be more conducive to the formation of strong tornadoes. Researchers are still look-ing into many questions about how the rear-flank down-draft may evolve in a given storm.

When the rear-flank downdraft strikes the ground as illustrated in Fig. 10.49, it may (under favorable shear con-ditions) interact with the forward-flank downdraft beneath

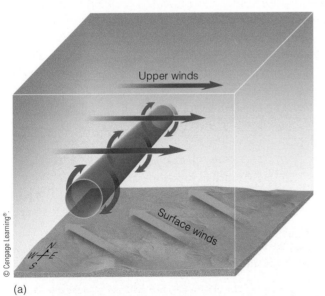

(a)

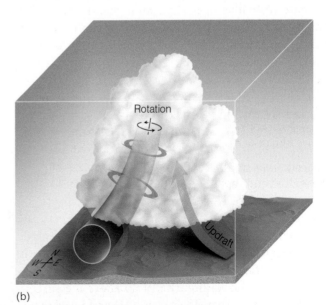

(b)

● **FIGURE 10.47** (a) A spinning vortex tube created by vertical wind shear. (b) The strong updraft in the developing thunderstorm carries the vortex tube into the thunderstorm producing a rotating air column that is oriented in the vertical plane.

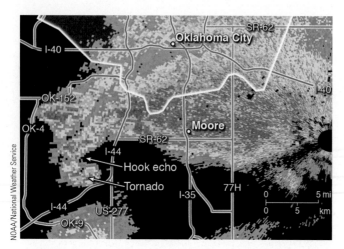

FIGURE 10.48 A tornado-spawning supercell thunderstorm over Oklahoma City during the afternoon of May 20, 2013, shows a hook echo west of Moore, Oklahoma, in its rainfall reflectivity pattern on a Doppler radar screen. The colors red and orange represent the heaviest precipitation. Compare this precipitation pattern with the precipitation pattern illustrated in Fig. 10.46 (A photo of this tornado can be found in Chapter 9, Fig. 9.13, on p. 258.)

the mesocyclone to initiate **tornadogenesis**—the formation of a tornado. A pre-existing boundary, such as an old gust front (outflow boundary), may also supply the surface air with horizontal spin that can be tilted and lifted into the storm by its updraft. At the surface of the mature supercell, the strong winds of the rear-flank downdraft now wrap around the updraft at the center of the mesocyclone. This situation may initiate additional spin that can be lifted into the mesocyclone. At this point, the lower half of the updraft begins to rise more slowly than the updraft aloft.

The rising updraft, which we can imagine as a column of air, now shrinks horizontally and stretches vertically. This *vertical stretching* of the spinning column of air causes the rising, spinning air to spin faster.[*] If the stretching process continues, the rapidly rotating air column may shrink into a narrow column of rapidly rotating air, a *tornado vortex*.

As air rushes upward and spins around the low-pressure core of the vortex, the air expands, cools, and, if sufficiently moist, condenses into a visible cloud, which is the *funnel cloud*. As the air beneath the funnel cloud is drawn into its core, the air cools rapidly and its moisture condenses, and the funnel cloud descends toward the surface. Upon reaching the ground, the tornado's circulation usually picks up dirt and debris, making it appear both dark and ominous. This ground-based debris may become visible before the funnel cloud is connected to it. While the air along the outside of the funnel is spiraling upward, Doppler radar reveals that, within the core of large violent tornadoes, the air is descending toward the extreme low pressure at the ground (which may be 100 mb lower than that of the surrounding air). As the air descends, it warms, causing the cloud droplets to evaporate. This process leaves the core free of clouds. Tornadoes usually develop in supercells near the right rear sector of the storm, on the southwestern side of a northeastward-moving storm, as shown in Fig. 10.49.

Many atmospheric situations can suppress tornado formation. For example, if the precipitation in the cloud is swept too far away from the updraft, or if too much precipitation wraps around the mesocyclone, the interactions necessary to produce the rear-flank downdraft are disrupted, and a tornado is not likely to form. Moreover, tornadoes

---

[*]As the rotating air column stretches vertically into a narrow column, its rotational speed increases, a situation called the *conservation of angular momentum*.

● **FIGURE 10.49** A classic mature tornadic supercell thunderstorm showing updrafts and downdrafts, along with surface air flowing counterclockwise and in toward the tornado. The flanking line is a line of cumulus clouds that form as surface air is lifted into the storm along the gust front.

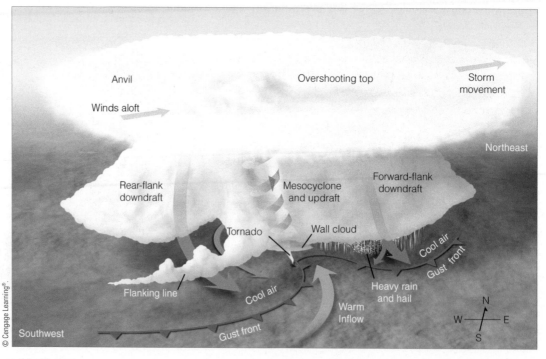

are not likely to form if the supercell is fed warm, moist air that is elevated above a deep layer of cooler surface air. And tornadoes usually will not form when the cool air of the rear-flank downdraft is too cold. Even among supercells with mesocyclones, only about 25 percent produce tornadoes. One of the main goals of tornado researchers is to better understand which supercells are most likely to generate tornadoes, especially the violent tornadoes that are the most destructive and deadly.

To an observer on the ground, the first sign that a supercell may give birth to a tornado is the sight of *rotating clouds* at the base of the storm.* If the area of rotating clouds lowers, it becomes the *wall cloud*. Notice in Fig. 10.49 that the tornado extends from within the wall cloud to Earth's surface. Sometimes the air is so dry that the swirling, rotating wind remains invisible until it reaches the ground and begins to pick up dust. Unfortunately, people sometimes have mistaken these "invisible tornadoes" for dust devils, only to find out (often too late) that they were not. Occasionally, the funnel is obscured by rain, dust, or darkness. When the tornado is not visible because it is surrounded by rain, it is referred to as being "*rain wrapped*." Even when not clearly visible, many tornadoes have a distinctive roar that can be heard as the tornado approaches. This sound, which has been described as "a roar like a thousand freight trains," appears to be loudest when the tornadic circulation is touching the surface. However, not all tornadoes make such sounds; when those that don't strike, they can be relatively silent killers.

Forecasters use a variety of information to help calculate the odds that a supercell might form on a given day or that a supercell might produce a tornado. The amount of instability in the atmosphere is represented by *Convective Available Potential Energy* (CAPE), which is a measure of how much energy is available to produce strong updrafts. The amount of wind shear can be calculated by comparing the wind speed and direction at various heights. And the amount of potential circulation within a storm can be estimated by calculating *storm-relative helicity*, which is a function of low-level wind shear. It measures how helical (corkscrew-like) the updraft will be in a growing thunderstorm.

Certainly, the likelihood that a thunderstorm will produce a tornado increases when the storm becomes a supercell, but not all supercells produce tornadoes. And not all tornadoes come from rotating thunderstorms.

NONSUPERCELL TORNADOES
Tornadoes that do not occur in association with a mid-level mesocyclone (or a pre-existing wall cloud) of a supercell are called

---

*Occasionally, people will call a sky dotted with mammatus clouds "a tornado sky." Mammatus clouds may appear with both severe and nonsevere thunderstorms as well as with a variety of other cloud types (see Chapter 4). Mammatus clouds are not funnel clouds, they do not rotate, and their appearance has no relationship to tornadoes.

nonsupercell tornadoes. These tornadoes may occur with intense multicell storms as well as with ordinary cell thunderstorms, even relatively weak ones. Some nonsupercell tornadoes extend from the base of a thunderstorm whereas others may begin on the ground and build upward in the absence of a condensation funnel.

Nonsupercell tornadoes may form along a gust front where the cool downdraft of the thunderstorm forces warm, humid air upward. Tornadoes that form along a gust front are commonly called **gustnadoes**. These relatively weak tornadoes normally are short-lived and rarely inflict significant damage. Gustnadoes typically form as a result of strongly converging winds along the edge of a rear-flank or forward downdraft. They are often seen as a rotating cloud of dust or debris rising above the surface (see ● Fig. 10.50).

Occasionally, rather weak, short-lived tornadoes will occur with rapidly building cumulus congestus clouds.

● **FIGURE 10.50** A gustnado that formed along a gust front swirls across the plains of eastern Nebraska.

NCAR/UCAR/NSF

● **FIGURE 10.51** A well-developed landspout moves over eastern Colorado.

Tornadoes such as these commonly form over northeastern Colorado and other parts of the High Plains. Because they look similar to waterspouts that form over water, and form in similar ways, they are sometimes called **landspouts*** (see ● Fig. 10.51).

---

*Landspouts occasionally form on the backside of a squall line where southerly winds ahead of a cold front and northwesterly winds behind it create swirling eddies that can be drawn into thunderstorms by their strong updrafts.

● Figure 10.52 illustrates how a landspout can form. Suppose, for example, that the winds at the surface converge along a boundary, as illustrated in Fig. 10.52a. (The wind may converge due to topographic irregularities or any number of other factors, including temperature and moisture variations.) Notice that along the boundary, the air is rising, condensing, and forming into a cumulus congestus cloud. Notice also that along the surface at the boundary there is horizontal rotation (spin) created by the wind blowing in opposite directions along the boundary. If the developing cloud should move over the region of rotating air (Fig. 10.52b), the spinning column may be drawn up into the cloud by the storm's updraft. In this case, the column of rising air typically narrows and its rotation intensifies as it stretches upward. As the spinning, rising air shrinks in diameter, it produces a tornadic structure, a *landspout,* similar to the one shown in Fig. 10.51. Landspouts usually dissipate when rain falls through the cloud and destroys the updraft. Although they are typically brief and affect only a small area, landspouts are still capable of causing serious damage. Tornadoes can form in this manner along many types of converging wind boundaries, including sea breezes and gust fronts. Nonsupercell tornadoes and funnel clouds can also form with thunderstorms when cold air aloft (associated with an upper-level trough) moves over a region. Common along the west coast of North America, these often short-lived tornadoes are sometimes called *cold-air funnels* (see ● Fig. 10.53).

WATERSPOUTS   A **waterspout** is a rotating column of air that is connected to a cumuliform cloud over a large body of water. The waterspout may be a tornado that formed over land and then traveled over water. In such a case, the waterspout is sometimes referred to as a *tornadic waterspout.* Such tornadoes can inflict major damage to ocean-going vessels, especially when the tornadoes are of

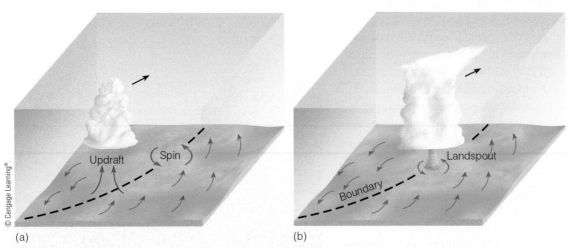

© Cengage Learning®

(a)   (b)

● **FIGURE 10.52** (a) Along the boundary of converging winds, the air rises and condenses into a cumulus congestus cloud. At the surface the converging winds along the boundary create a region of counterclockwise spin. (b) As the cloud moves over the area of rotation, the updraft draws the spinning air up into the cloud producing a nonsupercell tornado, or landspout. (Modified after Wakimoto and Wilson)

● **FIGURE 10.53** A funnel cloud—called a *cold-air funnel*—descends from a thunderstorm in California's Central Valley near Lodi.

the supercell variety. Strong waterspouts that form over water and then move over land also can cause considerable damage. For example, on August 30, 2009, an intense waterspout formed over the warm Gulf of Mexico, then moved onshore into Galveston, Texas, where it caused EF1 damage over several blocks and injured three people.

Waterspouts not associated with supercells that form over water, especially above warm, tropical coastal waters (such as in the vicinity of the Florida Keys, where almost 100 occur each month during the summer), are often referred to as "*fair weather*" *waterspouts.** These waterspouts are generally much smaller than an average tornado, as they have diameters usually between 3 and 100 meters. Fair weather waterspouts are also less intense, as their rotating winds are typically less than 45 knots. In addition, they tend to move more slowly than tornadoes and they only last for about 10 to 15 minutes, although some have existed for up to an hour.

Fair weather waterspouts tend to form in much the same way that landspouts do, when the air is conditionally unstable and cumulus clouds are developing. Some form with small thunderstorms, but most form with developing cumulus congestus clouds whose tops are frequently no higher than 3600 m (12,000 ft) and do not extend to the freezing level. Apparently, the warm, humid air near the water helps to create atmospheric instability, and the updraft beneath the resulting cloud helps initiate uplift of the surface air. Studies even suggest that gust fronts and

*"Fair weather" waterspouts can form over any large body of warm water. Hence, they occur frequently over the Great Lakes in summer.

converging sea breezes may play a role in the formation of some of the waterspouts that form over the Florida Keys. As with a landspout, a waterspout becomes more likely when a pre-existing boundary of converging air moves beneath a thunderstorm's updraft.

The waterspout funnel is similar to the tornado funnel in that both are clouds of condensed water vapor with converging winds that rise about a central core. Contrary to popular belief, the waterspout does not draw water up into its core; however, swirling spray can be lifted several meters when the waterspout funnel touches the water. A photograph of a particularly well-developed and intense waterspout is shown in ● Fig. 10.54.

Observing Tornadoes and Severe Weather

Starting as far back as the 1940s, organized networks of storm spotters kept an eye to the sky and reported the development of severe thunderstorms and tornadoes. Spotter networks are still important in notifying the National Weather Service of threatening conditions so that warnings can be promptly issued. In addition, laboratory models of tornadoes in chambers (called vortex chambers)

● **FIGURE 10.54** A powerful waterspout moves across Lake Tahoe, California. Compare this photo of a waterspout with the photo of a landspout in Fig. 10.51.

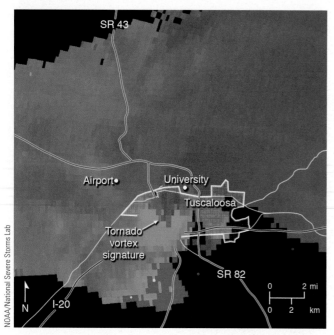

● **FIGURE 10.55** Doppler radar display of winds within a super-cell thunderstorm that moved through Tuscaloosa, Alabama, on April 27, 2011. The close packing of the horizontal winds blowing toward the radar (green shades) and those blowing away from the radar (red shades) indicate strong cyclonic rotation. A tornadic circulation (*tornado vortex signature*) exists where a small packet of red is adjacent to a small packet of green. (Doppler radar reflectivity for this tornado is shown in Figure 10.56.)

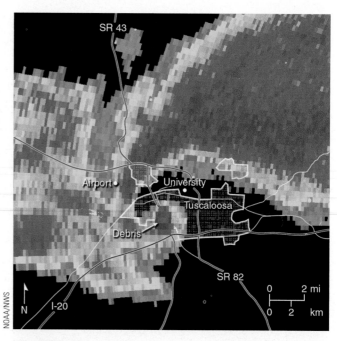

● **FIGURE 10.56** Doppler radar display showing precipitation inside a large supercell that takes on the shape of a hook. This *hook echo* is associated with a violent multi-vortex tornado that is moving through Tuscaloosa, Alabama, just south of the University of Alabama, on April 27, 2011. (Damage caused by this tornado is shown in Fig. 10.45 on p. 306.)

as well as high-resolution computer models also have provided insights into the three-dimensional processes at work in tornadic circulations.

In recent years, however, most of our knowledge about what goes on inside a tornado-generating thunderstorm has been gathered through the use of *Doppler radar*. Remember from Chapter 5 that a radar transmitter sends out microwave pulses and that when this energy strikes an object, a small fraction is scattered back to the antenna. Precipitation particles are large enough to bounce microwaves back to the antenna. Consequently, as we saw earlier, the colorful area on the radar screen in Fig. 10.48, p. 308, represents the amount of reflected microwave energy translated into precipitation intensity inside a supercell thunderstorm.

Doppler radar can do more than measure rainfall intensity; it can actually measure the speed at which precipitation is moving horizontally toward or away from the radar antenna. Because precipitation particles are carried by the wind, Doppler radar can peer into a severe storm and reveal its winds.

Doppler radar works on the principle that, as precipitation moves toward or away from the antenna, the returning radar pulse will change in frequency when compared to the transmitted frequency. A similar change occurs when the high-pitched sound (high frequency)

of an approaching noise source, such as a siren or train whistle, becomes lower in pitch (lower frequency) after it passes by the person hearing it. This change in frequency in sound waves or microwaves is called the *Doppler shift,* and this, of course, is where the Doppler radar gets its name.

To help distinguish the storm's air motions, wind velocities can be displayed in color. Winds blowing toward the radar antenna are usually displayed in green (or blue); winds blowing away from the antenna are usually shown in shades of red. Color contouring the wind field gives a good picture of how winds are changing within a storm and the possibility of a tornado (see ● Fig. 10.55).

Doppler radar can uncover many of the features of a severe thunderstorm. Mesocyclones have a distinct image (signature) on the radar display. Tornadoes also have a distinct signature on the radar screen, known as the *tornado vortex signature (TVS),* which shows up as a region of rapidly (or abruptly) changing wind directions within the mesocyclone, as shown in Fig. 10.55.

When Doppler radar displays precipitation intensity (reflectivity) inside a supercell thunderstorm, a signature of a mesocyclone (or tornado) may appear on the radar screen as a hook-shaped appendage, or *hook echo,* as shown in ● Fig 10.56.* The hook becomes visible as precipitation (and sometimes debris)

*See Fig. 10.44 on p. 306 for photo of this massive tornado.

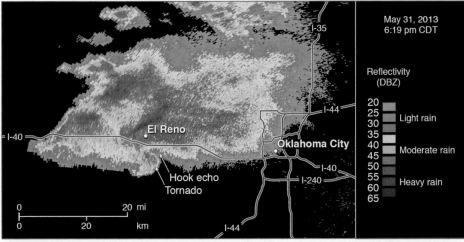

(a) Doppler radar (reflectivity)

May 31, 2013
6:19 pm CDT

Reflectivity
(DBZ)

20
25   Light rain
30
35
40   Moderate rain
45
50
55   Heavy rain
60
65

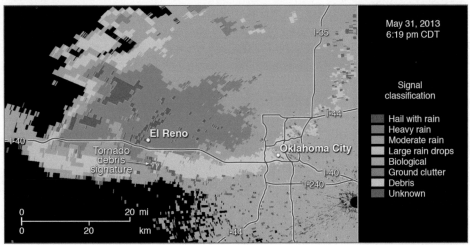

(b) Doppler radar, dual-polarization (hydrometeor classification)

May 31, 2013
6:19 pm CDT

Signal
classification

Hail with rain
Heavy rain
Moderate rain
Large rain drops
Biological
Ground clutter
Debris
Unknown

● FIGURE 10.57 (a) Doppler radar (reflectivity) shows precipitation intensity inside a large supercell thunderstorm on May 31, 2013, with heavy rain (red and yellow shades) falling west of Oklahoma City. A hook echo with a tornado exists to the south of El Reno, Oklahoma. (b) Doppler radar with dual-polarization technology more clearly depicts the different types of precipitation, as well as the tornado's debris, which shows up as a *tornado debris signature.*

swirls counterclockwise around the mesocyclone (or tornado). The doughnut-shaped dark red area at the end of the hook in Fig. 10.56 represents a massive multi-vortex tornado that is moving through Tuscaloosa, Alabama, on April 27, 2011, at about 5:10 p.m. (CST). The purple area in the center of the hook represents debris that, having been picked up by the tornado, is now swirling counterclockwise around it. Debris on a radar screen such as this is referred to as a *debris ball.* With the addition of *dual-polarization* technology to Doppler radar, debris can be observed even more clearly (see ● Fig. 10.57). Although the hook echo in Fig. 10.56 has a tornado embedded in it, it should be noted that not all hook echoes are associated with tornadoes and not all tornadoes show a distinctive hook echo on the radar screen.

Unfortunately, the resolution of the Doppler radar is not high enough to measure the actual wind speeds of most small tornadoes. However, a system called *Doppler lidar* uses a light beam (instead of microwaves) to measure the change in frequency of falling precipitation, cloud particles, and dust. Because it uses a shorter wavelength of radiation, it has a narrower beam and a higher resolution than does Doppler radar.

The network of more than 150 Doppler radar units deployed at selected weather stations within the continental United States is referred to as **NEXRAD** (an acronym for *NEX*t Generation Weather *RAD*ar). The NEXRAD system consists of the WSR-88D* Doppler radar and a set of computers that perform a variety of functions.

The computers take in data, display them on a monitor, and run computer programs called *algorithms,* which, in conjunction with other meteorological data, detect severe weather phenomena, such as storm cells, hail, mesocyclones, and tornadoes. Algorithms provide a great deal of information to the forecasters that allows them to make better decisions as to which thunderstorms are most likely to produce severe weather and possible flash flooding. In addition, the algorithms give advanced and improved warning of an approaching tornado. More reliable warnings, of course, cut down on the number of false alarms.

*The name WSR-88D stands for "Weather Surveillance Radar, 1988 Doppler."

● **FIGURE 10.58** A Doppler on Wheels mobile radar probes the thunderstorm that produced a tornado near La Grange, Wyoming, on June 5, 2009. The unit was one of several involved with the VORTEX2 field study.

Bob Henson

Because the Doppler radar shows horizontal air motion within a storm, it can help to identify the magnitude of other severe weather phenomena, such as gust fronts, derechoes, microbursts, and wind shears that are dangerous to aircraft. Certainly, as more and more information from Doppler radar becomes available, our understanding of the processes that generate severe thunderstorms and tornadoes will be enhanced, and hopefully there will be an even better tornado and severe storm warning system, resulting in fewer deaths and injuries.

As noted earlier, one recent advance in Doppler radar technology is the *polarimetric radar* (or *dual-polarization radar*), which transmits both a horizontal and a vertical radar pulse that, among other things, allows forecasters to better distinguish between very heavy rain and hail. The national NEXRAD radar network was upgraded in the early 2010s so that each of its Doppler radars now has dual-polarization capability. This should help improve the ability of forecasters to pinpoint areas where large hail, torrential rain, and flash flooding may occur. Researchers are also exploring the use of *phased array* radars, which allow a single radar unit to gather much more information than current Doppler radars by including many small transmitters and receivers on a flat plate.

Storm Chasing and Mobile Radar

Many people venture onto the Great Plains each spring to observe severe weather through "storm chasing," either informally or in organized tour groups. Most of this activity is unrelated to meteorological research. However, scientists have ventured into the field for more than 40 years in an attempt to directly observe tornadoes and collect observations as well as photos, films, and videos that might help unravel the secrets of tornado formation. One of the earliest accomplishments of tornado field research occurred on May 24, 1973, with the filming of a tornado near Union City, Oklahoma, at the same time that the tornado was being sampled by experimental Doppler radar. For the first time, scientists observed a tornado vortex signature (TVS) on radar, and they noted that the TVS appeared more than 20 minutes before a tornado actually formed.

The two most comprehensive field studies on tornadoes conducted to date are VORTEX (*V*erification of the *O*rigin of *R*otational *T*ornadoes *Ex*periment), which took place in 1994 and 1995, and VORTEX2, which occurred in 2009 and 2010. In both studies, scientists used an armada of observational vehicles and state-of-the-art equipment. The original VORTEX included a turboprop aircraft and dozens of weather stations attached to the tops of cars, as well as the debut of a series of Doppler radar units (Doppler on Wheels) mounted on trucks. VORTEX2 included a wider variety of mobile Doppler radar and lidar units, as well as compact surface weather stations and unmanned aircraft systems (see ● Fig. 10.58). To obtain as much information as possible, some instruments were placed directly in the path of an approaching storm, while others surrounded the storm. The data obtained from these studies is being combined with high-resolution computer models of tornadic supercells to provide new insights about the inner workings of supercells and tornadoes.

Storm chasing is a risky activity, even for veteran researchers. Three research-based storm chasers were killed in a violent tornado near El Reno, Oklahoma, on May 31, 2013, the same one shown on Doppler radar in Fig. 10.57a and 10.57b. With the research team positioned just ahead of it, the tornado suddenly changed direction, accelerated, and expanded to a record width of 4.2 km (2.6 mi), with multiple vortices spinning around the main circulation. One other person, a local resident, was killed while chasing the storm, and several other chase teams were struck by the tornado or by high winds nearby.

A mobile polarimetric Doppler radar operated by the University of Oklahoma detected in this storm near-surface winds estimated at 468 km/hr (291 mph). However, because EF ratings of tornadoes are based on damage rather than radar observations, and because the strongest winds of the tornado apparently did not hit any structures, the tornado was classified as an EF3. In recent years, experts have been working on a system that would expand the ways in which tornado wind speeds could be estimated and reported. The new system may incorporate mobile radar data, as well as information about the damage patterns that emerge when tornadoes pass through a heavily treed area.

SUMMARY

In this chapter, we examined thunderstorms and the atmospheric conditions that produce them. Thunderstorms are convective storms that produce lightning and thunder. Lightning is a discharge of electricity that occurs in mature thunderstorms. The lightning stroke momentarily heats the air to an incredibly high temperature. The rapidly expanding air produces a sound called thunder.

The ingredients for the isolated ordinary cell thunderstorm are humid surface air, plenty of sunlight to heat the ground, a conditionally unstable atmosphere, a "trigger" to start the air rising, and weak vertical wind shear. When these conditions prevail, and the air begins to rise, small cumulus clouds may grow into towering clouds and thunderstorms within 30 minutes.

When conditions are ripe for thunderstorm development, and moderate or strong vertical wind shear exists, the updraft in the thunderstorm may tilt and ride up and over the downdraft. As the forward edge of the downdraft (the gust front) pushes outward along the ground, the air is lifted and new cells form, producing a multicell thunderstorm. Some multicell storms form as a complex of thunderstorms, such as the squall line (which forms as a line of thunderstorms), and the Mesoscale Convective Complex (which forms as a cluster of storms). When convection in the multicell storm is strong, it may produce severe weather, such as strong damaging surface winds, hail, and flooding.

Supercell thunderstorms are intense thunderstorms with a single rotating updraft. The updraft and the downdraft in a supercell are nearly in balance, so that the storm may exist for many hours. Supercells are capable of producing severe weather, including strong damaging tornadoes.

Tornadoes are rapidly rotating columns of air with a circulation that reaches the ground. Tornadoes can form with supercells, as well as with less intense thunderstorms. Most tornadoes are less than a few hundred meters wide with wind speeds less than 100 knots, although violent tornadoes may have wind speeds that exceed 250 knots. A violent tornado may actually have smaller whirls (suction vortices) rotating within it. A normally small and less destructive cousin of the tornado is the "fair weather" waterspout that commonly forms above warm bodies of water. With the aid of Doppler radar, scientists are probing tornado-spawning thunderstorms, hoping to better predict tornadoes and to better understand where, when, and how they form.

KEY TERMS

The following terms are listed (with corresponding page numbers) in the order they appear in the text. Define each. Doing so will aid you in reviewing the material covered in this chapter.

ordinary cell (air mass) thunderstorms, 275
cumulus stage, 275
mature stage, 276
dissipating stage, 276
multicell thunderstorm, 277
overshooting top, 277
gust front, 277
straight-line winds, 278
shelf cloud, 278
roll cloud, 279
outflow boundary, 279
downburst, 279
microburst, 279
heat burst, 280
squall line, 280
bow echo, 281
derecho, 282
mesoscale convective complexes (MCCs), 282
supercell, 282
mesocyclone, 283
wall cloud, 283
flash floods, 286
lightning, 290
thunder, 290
sonic boom, 290

stepped leader, 292
return stroke, 292
dart leader, 293
heat lightning, 294
dry lightning, 294
St. Elmo's fire, 295
tornado, 296
funnel cloud, 296
Tornado Alley, 298
Dixie Alley, 298
suction vortices, 300
tornado watch, 301
tornado warning, 301
tornado emergency, 301
Fujita Scale, 302
Enhanced Fujita Scale (EF Scale), 303
tornado outbreak, 305
supercell tornadoes, 305
hook echo, 307
tornadogenesis, 308
nonsupercell tornadoes, 309
gustnadoes, 309
landspout, 310
waterspout, 310
NEXRAD, 313

QUESTIONS FOR REVIEW

1. What is a thunderstorm?
2. What atmospheric conditions are necessary for the development of ordinary cell (air mass) thunderstorms?
3. Describe the stages of development of an ordinary cell (air mass) thunderstorm.
4. How do downdrafts form in ordinary cell thunderstorms?
5. Why do ordinary cell thunderstorms most frequently form in the afternoon?
6. Explain why ordinary cell thunderstorms tend to dissipate much sooner than multicell storms.

7. How does the National Weather Service define a severe thunderstorm?

8. What atmospheric conditions are necessary for a multicell thunderstorm to form?

9. (a) How do gust fronts form?
   (b) What type of weather does a gust front bring when it passes?

10. (a) Describe how a microburst forms.
    (b) Why is the term *horizontal wind shear* often used in conjunction with a microburst?

11. How do derechoes form?

12. How does a squall line differ from a mesoscale convective complex (MCC)?

13. Give a possible explanation for the generation of a pre-frontal squall-line thunderstorm.

14. How do supercell thunderstorms differ from ordinary cell (air mass) thunderstorms?

15. Describe the atmospheric conditions at the surface and aloft that are necessary for the development of most supercell thunderstorms. (Include in your answer the role that the low-level jet plays in the rotating updraft.)

16. What is the difference between an HP supercell and an LP supercell?

17. When thunderstorms are *training*, what are they doing?

18. Where does the highest frequency of thunderstorms occur in the United States? Why there?

19. Why is large hail more common in Kansas than in Florida?

20. Describe one process by which thunderstorms become electrified.

21. How is thunder produced?

22. Explain how a cloud-to-ground lightning stroke develops.

23. Why is it unwise to seek shelter under an isolated tree during a thunderstorm? If caught out in the open, what should you do?

24. What is a tornado? Give some statistics about size, wind speed, and movement.

25. What is the primary difference between a tornado and a funnel cloud?

26. Why do tornadoes frequently move from southwest to northeast?

27. At what point in its life cycle would a tornado resemble a rope?

28. Why should you *not* open windows when a tornado is approaching?

29. Why is the central part of the United States more susceptible to tornadoes than any other region of the world?

30. How does a tornado *watch* differ from a tornado *warning*?

31. If you are in a single-story home (without a basement) during a tornado warning, what should you do?

32. Why was the original Fujita Scale for tornado damage replaced by the Enhanced Fujita Scale?

33. Supercell thunderstorms that produce tornadoes form in a region of strong vertical wind shear. Explain how the wind changes in speed and direction to produce this shear.

34. Explain how a nonsupercell tornado, such as a landspout, might form.

35. What atmospheric conditions lead to the formation of "fair weather" waterspouts?

36. Describe how Doppler radar measures the winds inside a severe thunderstorm.

37. How has Doppler radar helped in the prediction of severe weather?

QUESTIONS FOR THOUGHT AND EXPLORATION

1. Why does the bottom half of a dissipating thunderstorm usually "disappear" before the top?

2. Sinking air warms, yet thunderstorm downdrafts are usually cold. Why?

3. If you are confronted by a large tornado in an open field and there is no way that you can outrun it, your only recourse might be to run and lie down in a depression. If given the choice, when facing the tornado, would you run toward your left or toward your right as the tornado approaches? Explain your reasoning.

4. Suppose while you are standing on a high mountain ridge a thundercloud passes overhead. What would be the wisest thing to do—stand upright? lie down? or crouch? Explain.

5. Tornadoes apparently form in the region of a strong updraft, yet they descend from the base of a cloud. Why?

6. On a map of the United States, place the surface weather conditions (air masses, fronts, and so on)

as well as weather conditions aloft (jet stream, and so on) that are necessary for the formation of most supercell thunderstorms.

7. The number of strong tornadoes reported in the United States has held fairly steady over the last 60 years, while the number of weak tornadoes reported has doubled. What are at least two possible explanations for this trend? How would you investigate which of those explanations would be more likely?

8. Suppose you and several of your friends are invited to join a storm-chasing research expedition in the central United States. Each night, you have access to a computer in your hotel room. Which current weather and forecast maps would you use to estimate the likelihood of severe weather in your study area on the next day? Explain why you choose those maps.

9. A multi-vortex tornado with a rotational wind speed of 125 knots is moving from southwest to northeast at 30 knots. Assume the suction vortices within this tornado have rotational winds of 100 knots:

   (a) What is the maximum wind speed of this multi-vortex tornado?

   (b) If you are facing the approaching tornado, on which side (northeast, northwest, southwest, or southeast) would the strongest winds be found? the weakest winds? Explain both of your answers.

   (c) According to Table 10.2, p. 303, how would this tornado be classified on the Enhanced Fujita Scale?

GLOBAL **GEOSCIENCE** WATCH   Go to the Natural Disasters portal and search in the Academic Journals section for the report "Climatology of tornadoes associated with Gulf Coast-landfalling hurricanes" (*The Geographical Review*, July 2011). Based on information from the introductory and concluding sections, would you say that a hurricane that strikes Mobile, Alabama, would be more likely to produce a tornado in New Orleans or in Tallahassee, all else being equal?

ONLINE RESOURCES

 Visit www.cengagebrain.com to view additional resources, including video exercises, practice quizzes, an interactive eBook, and more.

Hurricanes

Contents

The indelible memory I have of covering the aftermath of Hurricane Katrina in New Orleans [August 2005] was how quickly and completely a modern, powered, connected, policed, orderly American city can descend into utter chaos. There was no 911 service; streets were impassable. No emergency rooms, no electricity, no stores, no communication system. The city had become pre-industrial, medieval. And no one was coming. No one was in charge. No one knew what to do. Once people were rescued from rooftops and deposited on dry land, they were left completely on their own. Reporters became, in some cases, first responders. We were the first individuals that some evacuees encountered when they trudged out of their underwater neighborhoods. They wanted food, water, diapers, medicine—they didn't want an interview. All we could do was get their stories. I recall an overwhelming sense of powerlessness that we could not do more to help people, instead of just making deadline.

John Burnett, National Public Radio

The introduction on the previous page describes the terrible conditions in New Orleans after Hurricane Katrina made landfall* in August 2005. Born over warm tropical waters and nurtured by a rich supply of water vapor, *hurricanes* can grow into ferocious storms that generate enormous waves, heavy rains, severe flooding, and winds that can exceed 150 knots. What exactly are hurricanes? How do they form? And why do they strike the east coast of the United States far more frequently than the west coast? These are some of the questions we will consider in this chapter.

Tropical Weather

In the broad belt around the earth known as the tropics—the region between 23½° north and 23½° south of the equator—the weather is much different than it is in the middle latitudes. In the tropics, the noon sun is always high in the sky, and so diurnal and seasonal changes in temperature are small. The daily heating of the surface and high humidity favor the development of cumulus clouds and afternoon thunderstorms. Most of these are individual thunderstorms that are not severe. Sometimes, however, they grow together into loosely organized systems called *non-squall clusters*. On other occasions, the thunderstorms will align into a row of vigorous convective cells or a *squall line*. The passage

*Landfall is the position along the coast where the center of a hurricane passes from ocean to land.

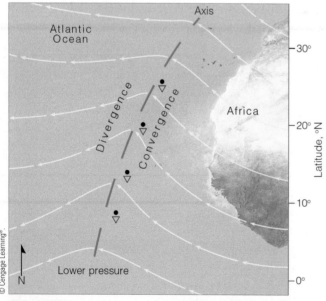

● **FIGURE 11.1** A tropical wave (also called an easterly wave) moving off the coast of Africa over the Atlantic. The wave is shown by the bending of streamlines—lines that show wind flow patterns. (The dashed green line is the axis of the trough.) The wave moves slowly westward, bringing fair weather on its western side and rainshowers on its eastern side.

of a squall line is usually noted by a sudden wind gust followed immediately by a heavy downpour. This deluge is then followed by several hours of relatively steady rainfall. Many of these tropical squall lines are similar to the middle-latitude squall lines described in Chapter 10.

Because it is warm all year long in the tropics, the weather is not characterized by four seasons (which, for the most part, are determined by temperature variations). Most of the tropics are marked instead by seasonal differences in precipitation. The greatest cloudiness and precipitation occur during the high-sun period, when the intertropical convergence zone moves into the region. Even during the dry season, precipitation can be irregular, as periods of heavy rain, lasting for several days, may follow an extreme dry spell.

The winds in the tropics generally blow from the east, northeast, or southeast—the trade winds. Because the variation in sea-level pressure is normally quite small, drawing isobars on a weather map provides little useful information. Instead of isobars, forecasters typically analyze **streamlines** that depict wind flow. Streamlines are useful because they show where surface air converges and diverges. Occasionally, the streamlines will be disturbed by a weak trough of low pressure called a **tropical wave,** or **easterly wave,** because it tends to move from east to west (see ● Fig. 11.1).

Tropical waves have wavelengths on the order of 2500 km (1550 mi) and travel from east to west at speeds of between 10 and 20 knots. Look at Fig. 11.1 and observe that, on the western side of the trough (green dashed line), where easterly and northeasterly surface winds diverge, sinking air produces generally fair weather. On its eastern side, southeasterly surface winds converge. The converging air rises, cools, and often condenses into showers and thunderstorms. Consequently, the main area of showers forms *behind* the trough. Occasionally, a tropical wave will intensify and grow into a hurricane.

Anatomy of a Hurricane

A **hurricane** is an intense storm of tropical origin, with sustained winds of at least 64 knots (74 mi/hr), and with considerably higher gusts, which forms over the warm northern Atlantic and eastern North Pacific oceans. Hurricanes are composed of intense convective clouds (thunderstorms) that sometimes grow to over 15 km (50,000 feet) in height. This same type of storm is given different names in different regions of the world. In the western North Pacific, it is called a **typhoon,** in India a *cyclone*, and in Australia a *tropical cyclone*. By international agreement, **tropical cyclone** is the general term for all hurricane-type storms that originate over tropical waters. For simplicity, we will refer to all of these storms as hurricanes.

● **FIGURE 11.2** Hurricane Igor over the North Atlantic Ocean, about 1370 km (850 mi) east of the Leeward Islands, as photographed from NASA's Aqua satellite on September 13, 2010. Because this storm is situated north of the equator, surface winds are blowing counterclockwise about its center (eye). The central pressure of the storm is 933 mb, with sustained winds of 130 knots (150 mi/hr) near its eye.

Eye

Eyewall

Spiral rain band

Rain-free area

NASA

● Figure 11.2 is a satellite image of Hurricane Igor, situated over the North Atlantic Ocean well east of the Caribbean Sea on September 13, 2010. The storm's thickest clouds cover an area approximately 500 km (310 mi) in diameter, which is about average for hurricanes. The relatively clear area at the center is its **eye.** Igor's eye is almost 40 km (25 mi) wide. Within the eye, winds are light and clouds are mainly broken. The surface air pressure is very low, around 933 mb (27.55 in.).* Notice that the clouds align themselves into spiraling bands (called *spiral rain bands*) that swirl in toward the storm's center, where they wrap themselves around the eye. Surface winds increase in speed as they blow counterclockwise and inward toward this center. (In the Southern Hemisphere, the winds blow clockwise around the center.) Adjacent to the eye is the **eyewall,** a ring of intense thunderstorms that whirl around the storm's center and may extend upward to almost 18 km (59,000 ft) above sea level. Within the eyewall, we find the heaviest precipitation and the strongest winds, which, in this storm, are 130 knots (150 mi/hr), with peak gusts of 160 knots (184 mi/hr).

If we were to venture from west to east (left to right) at the surface through the storm in Fig. 11.2, what might we experience? As we approach the hurricane, the sky becomes overcast with cirrostratus clouds; barometric pressure drops slowly at first, then more rapidly as we move closer to the center. Winds blow from the north

and northwest with ever-increasing speed as we near the eye. The high winds, which generate huge waves over 10 m (33 ft) high, are accompanied by heavy rainshowers. As we move into the eye, the winds slacken, rainfall ceases, and the sky brightens, as middle and high clouds appear overhead. The atmospheric pressure is now at its lowest point (933 mb), some 80 mb lower than the pressure measured on the outskirts of the storm. The brief respite ends as we enter the eastern region of the eyewall. Here, we are greeted by heavy rain and strong southerly winds. As we move away from the eyewall, the pressure rises, the winds diminish, the heavy rain lets up, and eventually the sky begins to clear.

This brief, imaginary venture raises many unanswered questions. Why, for example, is the surface pressure lowest at the center of the storm? And why is the weather clear almost immediately outside the storm area? To help us answer such questions, we need to look at a vertical view, a profile of the hurricane along a slice that runs through its center. A model that describes such a profile is given in ● Fig. 11.3.

The model shows that the hurricane is composed of an organized mass of thunderstorms* that are an integral part of the storm's circulation. Near the surface, moist tropical air flows in toward the hurricane's center. Adjacent to the eye, this air rises and the water vapor condenses into huge cumulonimbus clouds that produce heavy rainfall, as much as 15 cm (6 in.) per hour or more. Near the top of the clouds, the relatively dry air, having lost much of its moisture, begins to flow outward away from the center.

---

*An extreme low pressure of 870 mb (25.70 in.) was recorded in Typhoon Tip (while it was over the tropical Pacific Ocean) during October 1979, and Hurricane Wilma (while it was over the Gulf of Mexico) had a pressure reading of 872 mb (25.75 in.) during October 2005.

---

*These huge convective cumulonimbus clouds have surprisingly little lightning (and, hence, thunder) associated with them. Even so, for simplicity we will refer to these clouds as thunderstorms throughout this chapter.

● **FIGURE 11.3** A model that shows a vertical view of air motions and clouds in a typical hurricane in the Northern Hemisphere. The diagram is exaggerated in the vertical.

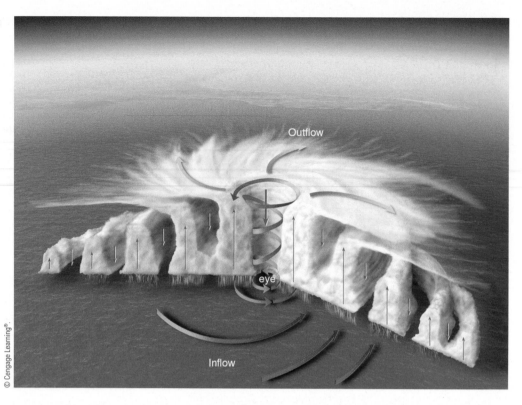

Outflow

eye

Inflow

This diverging air aloft actually produces a clockwise flow of air (anticyclonic in the Northern Hemisphere) several hundred kilometers from the eye. As this outflow reaches the storm's periphery, it begins to sink and warm, inducing clear skies. In the vigorous convective clouds of the eyewall, the air warms due to the release of large quantities of latent heat. This warming produces slightly higher pressures aloft, which initiate downward air motion within the eye. As the air sinks, it warms by compression. This process helps to account for the warm air and the absence of thunderstorms in the center of the storm (see ● Fig. 11.4).

● Figure 11.5 is a three-dimensional radar composite of Hurricane Katrina as it passes over the central area of the Gulf of Mexico. Compare Katrina's features with those of typical hurricanes illustrated in Fig. 11.2 and Fig. 11.3. Notice that the strongest radar echoes (heaviest rain) near the surface are located in the eyewall, adjacent to the eye.

Rainband

Eye

Eyewall

Rain rate (mm/hr)

0   10   20   30   40   50

NASA

● **FIGURE 11.5** A three-dimensional satellite view of Hurricane Katrina passing over the central Gulf of Mexico on August 28, 2005. The cutaway view shows concentric bands of heavy rain (red areas inside the clouds) encircling the eye. Notice that the heaviest rain (largest red area) occurs in the eyewall. The isolated tall cloud tower (in red) in the northern section of the eyewall indicates a cloud top of 16 km (52,000 ft) above the ocean surface. Such tall clouds in the eyewall often indicate that the storm is intensifying.

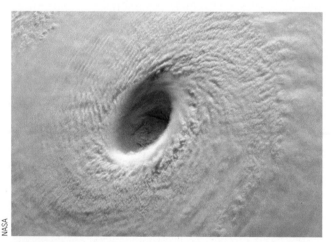

NASA

● **FIGURE 11.4** A close-up of the eye and eyewall of Typhoon Maysak in the Northwest Pacific on April 2, 2015, taken from the International Space Station. Notice that the thunderstorms of the eyewall completely encircle the eye, while the eye itself is almost cloud-free. The eye is about 30 km (19 mi) wide.

Hurricane Formation and Dissipation

We are now left with an important question: Where and how do hurricanes form? While there is no widespread agreement on how hurricanes actually form, it is known that certain necessary ingredients are required before a weak tropical disturbance will develop into a full-fledged hurricane.

THE RIGHT ENVIRONMENT   Hurricanes form over tropical waters where the winds are light, the humidity is high in a deep layer extending up through the troposphere, and the surface water temperature is warm, typically 26.5°C (80°F) or greater, over a vast area.* These conditions usually prevail over the tropical and subtropical North Atlantic and North Pacific oceans during the summer and early fall. These oceans usually reach their peak sea surface temperature in August or September because of the lag in seasonal temperature discussed in Chapter 2 (p. 49). The official hurricane season extends from May 15

---

*It was once thought that for hurricane formation, the ocean must be sufficiently warm through a depth of about 200 meters. It is now known that hurricanes can form in the eastern North Pacific when the warm layer of ocean water is only about 20 m (65 ft) deep.

to November 30 in the Northeast Pacific and from June 1 to November 30 in the North Atlantic, although a few hurricanes have formed outside these dates. ●Figure 11.6 shows the number of tropical storms and hurricanes that one might expect per century over the tropical Atlantic, based on data from 1870 to 2006. Notice that hurricane activity tends to pick up in August, peak in September, and then drop off rapidly.

For a mass of unorganized thunderstorms to develop into a hurricane, the surface winds must converge. In the Northern Hemisphere, converging air spins counterclockwise about an area of surface low pressure. Because this type of rotation will not develop on the equator where the Coriolis force is zero (see Chapter 6), hurricanes usually form at some distance from the equator, between about 5° and 20° latitude. (In fact, about two-thirds of all tropical cyclones form between 10° and 20° latitude.)

Hurricanes do not form spontaneously—they require a "trigger" to start the air converging. We know, for example, from Chapter 7 that surface winds converge along the intertropical convergence zone (ITCZ). Occasionally, when a wave forms along the ITCZ, an area of low pressure develops, convection becomes organized, and the system grows into a hurricane. Weak convergence also occurs on the eastern side of a tropical wave, where hurricanes sometimes form. In fact, many if not most Atlantic hurricanes can be traced to tropical waves that form over Africa. However, only a small fraction of all of the tropical disturbances that form over the course of a year ever grows into hurricanes.

Major Atlantic hurricanes tend to be more numerous when the western part of Africa is relatively wet. During

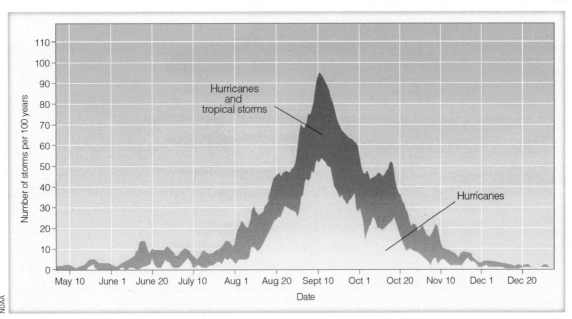

●**FIGURE 11.6** The total number of hurricanes and tropical storms (red shade) and hurricanes only (yellow shade) that one would expect every 100 years in the Atlantic Basin—the Atlantic Ocean, the Caribbean Sea, and the Gulf of Mexico—based on data extending from 1870 to 2006.

the wet years, tropical waves are stronger, better organized, and more likely to develop into strong Atlantic hurricanes.

Convergence of surface winds may also occur along a pre-existing atmospheric disturbance, such as a front that has moved into the tropics from middle latitudes. Although the temperature contrast between the air on both sides of the front is gone, converging winds may still be present so that thunderstorms can organize.

Even when all of the surface conditions appear nearly perfect for the formation of a hurricane (for example, warm water, humid air, converging winds, and the necessary trigger), the storm may not develop if the weather conditions aloft are not just right. For instance, in the region of the trade winds, and especially near latitude 20°, the air is often sinking in association with the subtropical high-pressure area. The sinking air warms and creates an inversion above the surface known as the **trade wind inversion.** When the inversion is strong, it can inhibit the formation of intense thunderstorms and hurricanes.

In addition, hurricanes do not form where the upper-level winds are strong. The resulting strong vertical wind shear tends to disrupt the organized pattern of convection and disperses heat and moisture, which are necessary for the growth of the storm.

The warmer-than-normal water over the eastern tropical Pacific during El Niño conditions favors the development of a greater number of hurricanes than average over the Northeast Pacific.* These same El Niño conditions help to generate strong vertical wind shear over the tropical North Atlantic, so fewer hurricanes than average tend to form there. When the water over the eastern tropical Pacific turns cooler than normal (La Niña conditions),

---

*Recall from Chapter 7, p. 199, that a major El Niño event is a condition where extensive ocean warming occurs over the eastern tropical Pacific.

vertical wind shear tends to weaken over the tropical North Atlantic, producing more favorable conditions for Atlantic hurricane development.

THE DEVELOPING STORM   The energy for a hurricane comes from the direct transfer of sensible heat and latent heat from the warm ocean surface. For a hurricane to form, a cluster of thunderstorms must become organized around a central area of surface low pressure. It appears that when a tropical wave moves at roughly the same speed as the surrounding upper-level flow, a protective zone of low wind shear and deep moisture may form that allows thunderstorms to gradually coalesce over several days. However, many questions remain about how such a cluster of thunderstorms may evolve into a hurricane.

One theory proposes that a hurricane forms in the following manner. Suppose that the trade wind inversion is weak and that thunderstorms start to organize along the ITCZ, or along a tropical wave. In the deep, moist, conditionally unstable environment, a huge amount of latent heat is released inside the clouds during condensation. The process warms the air aloft, causing the temperature near the cluster of thunderstorms to be much higher than the air temperature at the same level farther away. This warming of the air aloft causes a region of higher pressure to form in the upper troposphere (see ● Fig. 11.7), which, in turn, causes a horizontal pressure gradient aloft. The gradient induces the air aloft to move outward, away from the region of higher pressure in the anvils of the cumulonimbus clouds. This diverging air aloft, coupled with warming of the vertical air column, causes the surface pressure to drop and a small area of surface low pressure to form. The air now begins to spin counterclockwise (in the Northern Hemisphere) and in toward the region of surface

● **FIGURE 11.7** The top diagram shows an intensifying tropical cyclone. As latent heat is released inside the clouds, the warming of the air aloft creates an area of high pressure, which induces air to move outward, away from the high. The warming of the air lowers the air density, which in turn lowers the surface air pressure. As surface winds rush in toward the surface low, they extract sensible heat, latent heat, and moisture from the warm ocean. As the warm, moist air flows in toward the center of the storm, it is swept upward into the clouds of the eyewall. As warming continues, surface pressure lowers even more, the storm intensifies, and the winds blow even faster. This situation increases the transfer of heat and moisture from the ocean surface. The middle diagram illustrates how the air pressure drops rapidly as you approach the eye of the storm. The lower diagram shows how surface winds normally reach maximum strength in the region of the eyewall.

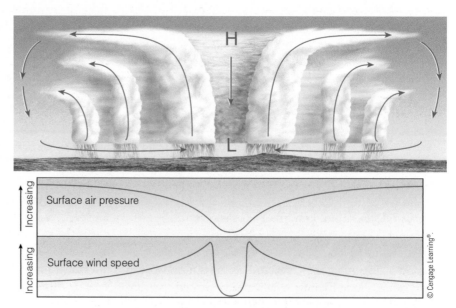

low pressure. As the air moves inward, its speed increases, just as figure skaters spin faster when they bring their arms in close to their bodies.

As the air moves over the warm water, small swirling eddies transfer heat energy from the ocean surface into the overlying air. The warmer the water and the greater the wind speed, the greater the transfer of sensible and latent heat into the air above. As the air sweeps in toward the center of lower pressure, the rate of heat transfer increases because the wind speed increases. Similarly, the higher wind speed causes greater evaporation rates, and the overlying air becomes nearly saturated. The turbulent eddies then transfer the warm, moist air upward, where the water vapor condenses to fuel new thunderstorms. As the surface air pressure lowers, wind speeds increase, more evaporation occurs at the ocean surface, and thunderstorms become more organized. The increasing surface winds generate large amounts of sea spray that partially evaporate, adding more moisture to the lower atmosphere. At the top of the thunderstorms, heat is lost by the clouds radiating infrared energy to space.

The driving force behind a hurricane is similar to that of a heat engine. In a heat engine, heat is taken in at a high temperature, converted into work, then ejected at a low temperature. In a hurricane, heat is taken in near the warm ocean surface, converted to kinetic energy (energy of motion or wind), and lost at its top through radiational cooling.

In a heat engine, the amount of work done is proportional to the difference in temperature between its input and output region. The maximum strength a hurricane can achieve is proportional to the difference in air temperature between the tropopause and the surface, and to the potential for evaporation from the sea surface. As a consequence, the warmer the ocean surface, the lower the minimum pressure of the storm, and the higher its winds. Because there is a limit to how intense the storm can become, peak wind gusts seldom exceed 200 knots (230 mi/hr).

After a hurricane forms, it may go through an internal cycle of intensification. In strong hurricanes, for example, the eyewall may become encircled by a second eyewall, as another band of strong thunderstorms forms perhaps 5 to 24 km (3 to 15 mi) out from the original eyewall. The growing outer eyewall cuts off the moisture supply to the original eyewall, causing it to dissipate. The dissipation of the original eyewall and the formation of a new one farther out from the eye is called **eyewall replacement.** As the replacement of the eyewall is taking place, the central pressure of the storm may rise, and its maximum winds may lessen. Eventually, however, the newly formed eyewall will usually contract toward the center of the storm as the hurricane re-intensifies, provided that the surface water remains warm and other conditions remain favorable.

THE STORM DIES OUT   If a hurricane remains over warm water, it may survive for a long time. For example, Hurricane Tina (1992) traveled for thousands of kilometers over deep, warm, tropical waters and maintained hurricane force winds for 24 days, making it one of the longest-lasting North Pacific hurricanes on record. Most hurricanes, however, maintain hurricane strength for less than a week.

Hurricanes weaken rapidly when they travel over colder water and lose their heat source. Studies indicate that if the water beneath the eyewall of the storm (the region of thunderstorms adjacent to the eye) cools by 2.5°C (4.5°F), the storm's energy source is cut off, and the storm will dissipate. Even a small drop in water temperature beneath the eyewall will noticeably weaken the storm. A hurricane can also weaken if the layer of warm water beneath the storm is shallow. In this situation, the strong winds of the storm generate powerful waves that produce turbulence in the ocean water under the storm. Such turbulence creates currents that bring to the surface cooler water from below. If the storm is moving slowly, it is more likely to lose intensity, because the eyewall remains over the cooler water for a longer period.

Hurricanes also dissipate rapidly when they move over a large landmass. Here, they not only lose their energy source but friction with the land surface causes surface winds to decrease and blow more directly into the storm, an effect that causes the hurricane's central pressure to rise. And a hurricane, or any tropical system for that matter, will rapidly dissipate should it move into a region of strong vertical wind shear.

Our understanding of hurricane behavior is far from complete. However, with the aid of computer model simulations, enhanced observations, and research projects, scientists are gaining new insight into how tropical cyclones form, intensify, and ultimately die.

HURRICANE STAGES OF DEVELOPMENT   Hurricanes go through a set of stages from birth to death. Initially, a *tropical disturbance* shows up as a mass of thunderstorms with only slight wind circulation. The tropical disturbance becomes a **tropical depression** when the winds increase to between 20 and 34 knots (23 and 39 mi/hr) and several closed isobars appear about its center on a surface weather map. When the isobars are packed together and the winds are between 35 and 63 knots (40 and 74 mi/hr), the tropical depression becomes a **tropical storm.** At this point, the storm gets a name. If the sustained winds reach 64 knots (74 mi/hr), the tropical storm is classified as a hurricane (see ● Figure 11.8). A tropical storm or hurricane will normally keep its name even after it weakens back to a tropical depression. If it moves to higher latitudes and begins to take on characteristics of a mid-latitude cyclone, it may be classified as a post-tropical cyclone

● FIGURE 11.8 Infrared satellite image from 8 a.m. (EDT) on July 10, 2012, showing three tropical systems over the eastern tropical Pacific, each in a different stage of its life cycle. From left to right are weakening Tropical Storm Daniel, formerly a hurricane, with sustained winds of 55 knots (63 mi/hr); Hurricane Emilia, near its peak strength with sustained winds of 120 knots (138 mi/hr); and a small center of low pressure (a tropical disturbance called "98E") that developed several days later into Hurricane Fabio.

(still keeping its original name). Sometimes a cyclonic storm over the ocean will have characteristics of both tropical and mid-latitude cyclones as it develops. If so, it would be classified as a subtropical storm and given a name as if it were a tropical storm.

INVESTIGATING THE STORM There are a variety of ways to obtain information about a developing hurricane and its environment. Visible, infrared, and enhanced infrared satellite images all provide a bird's-eye view of the storm, while sophisticated onboard radar instruments can actually peer into the storm and unveil its clouds as a three-dimensional image (see ● Fig. 11.9 and also Fig. 11.5 on p. 322). Some satellites are even equipped with onboard instruments capable of obtaining surface wind information in and around the storm (see ● Fig 11.10). A visible satellite image can be important in determining whether a developing hurricane will continue to strengthen. For example, the huge thunderstorms in the eyewall of the storm often produce a dense cirrus cloud shield that extends outward away from the eye, as illustrated in Fig. 11.3 on p. 322. If the storm in a visible satellite image has a well-defined eye and a dense cirrus cloud shield when it reaches hurricane strength, the storm will most likely continue to strengthen, as there appears to be insufficient wind shear to tear it apart.

Detailed information about a hurricane can also come from aircraft that fly directly into the storm. These so-called **hurricane hunters** carry instruments directly on the aircraft as well as instruments, such as the *dropsonde*, that are dropped from the aircraft into the storm. On its way down to the ocean surface, the dropsonde measures air temperature, humidity, and atmospheric pressure, which are transmitted back to the aircraft. Because the dropsonde is equipped with a Global Positioning System (GPS) that constantly monitors its changing position, it has the capability of providing wind information as well.

Another temperature-measuring device dropped from the aircraft is the *bathythermograph*, which falls into the ocean where it measures water temperature as it slowly descends beneath the surface. Such probes can also measure the speed of ocean currents and the salinity (saltiness) of the water, an important factor in determining water density.

BRIEF REVIEW

Before reading the next several sections, here is a review of some of the important points about hurricanes.

● Hurricanes are tropical cyclones, comprised of an organized mass of thunderstorms.

● Hurricanes have peak sustained winds about a central core (eye) that reach at least 64 knots (74 mi/hr), with considerably higher gusts.

● The strongest winds and the heaviest rainfall normally occur in the eyewall—a ring of intense convective clouds (thunderstorms) that surround the eye.

● Hurricanes form over warm tropical waters, where light surface winds converge, the humidity is high in a deep layer, and the winds aloft are weak.

● For a mass of thunderstorms to organize into a hurricane there must be some mechanism that triggers the formation, such as converging surface winds along the ITCZ, or a pre-existing atmospheric disturbance, such as a weak front from the middle latitudes or a tropical wave.

● Hurricanes derive their energy from the warm, tropical oceans and by evaporating water from the ocean's surface. Heat energy is converted to wind energy when the water vapor condenses and latent heat is released inside deep convective clouds.

● When hurricanes lose their source of warm water (either by moving over colder water or over a large landmass), they dissipate rapidly.

● The three primary stages in a developing hurricane are: tropical depression, tropical storm, and hurricane (tropical cyclone).

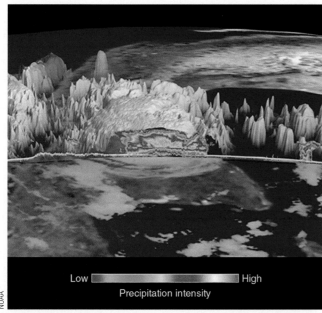

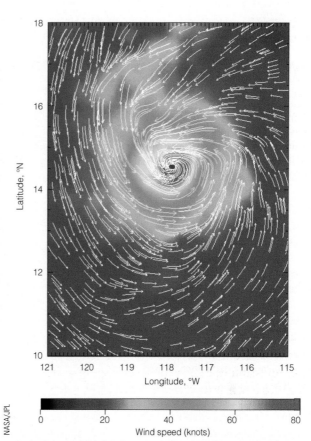

NOAA

NASA/JPL

● **FIGURE 11.9** Three-dimensional satellite image of Hurricane Karl over the Bay of Campeche on September 16, 2010, with precipitation intensity from the satellite *CloudSat*. Karl made landfall as a major hurricane along the coast of Mexico, northeast of Veracruz.

Up to this point, it is probably apparent that tropical cyclones called hurricanes are similar to middle-latitude cyclonic storms in that, at the surface, both have central cores of low pressure and winds that spiral counterclockwise (in the Northern Hemisphere) about their respective centers. However, there are many differences between the two systems, as described in Focus section 11.1.

HURRICANE MOVEMENT ● Figure 11.11 shows where most hurricanes are born and the general direction in which they move, whereas ● Figure 11.14 (p. 330) shows the actual paths taken by all hurricanes and tropical storms in the North Atlantic from 1980 to 2012. Notice that hurricanes that form over the warm,

● **FIGURE 11.10** Arrows show surface winds spinning counterclockwise around Hurricane Dora situated over the eastern tropical Pacific during August 1999. Colors indicate surface wind speeds. Notice that winds of 80 knots (92 mi/hr) are encircling the eye (the dark dot in the center). Wind speed and direction were obtained from the *QuikSCAT* satellite, which gathered wind data from 1999 to 2009.

tropical North Pacific and North Atlantic generally move toward the west or northwest. Steered by easterly winds, they move at an average speed of about 10 knots, often for a week or so. Gradually, they swing poleward around the subtropical high, and when they move far

● **FIGURE 11.11** Regions where tropical storms form (orange shading), the names given to storms, and the typical paths they take (red arrows).

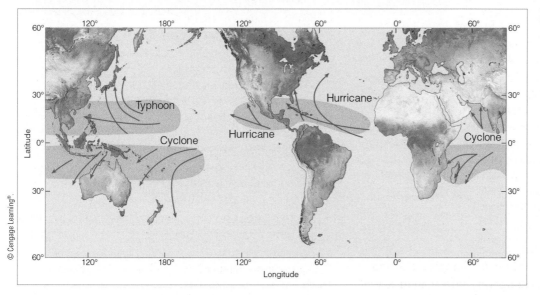

© Cengage Learning®

How Do Hurricanes Compare with Middle-Latitude Cyclones?

By now it should be apparent that a hurricane is much different from the mid-latitude cyclone that we discussed in Chapter 8. A hurricane derives its energy from the warm water and the latent heat of condensation, whereas the mid-latitude storm derives its energy from horizontal temperature contrasts. The vertical structure of a hurricane is such that its central column of air is warm from the surface upward; this is why hurricanes are called *warm-core lows*. A hurricane weakens with height, and the area of low pressure at the surface may actually become an area of high pressure above 12 km (40,000 ft). Mid-latitude cyclones, on the other hand, are *cold-core lows* that usually intensify with increasing height, with a cold upper-level low or trough often located above or to the west of the surface low.

A hurricane usually contains an eye where the air is sinking, while mid-latitude cyclones are characterized by centers of rising air. Hurricane winds are strongest near the surface, whereas the strongest winds of the mid-latitude cyclone are found aloft in the jet stream.

Further contrasts can be seen on a surface weather map. ● Figure 1 shows Hurricane Rita over the Gulf of Mexico and a mid-latitude cyclonic storm north of New England. Around the hurricane, the isobars are more circular, the pressure gradient is much steeper, and the winds are stronger. The hurricane has no fronts and is smaller (although Rita happens to be a large Category 5 hurricane). There are similarities between the two systems: Both are areas of surface low pressure, with winds moving counterclockwise about their respective centers. ▼ Table 1 summarizes the similarities and differences between the two systems.

It is interesting to note that some nor'easters (winter storms that move northeastward along the coastline of North America, characterized by heavy precipitation, high surf, and strong winds) may actually possess some of the characteristics of a hurricane. For example, a particularly powerful nor'easter during January 1989—one of the strongest on record—was observed to have a cloud-free eye, with surface winds in excess of 85 knots (98 mi/hr) spinning about a warm inner

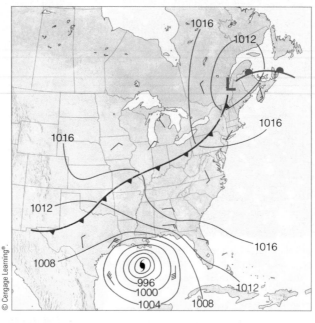

● **FIGURE 1** Surface weather map for the morning of September 23, 2005, showing Hurricane Rita over the Gulf of Mexico and a middle-latitude cyclonic storm system north of New England.

core. Occasionally, a hurricane will transition into a *post-tropical cyclone* such as a nor'easter. A particularly dramatic example is *Superstorm Sandy* in 2012 (see p. 341). Moreover, some *polar lows*—lows that develop over polar waters during winter—may exhibit many of the observed characteristics of a hurricane, such as a symmetric band of thunderstorms spiraling inward around a cloud-free eye, a warm-core area of low pressure, and strong winds near the storm's center. In fact, when surface winds within these polar storms

reach 58 knots (67 mi/hr), they are sometimes referred to as *Arctic hurricanes*.

Even though hurricanes weaken rapidly as they move inland, their circulation may draw in air with contrasting properties. If the hurricane links with an upper-level trough, it may actually become a mid-latitude cyclone, sometimes bringing very heavy rain over a wide area. Swept eastward by upper-level winds, the remnants of an Atlantic hurricane can become an intense mid-latitude autumn storm in Europe.

▼ **Table 1   Comparison of Hurricanes with Mid-Latitude Cyclonic Storms**

Conditions	TYPE OF STORM	
	Hurricane	Mid-Latitude Cyclone
Wind flow	Counterclockwise (NH)	Counterclockwise (NH)
	Clockwise (SH)	Clockwise (SH)
Strongest winds	Near surface; around eye	Aloft, in jet stream
Surface pressure	Lowest at center	Lowest at center
Vertical structure	Weakens with height; high pressure aloft; warm-core low	Strengthens with height, low pressure aloft; cold-core low
Air in center	Sinking	Rising
Weather fronts	No	Yes
Energy source	Warm water; release of latent heat	Horizontal temperature contrasts

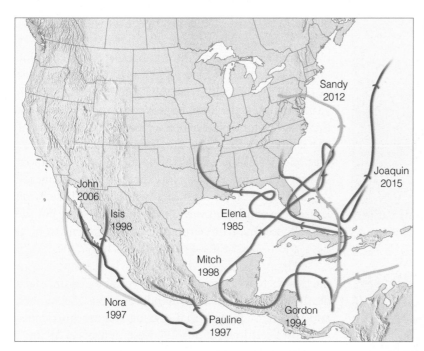

● FIGURE 11.12 Some erratic paths taken by hurricanes.

enough north, they become caught in the westerly flow, which curves them to the north or northeast. In the middle latitudes, a hurricane's forward speed normally increases, sometimes to more than 50 knots (58 mi/hr). The actual path of a hurricane (which appears to be determined by the structure of the storm and the storm's interaction with the environment) can vary considerably. Some take erratic paths and make odd turns that have occasionally caught weather forecasters by surprise (see Fig. 11.12), although modern forecasting techniques are likely to identify such twists and turns at least a day or two in advance. There have been many instances where a storm heading directly for land suddenly veered away and spared the region from almost certain disaster.

Look again at Fig 11.11 and notice that hurricanes apparently do not form over the South Atlantic and the eastern South Pacific, directly east and west of South America. Cooler water, vertical wind shear, and the unfavorable position of the ITCZ discourages hurricanes from developing in these regions. But guess what? For the first time since satellites began observing the south Atlantic, a hurricane formed off the coast of Brazil during March 2004. A visible satellite image of the storm is given in ● Fig. 11.13. So rare are tropical cyclones in this region that no government agency has an effective warning system for them, which is why the tropical cyclone was not given a name. It was informally dubbed *Catarina* because it struck the state of Santa Catarina. The storm caused more than

● FIGURE 11.13 An extremely rare tropical cyclone (with no official name) near 28°S latitude spins clockwise over the south Atlantic off the coast of Brazil during March 2004. Due to cool water and vertical wind shear, storms rarely form in this region of the Atlantic Ocean. In fact, this was the first hurricane-strength tropical cyclone ever officially reported there.

● FIGURE 11.14 Paths taken by hurricanes and tropical storms in the North Atlantic from 1980 to 2012.

(a) August

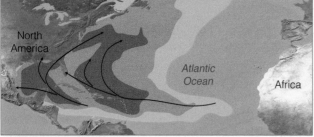

(b) September

Hurricane development
☐ Likely
☐ More likely
☐ Most likely
← Main tracks

(c) October

● FIGURE 11.15 Regions where Atlantic Basin hurricanes tend to form, and paths that are commonly taken, during the months of (a) August, (b) September, and (c) October. (Data from NOAA)

$350 million in damages and resulted in seven deaths. Six years later, in March 2010, another tropical storm (informally named Anita), formed east of Brazil, prompting the government of Brazil to begin maintaining a list of names for tropical cyclones. It is quite possible other such storms developed before the era of satellite imagery.

North Atlantic Hurricanes   Hurricanes that form over the tropical North Atlantic also move westward

or northwestward on a collision course with Central or North America. Most hurricanes, however, swing away from land and move northward, parallel to the coastline of the United States (see ● Fig. 11.14).* On average, one or two hurricanes will strike the United States coast each year, bringing high winds, huge waves, and torrential rain that may last for days. There is a great deal of year-to-year variability in United States hurricane landfalls: some years see none, while other years bring four or more. In addition, several tropical storms can be expected to make landfall on the United States coast in a typical year. These bring less-severe winds and waves than hurricanes do, but the rainfall accompanying a tropical storm—or even a tropical depression—can be just as heavy as in a hurricane.

● Figure 11.15 shows the regions where Atlantic Basin hurricanes tend to form and the typical paths they take during the active hurricane months of August, September, and October. Observe that, during August, hurricanes are most likely to form over the western tropical Atlantic, where they then either track westward into the Gulf of Mexico toward Texas, move northwestward into Florida, or follow a path parallel to the coast of the United States. In September, notice that the region where hurricanes are most likely to form stretches westward into the Gulf of Mexico and northward along the Atlantic seaboard. Typical hurricane paths take them into the central Gulf of Mexico or northeastward out over the Atlantic. Should an Atlantic hurricane track close to the coastline, it could make landfall anywhere from Florida to the mid-Atlantic states. In October, hurricanes are most likely to form in the western Caribbean and adjacent to the coast of North America, where they tend to take a more northerly trajectory. Note that Fig. 11.15 shows only the most common tracks observed over the long term. Any actual hurricane could behave much differently from these averages.

A hurricane moving northward over the Atlantic will normally survive as a hurricane for a much longer time than will its counterpart at the same latitude over the eastern Pacific. Why? An Atlantic hurricane moving northward usually stays over warmer water. In contrast, an eastern Pacific hurricane heading north will quickly move over much cooler water and, with its energy source cut off, will rapidly weaken.

Eastern Pacific Hurricanes   As we saw in an earlier section, many hurricanes form off the coast of Mexico over the Northeast Pacific. In fact, this area usually spawns about nine hurricanes each year, which is slightly more than the yearly average of six hurricanes born over the North Atlantic. Look back at Fig. 11.11 and notice that eastern North Pacific hurricanes normally move westward, away

*Sometimes hurricanes that remain over water and pose no threat to land are called "fish hurricanes" or "fish storms" because their greatest impact is on the fish in the open ocean.

from the coast. Because there is no landfall, little is heard about them. When one does move northwestward, it normally weakens rapidly over the cool water of the North Pacific. Occasionally, a Northeast Pacific hurricane will curve northward or even northeastward and slam into Mexico, causing destructive flooding. Hurricane Tico left 25,000 people homeless and caused an estimated $66 million in property damage after passing over Mazatlán, Mexico, in October 1983. The remains of Tico even produced record rains and flooding in Texas and Oklahoma. The record for top wind speed for any tropical cyclone on Earth in recent decades is held by Hurricane Patricia, whose sustained winds reached an estimated 185 knots (213 mi/hr) southwest of Manzanillo, Mexico, in October 2015. Fortunately, Patricia was a very small hurricane, and it weakened dramatically before landfall.

Even less frequently, a hurricane in the Northeast Pacific will stray far enough north to bring summer rains to southern California and Arizona, as did the remains of Hurricane Nora during September 1997. (Nora's path is shown in Fig. 11.12.) The only hurricane on record to reach the west coast of the United States with sustained hurricane-force winds slammed into extreme southern California near San Diego in October 1858.

The Hawaiian Islands, which are situated in the central North Pacific between about 20° and 23°N, appear to be in the direct path of many Northeast Pacific hurricanes and tropical storms. By the time most of these storms reach the islands, however, they have weakened considerably, and they typically pass harmlessly to the south or northeast. Occasionally a hurricane will form closer to Hawaii, in the Central Pacific, and such a hurricanes is somewhat more likely to threaten Hawaii. Two examples were Hurricane Iniki during September 1992 and Hurricane Iselle in August 2014. Iniki, the strongest hurricane known to strike Hawaii, made landfall on the island of Kauai as a Category 4 storm with sustained winds estimated at 122 knots (140 mi/hr). Iniki caused at least six deaths and inflicted close to $2 billion in destruction, a record for Hawaii. More than 14,000 structures on Kauai were damaged or destroyed. In 2014, Iselle became the strongest tropical cyclone on record to strike the Big Island of Hawaii, even though it had weakened to tropical storm strength before making landfall. Some areas received more than 30 cm (12 in.) of rain.

Naming Hurricanes and Tropical Storms

In an earlier section, we learned that hurricanes are given a name when they reach tropical storm strength. Before hurricanes and tropical storms were assigned names, they were identified according to their latitude and longitude.

This method was confusing, especially when two or more storms were present over the same ocean. When radio and television weathercasting became more important, the need to identify hurricanes for the public increased. To reduce the confusion, hurricanes were identified by letters of the alphabet. Starting in 1950, names such as Able and Baker were used for North Atlantic tropical storms. (These names correspond to the radio code words associated with each letter of the alphabet.) This method also led to confusion so, beginning in 1953, forecasters in the United States began using female names to identify Atlantic tropical storms and hurricanes. The list of names for each year was in alphabetical order, so that the name of the season's first storm began with the letter A, the second with B, and so on.

From 1953 to 1977, only female names were used. However, beginning in 1978, tropical storms in the eastern Pacific were alternately assigned female and male names, but not just English names, as Spanish and French ones were used too. This practice was adopted for North Atlantic hurricanes in 1979. The World Meteorological Organization now coordinates the naming systems for each tropical cyclone region around the world. The Northeast Pacific and Central Pacific each has its own list of storm names. In the Northwest Pacific, for example, most typhoons are named after birds, flowers, or other items rather than people. If a storm causes great damage, its name is retired for at least ten years. (None of the names retired from the North Atlantic have been reused as of this writing.)

▼ Table 11.1 gives the current list of names for North Atlantic tropical storms and hurricanes. The list of names for each year is recycled every six years, so the list for 2019 will be used again in 2025. If the number of named storms in any year should exceed the names on the list, which occurred for the first time in 2005, then tropical storms are assigned names from the Greek alphabet, such as Alpha, Beta, and Gamma. The last of the 27 named tropical systems in 2005 was Zeta, which formed on December 30. If it had developed just two days later, it would have been dubbed Alberto, the first name on the 2006 list.

Devastating Winds, the Storm Surge, and Flooding

When a hurricane is approaching from the south, its highest winds are usually on its eastern (right) side. The reason for this phenomenon is that the winds that push the storm along add to the winds on the east side and subtract from the winds on the west (left) side. The hurricane illustrated in ● Fig. 11.16 is moving northward along the east coast of the United States with winds of 100 knots swirling counterclockwise about its center. Because the storm is moving northward at about 25 knots, sustained winds on its

▼ Table 11.1 Names of Hurricanes and Tropical Storms in the North Atlantic*

2017	2018	2019	2020	2021	2022
Arlene	Alberto	Andrea	Arthur	Ana	Alex
Bret	Beryl	Barry	Bertha	Bill	Bonnie
Cindy	Chris	Chantal	Cristobal	Claudette	Colin
Don	Debby	Dorian	Dolly	Danny	Danielle
Emily	Ernesto	Erin	Edouard	Elsa	Earl
Franklin	Florence	Fernand	Fay	Fred	Fiona
Gert	Gordon	Gabrielle	Gonzalo	Grace	Gaston
Harvey	Helene	Humberto	Hanna	Henri	Hermine
Irma	Isaac	Imelda	Isaias	Ida	Ian
Jose	Joyce	Jerry	Josephine	Julian	Julia
Katia	Kirk	Karen	Kyle	Kate	Karl
Lee	Leslie	Lorenzo	Laura	Larry	Lisa
Maria	Michael	Melissa	Marco	Mindy	Matthew
Nate	Nadine	Nestor	Nana	Nicholas	Nicole
Ophelia	Oscar	Olga	Omar	Odette	Otto
Philippe	Patty	Pablo	Paulette	Peter	Paula
Rina	Rafael	Rebekah	Rene	Rose	Richard
Sean	Sara	Sebastien	Sally	Sam	Shary
Tammy	Tony	Tanya	Teddy	Teresa	Tobias
Vince	Valerie	Van	Vicky	Victor	Virginie
Whitney	William	Wendy	Wilfred	Wanda	Walter

*The list recycles every six years.

eastern side are about 125 knots, while on its western side, winds are only 75 knots.

The stronger winds on the storm's eastern side will likely cause the highest storm surge and most damage just east of the eye as the storm moves onshore. If the hurricane in Fig. 11.16 should suddenly change direction and move toward the west, its strongest winds, highest storm surge, and greatest potential for damage would now be just north of the eye.

Even though the hurricane in Fig. 11.16 is moving northward, there is a net transport of water directed from the east toward the coast. To understand this behavior, recall from Chapter 7 that as the wind blows over open water, the water beneath is set in motion. If we imagine the top layer of water to be broken into a series of layers, then we find each layer moving to the *right* of the layer above (in the Northern Hemisphere). This type of movement (bending) of water with depth (called the *Ekman Spiral*) causes a net transport of water (known as **Ekman transport**) to the right of the surface wind in the Northern Hemisphere. Hence, the north wind on the hurricane's left (western) side causes a net transport of water toward the shore. Here, the water piles up and rapidly inundates the region.

The high winds of a hurricane also generate large waves, sometimes 10 to 15 m (33 to 49 ft) high. These waves move outward, away from the storm, in the form of *swells* that carry the storm's energy to distant beaches. Consequently, the effects of a storm may be felt days before it arrives.

Although the hurricane's high winds inflict a great deal of damage, it is the huge waves, high seas, and *flooding* that normally cause most of the destruction. The flooding is also responsible for the loss of many lives. In fact, the majority of hurricane-related deaths during the past century has been due to flooding. The flooding is due, in part, to winds pushing water onto the shore and to the heavy rains, which may exceed 63 cm (25 in.) in 24 hours. Flooding is

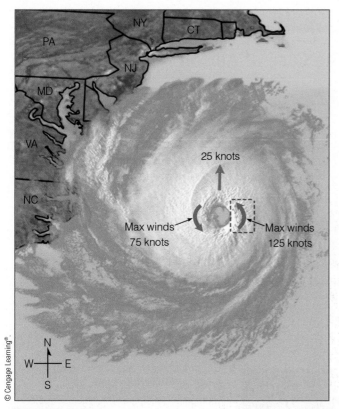

© Cengage Learning®.

● **FIGURE 11.16** A hurricane moving northward will have higher sustained winds on its eastern side than on its western side. The boxed area represents the region of strongest winds.

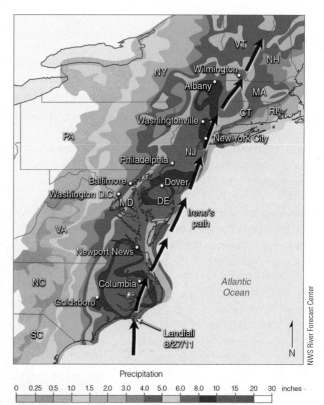

NWS River Forecast Center

● **FIGURE 11.18** Hurricane Irene's path (dark arrows) and estimated rainfall totals over the eastern United States from August 26 through August 29, 2011. Irene, the only hurricane to make landfall in the United States in 2011, was a massive but relatively weak storm that produced heavy rainfall and record flooding over sections of the Northeast.

also enhanced by the low pressure of the storm. The region of low pressure allows the ocean level to rise (perhaps half a meter), much like a soft drink rises up a straw as air is withdrawn. The combined effect of high water (which is usually well above the high-tide level), high winds, and the net Ekman transport toward the coast, produces the **storm surge**—an abnormal rise of several meters in the ocean level—which inundates low-lying areas and turns beachfront homes into piles of splinters (see ● Fig. 11.17). The storm surge is particularly damaging when it coincides with normal high tides.

The heavy rains produced by a hurricane can have a beneficial effect by providing much-needed rainfall

to drought-stricken regions. However, extreme flooding can occur well inland with strong hurricanes as well as with relatively weak storms, such as Hurricane Irene, which made landfall along the coast of North Carolina on August 26, 2011. Irene produced heavy rain and record flooding from North Carolina to Vermont (see ● Fig. 11.18). Flooding is not just associated with hurricane-strength tropical cyclones, as destructive floods can occur with tropical storms that do not reach hurricane strength. More on this topic is presented in Focus section 11.2.

● **FIGURE 11.17** When a storm surge moves in at high tide, it can inundate and destroy a wide swath of coastal lowlands.

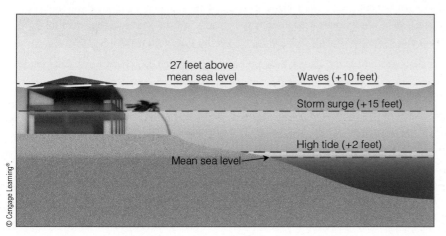

© Cengage Learning®.

Devastation from a Tropical Storm—The Case of Allison

Tropical storms that never become hurricanes can produce devastating floods. The most infamous example for the United States in recent years is Allison, which was the first tropical storm to have its name retired without ever reaching hurricane strength. In late May 2001, Allison began as a tropical wave that moved westward across the Atlantic. The wave continued its westward journey, and by the first of June it had moved across Central America and out over the Pacific Ocean. Here, it organized into a band of thunderstorms and a tropical depression. Upper-level winds guided the depression northward over the Gulf of Mexico, where the warm water fueled the circulation, and just east of Galveston, Texas, the depression became Tropical Storm Allison. Packing winds of 53 knots (61 mi/hr), Allison made landfall over the east end of Galveston Island on June 5. It drifted inland and weakened (see Fig. 2).

On the eastern side of the storm, heavy rain fell over parts of Texas and Louisiana. Some areas of southeast Texas received as much as 10 inches of rain in less than 5 hours. Homes, streets, and highways flooded as heavy rain continued to pound the area. But the worst was yet to come.

On June 7, as the upper-level winds began to change, the remnants of Allison drifted southwestward toward Houston. Heavy rain again fell over southeast Texas and Louisiana, where several tornadoes touched down. Over the Houston area, more than 20 inches of rain fell within a 12-hour period, submerging a vast part of the city. In six days the Port of Houston received a staggering 37 inches of rain.

The center of circulation drifted southward, moving off the Texas coast and out over the Gulf of Mexico on the evening of

● FIGURE 2 Visible satellite image showing the remains of Tropical Storm Allison centered over Texas on the morning of June 6, 2001. Heavy rain is falling from the thick clouds over Louisiana and eastern Texas.

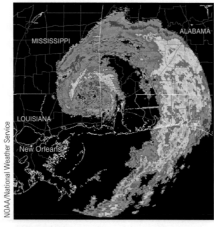

● FIGURE 3 Doppler radar display on June 11, 2001, showing bands of heavy rain swirling counterclockwise into the center of once–Tropical Storm Allison. The center of the storm, which is over Mississippi, has actually deepened and formed somewhat of an eye.

June 9. The flow aloft then guided the storm northeastward, where the storm made landfall again, but this time in southeastern Louisiana. Heavy rain continued to pound

Louisiana, creating one of the worst floods on record—a station in southern Louisiana reported a rainfall total of 30 inches. On June 11, a zone of maximum winds aloft (a jet streak) associated with the subtropical jet stream enhanced the outflow above the surface storm, and the remains of Tropical Storm Allison actually began to reintensify over land. As the storm entered Mississippi, its central pressure lowered, wind gusts reached 52 knots (60 mi/hr), and the center of circulation developed a weak-looking eye (see Fig. 3). As the system trekked eastward, it weakened and lost its eye, but continued to dump heavy rain over the southern Gulf States. Eventually, on June 14, the storm reached the Carolina coast.

Unfortunately, the storm slowed, then turned northward over North Carolina. Flooding became a major problem: Doppler radar estimated that up to 21 inches of rain had fallen over parts of the state. Severe weather broke out in Georgia and in the Carolinas, where some areas reported hail and downed trees due to gusty winds. The storm moved northeastward, parallel to the coast. A cold front moving in from the west eventually hooked up with the moisture from Allison. This situation caused heavy rain to fall over the mid-Atlantic states and southern New England. The storm finally accelerated to the northeast, away from the coast on June 18.

Allison, which never developed hurricane-strength winds, claimed the lives of 43 people, mainly due to flooding. The total damage from the storm totaled in the billions of dollars, with the Houston area alone sustaining over $2 billion in damage. If all the rain that fell from Allison could be placed in Texas, it would cover two-thirds of the state with water a foot deep.

CLASSIFYING HURRICANE STRENGTH In an effort to evaluate the possible damage a hurricane's sustained winds could do to a coastal area, the **Saffir-Simpson Scale** was developed. The scale numbers (which range from 1 to 5) are based on certain conditions at some time during the life of the storm. As the hurricane intensifies or weakens, the category, or scale number, is reassessed accordingly. Major hurricanes are classified as Category 3 and above. In the Northwest Pacific, a typhoon with sustained winds of at least 130 knots (150 mi/hr)—near the top end of Category 4 on the Saffir-Simpson Hurricane Wind Scale—is designated a **super typhoon** by the United States Joint Typhoon Warning Center.

The original *Saffir-Simpson Scale* used a hurricane's central pressure as a measure of the storm's wind strength. The modified **Saffir-Simpson Hurricane Wind Scale** (see ▼Table 11.2) does not use central pressure, because maximum winds today are accurately determined by modern observing techniques and are more directly relevant to a storm's impact. In addition, storm surge has been removed from the original scale because it has become clear that other factors besides wind speed play a major role in determining storm surge. For example, if two hurricanes have the same wind speed, the larger one will normally produce a greater surge. Landscape features along the coast, as well as underwater topography, are also critical in determining how high the storm surge will be and how far it will extend inland. Hurricane Andrew (1992) made landfall as a Category 5 storm, but its peak surge south of Miami was less than 17 feet. In contrast, Hurricane Ike (2008) made landfall as a Category 2, but it brought a surge as high as 20 feet to parts of Texas. ●Figure 11.19 illustrates how the storm surge could change along the coast as hurricanes with increasing intensity move onshore.

Because a hurricane storm surge can occur on top of natural high or low tides, it can be difficult to convey the actual high-water levels that coastal residents in the path of a hurricane might experience. The National Hurricane Center has recently been revising its procedures for communicating the risk of storm surge during hurricane

▼Table 11.2 **Saffir-Simpson Hurricane Wind Scale**

SCALE CATEGORY	WINDS (ONE MINUTE SUSTAINED)		SUMMARY*
	mi/hr	knots	
1	74–95	64–82	Very dangerous winds will produce some damage
2	96–110	83–95	Extremely dangerous winds will cause extensive damage
3	111–129	96–112	Devastating damage will occur
4	130–156	113–136	Catastrophic damage will occur
5	>156	>136	Catastrophic damage will occur

*The scale provides extensive information for each category on the potential harm to people and pets and potential damage to structures such as mobile homes, houses, apartments, and shopping centers.

threats. One of the newly designed tools is a map that incorporates both storm surge and tidal levels to show the maximum height of water above ground level that could occur during a given time span as a hurricane approaches.

●Figure 11.20 shows the number of hurricanes that have made landfall* along the coastline of the United States from 1900 through 2015. Out of a total of 191 hurricanes striking the American coastline, 69 (36 percent) were

Normal high tide

Category 1 [4-foot rise]

Category 3 [12-foot rise]

Category 5 [20-foot rise]

© Cengage Learning®.

● **FIGURE 11.19** The changing of the ocean level as hurricanes of different strength make landfall along the coast. The water level might rise about 4 feet on average with a typical Category 1 hurricane, but could rise to 20 feet (or more) with a Category 5 storm. Note that the exact surge for a particular hurricane and strength category can vary greatly from these numbers depending on the storm's size, the shape of the coastline where it strikes, and other factors. Tidal levels at landfall may also increase or reduce the actual water levels observed.

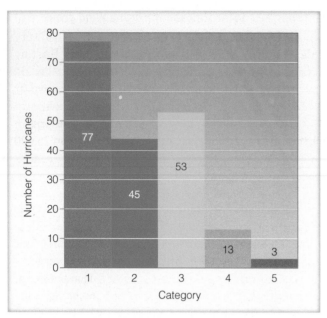

● **FIGURE 11.20** The number of hurricanes (by each category) that made landfall along the coastline of the United States from 1900 through 2015. All of the hurricanes struck the Gulf or Atlantic coasts. Categories 3, 4, and 5 are considered major hurricanes. No major hurricanes struck the United States between 2006 and 2015, the longest such period on record. (Note: Hurricane Sandy, 2012, was no longer classified as a hurricane when it struck New Jersey.)

major hurricanes, Category 3 or higher. Hence, along the Gulf and Atlantic coasts, on the average, about five hurricanes make landfall every three years, two of which are major hurricanes with winds in excess of 95 knots (110 mi/hr). However, the number of hurricanes in the North Atlantic can vary greatly from year to year and even from decade to decade.

HURRICANE-SPAWNED TORNADOES Although the high winds of a hurricane can devastate a region, considerable damage can also occur from hurricane-spawned tornadoes. About one-fourth of the hurricanes that strike the United States produce tornadoes. In fact, in 2004 six tropical systems produced just over 300 tornadoes in the southern and eastern United States. Most of these tornadoes develop in small supercell thunderstorms, which tend to occur within rain bands or along pre-existing boundaries that a hurricane may encounter as it moves inland. Tornadoes have been observed more than 500 km (310 mi) away from a hurricane's center. Moreover, tornadoes tend to form in the right front quadrant of an advancing hurricane, where vertical wind speed shear is greatest.

Researchers are still analyzing the factors that produce the strongest winds and damage within hurricane eyewalls, which are very difficult to observe directly. Studies of the 1992 catastrophic Hurricane Andrew (see p. 338) revealed that the most severe damage occurred close to the eyewall, where wind gusts of up to 154 knots (170 mi/hr)

were reported. It appears that the strongest winds lasted only a few seconds and struck in narrow bands less than 100 meters wide. Some research indicates that at least one tornado was involved. Other studies suggest the presence of tiny eddies, called *mini-whirls*, produced by very strong wind shear at low levels but perhaps not linked to a parent thunderstorm as a tornado would be. Some damage may be caused by strong downdrafts (microbursts) associated with the large, intense thunderstorms around the eyewall.

HURRICANE FATALITIES Up until 2005, the annual death toll from hurricanes in the United States, over a span of about 30 years, averaged fewer than 50 persons.* Most of these fatalities occurred because of flooding. The relatively low total was achieved in part because of the advanced warning provided by the National Weather Service and in part by the fact that only a few really intense storms had made landfall during this time. However, it became clear that massive loss of life was still possible in the United States when Hurricane Katrina slammed into Mississippi and Louisiana in 2005.

Even before it threatened the Gulf Coast, Hurricane Katrina inflicted more than $1 billion in damage and took 14 lives when it crossed South Florida. As Katrina moved into the Gulf of Mexico, intensified to Category 5 strength, and headed toward the upper Gulf Coast, evacuation orders were given to residents living in low-lying areas, including the city of New Orleans. Hundreds of thousands of people moved to higher ground but, unfortunately, many thousands of others either refused to leave their homes or had no means of leaving and were forced to ride out the storm. Tragically, more than 1500 people died in southeast Louisiana, many as a result of catastrophic flooding across the New Orleans area that occurred when several levees broke and parts of the city were inundated with water more than 20 feet deep. In addition, more than 200 people were killed along the Mississippi coast, where a giant storm surge of up to 27 feet pushed well inland.

The aftermath of an intense hurricane can be devastating in itself. The supply of fresh drinking water may be contaminated, and food may become scarce as grocery stores and markets are forced to close, sometimes for days or even weeks. Roads may be blocked by fallen trees and debris or by sand that was deposited during the storm surge. Electrical and telephone service may be disrupted or completely lost. And many people may be displaced from their damaged or destroyed homes. Even the cleanup efforts can prove deadly, as, in certain areas, poisonous snakes often find their way into various nooks and crannies of the debris.

*In other countries, the annual death toll was considerably higher. Estimates are that more than 3000 people died in Haiti from flooding and mud slides when Hurricane Jeanne moved through the Caribbean during September 2004.

Some Notable Hurricanes

GALVESTON, 1900 Before the era of satellites, radar, and hurricane warnings on radio and television, catastrophic loss of life was all too common in the United States when hurricanes made landfall. An estimated 8000 people (perhaps as many as 12,000) lost their lives when a hurricane slammed into Galveston, Texas, in September 1900 with a huge storm surge estimated at 15 feet high. This remains the deadliest natural disaster in United States history. Galveston had no seawall at the time, and the flood waters quickly swept into homes and businesses as the water pushed inland.

NEW ENGLAND, 1938 A powerful September hurricane slammed into the south shore of Long Island in 1938 as a strong Category 3 storm with a central pressure of 946 mb (27.94 in.) and a storm surge exceeding 12 feet, combined with additional water from rising tides. Pulled northward by a strong jet stream, this hurricane was dubbed the "Long Island Express" because of its extremely fast forward motion, estimated at close to 61 knots (70 mi/hr), the fastest hurricane motion on record. The rapid movement allowed less time for the storm to weaken over cooler waters, and it also led to especially high winds on the storm's eastern side. A record-high wind gust of 162 knots (186 mi/hr) was reported at Blue Hill Observatory near Boston. Because forecasters had expected the storm to move out to sea, residents had very little warning of the impending disaster. After blasting Long Island, the hurricane continued barreling northward, making a second landfall in Connecticut. Its intensity was little diminished, as a storm surge inundated downtown Providence, Rhode Island, and many nearby locations. In all, the hurricane damaged or destroyed more than 25,000 homes and took more than 600 lives, making it the deadliest hurricane in New England history.

CAMILLE, 1969 Hurricane Camille stands out as one of the most intense hurricanes to reach the coastline of the United States during the twentieth century (see ▼ Table 11.3). With a central pressure of 909 mb, tempestuous winds reaching 160 knots (184 mi/hr) and a storm surge more than 7 m (23 ft) above the normal high-tide level, Camille as a Category 5 storm unleashed its fury on Mississippi, destroying thousands of buildings. During its rampage, it caused an estimated $1.5 billion in property damage and took more than 200 lives.

HUGO, 1989 With maximum winds estimated at about 120 knots (138 mi/hr) and a central pressure near 934 mb, Hugo made landfall as a Category 4 hurricane near Charleston, South Carolina, about midnight on September 21 (see ● Fig. 11.21). The high winds and storm surge, which ranged between 2.5 and 6 m (8 and 20 ft),

▼ Table 11.3 **The Thirteen Most Intense Hurricanes (at Landfall) to Strike the United States from 1900 through 2015, Based on Central Pressure**

RANK	HURRICANE (MADE LANDFALL)	YEAR	CENTRAL PRESSURE (MILLIBARS/INCHES)	CATEGORY	DEATH TOLL
1	Florida (Keys)	1935	892/26.35	5	408
2	Camille (Mississippi)	1969	909/26.85	5	256
3	Andrew (South Florida)	1992	922/27.23	5	53
4	Katrina (Louisiana)	2005	920/27.17	3*	>1500
5	Florida (Keys)/South Texas	1919	927/27.37	4	>600**
6	Florida (Lake Okeechobee)	1928	929/27.43	4	>2000
7	Donna (Long Island, New York)	1960	930/27.46	4	50
8	Texas (Galveston)	1900	931/27.49	4	>8000
9	Louisiana (Grand Isle)	1909	931/27.49	4	350
10	Louisiana (New Orleans)	1915	931/27.49	4	275
11	Carla (South Texas)	1961	931/27.49	4	46
12	Hugo (South Carolina)	1989	934/27.58	4	49
13	Florida (Miami)	1926	935/27.61	4	243

*Although the central pressure in Katrina's eye was quite low, Katrina's maximum sustained winds of 110 knots at landfall made it a Category 3 storm.
**More than 500 of this total were lost at sea on ships. (The > symbol means "greater than.")

NOAA/National Weather Service

● **FIGURE 11.21** Satellite image of Hurricane Hugo approaching Charleston, South Carolina, on September 21, 1989.

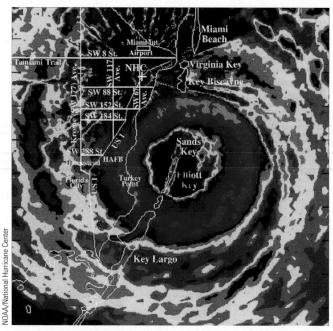

NOAA/National Hurricane Center

● **FIGURE 11.22** Radar image of Hurricane Andrew as it moves onshore over south Florida on the morning of August 24, 1992. The dark red and purple show where the heaviest rain is falling. Miami Beach is just to the north of the eye and the National Hurricane Center (NHC) is about 20 miles to the northeast of the eye.

hurled a thundering wall of water against the shore. Hugo knocked out power, flooded streets, and caused widespread destruction to coastal communities. The total damage in the United States attributed to Hugo was over $7 billion, with a death toll of 21 in the United States and 49 overall.

ANDREW, 1992   On August 21, 1992, as Tropical Storm Andrew churned westward across the Atlantic it began to weaken, prompting some forecasters to surmise that this tropical storm would never grow to hurricane strength. But Andrew moved into a region favorable for hurricane development. Even though it was outside the tropics near latitude 25°N, warm surface water and weak winds aloft allowed Andrew to intensify rapidly. And in just two days Andrew's winds increased from 45 knots (52 mi/hr) to 122 knots (140 mi/hr), turning an average tropical storm into one of the most intense hurricanes to strike Florida in more than 100 years (see Table 11.3). With sustained winds estimated at 140 knots (167 mi/hr) and a powerful storm surge, Andrew made landfall south of Miami on the morning of August 24 (see ● Fig. 11.22). The eye of the storm moved

over Homestead, Florida. Most of Andrew's destruction came not from the storm surge but from fierce winds that completely devastated the area (see ● Fig. 11.23). The hurricane roared westward across southern Florida, weakened slightly, then regained strength over the warm Gulf of Mexico. Surging northwestward, Andrew slammed into Louisiana as a Category 3 storm, with 100-knot (115 mi/hr) winds after midnight on August 26.

All told, Hurricane Andrew was one of the costliest natural disasters ever to hit the United States. It destroyed or damaged more than 200,000 homes and businesses, left more than 160,000 people homeless, caused over $30 billion in damages, and took 53 lives, including 41 in Florida.

● **FIGURE 11.23** A community in Homestead, Florida, devastated by Hurricane Andrew on August 24, 1992.

© Weather Catalog/Media Services

KATRINA AND RITA, 2005 Hurricane Katrina was the most damaging hurricane ever to hit the United States (although it is estimated that the 1926 Miami hurricane could be more than twice as costly if it struck today). Katrina was also by far the deadliest U.S. hurricane in more than 70 years. Forming over warm tropical water south of Nassau in the Bahamas, Katrina became a tropical storm on August 24, 2005, and a Category 1 hurricane just before making landfall in south Florida on August 25. (Katrina's path is shown in Focus section 11.3.) It moved southwestward across Florida and out over the eastern Gulf of Mexico. As Katrina moved westward, it passed over a deep band of warm water called the *Loop Current* that allowed Katrina to rapidly intensify. Within 12 hours, the hurricane increased from a Category 3 to a Category 5 storm with winds of 152 knots (175 mi/hr) and a central pressure of 902 mb.

Over the Gulf of Mexico, Katrina gradually turned northward toward Mississippi and Louisiana. As the powerful Category 5 hurricane moved slowly toward the coast, its rain bands near the center of the storm began to converge toward the storm's eye, thus cutting off moisture to the eyewall. As the old eyewall dissipated, a new one formed farther away in a phenomenon called *eyewall replacement*. The replacement of the eyewall weakened the storm such that Katrina made landfall near Buras, Louisiana, on August 29 as a Category 3 hurricane with sustained winds of 110 knots (127 mi/hr) and a central pressure of 920 mb. Despite its weakened winds, Katrina still carried an extremely high storm surge of between 6 m and 9 m (20 and 30 ft).

Katrina's strong winds and high storm surge on its eastern side devastated southern Mississippi, with Biloxi, Gulfport, and Pass Christian being particularly hard hit. The winds demolished all but the strongest structures, and

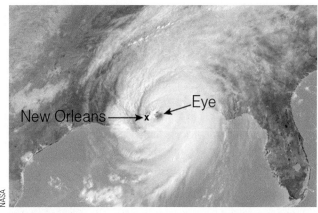

● **FIGURE 11.24** Hurricane Katrina just after making landfall along the Mississippi/Louisiana coast on the morning of August 29, 2005. Shown here, the storm is moving north with its eye due east of New Orleans, marked X on the image. At landfall, Katrina had sustained winds of 110 knots (127 mi/hr), a central pressure of 920 mb (27.17 in.), and a storm surge of more than 20 feet.

the huge storm surge scoured areas up to 6 km (10 mi) inland. The surge height of 27.8 feet measured at Pass Christian was the highest surge value ever recorded in the United States. The high winds and severe flooding from Katrina caused more than 200 deaths in Mississippi.

New Orleans and the surrounding parishes actually escaped the brunt of Katrina's winds, as the eye passed just to the east of the city (see ● Fig. 11.24). In addition, the storm surge in the New Orleans area was considerably less than in Mississippi. However, the combination of high winds, large waves, and storm surge led to disastrous breaches in the levee system that protected New Orleans from the Mississippi River, the Gulf of Mexico, and Lake Pontchartrain. When the levees gave way, water up to 20 feet deep invaded a large part of the city before thousands of people could escape (see ● Fig. 11.25). The immense

● **FIGURE 11.25** Floodwaters inundate New Orleans, Louisiana, during August 2005, after the winds and storm surge from Hurricane Katrina caused several levee breaks.

The Record-Setting Atlantic Hurricane Seasons of 2004 and 2005

Both 2004 and 2005 were active years for hurricane development over the tropical North Atlantic (see ● Fig. 4). During 2004, nine storms became full-fledged hurricanes. Out of the five hurricanes that made landfall in the United States, three (Charley, Frances, and Jeanne) plowed through Florida, and one (Ivan) came onshore just west of the Florida panhandle (see ● Fig. 5), making this the first time since record-keeping began in 1861 that four hurricanes have impacted the state of Florida in one year. Total damage in the United States from the five hurricanes exceeded $40 billion.

Then, in 2005, a record twenty-seven named storms developed (the most in a single season), of which fifteen (another record) reached hurricane strength. The 2005 Atlantic hurricane season also had four hurricanes (Emily, Katrina, Rita, and Wilma) reach Category 5 intensity for the first time since reliable record-keeping began. And Hurricane Wilma had the lowest central pressure ever measured in an Atlantic hurricane up to that point—882 mb (26.04 in.). Out of five hurricanes that made landfall in the United States, three (Dennis, Katrina, and Wilma) made landfall in hurricane-weary Florida and one (Ophelia)

skirted northward along Florida's east coast, giving Florida the dubious distinction of being the only state on record to experience eight hurricanes during the span of sixteen months (Fig. 4). To illustrate how unusual this event was, no major hurricanes (Category 3 or greater) made landfall in the United States between 2006 and 2015, and no hurricanes at all struck Florida during that period. Total damage in the United States from the five hurricanes that made landfall in 2005 exceeded $100 billion.

In both 2004 and 2005, very warm ocean water and weak vertical wind shear provided favorable conditions for hurricane development. In previous years, winds associated with a persistent upper-level trough over the eastern United States steered many tropical systems away from the coast before they could make landfall. However, in 2004 and in 2005, an area of high pressure replaced the trough, and winds tended to steer tropical cyclones on a more westerly track, toward the coastline of North America.

● **FIGURE 4** The paths of eight hurricanes that impacted Florida during 2004 and 2005. Notice that in 2004, hurricanes Frances and Jeanne made landfall at just about the same spot along Florida's southeast coast. The date under each hurricane's name indicates when the hurricane made landfall. After a very large loop in its path, Ivan made a second landfall in southwest Louisiana as a tropical depression on September 23.

● **FIGURE 5** Beach homes along the Gulf Coast at Orange Beach, Alabama, (a) before and (b) after Hurricane Ivan made landfall during September 2004. (Red arrows are for reference.)

devastation was compounded by the inability of rescuers to provide immediate assistance to many people, as floodwaters persisted for days in parts of the New Orleans area that lie below sea level. The death toll due to Hurricane Katrina eventually reached more than 1800, and the devastation wrought by the storm totaled more than $75 billion.

Less than a month later, powerful Hurricane Rita, another Category 5 storm with sustained winds of 152 knots (175 mi/hr), moved across the Gulf of Mexico, south of New Orleans. Strong, tropical storm–force easterly winds, along with another storm surge, caused some of the repaired levees in the New Orleans area to break again, flooding parts of the city that just days earlier had been pumped dry. Rita made landfall in southeast Texas, killing more than 100 people in the state. However, many of the deaths from Rita were caused by heat stress, as people evacuated from Rita's path in huge numbers (perhaps in part because of the recent Hurricane Katrina disaster), only to be caught in massive traffic jams in the middle of a major heat wave.

SANDY, 2012

Although it lost its hurricane status about three hours before making landfall, Hurricane Sandy (often referred to as *Superstorm Sandy*) brought the most deadly and damaging storm surge in more than 70 years to coastal areas of New Jersey and New York, including New York City. Sandy formed as a tropical storm late in the 2012 hurricane season, on October 22, in the western Caribbean. It struck Jamaica as a Category 1 hurricane, then intensified to Category 3–strength with winds of 100 knots (115 mi/hr) before striking Cuba. Afterward, Sandy weakened to a tropical storm and followed a typical track east of the Bahamas, moving toward the north-northwest and then to the northeast. Normally, such a storm would be expected to continue moving out to sea, especially so late in the hurricane season. However, some computer models provided more than five days' notice that Sandy might curve to the northwest and strike the eastern coastline of the United States. No other hurricane on record had made such a sharp left turn so far north, and the various models disagreed for several days about Sandy's forecast path. Eventually, however, they converged to accurately predict a very dangerous track. Sandy's entire track is shown in Fig. 11.12, on p. 329.

Sandy regained Category 1 strength on October 27, and the storm struck the coast just northeast of Atlantic City, New Jersey, on the evening of October 29 as a powerful *post-tropical cyclone*, with a central pressure of 945 mb (27.91 in.). Despite having been downgraded from hurricane status because it lacked a core of warm air, Sandy made landfall with sustained hurricane-force winds of 70 knots (80 mi/hr). Because Sandy was an extremely large storm, its winds were weaker than usual for a central pressure so low, but they were spread across a vast area (see ● Fig. 11.26). This situation helped generate an enormous storm surge that swept into New York Harbor and caused severe damage along much of the Long Island and New Jersey coast (see ● Fig. 11.27).

Sandy struck at close to high tide in the New York City area, which put roughly 5 feet of water on top of the 9-foot storm surge. Waters poured into the New York subway system and inundated much of lower Manhattan Island. More than 70 people died in the United States as a result of Sandy, and estimated damages totaled more than $50 billion, making it the second-costliest storm of tropical origin in the United States going back to 1900.

Because Sandy was not expected to make landfall as a hurricane, coastal residents had been alerted to the storm's serious dangers but had not been placed under a hurricane warning. In the aftermath of Sandy, the National Hurricane Center changed its policy so that hurricane warnings can now be issued and kept in place up to landfall even if a hurricane is expected to become a post-tropical cyclone before it strikes.

● **FIGURE 11.26** Hurricane Sandy over the Atlantic Ocean just after midnight (EDT) on October 29, 2012. Dark arrows show winds around the storm; color shading indicates wind speed (mi/hr). Heavy dashed lines show Sandy's path.

● **FIGURE 11.27** A home destroyed by Hurricane Sandy in Union Beach, New Jersey. Sandy devastated parts of coastal New Jersey and New York, causing more than 50 billion dollars in damages.

Ramin Talaie/Corbis via Getty Images

Devastating Tropical Cyclones around the World

Though horrifying enough, the statistics from even the worst hurricanes to strike the United States pale when compared to the stunning tolls that have occurred in other nations, especially those that have large populations in low-lying river deltas. More than 300,000 lives were taken as a killer tropical cyclone and storm surge ravaged the coast of Bangladesh with floodwaters in November 1970. In April 1991, a similar cyclone devastated the area with reported winds of 127 knots (146 mi/hr) and a storm surge of 7 m (23 ft). In all, the storm destroyed 1.4 million

houses and killed 140,000 people and one million cattle. And again in November 2007, Tropical Cyclone Sidr, a Category 4 storm with winds of 135 knots (155 mi/hr) moved into the region, killing more than 3000 people, damaging or destroying over one million houses, and flooding more than two million acres. Estimates are that Sidr adversely affected more than 8.5 million people. Unfortunately, the potential for a repeat of this type of disaster remains high in Bangladesh, because many people live along the relatively low, wide floodplain that slopes outward to the bay, and, historically, this region is in a path frequently taken by tropical cyclones.

In May 2008, Tropical Cyclone Nargis first took aim on Bangladesh (see ● Fig. 11.28), but then moved

● **FIGURE 11.28** Visible satellite image of Tropical Cyclone Nargis on May 2, 2008, as it begins to move eastward over the Bay of Bengal toward Myanmar (Burma), where its storm surge and floodwaters killed more than 140,000 people. (The red dashed lines show the path of Nargis.) Also on the image is the path of Tropical Cyclone Sidr (yellow arrows), which caused widespread destruction in Bangladesh during November 2007.

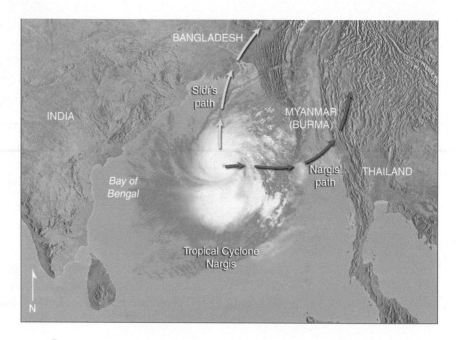

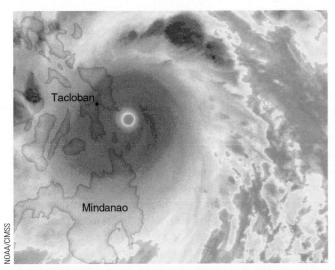

● **FIGURE 11.29** Infrared satellite image of Super Typhoon Haiyan as it approaches the central Philippines on November 7, 2013. The large central ring indicates very intense convection, with the tops of the clouds so high and cold that their temperatures range from –80°C to –90°C (–112°F to –130°F).

east, striking Myanmar (Burma), where strong tropical cyclones are much less common. Although the cyclone was accompanied by strong winds, it was the 16-foot storm surge and 8-foot waves that caused much of the damage. Nargis pushed floodwaters inland for at least 30 miles along the Irrawaddy Delta. No devastating tropical cyclone had been recorded in the delta in modern history, and millions of people were in flood-prone homes less than 10 feet above sea level. The cyclone killed at least 140,000 people as flooding washed away entire villages, in some places without leaving a single structure.

One of the strongest tropical cyclones to make landfall in modern times was Super Typhoon Haiyan, which struck the Philippines in November 2013. (In the Philippines, Haiyan was referred to as Yolanda.) Haiyan intensified over waters that were unusually warm well below the surface, so that the strong mixing from high winds did not bring up as much cooler water as usual. The typhoon was at its peak strength when it came ashore, with sustained winds estimated at 169 knots (195 mi/hr), the highest value for any landfalling tropical cyclone in the world (see ● Fig. 11.29). The powerful winds drove a

DID YOU KNOW?

The largest tropical cyclone on record was Super Typhoon Tip, which formed in the western Pacific on October 5, 1979. At its peak it had a circulation of tropical-storm–force winds that spanned 1350 miles, the distance from Key West, Florida, to Amarillo, Texas. On October 12, the storm's central pressure fell to 870 mb (25.69 in.), the lowest ever measured in any tropical system, and its winds reached an estimated 165 knots (190 mi/hr), a strength that ranks second only to that of Hurricane Patricia (2015). Fortunately, Typhoon Tip never made landfall.

massive storm surge into the island of Leyte. The city of Tacloban experienced the worst damage as a 5 m (17 ft) storm surge completely submerged the first level of many buildings. More than 6000 people died in the Philippines, with additional damage occurring when a weakened Haiyan moved on to bring high winds and heavy rains to China and Vietnam.

The deadliest known hurricane to strike the Western Hemisphere was the Great Hurricane of 1780, which claimed approximately 27,000 lives in the eastern Caribbean. The next-deadliest was Hurricane Mitch in October 1998. Mitch's high winds, huge waves (estimated maximum height 44 ft), and torrential rains destroyed vast regions of coastal Central America (for Mitch's path, see Fig. 11.12, p. 329). In the mountainous regions of Honduras and Nicaragua, rainfall totals from the storm may have reached 190 cm (75 in.). The heavy rains produced floods and deep mud slides that swept away entire villages, including structures and inhabitants. Mitch caused over $5 billion in damages, destroyed hundreds of thousands of homes, and killed more than 11,000 people. More than 3 million others were left homeless or were otherwise severely affected by this deadly storm.

Are major hurricanes on the increase worldwide? Will the intensity of hurricanes increase as the world warms? These questions are addressed in Focus section 11.4.

Hurricane Watches, and Warnings

With the aid of ship reports, satellites, radar, buoys, and reconnaissance aircraft, the location and intensity of tropical cyclones are pinpointed and their movements carefully monitored. Once a tropical depression forms, its motion and strength are predicted out to the following five days by the National Hurricane Center in Miami, Florida, or by the Central Pacific Hurricane Center in Honolulu, Hawaii. These forecasts are then updated every six hours (or more often if a hurricane is nearing shore). When a hurricane poses a direct threat to an area, a **hurricane watch** is issued, typically 24 to 48 hours before the storm arrives. When it becomes more certain that the storm will strike an area, a **hurricane warning** is issued (see ● Fig. 11.30). A map of surface wind speed probabilities, updated every few hours, provides the likelihood that winds of a particular speed (such as tropical storm force or hurricane force) will occur in a particular area over various time periods (see ● Fig. 11.31).

Hurricane warnings are designed to give people ample time to secure property and, if necessary, to evacuate the area. Even if a hurricane loses its tropical characteristics before striking land, the National Hurricane Center may opt to continue any existing hurricane warnings up until

Hurricanes in a Warmer World

In Focus section on 11.3, p. 340, we saw that 2005 was a record year for Atlantic hurricanes, with 27 named storms, 15 hurricanes, and 5 storms reaching Category 5 status on the Saffir-Simpson Hurricane Wind scale. It is clear, from ● Fig. 6 below, that the number of hurricanes reported in the Atlantic Basin has increased over the last several decades. Could the large numbers and high intensity of hurricanes in recent years be related to global warming?

We know that hurricanes are fueled by warm tropical water—the warmer the water, the more fuel available to drive the storm. A mere 0.6°C (1°F) increase in sea-surface temperature will increase the maximum winds of a hurricane by about 5 knots (almost 6 mi/hr) everything else being equal. Since the mid-1990s, sea-surface temperatures across the tropical North Atlantic have trended above the longer-term average. Some experts have attributed this phenomenon to a cyclic rise and fall in temperatures across this region, which apparently tends to peak about every 30 to 40 years. However, at least some of the warming appears to be related to an overall temperature increase throughout the world's oceans.

Between June and October 2005, sea-surface temperatures in the tropical Atlantic

were about 0.9°C (1.6°F) warmer than the long-term (1901–1970) average for that region. One study concluded that about half of the warming (about 0.4°C) stemmed from climate change caused by increasing concentrations of greenhouse gases in the atmosphere. Other studies indicate a global rise in the average intensity of the strongest tropical cyclones over the last several decades. These results suggest that warmer sea-surface temperatures may indeed be helping to boost the strength of hurricanes. One difficulty with these studies is that reliable and complete records of tropical cyclones have only been available since the 1970s, when observations from satellites became more extensive. As for the future, climate models predict that sea-surface temperatures in the tropics will rise by 0.6°C to 2.0°C by the end of this century, depending on the rate of greenhouse gas emissions. Should the upper end of these projections prove correct, a hurricane forming in today's atmosphere with maximum sustained winds of 130 knots (a strong Category 4 storm) could, in the warmer world, have maximum sustained winds of 140 knots (a Category 5 storm). Future hurricanes may also deposit more rain as they make landfall, because warmer temperatures globally are expected to increase evaporation from the

oceans and increase the amount of water vapor available for precipitation.

As sea-surface temperatures rise, will hurricanes become more frequent? Over the last 40 years, there has been no significant change in the number of tropical cyclones observed around the world, except for the increase in the number of Atlantic hurricanes noted earlier. Most climate models actually predict that the number of tropical cyclones will hold steady or decrease slightly by the end of this century, in part because of a projected weakening of the global tropical circulation that supports rising motion and hurricane formation. It is also possible that some oceans will warm more than others, which could lead to wind shear favoring the areas that are warming most rapidly and inhibiting hurricane formation in other areas.

Sophisticated instruments today allow scientists to peer into hurricanes and examine their structure and winds with much greater clarity than in the past. Any trends in hurricane frequency or intensity will likely become clearer when more reliable information on past tropical cyclone activity becomes available, especially from the ongoing investigation of sea sediment cores, which hold clues to past tropical cyclone occurrences.

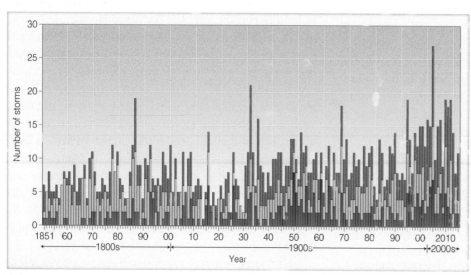

● **FIGURE 6** The total number of tropical storms and hurricanes (red bars), hurricanes only (yellow bars), and Category 3 hurricanes or greater (green bars) in the Atlantic Basin for the period 1851 through 2015. (NOAA)

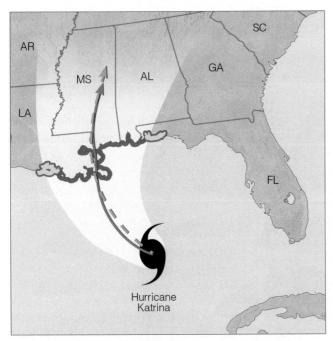

● **FIGURE 11.30** Hurricane Katrina over the Gulf of Mexico with sustained winds of 126 knots (145 mi/hr) on August 28, 2005, at 1 a.m. (CDT). The current movement of the storm is west-northwest at 8 mi/hr. The dashed orange line shows the hurricane's projected path; the solid blue line, the hurricane's actual path. Areas under a hurricane warning are in red. Those areas under a hurricane watch are in purple, while those areas under a tropical storm warning are in yellow. The light-shaded area denotes the track forecast cone, an estimate of the possible forecast error. On average, hurricanes will stay within the track forecast cone about 60% to 70% of the time.

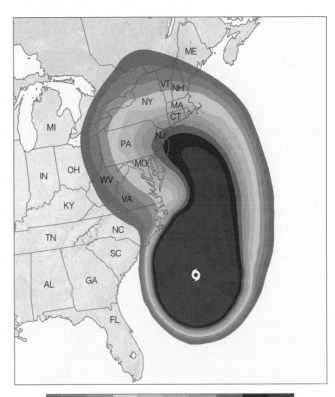

5% 10% 20% 30% 40% 50% 60% 70% 80% 90% 100%
Probability of tropical storm force winds

● **FIGURE 11.31** A map showing the likelihood that sustained winds from Hurricane Sandy will reach tropical storm force (34 knots or 39 mi/hr) at various locations, as predicted by NOAA's National Hurricane Center at 2 a.m. (EDT) on October 28, 2012, for the upcoming 120-hour period. The forecast was issued about 42 hours before Sandy made landfall on the evening of October 29 in New Jersey.

landfall. This new policy was implemented in 2013 as a result of Hurricane Sandy (see p. 341).

Because hurricane-force winds can extend a considerable distance on either side of where the storm is expected to make landfall, and because of uncertainty in the landfall forecast, hurricane warnings are issued for rather large swaths of coastline, about 500 km (310 mi) on average. Since the average swath of hurricane damage is normally about one-third this length, much of the area is "over-warned," and people in a warning area may feel that they are needlessly forced to evacuate. However, if the warning area were much narrower, it would increase the risk that hurricane damage might occur outside the warning area. Since hurricane impacts can cover a broad area, it is critical for people not to focus on the "skinny line" that denotes the best estimate of the track of the hurricane, but rather on the entire span of the hurricane warning. The *track forecast cone*, or "cone of uncertainty" (illustrated in Fig. 11.30) gives an estimate of the possible forecast error based on the five previous years of Atlantic tropical cyclone activity. Each hurricane will generally stay within the track forecast cone about 60 to 70 percent of the time.

Evacuation orders are given by local authorities, typically only for those low-lying coastal areas directly affected by the storm surge. People at higher elevations or farther from the coast are not usually requested to leave, in part because of the added traffic problems this would create, as was the case in Hurricane Rita (see p. 339). The time it takes to complete an evacuation puts a special emphasis on the timing and accuracy of the warning.

Hurricane Forecasting Techniques

As a potentially devastating storm such as Hurricane Katrina approaches land, will it intensify, maintain its strength, or weaken? Also, will it continue to move in the same direction and at the same speed? Such questions have challenged forecasters for decades. To forecast the intensity and movement of a hurricane, meteorologists use numerical weather prediction models, which are computer models that represent the hurricane and its environment in a greatly simplified manner.

Information from satellites, buoys, and reconnaissance aircraft (that deploy dropsondes into the eye of the storm) is fed into the models. The models then forecast the intensity and movement of the storm. There are a variety of forecast models, each one treating some aspect of the

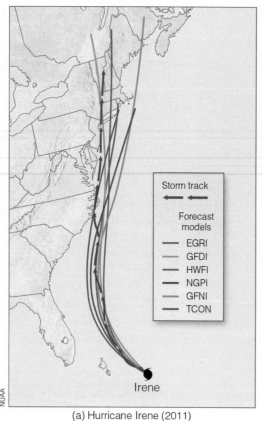

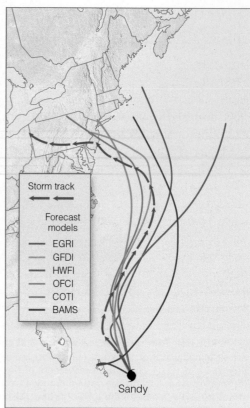

● **FIGURE 11.32** The paths projected by six numerical forecast models for (a) Hurricane Irene on August 24, 2011, three days before Irene struck the Outer Banks of North Carolina, and (b) Hurricane Sandy on October 25, 2012, four days before Sandy made landfall in New Jersey. The models are in close agreement for Hurricane Irene but vary greatly for Hurricane Sandy.

(a) Hurricane Irene (2011)

(b) Hurricane Sandy (2012)

atmosphere (such as evaporation of water from the ocean's surface) in a slightly different manner. Sometimes the models do not agree on where the storm will move and on how strong it will be. However, the best models in use today are more skillful than ever in their forecasting of hurricane movement.

The problem of different models forecasting different paths for the same hurricane has been addressed by using the method of *ensemble forecasting.* You may recall from Chapter 9 on p. 251 that an ensemble forecast is based on running several forecast models (or different simulations of the same model), each beginning with slightly different weather information. If the forecast models (or different versions of the same model) all agree that a hurricane will move in a particular direction, the forecaster will have confidence in making a forecast of the storm's movement. If, on the other hand, the models do not agree, then the forecaster will have to decide which model (or models) is most likely correct in forecasting the hurricane's track (see ● Fig. 11.32).

The use of ensemble forecasting along with better forecast models has helped raise the level of skill in forecasting hurricane paths. For example, in the 1970s, the projected position of a hurricane three days into the future was off by an average of around 700 km (435 mi). Today, the average error for the same forecast period has dropped to less than 240 km (150 mi).

Predicting hurricane intensity is a more challenging task. Forecasting of hurricane intensity showed little improvement for decades, but signs of progress began to appear by the early 2010s. To help predict hurricane intensity, forecasters traditionally used statistical models that compare the behavior of the present storm with that of similar tropical storms in the past. The results using these models have not been encouraging. More recently, forecasters have relied on *dynamical models,* some of which take into account the depth of warm ocean water in front of the storm's path to predict the storm's intensity. Recall from an earlier discussion (p. 325) that if the reservoir of warm water ahead of the storm is relatively shallow, ocean waves generated by the hurricane's wind will turbulently bring deeper, cooler water to the surface. The cooler water will cut off the storm's energy source, and the hurricane will weaken. On the other hand, should a deep layer of warm water exist ahead of the hurricane, cooler water will not be brought to the surface, and the storm will either maintain its strength or intensify, as long as other factors remain the same. So, knowing the depth of warm surface water ahead of the storm is important in predicting whether a hurricane will intensify or weaken. As new hurricane-prediction models with greater resolution are implemented, and as our understanding of the nature of hurricanes increases, forecasting hurricane intensification and movement should continue to improve.

Modifying Hurricanes

Because of the potential destruction and loss of lives that hurricanes can inflict, scientists have long thought about how the power of hurricanes might be blunted. During the 1960s, an experiment called Project STORMFURY carried out seeding on several hurricanes. The idea was to seed the clouds just outside the eyewall with just enough artificial ice nuclei so that the latent heat given off would stimulate cloud growth in this area of the storm, causing a new eyewall to replace the old one. This new, larger eyewall would have a weaker pressure gradient and weaker winds. (The process of eyewall replacement is described on p. 325.) Several storms were seeded in Project STORMFURY, with some encouraging results. However, it became evident that eyewall replacement also occurs naturally in many hurricanes that are not seeded, so it was impossible to know whether the seeding was responsible. No such attempts to modify hurricanes have taken place since the 1970s.

More recent studies using computer models have shown that tiny particles from air pollution may have the same effect as cloud seeding, helping the outer bands of a hurricane to strengthen at the expense of the inner core. However, if these particles made it into the inner core, they might actually cause strengthening, so trying this in an actual hurricane would be risky.

Another idea for weakening a hurricane is to place some form of oil (monomolecular film) on the water to retard the rate of evaporation and hence cut down on the release of latent heat inside the clouds. Some sailors, even in ancient times, would dump oil into the sea during stormy weather, claiming it reduced the winds around the ship. However, it would be much more difficult to maintain a coating of oil over the much larger area of ocean over which a hurricane might travel. One group of researchers has speculated that ocean spray has an effect on the winds of a hurricane. Their computer modeling suggests that the tiny spray reduces the friction between the wind and the sea surface. Consequently, with the same pressure gradient, the more ocean spray, the higher the winds. If this idea were to prove correct, limiting ocean spray from entering the air above might reduce the storm's winds. This concept has not yet been proven, however, so it is too soon to tell whether the ancient sailors were on to an idea that could actually help combat the terrible effects of hurricanes.

SUMMARY

Hurricanes are tropical cyclones with sustained winds of at least 64 knots (74 mi/hr) blowing counterclockwise about their centers in the Northern Hemisphere. A hurricane consists of a mass of organized thunderstorms that spiral in toward the extreme low pressure of the storm's eye. The most intense thunderstorms, the heaviest rain, and the highest winds occur outside the eye, in the region known as the eyewall. In the eye itself, the air is warm, winds are light, and skies may be broken or overcast.

Hurricanes (and all tropical cyclones) are born over warm tropical waters where the air is humid, surface winds converge, and thunderstorms become organized in a region of weak upper-level winds and weak vertical wind shear. Surface convergence may occur along the ITCZ, on the eastern side of a tropical wave, or along a front that has moved into the tropics from higher latitudes. If the disturbance becomes more organized, it becomes a tropical depression. If central pressures drop and surface winds increase, the depression becomes a tropical storm. At this point, the storm is given a name. Some tropical storms continue to intensify into full-fledged hurricanes, as long as they remain over warm water and are not disrupted either by strong vertical wind shear or by a large landmass.

The energy source that drives the hurricane comes primarily from the warm tropical oceans and from the release of latent heat. A hurricane is like a heat engine in that energy for the storm's growth is taken in at the surface in the form of sensible and latent heat, converted to kinetic energy in the form of winds, then lost at the cloud tops through radiational cooling.

The easterly winds in the tropics usually steer hurricanes westward. In the Northern Hemisphere, most storms then gradually swing northwestward around the subtropical high. If the storm moves into middle latitudes, the prevailing westerlies steer it northeastward. Because hurricanes derive their energy from the warm surface water and from the latent heat of condensation, they tend to dissipate rapidly when they move over cold water or over a large mass of land, where surface friction causes their winds to decrease and flow into their centers.

Although the high winds of a hurricane can inflict a great deal of damage, it is usually the huge waves and the flooding associated with the storm surge that cause the most destruction and loss of life. The Saffir-Simpson Hurricane Wind Scale was developed to estimate the potential destruction that a hurricane can cause. Computer forecast models have led to gradual improvement in our ability to predict the behavior of a hurricane several days in advance.

KEY TERMS

The following terms are listed (with corresponding page numbers) in the order they appear in the text. Define each. Doing so will aid you in reviewing the material covered in this chapter.

streamlines, 320
tropical wave (easterly wave), 320
hurricane, 320
typhoon, 320
tropical cyclone, 320
eye (of hurricane), 321
eyewall, 321
trade wind inversion, 324
eyewall replacement, 325
tropical depression, 325

tropical storm, 325
hurricane hunter, 326
Ekman transport, 332
storm surge, 333
Saffir-Simpson Scale, 334
super typhoon, 334
Saffir-Simpson Hurricane Wind Scale, 335
super typhoon, 334
hurricane watch, 343
hurricane warning, 343

QUESTIONS FOR REVIEW

1. What is a tropical (easterly) wave? How do these waves generally move in the Northern Hemisphere? Are showers found on the eastern or western side of the wave?
2. Why are streamlines, rather than isobars, used on surface weather maps in the tropics?
3. What is the name given to a hurricane-like storm that forms over the western North Pacific Ocean?
4. Describe the horizontal and vertical structure of a hurricane.
5. Why are skies often clear or partly cloudy in a hurricane's eye?
6. What conditions at the surface and aloft are necessary for hurricane development?
7. List three "triggers" that help in the initial stage of hurricane development.
8. (a) Hurricanes are sometimes described as a heat engine. What is the "fuel" that drives the hurricane?
   (b) What determines the maximum strength (the highest winds) that the storm can achieve?
9. Would it be possible for a hurricane to form over land? Explain.
10. If a hurricane in the Northern Hemisphere is moving westward at 10 knots, will the strongest winds be on its northern or southern side? Explain. If the same hurricane turns northward, will the strongest winds be on its eastern or western side?
11. What factors tend to weaken hurricanes?
12. Distinguish among a tropical depression, a tropical storm, and a hurricane.

13. In what ways is a hurricane different from a mid-latitude cyclone? In what ways are these two systems similar?

14. Why do most hurricanes move westward over tropical waters?

15. If the high winds of a hurricane are not responsible for inflicting the most damage, what is?

16. Most hurricane-related deaths are due to what?

17. Explain how a storm surge forms. How does it inflict damage in hurricane-prone areas?

18. Hurricanes are given names when the storm is in what stage of development?

19. When Hurricane Andrew moved over south Florida during August 1992, what was it that caused the relatively small areas of extreme damage?

20. As Hurricane Katrina moved toward the Louisiana coast, it underwent eyewall replacement. What actually happened to the eyewall during this process?

21. How do meteorologists forecast the intensity and paths of hurricanes?

22. How does a hurricane watch differ from a hurricane warning?

23. Why were some hurricanes once seeded with silver iodide, and why are they not seeded today?

24. Give two reasons why hurricanes are more likely to strike New Jersey than Oregon.

QUESTIONS FOR THOUGHT AND EXPLORATION

1. A hurricane just off the coast of northern Florida is moving northeastward, parallel to the eastern seaboard. Suppose that you live in North Carolina along the coast.

    (a) How will the surface winds in your area change direction as the hurricane's center passes due *east* of you? Illustrate your answer by making a sketch of the hurricane's movement and the wind flow around it.

    (b) If the hurricane passes east of you, the strongest winds would most likely be blowing from which direction? Explain your answer. (Assume that the storm does not weaken as it moves northeastward.)

    (c) The lowest sea-level pressure would most likely occur with which wind direction? Explain.

2. Give several reasons why a hurricane that once began to weaken can strengthen again.

3. Why are North Atlantic hurricanes more apt to form in October than in May?

4. Explain why the ocean surface water temperature is usually cooler after the passage of a hurricane. (Hint: The answer is not because the hurricane extracts heat from the water.)

5. Suppose this year five tropical storms develop into full-fledged hurricanes over the North Atlantic Ocean. Would the name of the third hurricane begin with the letter C? Explain.

6. You are in Darwin, Australia (on the north shore), and a hurricane approaches from the north. Where would the highest storm surge be, to the east or west? Explain

7. A friend tells you that his grandparents experienced a terrible hurricane named Frederic in 1975 while living on the Gulf Coast. How do you know there is something wrong with your friend's story even before you consult the climatological records?

GLOBAL **GEOSCIENCE** WATCH Go to the Statistics section of the Global Environment Watch: Meteorology portal and examine the charts "Cyclone Nargis: Total Affected Population by Township" and "Cyclone Nargis: Percent Affected Population by Township." What are the main differences between the two charts? How might you go about combining information from these two charts to assess which of the four districts were most and least affected?

ONLINE RESOURCES

 Visit www.cengagebrain.com to view additional resources, including video exercises, practice quizzes, an interactive eBook, and more.

CHAPTER 12

Global Climate

Contents

The climate is unbearable . . . At noon today the highest temperature measured was −33°C. We really feel that it is late in the season. The days are growing shorter, the sun is low and gives no warmth, katabatic winds blow continuously from the south with gales and drifting snow. The inner walls of the tent are like glazed parchment with several millimeters thick ice-armour. . . . Every night several centimeters of frost accumulate on the walls, and each time you inadvertently touch the tent cloth a shower of ice crystals falls down on your face and melts. In the night huge patches of frost from my breath spread around the opening of my sleeping bag and melt in the morning. The shoulder part of the sleeping bag facing the tent-side is permeated with frost and ice, and crackles when I roll up the bag . . . For several weeks now my fingers have been permanently tender with numb fingertips and blistering at the nails after repeated frostbites. All food is frozen to ice and it takes ages to thaw out everything before being able to eat. At the depot we could not cut the ham, but had to chop it in pieces with a spade. Then we threw ourselves hungrily at the chunks and chewed with the ice crackling between our teeth. You have to be careful with what you put in your mouth. The other day I put a piece of chocolate from an outer pocket directly in my mouth and promptly got frostbite with blistering of the palate.

Ove Wilson (quoted in David M. Gates, *Man and His Environment*)

Our opening comes from a report by Norwegian scientists on their encounter with one of nature's cruelest climates—that of Antarctica. Their experience illustrates the profound effect that climate can have on even ordinary events, such as eating a piece of chocolate. Though we may not always think about it, climate profoundly affects nearly everything in the middle latitudes, too. For instance, it influences our housing, clothing, the shape of landscapes, agriculture, how we feel and live, and even where we reside, as most people will choose to live on a sunny hillside rather than in a cold, dark, and foggy river basin. Entire civilizations have flourished in favorable climates and have moved away from, or perished in, unfavorable ones. We learned early in this text that *climate* is the average of the day-to-day weather over a long duration. But the concept of climate is much larger than this, for it encompasses, among other things, the daily and seasonal extremes of weather within specified areas.

When we speak of climate, then, we must be careful to specify the spatial location we are talking about. For example, the Chamber of Commerce of a rural town may boast that its community has mild winters with air temperatures seldom below freezing. This may be true several feet above the ground in an instrument shelter, but near the ground the temperature may drop below freezing on many winter nights.

Microclimate typically refers to the climate very near the ground in a small zone (sometimes as small as a few square feet) where the surface properties are consistent, such as a city park, an airport tarmac, or a forest clearing. Because a much greater extreme in daily air temperatures exists near the ground than several feet above, the microclimate for small plants is far more harsh than the thermometer in an instrument shelter would indicate.

When we examine the climate of a small area of Earth's surface, we are looking at the **mesoclimate**, which includes the conditions across regions such as forests, valleys, beaches, and towns. The climate of an even bigger area, such as a large state or a small nation, is called **macroclimate**. The climate extending over the entire Earth is often referred to as **global climate**.

In this chapter, we will concentrate on the larger scales of climate. We will begin with the factors that regulate global climate; then we will discuss how climates are classified. Finally, we will examine the different types of climate. (In Chapter 13, we will explore how both natural processes and human factors can result in climate change.)

A World with Many Climates

The world is rich in climatic types. From the teeming tropical jungles to frigid polar "wastelands," there seems to be an almost endless variety of climatic regions. The factors that produce the climate in any given place—the climatic controls—are the same that produce our day-to-day weather. Briefly, the controls are the

1. intensity of sunshine and its variation with latitude
2. distribution of land and water
3. ocean currents
4. prevailing winds
5. positions of high- and low-pressure areas
6. mountain barriers
7. altitude

Human settlements can also serve as a climate control, such as when a large forest is replaced with an open pasture. In the following sections we will focus on large-scale controls that are unrelated to human alteration of the environment. We can ascertain the effect these controls have on climate by observing the global patterns of two weather elements—temperature and precipitation.

GLOBAL TEMPERATURES   ● Figure 12.1 shows mean annual temperatures for the world. To eliminate the distorting effect of topography, the temperatures are corrected to sea level.* Notice that in both hemispheres the isotherms are oriented east-west, reflecting the fact that locations at the same latitude receive nearly the same amount of solar energy. In addition, the annual sunlight that each latitude receives decreases from low to high latitude; hence, annual temperatures tend to decrease from equatorial toward polar regions.**

The bending of the isotherms along the coastal margins is due in part to the unequal heating and cooling properties of land and water. Recall from Chapter 2 that oceans take longer than land areas to heat up in summer and cool down in winter, because water has a higher *specific heat capacity* than soil, meaning that it takes more energy to increase the temperature of water than it does to increase the temperature of soil by the same amount. The bending of the isotherms is also related to ocean currents and upwelling. For example, along the west coast of North and South America, ocean currents transport cool water equatorward. In addition to this, the wind in both regions blows toward the equator, parallel to the coast. This situation favors upwelling of cold water (see Chapter 7, p. 198), which cools the coastal margins. In the area of the eastern North Atlantic Ocean (north of 40°N), the poleward bending of the isotherms is due to the Gulf Stream and the North Atlantic Drift, which carry warm water northward.

---

*This correction is made by adding to each station above sea level an amount of temperature that would correspond to the normal (standard) temperature lapse rate of 6.5°C per 1000 m (3.6°F per 1000 ft).

**Average global temperatures for January and July are given in Figs. 3.14 and 3.15, respectively, on p. 66.

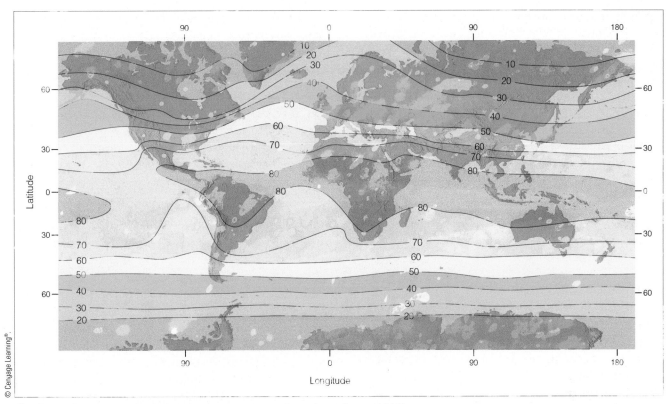

● FIGURE 12.1 Average annual sea-level temperatures throughout the world (°F).

The fact that landmasses heat up and cool off more quickly than do large bodies of water means that variations in temperature between summer and winter will be far greater over continental interiors than along the west coastal margins of continents. By the same token, the climates of interior continental regions will be more extreme, as they have (on the average) higher summer temperatures and lower winter temperatures than their west-coast counterparts. In fact, west coast climates are typically quite mild for their latitude.

When calculated across an entire year, the highest mean temperatures on Earth generally occur in the tropics, but the the hottest individual readings are observed in the subtropical deserts of the Northern Hemisphere. In summer, the subsiding air associated with the subtropical anticyclones produces generally clear skies and low humidity, and a high sun beating down upon a relatively barren landscape produces scorching heat.

The lowest annual mean temperatures occur over large landmasses at high latitudes. The coldest areas in the Northern Hemisphere are found in the interior of Siberia and Greenland, whereas the coldest area of the world is the Antarctic. During part of the year, the sun is below the horizon; when it is above the horizon, it is low in the sky and its rays do not effectively warm the surface. Consequently, most of Greenland and Antarctica remain in snow- and ice-covered year-round. The snow and ice

reflect perhaps 80 percent of the sunlight that reaches the surface. Much of the unreflected solar energy is used to transform the ice and snow into water vapor. The relatively dry air and the Antarctic's high elevation permit rapid radiational cooling during the dark winter months, producing extremely cold surface air.

The Antarctic continent covers the area around the South Pole and remains snow and ice-covered all year long, whereas the North Pole is surrounded by ocean. In summer a vast amount of arctic sea ice melts, allowing sunlight to be absorbed and mixed into the Arctic Ocean. This added heat keeps the average temperature of the Arctic higher than that of the Antarctic. The

DID YOU KNOW?

Even "summers" in the Antarctic can be brutal. In 1912, during the Antarctic summer, Robert Scott of Great Britain not only lost the race to the South Pole to Norway's Roald Amundsen, but perished in a blizzard trying to return to his starting point. Temperature data taken by Scott and his crew showed that the "summer" of 1912 was unusually cold, with air temperatures remaining below 30°F for nearly a month. These exceptionally low temperatures, which were colder than the team expected, eroded the men's health and created an increase in frictional drag on the sleds the men were pulling. Just before Scott's death, he wrote in his journal that "no one in the world would have expected the temperatures and surfaces which we encountered at this time of year."

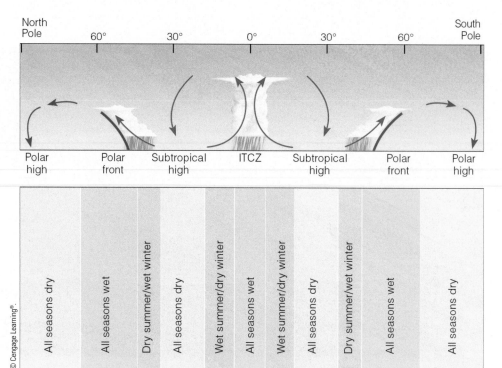

● FIGURE 12.2 A vertical cross section along a line running north to south illustrates the main global regions of rising and sinking air and how each region influences precipitation.

GLOBAL PRECIPITATION Appendix D, p. 470, shows the worldwide general pattern of annual precipitation, which varies from place to place. There are, however, certain regions that stand out as being wet or dry. For example, equatorial regions are typically wet, while the subtropics and the polar regions are relatively dry. The global distribution of precipitation is closely tied to the general circulation of winds described in Chapter 7 and to the distribution of mountain ranges and high plateaus.

● Figure 12.2 shows in simplified form how the general circulation influences the north-to-south distribution of precipitation to be expected on a uniformly water-covered Earth. Precipitation is most abundant where the air rises, least abundant where it sinks. Hence, one expects a great deal of precipitation in the tropics and along the polar front, and little near subtropical highs and at the poles. Let's look at this in more detail.

In tropical regions, the trade winds converge along the Intertropical Convergence Zone (ITCZ), producing rising air, towering clouds, and heavy precipitation all year long. Poleward of the equator, near latitude 30°, the sinking air of the subtropical highs produces a "dry belt" around the globe (although we will see that not all locations at this latitude have dry climates). The Sahara Desert of North Africa and the Mojave Desert of southwestern North America lie within this belt. Here, annual rainfall is exceedingly light and varies considerably from year to

year. Because the major wind belts and pressure systems shift with the season—northward in July and southward in January—the area between the rainy tropics and the dry subtropics is influenced by both the ITCZ and the subtropical highs.

In the cold air of the polar regions there is little moisture, so there is little precipitation. Winter storms drop light, powdery snow that remains on the ground for a long time because of the low evaporation rates. In summer, a ridge of high pressure tends to block storm systems that would otherwise travel into the area; hence, precipitation in polar regions is meager in all seasons.

There are exceptions to this idealized pattern. For example, in middle latitudes the migrating position of the subtropical anticyclones also has an effect on the west-to-east distribution of precipitation. The sinking air associated with these systems is more strongly developed on their eastern side. Hence, the air along the eastern side of an anticyclone tends to be more stable; it is also drier, as cooler air moves equatorward because of the circulating winds around these systems. In addition, along coastlines, cold upwelling water cools the surface air even more, adding to the air's stability. Consequently, in summer, when the Pacific high moves to a position centered off the California coast, a strong, stable subsidence inversion forms above coastal regions. With the strong inversion and the fact that the anticyclone tends to steer storms to the north, central and southern California areas experience little, if any, rainfall during the summer months.

On the western side of subtropical highs, the air is less stable and more moist, as warmer air moves

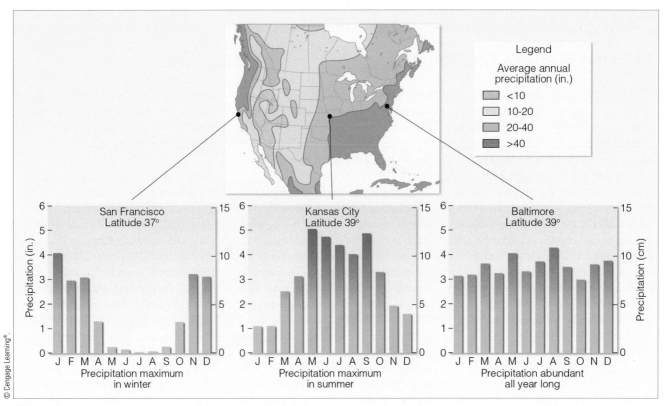

● **FIGURE 12.3** Average annual precipitation across North America, along with variation in annual precipitation for the reporting stations at major airports at three Northern Hemisphere cities (averaged for the period 1981–2010).

poleward. In summer, over the North Atlantic, the Bermuda high pumps moist tropical air northward from the Gulf of Mexico into the eastern two-thirds of the United States. The humid air is conditionally unstable to begin with, and by the time it moves over the heated ground, it becomes even more unstable. If conditions are right, the moist air will rise and condense into cumulus clouds, which may build into towering thunderstorms. Along the Gulf Coast, subtropical cities such as New Orleans, Louisiana, receive more than 127 cm (50 in.) of precipitation in a typical year (much more rain than in cities at the same latitude in other locations). In winter, the subtropical North Pacific high moves south, allowing storms traveling across the ocean to penetrate the western states, bringing much-needed rainfall to California after a long, dry summer. The Bermuda high also moves south in winter. Across much of the United States, intense winter storms develop and travel eastward, frequently dumping heavy precipitation as they go. Usually, however, the heaviest precipitation is concentrated in the eastern states, as moisture from the Gulf of Mexico moves northward ahead of these systems. Therefore, cities on the plains typically receive more rainfall in summer, and those on the west coast have maximum precipitation in winter. Cities in the Midwest and East usually have abundant precipitation all year long. ● Fig. 12.3 shows the average annual precipitation across

North America as well as the contrasts in seasonal precipitation for a west coast city (San Francisco), a central plains city (Kansas City), and an eastern city (Baltimore).

Mountain ranges disrupt the idealized pattern of global precipitation (1) by promoting convection (because their slopes are warmer than the surrounding air) and (2) by forcing air to rise along their windward slopes (*orographic uplift*). Consequently, the windward side of mountains tends to be "wet." As air descends and warms along the leeward side, there is less likelihood of clouds and precipitation. Thus, the leeward (downwind) side of mountains tends to be "dry." As Chapter 5 points out, a region on the leeward side of a mountain where precipitation is noticeably less is called a *rain shadow*.

A good example of the rain shadow effect occurs in the northwestern part of Washington State. Situated on the western side at the base of the Olympic Mountains, the Hoh rainforest annually receives an average of 340 cm (134 in.) of precipitation (see ● Fig. 12.4). On the eastern (leeward) side of this range, only about 100 km (62 mi) from the Hoh rainforest in the city of Port Townsend, the mean annual precipitation is less than 48 cm (19 in.), and irrigation is necessary to grow certain crops. ● Figure 12.5 shows a classic example of how topography produces several rain shadow effects. (Additional information on precipitation extremes is given in Focus section 12.1.)

Extreme Wet and Dry Regions

Most of the "rainiest" places in the world are located on the windward side of mountains. For example, Mount Waialeale on the island of Kauai, Hawaii, has the greatest annual average rainfall in the United States: 1164 cm (458 in.). Mawsynram, on the crest of the southern slopes of the Khasi Hills in northeastern India, is often considered the wettest place in the world as it receives an average of 1187 cm (467 in.) of rainfall each year, the majority of which falls during the summer monsoon, between April and October. Cherrapunji, which is only about 10 miles from Mawsynram, holds the greatest 12-month rainfall total of 2647 cm (1042 in.), and once received 249 cm (98 in.) in just 5 days.

Record rainfall amounts are often associated with tropical storms. On the island of La Réunion (about 650 km east of Madagascar in the Indian Ocean), a tropical cyclone dumped 114 cm (45 in.) of rain on

Foc-Foc in 12 hours. Heavy rains of short duration often occur with severe thunderstorms that move slowly or stall over a region. On July 4, 1956, 3 cm (1.2 in.) of rain fell from a thunderstorm on Unionville, Maryland, in one minute.

Snowfalls tend to be heavier where cool, moist air rises along the windward slopes of mountains. One of the snowiest places in North America is located at the Paradise Ranger Station in Mt. Rainier National Park, Washington. Situated at an elevation of 1646 m (5400 ft) above sea level, this station receives an average 1758 cm (692 in.) of snow annually, and holds the world's record 12-month snowfall total of 3109 cm (1224 in.), which fell between February 1971 and February 1972. A record seasonal snowfall total for North America of 2896 cm (1140 in.) fell on Mt. Baker Ski Area, Washington, during the winter of 1998–1999. The largest snow

accumulations on Earth are believed to occur in the Japanese Alps, where a snow depth of 1181 cm (465 in.) was reported on Mt. Ibuki on February 14, 1927.

As we noted earlier, the driest regions of the world lie in the frigid polar region, the leeward side of mountains, and in the belt of subtropical high pressure, between 15° and 30° latitude. Arica in northern Chile holds the world record for lowest annual rainfall—0.08 cm (0.03 in.) per year, averaged over a period of 59 years—and for the longest period without measurable rainfall (more than 14 years, from October 1903 through January 1918). In the United States, the driest regions are found in the Desert Southwest, the southern San Joaquin Valley of California, and Death Valley in Southern California, which averages only 5.9 cm (2.36 in.) of precipitation annually. ● Figure 1 gives additional information on world precipitation records.

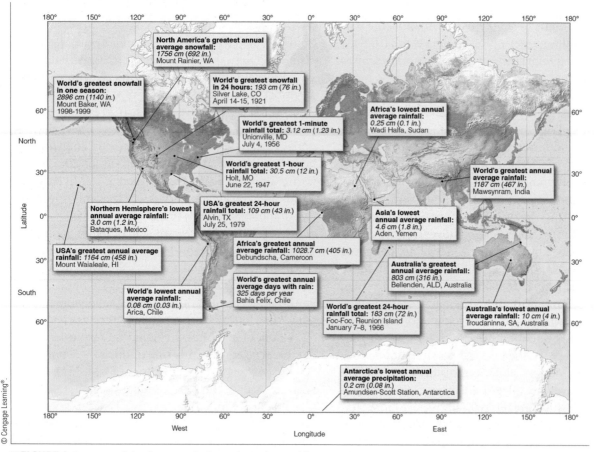

● FIGURE 1 Some precipitation records throughout the world.

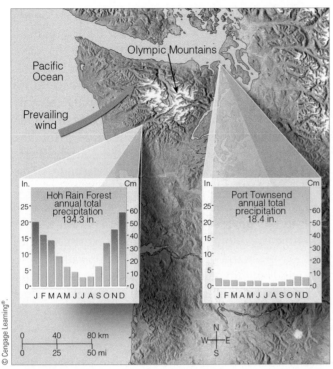

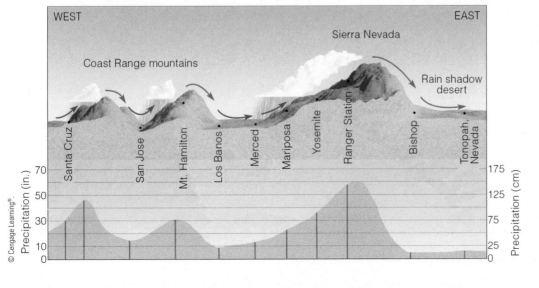

Hoh Rain Forest
annual total
precipitation
134.3 in.

Port Townsend
annual total
precipitation
18.4 in.

● **FIGURE 12.4** The effect of the Olympic Mountains in Washington State on average annual rainfall.

BRIEF REVIEW

Before going on to the section on climate classification, here is a brief review of some of the facts covered so far:

- Climate controls are the factors that govern the climate of any given region.

- The hottest places on Earth are generally found in the subtropical deserts of the Northern Hemisphere, where clear skies and sinking air, coupled with low humidity and a high summer sun beating down upon a relatively barren landscape, produce extreme heat. The warmest annual average temperatures are found in the tropics, where temperatures vary little through the year.

- The coldest places on Earth are generally found in the interior of high-latitude landmasses. The coldest areas of the Northern Hemisphere are found in the interior of Siberia and Greenland, whereas the coldest area of the world is the Antarctic.

- The wettest places in the world tend to be located on the windward side of mountains where warm, humid air rises upslope. On the downwind (leeward) side of a mountain there often exists a "dry" region, known as a *rain shadow*.

Climatic Classification— The Köppen System

The climatic controls interact to produce such a wide array of different climates that no two places experience exactly the same climate. However, the similarity of climates within a given area allows us to divide Earth into climatic regions.

A widely used classification of world climates based on the annual and monthly averages of temperature and precipitation was devised by the German scientist Waldimir Köppen (1846–1940). In the absence of adequate observing stations throughout the world, Köppen related the distribution and type of native vegetation to the various climates. In this way, climatic boundaries could be approximated where no climatological data were available. Initially published in 1918, the **Köppen classification system** has since been modified and refined by a number of researchers.

Köppen's scheme identifies five major climatic types, each designated by a capital letter:

A   *Tropical moist climates:* All months have an average temperature above 18°C (64°F). Since all months are warm, there is no real winter season.

B   *Dry climates:* Deficient precipitation most of the year. Potential evaporation and transpiration exceed precipitation.

C   *Moist mid-latitude climates with mild winters:* Warm-to-hot summers with mild winters. The

● **FIGURE 12.5** The effect of topography on average annual precipitation along a line running from the Pacific Ocean through central California into western Nevada.

average temperature of the coldest month is below 18°C (64°F) and above −3°C (27°F).

D *Moist mid-latitude climates with severe winters:* Warm summers and cold winters. The average temperature of the warmest month exceeds 10°C (50°F), and the coldest monthly average drops below −3°C (27°F).

E *Polar climates:* Extremely cold winters and summers. The average temperature of the warmest month is below 10°C (50°F). Since all months are cold, there is no real summer season.

In mountainous country, where rapid changes in elevation bring about sharp changes in climatic type, delineating the climatic regions is impossible. These regions are designated by the letter *H*, for *highland climates*.

● Figure 12.6 gives a simplified overview of the major climate types throughout the world, according to Köppen's system. Superimposed on the map are some of the climatic controls. These include the average annual positions of the semipermanent high- and low-pressure

areas, the average position of the Intertropical Convergence Zone in January and July, the major mountain ranges and deserts of the world, and some of the major ocean currents. Notice how the climatic controls impact the climate in different regions of the world. As we would expect, because of differences in the intensity and amount of solar energy, polar climates are found at high latitudes and tropical climates at low latitudes. Dry climates tend to be located on the downwind side of major mountain chains and near 30° latitude, where the subtropical highs (with their sinking air) are found. Climates with more moderate winters (C climates) tend to be equatorward of those with severe winters (D climates). Along the west coast of North America and Europe, warm ocean currents and prevailing westerly winds modify the climate such that coastal regions experience much milder winters than do regions farther inland.

Keep in mind that within the Köppen system each major climatic group contains subgroups that describe special regional characteristics, such as seasonal changes

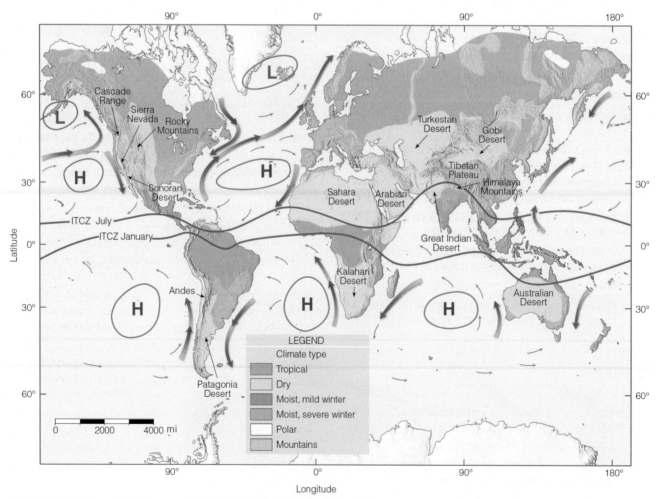

● **FIGURE 12.6** A simplified overview of the major climate types according to Köppen, along with some of the climatic controls. The large Hs and Ls on the map represent the average position of the semipermanent high- and low-pressure areas. The solid red lines show the average position of the Intertropical Convergence Zone (ITCZ) in January and July. The ocean currents in red are warm; those in blue are cold. The major mountain ranges and deserts of the world also are included.

in temperature and precipitation. The complete Köppen climatic classification system, including the criteria for the various subgroups, is given in Appendix E on p. 472.

Köppen's system has been criticized primarily because his boundaries (which relate vegetation to monthly temperature and precipitation values) do not correspond to the natural boundaries of each climatic zone. In addition, the Köppen system implies that there is a sharp boundary between climatic zones, when in reality there is a gradual transition. The Köppen system has been revised several times, most notably by the German climatologist Rudolf Geiger, who worked with Köppen on amending the climatic boundaries of certain regions. A popular modification of the Köppen system was developed by the American climatologist Glenn Trewartha, who redefined some of the climatic types and altered the climatic world map by putting more emphasis on the lengths of growing seasons and average summer temperatures.

Another system, created by American climatologist C. Warren Thornthwaite, uses an annual index that is related to monthly precipitation and to the potential monthly evapotranspiration,* since these are both factors critical for plant growth. The Thornthwaite system defines five major humidity provinces and their characteristic vegetations: rainforest, forest, grassland, steppe, and desert.

The Global Pattern of Climate

● Figure 12.7 gives a more detailed view of how the major climatic regions and subregions of the world are distributed, based mainly on the work of Köppen. We will first examine humid tropical climates in low latitudes and then we'll look at middle-latitude and polar climates. Bear in mind that each climatic region has many subregions of local climatic differences wrought by such factors as topography, elevation, and large bodies of water. Remember, too, that boundaries of climatic regions represent gradual transitions, not exact demarcations. Thus, the major climatic characteristics of a given region are best observed away from its periphery.

TROPICAL MOIST CLIMATES (GROUP A)

General characteristics: year-round warm temperatures (all months have a mean temperature above 18°C, or 64°F); abundant rainfall (typical annual average exceeds 150 cm, or 59 in.).

Extent: northward and southward from the equator to about latitude 15° to 25°.

Major types (based on seasonal distribution of rainfall): *tropical wet* (Af), *tropical monsoon* (Am), and *tropical wet* and *dry* (Aw).

At low elevations near the equator, in particular in the Amazon lowland of South America, the Congo River Basin of Africa, and the East Indies from Sumatra to New Guinea, high temperatures and abundant yearly rainfall combine to produce a dense, broadleaf, evergreen forest called a **tropical rainforest.** Here, many different plant species, each adapted to differing light intensity, present a crudely layered appearance of diverse vegetation. In the forest, little sunlight is able to penetrate through the thick crown cover to the ground. As a result, little plant growth is found on the forest floor. At the edge of the forest, however, or where a clearing has been made, abundant sunlight allows for the growth of tangled shrubs and vines, producing an almost impenetrable *jungle* (see ● Fig. 12.8, p. 360).

Within the **tropical wet climate** (Af),* seasonal temperature variations are small (normally less than 3°C) because the noon sun is always high and the number of daylight hours is relatively constant. However, there is a greater variation in temperature between day (average high about 32°C) and night (average low about 22°C) than there is between the warmest and coolest months. This is why people remark that winter comes to the tropics at night. The weather in a tropical wet climate is monotonous and sultry, with little change in temperature from one day to the next. Furthermore, almost every day, towering cumulus clouds form and produce heavy, localized showers by early afternoon. As evening approaches, the showers usually end and skies clear. Typical annual rainfall totals are greater than 150 cm (59 in.) and, in some cases, especially along the windward side of hills and mountains, the total may exceed 400 cm (157 in.).

The high humidity and cloud cover tend to keep maximum temperatures from reaching extremely high values. In fact, summer afternoon temperatures are normally higher in middle latitudes than here. Nighttime radiational cooling can produce saturation and, hence, a blanket of dew and—occasionally—fog covers the ground.

An example of a station with a tropical wet climate (Af) is Iquitos, Peru (see ● Fig. 12.9). Located near the equator (latitude 4°S), in the low basin of the upper Amazon River, Iquitos has an average annual temperature of 25°C (77°F), with an annual temperature range of only 2.2°C (4°F).** Notice also that monthly rainfall totals are more variable than monthly temperatures. This is due primarily to the migrating position of the Intertropical Convergence Zone (ITCZ) and its associated wind-flow patterns. Although

---

Evapotranspiration refers to the evaporation from soil and transpiration of plants.

---

*The tropical wet climate is also known as the *tropical rainforest climate.*

**Recall from Chapter 3 that annual temperature range is the difference in temperature between the average warmest month and the average coldest month.

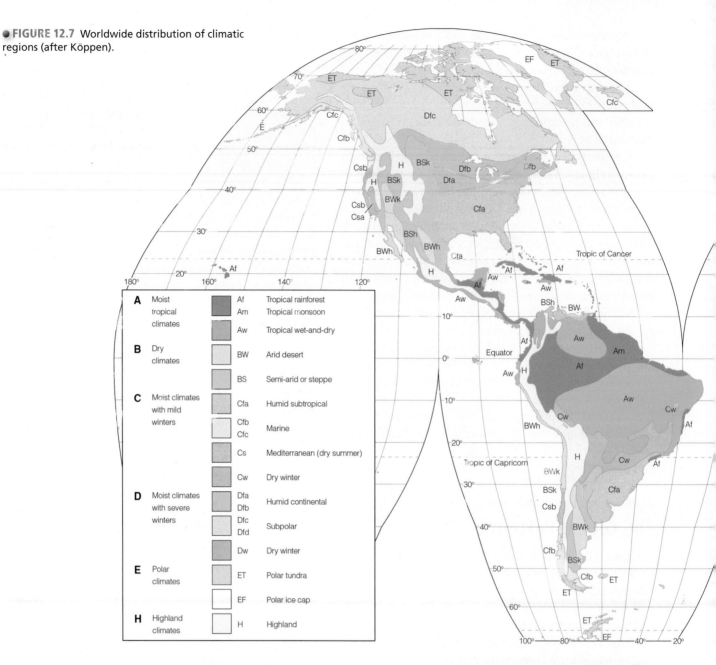

● FIGURE 12.7 Worldwide distribution of climatic regions (after Köppen).

A	Moist tropical climates		Af	Tropical rainforest
			Am	Tropical monsoon
			Aw	Tropical wet-and-dry
B	Dry climates		BW	Arid desert
			BS	Semi-arid or steppe
C	Moist climates with mild winters		Cfa	Humid subtropical
			Cfb / Cfc	Marine
			Cs	Mediterranean (dry summer)
			Cw	Dry winter
D	Moist climates with severe winters		Dfa / Dfb	Humid continental
			Dfc / Dfd	Subpolar
			Dw	Dry winter
E	Polar climates		ET	Polar tundra
			EF	Polar ice cap
H	Highland climates		H	Highland

● FIGURE 12.8 Tropical rainforest near Iquitos, Peru. (Climatic information for this region is presented in Fig. 12.9.)

Gregory G. Dimijian, M.D./Science Source

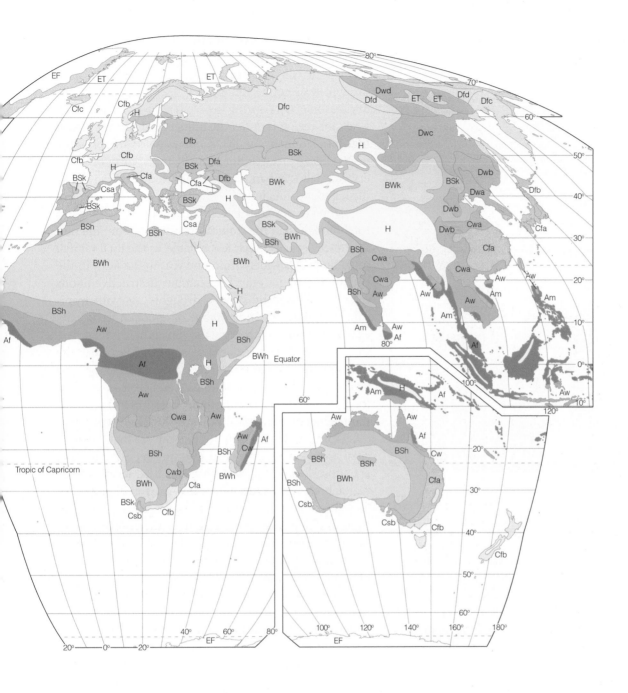

monthly precipitation totals vary considerably, the average for each month exceeds 6 cm, and consequently no month is considered deficient of rainfall.

Take a minute and look again at Fig. 12.8. From the photo, one might think that the soil beneath the forest's canopy would be excellent for agriculture, but actually, this is not true. As heavy rain falls on the soil, the water works its way downward, removing nutrients in a process called *leaching*. Strangely enough, many of the nutrients needed to sustain the lush forest actually come from dead trees that decompose. The roots of the living trees absorb this matter before the rains leach it away. When the forests are cleared for agricultural purposes, or for the timber, what is left is a thick red soil called **laterite.** When exposed to the intense sunlight of the tropics, the soil may harden

into a bricklike consistency, making cultivation almost impossible.

Köppen classified those tropical wet regions where the monthly precipitation totals drop below 6 cm for perhaps one or two months as **tropical monsoon climates** (Am). In the Am climate, yearly rainfall totals are similar to those of the tropical wet climate, usually exceeding 150 cm a year. Because the dry season is brief and copious rains fall throughout the rest of the year, there is sufficient soil moisture to maintain the tropical rainforest through the short dry period. Tropical monsoon climates can be seen in Fig. 12.7 along the coasts of Southeast Asia and southwestern India and in northeastern South America.

Poleward of the tropical wet region, total annual rainfall diminishes, and there is a gradual transition from the

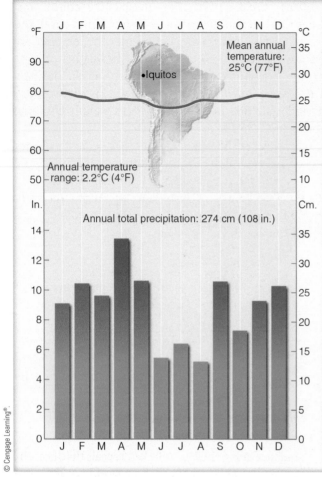

● **FIGURE 12.9** Temperature and precipitation data for Iquitos, Peru, latitude 4°S. A station with a tropical wet climate (Af). (This type of diagram is called a *climograph*. It shows monthly mean temperatures with a solid red line and monthly mean precipitation with bar graphs.)

tropical wet climate to the **tropical wet-and-dry climate** (Aw), where a distinct dry season prevails. Even though the annual precipitation usually exceeds 100 cm, the dry season, where the monthly rainfall is less than 6 cm (2.4 in.),

lasts for more than two months. Because tropical rainforests cannot survive this "drought," the jungle gradually gives way to tall, coarse **savanna grass,** scattered with low, drought-resistant deciduous trees (see ● Fig. 12.10). The dry season occurs during the winter, the low sun period, when the region is under the influence of the subtropical highs. In summer, the ITCZ moves poleward, bringing with it heavy precipitation, usually in the form of showers. Rainfall is enhanced by shallow lows moving slowly throughout the region.

Tropical wet-and-dry climates not only receive less total rainfall than the tropical wet climates, but the rain that does occur is much less reliable, as the total rainfall often fluctuates widely from one year to the next. In the course of a single year, for example, destructive floods may be followed by serious droughts. As with tropical wet regions, the daily range of temperature usually exceeds the annual range, but the climate here is much less monotonous. There is a cool season in winter when the maximum temperature averages 30°C to 32°C (86°F to 90°F). At night, the low humidity and clear skies allow for rapid radiational cooling and, by early morning, minimum temperatures typically drop to 20°C (68°F) or below.

From Fig. 12.7, p. 360-361, we can see that the principal areas having a tropical wet-and-dry climate (Aw) are those located in western Central America, in the region both north and south of the Amazon Basin (South America), in south-central and eastern Africa, in parts of India and Southeast Asia, and in northern Australia. In many areas (especially within India and Southeast Asia), the marked variation in precipitation is associated with the *monsoon*—the seasonal reversal of winds.

As we saw in Chapter 7, the Asian monsoon circulation stems in part from differential heating of landmasses and oceans. During winter in the Northern Hemisphere, winds blow outward, away from a cold, shallow high-pressure area centered over continental Siberia. These

● **FIGURE 12.10** Baobob and acacia trees illustrate typical trees of the East African grassland savanna, a region with a tropical wet-and-dry climate (Aw).

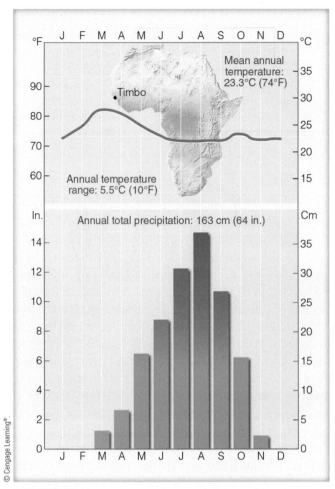

● **FIGURE 12.11** Climatic data for Timbo, Guinea, latitude 11°N. A station with a tropical wet-and-dry climate (Aw).

downslope, relatively dry northeasterly winds from the interior provide India and Southeast Asia with generally fair weather and the dry season. In summer, the wind-flow pattern reverses as air flows into a developing thermal low over the continental interior. The humid air from the water rises and condenses, resulting in heavy rain and the wet season. (A more detailed look at the winter and summer monsoon is shown in Fig. 7.20 on p. 186.)

An example of a station with a tropical wet-and-dry climate (Aw) is given in ● Fig. 12.11. Located at latitude 11°N in west Africa, Timbo, Guinea, receives an annual average 163 cm (64 in.) of rainfall. Notice that the rainy season is during the summer, when the ITCZ has migrated to its most northern position. Note also that practically no rain falls during the months of December, January, and February, when the region comes under the domination of the subtropical high-pressure area and its sinking air.

The monthly temperature patterns at Timbo are characteristic of most tropical wet-and-dry climates. As spring approaches, the noon sun is slightly higher, and the more intense sunshine produces greater surface heating and higher afternoon temperatures—usually above 32°C (90°F) and occasionally above 38°C

(100°F)—creating hot, dry, desert-like conditions. After this brief hot season, a persistent cloud cover and the evaporation of rain tends to lower the temperature during the summer. The warm, muggy weather of summer often resembles that of the tropical wet climate (Af). The rainy summer is followed by a warm, relatively dry period, with afternoon temperatures usually climbing above 30°C (86°F).

Poleward of the tropical wet-and-dry climate, the dry season becomes more severe. Clumps of trees are more isolated and the grasses dominate the landscape. When the potential annual water loss through evaporation and transpiration exceeds the annual water gain from precipitation, the climate is described as dry.

DRY CLIMATES (GROUP B)

General characteristics: deficient precipitation most of the year; potential evaporation and transpiration exceed precipitation.

Extent: the subtropical deserts extend from roughly 20° to 30° latitude in large continental regions of the middle latitudes, often surrounded by mountains.

Major types: arid (BW)—the "true desert"—and semi-arid (BS).

A glance at Fig. 12.7, p. 360-361, reveals that, according to Köppen, the dry regions of the world occupy more land area (about 26 percent) than any other major climatic type. These dry regions are deficient in water, meaning that potential annual loss of water through evaporation is greater than the annual water gained through precipitation. Thus, classifying a climate as dry depends not only on precipitation totals but also on temperature, which greatly influences evaporation. For example, 35 cm (14 in.) of precipitation in a hot climate will support only sparse vegetation, while the same amount of precipitation in much colder north-central Canada will support a conifer forest. A region with a low annual rainfall total is also more likely to be classified as dry if the majority of precipitation is concentrated during the warm summer months, when evaporation rates are greater.

Precipitation in a dry climate is both meager and irregular. Typically, the lower the average annual rainfall, the greater its variability. For example, a station that reports an annual rainfall of 5 cm (2 in.) may actually measure no rainfall for two years; then, in a single downpour, it may receive 10 cm (4 in.).

The major dry regions of the world can be divided into two primary categories. The first includes those sections of the subtropics (between latitudes 15° and 30°) where the sinking air of the subtropical anticyclones produces generally clear skies. The second is found in the continental areas of the middle latitudes. Here, far removed from

● **FIGURE 12.12** Rain streamers (virga) are common in dry climates, as falling rain evaporates into the drier air before ever reaching the ground.

● **FIGURE 12.13** Creosote bushes and cactus are typical of the vegetation found in the arid southwestern American deserts (BWh).

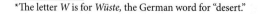

a source of moisture, areas are deprived of precipitation. Dryness in these areas is often accentuated by mountain ranges that produce a rain shadow effect.

Köppen divided dry climates into two types based on their degree of dryness: the *arid* (BW)* and the *semi-arid,* or *steppe* (BS). These two climatic types can be divided even further. For example, if the climate is hot and dry with a mean annual temperature above 18°C (64°F), it is either BWh or BSh (the *h* is for *heiss,* meaning "hot" in German). On the other hand, if the climate is cold (in winter, that is) and dry with a mean annual temperature below 18°C, then it is either BWk or BSk (where the *k* is for *kalt,* meaning "cold" in German).

The **arid climates** (BW) occupy about 12 percent of the world's land area. From Fig. 12.7, pp. 360-361, we can see that this climatic type is found along the west coast of South America and Africa and over much of the interior of Australia. Notice, also, that a swath of arid climate extends from northwest Africa all the way into central Asia. In North America, the arid climate extends from northern Mexico into the southern interior of the United States and northward along the leeward slopes of the Sierra Nevada. This region includes both the Sonoran and Mojave deserts and the Great Basin.

The southern desert region of North America is dry because it is dominated by the subtropical high most of the year, and cyclonic winter storm systems tend to weaken before they move into the area. The northern region is in the rain shadow of the Sierra Nevada. These regions are deficient in precipitation all year long, with many stations receiving less than 13 cm (5 in.) annually. As noted earlier, the rain that does fall is spotty, often in the form of scattered summer afternoon showers. Some of these showers can be downpours that change a parched gully into a raging torrent. More often than not, however, the rain evaporates into the dry air before ever reaching the ground, and the result is rain streamers (virga) dangling beneath the clouds (see ● Fig. 12.12).

Contrary to popular belief, few deserts are completely without vegetation. Although sparse, the vegetation that does exist must depend on the infrequent rains. Thus, most of the native plants are **xerophytes**—those capable of surviving prolonged periods of drought (see ● Fig. 12.13). Such vegetation includes various forms of cacti and short-lived plants that spring up during the rainy periods.

In low-latitude deserts (BWh), intense sunlight produces scorching heat on the parched landscape. Here, air temperatures can reach values as high as anywhere in the world. Maximum daytime readings during the summer can exceed 50°C (122°F), although 40°C to 45°C (104°F

---

*The letter *W* is for *Wüste,* the German word for "desert."

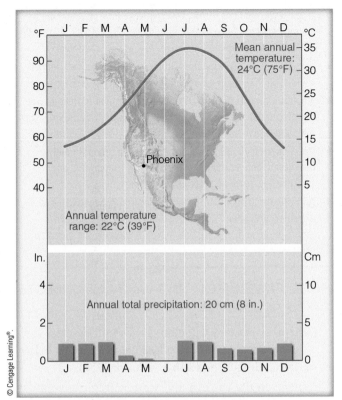

FIGURE 12.14 Climatic data for Phoenix, Arizona, latitude 33.5°N. A station with an arid climate (BWh).

to 113°F) are more common. In the middle of the day, the relative humidity is usually between 5 and 25 percent. At night, the air's relatively low water vapor content allows for rapid radiational cooling. Minimum temperatures often drop below 25°C (77°F). Thus, arid climates have large daily temperature ranges, often between 15°C and 25°C (27°F and 45°F) and occasionally greater.

During the winter, temperatures are more moderate, and minimums may, on occasion, drop below freezing. The variation in temperature from summer to winter produces large annual temperature ranges. We can see

this in the climate record for Phoenix, Arizona (see Fig. 12.14), a city in the southwestern United States with a BWh climate. Notice that the average annual temperature in Phoenix is 24°C (75°F), and that the average temperature of the warmest month (July) reaches a sizzling 35°C (95°F). As we would expect, rainfall is scant in all months. There is, however, a slight maximum in July and August. This is due to the summer monsoon, when more humid, southerly winds are likely to sweep over the region and develop into afternoon showers and thunderstorms.

In middle-latitude deserts (BWk), average annual temperatures are lower. Summers are typically warm to hot, with afternoon temperatures frequently reaching 40°C (104°F). Winters are usually extremely cold, with minimum temperatures sometimes dropping below −35°C (−31°F). Many of these deserts lie in the rain shadow of an extensive mountain chain, such as the Sierra Nevada and the Cascade mountains in North America, the Himalayan Mountains in Asia, and the Andes in South America. The meager precipitation that falls comes from an occasional summer shower or a passing mid-latitude cyclonic storm in winter.

Again, refer to Fig. 12.7, p. 360-361, and notice that around the margins of the arid regions, where rainfall amounts are greater, the climate gradually changes into **semi-arid** (BS). This region is called **steppe,** which typically has short bunchgrass, scattered low bushes, sparse trees, or sagebrush (see Fig. 12.15). In North America, this climatic region includes most of the western Great Plains,

DID YOU KNOW?

Phoenix, Arizona, a city with an arid climate, has a record 143 consecutive days without measured rainfall—from October 2005 to March 2006. And during the summer of 2011, Phoenix set a temperature record with 33 days of 110°F or greater.

FIGURE 12.15 Cumulus clouds forming over the steppe grasslands of western North America, a region with a semi-arid climate (BS).

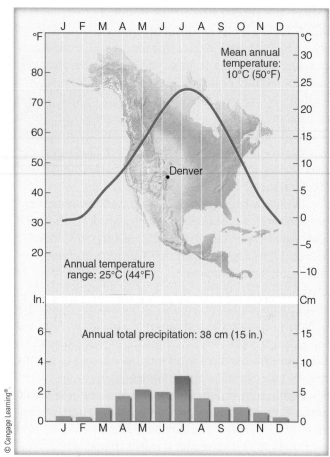

● **FIGURE 12.16** Climatic data for Denver, Colorado, latitude 40°N. A station with a semi-arid climate (BSk).

the southern coastal sections of California, and the northern valleys of the Great Basin. As in the arid region, northern areas experience lower winter temperatures and more frequent snowfalls. Annual precipitation is generally between 20 and 40 cm (8 and 16 in.). The climatic record for Denver, Colorado (see ● Fig. 12.16), exemplifies the semi-arid (BSk) climate.

As average rainfall amounts increase, the climate gradually changes to one that is more humid. Hence, the semi-arid (steppe) climate marks the transition between the arid and the humid climatic regions. (Before reading about moist climates, you may wish to read Focus section 12.2 about deserts that experience drizzle but little rainfall.)

BRIEF REVIEW

In this part of the chapter, we learned about the Köppen climate classification system and several climatic types. Here is a brief review of some important points we covered:

● The Köppen climate classification system includes the major climate types, defined by average temperature and precipitation experienced throughout the world at various times of the year.

● Tropical moist climates (Group A) are warm, with abundant rainfall. They experience no real winter, as each month has an average temperature of at least 18°C (64°F).

● Tropical wet climates (Group Af) experience little variation in average temperature and rainfall from month to month. All months are wet, which supports the growth of a tropical rainforest.

● Tropical wet and dry climates (Group Aw) feature a well-defined dry season that favors savanna grass and drought-resistant trees.

● Tropical monsoon climates (Group Am) are similar to tropical wet climates except that they experience a significant dry period lasting from one to two months.

● Dry climates (Group B) are deficient in precipitation throughout most of the year, with annual potential evaporation and transpiration exceeding precipitation.

● Arid climates (Group Bw)—the true deserts of the world—are extremely dry, whereas semi-arid climates (Group Bs) receive slightly more annual precipitation, which supports the growth of short bunchgrass in a region called steppe.

MOIST SUBTROPICAL MID-LATITUDE CLIMATES (GROUP C)

General characteristics: humid with mild winters (i.e., average temperature of the coldest month below 18°C, or 64°F, and above –3°C, or 27°F).

Extent: on the eastern and western regions of most continents, from about 25° to 40° latitude.

Major types: humid subtropical (Cfa), marine (Cfb), and dry-summer subtropical, or Mediterranean (Cs).

The Group C climates of the middle latitudes have distinct summer and winter seasons. Additionally, they have ample precipitation to keep them from being classified as dry. Although winters can be cold, and air temperatures can change appreciably from one day to the next, no month has a mean temperature below –3°C (27°F), for if it did, it would be classified as a D climate—one with severe winters.

The first C climate we will consider is the **humid subtropical climate** (Cfa).* Notice in Fig. 12.7, pp. 360-361, that Cfa climates are found principally along the east coasts of continents, roughly between 25° and 40° latitude. They dominate the southeastern section of the United States, as well as eastern China and southern Japan. In the Southern Hemisphere, they are found in southeastern South America and along the southeastern coasts of Africa and Australia. Many large cities in China and the United States experience a humid subtropical climate.

A trademark of the humid subtropical climate is its hot, muggy summers. This sultry summer weather occurs because Cfa climates are located on the western side of subtropical highs, where maritime tropical air from lower latitudes is swept poleward into these regions. Generally, summer dew point temperatures are high (often exceeding 23°C, or 73°F), and so is the relative humidity, even during the middle of the day. The high humidity combines with

---

*In the Cfa climate, the "f" means that all seasons are wet, and the "a" means that summers are long and hot. A more detailed explanation is given in Appendix E on p. 472.

A Desert with Clouds and Drizzle

We already know that not all deserts are hot. By the same token, not all deserts are sunny. In fact, some coastal deserts experience considerable cloudiness, especially low stratus and fog.

Amazingly, these coastal deserts are some of the driest places on Earth. They include the Atacama Desert of Chile and Peru, the coastal Sahara Desert of northwestern Africa, the Namib Desert of southwestern Africa, and a portion of the Sonoran Desert in Baja California (see ● Fig. 2). In the Atacama Desert, for example, some regions go without measurable rainfall for decades. Arica, in northern Chile, has an average annual rainfall of only 0.08 cm (0.03 in.).

This aridity is, in part, due to the fact that each region is adjacent to a large body of relatively cool water. Notice in Fig. 2 that these deserts are located along the western coastal margins of continents, where a subtropical high-pressure area causes prevailing winds to move cool water from higher latitudes along the coast. In addition, these winds help to accentuate the water's coldness by initiating *upwelling*— the rising of cold water from lower levels. The combination of these conditions tends to produce coastal water temperatures between 10°C and 15°C (50°F and 59°F), which is quite cool for such low latitudes. As surface air sweeps across the cold water, it is chilled to its dew point, often producing a blanket of fog and low clouds from which drizzle falls. The drizzle, however, accounts for very little rainfall. In most regions, it is only enough to dampen the streets with a mere trace of precipitation.

As the cool stable air moves inland, it warms, and the water droplets evaporate. Hence, most of the cloudiness and drizzle is found along the immediate coast. Although the relative humidity of this air is high, the dew-point temperature is comparatively low (often near that of the coastal surface water). Inland, further warming causes the air to rise. However, a stable subsidence inversion, associated with the subtropical highs, inhibits vertical motions by capping the rising air, causing it to drift back toward the ocean, where it sinks, completing a rather strong sea breeze circulation. The position of the subtropical highs, which tend to remain almost stationary, plays an additional role by preventing the Intertropical Convergence Zone with its rising, unstable air from entering the region.

And so we have a desert with clouds and drizzle—a desert that owes its existence in part to its proximity to rather cold ocean water and in part to the position and air motions of a subtropical high.

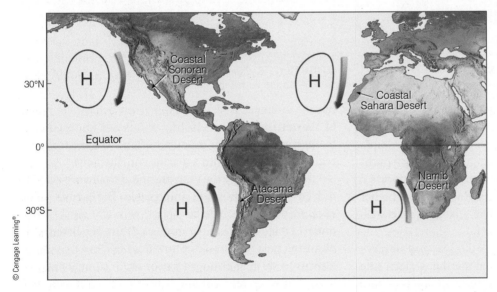

● **FIGURE 2** Location of coastal deserts (dark orange shade) that experience frequent fog, drizzle, and low clouds. (Blue arrows indicate prevailing winds and the movement of cool ocean currents.)

© Cengage Learning®

the high air temperature (usually above 32°C, or 90°F) to produce more oppressive conditions than are found in equatorial regions. Summer morning low temperatures often range between 21°C and 27°C (70°F and 81°F). Occasionally, a weak summer cool front will bring temporary relief from the sweltering conditions. However, devastating heat waves, sometimes lasting many weeks, can occur when an upper-level ridge moves over the area.

Winters tend to be relatively mild, especially in the lower latitudes, where air temperatures rarely dip much below freezing. Poleward regions experience winters that are colder and harsher. Here, frost, snow, and ice storms are more common, but heavy snowfalls are rare. Winter weather can be quite changeable, as almost summerlike conditions can give way to cold rain and wind in a matter of hours when a middle-latitude cyclonic storm and its accompanying fronts pass through the region.

Humid subtropical climates experience adequate and fairly well-distributed precipitation throughout the year, with typical annual averages between 80 and 165 cm (31 and 65 in.). In summer, when thunderstorms are common, much of the precipitation falls as afternoon showers.

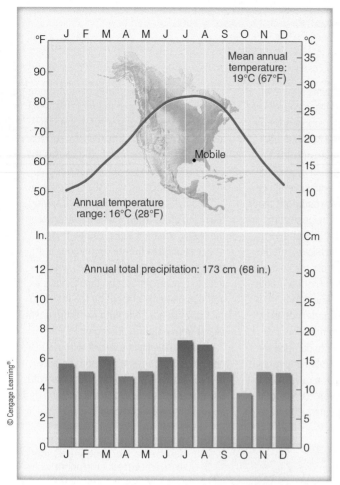

● FIGURE 12.17 Climatic data for Mobile, Alabama, latitude 30°N. A station with a humid subtropical climate (Cfa).

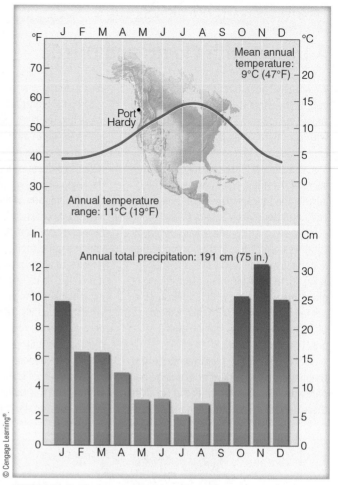

● FIGURE 12.18 Climatic data for Port Hardy, Canada, latitude 51°N. A station with a marine climate (Cfb).

Tropical storms entering the United States and China can substantially add to summer and autumn rainfall totals. Winter precipitation most often occurs with eastward-trekking middle-latitude cyclonic storms. In the southeastern United States, the abundant rainfall supports a thick pine forest that becomes mixed with oak at higher latitudes. The climate data for Mobile, Alabama, a city with a Cfa climate, is given in ● Fig. 12.17.

Glance back at Fig. 12.7, pp. 360-361, and observe that C climates extend poleward along the western side of most continents from about latitude 40° to 60°. These regions are dominated by prevailing winds from the ocean that moderate the climate, keeping winters considerably milder than stations located at the same latitude farther inland. In addition to this, summers are quite cool. When the summer season is both short and cool, the climate is designated as Cfc. Equatorward, where summers are longer (but still cool), the climate is classified as *west coast marine*, or simply **marine,** Cfb.*

Where mountains parallel the coastline, such as along the west coasts of North and South America, the

marine influence is restricted to narrow belts. Over much of Western Europe, prevailing winds are unobstructed by high mountains, so ocean air often sweeps across the region, providing it with a marine climate (Cfb).

During much of the year, marine climates are characterized by low clouds, fog, and drizzle. The ocean's influence produces adequate precipitation in all months, with much of it falling as light or moderate rain associated with maritime polar air masses. Snow does fall, but frequently it turns to slush after only a day or so. In some locations, topography greatly enhances precipitation totals. For example, along the west coast of North America, coastal mountains not only force air upward, enhancing precipitation, but they also slow a storm's eastward progress, which enables the storm to drop more precipitation on the area.

Along the northwest coast of North America, rainfall amounts decrease in summer. This phenomenon is caused by the northward migration of the subtropical Pacific high, which is located southwest of this region. The summer decrease in rainfall can be seen by examining the climatic record of Port Hardy (see ● Fig. 12.18), a station situated along the coast of Canada's Vancouver Island. The data illustrate another important characteristic of marine climates: the low annual temperature range for such a

*In the Cfb climate, the "b" means that summers are cooler than in those regions experiencing a Cfa climate. The temperature criteria for the various subregions are given in Appendix E, p. 472.

high-latitude station. The ocean's influence keeps daily temperature ranges low as well. In this climate type, it rains on many days and when it is not raining, skies are usually overcast. The heavy rains produce a dense forest of Douglas fir.

Moving equatorward of marine climates, the influence of the subtropical highs becomes greater, and the summer dry period more pronounced. Gradually, the climate changes from marine to one of **dry-summer subtropical** (Cs), which is also called **Mediterranean,** because it predominates along the coastal areas of the Mediterranean Sea. (Here the lowercase "s" stands for "summers dry.") Along the west coast of North America, Portland, Oregon, because it has rather dry summers, marks the transition between the marine climate and the dry-summer subtropical climate to the south.

The extreme summer aridity of the Mediterranean climate, which in California may exist for five months, is caused by the sinking air of the subtropical highs. In addition, these anticyclones divert summer storm systems poleward. During the winter, when the subtropical highs move equatorward, mid-latitude storms from the ocean frequent the region, bringing with them much-needed rainfall. Consequently, Mediterranean climates are characterized by mild, wet winters, and mild-to-hot, dry summers.

Where surface winds parallel the coast, upwelling of cold water helps keep the water itself and the air above it cool all summer long. In these coastal areas, which are often shrouded in low clouds and fog, the climate is called *coastal Mediterranean* (Csb). Here, summer daytime maximum temperatures usually reach about 21°C (70°F), while overnight lows often drop below 15°C (59°F). Inland, away from the ocean's influence, summers are hot and winters are a little colder than coastal areas. In this *interior Mediterranean climate* (Csa), summer afternoon temperatures usually climb above 34°C (93°F) and occasionally above 40°C (104°F).

● Figure 12.19 contrasts the coastal Mediterranean climate of San Francisco, California, with the interior Mediterranean climate of Sacramento, California. While Sacramento is only 130 km (80 mi) inland from San Francisco, Sacramento's average July temperature is 9°C (16°F) higher. As we would expect, Sacramento's annual temperature range is considerably greater, too. Although Sacramento and San Francisco both experience an occasional frost, snow in these areas is a rarity.

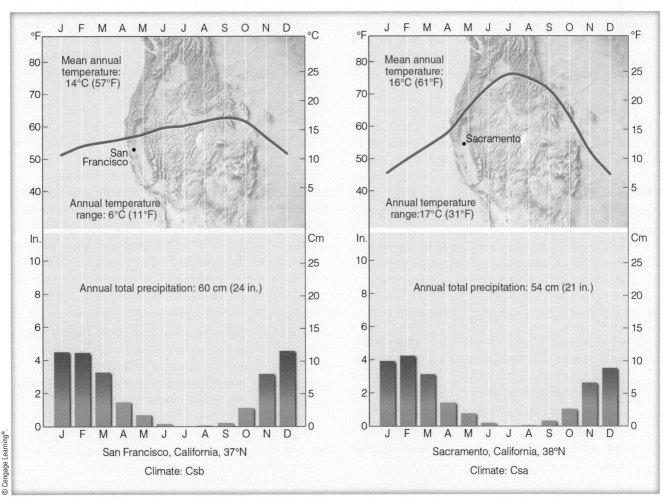

● **FIGURE 12.19** Comparison of a coastal Mediterranean climate, Csb (San Francisco, at left), with an interior Mediterranean climate, Csa (Sacramento, at right).

● **FIGURE 12.20** In the Mediterranean-type climates of North America, typical chaparral vegetation includes chamise, manzanita, and foothill pine.

In Mediterranean climates, yearly precipitation amounts range between 30 and 90 cm (11 and 35 in.). However, much more precipitation falls on surrounding hillsides and mountains. Because of the summer dryness, the land supports only a scrubby mixture of low-growing woody plants and trees called *chaparral* (see ● Fig. 12.20).

At this point, we should note that summers are not as dry along the Mediterranean Sea as they are along the west coast of North America. Moreover, coastal Mediterranean areas are also warmer, due to the lack of upwelling in the Mediterranean Sea.

Before leaving our discussion of C climates, note that when the dry season is in winter, the climate is classified as Cw. Over northern India and portions of China, the relatively dry winters are the result of northerly winds from continental regions circulating southward around the cold Siberian high. Many lower-latitude regions with a Cw climate would be tropical if they were not as high in elevation and, consequently, too cool to be designated as tropical.

When a moist climate turns dry, drought often results. What constitutes a drought and how is it measured? These are some of the questions addressed in Focus section 12.3.

MOIST CONTINENTAL CLIMATES (GROUP D)

General characteristics: warm-to-cool summers and cold winters (i.e., average temperature of warmest month exceeds 10°C, or 50°F, and the coldest monthly average drops below –3°C, or 27°F); winters are severe with snowstorms, blustery winds, bitter cold; climate controlled by large continent.

Extent: north of moist subtropical mid-latitude climates.

Major types: humid continental with hot summers (Dfa), humid continental with cool summers (Dfb), and subpolar (Dfc).

The D climates are controlled by large landmasses. Therefore, they are found only in the Northern Hemisphere. Look at the climate map, Fig. 12.7, pp. 360-361, and notice that D climates extend across North America and Eurasia, from about latitude 40°N to almost 70°N. In general, they are characterized by cold winters and warm-to-cool summers.

As we know, for a station to have a D climate, the average temperature of its coldest month must dip below –3°C (27°F). This is not an arbitrary number. Köppen found that, in Europe, this temperature marked the southern limit of persistent snow cover in winter.* Hence, D climates experience a great deal of winter snow that stays on the ground for extended periods. When the temperature drops to a point where every month has an average temperature below 10°C (50°F), the climate is classified as polar (E). Köppen found that an average monthly temperature of 10°C tended to represent the minimum temperature required for tree growth. So no matter how cold it gets in a D climate (and winters can get extremely cold), there is enough summer warmth to support the growth of trees.

There are two basic types of D climates: the **humid continental** (Dfa and Dfb) and the **subpolar** (Dfc). Humid continental climates are observed from about latitude 40°N to 50°N (60°N in Europe). In these areas, precipitation is adequate and fairly evenly distributed throughout the year, although interior stations experience maximum precipitation in summer. Annual precipitation totals usually range from 50 to 100 cm (20 to 40 in.). Native vegetation in the wetter regions includes forests of spruce, fir, pine, and oak. In autumn, nature's pageantry unveils itself as the leaves of deciduous trees turn brilliant shades of red, orange, and yellow (see ● Fig. 12.21).

Humid continental climates are subdivided on the basis of summer temperatures. Where summers are long and hot,** the climate is described as *humid continental with hot summers* (Dfa). Here, summers are often both hot and humid, especially in the southern regions. Midday temperatures often exceed 32°C (90°F) and occasionally 40°C (104°F). Summer nights are usually warm and humid as well. The frost-free season normally lasts from five to six months, long enough to grow a wide variety of crops. Winters tend to be windy, cold, and snowy. Farther north, where summers are shorter and not as hot, the climate is described as *humid continental with long cool summers* (Dfb). In Dfb climates, summers are not only cooler but much less humid. Temperatures may exceed 35°C (95°F)

---

*In North America, studies suggest that an average monthly temperature of 0°C (32°F) or below for the coldest month seems to correspond better to persistent winter snow cover.

**Hot means that the average temperature of the warmest month is above 22°C (72°F) and at least four months have a monthly mean temperature above 10°C (50°F). Again, a complete explanation for each subgroup is given in Appendix E, on p. 472.

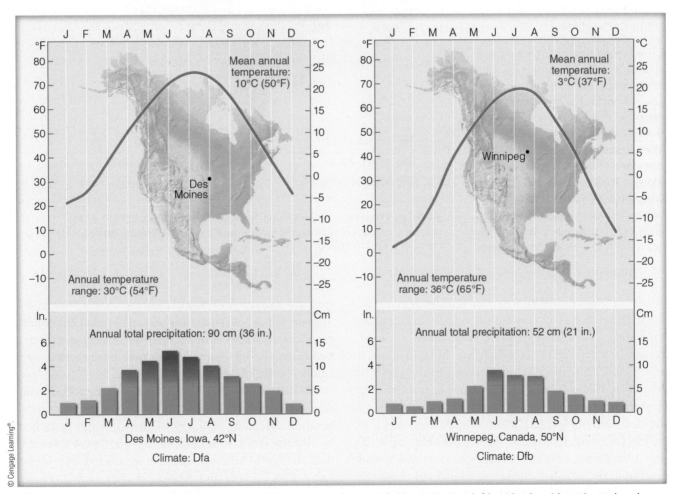

FIGURE 12.21 The leaves of deciduous trees burst into brilliant color during autumn over the countryside of Adirondack Park, New York, a region with a humid continental climate.

© Michio Hoshino/Minden Pictures

for a time, but extended hot spells lasting many weeks are rare. The frost-free season is shorter than in the Dfa climate, and normally lasts between three and five months. Winters are long, cold, and windy. It is not uncommon for temperatures to drop below –30°C (–22°F) and stay below –18°C (0°F) for days and sometimes weeks. Autumn is short, with winter often arriving right on the heels of summer. Spring, too, is short, as late spring snowstorms are common, especially in the more northern latitudes.

Figure 12.22 compares the Dfa climate of Des Moines, Iowa, with the Dfb climate of Winnipeg, Canada. Notice that both cities experience a large annual temperature

Mean annual temperature: 10°C (50°F)

Annual temperature range: 30°C (54°F)

Annual total precipitation: 90 cm (36 in.)

Des Moines, Iowa, 42°N

Climate: Dfa

Mean annual temperature: 3°C (37°F)

Annual temperature range: 36°C (65°F)

Annual total precipitation: 52 cm (21 in.)

Winnepeg, Canada, 50°N

Climate: Dfb

© Cengage Learning®.

FIGURE 12.22 Comparison of a humid continental hot summer climate, Dfa (Des Moines, at left), with a humid continental cool summer climate, Dfb (Winnipeg, at right).

When Does a Dry Spell Become a Drought?

When a region's average precipitation drops dramatically for an extended period of time, drought may result. The word *drought* refers to a period of abnormally dry weather that produces a number of negative consequences, such as crop damage or an adverse impact on a community's water supply. Keep in mind that drought is more than a dry spell. In the dry, summer subtropical (Csa) climate of California's Central Valley, it may not rain from May through September. This dry spell is normal for this region and, therefore, would not be considered a drought. However, if this summer dry spell were to occur in the humid subtropical (Cfa) climate of the southeastern United States, the lack of rain would constitute a drought and could be disastrous for many aspects of the community.

In an attempt to measure drought severity, Wayne Palmer, a scientist with the National Weather Service, developed the *Palmer Drought Severity Index (PDSI)*. The index takes into account average temperature and precipitation values to define drought severity. The index is most effective in assessing long-term drought that lasts several months or more. Drought conditions are indicated by a set of numbers that range from 0 (normal) to −4 (extreme drought). (See ▼ Table 1.) The index also assesses wet conditions with numbers that range from +2 (unusually moist) to +4 (extremely moist). The *Palmer Hydrological Drought Index (PHDI)* expands the PDSI by taking into account additional water (hydrological) information, such as a region's groundwater reserves and reservoir levels.

▼Table 1 **Palmer Drought Severity Index**

VALUE	DROUGHT	VALUE	MOISTURE
−4.0 or less	Extreme	+4.0 or greater	Extremely Moist
−3.0 to −3.9	Severe	+3.0 to +3.9	Very Moist
−2.0 to −2.9	Moderate	+2.0 to +2.9	Unusually Moist
−1.9 to +1.9	Normal	−1.9 to +1.9	Normal

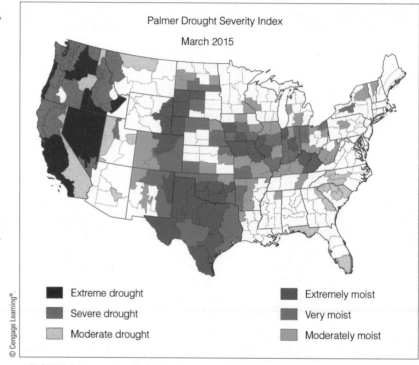

● **FIGURE 3** The Palmer Hydrological Drought Index for March 2015, showing long-term drought conditions and regions with sufficient moisture. (National Climatic Data Center, NOAA)

range. This is characteristic of climates located in the northern interior of continents. In fact, as we move poleward, the annual temperature range increases. In Des Moines, it is 30°C (54°F), while 950 km (590 mi) to the north in Winnipeg, it is 36°C (65°F). The summer precipitation maximum expected for these interior continental locations shows up well in Fig. 12.22. Most of the summer rain is in the form of isolated convective showers, although an occasional weak frontal system can produce more widespread precipitation, as can a cluster of thunderstorms—the *mesoscale convective complex* described in Chapter 10.

The weather in both climatic types can be quite changeable, especially in winter, when a brief warm spell is replaced by blustery winds and temperatures plummeting well below −30°C (−22°F).

When winters are severe and summers short and cool, with only one to three months having a mean temperature exceeding 10°C (50°F), the climate is described as *subpolar* (Dfc). From Fig. 12.7, p. 360-361, we can see that, in North America, this climate occurs in a broad belt across Canada and Alaska; in Eurasia, it stretches from Norway over much of Siberia. The exceedingly low

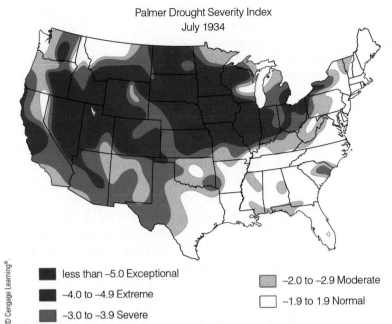

Palmer Drought Severity Index
July 1934

less than −5.0 Exceptional

−4.0 to −4.9 Extreme

−3.0 to −3.9 Severe

−2.0 to −2.9 Moderate

−1.9 to 1.9 Normal

● **FIGURE 4** Palmer Drought Severity Index for July 1934. (U.S. Department of Agriculture)

A drought apparent in the PDSI can take weeks or months to be reflected in the PHDI, because water supplies take time to drain down after dry conditions set in. Likewise, even after rains return to a drought-stricken region, it takes time for reservoirs and groundwater storage to be recharged, so the PHDI may lag the PSDI in showing recovery.

● Figure 3 shows the PHDI across the United States during March 2015. Notice that several parts of the West are in an extreme drought (dark red shade), including almost all of Nevada as well as the southern two-thirds of California, which at the time was in its fourth year of a destructive drought. Most of California's rains arrive in late autumn, winter, and early spring, yet drought conditions were persisting in early 2015 even after most of the rainy season had ended. Notice also in Fig. 3 that there are a few patches of green, especially over the northern Great Plains, indicating ample moisture.

Drought is not uncommon to North America. In fact, probably the worst weather-related disaster to hit the United States during the twentieth century was the great drought of the 1930s. That drought, which tragically coincided with the Great Depression, actually began in the late 1920s and continued into the late 1930s. It not only lasted a long time, but it also extended over a vast area (see ● Fig. 4).

The drought, coupled with poor farming practices, left the topsoil of the Great Plains ripe for wind erosion. As a result, wind storms lifted millions of tons of soil into the air, creating vast dust storms that buried farmhouses, reduced millions of acres to an unproductive wasteland, and financially ruined thousands of families. Because of the dust storms, the worst-affected region has been called the *Dust Bowl* and the 1930s are often referred to as "the Dust Bowl years." To worsen an already bad situation, the drought was accompanied by extreme summer heat that was most severe during the summers of 1934 and 1936.

How does the California drought of 2014–2015 compare with the drought year of 1934? We can see in Fig. 4 that during 1934 most of California was experiencing some form of drought. But notice that the 1934 drought in California was not nearly as extreme as that of 2015. However, in 1934 extreme dry conditions existed over a vast area of the upper Plains and far west, with many regions experiencing a Palmer Index of −4 or below. Unfortunately, this already disastrous drought became progressively worse. Millons of people were affected, many of whom eventually became destitute. Thousands migrated westward in search of employment, as depicted in the book and film *The Grapes of Wrath*.

temperatures of winter account for these areas being the primary source regions for continental polar and arctic air masses. Extremely cold winters coupled with cool summers produce large annual temperature ranges, as exemplified by the climate data in ● Fig. 12.23 for Fairbanks, Alaska.

Precipitation is comparatively light in the subpolar climates, especially in the interior regions, with most places receiving less than 50 cm (20 in.) annually. A good percentage of the precipitation falls when weak cyclonic storms move through the region in summer. The total snowfall is usually not large, but the cold air prevents melting, so snow stays on the ground for months at a time.

Because of the low temperatures, there is a low annual rate of evaporation that ensures adequate moisture to support the boreal* forests of conifers and birches known as **taiga** (see ● Fig. 12.24). Hence, the subpolar climate is known also as a *boreal climate* and as a *taiga climate*.

In the taiga region of northern Siberia and Asia, where the average temperature of the coldest month drops to a frigid −38°C (−36°F) or below, the climate is designated Dfd. Where the winters are considered dry, the climate is designated Dwd.

*The word *boreal* comes from the ancient Greek *Boreas*, meaning "wind from the north."

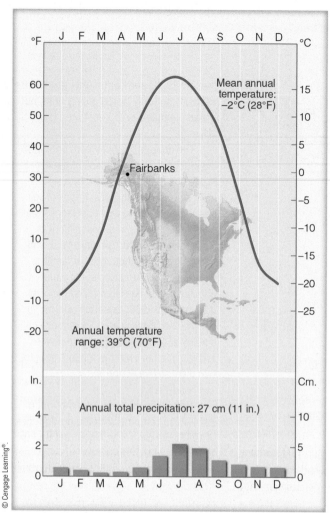

● FIGURE 12.23 Climatic data for Fairbanks, Alaska, latitude 65°N. A station with a subpolar climate (Dfc).

● FIGURE 12.24 Coniferous forests (taiga) such as this occur where winter temperatures are low and precipitation is abundant.

POLAR CLIMATES (GROUP E)

General characteristics: year-round low temperatures (i.e., average temperature of the warmest month is below 10°C, or 50°F).

Extent: northern coastal areas of North America and Eurasia; Greenland; and Antarctica.

Major types: polar tundra (ET) and polar ice caps (EF).

In the **polar tundra** (ET), the average temperature of the warmest month is below 10°C (50°F), but above freezing. (See ● Fig. 12.25, the climate data for Barrow, Alaska.) Here, the ground is permanently frozen, a condition known as **permafrost**. Permafrost can be less than 1 m (3 ft) or more than 1,000 m (3,300 ft) deep. Summer weather is usually just warm enough to thaw the topmost part of the soil, so during the summer the tundra turns swampy and muddy. Annual precipitation on the tundra is meager, with most stations receiving less than 20 cm (8 in.). In lower latitudes, this amount of precipitation would constitute a desert, but in the cold polar regions, evaporation rates are very low and moisture remains adequate. Because of the extremely short growing season, *tundra vegetation* consists of mosses, lichens, dwarf trees, and scattered woody vegetation, typically only several centimeters tall when fully grown (see ● Fig. 12.26).

Even though summer days are long, the sun is never very high above the horizon. Additionally, some of the sunlight that reaches the surface is reflected by snow and ice, while some is used to melt the frozen soil. Consequently, in spite of the long hours of daylight, summers are quite cool. Even so, there are still large annual temperature ranges between the cool summers and the extremely cold winters.

When the average temperature for every month drops below freezing, plant growth is impossible, and the region

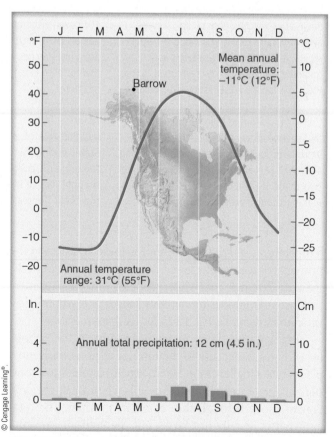

● FIGURE 12.25 Climatic data for Barrow, Alaska, latitude 71°N. A station with a polar tundra climate (ET).

● FIGURE 12.26 Tundra vegetation in Alaska. This type of tundra is composed mostly of sedges and dwarfed wildflowers that bloom during the brief growing season.

is perpetually covered with snow and ice. This climatic type is known as **polar ice cap** (EF). It occupies the interior ice sheets of Greenland and Antarctica, where the depth of ice in some places measures thousands of meters. In this region, temperatures are never much above freezing, even

● FIGURE 12.27 Climatic data for Eismitte, Greenland, latitude 71°N. Located in the interior of Greenland at an elevation of almost 10,000 feet above sea level, Eismitte has a polar ice cap climate (EF).

during the middle of "summer." The coldest places in the world are located here. Precipitation is extremely meager with many places receiving less than 10 cm (4 in.) annually. Most precipitation falls as snow during the "warmer" summer. Strong downslope katabatic winds frequently whip the snow about, adding to the climate's harshness. The data in ● Fig. 12.27 for Eismitte, Greenland, illustrate the severity of an EF climate.

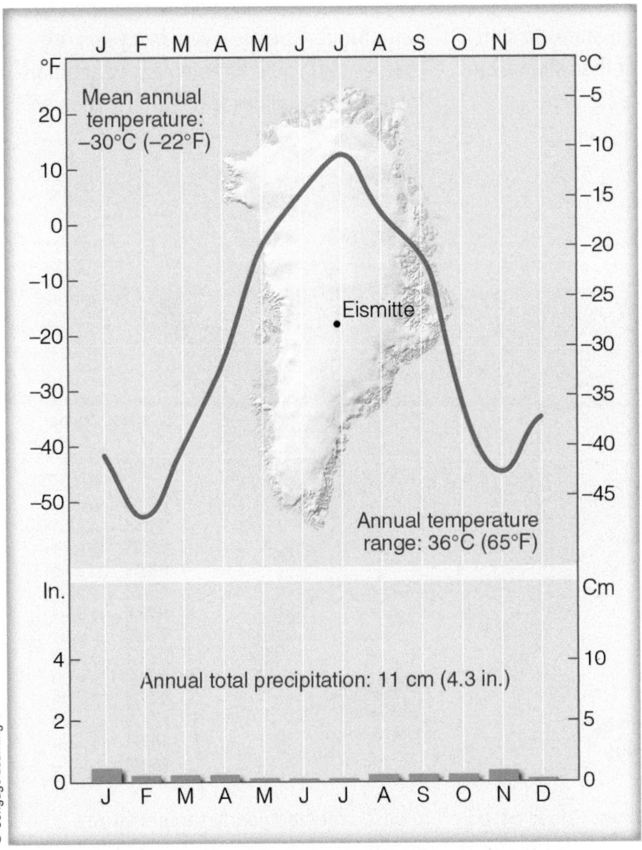

In Chapter 13 we will look at how Earth's climate is changing and how it has warmed over the past 100-plus years. Such changes may have an effect on the Köppen climate boundaries, but any boundary change is often so small that it is difficult to assess. One possible way to observe these changes, however, is to see how changing temperatures are influencing regions where plants grow. In a warmer world, are the plant hardiness zones shifting poleward? Focus section 12.4 addresses this question.

HIGHLAND CLIMATES (GROUP H)   You do not have to visit the polar regions to experience a polar climate. Because temperature decreases with altitude, climatic changes experienced when climbing 300 m (1000 ft) in elevation are about equivalent in high latitudes to horizontal changes experienced when traveling 300 km (186 mi) northward. (This distance is equal to about 3° latitude.) Moreover, as air rises along the windward side of a mountain, precipitation amounts usually increase the higher you go. When you are ascending a high mountain, such as Mt. Everest, you can travel through many climatic regions in a relatively short distance. Thus, **highland climates** often show a great deal of variation in temperature, precipitation, and vegetation over a relatively short vertical change in elevation.

Are Plant Hardiness Zones Shifting Northward?

Millions of Americans pay close attention to climate in order to help them decide what plants to grow. Temperatures across the United States have been warming in recent decades. In fact, 2012 was the warmest year in more than a century of record keeping for the 48 contiguous states, beating 1934 by a full degree Fahrenheit. Moreover, winters are warming faster than summers, which could mean that plants are less likely to be killed by severe cold. The winter of 2015–2016 was the warmest on record for the contiguous United States. These changes may not be enough to cause a drastic change to the Köppen climate classification system described in this chapter, but we might expect that climate boundaries are shifting in some locations. One way to assess this situation is to look at the temperatures that plants need to grow and thrive.

One of the main sources of guidance for farmers, gardeners, and landscapers is the plant hardiness map produced by the United States Department of Agriculture (USDA). This map (see ● Fig. 5) places each part of the United States, including Puerto Rico, into one of 13 numbered zones based on the average annual minimum temperature (the coldest reading one might expect to observe in a typical year). Each zone corresponds to a range of 10°F. For example, Chicago is in zone 5, which means that the average annual minimum temperature is between -20 and -11°F. Each major zone is divided into subzones with intervals of 5°F. (For clarity, Fig. 5 does not show subzones.)

There are other important ways to measure the risk posed by cold weather to plants, such as the average dates of the first frost and freeze in autumn or the last frost and freeze in spring. However, the average annual minimum temperature serves as a useful single index for determining which plants might do well in a given location. This guideline is especially true for perennial plants such as trees and shrubs that must survive each winter's cold.

In 2012, the USDA issued the latest version of its map (see Fig. 5), based on temperatures observed from 1976 to 2005.

This updated edition replaced a previous version of the map, released in 1990, that was based on data from 1974 to 1986. Not surprisingly—as the planet is undergoing long-term warming related to greenhouse gases—the new map showed that many parts of the United States had shifted into warmer zones of plant hardiness. In fact, more than half of 34 cities highlighted on the 1990 map ended up in a new zone in the 2012 map.

The 1990 map drew on data spanning only 13 years, which is smaller than the 30-year period typically used in climatology. Therefore, the USDA stated that the new map was not intended to serve as an illustration of long-term climate change. However, the maps are consistent with other analyses showing an overall trend toward warmer temperatures in the United States since the 1970s. Conditions may even be leaping ahead of the revised map. One study concluded that there had been enough warming from 2005 to 2012 for about a third of the United States to shift another half zone warmer than shown on the most recent USDA map.

The winter of 2015–2016 brought many dramatic cases of mildness. December 25 was the warmest Christmas Day on record for many East Coast cities, from Portland, Maine, which hit 62°F (17°C), to Jacksonville, Florida, which soared to 82°F (28°C). The famed cherry blossoms in Washington, D.C., reached their 2016 peak bloom on March 25, a few days ahead of average. In fact, only six years between 1995 and 2016 saw the peak bloom occur later than the long-term average of April 4.

Even in a warming climate, however, cold extremes can still occur. During the very mild winter of 2015–2016, Boston warmed to 50°F on a total of 35 days, the most recorded in any winter. Yet on February 14, the city dipped to a frigid –9°F (–23°C), its coldest reading in more than 50 years. One of the key questions in climate research is whether a warming climate will tend to make annual minimum temperatures reliably warmer, or whether the general warming will be interspersed with frigid extremes every few years, extremes that could threaten any vegetation planted with milder temperatures in mind.

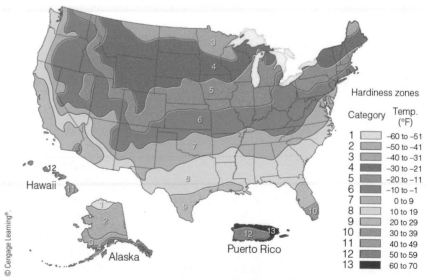

● FIGURE 5 Plant Hardiness Zones based on lowest expected yearly temperatures. (Adapted from the official Plant Hardiness Zone Map produced jointly by the USDA Agriculture Research Service and Oregon State University.)

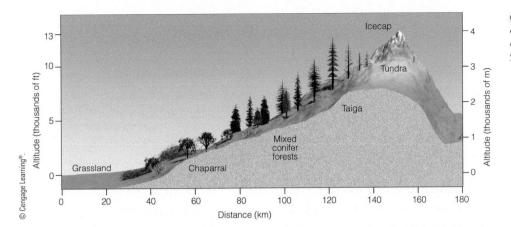

● **FIGURE 12.28** Vertical view of changing vegetation and climate due to elevation in the central Sierra Nevada.

● Figure 12.28 shows how the climate and vegetation change along the western slopes of the central Sierra Nevada. (See Fig. 12.5, p. 357, for the precipitation patterns for this region.) Notice that, at the base of the mountains, the climate and vegetation represent semi-arid conditions, while in the foothills the climate becomes Mediterranean and the vegetation changes to chaparral. Higher up, thick fir and pine forests prevail. At still higher elevations, the climate is subpolar and the taiga gives way to dwarf trees and tundra vegetation. Near the summit there are permanent patches of ice and snow, with some small glaciers nestled in protected areas. Hence, in less than 13,000 vertical feet, the climate has changed from semi-arid to polar.

SUMMARY

In this chapter, we examined global temperature and precipitation patterns, as well as the various climatic regions throughout the world. Tropical climates are found in low latitudes, where the noon sun is always high, day and night are nearly equal in length, every month is warm, and no real winter season exists. Some of the rainiest places in the world exist in the tropics, especially where warm, humid air rises upslope along mountain ranges.

Dry climates prevail where potential evaporation and transpiration exceed precipitation. Some deserts, such as the Sahara, are mainly the result of sinking air associated with the subtropical highs, while others, due to the rain shadow effect, are found on the leeward side of mountains. Many deserts form in response to both of these effects.

Middle latitudes are characterized by a distinct winter and summer season. Winters tend to be milder in lower latitudes and more severe in higher latitudes. Along the east coast of some continents, summers tend to be hot and humid as moist air sweeps poleward around the subtropical highs. The air often rises and condenses into afternoon thunderstorms in this humid subtropical climate. The west coasts of many continents tend to be drier, especially in summer, as the combination of cool ocean water and sinking air of the subtropical highs inhibit to a large degree the formation of cumuliform clouds.

In the middle of large continents, such as North America and Eurasia, summers are usually wetter than winters. Winter temperatures are generally lower than those experienced in coastal regions. As one moves northward, summers become shorter and winters longer and colder. Polar climates prevail at high latitudes, where winters are severe and there is no real summer. When ascending a high mountain, one can travel through many climatic zones in a relatively short distance.

KEY TERMS

The following terms are listed (with corresponding page number) in the order they appear in the text. Define each. Doing so will aid you in reviewing the material covered in this chapter.

microclimate, 352
mesoclimate, 352
macroclimate, 352
global climate, 352
climatic controls, 352
Köppen classification system, 357
tropical rainforest, 359
tropical wet climate, 359
laterite, 361
tropical monsoon climate, 361
tropical wet-and-dry climate, 362
savanna grass, 362
arid climate, 364
xerophytes, 364
semi-arid climate, 365
steppe, 365
humid subtropical climate, 366
marine climate, 368
dry-summer subtropical (Mediterranean) climate, 369
humid continental climate, 369
subpolar climate, 370
taiga, 373
polar tundra climate, 374
permafrost, 374
polar ice cap climate, 375
highland climate, 375

QUESTIONS FOR REVIEW

1. What factors determine the global pattern of precipitation?
2. Explain why, in North America, there is typically a precipitation maximum along the West Coast in winter, a maximum on the Central Plains in summer, and a fairly even distribution between summer and winter along the East Coast.
3. What climate information did Köppen use in classifying climates?
4. How did Köppen define a tropical climate? How did he define a polar climate?
5. According to Köppen's climatic system (Fig. 12.7, pp. 360-361), what major climatic type is most abundant: (a) in North America; (b) in South America; (c) throughout the world?
6. What is the primary factor that makes a dry climate "dry"?
7. In which climatic region would each of the following be observed: tropical rainforest, xerophytes, steppe, taiga, tundra, and savanna?
8. What are the controlling factors (the major climatic controls) that produce the following climatic regions? (a) tropical wet and dry; (b) Mediterranean; (c) marine; (d) humid subtropical; (e) subpolar; (f) polar ice cap
9. Why are marine climates (Cs) usually found on the west coast of continents?
10. Why are large annual temperature ranges characteristic of D-type climates?
11. Why are D climates found in the Northern Hemisphere but not in the Southern Hemisphere?
12. Explain why a tropical rainforest climate will support a tropical rainforest, while a tropical wet-and-dry climate will not.

13. What is the primary distinction between a Cfa and a Dfa climate?

14. Explain how arid deserts can be found adjacent to oceans.

15. Why did Köppen use the 10°C (50°F) average temperature for July to distinguish between D and E climates?

16. What accounts for the existence of a BWk climate in the western Great Basin of North America?

17. Barrow, Alaska, receives a mere 11 cm (about 4.3 in.) of precipitation annually. Explain why its climate is not classified as arid or semi-arid.

18. Explain why subpolar climates are also known as boreal climates and taiga climates.

QUESTIONS FOR THOUGHT AND EXPLORATION

1. Why do cities east of the Rockies, such as Denver, Colorado, get much more precipitation than cities east of the Sierra Nevada, such as Reno, Nevada?

2. According to the Köppen system of climate classification, which type of climate is found in your area?

3. Los Angeles, Seattle, and Boston are all coastal cities, yet Boston has a continental rather than a marine climate. Explain why.

4. Why are many structures in polar regions built on pilings?

5. Why are summer afternoon temperatures in a humid subtropical climate (Cfa) often higher than in a tropical wet climate (Af)?

6. Why are humid subtropical climates (Cfa) found in regions bounded by 20° and 40° (N or S) latitudes, and nowhere else?

7. In which of the following climate types is virga likely to occur most frequently: humid continental, arid desert, or polar tundra? Explain why.

8. As shown in Figure 12.19, p. 369, San Francisco and Sacramento, California, have similar mean annual temperatures but different annual temperature ranges. What factors control the annual temperature ranges at these two locations?

9. Why is there a contrast in climate types on either side of the Rocky Mountains, but not on either side of the Appalachian Mountains?

10. Sketch graphs of annual variation of temperature and precipitation for a coastal location, and also for a location in the center of a large continent. Explain any differences in your graphs.

11. On a blank map of the world, roughly outline where Köppen's major climatic regions are located.

12. Over the past 100 years or so, Earth has warmed by around 1°C (1.8°F). If additional warming should occur as expected over the next 100 years, explain how this rise in temperature might influence the boundary between C and D climates. How might the warming influence the boundary between D and E climates?

GLOBAL **GEOSCIENCE** WATCH   Within the Global Environment Watch: Meteorology portal, go to the Website and Blogs section and choose websites from three different state climate centers. How does each center go about describing the average climate of its state? Does it use only the variables outlined in the Köppen classification system, or are other climatological factors included?

ONLINE RESOURCES

 Visit www.cengagebrain.com to view additional resources, including video exercises, practice quizzes, an interactive eBook, and more.

CHAPTER 13

Earth's Changing Climate

Contents

A change in our climate however is taking place very sensibly. Both heats and colds are becoming much more moderate within the memory even of the middle-aged. Snows are less frequent and less deep. They do not often lie, below the mountains, more than one, two, or three days, and very rarely a week. They are remembered to have been formerly frequent, deep, and of long continuance. The elderly inform me the earth used to be covered with snow about three months in every year. The rivers, which then seldom failed to freeze over in the course of the winter, scarcely ever do now. This change has produced an unfortunate fluctuation between heat and cold, in the spring of the year, which is very fatal to fruits. In an interval of twenty-eight years, there was no instance of fruit killed by the frost in the neighborhood of Monticello. The accumulated snows of the winter remaining to be dissolved all together in the spring, produced those overflowings of our rivers, so frequent then, and so rare now.

Thomas Jefferson, *Notes on the State of Virginia, 1781*

The opening passage of this chapter describes how our third American president, Thomas Jefferson, perceived the evolution of climate in the state of Virginia and in his own backyard. Some changes in climate are due to natural processes, whereas others are related to human activity. One of the great environmental concerns of our time is the **climate change** now unfolding as a result of greenhouse gases being added to our atmosphere. Glaciers are melting, sea level is rising, precipitation is becoming more intense in many areas, and global temperature is increasing each decade. Extensive research has shown that the primary cause of these changes over the last few decades is human (anthropogenic) activity—the burning of fossil fuels.

We know that climate has changed in the past, and nothing suggests that it will not continue to change, both globally and locally. As the urban environment grows, its climate differs from that of the region around it. Sometimes the difference is striking, as when city nights are warmer than the nights of the outlying rural areas. Other times, the difference is subtle, as when a layer of smoke and haze covers a city. Climate variations such as a persistent drought or a delay in the annual monsoon rains can adversely affect the lives of millions. Even small changes can be problematic when averaged over many years, as when grasslands once used for grazing gradually become uninhabited deserts. In this chapter, we will first look at the evidence for climate change in the past; then we will investigate the causes of climate change from both natural processes and human activity.

Reconstructing Past Climates

A mere 18,000 years ago, Earth was in the grip of a cold spell, with *alpine glaciers* extending their icy fingers down river valleys and huge ice sheets (*continental glaciers*) covering vast areas of North America and Europe (see • Fig. 13.1). The ice at that time measured several kilometers thick and extended as far south as New York and the Ohio River Valley. The glaciers advanced and retreated more than 20 times during the last 2.5 million years. In the warmer periods, between glacier advances, average global temperatures were similar to those at present, even slightly higher in some cases. The advance and retreat of glaciers is closely related to variations in how Earth orbits the sun. Research in this area suggests we are thousands of years away from the next return of the glaciers.

Presently, glaciers cover less than 10 percent of Earth's land surface. The total volume of ice over the face of Earth amounts to about 25 million cubic kilometers. Most of this ice is in the Greenland and Antarctic ice sheets, and its accumulation over time has allowed scientists to measure past climatic changes. If global temperatures were to rise enough so that all of this ice melted, the level of the oceans would rise about 66 m (217 ft) (see • Fig. 13.2). Imagine the catastrophic results: Many major cities (such as New York, Shanghai, Tokyo, and London) would be inundated. Even a rise in global temperature of several degrees Celsius may be enough to raise sea level by a meter or more this century, flooding coastal lowlands and increasing the impact of storm surges.

Geological evidence left behind by advancing and retreating glaciers suggests that global climate has undergone slow but continuous changes. To reconstruct past climates, scientists must examine and then carefully piece together all the available evidence. Unfortunately, the evidence only gives a general understanding of what past climates were like. For example, fossil pollen of a tundra plant collected in a layer of sediment in New England and dated to be 12,000 years old suggests that the climate of that region was much colder than it is today.

Other evidence of global climatic change comes from core samples taken from ocean floor sediments and ice from Greenland and Antarctica. A multiuniversity research project known as CLIMAP (Climate: Long-Range Investigation Mapping and Prediction) has studied the past million years of global climate. Thousands of meters of ocean sediment obtained with a hollow-centered drill were analyzed. This sediment contained the remains of calcium carbonate shells of organisms that once lived near the surface. Because certain organisms can only live within a narrow range of temperature, the distribution and type of organisms within the sediment indicate the temperature of the surface water.

In addition, the oxygen-isotope* ratio of these shells provided information about the sequence of glacier advances. How? Most of the oxygen in sea water is composed of 8 protons and 8 neutrons in its nucleus, giving it an atomic weight of 16. However, about one out of every thousand oxygen atoms contains an extra 2 neutrons, giving it an atomic weight of 18. When ocean water evaporates, the heavy oxygen-18 tends to be left behind. Consequently, during periods of glacier advance, the oceans, which, therefore, contain less water, have a higher concentration of oxygen-18. Since the shells of marine organisms are constructed from the oxygen atoms existing in ocean water, determining the ratio of oxygen-18 to oxygen-16 within these shells yields information about how the climate may have varied in the past. A higher ratio of oxygen-18 to oxygen-16 in the sediment record suggests a colder climate, whereas a lower ratio suggests a warmer climate. Using data such as these, the CLIMAP project was able to reconstruct Earth's surface ocean temperature for various times during the past (see • Fig. 13.3).

Vertical ice cores extracted from ice sheets in Antarctica and Greenland provide additional information on

---

*Isotopes are atoms whose nuclei have the same number of protons but different numbers of neutrons.

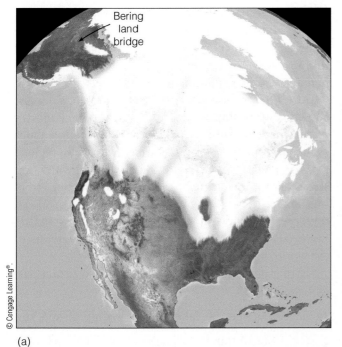

(a)

(b)

● **FIGURE 13.1** Extent of glaciation about 18,000 years ago over (a) North America and over (b) western Europe.

● **FIGURE 13.2** If all the ice locked up in glaciers and ice sheets were to melt, estimates are that this coastal area of south Florida would be under more than 61 m (200 ft) of water. Even a relatively small one-meter rise in sea level would threaten millions of people with rising seas. In fact, latest research suggests sea level could rise one meter or more by the end of this century due in part to the rapid melting of ice in Greenland and Antarctica, with additional rises expected thereafter.

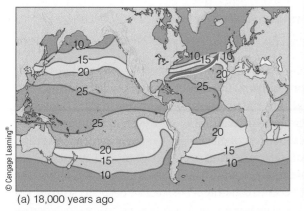

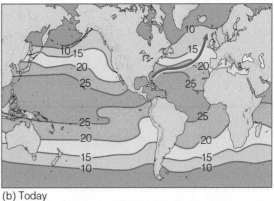

(a) 18,000 years ago

(b) Today

● **FIGURE 13.3** (a) Sea surface isotherms (°C) during August 18,000 years ago and (b) during August today. During the Ice Age (diagram a) the Gulf Stream (heavy red arrow) approached Europe at a lower latitude, depriving northern Europe of its warmth and producing a strong north-to-south ocean surface temperature gradient.

past temperature patterns. Glaciers form over land where temperatures are sufficiently low so that, during the course of a year, more snow falls than will melt. Successive snow accumulations over many years compact the snow, which slowly recrystallizes into ice. Since ice is composed of hydrogen and oxygen, examining the oxygen-isotope ratio in ancient cores provides a past record of temperature trends. Generally, the colder the air when the snow fell, the richer the concentration of oxygen-16 in the core. Moreover, bubbles of ancient air trapped in the ice can be analyzed to determine the past composition of the atmosphere (illustrated later in this chapter; see Fig. 13.13, p. 393).

Ice cores also record the causes of climate changes. One such cause is deduced from layers of sulfuric acid in the ice. The sulfuric acid originally came from large volcanic explosions that injected huge quantities of sulfur into the stratosphere. The resulting sulfate *aerosols* eventually fell to Earth in polar regions as acid snow, which was preserved in the ice sheets. (As we discussed in Chapter 1, aerosols are tiny airborne particles or droplets.) The Greenland ice cores also provide a continuous record of sulfur from human sources. Moreover, ice cores at both poles are being analyzed for many chemicals that provide records of biological and physical changes in the climate system, such as a beryllium isotope (^{10}Be) that indicates solar activity. Various types of dust collected in the cores indicate whether the climate was arid or wet.

Still other evidence of climatic change comes from the study of annual growth rings of trees, called **dendrochronology**. As a tree grows, it produces a layer of wood cells under its bark. Each year's growth appears as a ring, which can be seen in a cross section of the trunk. The change in the thickness of the rings, especially those rings that formed late in the tree's life (the outer rings), indicates climatic changes that may have taken place from one year to the next (see ● Fig. 13.4). The presence of frost rings during particularly cold periods and the chemistry of the wood itself provide additional information about a changing climate. Tree rings are only useful in regions that experience an annual cycle and in trees that are stressed by temperature or moisture during their growing season. The growth of tree rings has been correlated with precipitation and temperature patterns extending hundreds of years into the past in various regions of the world.

Other data that have been used to reconstruct past climates include:

1. records of natural lake-bottom sediment and soil deposits
2. the study of pollen in deep ice caves, soil deposits, and sea sediments
3. certain geologic evidence (ancient coal beds, sand dunes, and fossils), and the change in the water level of closed-basin lakes
4. documents concerning droughts, floods, crop yields, rain, snow, and dates of lakes freezing and trees blossoming
5. oxygen-isotope ratios of corals
6. the dates of calcium carbonate layers of stalactites in caves
7. borehole temperature profiles, which can be inverted to give records of past temperature change at the surface
8. deuterium (heavy hydrogen) ratios in ice cores, which indicate temperature changes

Even with all of this knowledge, our picture of past climates is still incomplete. With this shortcoming in mind, we will examine what the information gained about past climates does reveal.

Climate Throughout the Ages

Throughout much of Earth's history, the global climate was probably much warmer than it is today. During most of this time, the polar regions were free of ice. These comparatively warm conditions, however, were interrupted by several periods of glaciation. Geologic evidence suggests that one glacial period occurred about 700 m.y.a. (stands for "million years ago") and another about 300 m.y.a. The most recent one—the *Pleistocene epoch* or, simply, the **Ice Age**—began about 2.6 m.y.a. Often, each advance and retreat of glaciers within the Pleistocene epoch is simply referred to as an ice age (using lowercase letters).

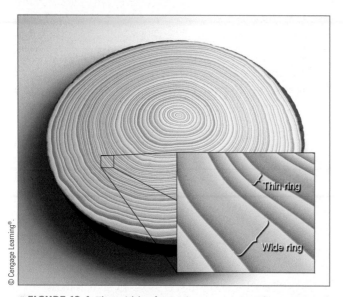

© Cengage Learning®.

● **FIGURE 13.4** The width of tree rings represents the amount of time each year that conditions were favorable for tree growth. The width of tree rings and the ratio of the light early wood to the darker later wood is controlled by a number of factors. In dry regions, the dominant factor may be precipitation, whereas at high latitudes or high elevations, it may be temperature.

Let's summarize the climatic conditions that led up to the Pleistocene.

About 65 m.y.a., Earth was warmer than it is now; polar ice caps did not exist. Beginning about 55 m.y.a., Earth entered a long cooling trend. After millions of years, polar ice appeared. As average temperatures continued to lower, the ice grew thicker, and by about 10 m.y.a., a deep blanket of ice covered the Antarctic. Meanwhile, snow and ice began to accumulate in high mountain valleys of the Northern Hemisphere, and alpine glaciers soon appeared.

About 2.5 m.y.a., continental glaciers appeared in the Northern Hemisphere, marking the beginning of the Pleistocene epoch—the time of the ice ages mentioned above. The Pleistocene, however, was not a period of continuous glaciation but a time when glaciers alternately advanced and retreated (melted back) over large portions of North America and Europe. Between the glacial advances were warmer periods called **interglacial periods,** which lasted for 10,000 years or more. The most recent was the *Eemian interglacial period*, which lasted from about 130,000 to about 114,000 y.a. (stands for "years ago"). A recently obtained ice core indicates that summer temperatures over Greenland may have averaged as much as 8°C (14°F) above today's values during the peak warmth of this period.

The most recent North American glaciers reached their maximum thickness and extent about 18,000–22,000 years ago y.a. At that time, average temperatures in Greenland were about 10°C (18°F) lower than at present and tropical average temperatures were about 4°C (7°F) lower than they are today. Because a great deal of water was in the form of ice over land, sea level was perhaps 120 m (395 ft) lower than it is now. The lower sea level exposed vast areas of land, such as the *Bering land bridge* (a strip of land that connected Siberia to Alaska as shown in Fig. 13.1a), which allowed human and animal migration from Asia to North America.

The ice began to retreat about 14,700 y.a. as surface temperatures slowly rose, producing a warm spell called the *Bölling-Alleröd* period (see ● Fig. 13.5). Then, about 13,000 y.a., the average temperature suddenly dropped and northeastern North America and northern Europe reverted back to glacial conditions. By about 10,000 y.a., the cold spell (known as the **Younger Dryas***) had ended abruptly, and temperatures had risen rapidly in many areas. Between about 9000 and 6000 y.a., the continental ice sheets over North America disappeared, and summer temperatures in the higher latitudes of the Northern Hemisphere were, on average, several degrees C above today's readings. This northern-latitude warm spell during the current interglacial period, or *Holocene epoch,* is sometimes called the **mid-Holocene maximum.** However, temperatures across much of the rest of the globe

*This exceptionally cold spell is named after the *Dryas,* an Arctic flower.

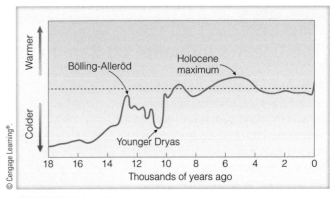

● **FIGURE 13.5** Relative air temperature variations (warmer and cooler periods) during the past 18,000 years. These data, which represent temperature records compiled from a variety of sources, only give an approximation of temperature changes. Some regions of the world experienced a cooling and other regions a warming that either preceded or lagged behind the temperature variations shown in the diagram.

appear to have been close to, or even slightly below, their current averages. About 5000 y.a., a cooling trend set in across northern latitudes, during which extensive alpine glaciers returned, but not continental glaciers.

It is interesting to note that ice core data from Greenland reveal that rapid shifts in climate (from ice age conditions to a much warmer state) took place in as little as a decade over central Greenland around the end of the Younger Dryas. The data also reveal that similar rapid shifts in climate occurred several times toward the end of the ice age. What could cause such rapid changes in temperature? One possible explanation is given in Focus section 13.1.

TEMPERATURE TRENDS DURING THE PAST 1000 YEARS ● Figure 13.6 shows how the average surface air temperature changed in the Northern Hemisphere during the last 1000 years.* The data needed to reconstruct the temperature profile in Fig. 13.6 comes from a variety of sources, including tree rings, corals, ice cores, historical records, and thermometers. Notice that about 1000 y.a., the Northern Hemisphere was about 1°C cooler than its average over the last couple of decades. However, certain regions in the Northern Hemisphere were warmer than others. For example, during this time vineyards flourished

*The National Academy of Science published a report comparing the work in Fig. 13.6 to many other reconstructions of temperatures during the past 1000 years, and found that they all gave basically the same results.

The Ocean's Influence on Rapid Climate Change

During the last glacial period, the climate around Greenland (and probably other areas of the world, such as northern Europe) underwent shifts, from ice-age temperatures to much warmer conditions in a matter of years. What could bring about such large fluctuations in temperature over such a short period of time? It now appears that a vast circulation of ocean water, known as the *conveyor belt,* plays a major role in the climate system.

● Figure 1 illustrates the movement of the ocean conveyor belt, or *thermohaline circulation.** The conveyor-like circulation begins in the north Atlantic near Greenland and Iceland, where salty surface water is cooled through contact with cold Arctic air masses. The cold, dense water sinks and flows southward through the deep Atlantic Ocean, around Africa, and into the Indian and Pacific Oceans.

In the North Atlantic, the sinking of cold water draws warm water northward from lower latitudes. As this water flows northward, evaporation increases the water's salinity (dissolved salt content) and density. When this salty, dense water reaches the far regions of the North Atlantic, it gradually sinks to great depths. This warm part of the conveyor delivers an incredible amount of tropical heat to the northern Atlantic. During the winter, this heat is transferred to the overlying atmosphere, and evaporation moistens the air. Strong westerly winds then carry this warmth and moisture into northern and western Europe, where it causes winters to be much warmer and wetter than one would normally expect for this latitude.

Ocean sediment records along with ice-core records from Greenland suggest

*Thermohaline circulations are ocean circulations produced by differences in temperature and/or salinity. Changes in ocean water temperature or salinity create changes in water density.

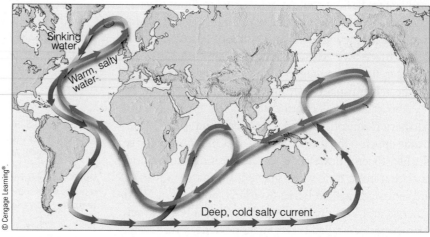

© Cengage Learning®

● FIGURE 1 The ocean conveyor belt. In the North Atlantic, cold, salty water sinks, drawing warm water northward from lower latitudes. The warm water provides warmth and moisture for the air above, which is then swept into northern Europe by westerly winds that keep the climate of that region milder than one would normally expect. When the conveyor belt stops, winters apparently turn much colder over northern Europe.

that the giant conveyor belt has switched on and off during the last glacial period. Such events have apparently coincided with rapid changes in climate. For example, when the conveyor belt is strong, winters in northern Europe tend to be wet and relatively mild. However, when the conveyor belt is weak or stops altogether, winters in northern Europe appear to turn much colder. This switching from a period of milder winters to one of severe cold shows up many times in the climate record. One such event—the Younger Dryas—illustrates how quickly climate can change and how western and northern Europe's climate can cool within a matter of decades, then quickly return back to milder conditions.

Apparently, one mechanism that can switch the conveyor belt off is a massive influx of freshwater. For example, about 13,000 years ago, freshwater from a huge glacial lake may have started to flow down the St. Lawrence River and into the North Atlantic. This massive inflow of freshwater may have reduced the salinity (and, hence,

density) of the surface water to the point that it stopped sinking. The conveyor may have shut down for about 1000 years during which time severe cold engulfed much of northern Europe in the Younger Dryas event. The conveyor belt possibly started up again when freshwater began to drain down the Mississippi rather than into the North Atlantic. It was during this time that milder conditions returned to northern Europe.

Will increasing levels of CO_2 have an effect on the conveyor belt? Some climate models predict that as CO_2 levels increase, more precipitation will fall over the North Atlantic. This situation reduces the density of the sea water and slows down the conveyor belt. Computer models predict that the conveyor belt could slow by about 20 to 45 percent over the twenty-first century, depending on how quickly carbon dioxide emissions increase. This slowdown of the conveyor belt would tend to keep Europe from warming as rapidly as the rest of the world. However, the conveyor belt is not expected to shut down entirely.

and wine was produced in England, indicating warm, dry summers and the absence of cold springs. This relatively warm, tranquil period of several hundred years over western Europe is sometimes referred to in that region as the *Medieval Climatic Optimum.* It was during the early part of

the millennium that Vikings colonized Iceland and Greenland and traveled to North America.

Notice in Fig. 13.6 that the temperature curve indicates a relatively warm period during the eleventh to fourteenth centuries—relatively warm, but still cooler

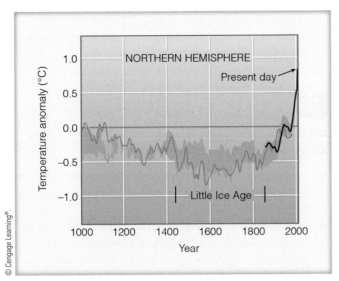

● **FIGURE 13.6** The average temperature variations over the Northern Hemisphere for the last 1000 years relative to the 1961 to 1990 average (zero line). The blue line represents air temperatures constructed from tree rings, corals, ice cores, and pollen. Yearly temperature data measured by thermometers are in black. This reconstruction has been compared to other similar reconstructions. The area shaded orange represents where these reconstructions overlap the data by 50 percent or more. (*Source:* Adapted from Susan Solomon, ed., Climate Change 2007–The Physical Science Basis: Working Group 1 Contribution to the Fourth Assessment report of the IPCC. Vol. 4 Cambridge University Press, 2007. Present-day data courtesy of NOAA.)

than the twentieth century. During this time, the relatively mild climate of Western Europe began to show large variations. For several hundred years the climate grew stormy. Both great floods and great droughts occurred. Extremely cold winters were followed by relatively warm ones. During the cold spells, the English vineyards and the Viking settlements suffered. Europe experienced several famines during the 1300s. The ice

pack of Greenland began advancing, and the Viking colony in Greenland perished.*

Again look at Fig. 13.6 and observe that the Northern Hemisphere experienced a slight cooling during the fifteenth to nineteenth centuries. This cooling was significant enough in certain areas to allow alpine glaciers to increase in size and advance down river canyons. Though the cooling averaged less than 1°C across the hemisphere, it had a major impact, especially in many areas of Europe, where winters were long and severe; summers, short and wet. The vineyards in England actually vanished, and farming became impossible in the more northern latitudes. There is no evidence that this cold spell existed worldwide. However, over Europe, this cold period has come to be known as the **Little Ice Age.**

TEMPERATURE TRENDS DURING THE PAST 100-PLUS YEARS In the early 1900s, the average global surface temperature began to rise (see ● Fig. 13.7). Notice that, from about 1900 to 1945, the average temperature rose nearly 0.5°C. Following the warmer period, Earth began to cool slightly over the next 25 years or so. In the late 1960s and 1970s, the cooling trend ended over most of the Northern Hemisphere, and by the mid-1970s, a warming trend set in that has continued into the twenty-first century. The increase in average temperature experienced over the Northern Hemisphere during the twentieth century is likely to have been the largest of any century during the past 1000 years. The year 2015 was the warmest up to that point in global records going back to 1880, and research suggests it was the warmest in more than 1000 years.

---

*Although climate change played a role in the demise of the Viking colony in northern Greenland, it was also their inability to adapt to the climate and to learn hunting and farming techniques from the Inuit that led to their downfall.

● **FIGURE 13.7** The orange and blue bars represent the annual average temperature variations over the globe (land and sea) from 1880 through 2015. Temperature changes are compared to the average surface temperature from 1951–1980. The dark solid line shows the five-year average temperature change.

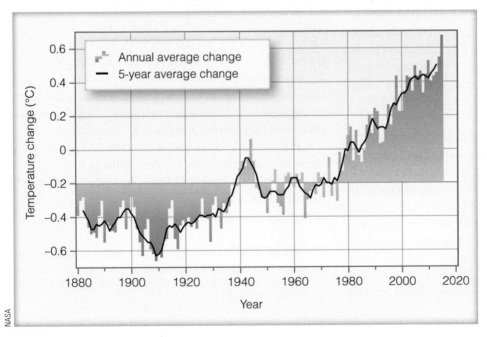

The average warming experienced over the globe, however, has not been uniform. The greatest warming has occurred in the Arctic and over the mid-latitude continents in winter and spring, whereas a few areas have not warmed in recent decades, such as areas of the oceans in the Southern Hemisphere and parts of Antarctica. The United States has experienced less warming than some other parts of the world. Moreover, most of the warming has occurred at night—a situation that has lengthened the frost-free seasons in many mid- and high-latitude regions. (In recent decades, the warming has been more equally distributed between day and night.)

The changes in air temperature shown in Fig. 13.7 are derived from three main sources: air temperatures over land, air temperatures over ocean, and sea surface temperatures. There are, however, uncertainties in the temperature record. For example, during this time period recording stations have moved, and techniques for measuring temperature have varied. Also, marine observing stations are scarce. In addition, urbanization (especially in developed nations) tends to artificially raise average temperatures as cities grow (called the *urban heat island effect*). Taking all of this information into account, along with improved information about sea surface temperatures, the **global warming** trend occurring between the early 1880s and the mid-2010s measured about 1.0°C (1.8°F). Look at Fig. 13.7 and observe that although the global temperature trend was relatively flat during the decade from 2000 to 2009, that decade was still warmer than the decade of the 1990s, which in turn was warmer than the decade of the 1980s.

A global increase in temperature of 1°C over about 100 years may seem small, but global temperatures probably have not varied by more than 2°C during the past 10,000 years. Consequently, an increase of 1°C becomes significant when compared with temperature changes over thousands of years.

Up to this point we have examined the temperature record of Earth's surface and observed that Earth has been in a warming trend for more than 100 years. A common question regarding this *global warming* is whether the warming trend is due to natural variations in the climate system, human activities, or a combination of the two. As we will see later in this chapter, climate scientists have concluded that most of the recent warming is due to an enhanced greenhouse effect caused by increasing levels of greenhouse gases, such as CO_2, mainly as a result of fossil

fuel burning (oil, gas, and coal).* If human activities are primarily responsible for this global warming, what was it that led to global warming trends of the past, before human beings walked on the surface of this planet?

BRIEF REVIEW

Before going on to the next section, here is a brief review of some of the facts and concepts we covered so far:

- Earth's climate is constantly undergoing change. Evidence suggests that throughout much of Earth's history, Earth's climate was much warmer than it is today.
- The most recent glacial period (or Ice Age) began about 2.5 m.y.a. During this time, glacial advances were interrupted by warmer periods called *interglacial periods*. In North America, continental glaciers reached their maximum thickness and extent about 18,000 to 22,000 y.a. and disappeared completely from North America by about 6000 y.a.
- The Younger Dryas event represents a time about 12,000 years ago when northeastern North America and northern Europe reverted back to glacier conditions.
- From the late 1800s to the early 2010s, Earth's surface temperature increased by about 1.0°C. The rate of global warming has increased over the last several decades.

Climate Change Caused by Natural Events

There are three "external" causes of climate change. They are:

1. changes in incoming solar radiation
2. changes in the composition of the atmosphere
3. changes in Earth's surface

Natural phenomena can cause climate to change by all three mechanisms, whereas human activities can change climate by both the second and third mechanisms. In addition to these external causes, there are "internal" causes of climate change, such as changes in the circulation patterns of the ocean and atmosphere, which redistribute energy within the climate system rather than altering the total amount of energy it holds.

Part of the complexity of the climate system is the intricate interrelationship of the elements involved. For example, if temperature changes, many other elements may be altered as well. The interactions among the atmosphere, the oceans, and the ice are extremely complex and

---

*Earth's atmospheric greenhouse effect is due mainly to the absorption and emission of infrared radiation by gases, such as water vapor, CO_2, methane, nitrous oxide, and chlorofluorocarbons. Refer back to Chapter 2 for additional information on this topic.

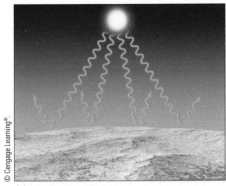
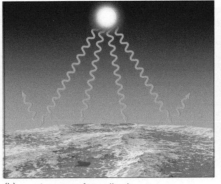
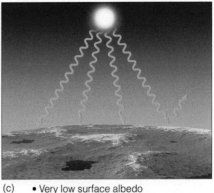

(a) • High surface albedo
• Low absorption of sunlight
• Gradual surface warming

(b) • Lower surface albedo
• Higher absorption of sunlight
• Surface warming increases

(c) • Very low surface albedo
• Much higher absorption of sunlight
• Surface warming enhanced

● **FIGURE 13.8** On a warming planet, the snow-albedo positive feedback would enhance the warming. (a) In polar regions snow reflects much of the sun's energy back to space. (b) If the air temperature were to gradually increase, some of the snow would melt, less sunlight would be reflected, and more sunlight would reach the ground, warming it more quickly. (c) The warm surface would enhance the snowmelt which, in turn, would accelerate the rise in temperature.

the number of possible interactions among these systems is enormous. No climatic element within the system is isolated from the others, which is why the complete picture of Earth's changing climate is not totally understood. With this in mind, we will first investigate how feedback systems work; then we will consider some of the current theories as to why Earth's climate changes naturally.

CLIMATE CHANGE: FEEDBACK MECHANISMS In Chapter 2, we learned that the Earth-atmosphere system is in a delicate balance between incoming and outgoing energy. If this balance is upset, even slightly, global climate can undergo a series of complicated changes.

Let's assume that the Earth-atmosphere system has been disturbed to the point that Earth has entered a slow warming trend. Over the years the temperature slowly rises, and water from the oceans rapidly evaporates into the warmer air. The increased quantity of water vapor absorbs more of Earth's infrared energy, thus strengthening the atmospheric greenhouse effect.

This strengthening of the greenhouse effect raises the air temperature even more, which, in turn, allows more water vapor to evaporate into the atmosphere. The greenhouse effect becomes even stronger, and the air temperature rises even more. This situation is known as the **water vapor–greenhouse feedback.** It represents a **positive feedback mechanism** because the initial increase in temperature is reinforced by the other processes. If this feedback were left unchecked, Earth's temperature would increase until the oceans completely evaporated. Such a chain reaction is called a *runaway greenhouse effect.*

Another positive feedback mechanism is the **snow-albedo feedback,** in which an increase in global surface air temperature might cause snow and ice to melt in polar latitudes. This melting would reduce the albedo

(reflectivity) of the surface, allowing more solar energy to reach the surface, which would further raise the temperature (see ● Fig. 13.8).

All feedback mechanisms work simultaneously and in both directions. Consequently, the snow-albedo feedback produces a positive feedback on a cooling planet as well. Suppose, for example, Earth were in a slow cooling trend. Lower temperatures might allow for a greater snow cover in middle and high latitudes, which would increase the albedo of the surface so that much of the incoming sunlight would be reflected back to space. Lower temperatures might further increase the snow cover, causing the air temperature to lower even more. If left unchecked, the snow-albedo positive feedback would produce a *runaway ice age,* which is highly unlikely on Earth because other feedback mechanisms in the atmospheric system would be working to moderate the magnitude of the cooling.

Along with positive feedback mechanisms, there are **negative feedback mechanisms**—those that tend to weaken the interactions among the variables rather than reinforce them. For example, a warming planet emits more infrared radiation.* If Earth's climate system were in a runaway greenhouse effect, the increase in radiant energy from the surface would greatly slow the rise in temperature and help to stabilize the climate. The increase in radiant energy from the surface as the planet warms is the strongest negative feedback in the climate system, and it greatly lessens the possibility of a runaway greenhouse effect. Consequently, although human-produced enhancement of the greenhouse

---

*Recall from Chapter 2, p. 34, that the outgoing infrared radiation from the surface increases at a rate proportional to the fourth power of the surface's absolute temperature. This relationship is called the Stefan-Boltzmann law. In effect, doubling the absolute temperature of Earth's surface would result in 16 times more energy emitted.

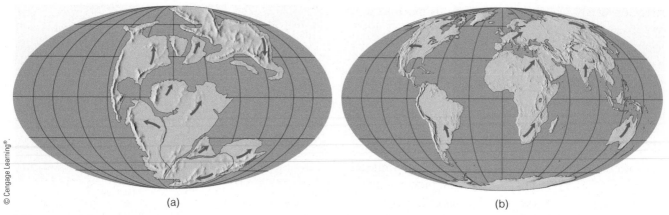

● FIGURE 13.9 Geographical distribution of (a) landmasses about 150 million years ago, and (b) today. Arrows show the relative direction of continental movement.

effect could have critical impacts, there is no evidence that a runaway greenhouse effect has ever occurred on Earth, and no indication that it will occur in the future.

Another negative feedback in Earth's climate system is the **chemical weathering–CO$_2$ feedback**. Chemical weathering is a process by which carbon dioxide is removed from the atmosphere as silicate minerals in rocks decompose in the presence of moisture. In this feedback, as chemical weathering increases, the amount of CO$_2$ in the atmosphere decreases. Chemical weathering (and the removal of CO$_2$ from the atmosphere) will generally take place more rapidly on a warmer planet, as chemical reactions speed up and greater evaporation from the oceans leads to more precipitation over the continents. As CO$_2$ leaves the atmosphere more quickly, CO$_2$ levels drop and Earth's climate begins to cool and stabilize. As temperatures dip, less water evaporates from the oceans, chemical weathering decreases, and the removal of CO$_2$ from the atmosphere diminishes.

In summary, the Earth–atmosphere system has a number of processes called *feedback mechanisms* that can either intensify or weaken tendencies toward climate change. Although we do not worry about a runaway greenhouse effect or an ice-covered Earth anytime in the future, there is concern that large positive feedback mechanisms may be working in the climate system to produce accelerated melting of ice in polar regions, especially in Greenland.

CLIMATE CHANGE: PLATE TECTONICS AND MOUNTAIN BUILDING Earlier, we saw that one of the external causes of climate change is a change in the surface of Earth. During the geologic past, Earth's surface has undergone extensive modifications. One involves the slow shifting of the continents and the ocean floors. This motion is explained in the widely accepted **theory of plate tectonics**. According to this theory, Earth's outer shell is composed of huge plates that fit together like pieces of a jigsaw puzzle. The plates, which slide over a partially molten zone below them, move in relation to one another. Continents

are embedded in the plates and move along like luggage riding piggyback on a conveyor belt. The rate of motion is extremely slow, only a few centimeters per year.

Besides providing insights into many geological processes, the theory of plate tectonics also helps to explain past climates. For example, we find glacial features near sea level in Africa today, suggesting that the area underwent a period of glaciation hundreds of millions of years ago. Were temperatures at low elevations near the equator ever cold enough to produce ice sheets? Probably not. The ice sheets formed when the African landmass was located at a much higher latitude. Over the many millions of years since then, the land has slowly moved to its present position. Thinking along the same line, we can see how the fossil remains of tropical vegetation can be found under layers of ice in polar regions today.

According to the theory of plate tectonics, the now-existing continents were at one time joined together in a single huge continent, which broke apart. Its pieces slowly moved across the face of the globe, thus changing the distribution of continents and ocean basins, as illustrated in ● Fig. 13.9. When landmasses are concentrated in middle and high latitudes (as they are today), it appears that ice sheets are more likely to form. During these times, there is a greater likelihood that more sunlight will be reflected back into space from the snow that falls over the continent in winter. Less sunlight absorbed by the surface lowers the air temperature, which allows for a greater snow cover, and, over thousands of years, the formation of continental glaciers. The amplified cooling that takes place over the snow-covered land is the snow-albedo feedback mentioned earlier.

Some periods of climatic change, taking place over millions of years, may be related to the rate at which the plates move and, hence, related to the amount of CO$_2$ in the air. For example, during times of rapid movement, an increase in volcanic activity vents massive quantities of CO$_2$ into the atmosphere, which enhances the atmospheric greenhouse effect, causing global temperatures to rise. Conversely, when plate movement is much slower, less volcanic activity puts less CO_2 into the atmosphere, which could result in a weaker greenhouse effect and a cooler planet. (It is important to note that modern-day volcanic activity emits less than 1 percent of the CO_2 produced by human activity.)

Mountains can cause the climate to change in a variety of ways. A chain of volcanic mountains forming perpendicular to the mean wind flow may disrupt the air-flow over mountains, altering the climate both upwind and downwind of them. By the same token, mountain building that occurs when two continental plates collide (like that which presumably formed the Himalayan mountains and Tibetan highlands) can have a marked influence on global circulation patterns and, hence, on the climate of an entire hemisphere. It has been theorized that the uplift of the Himalayas and Tibetan plateau helped to increase chemical weathering, thus pulling more carbon dioxide out of the air and lowering global temperature—perhaps enough to help trigger the onset of the Ice Age.

Up to now, we have examined how climatic variations can take place over millions of years due to the movement of continents and the associated restructuring of landmasses. We will now turn our attention to variations in Earth's orbit that may account for climatic fluctuations that take place on a time scale of tens of thousands of years.

CLIMATE CHANGE: VARIATIONS IN THE EARTH'S ORBIT

Another external cause of climate change involves a change in the amount of solar radiation that reaches Earth. A theory ascribing climatic changes to variations in Earth's orbit is the **Milankovitch theory**, named for the astronomer Milutin Milankovitch, who first proposed the idea in the 1930s. The basic premise of this theory is that, as Earth travels through space, three separate cyclic movements combine to produce variations in the amount, location, and timing of solar energy that falls on Earth.

The first cycle deals with changes in the shape **(eccentricity)** of Earth's orbit as Earth revolves about the sun. Notice in ● Fig. 13.10 that Earth's orbit changes from being elliptical (dashed line) to being nearly circular (solid line). To go from more circular to elliptical and back again takes about 100,000 years. The greater the eccentricity of the orbit (that is, the more elliptical the orbit), the greater the variation in solar energy received by Earth between its closest and farthest approach to the sun.

Presently, we are in a period of low eccentricity, which means that our annual orbit around the sun is more circular. Moreover, Earth is closer to the sun in January and farther away in July (see Chapter 2, p. 44). The difference in distance (which only amounts to about 3 percent) is responsible for a nearly 7 percent increase in the solar

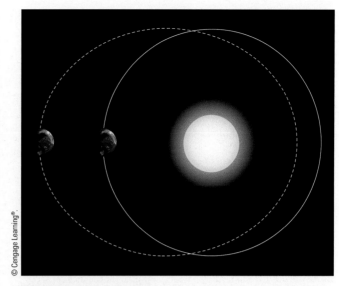

● **FIGURE 13.10** For Earth's orbit to stretch from nearly a circle (solid line) to an elliptical orbit (dashed line) and back again takes nearly 100,000 years. (Diagram is highly exaggerated and is not to scale.)

© Cengage Learning®.

energy received at the top of the atmosphere from July to January. When the difference in distance is 9 percent (a highly eccentric orbit), the difference in solar energy received between July and January is on the order of 20 percent. In addition, the more eccentric orbit changes the length of seasons in each hemisphere by changing the length of time between the vernal and autumnal equinoxes. Although rather large percentage changes in solar energy can occur between summer and winter, the globally and annually averaged change in solar energy received by Earth (due to orbital changes) hardly varies at all. It is the *distribution* of incoming solar energy that changes, not the totals.

The second cycle takes into account the fact that, as Earth rotates on its axis, it wobbles like a spinning top. This wobble, known as the **precession** of Earth's axis, occurs in a cycle of about 23,000 years. Presently, Earth is closer to the sun in January and farther away in July. Due to precession, the reverse will be true in about 11,000 years (see ● Fig. 13.11). In about 23,000 years we will be back to where we are today. This means, of course, that if everything else remains the same, 11,000 years from now seasonal variations in the Northern Hemisphere should be greater than at present. The opposite would be true for the Southern Hemisphere, although the effects there in both directions would be somewhat muted because of the greater proportion of ocean to land in the Southern Hemisphere.

The third cycle takes about 41,000 years to complete and relates to the changes in tilt **(obliquity)** with respect to Earth's orbit. Presently, Earth's orbital tilt is 23½°, but during the 41,000-year cycle the tilt varies from about 22° to 24½° (see ● Fig. 13.12). The smaller the tilt, the less seasonal variation there is between summer and winter in middle and high latitudes; thus, winters tend to be milder

● FIGURE 13.11 (a) Like a spinning top, Earth's axis of rotation slowly moves and traces out the path of a cone in space. (b) Presently Earth is closer to the sun in January, when the Northern Hemisphere experiences winter. (c) In about 11,000 years, due to precession, Earth will be closer to the sun in July, when the Northern Hemisphere experiences summer.

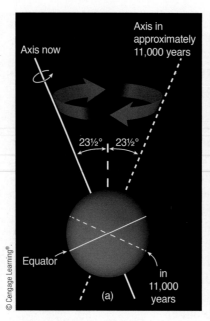

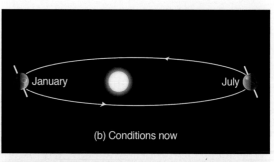

(b) Conditions now

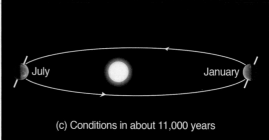

(c) Conditions in about 11,000 years

and summers cooler. (Recall from Chapter 2 that this tilt of Earth is the true "reason for the seasons.")

Over the extensive land areas located at high latitudes of the Northern Hemisphere, ice sheets are more likely to

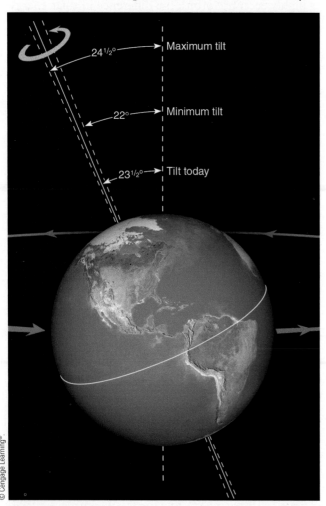

● FIGURE 13.12 Earth currently revolves around the sun while tilted on its axis by an angle of 23½°. During a period of 41,000 years, this angle of tilt ranges from about 22° to 24½°.

form when less solar radiation reaches the surface in summer. Less sunlight promotes lower summer temperatures. During the cooler summer, snow from the previous winter may not totally melt. The summer cooling predominates because even though there is an increase in winter insolation, very little sunlight ever reaches the far north in winter. The accumulation of snow over many years increases the albedo of the surface. Less sunlight reaches the surface, summer temperatures continue to fall, more snow accumulates, and continental ice sheets gradually form. At this point, it is interesting to note that when all of the Milankovich cycles are taken into account, the present trend should be toward *cooler summers* over high latitudes of the Northern Hemisphere.

In summary, the Milankovitch cycles that combine to produce variations in solar radiation received at Earth's surface include:

1. changes in the shape *(eccentricity)* of Earth's orbit about the sun
2. *precession* of Earth's axis of rotation, or wobbling
3. changes in the tilt *(obliquity)* of Earth's axis

In the 1970s, scientists of the CLIMAP project (described on p. 382) found strong evidence in deep-ocean sediments that variations in climate during the past several hundred thousand years were closely associated with the Milankovitch cycles. More recent studies have strengthened this premise. For example, analyses show that during the past 800,000 years, the extent and depth of ice sheets have peaked about every 100,000 years. This timing corresponds to the 100,000-year cycling of the eccentricity in Earth's orbit discussed earlier. Superimposed on this are smaller ice advances that show up at intervals of about 41,000 years and 23,000 years; as we saw above, Earth's obliquity varies on a 41,000-year cycle. So it does appear that the

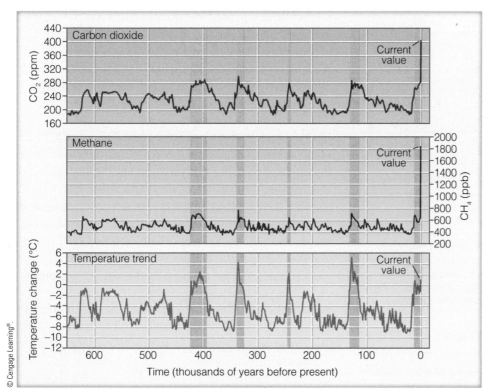

© Cengage Learning®.

● FIGURE 13.13 Variations of carbon dioxide (top, ppm), methane (middle, ppb), and temperatures (bottom, °C) during the past 650,000 years. Concentrations of gases are derived from air bubbles trapped within the ice sheets of Antarctica and extracted from ice cores. Temperatures are derived from the analysis of oxygen isotopes. The shaded orange band indicates current and most recent interglacial warm periods. (Note: *ppm* represents "parts per million by volume," and *ppb* represents "parts per billion by volume".) (Adapted from the Technical Summary by the Working Group 1 contribution to the Fourth Assessment Report to the IPCC, 2007.)

Milankovitch cycles play a role in the frequency of glaciation and the severity of natural climatic variation.

But orbital changes alone are probably not totally responsible for ice buildup and retreat. Evidence (from trapped air bubbles in the ice sheets of Greenland and Antarctica representing thousands of years of snow accumulation) reveals that CO_2 concentrations were about 30 percent lower during colder glacial periods than during warmer interglacial periods. This knowledge suggests that relatively small amounts of CO_2 in the atmosphere may have had the effect of amplifying the cooling initiated by the orbital changes. Likewise, increasing CO_2 levels at the end of the glacial period may have accounted for the rapid melting of the ice sheets.* Analysis of air bubbles in Antarctic ice cores reveals that methane, another important greenhouse gas, follows a pattern similar to that of CO_2 (see ● Fig. 13.13).

Interestingly, research shows that temperature changes thousands of years ago actually *preceded* the CO_2 changes. This observation supports the idea that CO_2 is a positive feedback in the climate system, where higher temperatures lead to higher CO_2 levels and lower temperatures to lower CO_2 levels. Consequently, CO_2 can be viewed as an internal, natural part of Earth's climate system.

Just why atmospheric CO_2 levels have varied as glaciers expanded and contracted is not clear, but it may be due to changes in biological activity taking place in the oceans. Moreover, as oceans cool, they remove more CO_2 from the atmosphere. Perhaps, also, changing levels of CO_2 indicate a shift in ocean circulation patterns. Such shifts, brought on by changes in precipitation and evaporation rates, may alter the distribution of heat energy around the world. Alteration wrought in this manner could, in turn, affect the global circulation of winds, which may explain why alpine glaciers in the Southern Hemisphere expanded and contracted in tune with Northern Hemisphere glaciers during the last ice age, even though the Southern Hemisphere (according to the Milankovitch cycles) was not in an orbital position for glaciation.

Still other factors may work in conjunction with Earth's orbital changes to explain the temperature variations between glacial and interglacial periods. Some of these are:

1. the amount of dust and other aerosols in the atmosphere
2. the reflectivity of the ice sheets
3. the concentration of other greenhouse gases
4. the changing characteristics of clouds
5. the rebounding of land, having been depressed by ice

Hence, the Milankovitch cycles, in association with other natural factors, may explain the advance and retreat of ice over periods of 10,000 to 100,000 years. But what caused the ice ages to begin in the first place? And why were periods of glaciation so infrequent during earlier geologic times? The Milankovitch theory does not attempt to answer these questions, although it is likely that changes in Earth's geography related to plate tectonics are involved.

*During peak CO_2 levels, its concentration was less than 300 ppm, which is substantially lower than its concentration of about 405 ppm in today's atmosphere.

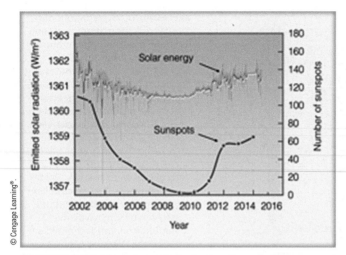

● **FIGURE 13.14** Changes in total solar irradiance (solar energy output) (red line) in watts per square meter (red line) measured by the NASA-sponsored SORCE satellite from 2003 to 2012. Gray line represents the number of sunspots observed each year.

CLIMATE CHANGE: VARIATIONS IN SOLAR OUTPUT

Solar energy measurements made by sophisticated instruments aboard satellites show that the sun's energy output (called *brightness*) varies slightly—by a fraction of one percent—with sunspot activity.

Sunspots are huge magnetic storms on the sun that show up as cooler (darker) regions on the sun's surface. They occur in cycles, with the number and size reaching a maximum approximately every 11 years. During periods of maximum sunspots, the sun emits more energy (about 0.1 percent more) than during periods of sunspot minimums (see ● Fig. 13.14). Evidently, the greater number of bright areas *(faculae)* around the sunspots radiate more energy, which offsets the effect of the dark spots.

It appears that the 11-year sunspot cycle has not always prevailed. Between 1645 and 1715, during the period known as the **Maunder minimum,*** very few sunspots were observed. This minimum coincided with part of the Little Ice Age (recall that this was a cool spell in the temperature record experienced mainly over Europe). Some scientists suggest that a reduction in the sun's energy output may have played a role in exacerbating this cool spell, although the cooling started many years before the Maunder minimum began. Fluctuations in solar output may account for small climatic changes over time scales of decades and centuries. Many theories have been proposed linking solar variations to climate change, but none has been proven. Currently, instruments aboard satellites and solar telescopes on Earth are monitoring the sun to observe how much its energy output varies. To date, these measurements show that solar output has only changed a fraction of one percent over several decades, which is not enough time to explain recent worldwide warming. In fact, the solar cycle that peaked in the early 2010s was the weakest in more than 100 years, yet global temperatures continued to climb. Because high-quality data sources have only been available for several decades, it may be some time before we fully understand the relationship between solar activity and climate change on Earth.

CLIMATE CHANGE: ATMOSPHERIC PARTICLES

Microscopic liquid and solid particles (*aerosols*) that enter the atmosphere from both natural and human-induced sources can have an effect on climate. The effect these particles have on the climate is exceedingly complex and depends upon a number of factors, such as the particle's size, shape, color, chemical composition, and vertical distribution above the surface. In this section, we will examine those particles that enter the atmosphere through natural means.

Particles Near the Surface Particles can enter the atmosphere in a variety of natural ways. For example, wildfires can produce copious amounts of tiny smoke particles, and dust storms can sweep tons of fine particles into the atmosphere. Smoldering volcanoes can release significant quantities of sulfur-rich aerosols into the lower atmosphere. And even the oceans are a major source of natural sulfur aerosols, as tiny drifting aquatic plants—phytoplankton—produce a form of sulfur (dimethylsulphide, DMS) that slowly diffuses into the atmosphere, where it combines with oxygen to form sulfur dioxide, which in turn converts to sulfate aerosols. Although the effect these particles have on the climate system is complex, the overall effect they have is to *cool the surface* by preventing sunlight from reaching the surface.

Volcanic Eruptions Volcanic eruptions can have a major impact on climate. During volcanic eruptions, fine particles of ash and dust (as well as gases) can be ejected into the atmosphere (see ● Fig 13.15). Scientists agree that the volcanic eruptions having the greatest impact on climate are those rich in sulfur gases. These gases, when ejected into the stratosphere,* combine with water vapor in the presence of sunlight to produce tiny, reflective sulfuric acid particles that grow in size, forming a dense layer of haze. The haze may reside in the stratosphere for several years, absorbing and reflecting back to space a portion of the sun's incoming energy. The reflection of incoming sunlight by the haze tends to cool the air at Earth's surface, especially in the hemisphere where the eruption occurs.

Two of the largest volcanic eruptions of the twentieth century in terms of their sulfur-rich veil were that of El Chichón in Mexico during April 1982, and Mount Pinatubo in the Philippines during June 1991. The eruption of Mount Pinatubo in 1991 was many times greater than that

---

*This period is named after E. W. Maunder, the British solar astronomer who first discovered the low sunspot period sometime in the late 1880s.

*You may recall from Chapter 1 that the stratosphere is a stable layer of air above the troposphere, typically about 11 to 50 km (7 to 31 mi) above Earth's surface.

USGS

● **FIGURE 13.15** Large volcanic eruptions rich in sulfur can affect climate. As sulfur gases in the stratosphere transform into tiny reflective sulfuric acid particles, they prevent a portion of the sun's energy from reaching the surface. Here, the Philippine volcano Mount Pinatubo erupts during June 1991.

of Mount St. Helens in the Pacific Northwest in 1980. In fact, the largest eruption of Mount St. Helens was a lateral explosion that pulverized a portion of the volcano's north slope. The ensuing dust and ash (and very little sulfur) had virtually no effect on global climate, as the volcanic material was confined mostly to the lower atmosphere and fell out quite rapidly over a large area of the northwestern United States.

Mount Pinatubo ejected an estimated 20 million tons of sulfur dioxide (more than twice that of El Chichón), which gradually worked its way around the globe. For major eruptions such as this one, mathematical models predict that average hemispheric temperatures can drop by about 0.2° to 0.5°C or more for one to three years after the eruption. Model predictions agreed with temperature changes brought on by the Pinatubo eruption, as in early 1992 the mean global surface temperature had decreased by about 0.5°C (see ● Fig. 13.16). The cooling might even have been greater had the eruption not coincided with a major El Niño event in 1991–1992 (see Chapter 7, p. 199, for more information on El Niño). In spite of El Niño, the eruption of Mount Pinatubo produced the two coolest years of the decade, 1991 and 1992. Even more-modest volcanoes may have a climate impact if they erupt frequently. Scientists, for example, have found that a series of eruptions from dozens of smaller volcanoes may have impacted global warming. This impact may have reduced global warming by 25 percent from 2000 to 2009 from what otherwise would have been expected.

Climate change is not just about temperature. In fact, precipitation may be the most important weather element associated with climate change in terms of an impact on humans. The Pinatubo eruption of 1991 had a large impact on the world's hydrologic cycle, even causing

drought over certain areas owing to the effects (described in the next section) that volcanic emissions can exert on cloud and precipitation formation. As we have just seen, volcanic eruptions rich in sulfur tend to cool Earth's surface. An infamous cold spell often linked to volcanic activity occurred during the year 1816, which has come to be known as "the year without a summer." In Europe that year, bad weather contributed to a poor wheat crop, and famine spread across the land. In North America, unusual blasts of cold polar air moved through Canada and the northeastern United States between May and September. The cold spell brought heavy snow in June and killing frosts in July and August. In the warmer days that followed each cold snap, farmers replanted, only to have another cold outbreak damage the planting. The unusually cold summer was followed by a bitterly cold winter. The cold

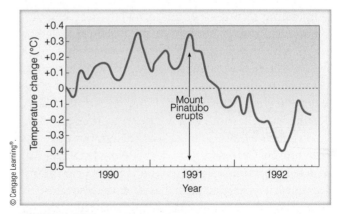

© Cengage Learning®.

● **FIGURE 13.16** Changes in average global air temperature from 1990 to 1992. After the eruption of Mount Pinatubo in June 1991, the average global temperature by July 1992 decreased by almost 0.5°C (0.9°F) from the 1981 to 1990 average (dashed line). (Data courtesy of John Christy, University of Alabama, Huntsville, and R. Spencer, NASA Marshall Space Flight Center.)

weather over eastern North America and western Europe followed the massive eruption in 1815 of Mount Tambora in Indonesia. In addition, a smaller volcanic eruption had occurred in 1809, from which the climate system may not have fully recovered when Tambora erupted in 1815.

In an attempt to correlate sulfur-rich volcanic eruptions with long-term trends in global climate, scientists are measuring the acidity of annual ice layers in Greenland and Antarctica. Generally, the greater the concentration of sulfuric acid particles in the atmosphere, the greater the acidity of the ice layer. Relatively acidic ice has been uncovered from about 1350 to about 1700, a time that corresponds to the early part of the *Little Ice Age*. Such findings suggest that sulfur-rich volcanic eruptions may have played an important role in triggering this comparatively cool period and, perhaps, other cool periods during the geologic past. Moreover, recent core samples taken from the northern Pacific Ocean reveal that volcanic eruptions in the northern Pacific were at least 10 times larger 2.5 million years ago (a time when Northern Hemisphere glaciation began) than previous volcanic events recorded elsewhere in the sediment.

Particles in the Upper Atmosphere and Mass Extinction   Could the injection of particles high into the atmosphere by a catastrophic event cause climate change and the demise of living creatures? About 65 million years ago, the dinosaurs, along with an estimated 70 to 75 percent of all plant and animal species on Earth, died in a mass extinction. What could have caused such a devastating event?

About 65 m.y.a., a giant meteorite estimated to measure some 10 km (6 mi) in diameter slammed into Earth near the Yucatán Peninsula of Mexico at about 44,000 mi/hr (see ● Fig. 13.17). One widely accepted theory proposes that the impact sent billions of tons of dust and debris into the upper atmosphere, where such particles would have circled the globe for months, greatly reducing the sunlight reaching Earth's surface. Reduced sunlight would have disrupted photosynthesis in plants which, in turn, could lead to a breakdown in the planet's food chain. Lack of food, as well as much cooler conditions brought on by the dust, must have had an adverse effect on life, especially large plant-eating dinosaurs.

Evidence for this catastrophic collision comes from the geologic record, which shows a thin layer of sediment deposited worldwide, about the time the dinosaurs disappeared. The sediment contains *iridium*, an element that is rare on Earth but common in certain types of meteorites.

As evidence has accumulated over the last few years, most scientists now consider the Yucatán meteorite strike to be the main explanation for the extinction of the dinosaurs. However, other factors may have played a role as well, including huge volcanic eruptions that occurred in India at about the same time.

BRIEF REVIEW

Up to this point, we have examined a number of ways Earth's climate can change by natural means. Before going on to the next section, which covers climate change brought on by human activities, here is a brief review of some of the facts and concepts we covered so far:

- The external causes of climate change include: (1) changes in incoming solar radiation; (2) changes in the composition of the atmosphere; (3) changes in the surface of Earth.

- The shifting of continents, along with volcanic activity and mountain building, are possible causes of natural climate change.

- The Milankovitch theory (in association with other natural forces) proposes that alternating glacial and interglacial episodes during the past 2.5 million years are the result of small variations in the tilt of Earth's axis and in the geometry of Earth's orbit around the sun.

- Trapped air bubbles in the ice sheets of Greenland and Antarctica reveal that CO_2 levels and methane levels were lower during colder glacial periods and higher during warmer interglacial periods. But even when the levels were higher, they still were much lower than they are today.

- Fluctuations in solar output (brightness) may account for small climatic changes over periods of decades to centuries.

- Volcanic eruptions, rich in sulfur, may be responsible for cooler periods in the geologic past.

● **FIGURE 13.17** Artist's interpretation of a giant meteorite striking Earth's surface about 65 million years ago, creating a huge explosion that sent billions of tons of dust and debris into the upper atmosphere where it blocked out sunlight and caused climate change on Earth.

Climate Change Caused by Human (Anthropogenic) Activities

Earlier in this chapter we saw how variations in atmospheric carbon dioxide may have contributed to changes in global climate spanning thousands and even millions of years. Today, we are modifying the atmosphere by injecting into it vast quantities of particles and greenhouse gases without fully understanding the long-term consequences. In this section, we will first look at how particles emitted into the lower atmosphere by human activities may be affecting climate. Then we will examine how CO_2 and other trace gases are enhancing Earth's greenhouse effect, producing global warming.

CLIMATE CHANGE: AEROSOLS INJECTED INTO THE LOWER ATMOSPHERE

In a previous section we learned that tiny solid and liquid particles (aerosols) can enter the atmosphere from both human-induced and natural sources. The human-induced sources include emissions from factories, autos, trucks, aircraft, power plants, and home furnaces and fireplaces, to name a few. Many aerosols are not injected directly into the atmosphere, but form when gases convert to particles. Some particles, such as sulfates and nitrates, mainly reflect incoming sunlight, whereas others, such as soot, readily absorb sunlight. Many of the particles that reduce the amount of sunlight reaching Earth's surface tend to cause a *net cooling* of the surface air during the day.

In recent years, the effect of highly reflective **sulfate aerosols** on climate has been extensively researched. Earlier we learned that sulfate aerosols can come from natural sources, such as the oceans. However, the majority of these sulfate particles in the lower atmosphere are directly related to human activities and come primarily from the combustion of sulfur-containing fossil fuels.

Sulfur pollution, which has more than doubled globally since preindustrial times, enters the atmosphere mainly as sulfur dioxide gas. There, it transforms into tiny sulfate droplets or particles. Since these aerosols usually remain in the lower atmosphere for only a few days, they do not have time to spread around the globe. Hence, they are not well mixed and their effect is felt mostly over the Northern Hemisphere, especially over polluted regions.

Sulfate aerosols not only reflect incoming sunlight back to space, but they also serve as cloud condensation nuclei, tiny particles on which cloud droplets form. Thus, they have the potential for altering the physical characteristics of clouds. For example, if the number of sulfate aerosols and, hence, condensation nuclei inside a cloud should increase, the cloud would have to share its available moisture with the added nuclei, a situation that should produce many more (but smaller) cloud droplets. The greater number of droplets would reflect more sunlight and have the effect of *brightening* the cloud and reducing the amount of sunlight that reaches the surface. This process could also reduce the chance that precipitation could fall.

In summary, sulfate aerosols reflect incoming sunlight, which tends to lower Earth's surface temperature during the day. Sulfate aerosols may also modify clouds by increasing their reflectivity. Because sulfate pollution increased dramatically over industrialized areas of the Northern Hemisphere from the 1940s into the 1970s, it may help explain why global temperatures showed little warming during this period, and perhaps why warming intensified in the 1980s and 1990s as sulfate pollution decreased. The effects of aerosols in the lower atmosphere on climate are a topic of active research. Information on the possible effect on climate from particles injected into the atmosphere during a nuclear war is given in Focus section 13.2.

CLIMATE CHANGE: GREENHOUSE GASES

We learned in Chapter 2, p. 37, that carbon dioxide is a greenhouse gas that strongly absorbs infrared radiation and plays a major role in the warming of the lower atmosphere. Everything else being equal, the more CO_2 in the atmosphere, the warmer the surface air. We also know that CO_2 has been increasing steadily in the atmosphere, primarily due to human activities, such as the burning of fossil fuels like coal, oil, and natural gas (see Fig. 1.18, p. 17). Deforestation is adding to this increase. Through the process of photosynthesis, the leaves of trees remove CO_2 from the atmosphere. The CO_2 is then stored in leaves, branches, and roots. When the trees are cut and burned, or left to rot, the CO_2 goes back into the atmosphere.

Presently, the annual average of CO_2 is about 405 ppm, and the concentration is increasing by about 2 to 3 ppm per year. In recent studies, scientists have examined scenarios that show the atmospheric concentration of CO_2 to be anywhere from 421 to 1313 ppm by the end of this century. The actual number will depend on how much CO_2 is emitted by human activity in the coming decades, and on how natural processes interact with the increase in CO_2.

To complicate the picture, increasing concentrations of other greenhouse gases—such as methane (CH_4), nitrous oxide (N_2O), and chlorofluorocarbons (CFCs)—all readily absorb infrared radiation.* Overall, these gases tend to be much less prevalent than CO_2 in the atmosphere, but many of them are more powerful per molecule in their heat-trapping effect. Collectively, these gases enhance the atmospheric greenhouse effect by a substantial amount—almost half of the effect now produced by CO_2. Before we

---

*Refer back to Chapter 1 and Table 1.1 on p. 16 for additional information on the average concentration of these greenhouse gases.

Nuclear Winter—Climate Change Induced By Nuclear War

Many studies indicate that a nuclear war brought on by either human carelessness or negligence would drastically modify Earth's climate, instigating climate change unprecedented in recorded human history.

Researchers assume that a nuclear war would raise an enormous pall of thick, sooty smoke from massive fires that would burn for days, even weeks, following an attack. The smoke would drift higher into the atmosphere, where it would be caught in the upper-level westerlies and circle the middle latitudes of the Northern Hemisphere. Unlike soil dust, which mainly scatters and reflects incoming solar radiation, soot particles readily absorb sunlight. Hence, months, or perhaps years, after the war, sunlight would virtually be unable to penetrate the smoke layer, bringing darkness or, at best, twilight at midday.

Such reduction in solar energy would cause surface air temperatures over landmasses to drop below freezing, even during the summer, resulting in extensive damage to plants and crops and the death of millions (or possibly billions) of people. The dark, cold, and gloomy conditions that would be brought on by nuclear war are often referred to as *nuclear winter.*

As the lower troposphere cools, the solar energy absorbed by the smoke particles in the upper troposphere would cause this region to warm. The end result would be a strong, stable temperature inversion extending from the surface up into the higher atmosphere. A strong inversion would lead to a number of adverse effects, such as suppressing convection, altering precipitation processes, and causing major changes in the general wind patterns.

The heating of the upper part of the smoke cloud would cause it to rise upward into the stratosphere, where it would then drift around the world. Thus, about one-third of the smoke would remain in the atmosphere for up to a decade. The other two-thirds would be washed out in a month or so by precipitation. This smoke lofting, combined with persisting sea ice formed by the initial cooling, would produce climatic change that would remain for more than a decade.

Virtually all research on nuclear winter, including models and analog studies, confirms this gloomy scenario. Observations of forest fires show lower temperatures under the smoke, confirming part of the theory. Studies now show that even a regional nuclear conflict could not only produce widespread, dramatic cooling, but it could also trigger chemical reactions that would produce major global depletion of stratospheric ozone. The implications of nuclear winter are clear: A nuclear war would drastically alter global climate and would devastate our living environment.

examine how increasing levels of greenhouse gases are producing climate change, we will look at how humans may be altering the climate by changing the landscape.

CLIMATE CHANGE: LAND USE CHANGES All climate models predict that, as human populations continue to spew greenhouse gases into the air, the climate will change and Earth's surface will warm. But are humans changing the climate by other activities as well? Modifications of Earth's surface taking place right now could potentially be influencing the immediate climate of certain regions. For example, studies show that about half the rainfall in the Amazon River Basin is returned to the atmosphere through evaporation and through transpiration from the leaves of trees. Consequently, as is happening now, clearing large areas of tropical rainforests in South America to create open areas for farms and cattle ranges will most likely cause a decrease in evaporative cooling. This decrease, in turn, could lead to a warming in that area of at least several degrees Celsius. In turn, the reflectivity of the deforested area will change. Similar changes in albedo result from the overgrazing and excessive cultivation of grasslands in semi-arid regions, causing an increase in desert conditions (a process known as **desertification**).

Billions of acres of the world's range and cropland, along with the welfare of millions of people, have been affected by desertification in recent decades. One of the main causes is believed to be overgrazing, although overcultivation, poor irrigation practices, and deforestation also play a role. The effect desertification will have on climate, as surface albedos increase and more dust is swept into the air, is uncertain. Some studies have proposed that allowing livestock to graze over larger areas, as their wild predecessors did, could reduce the risk of desertification by minimizing the effects of more intensive, localized grazing. Desertification may also be reversible in some areas if a long period of rainfall deficit is followed by adequate rainfall. A multiyear dry period across the Sahel region of North Africa (a region bounded on the north by the Sahara Desert and to the south by grasslands) led to devastating drought and famine during the 1970s and 1980s. However, the annual rainfall increased after this period, and vegetation has now returned to many of the affected areas during wet years.

Some scientists have concluded that humans were altering climate way before modern civilizations came along. A growing amount of research indicates that pre-industrial farming had been boosting methane and carbon dioxide in the atmosphere for up to 8000 years before the more rapid increase brought about by industrialization. One of the scientists involved in this research, retired professor William Ruddiman (University of Virginia), has theorized that, without the pre-industrial farming, we would have entered a naturally occurring ice age—an idea that remains controversial. Ruddiman even suggests that the Little Ice Age of the fifteenth through the nineteenth centuries in Europe was caused by plagues that killed millions of people and thereby resulted in a reduction in farming.

The reasoning behind this idea goes something like this: As forests are cleared for farming, levels of CO_2 and methane increase, producing a strong greenhouse effect and a rise in surface air temperature. When catastrophic plagues strike—the bubonic plague, for instance—high mortality rates cause farms to be abandoned. As forests begin to take over the untended land, levels of CO_2 and methane drop, causing a reduction in the greenhouse effect and a corresponding drop in air temperature. After the plague abates, farming eventually returns, forests are cleared, levels of greenhouse gases go up, and surface air temperatures rise. (It should be emphasized that the rapid increase in greenhouse gases over the last few decades is primarily due to fossil fuel use, rather than land-use processes such as those described above.)

Climate Change: Global Warming

We have seen several times in this chapter that Earth's atmosphere is in a warming trend that began during the late nineteenth century. This warming trend is real, as the average global surface air temperature since the late 1800s has risen by about 1.0°C (1.8°F). Moreover, the global average for each decade since the 1980s has been warmer than that of the preceding decade. There are many signs of increasing global warmth other than temperature readings. For example, the amount of water locked in the world's glaciers and ice sheets is steadily decreasing, and sea level is steadily rising. Global warming might even be apparent where you live. The growing season, for example, may be extended, or the changing of the leaf color in autumn may be observed later than in the past.* Global warming in any given year, however, is small, and only becomes significant when averaged over many years, such as decades. So it is important not to think of global warming based on a specific weather event. A few facts illustrate the point. In 2014, a January cold wave across eastern North America sent temperatures plummeting. Atop Mount Mitchell in North Carolina, a low temperature of −31°C (−24°F) was the second coldest reading ever observed there. Binghamton, New York, dipped below 0°F on ten days, the most ever recorded there in January. Yet at

the same time, it was the third warmest January on record in California, and globally, January 2014 was the fourth warmest on record. (Note that the United States only represents about 2 percent of the entire surface area of the planet, so one cannot use United States conditions alone as an index of how much the entire world is warming.) For more on the relationship between climate change and extreme weather, see Focus section 13.3.

RECENT GLOBAL WARMING: PERSPECTIVE   Is the warming trend experienced over the past 100-plus years due to increasing greenhouse gases and an enhanced greenhouse effect? Before we can address this question, we need to review a few concepts we learned in Chapter 2.

Radiative Forcing Agents   We know from Chapter 2 that our world without water vapor, CO_2, and other greenhouse gases would be a colder world—about 33°C (59°F) colder than at present. With an average surface temperature of about −18°C (0°F), much of the planet would be uninhabitable. In Chapter 2, we also learned that when the rate of the incoming solar energy balances the rate of outgoing infrared energy from Earth's surface and atmosphere, the Earth-atmosphere system is in a state of *radiative equilibrium*. Increasing concentrations of greenhouse gases can disturb this equilibrium and are, therefore, referred to as **radiative forcing agents.** The **radiative forcing*** provided by extra CO_2 and other greenhouse gases increased by about 3 watts per square meter (W/m²) over the past several hundred years, with the most rapid increase occurring over the last several decades. At the same time, increasing amounts of sun-blocking aerosols emitted by human activity (such as sulfates and other pollutants), together with their effects on cloudiness, have led to a decrease in radiative forcing estimated at roughly 1 W/m², which counteracts part of the greenhouse forcing.

Overall, it is very likely that most of the warming during the last few decades comes from increasing levels of greenhouse gases. But what part does natural climate variability play in global warming? And with atmospheric concentrations of CO_2 increasing by more than 30 percent since the early 1900s, why has the observed increase in global temperature been relatively small?

We already know that the climate may change due to natural events. For example, changes in the sun's energy output (called *solar irradiance*) and volcanic eruptions rich in sulfur are, as we have discussed, two major natural radiative forcing agents. The most recent studies show that since the middle 1700s, changes in the sun's energy output have increased the total radiative forcing on the

---

*Focus section 12.4 in Chapter 12 on p. 376 looks at the influence of climate change on plant hardiness zones.

*Radiative forcing is interpreted as an increase (positive) or a decrease (negative) in net radiant energy observed over an area in the middle of the tropopause. All factors being equal, an increase in *radiative forcing* may induce surface *warming*, whereas a *decrease* may induce surface *cooling*.

Climate Change and Extreme Weather

Americans dealt with a succession of extreme weather events in 2011 that seemed never-ending. Two major winter storms struck the Eastern Seaboard in January, leaving New York City with its snowiest January on record. A February storm paralyzed Chicago with 21.2 inches and brought several cities their heaviest snow in history (see ● Fig. 2). The onslaught shifted in the spring, as a catastrophic four-day series of tornadoes swept through the southeast United States. The worst day by far was April 27, which brought the nation's most prolific 24-hour outbreak of twisters on record, and the deadliest in more than 80 years. At least 322 people were killed and an estimated $10 billion in damage occurred. Less than a month later, a violent tornado ripped through Joplin, Missouri, killing an estimated 159 people—the largest toll from a single tornado since 1947.

Disastrous weather continued into the late spring and summer of 2011. Record floods poured across the Missouri and Mississippi river valleys, and unprecedented drought and heat struck the Southern Plains. Texas recorded the hottest summer in history for any state, with a statewide average temperature of 86.8°F. Tropical cyclones largely bypassed the nation, but Hurricane Irene brought destructive floods to New England in August, followed by the Northeast's heaviest October snows in more than a century.

Disasters such as hurricanes, tornadoes, wildfires, and floods often have people wondering whether climate change might be boosting the occurrence of such extremes. No single weather event is "caused" by a changing climate, but research shows that human-produced greenhouse gases are making some types of extreme weather more likely. For example, heat waves are expected to become more frequent and intense in coming decades. There is also evidence that precipitation is becoming increasingly

concentrated in very heavy rain and snow events in many parts of the world, including the United States. A warming climate allows more water to evaporate into the atmosphere from the oceans, which can help intensify rain and snow where it is falling. However, warmer temperatures also help draw moisture from already dry land. Computer models suggest that both precipitation and drought will continue to intensify during this century.

Some other kinds of extreme weather are less directly connected to increased greenhouse gases. For example, tornadoes are very localized events that develop as various ingredients come together on a larger scale to create severe thunderstorms. The number of tornado reports in the United States has roughly doubled since the 1950s, but this phenomenon is mainly due to the growing number of storm spotters and chasers, as well as better post-storm surveys. There has been no significant trend in the number of the strongest tornadoes (those ranked EF3 or greater on the Enhanced Fujita Scale). Some research has pointed to a potential increase in severe thunderstorms across parts of the southern and eastern United States as the climate warms and average instability increases. In order for the most violent tornadoes to occur, this high instability must also be accompanied by substantial vertical wind shear. As the nation warms, vertical wind shear is expected to decrease on average. However, we can still expect some periods when adequate vertical wind shear is present along with high instability. Thus, it now appears that severe weather episodes may become more variable, with the possibility of fewer tornado outbreaks overall but more intense outbreaks when enough wind shear is present. In the United States, the terrible tornado year of 2011, which caused more than 500 deaths and billions of dollars in damage, was followed by the relatively

tranquil tornado season of 2012, which produced the fewest tornadoes of any year in more than 60 years of recordkeeping.

A growing area of research focuses on *detection* and *attribution*. These studies attempt to identify a change in climate, and determine how much can be attributed to human-produced greenhouse gases. With the help of new statistical and numerical modeling tools, scientists are now beginning to estimate how the odds for a given weather event might have been boosted by our warming planet, or what part of a given event may be related to human-produced greenhouse gases. For example, one study in 2015 used models to estimate that climate change was responsible for 8 to 27 percent of the summertime soil-moisture deficit observed in California during the drought years of 2012, 2013, and 2014.

● **FIGURE 2** A man carefully maneuvers between stranded cars on Chicago's Lake Shore Drive after a huge snowstorm during February 2011.

REUTERS/John Gress/Landov

climate system by only a small amount, perhaps about 0.05 W/m². On the other hand, volcanic eruptions that inject sulfur-rich particles into the stratosphere produce a negative forcing, which lasts for a few years after the eruption. Several major eruptions occurred between 1880 and 1920,

as well as between 1960 and 1991. In addition, a number of smaller eruptions took place between 2000 and 2011. Because of the latter, the combined change in radiative forcing due to both volcanic activity and solar activity from 1988 to 2011 appears to have been slightly *negative*, which means that the net effect is that of *cooling* Earth's surface. Thus, natural factors may have actually reduced some of the global warming that we otherwise would have been expecting since 1998.

Climate Models and Recent Temperature Trends

We know that Earth's average surface temperature has increased by about 1.0°C (1.8°F) since the late nineteenth century. How does this observed temperature change compare with temperature changes derived from climate models using different forcing agents? Before we look at what climate models reveal, it is important to realize that the interactions between Earth and its atmosphere are so complex that it is difficult to unequivocally *prove* that Earth's present warming trend is due entirely to increasing concentrations of greenhouse gases. The problem is that any human-induced signal of climate change is superimposed on a background of natural climatic variations ("noise"), such as the El Niño–Southern Oscillation (ENSO) phenomenon (discussed in Chapter 7). And, in the temperature observations, it can be difficult to separate the signal of climate change from the noise of natural climate variability. However, today's more sophisticated climate models are much better at filtering out this noise while at the same time taking into account those forcing agents that are both natural and human-induced.

● Figure 13.18a shows the changes predicted in global surface air temperature from 1900 to 2005 by different climate models, which, as you recall, are mathematical models that simulate climate. Here the models use only natural forcing agents, such as solar energy and volcanic eruptions. Notice that the models' projected temperature (blue line) does not follow the observed trend in surface air temperature (gray line). In fact, the models project that temperatures now would be roughly similar to those in the early 1900s. Figure 13.18b shows how the models project changes in global surface air temperature from 1900 to 2005 when *both* natural forcing agents and human forcing agents, such as greenhouse gases and sulfur aerosols, are added to the models. Notice how the projected temperature change (red line) now closely follows the observed temperature trend (gray line). It is climate studies using computer models such as these that have led scientists to conclude that *most* of the warming since the middle of the twentieth century is very likely the result of increasing levels of greenhouse gases.

The Intergovernmental Panel on Climate Change (IPCC), a committee of more than 2000 leading earth scientists from around the globe, has produced the world's most comprehensive reports on climate change for more than 25 years. The IPCC published in-depth climate assessments in 1990, 1995, 2001, 2007, and again in 2013. The 2013 Fifth Assessment Report of the IPCC states that:

> It is extremely likely that human influence has been the dominant cause of the observed warming since the mid–20th century. [In the report, "extremely likely" means a probability of at least 95 percent.]

FUTURE CLIMATE CHANGE: PROJECTIONS Climate models project that, by the end of this century, increasing concentrations of greenhouse gases will result in an additional global warming that could be as much as several degrees Celsius. The newest, most sophisticated models take into account a number of important relationships, including the interactions between the oceans and the atmosphere, the processes by which CO_2 is removed from

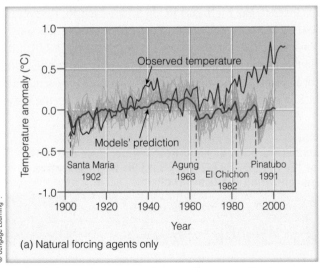

(a) Natural forcing agents only

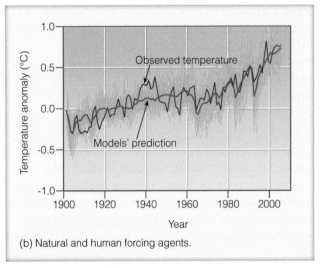

(b) Natural and human forcing agents.

● **FIGURE 13.18** (a) Projected global surface air temperature changes from 1900 to 2005 using only natural forcing agents (dark blue line) compared to observed global surface air temperature changes (gray line). Light blue lines show range of model simulations. (Names and dates of major volcanic eruptions are given at the bottom of the graph.) (b) Projected global surface air temperature changes using both natural and human forcing agents (dark red line) compared to observed global surface air temperature changes (gray line). Orange lines show range of model simulations. (Temperature changes in both (a) and (b) are relative to the period 1901 to 1950.) (Adapted from the Technical Summary by the Working Group 1 contribution to the Fourth Assessment Report to the IPCC, 2007.)

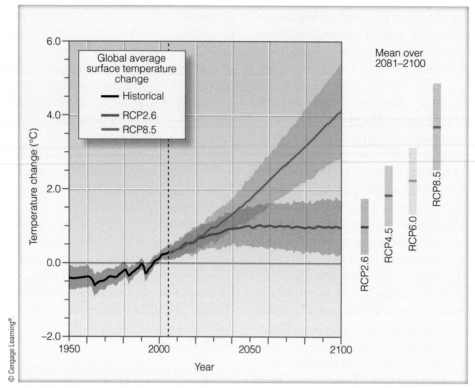

●**FIGURE 13.19** Global average projected surface air temperature changes (°C) above the 1986–2005 average (dark gray zero line) for the years 2005 to 2100. Temperature changes inside the graph and to the right of the graph are based on multi-climate models with different scenarios. Each scenario describes how the average temperature will change based on different pathways for the rate of change in greenhouse gases and various forcing agents. The black line shows global temperature change from 1950 to 2005. The two tinted curves from 2005 to 2100 show the range of likely temperature changes over time for two RCPs (representative concentration pathways), RCP2.6 (purple) and RCP 8.5 (red). The vertical tinted bars on the right side of the figure indicate the likely range of temperature change by 2081–2100 for all four RCPs. The thick solid bar within each projection gives the best estimate for temperature change for each RCP. (*Source: Climate Change 2013: The Physical Science Basis,* Summary for Policymakers, by the Working Group 1 contribution to the Fifth Assessment Report to the IPCC, 2013. Reprinted by permission of the Intergovernmental Panel on Climate Change.)

the atmosphere, and the cooling effect produced by sulfate aerosols in the lower atmosphere. The models also predict that, as the air warms, additional water will evaporate from the ocean surfaces and enter the atmosphere as water vapor. The added water vapor (which is the most abundant greenhouse gas) will produce a feedback on the climate system by enhancing the atmospheric greenhouse effect and accelerating the temperature rise. (This phenomenon, the *water vapor–greenhouse feedback,* is described on p. 39.) Without this feedback produced by the added water vapor, the models predict that the warming would be much less.

● Figure 13.19 shows climate models' projected warming during this century from increasing levels of greenhouse gases and various forcing agents. Four sets of model simulations were produced, as shown by the four bands on the right side of the images. Each set of simulations used a different representative concentration pathway (RCP), which describes how the total radiative forcing might change over this century (see ▼ Table 13.1). To produce the low-end RCP2.6, greenhouse gas emissions would have to decline drastically over the next several

decades. If emissions continue to increase rapidly, we will be closer to the high-end RCP8.5. For each RCP, several dozen climate models were used to create a range of projections. The shaded area around the blue trace (RCP2.6) and the red trace (RCP8.5) tell us that the climate models do not all project the same amount of warming for a particular RCP. This is because each climate model handles the very

▼**Table 13.1** **The Projected Average Surface Air Temperature Ranges and Best Temperature Estimates for the Period 2081–2100, Using Six Representative Concentration Pathways (RCPs)***

NAME OF PATHWAY	LIKELY TEMPERATURE RANGE, °C	MEAN ESTIMATED TEMPERATURE CHANGE, °C
RCP2.6	0.3–1.7	1.0
RCP4.5	1.1–2.6	1.8
RCP6.0	1.4–3.1	2.2
RCP8.5	2.6–4.8	3.7

complex interactions of the Earth–atmosphere system in a somewhat different way. Each model has its strengths and weaknesses, so by combining the results of several models, scientists can gain a more complete sense of what the future may hold. The other uncertainty, of course, is how much greenhouse gas will be emitted this century, as shown in the sharply different results for each of the RCPs.

The IPCC in its 2013 report concluded that increases in CO_2 would likely produce surface warming in the range of 0.3°C to 4.8°C from the period 1986–2005 to 2081–2100. The actual amount of warming will depend largely on the rate of fossil fuel burning through the century. If, during this century, the surface temperature should increase by more than 2°C, the warming would be more than twice that experienced during the twentieth century. If this happens, it is likely that the warming over this, the twenty-first century, will probably be greater than any warming during the past 10,000 years. (For additional information on the climate models that are used to predict changes in future surface air temperatures, read Focus section 13.4.)

Uncertainties About Greenhouse Gases Although carbon dioxide continues to increase in the atmosphere, and model projections agree that warming can be expected this century, some uncertainties remain over how water and land will influence the rate of CO_2 increase. For example, the oceans and the vegetation on land now absorb about half of the CO_2 emitted by human sources, although the exact proportion varies from year to year. The microscopic plants (phytoplankton) dwelling in the oceans extract CO_2 from the atmosphere during photosynthesis and store some of it below the oceans' surface, where they die. Could a warming Earth trigger a large blooming of these microscopic plants, in effect reducing the rate at which atmospheric CO_2 is increasing?

Current models show that warming the planet tends to *reduce* both ocean and land intake of CO_2. Therefore, if human-induced CO_2 emissions continue to increase at their present rate, more of that CO_2 should remain in the atmosphere, further enhancing global warming. An example of how rising temperatures can play a role in altering the way landmasses absorb and emit CO_2 is found in the Alaskan tundra. There, temperatures in recent years have risen to the point where more frozen soil melts in summer than it used to. Accordingly, during the warmer months, deep layers of exposed decaying peat moss release CO_2 into the atmosphere. Until recently, this region absorbed more CO_2 than it released. Now, however, much of the tundra acts as a producing source of CO_2. Moreover, recent analyses suggest that warmer temperatures will further increase the net amount of CO_2 released from Earth's tundra.

Deforestation accounts for about 10 to 15 percent of the observed increase in atmospheric CO_2. That percentage

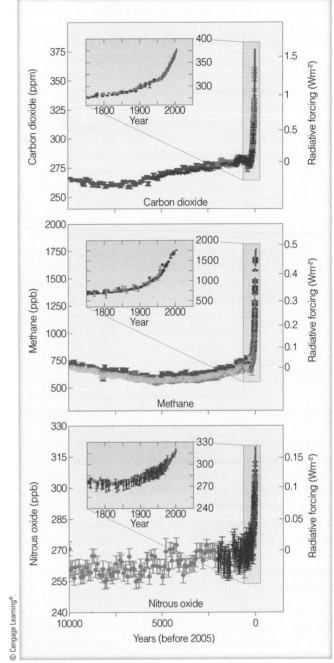

● **FIGURE 13.20** Changes in the greenhouse gases carbon dioxide, methane, and nitrous oxide indicated from ice core and modern data. (*Source:* Adapted from Climate Change 2007, *The Physical Science Basis*, by the Working Group 1 contribution to the Fourth Assessment Report to the IPCC, 2007. Reprinted by permission of the Intergovernmental Panel on Climate Change.)

is lower than it was in the 1990s, in part because of progress in limiting deforestation. Hence, changes in land use can influence levels of CO_2 concentrations, especially if the practice of deforestation is replaced by reforestation.

Perhaps the single biggest uncertainty in climate change over coming decades is the rate at which human activities will add greenhouse gases to the atmosphere.
● Fig. 13.20 indicates the dramatic rise in CO_2 levels

Climate Models—A Quick Look

Climate models that simulate the physical processes of the atmosphere (and the oceans) are called *general circulation models*, or *GCMs* for short. When an atmospheric component of a GCM is linked to an ocean component, the model is said to be "coupled," and the model is called an *Atmosphere–Ocean General Circulation Model*, or *AOGCM*. General Circulation Models use mathematics and the laws of physics to describe the general behavior of the atmosphere. To reduce some of the atmosphere's complexities, the models make simplified assumptions about the atmosphere and describe the atmosphere in more simplified physical terms. They also reduce many of the small-scale atmospheric processes (such as those due to clouds) into a single approximation, or parameter, which is known as *parameterization*.

The GCMs represent the atmosphere by dividing it up into grid squares, perhaps 100 kilometers on a side. The size of the squares has been getting steadily smaller over the years as computing power increases and model resolution improves. General circulation models simulate the behavior of the real atmosphere and describe the major circulation features as well as the seasonal and latitudinal temperature patterns.

Today's coupled models are extremely sophisticated, taking into account land and vegetation processes, atmospheric chemistry, carbon cycles in the ocean and on land, ice and snow cover, and aerosols (see Fig. 3). These models are increasingly referred to as Earth system models because they include not only an evaluation of oceans and atmosphere, but other parts of our global ecosystem.

A General Circulation Model is first run for a few decades to make sure that the model simulates the real atmosphere. Then, to see how some variables (such as increasing levels of CO_2) might influence the atmosphere, the model is repeatedly run with increasing concentrations of CO_2. In this manner, the GCMs reveal how the atmosphere and its circulation might change with time, due to increasing levels of greenhouse gases. When the models are run with different scenarios (that is, varying concentrations of greenhouse gases and different forcing agents), the end result is usually a variation in the predicted temperature (such as those temperature projections shown in Fig. 13.18 on p. 401 and in Fig. 13.19 on p. 402).

Just as weather forecasters examine the results from several different models, each of which handles the atmosphere in a slightly different way, climate scientists make use of a variety of models from more than a dozen research centers around the globe. In addition, the researchers must consider natural climate variability. If a particular experiment is repeated using the same model but with slightly different initial conditions, the results may be somewhat different. Typically, the large-scale trends in global climate over decades are consistent from model to model and run to run, but there may be regional variations. Climate researchers, therefore, rely on *ensembles*, containing dozens of model runs, in order to see the full range of possibilities that may occur over the coming decades.

General circulation models are not perfect: They have imperfect parameterization of all processes, and they cannot resolve processes that unfold over distances smaller than the grid spacing. However, as mentioned previously, today's models are extremely sophisticated, and serve as the most reliable tools available for estimating climate change. As computers continue to become even more powerful, researchers are experimenting with models that blend global coverage with some aspects of higher-resolution models over specified areas. Another option is to use higher resolution when carrying out a model run for particular periods of time, such as a single decade later in the century. Simulations like these could lead to more accurate and detailed depictions of the weather and climate features of particular interest that might emerge in future climates.

FIGURE 3 Components of an Earth system model.

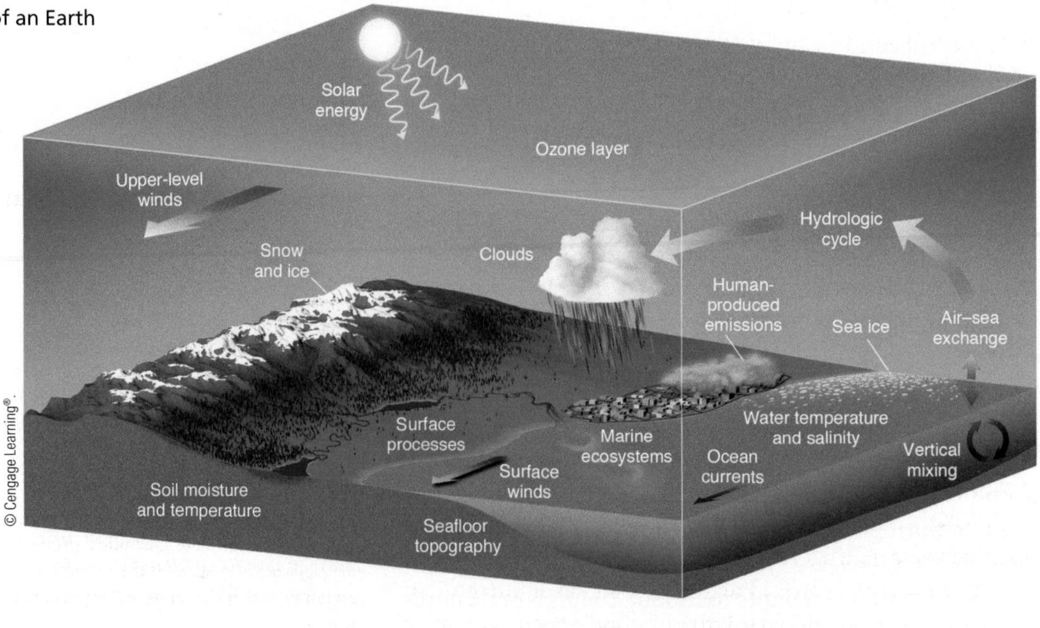

during the twentieth century. In the year 1990, carbon dioxide levels were increasing by about 1.5 ppm/year, whereas today they are increasing by about 2 ppm/year. If this trend continues, CO_2 concentrations could easily exceed 550 ppm by the end of this, the twenty-first century. In Fig 13.20 notice that the atmospheric concentration of methane has increased dramatically over the last 250 years, and it is still increasing, although with some variability in the last 20 years. Also notice that atmospheric concentrations of nitrous oxide have risen quickly, and its concentration is still rising.

Since the mid-1990s, the atmospheric concentration of a group of greenhouse gases called *chlorofluorocarbons* (halocarbons) has been decreasing. However, the substitute compounds for chlorofluorocarbons, which are also greenhouse gases, have been increasing. Moreover, the total amount of surface ozone has probably increased by more than 30 percent since 1750. This increase is counterbalanced by the fact that the majority of ozone is found in the stratosphere, where its maximum concentration is typically less than 12 ppm. Although ozone is a greenhouse gas, it plays a very *minor* role in the enhancement of the greenhouse effect because its concentration near Earth's surface is typically less than 0.04 ppm. The concentration of ozone varies greatly from region to region, and depends on the production of photochemical smog. The increase in surface ozone has probably led to a very small increase in radiative forcing. Before going on to the next section, you may wish to read Focus section 13.5, which details some of the misconceptions that have arisen about global warming, ozone, and the ozone hole.

The Question of Clouds As the atmosphere warms and more water vapor is added to the air, global cloudiness might increase as well. How, then, would clouds—coming as they do in a variety of shapes and sizes and at different altitudes—affect the climate system?

Clouds reflect incoming sunlight back to space, a process that tends to cool the climate, but clouds also emit infrared radiation to Earth, which tends to warm it. Just how the climate will respond to changes in cloudiness will depend on the type of clouds that form, their height above the surface, and their physical properties, such as liquid water (or ice) content, depth, and droplet size distribution. For example, high, thin cirriform clouds (composed mostly of ice) appear to promote a net warming effect: They allow a good deal of sunlight to pass through (which warms Earth's surface), yet because they are cold, they warm the atmosphere around them by absorbing more infrared radiation from Earth than they emit upward. Low stratified clouds, on the other hand, tend to promote a net cooling effect. Composed mostly of water droplets, they reflect much of the sun's incoming energy, which cools Earth's surface, and because their tops are relatively warm, they radiate to space much of the infrared energy they receive from Earth. Satellite data confirm that, overall, the current global mixture of clouds has a *net cooling*

Anthony John West/CORBIS

● **FIGURE 13.21** Jet contrails can have an effect on climate by reflecting incoming sunlight and by emitting infrared energy to the surface.

effect on our planet, which means that, without any clouds, our atmosphere would be warmer.

Additional clouds in a warmer world would not necessarily have a net cooling effect, however. Their influence on the average surface air temperature would depend on their extent and on whether low or high clouds dominate the climate scene. Consequently, the feedback from clouds could potentially enhance or reduce the warming produced by increasing greenhouse gases. Most models show that as the surface air warms, there will be an increase in the typical altitude of cirrus clouds. This could cause an overall positive feedback. In other words, warmer temperatures may change the global arrangement of clouds in a way that leads to further warming.*

There is another cloud-related factor to consider. Jet aircraft influence climate by producing contrails (condensation trails) high in the troposphere, generally above about 20,000 feet (see ● Fig. 13.21). Most contrails form as a cirrus-like trail behind the aircraft. Some disappear quickly, whereas others persist over time, occasionally stretching across the sky as streamers of cirriform clouds that coalesce into a white canopy. Contrails can affect climate by enhancing cirriform cloudiness and by adding ice crystals to existing cirriform clouds, thus changing their albedo. Because contrails reflect sunlight and absorb infrared energy, they have the ability to alter the temperature near the ground. Overall, contrails have a net warming effect on the planet's radiative balance, so it is possible that any increase in global air travel may lead to additional warming.

---

*In addition to the amount and distribution of clouds, the way in which climate models calculate the optical properties of a cloud (such as albedo) can have a large influence on the model's calculations. Also, there is much uncertainty as to how clouds will interact with aerosols, and what the net effect will be.

The Impact of Ozone on the Greenhouse Effect and Climate Change

Ozone is indeed a greenhouse gas, but its influence on the greenhouse effect is just minor. Why? Because of the following two conditions:

1. The concentration of atmospheric ozone is extremely small. Near Earth's surface, ozone averages only about 0.04 ppm, and in the stratosphere where it is more concentrated, its average value is only between 5 and 12 ppm. By comparison, the average value of carbon dioxide in our atmosphere is about 405 ppm.
2. Ozone only absorbs infrared energy in a very narrow band, near 10 μm. Look at Fig. 2.11 on p. 37 and observe that both water vapor and carbon dioxide are much more prolific absorbers of infrared energy than is ozone.

Accordingly, given these two facts, any small change in ozone concentration would have negligible impact on the greenhouse effect and on climate change.

The Impact of the Ozone Hole on Climate Change

How does the ozone hole affect climate change? You may recall that we briefly looked at the ozone hole in Chapter 1 on p. 18.* There we saw that over springtime Antarctica ozone concentrations in the stratosphere plummet, in some years leaving virtually no protective ozone above this region. Now, ozone readily absorbs incoming ultraviolet solar (UV) radiation at wavelengths below about 0.3 μm. So, does this fact mean that the formation of the ozone hole enhances global warming by allowing more UV radiation to reach the surface and warm it?

We know from Chapter 2 (p. 36) that the sun emits only a small fraction of its total energy output at ultraviolet wavelengths. Although UV waves do carry more

*The ozone hole is covered in more detail in Chapter 14 on p. 425.

energy than visible waves, there are too few of them to produce much warming. Those UV waves that do reach the surface mostly impinge upon snow and ice, ensuring that virtually no surface warming occurs.

It's interesting to note that the main temperature-related effect of ozone depletion is in the lower stratosphere, which has been *cooling*. Temperatures at this height (above about 20 km, or 12 mi) have dropped to record lows in recent years, in large part due to the loss of ozone.

Therefore, the depletion of ozone over Antarctica during its spring (that is, the ozone hole) does not enhance global warming at Earth's surface. We then must take care not to link the ozone hole with global warming. These are two distinctly different atmospheric conditions virtually unrelated—basically a case of apples and oranges.

The Impact of Oceans The oceans are a critical part of Earth's climate system, yet the exact effect they will have on climate change is not fully understood. For example, the oceans have a large capacity for storing heat energy. In fact, more than 90 percent of the energy trapped by increased greenhouse gases in recent decades has gone not into the atmosphere but into the ocean. Only a slight change in the rate of oceanic heat storage can thus have a big impact on atmospheric warming. Variations in this heat storage, perhaps related to ocean circulation patterns, could help explain why some decades show more atmospheric warming than others. The vast amount of heat stored by the ocean also means that some global warming would continue to occur even if fossil-fuel emissions were completely stopped. It appears that increased heat storage in the Pacific Ocean since the late 1990s played a role in slowing down atmospheric warming for more than a decade. However, the record global temperatures observed in 2014, 2015, and 2016, along with the strong 2015–16 El Niño event, suggest that more of the heat stored in the oceans has again been making its way into the atmosphere.

CONSEQUENCES OF CLIMATE CHANGE: THE POSSIBILITIES

If the world continues to warm as predicted by climate models, where will most of the warming take place? Climate models predict that land areas will warm more rapidly than the global average, particularly in the

northern high latitudes in winter (see ● Fig. 13.22a). We can see in Fig. 13.22b that the greatest surface warming for the period 2001 to 2006 occurred over landmasses in the high latitudes of the Northern Hemisphere, especially over Canada and Russia. These observations of global average temperature change indicate that climate models are on target with their warming projections.

As high-latitude regions of the Northern Hemisphere continue to warm, modification of the land may actually enhance the warming. For example, the dark-green boreal forests* of the high latitudes absorb up to three times as much solar energy as does the snow-covered tundra. Consequently, the winter temperatures in subarctic regions are, on the average, much higher than they would be without trees. If warming allows the boreal forests to expand into the tundra, the forests may accelerate the warming in that region. As the temperature rises, organic matter in the soil should decompose at a faster rate, adding more CO_2 to the air, which might accelerate the warming even more. Trees that grow in a climate zone defined by temperature may become especially hard hit as rising temperatures place them in an inhospitable environment. In a weakened state, they may become more susceptible to insects and disease. These changes in temperature will

*The boreal forest consists of woodlands (northern part) and conifers and some hardwoods (southern part). Its northern boundary is next to the tundra along the Arctic tree line.

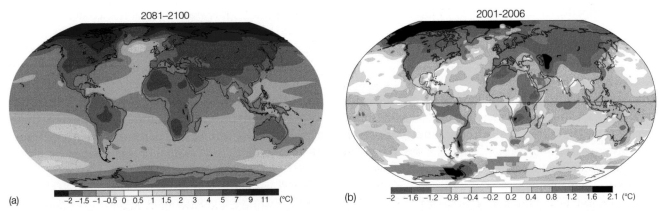

2081–2100

2001-2006

(a) -2 –1.5 –1 –0.5 0 0.5 1 1.5 2 3 4 5 7 9 11 (°C)

(b) -2 –1.6 –1.2 –0.8 –0.4 –0.2 0.2 0.4 0.8 1.2 1.6 2.1 (°C)

● **FIGURE 13.22** (a) Projected surface air temperature changes averaged for the decade 2081-2100 (using the RCP8.5 scenario) compared to the average surface temperature for the period 1986–2005. The largest increase in air temperature is projected to be over landmasses and in the Arctic region. (b) The average change in surface air temperature for the period 2001–2006 compared to the average for the years 1951–1980. The greatest warming was over the Arctic region and the high-latitude landmasses of the Northern Hemisphere. (Diagram [a] *Source: Climate Change 2007: The Physical Science Basis.* Working Group I Contribution to the Fourth Assessment Report of the Intergovernmental Panel on Climate Change, Figure SPM.6; Figure SPM.7. Cambridge University Press. Diagram [b] Courtesy NASA.)

also affect people in many ways, of course, including direct effects on human health. For example, with heat waves expected to become more frequent and intense, heat-related deaths are expected to increase, although there could be some compensating decrease in cold-related illnesses.

Precipitation Changes in precipitation and drought may be just as important as changes in temperature over the coming decades. As with temperature, changes in precipitation will not be evenly distributed, as some areas will tend to get more precipitation and others less. Since the middle of the twentieth century, precipitation has generally increased over the middle- and high-latitude land areas of the Northern Hemisphere, while decreasing over some subtropical land areas. In many areas, there has also been an increase in the intensity of the heaviest precipitation events during the last 50 years or so. Notice in ● Fig. 13.23 that the models project a further increase in average precipitation over high latitudes of the Northern Hemisphere and a continued decrease over parts of the subtropics. A decrease in precipitation in this region could have an adverse effect by placing added stress on agriculture.

Some models suggest that changes in global patterns of precipitation might lead to more extreme floods and more severe drought. Even in places where average annual precipitation does not change, it is possible that rainfall and snowfall will be focused in increasingly heavy wet spells, with longer dry periods in between. In addition, warming temperatures will tend to cause soil to dry out more quickly, exacerbating the impact of drought when it occurs. In mountainous regions of western North America, where much of the precipitation falls in winter, a greater fraction of precipitation might fall mainly as rain, causing a decrease in the snowmelt runoff that fills the reservoirs during the spring. In California, the reduction in water

storage because of increasing temperatures and decreasing runoff could threaten the state's agriculture over the long term.

Sea Level Rise Another consequence of climate change is an increase in sea level, as land-based ice sheets and glaciers recede and the oceans continue to expand as they slowly warm. From 1900 to 2010, globally averaged sea level rose about 19 cm (7.5 in.), with the pace accelerating from the 1990s onward. About half of that was a result of melting glaciers and ice sheets, with the other half produced by the expansion of oceans as they warm. Although we often think of sea level as being a fixed height

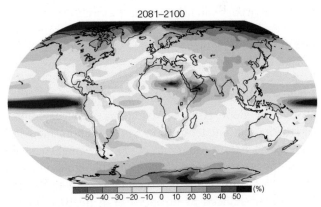

2081–2100

–50 –40 –30 –20 –10 0 10 20 30 40 50 (%)

● **FIGURE 13.23** Projected relative changes in annual mean precipitation (in percent) for the period 2081–2100 (using the RCP8.5 scenario) compared to the average for the period 1986–2005. The stippled areas represent regions where the average change from multiple models is large compared to natural variability and more than 90 percent of the models agree as to whether precipitation will increase or decrease; hatched regions indicate the average change from multiple models is small compared to natural variability. (Source: Climate Change 2013: The Physical Science Basis. Contribution of Working Group I to the Fifth Assessment Report of the Intergovernmental Panel on Climate Change, 2013. Reprinted by permission of the Intergovernmental Panel on Climate Change.)

around the world, it can actually vary by more than 30 cm (12 in.) based on natural processes. These include prevailing winds that can pile up water in the western tropical Pacific and along the east coasts of Asia and North America. In some areas, such as Scandinavia, the land is still rebounding from the weight of ice sheets thousands of years ago. Other coastal areas are plagued with subsidence, as the land slowly settles from natural processes or the withdrawal of underground fluids. Future changes in ocean circulation will lead to greater rises in sea level in some areas than in others.

The amount of additional rise in sea level this century will also depend on the future rate of greenhouse gas emissions, how much the air and water temperature increase in response to these greenhouse gases, and how quickly the vast ice sheets of Greenland and Antarctica melt. In 2013, the IPCC projected that the twenty-first century rise in sea level will likely be somewhere between 26 and 82 cm (10 to 32 in.). The increase could be even larger if some of the ice cliffs along the edge of Antarctica were to fracture and collapse into the sea, which happened during previous warm periods in Earth's history. Should this happen in the coming decades, recent modeling has found that sea level could rise by as much as 150 cm (59 in.) this century, with even larger increases afterward.

Millions of people in coastal communities around the world will be increasingly affected by sea level rise. Rising ocean levels and warming oceans could also have a damaging influence on many marine ecosystems, such as coral reefs. In addition, coastal groundwater supplies can become contaminated with saltwater. And as we saw in Chapter 11, as sea surface temperatures increase, the average intensity of hurricanes will likely increase as well, other factors being equal. (For more information on hurricanes and global warming, look again at Focus section 11.4, "Hurricanes in a Warmer World," on p. 344.)

Effects in Polar Regions

In polar regions, as elsewhere around the globe, rising temperatures produce complex interactions among temperature, precipitation, and wind patterns. Hence, in Antarctica, more snow might actually fall in the warmer (but still cold) air. This situation could allow snow to build up across the interior, although it may be counterbalanced by an increase in melting already taking place along the Antarctic coastline. Over Greenland, which is experiencing rapid melting of ice and snow, any increase in precipitation will likely be offset by rapid

DID YOU KNOW?

In our warmer world, many freshwater lakes in northern latitudes are freezing later in the fall and thawing earlier in the spring than they did in years past. Wisconsin's Lake Mendota, for example, now averages about 40 fewer days with ice than it did 150 years ago.

melting, and so it is expected that the ice sheet will continue to shrink. In the Arctic Ocean, warming has caused sea ice to shrink and thin dramatically since the 1990s. (Sea ice is formed by the freezing of sea water.) During the summer of 2012, the extent of sea ice was reduced to new record lows (see ● Fig. 13.24). If the warming in this region continues at its present rate, summer sea ice may, at times, shrink to cover less than 10 percent of the Arctic Ocean by the middle of this century, or even sooner.

Effects on Ecosystems

Increasing levels of CO_2 in a warmer world could have many other consequences. For example, greater amounts of CO_2 act as a "fertilizer" for some plants, accelerating their growth, although this process can slow over time if water, nitrogen, and other nutrients are not plentiful enough to sustain the growth. In some ecosystems, certain plant species could become so dominant that others are eliminated. In tropical areas, where many developing nations are located, the effects of climate change may actually decrease crop yield, whereas higher latitudes might benefit from a longer growing season and an earlier snowmelt. Extremely cold winters might become less numerous with fewer bitter cold spells. However, wildfires may continue to become more prevalent during dry spells in forested high-latitude areas. (The city of Fort McMurray in northern Canada was engulfed by a huge wildfire in May 2016, much earlier in the year than wildfire is normally observed in the region.) Thus, while there will be some "winners" and some "losers," the most recent analyses suggest that the impact of climate change on agriculture and ecosystems may become increasingly negative by later in this century.

CLIMATE CHANGE: EFFORTS TO CURB The most obvious way to curb global warming is to reduce greenhouse gas emissions by reducing the use of fossil fuels. Burning natural gas produces less carbon dioxide than burning oil and coal, and the rapid growth of natural gas use apparently helped lead to a temporary drop in CO_2 emissions in the United States in the early 2010s. However, natural gas production also leads to emissions of methane, a powerful greenhouse gas, as a byproduct. Researchers are now investigating this phenomenon and the extent to which it may be counteracting the benefits of reduced CO_2 emissions. Increasing the use of alternative energy sources can also play a major role in curbing global warming. Technologies such as solar and wind power—the two fastest-growing energy sources worldwide—produce virtually no greenhouse gases other than those required to build and maintain the facilities.

Diplomatic Efforts

In an attempt to mitigate the impact humans have on the climate system, representatives from 160 countries met at Kyoto, Japan, in 1997 to work out a formal agreement to limit greenhouse gas

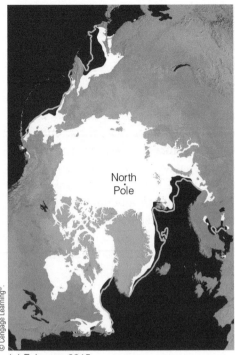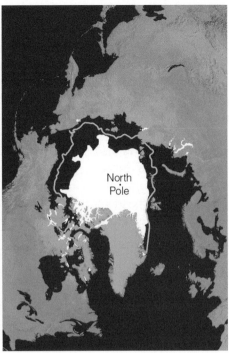

(a) February 2015   (b) September 2015

● FIGURE 13.24 The extent of Arctic sea ice in (a) March 2015, when the ice cover was at or near its maximum for the year, and in (b) September 2015, when the ice cover was near or at its minimum for the year. The orange line in (a) represents the median maximum of the ice cover for the period 1979–2000. The orange line in (b) represents the median minimum extent of the ice cover for the period 1979–2000.

© Cengage Learning®.

emissions in industrialized nations. The international agreement—called the *Kyoto Protocol*—was adopted in 1997 and put into force in February 2005. The protocol set mandatory targets for reducing greenhouse gas emissions in countries that adopted the plan. The overall goal was to reduce greenhouse gas emissions in developed countries by at least 5 percent below existing 1990 levels during the 5-year period of 2008 through 2012. For the industrialized nations that participated in the Kyoto Protocol, emissions dropped by more than 22 percent. However, the United States did not ratify the protocol, and many developing nations such as China were not required to carry out emission reductions, since they had been responsible for only a small part of the accumulated CO_2 up to that point. As a result, the global total of greenhouse gas emissions actually *increased* by more than 25 percent from 1990 to 2012.

The Kyoto Protocol has been followed by the Paris Agreement, which was adopted by 195 countries in 2015. Under this agreement, each nation has set voluntary targets for reducing emissions and will report their progress on a regular basis. In addition, several cities and countries, including Costa Rica, Iceland, and Norway, have pledged to become *carbon neutral*—meaning that all of their greenhouse gas emissions would be offset by activities such as planting trees, so that the country ends up with no net emissions. Many global businesses are also striving to become carbon neutral.

In the United States, many cities and states have implemented their own climate change policies. For example, California has set targets for reducing greenhouse gas emissions to 1990 levels by the year 2020, with additional reductions of 40 percent below 1990 levels by 2030 and 80 percent by 2050. In addition, the mayors of more than 1000 towns and cities in the United States have pledged to reduce the levels of carbon emissions in their municipalities below 1990 levels.

Geoengineering In recent years, as the difficulty of reducing global emissions has become more evident, the idea of using technology to mitigate climate change has gained interest among researchers and policy makers. Called **geoengineering**, the idea is to use global-scale technological fixes to counter climate change by either (1) removing greenhouse gases from the atmosphere or by (2) changing the amount of sunlight that reaches Earth. Several geoengineering ideas that remove CO_2 from the atmosphere include fertilizing the oceans with plants that absorb CO_2, sprinkling iron-rich particles over portions of the ocean to promote the growth of carbon-absorbing phytoplankton, and placing large drifting vertical pipes into the ocean so that wave activity will pull up nutrient-rich water from below to promote algae blooms.

One significant drawback to all of these proposals is that they do nothing to reduce the gradual acidification of oceans, which occurs as the oceans absorb carbon dioxide. Acidification poses a serious threat to shellfish, coral, and many other forms of marine life that rely on calcification. One proposal that takes this problem into account is to extract CO_2 from the atmosphere with "synthetic trees" made of recyclable chemicals that react with CO_2 in the air.

To prevent sunlight from reaching Earth's surface, one idea proposes placing an array of reflecting mirrors in space high above Earth. Another proposal suggests injecting highly reflective sulfate aerosols into the stratosphere.

In one study, scientists using climate models placed tons of sulfate aerosols—on the order of those lofted by Mount Pinatubo in 1991—into the stratosphere at various intervals. The study concluded that injecting these sulfate aerosols every one to four years in conjunction with reducing greenhouse gases could provide a "grace period" of up to 20 years before major cutbacks in greenhouse gas emissions would be required.

All of these geoengineering proposals may have unforeseen or unwanted consequences. Injecting the stratosphere with sulfate particles, for example, might alter the temperature of the upper atmosphere and affect the fragile ozone layer. Carrying out large-scale geoengineering would be quite expensive, and it would also call for global agreement on techniques and procedures, since the results could affect the entire planet. As we saw with the Kyoto Protocol, such global agreement can be very difficult to obtain. In short, the science of geoengineering is intriguing, but it poses costly complex political, financial, and technological challenges.

CLIMATE CHANGE: A FINAL NOTE Cutting down on the emissions of greenhouse gases and pollutants has several benefits. A reduction in greenhouse gas emissions would slow down the enhancement of Earth's greenhouse effect and reduce global warming. The simultaneous reduction in air pollutants produced by fossil fuel use would help limit haze, slow the production of photochemical smog, and reduce the impact on human health. Even if the greenhouse warming were to end up toward the lower end of what modern climate models project, these measures would certainly benefit humanity.

SUMMARY

In this chapter, we considered some of the many ways Earth's climate can be changed. First, we saw that Earth's climate has undergone considerable change during the geologic past. Some of the evidence for a changing climate comes from tree rings (dendrochronology), chemical analysis of oxygen isotopes in ice cores and fossil shells, and geologic evidence left behind by advancing and retreating glaciers. The evidence from these suggests that, throughout much of the geologic past (long before humanity arrived on the scene), Earth was warmer than it is today. There were cooler periods, however, during which glaciers advanced over large sections of North America and Europe.

We examined some of the possible causes of climate change, noting that the problem is extremely complex, as a change in one variable in the climate system almost immediately changes other variables. Climate changes can be brought on by both natural events and by human (anthropogenic) activities. One theory of natural cause of climate change suggests that the shifting of the continents, along with volcanic activity and mountain building, may account for variations in climate that take place over millions of years.

Another theory of natural cause of climate change is the Milankovitch theory, which proposes that alternating glacial and interglacial episodes during the past 2.5 million years are the result of small variations in the tilt of Earth's axis and in the geometry of Earth's orbit around the sun. Climate change may also be brought on naturally by volcanic eruptions rich in sulfur and by variations in the sun's energy output.

Human activities, such as emitting vast quantities of greenhouse gases into the atmosphere, can produce climate changes worldwide. Global average temperatures since the late nineteenth century have risen by about 1.0°C. Many studies have found that increasing concentrations of greenhouse gases are the primary cause of this warming. Sophisticated climate models project that, as levels of CO_2 and other greenhouse gases continue to increase, Earth's surface will warm substantially by the end of this century. The models also predict that, as Earth warms, there will be a global increase in atmospheric water vapor, more extreme precipitation events, a worsening of drought impacts, a more rapid melting of sea ice, and a rise in sea level.

KEY TERMS

The following terms are listed (with corresponding page numbers) in the order they appear in the text. Define each. Doing so will aid you in reviewing the material covered in this chapter.

climate change, 382
dendrochronology, 384
Ice Age, 384
interglacial periods, 385
Younger Dryas (event), 385
mid-Holocene maximum, 385
Little Ice Age, 387
global warming, 388
water vapor–greenhouse feedback, 389
positive feedback mechanism, 389
snow-albedo feedback, 389
negative feedback mechanism, 389
chemical weathering–CO_2 feedback, 390
theory of plate tectonics, 390
Milankovitch theory, 391
eccentricity, 391
precession, 391
obliquity, 391
Maunder minimum, 394
sulfate aerosols, 397
desertification, 398
radiative forcing agents, 399
radiative forcing, 399
geoengineering, 409

QUESTIONS FOR REVIEW

1. What methods do scientists use to determine climate conditions that have occurred in the past?
2. Explain how the changing climate influenced the formation of the Bering land bridge.
3. How does today's average global temperature compare with the average temperature during most of the past 1000 years?
4. What is the Younger Dryas episode? When did it occur?
5. How does a positive feedback mechanism differ from a negative feedback mechanism? Is the water vapor–greenhouse feedback considered positive or negative? Explain.

6. Explain why the chemical weathering–CO_2 feedback is a negative feedback on Earth's climate system.

7. How does the theory of plate tectonics explain climate change over periods of millions of years?

8. Describe the Milankovitch theory of climatic change by explaining how each of the three cycles alters the amount of solar energy reaching Earth.

9. Given the analysis of air bubbles trapped in polar ice during the past 800,000 years, were CO_2 levels generally higher or lower during warmer periods? Were methane levels higher or lower at this time?

10. How do sulfate aerosols in the lower atmosphere affect surface air temperatures during the day?

11. Describe the scenario of nuclear winter.

12. Do volcanic eruptions rich in sulfur tend to warm or cool Earth's surface? Explain.

13. Explain how variations in the sun's energy output might influence global climate.

14. Climate models predict that increasing levels of CO_2 will cause the mean global surface temperature to rise significantly by the year 2100. What other greenhouse gas must also increase in concentration in order for the amount of predicted temperature rise to occur?

15. Describe some of the natural and human-induced radiative forcing agents and their effect on climate.

16. List five ways natural events can cause climate change.

17. List three ways human (anthropogenic) activities can cause climate change.

18. Describe how clouds influence the climate system.

19. In Fig. 13.18a, p. 401, explain why the actual rise in surface air temperature (gray line) is much greater than the projected rise in temperature due to natural forcing agents.

20. Why have climate scientists concluded that most of the warming experienced during the last 50 years has been due to increasing concentrations of greenhouse gases?

21. List some of the potential consequences of climate change on the atmosphere and its inhabitants.

22. Is CO_2 the only greenhouse gas we should be concerned with for climate change? If not, what are the other gases?

QUESTIONS FOR THOUGHT AND EXPLORATION

1. Ice cores extracted from Greenland and Antarctica have yielded valuable information on climate changes during the past 800,000 years. What do you feel might be some of the limitations in using ice core information to evaluate past climate changes?

2. When glaciation was at a maximum (about 18,000 years ago), was global precipitation greater or less than at present? Explain your reasoning.

3. Consider the following climate change scenario. Warming global temperatures increase saturation vapor pressures over the ocean. As more water evaporates, increasing quantities of water vapor build up in the troposphere. More clouds form as the water vapor condenses. The clouds increase the albedo, resulting in decreased amounts of solar radiation reaching Earth's surface. Is this scenario plausible? What type(s) of feedback(s) is/are involved? What type of clouds (high or low)?

4. Are ice ages in the Northern Hemisphere more likely when the tilt of Earth is at a maximum or a minimum? Explain.

5. Are ice ages in the Northern Hemisphere more likely when the sun is closest to Earth during summer or during winter? Explain.

6. Why did periods of glacial advance in the higher latitudes of the Northern Hemisphere tend to occur with colder summers, but not necessarily with colder winters?

GLOBAL GEOSCIENCE WATCH Go to the Reference section of the Global Environment Watch: Climate Change portal and search for the report "Climate Modeling" from *Environmental Science: In Context* (or other reference works in this portal). What would a climate model and a weather forecasting model have in common? How might they be different? What aspects of the average climate expected by the 2050s might be most useful for city planners or farmers to know?

ONLINE RESOURCES

 Visit www.cengagebrain.com to view additional resources, including video exercises, practice quizzes, an interactive eBook, and more.

CHAPTER 14

Air Pollution

Contents

Air pollution makes the earth a less pleasant place to live. It reduces the beauty of nature. This blight is particularly noticed in mountain areas. Views that once made the pulse beat faster because of the spectacular panorama of mountains and valleys are more often becoming shrouded in smoke. When once you almost always could see giant boulders sharply etched in the sky and the tapered arrowheads of spired pines, you now often see a fuzzy picture of brown and green. The polluted air acts like a translucent screen pulled down by an unhappy God.

Louis J. Battan, *The Unclean Sky*

415

Every deep breath fills our lungs. Most of what we inhale is gaseous nitrogen and oxygen. However, we may also inhale, in minute quantities, other gases and particles, some of which could be considered pollutants. These contaminants come from car exhaust, chimneys, factories, power plants, and other sources related to human activities.

Virtually every large city has to contend in some way with air pollution, which clouds the sky, injures plants, and damages property. Some pollutants merely have a noxious odor, whereas others can cause severe health problems. The cost is high. In the United States, for example, outdoor air pollution takes its toll in health care and lost work productivity at an annual expense that runs into billions of dollars. Estimates are that, worldwide, nearly 1 billion people in urban environments are continuously being exposed to health hazards from air pollutants. Poor air quality indoors also takes an enormous toll around the globe, especially from the use of indoor cookstoves in developing countries. As many as one out of eight deaths worldwide—several million each year—are connected to indoor and outdoor air pollution.

This chapter takes a look at this serious contemporary concern. We begin by briefly examining the history of problems in this area, and then go on to explore the types and sources of air pollution, as well as the weather that can produce an unhealthy accumulation of pollutants. Finally, we investigate how air pollution influences the urban environment and also how it brings about unwanted acid precipitation.

A Brief History of Air Pollution

Strictly speaking, air pollution is not a new problem. More than likely it began when humans invented fire whose smoke choked the inhabitants of poorly ventilated caves. In fact, very early accounts of air pollution characterized the phenomenon as "smoke problems," the major cause being people burning wood and coal to keep warm. (Indoor air pollution is now less of a problem in industrialized countries, although cookstoves are major sources of both indoor and outdoor air pollution in many parts of the developing world.)

To alleviate the smoke problem in old England, King Edward I issued a proclamation in 1273 forbidding the use of sea coal, an impure form of coal that produced a great deal of soot and sulfur dioxide when burned. One person was reputedly executed for violating this decree. In spite of such restrictions, the use of coal as heating fuel grew during the fifteenth and sixteenth centuries.

As industrialization increased, the smoke problem worsened. In 1661, the prominent scientist John Evelyn wrote an essay deploring London's filthy air. And by the 1850s, London had become notorious for its "pea soup" fog, a thick mixture of smoke and fog that hung over the city. These fogs could be dangerous. In 1873, one was responsible for as many as 700 deaths. Another, in 1911, claimed the lives of 1150 Londoners. To describe this chronic atmospheric event, a physician, Harold Des Voeux, coined (in 1905) the word *smog*, meaning a combination of smoke and fog.

Little was done to control the burning of coal as time went by, primarily because it was extremely difficult to counter the basic attitude of the powerful industrialists: "Where there's muck, there's money." London's acute smog problem intensified. Then, during the first week of December 1952, a major disaster struck. The winds died down over London and the fog and smoke became so thick that people walking along the street literally could not see where they were going (see ● Fig. 14.1). This particularly disastrous smog lasted 5 days and took at least 4000 lives, prompting Parliament to pass a Clean Air Act in 1956. Additional air pollution incidents occurred in England

● FIGURE 14.1 The fog and smoke were so dense in London during December 1952 that visibilities were often restricted to less than 100 feet and streetlights had to be turned on during the middle of the day.

● FIGURE 14.2 Pupils cover their noses after school in heavy smog on December 23, 2015, in Binzhou, China.

during 1956, 1957, and 1962, but due to the strong legislative measures taken against air pollution, London's air today is much cleaner, and "pea soup" fogs are a thing of the past.

Air pollution episodes were by no means limited to Great Britain. During the winter of 1930, for instance, Belgium's highly industrialized Meuse Valley experienced an air pollution tragedy when smoke and other contaminants accumulated in a narrow steep-sided valley. The tremendous buildup of pollutants caused about 600 people to become ill, and ultimately 63 died. Not only did humans suffer, but cattle, birds, and rats also fell victim to the deplorable conditions.

The industrial revolution brought air pollution to the United States, as homes and coal-burning industries belched smoke, soot, and other undesirable emissions into the air. Soon, large industrial cities, such as St. Louis and Pittsburgh (which became known as the "Smoky City"), began to feel the effects of the ever-increasing use of coal. As early as 1911, studies documented the irritating effect of smoke particles on the human respiratory system and the "depressing and devitalizing" effects of the constant darkness brought on by giant, black clouds of smoke. By 1940, the air over some cities had become so polluted that automobile headlights had to be turned on during the day.

The first major documented air pollution disaster in the United States occurred at Donora, Pennsylvania, during October 1948, when industrial pollution became trapped in the Monongahela River Valley. During the ordeal, which lasted 5 days, more than 20 people died and thousands became ill.* Several times during the 1960s, air pollution levels became dangerously high over New York City. Meanwhile, on the West Coast, in cities such as Los Angeles, the ever-increasing number of automobiles, coupled with large petroleum processing plants, were

instrumental in generating a different type of pollutant—*photochemical smog*—that forms in sunny weather and irritates the eyes. Toward the end of World War II, Los Angeles had its first (of many) smog alerts.

Air pollution episodes in Los Angeles, New York, and other large American cities led to the establishment of much stronger emission standards for industry and automobiles. The Clean Air Act of 1970, for example, empowered the federal government to set emission standards that each state was required to enforce. The Clean Air Act was revised in 1977 and updated by Congress in 1990 to include even stricter emission requirements for autos and industry. The new version of the Act also includes incentives to encourage companies to lower emissions of those pollutants contributing to the current problem of acid rain. Moreover, amendments to the Act have identified 189 toxic air pollutants for regulation, and in 2001, the United States Supreme Court, in a unanimous ruling, made it clear that cost need not be taken into account when setting clean air standards.

As scientific findings accumulate and health standards improve, the range of substances that are considered to be pollutants continues to grow. In 2007 the United States Supreme Court ruled that the greenhouse gas carbon dioxide (CO_2) is a pollutant covered by the Clean Air Act. Because of this ruling, the Environmental Protection Agency, in 2010, began to regulate CO_2 as a pollutant on the premise that, as a greenhouse gas, CO_2 causes a risk to public health. The Clean Power Plan, issued by EPA in 2015, is designed to reduce CO_2 as well as other emissions from power plants. Meanwhile, some of the most urgent air pollution problems on Earth are now found in developing countries such as China and India, where rapidly growing populations can be subject to massive levels of emissions from coal plants, vehicles, and other sources (see ● Fig. 14.2).

*Additional information about the Donora air pollution disaster is given in Focus section 14.5 on p. 434.

Indoor Air Pollution

When people think of air pollution, most think of outside air where automobiles, factories, and power plants spew countless tons of contaminants into the air. But, surprisingly, the air we breathe inside our homes and other structures can be between 5 and 100 times more polluted than the air we breathe outdoors. The most dangerous indoor pollution occurs in developing countries, where cookstoves that use wood or dung spew out large amounts of soot. Millions of deaths are caused each year by this indoor pollution. However, even the homes in affluent, industrialized countries can harbor a variety of dangerous indoor pollutants (see ● Fig.1).

The Environmental Protection Agency (EPA) has identified many sources of indoor air pollution, ranging from building materials, pressed wood products, furnishings, and home cleaning products, to pesticides, adhesives, and personal care products. In addition, heating sources (such as unvented kerosene heaters, wood stoves, and fireplaces) can release a variety of pollutants into a home. The pollution impact of any heating source depends upon several factors, such as how old the source is, the level of maintenance it receives, as well as its location and access to ventilation. For example, a gas cooking stove or heating stove with improper fittings and adjustments can cause a significant emission of carbon monoxide (CO). New carpets and padding (as well as the adhesives used in their installation) can emit volatile organic compounds.

Some pollution sources, such as building materials and foam insulation, produce a constant stream of pollutants, whereas activities such as tobacco smoking emit pollutants into the air on an intermittent basis. In certain instances, outside pollution is brought indoors. Some pollutants enter homes and other structures through cracks and holes in foundations and basements, as is the case with radon.

© C. Donald Ahrens

● FIGURE 1 With candles burning, a fire in the fireplace, and stain-resistant material on rugs and carpets, there are probably more air pollutants in this living room than there are in a similar-size volume of air outdoors.

Radon is a colorless, odorless gas—a natural radioactive compound—that forms as the uranium in soil and rock breaks down. Radon is found everywhere on Earth and only becomes a problem when it leaks out of the soil and becomes trapped inside homes and buildings. The radon gas that seeps in through cracks and other openings in a building can accumulate to levels that create a serious health threat. Radon concentrations vary greatly from structure to structure and can only be measured by devices known as *radon detectors*. Inside a home, radon decays into *polonium*, a solid substance that attaches itself to dust in the air. The tiny dust particles can be inhaled deep into the lungs, where they attach to lung tissue. As the polonium decays, it damages the lung tissue, sometimes producing mutated cells that may develop into lung cancer.

Another major chemical pollutant found inside our homes is *formaldehyde*. It is a colorless, pungent-smelling gas, used widely to manufacture building materials, such as particle board and plywood paneling, insulation, and other household products. Exposure to formaldehyde can cause watery eyes, burning sensations in the nose and throat, breathing difficulties, and nausea. It can also trigger attacks in individuals suffering from asthma. More than 50,000 people who lived in mobile homes provided by the federal government following Hurricanes Katrina and Rita in 2005 were eligible for payments totaling more than $40 million following a class-action lawsuit related to formaldehyde emissions.

Another polluter of our indoor air is *asbestos*, a mineral fiber once used in insulation and as a fire retardant. Manufacturers in the United States have greatly reduced the use of asbestos; it is banned in many nations, but still used for construction in some developing countries. In the United States, much asbestos still remains in furnace and pipe insulation, texturing materials, and floor tiles of older buildings. The most lethal fibers of asbestos are invisible. When inhaled, these tiny particles can accumulate and remain deep in the lungs for extended periods of time, where they damage tissue and potentially cause cancer or *asbestosis*, a permanent scarring of the lungs that can be fatal.

Smoking tobacco indoors can also create an extremely dangerous health situation. Environmental tobacco smoke is a complex mixture of more than 4700 different compounds. These pollutants enter the body as particles and as gases, such as carbon monoxide and hydrogen cyanide. Exposure to tobacco smoke greatly increases the risk of developing lung cancer in both smokers and nonsmokers. (Studies show that the nonsmoking spouses of smokers experience a 30 percent increase in the occurrence of lung cancer.) The small children of smokers are also more likely to fall victim to such illnesses as bronchitis and pneumonia. Heart disease is closely linked with exposure to tobacco smoke, as is premature aging of the skin.

Most of the pollutants we will examine in the next several sections are those more typically found outside. But as we have already mentioned, air pollution can be a major health problem inside a structure. Focus section 14.1 addresses some of the health risks of indoor air pollution.

Types and Sources of Air Pollutants

Air pollutants are airborne substances (either solids, liquids, or gases) that occur in concentrations high enough to threaten the health of people and animals, to harm

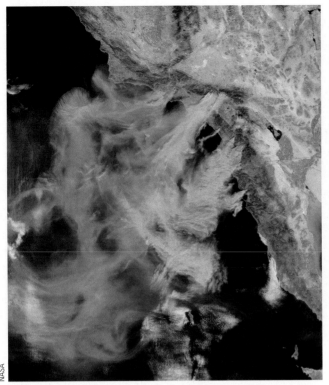

FIGURE 14.3 Strong northeasterly Santa Ana winds on October 28, 2003, blew the smoke from massive wildfires across southern California out over the Pacific Ocean.

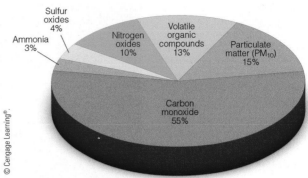

FIGURE 14.4 Estimates of emissions of the primary air pollutants in the United States on a per-weight basis as of 2013. (Data courtesy of United States Environmental Protection Agency.)

vegetation and structures, or to toxify a given environment. Air pollutants come from both natural sources and human activities. Examples include dust and soot from Earth's surface picked up and carried aloft by wind, tons of ash and dust belched into our atmosphere by volcanoes, and vast quantities of drifting smoke produced by forest fires (see Fig. 14.3).

Human-induced pollution enters the atmosphere from both *fixed sources* and *mobile sources.* Fixed sources encompass industrial complexes, power plants, homes, office buildings, and so forth; mobile sources include motor vehicles, ships, and jet aircraft. Certain pollutants are called **primary air pollutants** because they enter the atmosphere directly—from smokestacks and tailpipes, for example. Other pollutants, known as **secondary air pollutants,** form only when a chemical reaction occurs between a primary pollutant and some other component of air, such as water vapor or another pollutant.

Figure 14.4 shows that carbon monoxide is the most abundant primary air pollutant in the United States. The primary source for all pollutants is transportation (motor vehicles, and so on), with fuel combustion from stationary (fixed) sources coming in a distant second. Although hundreds of pollutants are found in our atmosphere, most fall into five groups, recognized as "criteria pollutants" under the Clean Air Act, which are summarized in the following section. Note in Figure 14.4 that emissions are characterized in terms of their total weight. Because different pollutants have different lifetimes in the atmosphere,

they can also be measured by the number of parts per million (ppm) or parts per billion (ppb).

PRINCIPAL AIR POLLUTANTS The term **particulate matter** represents a group of solid particles and liquid droplets that are small enough to remain suspended in the air. Collectively known as *aerosols,* this grouping includes solid particles that may irritate people but are not usually poisonous, such as soot (tiny solid carbon particles), dust, smoke, and pollen. Some of the more dangerous substances include asbestos fibers and arsenic. Tiny liquid droplets of sulfuric acid, polychlorinated biphenyls (PCBs), oil, and various pesticides are also placed into this category.

Because it often dramatically reduces visibility in urban environments, particulate matter pollution is the most noticeable (see Fig. 14.5). Some particulate matter

FIGURE 14.5 A thick layer of particulate matter (mostly smoke) and haze covers Santiago, Chile.

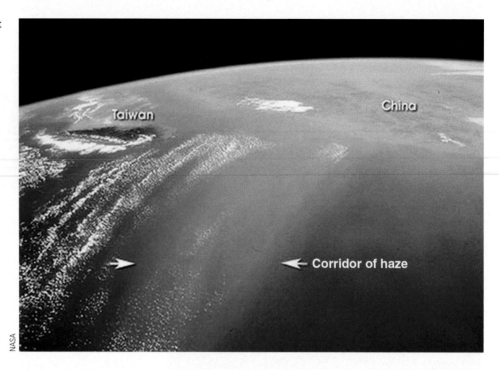

● **FIGURE 14.6** A thick haze about 200 km wide and about 600 km long covers a portion of the East China Sea. The haze is probably a mixture of industrial air pollution, dust, and smoke.

Taiwan

China

← Corridor of haze

observed in urban air includes iron, copper, nickel, and lead. This type of pollution can immediately influence the human respiratory system. Once inside the lungs, it can make breathing difficult, particularly for those suffering from chronic respiratory disorders. Lead particles especially are dangerous as they tend to fall out of the atmosphere and become absorbed into the body through ingestion of contaminated food and water supplies. Fortunately, lead emissions in the United States have been virtually eliminated (look ahead to Fig. 14.11 on p. 426 for data on lead emissions). However, particulate pollution remains a serious problem. Researchers have found that particulate pollution poses a major risk not only to our respiratory system but also to our cardiovascular system. The smallest particulates—those less than 2.5 microns (0.0001 in.) in diameter—are tiny enough to pass from the lungs into the bloodstream, where they can increase the risk of heart attack, arryhthmia, congestive heart failure, and stroke. One study estimated that particulate pollution may be responsible for as many as 10,000 heart disease fatalities per year in the United States. Recent estimates are that more than half of the global deaths associated with outdoor air pollution are actually caused by cardiovascular problems, not respiratory ailments.

Roughly 18 million metric tons of particulate matter is emitted over the United States each year. One major problem with particulate pollution is that it can remain in the atmosphere for some time. Larger, heavier particles with diameters greater than about 10 μm* (0.01 mm) tend to settle to the ground in about a day or so after being emitted. Finer, lighter particles with diameters less than 1 μm (0.001 mm) can remain suspended in the lower atmosphere for several weeks.

Finer particles with diameters smaller than 10 μm are referred to as PM_{10}. These particles pose the greatest health risk, as they are small enough to penetrate the lung's natural defense mechanisms. Moreover, winds can carry these fine particles great distances before they finally reach the surface. In fact, suspended particles from sources in Europe and Asia are believed responsible for the brownish cloud layer called *Arctic haze* that forms over the Arctic each spring. And strong winds over northern China can pick up dust particles and sweep them eastward, where they may settle on North America. This *Asian dust*, which has increased dramatically in recent decades, can reduce visibility, produce spectacular sunrises and sunsets, and coat everything with a thin veneer of particles (see ● Fig. 14.6).

Studies show that particulate matter with diameters less than 2.5 μm, called $PM_{2.5}$, are especially dangerous. For one thing, they can penetrate farther into the lungs. Moreover, these tiny particles frequently consist of toxic or carcinogenic (cancer-causing) combustion products. Of recent concern are the $PM_{2.5}$ particles found in diesel soot. Relatively high amounts of these particles have been measured inside school buses with high amounts also observed downwind of traffic corridors and truck terminals. Another important source of tiny particulates is wildfire. More than 10 percent of the total $PM_{2.5}$ emissions across the United States in recent years have been attributed to emissions from wildland fires.

Rain and snow remove many of these particles from the air; even the minute particles are removed by ice crystals and cloud droplets. In fact, numerical simulations

*Recall that one micrometer (μm) is one-millionth of a meter. (The thickness of this page in print is about 100 micrometers.)

of air pollution suggest that the predominant removal mechanism occurs when these particles act as nuclei for cloud droplets and ice crystals. As we learned in Chapter 13, a long-lasting accumulation of suspended particles (especially those rich in sulfur) is not only aesthetically unappealing but it also has the potential for affecting the climate, as some particles reflect incoming sunlight and prevent a portion of the sun's energy from reaching the surface. Thus, as we saw in Chapter 13, those *sulfur-rich aerosols* have the net effect of cooling Earth's surface.

Many suspended particles are hygroscopic, meaning that water vapor readily condenses onto them. As a thin film of water forms on the particles, they grow in size. When they reach a diameter between 0.1 and 1.0 μm, these *wet haze* particles effectively scatter incoming sunlight, giving the sky a milky white appearance. The particles are usually sulfate or nitrate particulate matter from combustion processes, such as those produced by diesel engines and power plants. The hazy air mass may become quite thick, and on humid summer days it often becomes well-defined, as illustrated in ● Figure 14.7.

Carbon monoxide (CO), a major pollutant of city air, is a colorless, odorless, poisonous gas that forms during the incomplete combustion of carbon-containing fuels. As we saw earlier, carbon monoxide is the most plentiful of the primary pollutants (Fig. 14.4a).

The Environmental Protection Agency (EPA) estimates that more than 60 million metric tons of carbon monoxide enter the air annually over the United States through human activity alone, with about one-third of that coming from highway vehicles. Due to stricter air quality standards and the use of emission-control devices, carbon monoxide levels unrelated to wildfire decreased by about 50 percent from the early 1970s to the early 2000s, and they have dropped by more than 30 percent since then.

It is fortunate that carbon monoxide is quickly removed from the atmosphere by microorganisms in the soil, because even in small amounts this gas is dangerous. Hence, it poses a serious problem in poorly ventilated areas, such as highway tunnels and underground parking garages. Because carbon monoxide cannot be seen or smelled, it can kill without warning. Here's how: Normally, your cells obtain oxygen through a blood pigment called *hemoglobin,* which picks up oxygen from the lungs, combines with it, and carries it throughout your body. Unfortunately, human hemoglobin prefers carbon monoxide to oxygen, so if there is too much carbon monoxide in the air you breathe, your brain will soon be starved of oxygen, and headache, fatigue, drowsiness, and even death may result.*

Sulfur dioxide (SO$_2$) is a colorless gas that comes primarily from the burning of sulfur-containing fossil fuels (such as coal and oil). Its primary sources include power plants, heating devices, smelters, petroleum refineries, and paper mills. However, it can also enter the atmosphere naturally during volcanic eruptions and as sulfate particles from ocean spray.

Sulfur dioxide readily oxidizes (combines with oxygen) to form the secondary pollutants *sulfur trioxide* (SO$_3$) and, in moist air, highly corrosive *sulfuric acid* (H$_2$SO$_4$). Winds can carry these particles great distances before they reach Earth as undesirable contaminants. When inhaled into the lungs, high concentrations of sulfur dioxide aggravate respiratory problems, such as asthma, bronchitis, and emphysema. Sulfur dioxide in large quantities can cause injury to certain plants, such as lettuce and spinach, sometimes producing bleached marks on their leaves and reducing their yield.

Volatile organic compounds (VOCs) represent a class of organic compounds that are mainly **hydrocarbons**—individual organic compounds composed of hydrogen

---

*Should you become trapped in your car during a snowstorm, and you have your engine and heater running to keep warm, roll down the window just a little. This action will allow the escape of any carbon monoxide that may have entered the car through leaks in the exhaust system.

● **FIGURE 14.7** Cumulus clouds and a thunderstorm rise above the thick layer of haze that frequently covers the eastern half of the United States on humid summer days.

© C. Donald Ahrens

and carbon. At room temperature they occur as solids, liquids, and gases. Although thousands of such compounds are known to exist, methane (which occurs naturally and poses no known dangers to health) is the most abundant. Other volatile organic compounds include benzene, formaldehyde, and some chlorofluorocarbons. The Environmental Protection Agency estimates that more than 10 million metric tons of VOCs are emitted into the air over the United States each year, with about 15 percent of the total coming from vehicles used for transportation and about 20 percent from wildfires. The nation's VOC levels unrelated to wildfire dropped by roughly 60 percent from 1980 to 2010.

Certain VOCs, such as benzene (an industrial solvent) and benzo-a-pyrene (a byproduct of burning wood, smoking, and barbecuing), are known to be carcinogens—cancer-causing agents. Although many VOCs are not intrinsically harmful, some will react with nitrogen oxides in the presence of sunlight to produce secondary pollutants, which are harmful to human health.

Nitrogen oxides are gases that form when some of the nitrogen in the air reacts with oxygen during the high-temperature combustion of fuel. The two primary nitrogen pollutants are **nitrogen dioxide (NO_2)** and **nitric oxide (NO),** which, together, are commonly referred to as NO_x—or, simply, *oxides of nitrogen.*

Although both nitric oxide and nitrogen dioxide are produced by natural bacterial action, their concentration in urban environments is between 10 and 100 times greater than in nonurban areas. In moist air, nitrogen dioxide reacts with water vapor to form corrosive nitric acid (HNO_3), a substance that adds to the problem of acid rain, which we will address later.

The primary sources of nitrogen oxides are motor vehicles, power plants, and waste disposal systems. High concentrations are believed to contribute to heart and lung problems, as well as to lowering the body's resistance to respiratory infections. Studies on test animals suggest that nitrogen oxides may encourage the spread of cancer. Moreover, nitrogen oxides are highly reactive gases that play a key role in producing ozone and other ingredients of photochemical smog.

OZONE IN THE TROPOSPHERE   As mentioned earlier, the word **smog** originally meant the combining of smoke and fog. Today, however, the word mainly refers to the type of smog that forms in large cities, such as Los Angeles.

Because this type of smog forms when chemical reactions take place in the presence of sunlight (called *photochemical re-actions*), it is termed **photochemical smog,** sometimes called *Los Angeles-type smog.* When the smog is composed of sulfurous smoke and foggy air, it is sometimes called *London-type smog.*

The main component of photochemical smog is the gas **ozone (O_3).** Ozone is an invisible but noxious substance with an unpleasant odor that irritates eyes and the mucous membranes of the respiratory system, aggravating chronic diseases, such as asthma and bronchitis. Even in healthy people, exposure to relatively low concentrations of ozone for six or seven hours during periods of moderate exercise can significantly reduce lung function. This situation often is accompanied by symptoms such as chest pain, nausea, coughing, and pulmonary congestion. Ozone also attacks rubber, retards tree growth, and damages crops. Each year, in the United States alone, ozone is responsible for crop yield losses of several billion dollars.

We will see later that ozone forms naturally in the stratosphere through the combining of molecular oxygen and atomic oxygen. There, *stratospheric ozone* provides a protective shield against the sun's harmful ultraviolet rays. However, near the surface, in polluted air, ozone—often referred to as *tropospheric (or ground-level) ozone*—is a secondary pollutant that is not emitted directly into the air. Rather, it forms from a complex series of chemical reactions involving other pollutants, such as nitrogen oxides and volatile organic compounds (hydrocarbons). Because sunlight is required to produce ozone, concentrations of tropospheric ozone are normally higher during the afternoons and during the summer months, when sunlight is more intense. (For more information on the formation of ozone in polluted air, read Focus section 14.2.)

Ozone in the Stratosphere   Recall from Chapter 1 that the stratosphere is a region of the atmosphere that lies above the troposphere between about 10 and 50 km (6 and 31 mi) above Earth's surface. The atmosphere is stable in the stratosphere because of a strong temperature inversion—a situation in which the air temperature increases rapidly with height (look back at Fig. 1.23, p. 21). The inversion is due, in part, to the gas ozone, which absorbs ultraviolet radiation at wavelengths of less than about 0.3 micrometers.

In the stratosphere, above middle latitudes, notice in ● Fig. 14.8 that ozone, as a fraction of the entire atmosphere, is most dense at an altitude near 25 km (16 mi). Even at this altitude, its concentration is quite small, as there are only about 12 ozone molecules for every million air molecules (12 ppm).* Although thin, this layer of ozone is significant, for it shields Earth's inhabitants

---

*With a concentration of ozone of only 12 parts per million in the stratosphere, the composition of air here is about the same as it is near Earth's surface—mainly 78 percent nitrogen and 21 percent oxygen.

The Formation of Ground-Level Ozone in Polluted Air

The process leading to the production of tropospheric ozone (ozone near ground level) involves a variety of components of polluted air. This process unfolds along the following lines: Sunlight (with wavelengths shorter than about 0.41 μm) dissociates nitrogen dioxide into nitric oxide and atomic oxygen, which may be expressed by

$$NO_2 + \text{solar radiation} \rightarrow NO + O.$$

The atomic oxygen then combines with molecular oxygen (in the presence of a third molecule, M), to form ozone, as

$$O_2 + O + M \rightarrow O_3 + M.$$

The ozone is then destroyed by combining with nitric oxide; thus

$$O_3 + NO \rightarrow NO_2 + O_2.$$

If sunlight is present, however, the newly formed nitrogen dioxide will break down into nitric oxide and atomic oxygen. The atomic oxygen then combines with molecular oxygen to form ozone again.

As a result of these reactions, large concentrations of ozone can form in polluted air only if some of the nitric oxide (NO) reacts with other gases *without removing ozone in the process.* This process can take place in polluted air as unburned or partially burned hydrocarbons (released into the air by automobiles and industry) react with a variety of gases to form reactive molecules. These molecules then combine with nitric oxide (NO) to produce nitrogen dioxide (NO_2) and other products. In this manner, nitric oxide can react with hydrocarbons to form nitrogen dioxide *without removing ozone.* Hence, certain hydrocarbons in polluted air allow ozone concentrations to increase by preventing nitric oxide from destroying the ozone as rapidly as it forms.

The hydrocarbons (VOCs) also react with oxygen and nitrogen dioxide to produce other undesirable contaminants, such as *PAN* (peroxyacetyl nitrate)—a pollutant that irritates eyes and is extremely harmful to vegetation—and organic compounds. Ozone, PAN, and small amounts of other oxidating pollutants are the ingredients of photochemical smog. Instead of being specified individually, these pollutants are sometimes grouped under a single heading called *photochemical oxidants.**

Hydrocarbons (VOCs) do occur naturally in the atmosphere, as they are given off by vegetation. Oxides of nitrogen drifting downwind from urban areas can react with these natural hydrocarbons and produce smog in relatively uninhabited areas. This phenomenon has been observed downwind of cities such as Los Angeles, London, and New York. Some regions have so much natural (background) hydrocarbon that it may be difficult to reduce ozone levels as much as desired.

In spite of vast efforts to control ozone levels in some major metropolitan areas, results have been generally disappointing because ozone, as we have seen, is a secondary pollutant that forms from chemical reactions involving other pollutants. Ozone production should decrease in most areas when emissions of *both* nitrogen oxides and hydrocarbons (VOCs) are reduced. However, the reduction of only one of these pollutants will not necessarily diminish ozone production because the oxides of nitrogen act as a catalyst for producing ozone in the presence of hydrocarbons.

*An *oxidant* is a substance (such as ozone) whose oxygen combines chemically with another substance.

---

from harmful amounts of ultraviolet solar radiation, which at wavelengths below 0.3 micrometers has enough energy to cause skin cancer in humans.

In recent decades, much attention has been placed on reducing the depletion of stratospheric ozone because of the many risks such a loss poses. These include:

- An increase in the number of cases of skin cancer·
- A sharp increase in eye cataracts and sunburns.
- Suppression of the human immune system.
- An adverse impact on crops and animals due to an increase in ultraviolet radiation.
- A reduction in the growth of ocean phytoplankton.
- A cooling of the stratosphere that could alter stratospheric wind patterns, possibly affecting the destruction of ozone.

Ozone (O_3) forms naturally in the stratosphere when atomic oxygen (O) combines with molecular oxygen (O_2) in the presence of another molecule. Although it forms mainly above 25 kilometers, ozone gradually drifts downward by mixing processes, producing a peak concentration in middle latitudes near 25 kilometers. (In polar regions, its maximum concentration is found at lower levels.) Ozone is broken down into molecular and atomic oxygen when it absorbs ultraviolet (UV) solar radiation with wavelengths between 0.2 and 0.3 micrometers (see ●Fig. 14.9). Thus

$$O_3 + UV \rightarrow O_2 + O.$$

The idea that human activities could alter the amount of stratospheric ozone was first explored in the early 1970s, at a time when Congress was pondering whether or not the United States should build a supersonic jet transport similar to the French-British Concorde. Among the gases emitted from the engines of this type of aircraft are nitrogen oxides. Although the United States' supersonic aircraft was being designed to fly in the stratosphere below the level of maximum ozone, it was feared that the nitrogen oxides would eventually have an adverse effect on the ozone. This factor was one of many considered when Congress decided to halt the development of the United States' version of the supersonic transport in 1971.

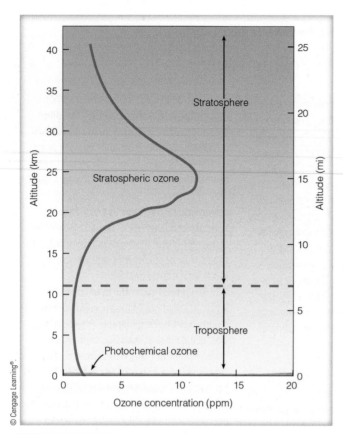

● FIGURE 14.8 The average distribution of ozone above Earth's surface in the middle latitudes.

Concerns about ozone loss gained more prominence later in the 1970s, as scientists found that *chlorofluorocarbons* (CFCs) posed a serious threat to stratospheric ozone. At the time, chlorofluorocarbons were the most widely

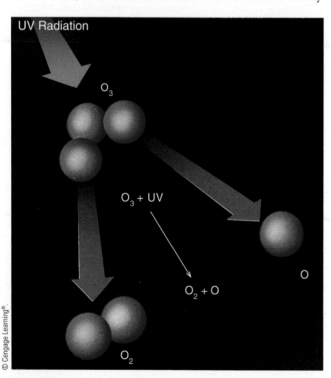

● FIGURE 14.9 An ozone molecule absorbing ultraviolet radiation can become molecular oxygen (O_2) and atomic oxygen (O).

used propellants in spray cans, such as deodorants and hairsprays. In the troposphere, these gases are quite safe, being nonflammable, nontoxic, and unable to chemically combine with other substances.* Hence, these gases slowly diffuse upward without being destroyed. They enter the stratosphere

1. near breaks in the tropopause; especially in the vicinity of jet streams, and
2. in building thunderstorms, especially those that develop in the tropics along the Intertropical Convergence Zone and penetrate the lower stratosphere.

Once CFC molecules reach the middle stratosphere, ultraviolet energy that is normally absorbed by ozone breaks them up, releasing atomic *chlorine* in the process—and chlorine rapidly destroys ozone. In fact, estimates are that a single chlorine atom removes as many as 100,000 ozone molecules before it is taken out of action by combining with other substances.

Since the average lifetime of a CFC molecule is between 50 and 100 years, any increase in the concentration of CFCs is long lasting and a genuine threat to the concentration of ozone. In the late 1970s, the use of CFCs in aerosol cans was banned in the United States and a number of other countries. However, CFCs were still allowed to be used in refrigerators and air-conditioning units. It soon became apparent that CFCs continued to pose a threat to stratospheric ozone, especially with the discovery in the mid-1980s of major ozone depletion during the springtime above Antarctica, where several factors combine to make severe ozone loss possible. This sharp drop in ozone is known as the ozone hole. (More information on the ozone hole is provided in Focus section 14.3, p. 425.)

The discovery of the ozone hole added urgency to the development of an international agreement called the *Montreal Protocol*, which was signed in 1987. This agreement established a timetable for reducing CFC emissions and the use of bromine compounds (halons), which destroy ozone at a rate more than 50 times greater than that of chlorine compounds.** One sign of the success of the Montreal Protocol is the decrease in atmospheric concentrations since the 1990s of most ozone-depleting gases. Although the use of CFCs has has been virtually eliminated, there are still millions of kilograms in the troposphere that will continue to slowly diffuse upward. Thus, we will continue to see signs of ozone depletion for years to come.

A United Nations assessment in 2014 found that the total amount of atmospheric ozone was relatively unchanged since 2000 outside polar regions, after having

---

* Recall that CFCs do act as strong greenhouse gases in the troposphere.

**There are many chemical reactions that involve chlorine and bromine and the destruction of ozone in the stratosphere.

The Ozone Hole

In 1974, two chemists from the University of California at Irvine—F. Sherwood Rowland and Mario J. Molina—warned that increasing levels of CFCs would eventually deplete stratospheric ozone on a global scale. Their studies suggested that ozone depletion would occur gradually and would perhaps not be detectable for many years to come. It was surprising, then, when in 1985 British researchers identified a year-to-year decline in stratospheric ozone over Antarctica. Their findings, corroborated later by satellites and balloon-borne instruments, showed that since the late 1970s ozone concentrations had diminished each year during the months of September and October. This decrease in stratospheric ozone over springtime Antarctica is known as the ozone hole. In years of severe depletion, such as in 2006, the ozone hole covers almost twice the area of the Antarctic continent (see ● Fig. 2).

To understand the causes behind the ozone hole, scientists in 1986 organized the first *National Ozone Expedition*, NOZE-1, which set up a fully instrumented observing station near McMurdo Sound, Antarctica. During 1987, with the aid of instrumented aircraft, NOZE-2 got under way. The findings from these research programs helped scientists put together the pieces of the ozone puzzle.

The stratosphere above Antarctica has one of the world's highest ozone concentrations. Most of this ozone forms over the tropics and is brought to the Antarctic by stratospheric winds. During September and October (spring in the Southern Hemisphere), a belt of stratospheric winds called the *polar vortex* encircles the Antarctic region near 66°S latitude, essentially isolating the cold Antarctic stratospheric air from the warmer air of the middle latitudes. During the long dark Antarctic winter, temperatures inside the vortex can drop to −85°C (−121°F). This

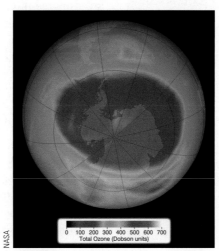

● **FIGURE 2** Ozone distribution over the South Pole on September 11, 2014, as measured by ozone monitoring equipment on NASA's *Aura* satellite. Notice that the lowest ozone concentration or ozone hole (purple shades) covers most of Antarctica. The color scale at the bottom of the image shows total ozone values in Dobson units (DU). A Dobson unit is the physical thickness of the ozone layer if it were brought to Earth's surface (500 DU equals 5 mm).

frigid air allows for the formation of *polar stratospheric clouds*. These ice clouds are critical in facilitating chemical interactions among nitrogen, hydrogen, and chlorine atoms, the end product of which is the destruction of ozone.

In 1986, the NOZE-1 study detected unusually high levels of chlorine compounds in the stratosphere, and, in 1987, the instrumented aircraft of NOZE-2 measured enormous increases in chlorine compounds when it entered the polar vortex. These findings, in conjunction with other chemical discoveries, allowed scientists to pinpoint *chlorine* from CFCs as the main cause of the ozone hole.

Even with a decline in ozone-destroying chemicals, the largest Antarctic ozone hole observed to date occurred during September

2006 (Fig. 2). Yearly variations in the size and depth of the ozone hole are attributed mainly to changes in polar stratospheric temperatures.

Because ozone protects Earth's inhabitants from the sun's harmful UV rays, the springtime drop in ozone over the Southern Hemisphere is serious enough that government agencies in New Zealand and Australia warn their citizens to protect themselves from the sun's rays when the ozone hole forms each year.

In the Northern Hemisphere's polar Arctic, airborne instruments and satellites during the late 1980s and 1990s measured high levels of ozone-destroying chlorine compounds in the stratosphere. But observations could not detect an ozone hole like the one that forms over the Antarctic. (Compare Fig. 2 with Fig. 14.9.) Why would the Arctic not have an ozone hole like its southern counterpart? For one thing, the circulation of stratospheric air over the Arctic (an ocean surrounded by continents) differs from that over the Antarctic (a continent ringed by oceans). The Arctic stratosphere is normally too warm for the widespread development of clouds that help activate chlorine molecules. However, in 1997 and again in 2011, a very cold Arctic stratosphere teamed up with ozone-destroying chemicals to cause significant ozone depletion.

During the past 40 years or so, we have learned much about stratospheric ozone and the ozone hole. Ozone-depleting substances are now regulated, and emissions of these substances are essentially zero. The ozone hole is still there—stronger in some years, weaker in others. The latest research anticipates that mid-latitude and Arctic ozone depletion will end by the middle of this century, with the Antarctic ozone hole virtually extinct by later in the century.

declined by about 2.5 percent during the 1980s and early 1990s. Studies suggest that stratospheric ozone concentrations above the United States have been running about 3 to 5 percent below their pre-1980 averages. Indications are that ozone concentrations in the stratosphere will return to pre-1980 levels before the mid-twenty-first

century, except over Antarctica, where the ozone concentrations will likely remain low until about 2070.

Natural processes can exacerbate ozone depletion for periods of time. Satellite measurements in 1992 and 1993 revealed that ozone concentrations had dropped to record low levels over much of the globe. The decrease appears to

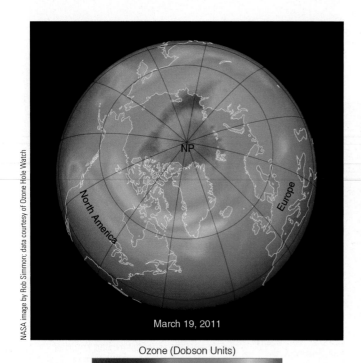

NASA image by Rob Simmon; data courtesy of Ozone Hole Watch

March 19, 2011

Ozone (Dobson Units)

110   220   330   440   550

● FIGURE 14.10 Ozone distribution over the Arctic for March 19, 2011, as measured by the *Ozone Monitoring Instrument* (OMI) on NASA's *Aura* satellite. Total ozone amounts are in Dobson units (DU). A Dobson unit is the physical thickness of the ozone layer if it were brought to Earth's surface (500 DU equals 5 mm). Notice low readings over the Arctic (blue shade) and that these readings are not nearly as low as readings over the Antarctic as shown in Fig. 2, p. 425.

stem not only from ozone-destroying chemicals but also from the 1991 volcanic eruption of Mt. Pinatubo, which sent tons of sulfur dioxide gas into the stratosphere where it formed tiny droplets of sulfuric acid. These droplets not only enhance the ozone destructiveness of the chlorine chemicals but also alter the circulation of air in the stratosphere, making it more favorable for ozone depletion. During the mid-1990s, wintertime ozone levels dropped well below normal over much of the Northern Hemisphere. By 1997, springtime ozone levels over the Arctic had dropped dramatically. This decrease apparently was due to ozone-destroying pollution along with natural cold stratospheric weather patterns that favored ozone reduction.

During the early years of this century, springtime ozone concentrations over the Arctic have varied greatly from year to year. For example, 2010 had relatively high stratospheric ozone levels, whereas in 2011 ozone levels approached the lowest readings ever observed there (see ● Fig. 14.10). Although the ozone layer is in a long-term state of recovery, it remains vulnerable to large annual variations, largely because of the year-to-year influence of stratospheric temperatures on the effects of ozone-destroying chemicals.

There are now two major substitutes for CFCs: *hydrochlorofluorocarbons* (HCFCs) and *hydrofluorocarbons*

(HFCs). The HCFCs contain fewer chlorine atoms per molecule than CFCs and, therefore, pose much less danger to the ozone layer, whereas HFCs contain no chlorine at all. However, like CFCs, these two groups of substitutes are also greenhouse gases that can enhance global warming. Because of this, the Montreal Protocol now calls for HCFCs to be phased out by 2030, and several nations have proposed phasing out HFCs as well. (Diminishing the potential severity of climate change by reducing the amount of greenhouse gas added to the atmosphere has proven to be an important and beneficial side effect of the Montreal Protocol.)

AIR POLLUTION: TRENDS AND PATTERNS   Over the past few decades, major strides have been made in the United States to improve the quality of the air we breathe. ● Figure 14.11 shows the estimated emission trends over the United States for the primary pollutants. Notice that

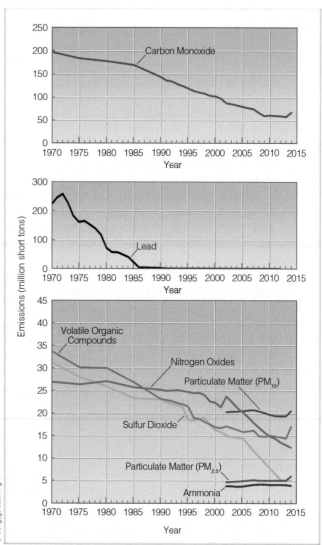

● FIGURE 14.11 Emission estimates of six pollutants in the United States from 1970 through 2014. Emissions related to wildfires are not included. (Data courtesy of United States Environmental Protection Agency.)

since the Clean Air Act of 1970, and including subsequent revisions to the Act, emissions of most pollutants have fallen off substantially, with lead showing the greatest reduction, primarily because of the gradual elimination of leaded gasoline.

Although the situation has improved, we can see from Fig. 14.10 that much more needs to be done. Large quantities of pollutants still spew into our air. In fact, many areas of the country do not conform to the standards for air quality set by the Clean Air Act of 1990. A large part of the problem of pollution control lies in the fact that, even with stricter emission laws, increasing numbers of vehicles and other sources (including roughly 250 million passenger vehicles in the United States) can overwhelm control efforts.

The National Ambient Air Quality Standards are maintained by the Environmental Protection Agency (EPA). *Primary standards* are set to protect human health, whereas *secondary standards* protect human welfare, as measured by the effects of air pollution on visibility, crops, and buildings. Areas that do not meet air quality standards are called *nonattainment areas*. The EPA also issues specific rules to address a variety of local, state, and regional air pollution concerns. For example, the Cross-State Air Pollution Rule restricts power plant emissions that are carried from one state into another. Even with stronger emission laws, estimates are that presently more than 160 million Americans are breathing air that at times does not meet at least one of the EPA primary standards.

To indicate the air quality in a particular region, the EPA developed the **air quality index (AQI)**. The index includes the pollutants carbon monoxide, sulfur dioxide, nitrogen dioxide, particulate matter, and ozone. On any given day, the pollutant measuring the highest value is the one used in the index. The pollutant's measurement is then converted to a number that ranges from 0 to 500 (see ▼ Table 14.1). When the pollutant's value is the same as the primary ambient air quality standard, the pollutant is assigned an AQI number of 100. A pollutant is considered unhealthful when its AQI value exceeds 100. ● Figure 14.12 shows the number of unhealthful days across the United States during 2008. When the AQI value is between 51 and 100, the air quality is described as "moderate." Although these levels may not be harmful to humans during a 24-hour period, they may exceed long-term standards. Notice that the AQI is color-coded, with

▼Table 14.1 **The Air Quality Index (AQI)**

AQI VALUE	AIR QUALITY	GENERAL HEALTH EFFECTS	RECOMMENDED ACTIONS
0–50	Good	None	None
51–100	Moderate	There may be a moderate health concern for a very small number of individuals. People unusually sensitive to ozone may experience respiratory symptoms.	When O_3 AQI values are in this range, unusually sensitive people should consider limiting prolonged outdoor exposure.
101–150	Unhealthy for sensitive groups	Mild aggravation of symptoms in susceptible persons.	Active people with respiratory or heart disease should limit prolonged outdoor exertion.
151–200	Unhealthy	Aggravation of symptoms in susceptible persons, with irritation symptoms in the healthy population.	Active children and adults with respiratory or heart disease should avoid extended outdoor activities; everyone else, especially children, should limit prolonged outdoor exertion.
201–300	Very unhealthy	Significant aggravation of symptoms and decreased exercise tolerance in persons with heart or lung disease, with widespread symptoms in the healthy population.	Active children and adults with existing heart or lung disease should avoid outdoor activities and exertion. Everyone else, especially children, should limit outdoor exertion.
301–500	Hazardous	Significant aggravation of symptoms. Premature onset of certain diseases. Premature death may occur in ill or elderly people. Healthy people may experience a decrease in exercise tolerance.	Everyone should avoid all outdoor exertion and minimize physical outdoor activities. Elderly and persons with existing heart or lung disease should stay indoors

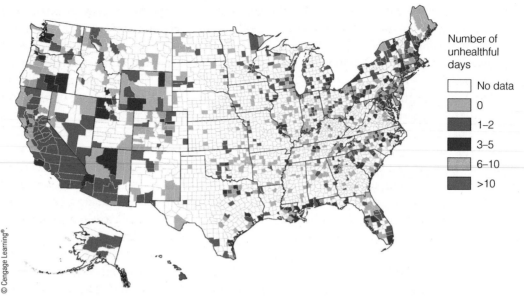

FIGURE 14.12 The number of unhealthful days (by county) across the United States for any one of the five "criteria" pollutants (CO, SO_2, NO_2, O_3, and particulate matter) during 2013. (Data courtesy of United States Environmental Protection Agency.)

Number of unhealthful days

☐ No data
0
1–2
3–5
6–10
>10

© Cengage Learning®.

each color corresponding to an AQI level of health concern. The color green indicates "good" air quality; the color red, "unhealthy" air; and maroon, "hazardous" air. Table 14.1 also shows the health effects and the precautions that should be taken when the AQI value reaches a certain level.

Higher emission standards, along with cleaner fuels (such as natural gas) and expanded use of renewable energy, have made the air over our large cities cleaner today than it was years ago. In fact, total emissions of toxic chemicals spewed into the skies over the United States have been declining steadily since the EPA began its inventory of these chemicals in 1987. But particulates and ground-level ozone are still pervasive problems. Because ozone is a secondary pollutant, its formation is controlled by the concentrations of other pollutants, namely nitrogen oxides and hydrocarbons (VOCs). Moreover, weather conditions play a vital role in ozone formation, as ozone reaches its highest concentrations in sunny, hot weather

when surface winds are light and a stagnant high-pressure area covers the region.

As a result of these factors, year-to-year ozone trends are quite variable, although the Los Angeles area has shown an overall decline in the number of unhealthful days due to ozone (see ● Fig. 14.13). In fact, the year 2015 saw a record-low number of unhealthful days in the Los Angeles area for ozone as well as particulates.

BRIEF REVIEW

Before going on to the next several sections, here is a brief review of some of the important points presented so far.

● Near the surface, primary air pollutants (such as particulate matter, CO, SO_2, NO, NO_2, and VOCs) enter the atmosphere directly, whereas secondary air pollutants (such as O_3) form when a chemical reaction takes place between a primary pollutant and some other component of air.

FIGURE 14.13 The number of days ozone exceeded the 8-hour federal standard established in 1997 (0.08 ppm) and maximum 8-hour ozone concentration (ppm) for Los Angeles and surrounding areas in the South Coast air basin. (Courtesy of South Coast Air Pollution District.)

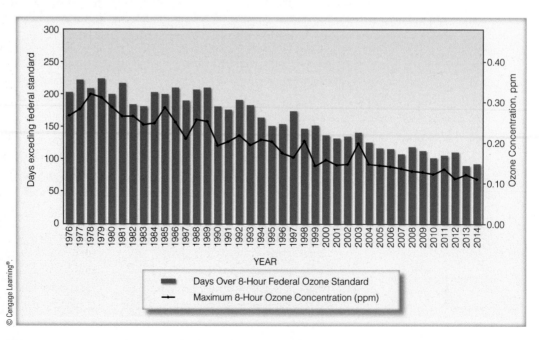

© Cengage Learning®.

Days Over 8-Hour Federal Ozone Standard
Maximum 8-Hour Ozone Concentration (ppm)

(a)

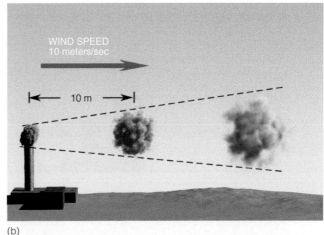

(b)

●**FIGURE 14.14** If each chimney emits a puff of smoke every second, then where the wind speed is low (a), the smoke puffs are closer together and more concentrated. Where the wind speed is greater (b), the smoke puffs are farther apart and more diluted as turbulent eddies mix the smoke with the surrounding air.

- The word "smog" (coined in London in the early 1900s) originally meant the combining of smoke and fog. Today, the word mainly refers to photochemical smog—pollutants that form in the presence of sunlight.

- Stratospheric ozone forms naturally in the stratosphere and provides a protective shield against the sun's harmful ultraviolet rays. Tropospheric (ground-level) ozone that forms in polluted air is a health hazard and is the primary ingredient of photochemical smog.

- Human-induced chemicals, such as chlorofluorocarbons (CFCs), have been altering the amount of ozone in the stratosphere by releasing chlorine, which rapidly destroys ozone.

- Even though the emissions of most pollutants have declined across the United States since 1970, millions of Americans are breathing air that does not meet air quality standards.

Factors That Affect Air Pollution

If you live in a region that periodically experiences photochemical smog, you may have noticed that these episodes often occur with clear skies, light winds, and generally warm, sunny weather. Although this may be "typical" air pollution weather, it by no means represents the only weather conditions necessary to produce high concentrations of pollutants, as we will see in the following sections.

THE ROLE OF THE WIND Wind speed plays a role in diluting pollution. When vast quantities of pollutants are spewed into the air, the wind speed determines how quickly the pollutants mix with the surrounding air and, of course, how fast they move away from their source. Strong winds tend to lower the concentration of pollutants by spreading them apart as they move downwind. This process of spreading is called **dispersion**. Moreover, the stronger the wind, the more turbulent the air. Turbulent air produces swirling eddies that dilute the pollutants by mixing them with the cleaner surrounding air. Hence, when the wind

dies down, pollutants are not readily dispersed, and they tend to become more concentrated (see ●Fig. 14.14).

THE ROLE OF STABILITY AND INVERSIONS Recall from Chapter 5 that atmospheric stability determines the extent to which air will rise. Remember also that an unstable atmosphere favors vertical air currents, whereas a stable atmosphere strongly resists upward vertical motions. Smoke emitted into a stable atmosphere thus tends to spread horizontally rather than to mix vertically.

The stability of the atmosphere is determined by the way the air temperature changes with height (the lapse rate). When the measured air temperature decreases rapidly as we move up into the atmosphere, the atmosphere tends to be more unstable, and pollutants tend to be mixed vertically, as illustrated in ●Fig. 14.15a. If, however, the measured air temperature either decreases quite slowly as we ascend, or actually increases with height (remember that this is called an *inversion*), the atmosphere is stable. An inversion represents an extremely stable atmosphere where warm air lies above cool air (see Fig. 14.15b). Any air parcel that attempts to rise into the inversion will, at some point, be cooler and heavier (more dense) than the warmer air surrounding it. Hence, the inversion acts like a lid on vertical air motions.

The inversion depicted in Fig. 14.15b is called a **radiation** (or **surface**) **inversion**. This type of inversion typically forms during the night and early morning hours when the sky is clear and the winds are light. As we saw in Chapter 3, radiation inversions also tend to be well developed during the long nights of winter.

In Fig. 14.15b, notice that within the stable inversion, the smoke from the shorter stacks does not rise very high, but spreads out, contaminating the area around it. In the relatively unstable air above the inversion, smoke from the taller stack is able to rise and become dispersed. Since radiation inversions are often rather shallow, it should be apparent

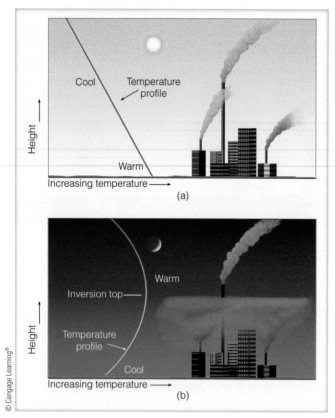

●FIGURE 14.15 (a) During the afternoon, when the atmosphere is most unstable, pollutants rise, mix, and disperse downwind. (b) At night when a radiation inversion exists, pollutants from the shorter stacks are trapped within the inversion, while pollutants from the taller stack, above the inversion, are able to rise and disperse downwind.

why taller chimneys have been replacing many shorter ones. In fact, taller chimneys disperse pollutants better than shorter ones even in the absence of a surface inversion because the taller chimneys are able to mix pollutants throughout a greater volume of air. Although these taller stacks do improve the air quality in their immediate area, they may also contribute to the acid rain problem by allowing the pollutants to be swept great distances downwind.

As the sun rises and the surface warms, the radiation inversion normally weakens and disappears before noon. By afternoon, the atmosphere is sufficiently unstable so that, with adequate winds, pollutants are able to disperse vertically (Fig. 14.15a). The changing atmospheric stability, from stable in the early morning to conditionally unstable in the afternoon, can have a profound effect on

DID YOU KNOW?

An unusually strong and long-lived inversion covered much of the Great Basin for several weeks in January 2013. On the morning of January 23, the temperature was a frigid 4°F in dense smog at Salt Lake City (Utah) International Airport (elevation 1288 m or 4226 ft). Almost a kilometer higher up, in the mountains just to the east of town, Park City (elevation 2100 m or 6900 ft) was experiencing sunshine and fresh air, with a temperature of around 45°F.

the daily concentrations of pollution in certain regions. For example, on a busy city street corner, carbon monoxide levels can be considerably higher in the early morning than in the early afternoon (with the same flow of traffic). Changes in atmospheric stability can also cause smoke plumes from chimneys to change during the course of a day. (Some of these changes are described in Focus section 14.4.)

Radiation inversions normally last just a few hours. On the other hand, **subsidence inversions** may persist for several days or longer. These are the ones commonly associated with major air pollution episodes. They form as the air above a deep anticyclone slowly sinks (subsides) and warms.*

A typical temperature profile of a subsidence inversion that forms along the California coast in summer is shown in ●Fig. 14.16. Notice that in the relatively unstable air beneath the inversion, the pollutants are able to mix vertically up to the inversion base. The stable air of the inversion, however, inhibits vertical mixing and acts like a lid on the pollution below, preventing it from entering into the inversion.

In Fig. 14.16, the region of relatively unstable (well-mixed) air that extends from the surface to the base of the inversion is referred to as the **mixing layer.** The vertical extent of the mixing layer is called the **mixing depth.** Observe that if the inversion rises, the mixing depth increases and the pollutants will be dispersed throughout a greater volume of air; if the inversion lowers, the mixing depth decreases and the pollutants will become more concentrated, sometimes reaching unhealthy levels. Since the atmosphere tends to be most unstable in the afternoon and most stable in the early morning, we typically find the greatest mixing depth in the afternoon and the shallowest one (if one exists at all) in the early morning. Consequently, during the day, the top of the mixing layer may be clearly visible (see ●Fig. 14.17). Moreover, during take-off or landing on daylight flights out of large urban areas, the top of the mixing layer may sometimes be observed.

The position of the semipermanent Pacific high off the coast of California contributes greatly to the air pollution in that region. The Pacific high promotes sinking air, which warms the air aloft. Surface winds around the high promote upwelling of ocean water. Upwelling—the rising of cold water from below—makes the surface water cool, which, in turn, cools the air above. Warm air aloft coupled with cool, surface (marine) air, together produce a strong and persistent subsidence inversion—one that exists 80 to 90 percent of the time over the city of Los Angeles between June and October, the smoggy months. (Look at the strong subsidence inversion portrayed in Fig. 14.16.) The pollutants trapped within the cool marine air are occasionally

---

*Remember from Chapter 2 that sinking air always warms because it is being compressed by the surrounding air.

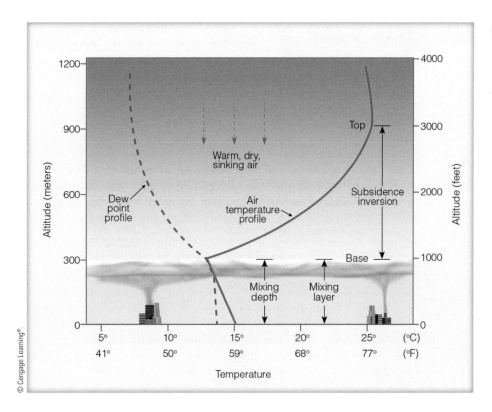

FIGURE 14.16 A strong subsidence inversion along the coast of California. The base of the stable inversion acts as a cap or lid on the pollutants below by preventing them from escaping into the warmer air above. If the inversion lowers, the mixing depth decreases, and the pollutants are concentrated within a smaller volume.

swept eastward by a sea breeze. This action carries smog from the coastal regions into the interior valleys producing a *smog front* (see Fig. 14.18).

THE ROLE OF TOPOGRAPHY The shape of the landscape (topography) plays an important part in trapping pollutants. We know from Chapter 3 that, at night, cold air tends to drain downhill, where it settles into low-lying basins and valleys. The cold air can strengthen a pre-existing surface inversion, and it can carry pollutants downhill from the surrounding hillsides (see Fig. 14.19).

Valleys prone to pollution are those completely enclosed by mountains and hills. The surrounding mountains tend to block the prevailing wind. With light winds, and a shallow mixing layer, the poorly ventilated cold valley air can only slosh back and forth like a murky bowl of soup.

Valleys susceptible to stagnant air exist in just about all mountainous regions. Air pollution concentrations in these valleys tend to be greatest during the colder months. During the warmer months, daytime heating can warm the sides of the valley to the point that upslope valley winds vent the pollutants upward, as if in a chimney.

FIGURE 14.17 A thick layer of polluted air is trapped in the valley. The top of the polluted air marks the base of a subsidence inversion and the top of the mixing layer.

FIGURE 14.18 The leading edge of cool, marine air carries pollutants into Riverside, California, as an advancing smog front.

Smokestack Plumes

We know that the stability of the air (especially near the surface) changes during the course of a day. These changes can influence the pollution near the ground as well as the behavior of smoke leaving a chimney. ● Figure 3 illustrates different smoke plumes that can develop with adequate wind, but different types of stability.

In Fig. 3a, it is early morning, the winds are light, and a radiation inversion extends from the surface to well above the height of the smokestack. In this stable environment, there is little up-and-down motion, so the smoke spreads horizontally rather than vertically. When viewed from above, the smoke plume resembles the shape of a fan, and it is referred to as a *fanning smoke plume*.

Later in the morning, the surface air warms quickly and destabilizes as the radiation inversion gradually disappears from the surface upward (Fig. 3b). However, the air above the chimney is still stable, as indicated by the presence of the inversion, so vertical motions are confined to the region near the surface. Because of this, the smoke mixes downwind, increasing the concentration of pollution at the surface—sometimes to dangerously high levels. This effect is called *fumigation*.

If daytime heating of the ground continues, the depth of atmospheric instability increases. Notice in Fig. 3c, that the inversion

has completely disappeared. Light-to-moderate winds combine with rising and sinking air to cause the smoke to move up and down in a wavy pattern, producing a *looping smoke plume*. Thus, the behavior of smoke plumes provides a clue to the stability of the atmosphere, which yields important information about the dispersion of pollutants.

Of course, other factors influence the dispersion of pollutants from a chimney, including the pollutants' temperatures and exit velocities, wind speed and direction, and, as we saw in an earlier section, the chimney's height. Overall, taller chimneys, greater wind speeds, and higher exit velocities result in a lower concentration of pollutants.

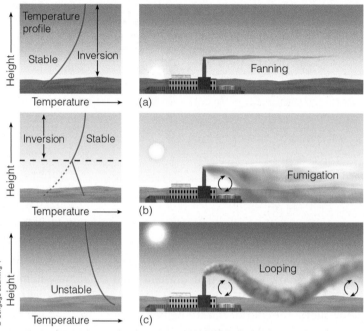

● **FIGURE 3** As the vertical temperature profile changes during the course of a day (a through c), the pattern of smoke emitted from the stack changes as well.

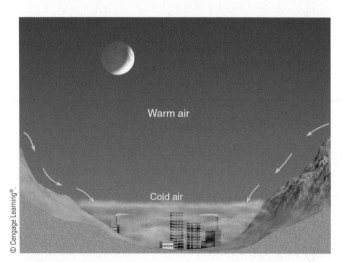

● **FIGURE 14.19** At night, cold air and pollutants drain downhill and settle in low-lying valleys.

The pollution problem in several large cities is, at least partly, due to topography. For example, the city of Los Angeles is surrounded on three sides by hills and mountain.* Cool marine air from off the ocean moves inland and pushes against the hills, which tend to block the air's eastward progress. Unable to rise, the cool air settles in the basin, trapping pollutants from industry and millions of autos. Baked by sunlight, the pollutants become the infamous photochemical smog shown in ● Fig. 14.20. By the same token, the "mile high" city of Denver, Colorado, sits in a broad shallow basin that frequently traps both cold air and pollutants. The much larger Great Basin that spans Utah and neighboring states is also prone to trapping cold

*The city of Los Angeles with its surrounding topography is shown in Fig 7.13 on p. 183.

Jon Hicks/Getty Images

● **FIGURE 14.20** Thick smog, such as shown in this photo of Los Angeles taken during May 2015, has become less common over the region in recent years.

air, occasionally for weeks at a time, which can put cities in the region at risk of significant pollution episodes.

SEVERE AIR POLLUTION POTENTIAL The greatest potential for an episode of severe air pollution occurs when all of the factors mentioned in the previous sections come together simultaneously. Ingredients for a major buildup of atmospheric pollution are:

- many sources of air pollution (preferably clustered close together)
- a deep high-pressure area that becomes stationary over a region
- light surface winds that are unable to disperse the pollutants
- a strong subsidence inversion produced by the sinking of air aloft
- a shallow mixing layer with poor ventilation
- a valley where the pollutants can accumulate
- clear skies so that radiational cooling at night will produce a surface inversion, which can cause an even greater buildup of pollutants near the ground
- and, for photochemical smog, adequate sunlight to produce secondary pollutants, such as ozone

Light winds and poor vertical mixing can produce a condition known as **atmospheric stagnation**. The atmospheric conditions resulting in air stagnation usually occur during particular weather patterns. For example, a region under the domination of a surface high-pressure area or ridge often experiences clear skies, light winds, and a

subsidence inversion. Moreover, where warm air rides up over cold surface air, such as ahead of an advancing warm front, stable atmospheric conditions promote the trapping of pollutants near the surface. On the other hand, the strong and gusty winds and generally less-stable air behind a cold front usually results in good dispersion (see ● Fig. 14.21).

If air stagnation conditions persist for several days or longer where there are ample pollutant sources, the buildup of pollution can lead to some of the worst air pollution disasters on record, such as the one in the valley city of Donora, Pennsylvania, where in 1948 seventeen people died within fourteen hours. (Additional information on the Donora disaster is found in Focus section 14.5.)

Air Pollution and the Urban Environment

For more than 100 years, it has been known that cities are generally warmer than surrounding rural areas. This region of city warmth, known as the **urban heat island,** can influence the concentration of air pollution. However, before we look at its influence, let's see how the heat island actually forms.

The urban heat island is formed by industrial and urban development. In rural areas, a large part of the incoming solar energy evaporates water from vegetation and soil. In cities, where less vegetation and exposed soil exists, the majority of the sun's energy is absorbed by urban structures and asphalt. Hence, during warm daylight hours, less evaporative cooling in cities allows surface temperatures to rise higher than in rural areas.*

---

*The cause of the urban heat island is quite involved. Depending on the location, time of year, and time of day, any or all of the following differences between cities and their surroundings can be important: albedo (reflectivity of the surface), surface roughness, emissions of heat, emissions of moisture, and emissions of particles that affect net radiation and the growth of cloud droplets.

Five Days In Donora—An Air Pollution Episode

On Tuesday morning, October 26, 1948, a cold surface anticyclone moved over the eastern half of the United States. There was nothing unusual about this high-pressure area; with a central pressure of only 1025 mb (30.27 in.), it was not exceptionally strong (see ● Fig. 4). Aloft, however, a large blocking-type ridge formed over the region, and the jet stream, which moves the surface pressure features along, was far to the west. Consequently, the surface anticyclone became entrenched over Pennsylvania and remained nearly stationary for five days.

The widely spaced isobars around the high-pressure system produced a weak pressure gradient and generally light winds throughout the area. These light winds, coupled with the gradual sinking of air from aloft, set the stage for a disastrous air pollution episode.

On Tuesday morning, radiation fog gradually settled over the moist ground in Donora, a small town nestled in the Monongahela Valley of western Pennsylvania. Because Donora rests on bottomland (Fig. 4), surrounded by rolling hills, its residents were accustomed to fog, but not to what was to follow.

The strong radiational cooling that formed the fog, along with the sinking air of the anticyclone, combined to produce a strong temperature inversion. Light, downslope winds spread cool air and contaminants over Donora from the community's steel mill, zinc smelter, and sulfuric acid plant.

The fog with its burden of pollutants lingered into Wednesday. Cool drainage winds during the night strengthened the inversion and added more effluents to the already filthy air. The dense fog layer blocked sunlight from reaching the ground. With essentially no surface heating, the mixing

depth lowered and the pollution became more concentrated. Unable to mix and disperse either horizontally or vertically, the dirty air became confined to a shallow, stagnant layer.

Meanwhile, the factories continued to belch impurities into the air (primarily sulfur dioxide and particulate matter) from stacks no higher than 40 m (130 ft) tall. The fog gradually thickened into a moist clot of smoke and water droplets. By Thursday, the visibility had decreased to the point where one could barely see across the street. At the same time, the air had a penetrating, almost sickening, smell of sulfur dioxide. At this point, a large percentage of the population became ill.

The episode reached a climax on Saturday, as 17 deaths were reported. As

the death rate mounted, alarm swept through the town. An emergency meeting was called between city officials and factory representatives to see what could be done to cut down on the emission of pollutants.

The light winds and unbreathable air persisted until, on Sunday, an approaching storm generated enough wind to vertically mix the air and disperse the pollutants. A welcome rain then cleaned the air further. All told, the episode had claimed the lives of 22 people. During the five-day period, about half of the area's 14,000 inhabitants experienced some ill effects from the pollution. Most of those affected were older people with a history of cardiac or respiratory disorders.

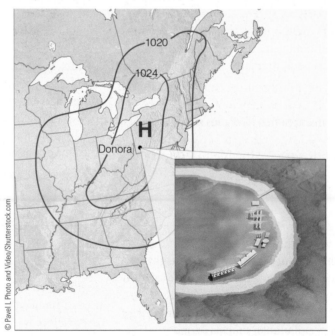

● FIGURE 4 Surface weather map that shows a stagnant anticyclone over the eastern United States on October 26, 1948. The insert map shows a schematic depiction of the town of Donora on the Monongahela River.

At night, the solar energy (stored as vast quantities of heat in city buildings and roads) is slowly released into the city air. Additional city heat is given off at night (and during the day) by vehicles and factories, as well as by industrial and domestic heating and cooling units. The release of heat energy is retarded by the tall city walls that do not allow infrared radiation to escape as readily as do the relatively level surfaces of the surrounding countryside.

The slow release of heat tends to keep nighttime city temperatures higher than those of the faster cooling rural areas. Overall, the heat island is strongest

1. at night when compensating sunlight is absent,
2. during the winter when nights are longer and there is more heat generated in the city,
3. when the region is dominated by a high-pressure area with light winds, clear skies, and less humid air.

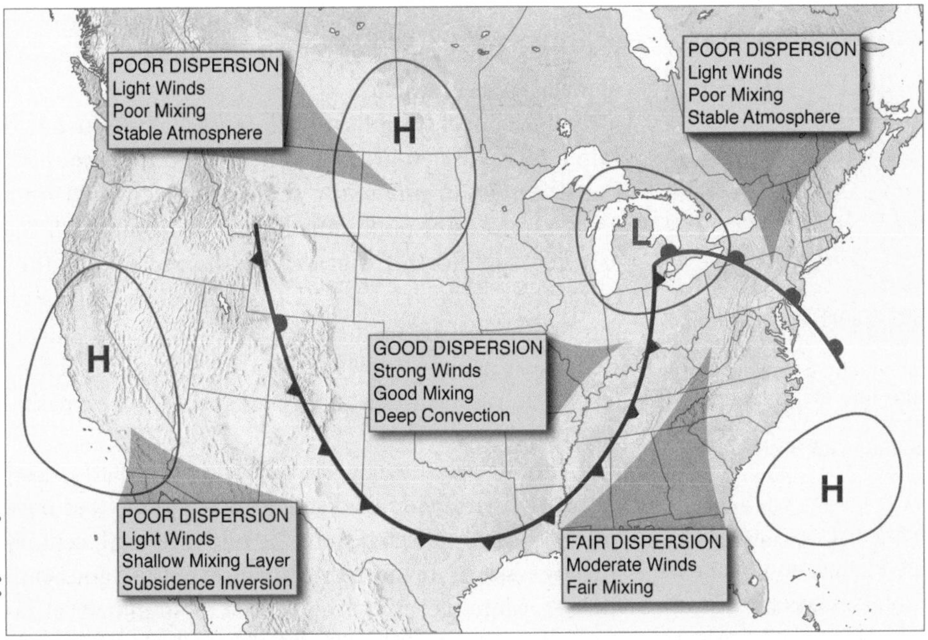

● FIGURE 14.21 Weather patterns associated with poor, fair, and good air pollution dispersion.

Over time, increasing numbers (and sizes) of urban heat islands affect climatological temperature records, producing artificial warming in climatic records taken in cities. Scientists take this warming into account when interpreting climate change over the past 100-plus years.

The constant outpouring of pollutants into the environment may influence the climate of a city. Certain particles reflect solar radiation, thereby reducing the sunlight that reaches the surface. Some particles serve as nuclei upon which water and ice form. Water vapor condenses onto these particles when the relative humidity is as low as 70 percent, forming haze that greatly reduces visibility. Moreover, the added nuclei increase the frequency of city fog.*

Many studies over the last few decades indicate that precipitation can be greater in urban areas than in surrounding countrysides. This phenomenon may be due in part to the increased roughness of city terrain, the result of large structures that cause surface air to slow and gradually converge. This pile-up of air over the city then slowly rises. At the same time, city heat warms the surface air, making it more unstable, which enhances rising air motions, which, in turn, aids in forming clouds and thunderstorms. This process helps explain why precipitation tends to be heavier in and near urban areas, and sometimes just downwind from them. Enhanced precipitation has been found over the years in areas near Chicago, Illinois; St. Louis, Missouri; Paris, France; Atlanta, Georgia; Beijing, China; and other locations. The effects are generally more evident with showers and thunderstorms rather than steady rains.

▼ Table 14.2 summarizes the environmental influence of cities by contrasting the urban environment with the rural.

The large concentrations of particulates in urban areas can also help boost rainfall within and downwind of cities in some cases. The main exception is for cities located in dry regions, where large numbers of aerosols can actually decrease rainfall, as the particulates (nuclei) compete for the available moisture—similar to what happens when a cloud is overseeded, discussed in Chapter 5. On clear still nights when the heat island is pronounced, a small thermal low-pressure area may form over the city. Sometimes a light breeze—which may be called a **country breeze**—blows from the countryside into the city. If there are major industrial areas along the city's outskirts, pollutants are carried

▼ Table 14.2 Contrast of the Urban and Rural Environment (Average Conditions)*

CONSTITUENTS	URBAN AREA (CONTRASTED TO RURAL AREA)
Mean pollution level	higher
Mean sunshine reaching the surface	lower
Mean temperature	higher
Mean relative humidity	lower
Mean visibility	lower
Mean wind speed	lower
Mean precipitation	higher
Mean amount of cloudiness	higher
Mean thunderstorm (frequency)	higher

*The impact that tiny liquid and solid particles (aerosols) may have on a larger scale is complex and depends upon a number of factors that are addressed in Chapter 13.

*Values are omitted because they vary greatly depending upon city, size, type of industry, and season of the year.

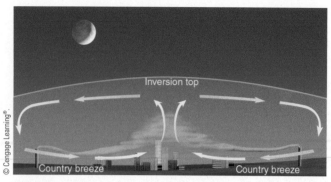

● FIGURE 14.22 On a clear, relatively calm night, a weak country breeze carries pollutants from the outskirts into the city, where they concentrate and rise due to the warmth of the city's urban heat island. This effect may produce a pollution (or dust) dome from the suburbs to the center of town.

into the heart of town, where they become even more concentrated. Such an event is especially likely if an inversion inhibits vertical mixing and dispersion (see ● Fig. 14.22).

Before we look at the problem of acid rain, it is important to note that, on average, more people in the United States die from excessive heat during heat waves than from any other extreme weather event. But recent studies suggest that perhaps a fair percentage of these deaths actually occur due to exposure to high levels of air pollution. Focus section 14.6, "Heat Waves and Air Pollution: A Deadly Team," examines this possibility in more detail.

Acid Deposition

Air pollution emitted from industrial areas, especially products of combustion, such as oxides of sulfur and nitrogen, can be carried many kilometers downwind. Either these particles and gases slowly settle to the ground in dry form (*dry deposition*) or they are removed from the air during the formation of cloud particles and then carried to the ground in rain and snow (*wet deposition*). **Acid rain** and *acid precipitation* are common terms used to describe wet deposition, while **acid deposition** encompasses both dry and wet acidic substances. How, then, do these substances become acidic?

Emissions of sulfur dioxide (SO_2) and oxides of nitrogen may settle on the local landscape, where they transform into acids as they interact with water, especially during the formation of dew or frost. The remaining airborne particles may transform into tiny dilute drops of sulfuric acid (H_2SO_4) and nitric acid (HNO_3) during a complex series of chemical reactions involving sunlight, water vapor, and other gases. These acid particles may then fall slowly to the surface, or they may adhere to cloud droplets or to fog droplets, producing **acid fog.** They may even act as nuclei on which the cloud droplets begin to grow. When precipitation occurs in the cloud, it carries the acids to the ground. Because of this, precipitation grew increasingly acidic during

the mid- to late twentieth century, with the increase of air pollution in many parts of the world, especially downwind of major industrial areas.

High concentrations of pollutants that produce acid rain can be carried great distances from their sources. For example, in one study scientists discovered high concentrations of pollutants hundreds of miles off the east coast of North America. It is suspected that they came from industrial East Coast cities. Even though most pollutants are washed from the atmosphere during storms, some may be swept over the Atlantic, reaching places such as Bermuda and Ireland. Acid rain knows no national boundaries.

Although studies suggest that acid precipitation may be nearly worldwide in distribution, regions that have been noticeably affected since the mid-twentieth century include eastern North America, central Europe, and Scandinavia. More recently, as a result of rapid industrialization, China and other parts of Asia have been affected. In some places, acid precipitation occurs naturally, such as in northern Canada, where natural fires in exposed coal beds produce tremendous quantities of sulfur dioxide. By the same token, acid fog can form by natural means.

Precipitation is naturally somewhat acidic. The carbon dioxide occurring naturally in the air dissolves in precipitation, making it slightly acidic with a pH between 5.0 and 5.6. Consequently, precipitation is considered acidic when

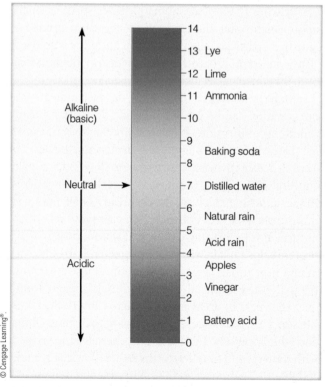

● FIGURE 14.23 The pH scale ranges from 0 to 14, with a value of 7 considered neutral. Values greater than 7 are alkaline and below 7 are acidic. The scale is logarithmic, which means that rain with pH 3 is 10 times more acidic than rain with pH 4 and 100 times more acidic than rain with pH 5.

Heat Waves and Air Pollution: A Deadly Team

Some of the worst weather calamities of the twenty-first century have been triggered by intense, prolonged episodes of heat. By itself, the cumulative impact of high temperatures can be enough to kill people, especially those in poor health who have little or no way to cool down. But air pollution can also play a significant role in the deadly nature of heat waves.

The numbers involved are staggering. Mortality specialists have found that more than 40,000 people died as a result of a heat wave that gripped Europe during the summer of 2003. Likewise, the record-breaking heat that struck Moscow in 2010 took more than 10,000 lives. Because a heat wave unfolds over days and weeks, death tolls such as these are calculated by taking the total number of fatalities observed and adjusting that total based on the number of deaths that would normally have been expected at a given time of year.

Like a gang of criminals, high temperatures and air pollution often appear at the same time and place, so it can be difficult to tell whether the villains acted individually or collectively. The stationary domes of upper-level high pressure that can lead to heat waves also tend to produce multiple days of sunshine and stagnant air, conditions ideal for the formation of ground-level ozone and the accumulation of small particulates in the atmosphere. These, in turn, are directly linked to a variety of dangerous, potentially fatal health conditions, including cardiovascular and respiratory problems.

Several studies of the European heat wave of 2003 examined the causes of death and the potential role of pollutants. One team found that between 13 and 30 percent of the roughly 1000 deaths in Switzerland associated with the heat wave were actually caused by ozone rather than by the heat itself. Another study attributed between 21 and 38 percent of more than 2000 heat wave fatalities in the United Kingdom to the effects of ozone and particles with diameters smaller than 10 micrometers (PM_{10}). One reason why it is difficult to narrow the ranges reported in studies like these is that researchers are not yet sure of the extent to which heat and pollution have a synergistic effect, as opposed to acting independently.

To get at this question, one analysis in the United States drew on data from the massive National Morbidity, Mortality, and Air Pollution Study (NMMAPS), which combined health- and weather-related data for about 100 urban areas from 1987 through 2000. Using NMMAPS, the researchers found that, as a rule, ozone did tend to boost the link between hot weather and deaths related to heart disease. For example, a 10°C (18°F) increase in temperature produced about 1 percent more cardiovascular deaths when ozone concentrations were at their lowest, but the increase was about 8 percent when ozone levels were highest.

The role of air pollution in heat-related health problems could be magnified in coming decades. Climate change simulations indicate that heat waves may increase in number, duration, and strength across many parts of the world. At the same time, an ever greater proportion of the world's population is clustering in urban areas. In the absence of stringent air pollution control, this trend could increase both the amount of air pollution in a given area and the number of people vulnerable to its health effects.

June 10, 2010 August 7, 2010

● **FIGURE 5** The severe effects of Moscow's 2010 heat wave on air quality can be seen in these two photos of the same street, one taken on June 17 (before the heat wave began) and the other on August 7 (near the peak of the heat wave).

its pH is below about 5.0 (see ●Fig. 14.23). In the northeastern United States, where emissions of sulfur dioxide are primarily responsible for the acid precipitation, typical pH values range between 4.0 and 4.7 (see ●Fig. 14.24). But acid precipitation is not confined to the Northeast; it is also found in the southeastern states, too. Along the West Coast, the main cause of acid deposition appears to be the oxides of nitrogen released in automobile exhaust. In Los Angeles, acid fog is a more serious problem than acid rain, especially along the coast, where fog is most prevalent. The fog's pH is usually between 4.4 and 4.8, although pH values of 3.0 and below have been measured.

● FIGURE 14.24 Values of pH in precipitation over the United States during 2012. (National Atmospheric Deposition Program)

pH

■ >5.3
■ 5.1 to 5.3
■ 4.8 to 5.0
■ 4.5 to 4.7
■ <4.5

© Cengage Learning®.

High concentrations of acid deposition can damage plants and water resources. Freshwater ecosystems seem to be particularly sensitive to changes in acidity. Concern has centered mainly on areas where interactions with alkaline soil are unable to neutralize the acidic inputs. Studies indicate that thousands of lakes in the United States and Canada became so acidified in the late twentieth century that entire fish populations may have been adversely affected. Many trees in Germany showed signs

of a blight that was due, in part, to acid deposition. Apparently, acidic particles raining down on the forest floor for decades have caused a chemical imbalance in the soil that, in turn, causes serious deficiencies in certain elements necessary for tree growth. The trees were thus weakened and become susceptible to insects and drought. The impact on large forests was called Waldsterben, or "forest death." The same processes may have affected North American forests, but at a much slower pace, with symptoms apparent in many forests at higher elevations from southeastern Canada to South Carolina (see ● Fig. 14.25). Moreover, acid precipitation has also been a problem in the mountainous West where high mountain lakes and forests seem to have been most affected.

The effects of acid rain are not only felt in rural areas. Acid deposition has eroded the foundations of structures in many cities throughout the world, as well as damaging priceless outdoor statues, sculptures, and fountains. The cost of this damage to building surfaces, monuments, and other structures has been estimated to run into the billions of dollars each year.

Observations and computer models in the 1980s greatly clarified the threat posed by acid precipitation and the processes behind it. As a result, the Clean Air Act of 1990 imposed a reduction in the United States' emissions of sulfur dioxide and nitrogen dioxide. Many other countries implemented similar measures. The United States also implemented a "cap and trade" system by which companies that reduced their emissions had the right to sell allowances to other companies that had yet to do so. Such measures have helped produce large decreases in emissions (in many cases, more than 50 percent), and the threat of acid precipitation has lessened somewhat, although acidification is still a problem for many streams and lakes.

© Frederic Georgia/Science Source

● FIGURE 14.25 The effects of acid fog in the Great Smoky Mountains of Tennessee.

SUMMARY

In this chapter, we found that air pollution has plagued humanity for centuries. Air pollution problems began when people tried to keep warm by burning wood and coal. These problems worsened during the industrial revolution as coal became the primary fuel for both homes and industry. Even though many American cities do not meet all of the air quality standards set by the federal Clean Air Act of 1990, the air over our large cities is cleaner today than it was 50 years ago owing to stricter emission standards and cleaner fuels.

We examined the types and sources of air pollution and found that primary air pollutants enter the atmosphere directly, whereas secondary pollutants form by chemical reactions that involve other pollutants. The secondary pollutant ozone is the main ingredient of photochemical smog—a smog that irritates the eyes and forms in the presence of sunlight. In polluted air near the surface, ozone forms during a series of chemical reactions involving nitrogen oxides and hydrocarbons (VOCs). In the stratosphere, ozone is a naturally occurring gas that protects us from the sun's harmful ultraviolet rays. We learned that human-induced gases, such as chlorofluorocarbons, work their way into the stratosphere where they release atomic chlorine that rapidly destroys ozone, especially in polar regions.

We looked at the air quality index and found that a number of areas across the United States still have days considered unhealthy according to the standards set by the United States Environmental Protection Agency. We also looked at the main factors affecting air pollution and found that most air pollution episodes occur when winds are light, skies are clear, the mixing layer is shallow, the atmosphere is stable, and a strong inversion exists. These conditions usually prevail when a high-pressure area stalls over a region.

We observed that, on the average, urban environments tend to be warmer and more polluted than the rural areas that surround them. We saw that pollution from industrial areas can modify environments downwind from them, as oxides of sulfur and nitrogen are swept into the air, where they may transform into acids that fall to the surface. Although acid deposition remains a serious problem in some parts of the world, other areas have seen significant improvement by reducing the pollutants that cause acid precipitation.

KEY TERMS

The following terms are listed (with corresponding page numbers) in the order they appear in the text. Define each. Doing so will aid you in reviewing the material covered in this chapter.

air pollutants, 418
primary air pollutants, 419
secondary air pollutants, 419
particulate matter, 419
carbon monoxide (CO), 421
sulfur dioxide (SO_2), 421
volatile organic compounds (VOCs), 421
hydrocarbons, 421
nitrogen dioxide (NO_2), 422
nitric oxide (NO), 422
smog, 422
photochemical smog, 422
ozone (O_3), 422
ozone hole, 424
air quality index (AQI), 427
dispersion (of pollutants), 429
radiation (surface) inversion, 429
subsidence inversion, 430
mixing layer, 430
mixing depth, 430
atmospheric stagnation, 433
urban heat island, 433
country breeze, 435
acid rain, 436
acid deposition, 436
acid fog, 436

QUESTIONS FOR REVIEW

1. What are some of the main sources of air pollution?
2. How do primary air pollutants differ from secondary air pollutants?
3. List a few of the substances that fall into the category of particulate matter.
4. How does PM_{10} particulate matter differ from that called $PM_{.5}$? Which poses the greater risk to human health?
5. How is particulate matter removed from the atmosphere?

6. Describe the primary sources and some of the health problems associated with each of the following pollutants:
   (a) carbon monoxide (CO)
   (b) sulfur dioxide (SO_2)
   (c) volatile organic compounds (VOCs)
   (d) nitrogen oxides

7. How does London-type smog differ from Los Angeles–type smog?

8. What is photochemical smog? How does it form? What is the main component of photochemical smog?

9. Why is photochemical smog more prevalent during the summer and early fall than during the middle of winter?

10. Why is stratospheric ozone beneficial to life on Earth, whereas tropospheric (ground-level) ozone is not?

11. If most of the ozone in the stratosphere were destroyed, what possible effects might this have on Earth's inhabitants?

12. As shown in Fig. 14.11 (p. 426), there is a steady drop in the concentration of several pollutants after 1970. What is the reason for this decrease?

13. (a) On the AQI scale, when is a pollutant considered unhealthful?
    (b) On the AQI scale, how would air be described if it had an AQI value of 250 for ozone?
    (c) What would be the general health effects with an AQI value of 250 for ozone? What precautions should a person take given this value?

14. Why is a light wind, rather than a strong wind, more conducive to high concentrations of air pollution?

15. How does atmospheric stability influence the accumulation of air pollutants near the surface?

16. Why is it that polluted air and inversions seem to go hand in hand?

17. Major air pollution episodes are mainly associated with radiation inversions or subsidence inversions. Explain why.

18. Give several reasons why taller smokestacks are better than shorter ones at improving the air quality in their immediate area.

19. How does the mixing depth normally change during the course of a day? As the mixing depth changes, how does it affect the concentration of pollution near the surface?

20. For least-polluting conditions, what would be the best time of day for a farmer to burn agricultural debris? Explain your reasoning.

21. Explain why most severe episodes of air pollution are associated with slow moving or stagnant high-pressure areas.

22. How does topography influence the concentration of pollutants in cities such as Los Angeles and Denver? In mountainous terrain?

23. List the factors that can lead to a major buildup of atmospheric pollution.

24. What type of weather pattern is best for dispersion of pollutants?

25. What is an urban heat island? Is it more strongly developed at night or during the day? Explain.

26. What causes the "country breeze"? Why is it usually more developed at night than during the day? Would it be more easily developed in summer or winter? Explain.

27. How can pollution play a role in influencing the precipitation downwind of certain large industrial complexes?

28. What is acid deposition? Why is acid deposition considered a serious problem in many regions of the world? How does precipitation become acidic?

QUESTIONS FOR THOUGHT AND EXPLORATION

1. Would you expect a fumigation-type smoke plume on a warm, sunny afternoon? Explain.

2. Give a few reasons why, in industrial areas, nighttime pollution levels might be higher than daytime levels.

3. Explain this apparent paradox: High levels of tropospheric (ground-level) ozone are "bad" and we try to reduce them, whereas high levels of stratospheric ozone are "good" and we try to maintain them.

4. A large industrial smokestack located within an urban area emits vast quantities of sulfur dioxide and nitrogen dioxide. Following criticism from local residents that emissions from the stack are contributing to poor air quality in the area, the management raises the height of the stack from 10 m (33 ft) to 100 m (330 ft). Will this increase in stack height change any of the existing air quality

problems? Will it create any new problems? Explain.

5. If the sulfuric acid and nitric acid in rainwater can adversely affect soil, trees, and fish, why doesn't this same acid adversely affect people when they walk in the rain?

6. Which do you feel is likely to be more acidic: acid rain or acid fog? Explain your reasoning.

7. Keep a log of the daily AQI readings in your area and note the pollutants listed in the index. Also, keep a record of the daily weather conditions, such as cloud cover, high temperature for the day, and average wind direction and speed. See if there is any relationship between these weather conditions and high AQI readings for certain pollutants.

GLOBAL **GEOSCIENCE** WATCH   Go to the Global Environment Watch: Air Pollution portal. Within the Document section, open the document "Emissions Trading," from *Climate Change: In Context*. When was the first U.S. cap-and-trade system created by legislation? What pollutant was it designed to control? What could explain why it has proven more difficult to institute a cap-and-trade system for carbon dioxide?

ONLINE RESOURCES

 Visit www.cengagebrain.com to view additional resources, including video exercises, practice quizzes, an interactive eBook, and more.

CHAPTER 15

Light, Color, and Atmospheric Optics

Contents

The sky is clear, the weather cold, and the year, 1818. Near Baffin Island in Canada, a ship with full sails enters unknown waters. On board are the English brothers James and John Ross, who are hoping to find the elusive "Northwest Passage," the waterway linking the Atlantic and Pacific oceans. On this morning, however, their hopes would be dashed, for directly in front of the vessel, blocking their path, is a huge towering mountain range. Disappointed, they turn back and report that the Northwest Passage does not exist.

Almost ninety years later, Rear Admiral Robert Peary of the United States Navy met the same barrier and called it "Crocker Land." What type of treasures did this mountain conceal—gold, silver, precious gems? The curiosity of explorers from all over the world had been aroused. Speculation was the rule, until, in 1913, the American Museum of Natural History commissioned Donald MacMillan to lead an expedition to solve the mystery of Crocker Land. At first, the journey was disappointing. Where Peary had seen mountains, MacMillan saw only vast stretches of open water. Finally, ahead of his ship was Crocker Land, but it was more than two hundred miles west from where Peary had encountered it. MacMillan sailed on as far as possible. Then he dropped anchor and set out on foot with a small crew of men.

As the team moved toward the mountains, the mountains seemed to move away from them. If they stood still, the mountains stood still; if they started walking, the mountains receded again. Puzzled, they trekked onward over the glittering snowfields until huge mountains surrounded them on three sides. But in the next instant the sun disappeared below the horizon and, as if by magic, the mountains dissolved into the cold arctic twilight. Dumbfounded, the men looked around only to see ice in all directions—not a mountain was in sight. There they were, the victims of one of nature's greatest practical jokes, for Crocker Land was a mirage.

The sky is full of visual events. Optical illusions (mirages) can appear as towering mountains or wet roadways. In clear weather, the sky can appear blue, while the horizon appears milky white. Sunrises and sunsets can fill the sky with brilliant shades of pink, red, orange, and purple. At night, away from city lights, the sky is black, except for the light from the stars, planets, and the moon. The moon's size and color seem to vary during the night, and the stars twinkle. To understand what we see in the sky, we will take a closer look at sunlight, examining how it interacts with the atmosphere to produce an array of atmospheric visuals.

White and Colors

We know from Chapter 2 that nearly half of the solar radiation that reaches the atmosphere is in the form of visible light. As sunlight enters the atmosphere, it is either absorbed, reflected, scattered, or transmitted on through. How objects at the surface respond to this energy depends on their general nature (color, density, composition) and the wavelength of light that strikes them. How do we see? Why do we see various colors? What kind of visual effects do we observe because of the interaction between light and matter? In particular, what can we *see* when light interacts with our atmosphere?

We perceive light because radiant energy from the sun travels outward in the form of electromagnetic waves. When these waves reach the human eye, they stimulate antenna-like nerve endings in the retina. These antennae are of two types—*rods* and *cones*. The rods respond to all wavelengths of visible light and give us the ability to distinguish light from dark. If people possessed rod-type receptors only, then only black and white vision would be possible. The cones respond to specific wavelengths of visible light. The cones fire an impulse through the nervous system to the brain, and we perceive this impulse as the sensation of color. (Color blindness is caused by missing or malfunctioning cones.) Wavelengths of radiation shorter than those of visible light do not stimulate color vision in humans.

White light is perceived when all visible wavelengths strike the cones of the eye with nearly equal intensity.* Because the sun radiates almost half of its energy as visible light, all visible wavelengths from the midday sun reach the cones, and the sun usually appears nearly white. A star that is cooler than our sun radiates most of its energy at slightly longer wavelengths; therefore, it appears redder. On the other hand, a star much hotter than our sun radiates more energy at shorter wavelengths and thus appears

bluer. A star whose temperature is about the same as the sun's appears white.

Objects that are not hot enough to produce radiation at visible wavelengths can still have color. Everyday objects we see as red are those that absorb all visible radiation except red. The red light is reflected from the object to our eyes. Blue objects have blue light returning from them, since they absorb all visible wavelengths except blue. Some surfaces absorb all visible wavelengths and reflect no light at all. Since no radiation strikes the rods or cones, these surfaces appear black. Therefore, when we see colors, we know that light must be reaching our eyes.

Clouds and Scattered Light

It is interesting to watch the underside of a puffy, growing cumulus cloud change color from white to dark gray or black. When we see this change happen, our first thought is usually, "It's going to rain." Why is the cloud initially white? Why does it change color? To answer these questions, we need to investigate in more detail the concept of *scattering*.

When sunlight bounces off a surface at the same angle at which it strikes the surface, we say that the light is **reflected,** and call this phenomenon *reflection*. There are various constituents of the atmosphere, however, that tend to deflect sunlight from its path and send it out in all directions. We know from Chapter 2 that radiation reflected in this way is said to be **scattered.** (Scattered light is also called *diffuse light*.) During the scattering process, no energy is gained or lost and, therefore, no temperature changes occur. In the atmosphere, scattering is usually caused by very small objects, such as air molecules, fine particles of dust, water molecules, and some pollutants. Just as the ball in a pinball machine bounces off the pins in many directions, so solar radiation is knocked about by small particles in the atmosphere.

Typical cloud droplets are large enough to effectively scatter all wavelengths of visible radiation more or less equally (see ● Fig. 15.1). Clouds, even small ones, are optically thick, meaning that very little unscattered light gets through them. These same clouds are poor absorbers of sunlight. Hence, when we look at a cloud, it appears white because countless cloud droplets scatter all wavelengths of visible sunlight in all directions (see ● Fig. 15.2).

As a cloud grows larger and taller, more sunlight is reflected from it and less light can penetrate all the way through it (see ● Fig. 15.3). In fact, relatively little light penetrates a cloud whose thickness is 1000 m (3300 ft). Since little sunlight reaches the underside of the cloud, little light is scattered, and the cloud base appears dark. At the same time, if droplets near the cloud base grow larger, they become less effective scatterers and better absorbers. As a

*Recall from Chapter 2 that visible white light is a combination of waves with different wavelengths. The wavelengths of visible light in decreasing order are: red (longest), orange, yellow, green, blue, and violet (shortest).

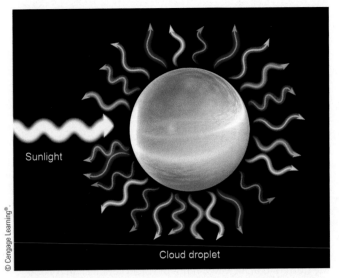

● **FIGURE 15.1** Cloud droplets scatter all wavelengths of visible white light about equally. The different colors represent different wavelengths of visible light.

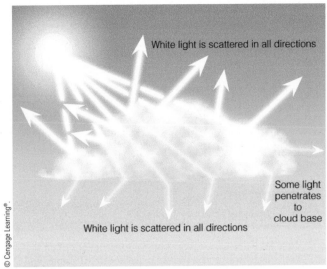

● **FIGURE 15.2** Since tiny cloud droplets scatter visible light in all directions, light from many billions of droplets turns a cloud white.

result, the meager amount of visible light that does reach this part of the cloud is absorbed rather than scattered, which makes the cloud appear even darker. These same cloud droplets may even grow large and heavy enough to fall to the earth as rain. From a casual observation of clouds, we know that dark, threatening ones frequently produce rain. Now we know why they appear so dark.

BLUE SKIES AND HAZY DAYS The sky appears blue because light that stimulates the sensation of blue color is reaching the retina of the eye. How does this happen?

Individual air molecules are much smaller than cloud droplets—their diameters are small even when compared with the wavelength of visible light. Each air molecule of

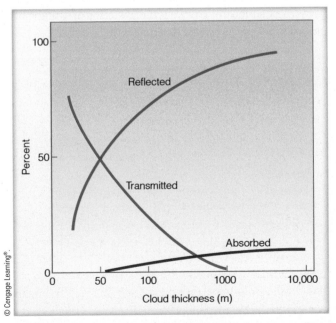

● **FIGURE 15.3** Average percent of radiation reflected, absorbed, and transmitted by clouds of various thickness.

oxygen and nitrogen is a *selective scatterer* in that each scatters shorter waves of visible light much more effectively than longer waves. This selective scattering is also known as *Rayleigh scattering.*

As sunlight enters the atmosphere, the shorter visible wavelengths of violet, blue, and green are scattered more by atmospheric gases than are the longer wavelengths of yellow, orange, and especially red. (Violet light is scattered about 16 times more than red light.) As we view the sky, the scattered waves of violet, blue, and green strike the eye from all directions. Because our eyes are more sensitive to blue light, these waves, viewed together, produce the sensation of blue coming from all around us (see ● Fig. 15.4). Therefore, when we look at the sky it appears blue (see ● Fig. 15.5). (Earth, by the way, is not the only planet with a colorful sky. On Mars, dust in the air turns the sky red at midday and purple at sunset.)

The selective scattering of blue light by air molecules and very small particles can make distant mountains appear blue, such as happens with the Blue Ridge Mountains of Virginia (see ● Fig. 15.6) and the Blue Mountains of Australia. In some places, a *blue haze* may cover the landscape, even in areas far removed from human contamination. Although its cause is still controversial, the blue haze

DID YOU KNOW?

Ever see a green thunderstorm? Severe thunderstorms that form over the Great Plains often appear to have green areas. The green color may be due to reddish sunlight (especially at sunset) penetrating the storm, then being scattered by cloud particles composed of water and ice. With much of the red light removed, the scattered light casts the underside of the cloud as a faint greenish hue. This phenomenon appears to occur most often when thunderstorms have very large amounts of moisture, sometimes including hail.

● **FIGURE 15.4** The sky appears blue because billions of air molecules selectively scatter the shorter wavelengths of visible light more effectively than the longer ones. This causes us to see blue light coming from all directions.

● **FIGURE 15.5** Blue skies and white clouds. The selective scattering of blue light by air molecules produces the blue sky, while the scattering of all wavelengths of visible light by liquid cloud droplets produces the white clouds.

appears to be the result of a particular process. Extremely tiny particles (hydrocarbons called *terpenes*) are released by vegetation to combine chemically with small amounts of ozone, a reaction that produces the extremely tiny particles that selectively scatter blue light.

When small particles, such as fine dust and salt, become suspended in the atmosphere, the color of the sky begins to change from blue to milky white. Although these particles are small, they are large enough to scatter all wavelengths of visible light fairly evenly in all directions. (This type of scattering is called *Mie scattering.*) When our eyes are bombarded by all wavelengths of visible light, the sky appears milky white, the visibility lowers, and we call the day "hazy." If the humidity is high enough, soluble particles (nuclei) will "pick up" water vapor and grow into haze particles. Thus, the color of the sky gives us a hint about how much material is suspended in the air: the more particles, the more scattering, and the

whiter the sky becomes. On top of a high mountain, when we are above many of these haze particles, the sky usually appears a deep blue.

Haze can scatter light from a rising or setting sun in a way that produces bright light beams, or **crepuscular rays,** radiating across the sky (see ● Fig. 15.7). When the bright light beams appear to converge toward the part of the horizon opposite from the sun, the beams of light are called **anticrepuscular rays** (see ● Fig. 15.8). A similar effect occurs when the sun shines through a layer of clouds (see ● Fig. 15.9). Dust, tiny water droplets, or haze in the air beneath the clouds scatter sunlight, making that region of the sky appear bright with rays. Because these rays seem to reach downward from clouds, some people will remark that the "sun is drawing up water." In England, this same phenomenon is referred to as "Jacob's ladder." No matter what name these sunbeams go by, it is the scattering of sunlight by particles in the atmosphere that makes them visible.

● **FIGURE 15.6** The Blue Ridge Mountains in Virginia. The blue haze is caused by the scattering of blue light by extremely small particles—smaller than the wavelengths of visible light. Notice that the scattered blue light causes the most distant mountains to become almost indistinguishable from the sky.

● **FIGURE 15.7** The bright light beams radiating across the sky from behind the clouds are crepuscular rays.

● **FIGURE 15.8** Anticrepuscular rays photographed near Boulder, Colorado. The rays appear to converge toward the horizon opposite from a setting sun, which is to the back of the photographer.

● **FIGURE 15.9** The scattering of sunlight by dust and haze produces these white bands of crepuscular rays that appear to radiate downward toward the surface.

Red Suns and Blue Moons

At midday, the sun seems a brilliant white, while at sunset it usually appears to be yellow, orange, or red. At noon, when the sun is high in the sky, light from the sun is most intense—all wavelengths of visible light are able to reach the eye with about equal intensity, and the sun appears white. (Looking directly at the sun, especially during this time of day, can cause irreparable damage to the eye. Normally, we get only glimpses or impressions of the sun out of the corner of our eye.)

Near sunrise or sunset, however, the rays coming directly from the sun strike the atmosphere at a low angle. These rays pass through much more atmosphere than at any other time during the day. (When the sun is 4° above the horizon, sunlight must pass through an atmosphere

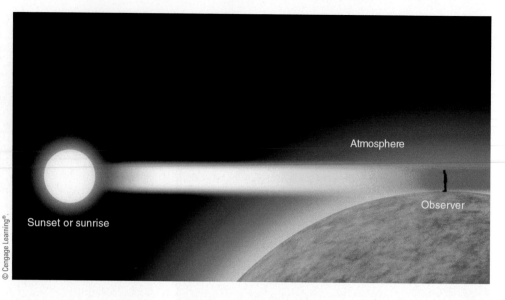

© Cengage Learning®.

FIGURE 15.10 Because of the selective scattering of radiant energy by a thick section of atmosphere, the sun at sunrise and sunset appears either yellow, orange, or red. The more particles in the atmosphere, the more scattering of sunlight, and the redder the sun appears.

Atmosphere

Sunset or sunrise

Observer

more than 12 times thicker than when the sun is directly overhead.) By the time sunlight has penetrated this large amount of air, most of the shorter waves of visible light have been scattered away by the air molecules. Just about the only waves from a setting sun that make it through the atmosphere on a fairly direct path are the yellow, orange, and red. Upon reaching the eye, these waves produce a sunset that is bright yellow-orange.

Such sunsets occur only when the atmosphere is fairly clean, such as after a recent rain. If the atmosphere contains many fine particles with diameters a little larger than those of air molecules, slightly longer (yellow) waves also will be scattered away. Only orange and red waves would

penetrate through to the eye, and the sun would appear red-orange (see Fig. 15.10). When the atmosphere becomes loaded with particles, only the longest red wavelengths are able to penetrate the atmosphere, and we see a red sun.

Natural events may produce red sunrises and sunsets over the oceans. For example, the scattering characteristics of small suspended salt particles and water molecules are responsible for the brilliant red suns that can be observed from a beach (see Fig. 15.11). Volcanic eruptions rich in sulfur can produce red sunsets, too. Such red sunsets are actually produced by a highly reflective cloud of sulfuric acid droplets, formed from sulfur dioxide gas injected into the stratosphere during powerful eruptions. Two examples of these eruptions are the Mexican volcano El Chichón in 1982 and the Philippine volcano Mt. Pinatubo in 1991. Fine particles from them, moved by the winds aloft, circled the globe, producing beautiful sunrises and sunsets for months and even years after the eruptions. These same volcanic particles in the stratosphere can turn the sky red after sunset, as some of the red light from the setting sun bounces off the bottom of the particles back to Earth's surface. Generally, these volcanic red sunsets occur about an hour after the actual sunset (see Fig. 15.12).

Occasionally, the atmosphere becomes so laden with dust, smoke, and pollutants that even red waves are unable to pierce the filthy air. An eerie effect then occurs. Because no visible waves enter the eye, the sun literally disappears before it reaches the horizon.

The scattering of light by large quantities of atmospheric particles can cause some rather unusual sights. If the volcanic ash, dust, smoke particles, or pollutants are roughly uniform in size, they can selectively scatter the sun's rays. Even at noon, various colored suns have appeared—orange suns, green suns, and even blue suns—although green and blue suns are extremely unusual. For blue suns to appear, the size of the suspended particles

© C. Donald Ahrens

FIGURE 15.11 Red sunset near the coast of Iceland. The reflection of sunlight off the slightly rough water is producing a glitter path.

● **FIGURE 15.12** Bright red sky over California produced by the sulfur-rich particles from the volcano Mt. Pinatubo during September 1992. The photo was taken about an hour after sunset.

must be similar to the primary wavelengths of visible light. When these particles are present they tend to scatter red light more than blue, which causes a bluing of the sun and a reddening of the sky. Although rare, the same phenomenon can happen to moonlight, making the moon appear blue; thus, the expression "once in a blue moon." (The term "blue moon" also refers to the second full moon in a calendar month that has two full moons).

In summary, the scattering of light by small particles in the atmosphere causes many familiar effects: white clouds, blue skies, hazy skies, crepuscular rays, and colorful sunsets. In the absence of any scattering, we would simply see a white sun against a black sky—not an attractive alternative.

Twinkling, Twilight, and the Green Flash

Light that passes through a substance is said to be *transmitted*. Upon entering a denser substance, transmitted light slows in speed. If it enters the substance at an angle, the light's path also bends. This bending is called **refraction.** The amount of refraction depends primarily on two factors: the density of the material and the angle at which the light enters the material.

Refraction can be demonstrated in a darkened room by shining a flashlight into a beaker of water (see ● Fig. 15.13). If the light is held directly above the water so

that the beam strikes the surface of the water straight on, no bending occurs. But if the light enters the water at some angle, it bends toward the *normal,* which is the dashed line in Fig. 15.13 running perpendicular to the air-water boundary. (The normal is simply a line that intersects any surface at a right angle. We use it as a reference to see how much bending occurs as light enters and leaves various substances.) A small mirror on the bottom of the beaker reflects the light upward. This reflected light bends away from the normal as it re-enters the air. We can summarize these observations as follows: *Light that travels from a less-dense to a more-dense medium loses speed and bends toward the normal, whereas light that enters a less-dense*

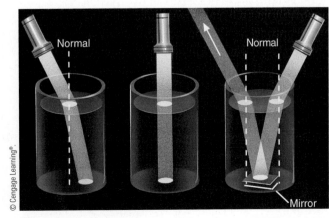

● **FIGURE 15.13** The behavior of light as it enters and leaves a more-dense substance, such as water.

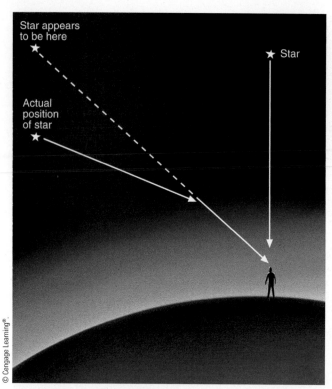

● **FIGURE 15.14** Due to the bending of starlight by the atmosphere, stars not directly overhead appear to be higher than they really are. (The angle between positions is highly exaggerated.)

medium increases in speed and bends away from the normal.

The refraction of light within the atmosphere causes a variety of visual effects. At night, for example, the light from the stars that we see directly above us is not bent, but starlight that enters Earth's atmosphere at an angle is bent. In fact, a star whose light enters the atmosphere just above the horizon has more atmosphere to penetrate and is thus refracted the most. As we can see in ● Fig. 15.14 the bending is toward the normal as the light enters the more-dense atmosphere. By the time this "bent" starlight reaches our eyes, the star appears to be higher than it actually is because our eyes cannot detect that the light path is bent. We see light coming from a

particular direction and interpret the star to be in that direction. So, the next time you take a midnight stroll, point to any star near the horizon and remember: This point is only where the star appears to be. To point to the star's true position, you would have to lower your arm just a tiny bit.

As starlight enters the atmosphere, it often passes through regions of differing air density. Each of these regions deflects and bends the tiny beam of starlight, constantly changing the apparent position of the star. This causes the star to appear to *twinkle* or flicker, a condition known as **scintillation.** Planets, being much closer to us, appear larger, and usually do not twinkle because their size is greater than the angle at which their light deviates as it penetrates the atmosphere. Planets sometimes twinkle, however, when they are near the horizon, where the bending of their light is greatest.

The refraction of light by the atmosphere has some other interesting consequences. For example, the atmosphere gradually bends the rays from a rising or setting sun or moon. Because light rays from the lower part of the sun (or moon) are bent more than those from the upper part, the sun appears to flatten out on the horizon, taking on an elliptical shape. (The sun in Fig. 15.16 shows this effect.) Also, since light is bent most on the horizon, the sun and moon both appear to be higher than they really are, so they both rise about two minutes earlier and set about two minutes later than they would if there were no atmosphere (see ● Fig. 15.15).

You may have noticed that on clear days the sky is often bright for some time after the sun sets. The atmosphere refracts and scatters sunlight to our eyes, even though the sun itself has disappeared from our view (look back at Fig. 15.12). **Twilight** is the name given to the time after sunset (and immediately before sunrise) when the sky remains illuminated, allowing outdoor activities to continue without artificial lighting.

The length of twilight depends on season and latitude. During the summer in middle latitudes, twilight adds about 30 minutes of light to each morning and evening.

● **FIGURE 15.15** The bending of sunlight by the atmosphere causes the sun to rise about two minutes earlier, and set about two minutes later, than it would otherwise.

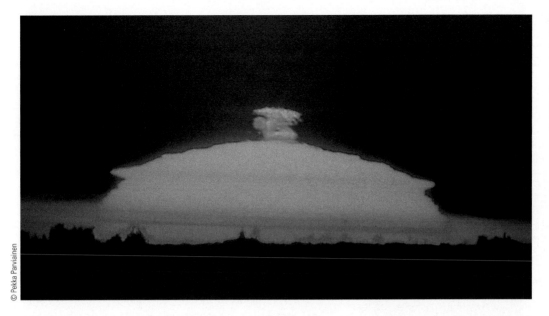

● FIGURE 15.16 The very light green on the upper rim of the sun is the green flash. Also, observe how the atmosphere makes the sun appear to flatten on the horizon into an elliptical shape.

The duration of twilight increases with increasing latitude, especially in summer. At high latitudes during the summer, morning and evening twilight may converge, producing a *white night*—a night-long twilight. At low latitudes, twilight tends to be shorter, with more abrupt transitions between daylight and darkness.

In general, without the atmosphere, there would be no refraction or scattering, and the sun would rise later and set earlier than it now does. Instead of twilight, darkness would arrive immediately at the sun's disappearance below the horizon. Imagine the number of evening softball games that would be called because of instant darkness.

Occasionally, a flash of green light—called the **green flash**—can be seen near the upper rim of a rising or setting sun (see ● Fig. 15.16). Remember from our earlier discussion that when the sun is near the horizon, its light must penetrate a thick section of atmosphere. This thick atmosphere refracts sunlight, with purple and blue light bending the most, and red light the least. Because of this bending, more blue light should appear along the top of the sun. But because the atmosphere selectively scatters blue light, very little reaches us, and we see green light instead.

Usually, the green light is too faint for the human eye to see. However, under certain atmospheric conditions, such as when the surface air is very hot or when an upper-level inversion exists, the atmosphere magnifies the green light. When this happens, a momentary flash of green light appears, often just before the sun disappears from view. The flash usually lasts about a second, although in polar regions it can last longer. Here, the sun slowly changes in elevation and the flash may exist for many minutes. Members of Admiral Byrd's expedition in the south polar region reported seeing the green flash for 35 minutes in September as the sun slowly rose above the horizon, marking the end of the long winter.

BRIEF REVIEW

Up to this point, we have examined how light can interact with our atmosphere. Before going on, here is a review of some of the important concepts and facts we have covered:

- When light is scattered, it is sent in all directions—forward, sideways, and backward.
- White clouds, blue skies, hazy skies, crepuscular rays, anticrepuscular rays, and colorful sunsets are all the result of sunlight being scattered.
- The bending of light as it travels through regions of differing density is called *refraction*.
- As light travels from a less-dense substance (such as outer space) and enters a more-dense substance at an angle (such as our atmosphere), the light bends downward, toward the normal. This effect causes stars, the moon, and the sun to appear just a tiny bit higher than they actually are.

The Mirage: Seeing Is Not Believing

In the atmosphere, when an object appears to be displaced from its true position, we call this phenomenon a **mirage.** A mirage is not a figment of the imagination—our minds are not playing tricks on us, but the atmosphere is.

Atmospheric mirages are created by light passing through and being bent (refracted) by air layers of different densities. Such changes in air density are usually caused by sharp changes in air temperature. The greater

● **FIGURE 15.17** The road in the photo appears wet because blue skylight is bending up into the camera as the light passes through air of different densities.

the rate of temperature change, the more the light rays are bent. For example, on a warm, sunny day, black road surfaces absorb a great deal of solar energy and become very hot. Air in contact with these hot surfaces warms by conduction and, because air is a poor thermal conductor, we find much cooler air only a few meters higher. On hot days, these road surfaces often appear wet (see ● Fig. 15.17). Such "puddles" disappear as we approach them, and advancing cars seem to swim in them. Yet we know the road is dry. The apparent wet pavement above a road is the result of blue skylight refracting up into our eyes as it travels through air of different densities. A similar type of mirage occurs in deserts during the hot summer. Many thirsty travelers have been disappointed to find that what appeared to be a water hole was in actuality hot desert sand.

Sometimes, these "watery" surfaces appear to *shimmer*. The shimmering results as rising and sinking air near the ground constantly change the air density. As light moves through these regions, its path also changes, causing the shimmering effect.

When the air near the ground is much warmer than the air above, objects may not only appear to be lower than they really are, but also they often appear to be inverted. These mirages are called **inferior** (lower) **mirages.** The

tree in ● Fig. 15.18 certainly doesn't grow upside down. So why does it look that way? It appears to be inverted because light reflected from the top of the tree moves outward in all directions. Rays that enter the hot, less-dense air above the sand are refracted upward, entering the eye from below. The brain is fooled into thinking that these rays come from below the ground, which makes the tree appear upside down. However, some light from the top of the tree travels directly toward the eye through air of nearly constant density and, therefore, bends very little. These rays reach the eye "straight on," and the tree appears upright. Off in the distance, then, we see a tree and its upside-down image beneath it. (Some of the trees in Fig. 15.17 show this effect.)

The atmosphere can play optical jokes on us in extremely cold areas, too. In polar regions, air next to a snow surface can be much colder than the air many meters above. Because the air in this cold layer is very dense, light from distant objects entering it bends toward the normal in such a way that the objects can appear to be shifted upward. This phenomenon is called a **superior** (upward) **mirage.** ● Figure 15.19 shows the conditions favorable for a superior mirage. (A special type of mirage, the *Fata Morgana,* is described in Focus section 15.1.)

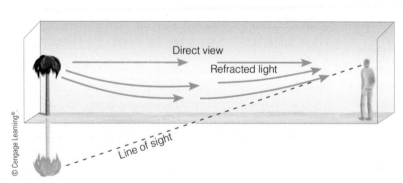

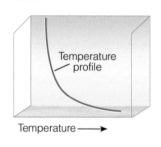

● **FIGURE 15.18** Inferior mirage over hot desert sand.

452 CHAPTER 15

The *Fata Morgana*

A special type of superior mirage is the *Fata Morgana*, a mirage that transforms a fairly uniform horizon into one of vertical walls and columns with spires (see ● Fig. 1). According to legend, *Fata Morgana* (Italian for "fairy Morgan") was the half-sister of King Arthur. Morgan, who was said to live in a crystal palace beneath the water, had magical powers that could build fantastic castles out of thin air. Looking across the Straits of Messina (between Italy and Sicily), residents of Reggio, Italy, on occasion would see buildings, castles, and sometimes whole cities appear, only to vanish again in minutes. The *Fata Morgana* is observed where the air temperature increases with height above the surface, slowly at first, then more rapidly, then slowly again. Consequently, mirages like the *Fata Morgana* are frequently seen where warm air rests above a cold surface, such as above large bodies of water and in polar regions.

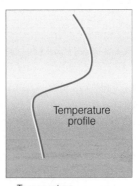
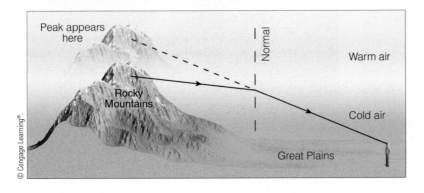

● **FIGURE 1** The *Fata Morgana* mirage over water. The mirage is the result of refraction—light from small islands and ships is bent in such a way as to make them appear to rise vertically above the water.

● **FIGURE 15.19** The formation of a superior mirage. When cold air lies close to the surface with warm air aloft, light from distant mountains is refracted toward the normal as it enters the cold air. This causes an observer on the ground to see mountains higher and closer than they really are.

Halos, Sundogs, and Sun Pillars

A ring of light encircling and extending outward from the sun or moon is called a **halo.** Such a display is produced when sunlight or moonlight is refracted as it passes through ice crystals. Hence, the presence of a halo indicates that *cirriform clouds* are present. (Refer to Chapter 6 for a more complete description of cirriform clouds.)

The most common type of halo is the 22° halo—a ring of light 22° from the sun or moon.* Such a halo forms when tiny suspended column-type ice crystals (with diameters less than 20 micrometers) become randomly oriented as air molecules constantly bump against them. The refraction of light rays through these crystals forms

---

*Extend your arm and spread your fingers apart. An angle of 22° is about the distance from the tip of the thumb to the tip of the little finger.

● **FIGURE 15.20** A 22° halo around the sun, produced by the refraction of sunlight through ice crystals.

● **FIGURE 15.22** Halo with an upper tangent arc.

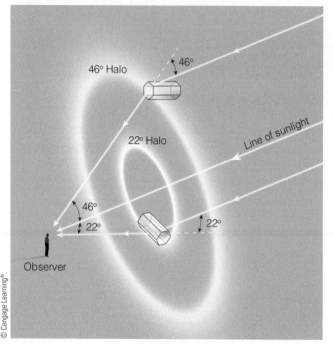

● **FIGURE 15.21** The formation of a 22° and a 46° halo with column-type ice crystals.

● **FIGURE 15.23** Refraction and dispersion of light through a glass prism.

a halo like the one shown in ● Fig. 15.20. Less common is the 46° halo, which forms in a similar fashion to the 22° halo (see ● Fig. 15.21). With the 46° halo, however, the light is refracted through column-type ice crystals that have diameters in a narrow range between about 15 and 25 micrometers.

Occasionally, a bright arc of light can be seen at the top of a 22° halo (see ● Fig. 15.22). Since the arc is tangent to the halo, it is called a **tangent arc.** Apparently, the arc forms as large, six-sided (hexagonal), pencil-shaped ice crystals fall with their long axes horizontal to the ground. Refraction of sunlight through the ice crystals produces the bright arc of light.

A halo is usually seen as a bright, white ring, but some refraction effects can cause it to have color. To understand this, we must first examine refraction more closely.

When white light passes through a glass prism, it is refracted and split into a spectrum of visible colors (see ● Fig. 15.23). Each wavelength of light is slowed by the glass, but each is slowed a little differently. Because longer wavelengths (red) slow the least and

DID YOU KNOW?

On December 14, 1890, a mirage—lasting several hours—reportedly gave the residents of Saint Vincent, Minnesota, a clear view of cattle nearly 8 miles away.

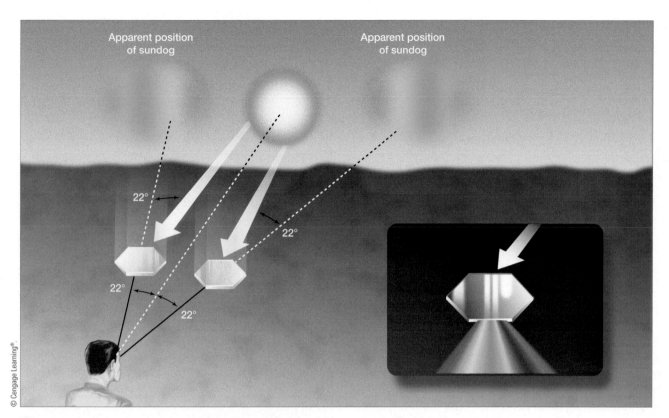

● **FIGURE 15.24** Platelike ice crystals falling with their flat surfaces parallel to Earth produce sundogs.

shorter wavelengths (violet) slow the most, red light bends the least, and violet light bends the most. The breaking up of white light by "selective" refraction is called **dispersion.** As light passes through ice crystals, dispersion causes red light to be on the inside of the halo and blue light on the outside.

When hexagonal, platelike ice crystals are present in the air, they tend to fall slowly and orient themselves horizontally (see ● Fig. 15.24). (The horizontal orientation of these

ice crystals prevents a ring halo.) In this position, the ice crystals act as small prisms, refracting and dispersing sunlight that passes through them. If the sun is near the horizon in such a configuration that it, ice crystals, and observer are all in the same horizontal plane, the observer will see a pair of brightly colored spots, one on either side of the sun. These colored spots are called **sundogs,** *mock suns,* **or parhelia**— meaning "with the sun" (see ● Fig. 15.25). The colors usually grade from red (bent least) on the inside closest to the sun to

● **FIGURE 15.25** The bright areas on each side of the sun are sundogs.

● **FIGURE 15.26** A brilliant red sun pillar extending upward above the sun, produced by the reflection of sunlight off ice crystals.

NASA

blue (bent more) on the outside. The refraction of sunlight through ice crystals can produce a variety of optical visuals, sometimes turning white cirriform clouds into a display of vibrant colors.

Whereas sundogs, tangent arcs, and halos are caused by *refraction* of sunlight *through* ice crystals, **sun pillars** are caused by *reflection* of sunlight *off* ice crystals. Sun pillars appear most often at sunrise or sunset as a vertical shaft of light extending upward or downward from the sun (see ● Fig. 15.26). Pillars may form as hexagonal platelike ice crystals fall with their flat bases oriented horizontally.

As the tiny crystals fall in still air, they tilt from side to side like a falling leaf. This motion allows sunlight to reflect off the tipped surfaces of the crystals, producing a relatively bright area in the sky above or below the sun. Pillars may also form as sunlight reflects off hexagonal, pencil-shaped ice crystals that fall with their long axes oriented horizontally. As these crystals fall, they can rotate about their horizontal axes, producing many orientations that reflect sunlight. Look for sun pillars when the sun is low on the horizon and cirriform (ice crystal) clouds are present. ● Figure 15.27 summarizes some of the optical phenomena that form when cirriform clouds are present.

Rainbows

Now we come to one of the most spectacular light shows observed on Earth—the rainbow. **Rainbows** occur when rain is falling in one part of the sky while the sun is shining in another. (Rainbows also may form from the sprays of waterfalls and water sprinklers.) To see a rainbow, we must face the falling rain with the sun at our backs. Look closely at ● Fig. 15.28 and note that when we see a rainbow in the evening, we are facing east toward the rainshower. Behind us—in the west—it is usually clear, or at least clear enough for sunlight to reach the showers. Because clouds tend to move from west to east in middle latitudes, the clear skies in the west suggest that the showers will give way to clearing. However, when we see a rainbow in the morning, we are facing west, toward the rainshower. It is a good bet that the clouds and showers will move toward us and it will rain soon. These observations explain why the following weather rhyme became popular:

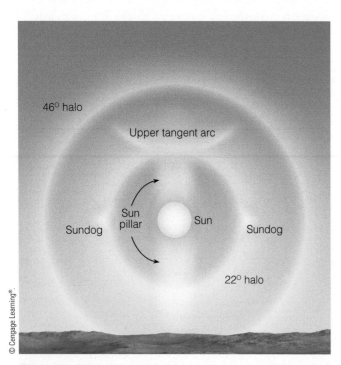

© Cengage Learning®

● **FIGURE 15.27** Optical phenomena that form when cirriform ice crystal clouds are present.

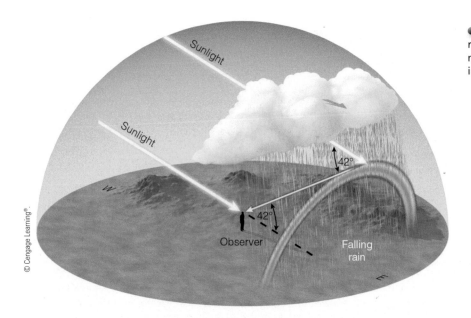

● FIGURE 15.28 When you observe a rainbow, the sun is always to your back. In middle latitudes, a rainbow in the evening indicates that clearing weather is ahead.

Rainbow in the morning, sailors take warning.
Rainbow at night, sailor's delight.*

When we look at a rainbow we are looking at sunlight that has entered the falling drops, and, in effect, has been redirected back toward our eyes. Exactly how this process happens requires some discussion.

As sunlight enters a raindrop, it slows and bends (see ● Fig. 15.29). Although most of this light passes right on through the drop and we do not see it, some of it strikes

---

*This rhyme is often used with the words "red sky" in the place of "rainbow." The red sky makes sense when we consider that it is the result of red light from a rising or setting sun being reflected from the underside of clouds above us. In the morning, a red sky indicates that it is clear to the east and cloudy to the west. A red sky in the evening suggests the opposite.

the backside of the drop at such an angle that it is reflected within the drop. The angle at which this occurs is called the *critical angle,* and for water, this angle is 48°. Notice in Fig 15.29 that light that strikes the back of a raindrop at an angle exceeding the critical angle bounces off the back of the drop and is *internally reflected* toward our eyes. Because each light ray bends differently from the rest, each ray emerges from the drop at a slightly different angle as illusrated in Fig 15.29.

Observe in ● Fig 15.30 that inside a raindrop violet light is refracted the most and red light the least. This

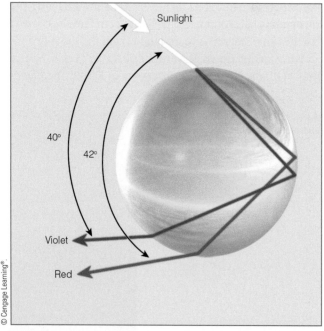

● FIGURE 15.30 Sunlight internally reflected and dispersed by a raindrop. Light rays are internally reflected only when they strike the backside of the drop at an angle greater than the critical angle for water. Refraction of the light as it enters the drop causes the point of reflection (on the back of the drop) to be different for each color. Hence, the colors are separated from each other when the light emerges from the raindrop.

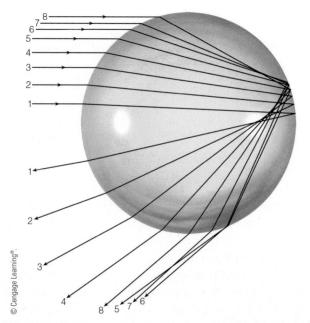

● FIGURE 15.29 Rays of sunlight entering a raindrop bounce off the back of the drop and are redirected toward our eyes.

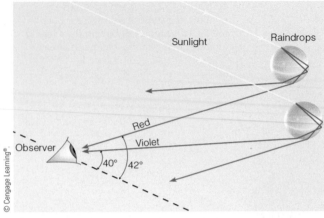

● **FIGURE 15.31** The formation of a primary rainbow. The observer sees red light from the upper drop and violet light from the lower drop.

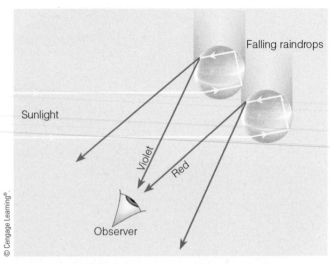

● **FIGURE 15.33** Two internal reflections are responsible for the weaker, secondary rainbow. Notice that the eye sees violet light from the upper drop and red light from the lower drop.

situation causes red light to emerge from the drop at an angle of 42° from the beam of sunlight, whereas violet light emerges at an angle of 40°. The light leaving the drop is, therefore, dispersed into a spectrum of colors from red to violet. Since we see only a single color from each drop, it takes myriads of raindrops (each refracting and reflecting light back to our eyes at slightly different angles) to produce the brilliant colors of a *primary rainbow*. If raindrops are not spread widely enough across the sky, then the rainbow may not appear "complete."

Figure 15.30 might lead us erroneously to believe that red light should be at the bottom of the bow and violet at the top. A more careful observation of the behavior of light leaving two drops (● Fig 15.31) shows us why the reverse is true. When violet light from the *lower drop* reaches an observer's eye, red light from the same drop is arriving farther down, toward the observer's waist. Notice that red light reaches the observer's eye from the *higher*

drop. Because the color red comes from higher drops and the color violet from lower drops, the colors of a primary rainbow change from red on the outside (top) to violet on the inside (bottom).

Sometimes a larger, second (secondary) rainbow with its colors reversed can be seen above the primary bow (see ● Fig. 15.32). Usually this secondary bow is much fainter than the primary one. The *secondary bow* is caused when sunlight enters the raindrops at an angle that allows the light to make two internal reflections in each drop. Each reflection weakens the light intensity and makes the bow dimmer. ● Figure 15.33 shows that the color reversals—with red now at the bottom and violet on top—arise from the way the light emerges from each drop after going through two internal reflections.

● **FIGURE 15.32** A primary and a secondary rainbow.

● **FIGURE 15.34** The corona around the moon results from the diffraction of light by tiny liquid cloud droplets of uniform size.

● **FIGURE 15.35** Corona around the sun, photographed in Colorado. This type of corona, called *Bishop's ring*, is the result of diffraction of sunlight by tiny volcanic particles emitted from the volcano El Chichón in 1982.

As you look at a rainbow, keep in mind that only one ray of light is able to enter your eye from each drop. Every time you move, whether it be up, down, or sideways, the rainbow moves with you. The reason for this is that, with every movement, light from different raindrops enters your eye. The bow you see is not exactly the same rainbow that the person standing next to you sees. In effect, each of us has a personal rainbow to ponder and enjoy!

Coronas and Cloud Iridescence

When the moon is seen through a thin veil of clouds composed of tiny spherical water droplets, a bright ring of light, called a **corona** (meaning "crown"), may appear to rest on the moon (see ● Fig. 15.34). The same effect can occur with the sun, but, due to the sun's brightness, it is usually difficult to see.

The corona is a result of **diffraction** of light—the bending of light as it passes *around* objects. To understand the corona, imagine water waves moving around a small stone in a pond. As the waves spread around the stone, the trough of one wave may meet the crest of another wave. This situation causes the waves to cancel each other, thus producing calm water. Where two crests come together, they produce a much larger wave. The same thing happens when light passes around tiny cloud droplets. Where light waves come together, we see bright light; where they cancel each other, we see darkness. Sometimes, the corona appears white, with alternating bands of light and dark. On other occasions, the rings have color (see ● Fig. 15.35).

The colors appear when the cloud droplets (or any kind of small particles, such as volcanic ash) are of uniform size.

Because the amount of bending due to diffraction depends upon the wavelength of light, the shorter-wavelength blue light appears on the inside of a ring, whereas the longer-wavelength red light appears on the outside. These colors may repeat over and over, becoming fainter as each ring is farther from the moon or sun. Clouds that have recently formed (such as thin altostratus and altocumulus) are the best corona producers.

When various sizes of droplets exist within a cloud, the corona becomes distorted and irregular. Sometimes the cloud exhibits patches of color, often pastel shades of pink, blue, or green. These bright areas produced by diffraction are called **iridescence** (see ● Fig. 15.36). Cloud iridescence is most often seen within 20° of the sun and is often associated with thin clouds, such as cirrocumulus and altocumulus. (Additional optical phenomena—the glory and the *Heiligenschein*—are described in Focus section 15.2.)

● **FIGURE 15.36** Cloud iridescence.

Glories and The *Heiligenschein*

When an aircraft flies above a cloud layer composed of tiny water droplets, a set of colored rings called the *glory* may appear around the shadow of the aircraft (see ●Fig. 2). The same effect can happen when you stand with your back to the sun and look into a cloud or fog bank, as a bright ring of light may be seen around the shadow of your head. In this case, the glory is called the *brocken bow,* after the Brocken Mountains in Germany, where it is particularly common.

For the glory and the brocken bow to occur, the sun must be to your back, so that sunlight can be returned to your eye from the water droplets. Sunlight that enters the small water droplet along its edge is refracted, then reflected off the backside of the droplet. The light then exits at the other side of the droplet, being refracted once again (see ●Fig. 3). The colorful rings may be due to the various angles at which different colors leave the droplet.

On a clear morning with dew on the grass, stand facing the dew with your back to the sun and observe that, around the shadow of your head, is a bright area—the *Heiligenschein* (German for "halo"). The *Heiligenschein* forms when sunlight that falls on nearly spherical dew drops is focused and reflected back toward the sun along nearly the same path that it took originally. The light, however, does not travel along the exact path; it actually spreads out just enough to be seen as bright white light around the shadow of your head on a dew-covered lawn (see ●Fig. 4).

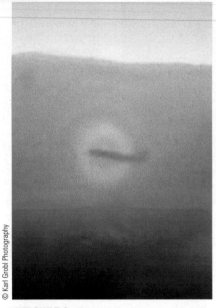

© Karl Grobl Photography

●**FIGURE 2** The series of rings surrounding the shadow of the aircraft is called the *glory*.

© C. Donald Ahrens

●**FIGURE 4** The *Heiligenschein* is the ring of light around the shadow of the observer's head.

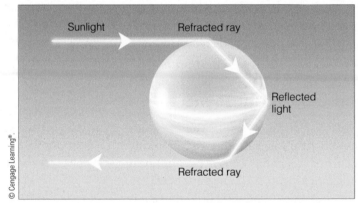

© Cengage Learning®.

Sunlight   Refracted ray

Reflected light

Refracted ray

●**FIGURE 3** Light that produces the glory follows this path in a water droplet.

SUMMARY

The scattering of sunlight in the atmosphere can produce a variety of atmospheric visuals, from hazy days and blue skies to crepuscular rays and blue moons. Refraction of light by the atmosphere causes stars near the horizon to appear higher than they really are. It also causes the sun and moon to appear to rise earlier and set later than they otherwise would. Under certain atmospheric conditions, the amplification of green light near the upper rim of a rising or setting sun produces the elusive green flash.

Mirages form when refraction of light displaces objects from their true positions. Inferior mirages cause objects to appear lower than they really are, while superior mirages displace objects upward.

Halos and sundogs form from the refraction of light through ice crystals. Sun pillars are the result of sunlight reflecting off gently falling ice crystals. Diffraction of light produces coronas and cloud iridescence. The refraction, reflection, and dispersion of light in raindrops create a rainbow. If you want to see a rainbow, the sun must be to your back, and rain must be falling in front of you.

KEY TERMS

The following terms are listed (with corresponding page numbers) in the order they appear in the text. Define each. Doing so will aid you in reviewing the material covered in this chapter.

reflected (light), 444
scattered (light), 444
crepuscular rays, 446
anticrepuscular rays, 446
refraction (of light), 449
scintillation, 450
twilight, 450
green flash, 451
mirage, 451
inferior mirage, 452
superior mirage, 452
halo, 453
tangent arc, 454
dispersion (of light), 455
sundog (or parhelia), 455
sun pillars, 456
rainbow, 456
corona, 459
diffraction (of light), 459
iridescence, 459

QUESTIONS FOR REVIEW

1. Why are cumulus clouds normally white?
2. Why do the undersides of building cumulus clouds frequently change color from white to dark gray or even black?
3. Explain why the sky is blue during the day and black at night.
4. What can make a setting (or rising) sun appear red?
5. If Earth had no atmosphere, what would be the color of the daytime sky?
6. Explain why the horizon sky appears white on a hazy day.
7. What process (refraction or scattering) produces crepuscular rays?
8. Why do stars "twinkle"?
9. How does refraction of light differ from reflection of light?
10. How long does twilight last on the moon? (Hint: The moon has no atmosphere.)
11. At what time of day would you expect to observe the green flash?
12. How does light bend as it enters a more-dense substance at an angle? How does it bend upon leaving the more-dense substance? Make a sketch to illustrate your answer.
13. On a clear, dry, warm day, why do dark road surfaces frequently appear wet?
14. What atmospheric conditions are necessary for an inferior mirage? A superior mirage?
    (a) Describe how a halo forms.
    (b) How is the formation of a halo different from that of a sundog?
16. Would you expect to see a ringed halo if the sky contains a few wispy cirrus clouds? Explain.
17. What process is believed to be mainly responsible for the formation of sun pillars: refraction, reflection, or scattering?
18. Explain why this rhyme makes sense:

    Rainbow in the morning, runner take warning.
    Rainbow at night (evening), runner's delight.

19. Why can a rainbow only be observed if the sun is at the observer's back?
20. Why are secondary rainbows much dimmer than primary rainbows?
21. How would you distinguish a corona from a halo?
22. What process is primarily responsible for the formation of cloud iridescence: reflection, refraction, or diffraction of light?

QUESTIONS FOR THOUGHT AND EXPLORATION

1. Why is it often difficult to see the road while driving on a foggy night with your high beam lights on?
2. Explain why the notion that "the sky is blue because of reflected light from the oceans" is false.
3. Why does smoke rising from a cigarette appear blue, yet appears white when blown from the mouth?
4. If there were no atmosphere surrounding Earth, what color would the sky be at sunrise? At sunset? What color would the sun be at noon? At sunrise? At sunset?
5. Explain why, on a cloudless day, the sky will usually appear milky white before it rains and a deeper blue after it rains.
6. Why are rainbows seldom observed at noon?
7. During the day, clouds are white and the sky is blue. Why then, during a full moon, do cumulus clouds appear faintly white, while the sky does not appear blue?
8. During Ernest Shackleton's last expedition to Antarctica, on May 8, 1915, seven days after the sun had set for the winter, he saw the sun reappear. Explain how this event—called the *Novaya Zemlya effect*—can occur.
9. Choose a 3-day period in which to observe the sky 5 times each day. Record in a notebook the number of times you see halos, crepuscular rays, coronas, cloud iridescence, sundogs, rainbows, and other phenomena.

GLOBAL **GEOSCIENCE** WATCH   Within the Global Environmental Watch: Solar Energy portal, find and open the document "Solar Energy," from *Climate Change: In Context*. Among three behaviors of light discussed in this chapter—reflection, refraction, and absorption—which are most important in the two main types of equipment designed to collect solar energy? How many 100-watt light bulbs could be powered if every bit of energy reaching Earth from the sun were harvested for the task?

ONLINE RESOURCES

Visit www.cengagebrain.com to view additional resources, including video exercises, practice quizzes, an interactive eBook, and more.

Units, Conversions, Abbreviations, and Equations

LENGTH

1 kilometer (km)	=	1000 m
	=	3281 ft
	=	0.62 mi
1 statute mile (mi)	=	5280 ft
	=	1609 mi
	=	1.61 km
1 nautical mile (nm)	=	1.15 mi
	=	1.85 km
	=	0.87 nm
1 meter (m)	=	100 cm
	=	3.28 ft
	=	39.37 in.
1 foot (ft)	=	12 in.
	=	30.48 cm
	=	0.305 m
1 centimeter (cm)	=	0.39 in.
	=	0.01 m
	=	10 mm
1 inch (in.)	=	2.54 cm
	=	0.08 ft
1 millimeter (mm)	=	0.1 cm
	=	0.001 m
	=	0.039 in.
1 micrometer (µm)	=	0.0001 cm
	=	0.000001 m
1 degree latitude	=	111 km
	=	60 nautical mi
	=	69 statute mi

AREA

1 square centimeter (cm²)	=	0.15 in.²
1 square inch (in.²)	=	6.45 cm²
1 square meter (m²)	=	10.76 ft²
1 square foot (ft²)	=	0.09 m²

VOLUME

1 cubic centimeter (cm³)	=	0.06 in.³
1 cubic inch (in.³)	=	16.39 cm³
1 liter (l)	=	1000 cm³
	=	0.264 gallon (gal) U.S.

SPEED

1 knot	=	1 nautical mi/hr
	=	1.15 statute mi/hr
	=	0.51 m/sec
	=	1.85 km/hr
1 mile per hour (mi/hr)	=	0.87 knot
	=	0.45 m/sec
	=	1.61 km/hr
1 meter per second (m/sec)	=	0.54 knot
	=	0.62 mi/hr
	=	0.28 m/sec
1 meter per second (m/sec)	=	1.94 knots
	=	2.24 mi/hr
	=	3.60 km/hr

FORCE

1 dyne	=	1 gram centimeter per second per second
	=	2.2481×10^{-6} pound (lb)
1 newton (N)	=	1 kilogram meter per second per second
	=	10^5 dynes
	=	0.2248 lb

MASS

1 gram (g)	=	0.035 ounce
	=	0.002 lb
1 kilogram (kg)	=	1000 g
	=	2.2 lb

ENERGY

1 erg	=	1 dyne per cm
	=	2.388×10^{-8} cal
1 joule (J)	=	1 newton meter
	=	0.239 cal
	=	10^7 erg
1 calorie (cal)	=	4.186 J
	=	4.186×10^7 erg

PRESSURE

1 millibar (mb)	=	1000 dynes/cm^2
	=	0.75 millimeter of mercury (mm Hg)
	=	0.02953 inches of mercury (in. Hg)
	=	0.01450 pounds per square inch (lb/in.^{2})
	=	100 pascals (Pa)

1 standard atmosphere	=	1013.25 mb
	=	760 mm Hg
	=	29.92 in. Hg
	=	14.7 lb/in.^{2}
1 inch of mercury	=	33.865 mb
1 millimeter of mercury	=	1.3332 mb
1 pascal	=	0.01 mb
	=	1 N/m^2
1 hectopascal (hPa)	=	1 mb
1 kilopascal (kPa)	=	10 mb

POWER

1 watt (W)	=	1 J/sec
	=	14.3353 cal/min
1 cal/min	=	0.06973 W
1 horse power (hp)	=	746 W

POWERS OF TEN

PREFIX

nano	one-billionth	=	10^{-9}	=	0.000000001
micro	one-millionth	=	10^{-6}	=	0.000001
milli	one-thousandth	=	10^{-3}	=	0.001
centi	one-hundredth	=	10^{-2}	=	0.01
deci	one-tenth	=	10^{-1}	=	0.1
hecto	one hundred	=	10^{2}	=	100
kilo	one thousand	=	10^{3}	=	1000
mega	one million	=	10^{6}	=	1,000,000
giga	one billion	=	10^{9}	=	1,000,000,000

TEMPERATURE

$$°C = \tfrac{5}{9} (°F - 32)$$

To convert degrees Fahrenheit (°F) to degrees Celsius (°C): Subtract 32 degrees from °F, then divide by 1.8.

To convert degrees Celsius (°C) to degrees Fahrenheit (°F): Multiply °C by 1.8, then add 32 degrees.

To convert degrees Celsius (°C) to Kelvins (K): Add 273 to Celsius temperature, as

$$K = °C + 273.$$

▼ Table A.1 Temperature Conversions

°F	°C	°F	°C	°F	°C	°F	°C	°F	°C	°F	°C	°F	°C	°F	°C
−40	−40	−20	−28.9	0	−17.8	20	−6.7	40	4.4	60	15.6	80	26.7	100	37.8
−39	−39.4	−19	−28.3	1	−17.2	21	−6.1	41	5.0	61	16.1	81	27.2	101	38.3
−38	−38.9	−18	−27.8	2	−16.7	22	−5.6	42	5.6	62	16.7	82	27.8	102	38.9
−37	−38.3	−17	−27.2	3	−16.1	23	−5.0	43	6.1	63	17.2	83	28.3	103	39.4
−36	−37.8	−16	−26.7	4	−15.6	24	−4.4	44	6.7	64	17.8	84	28.9	104	40.0
−35	−37.2	−15	−26.1	5	−15.0	25	−3.9	45	7.2	65	18.3	85	29.4	105	40.6
−34	−36.7	−14	−25.6	6	−14.4	26	−3.3	46	7.8	66	18.9	86	30.0	106	41.1
−33	−36.1	−13	−25.0	7	−13.9	27	−2.8	47	8.3	67	19.4	87	30.6	107	41.7
−32	−35.6	−12	−24.4	8	−13.3	28	−2.2	48	8.9	68	20.0	88	31.1	108	42.2
−31	−35.0	−11	−23.9	9	−12.8	29	−1.7	49	9.4	69	20.6	89	31.7	109	42.8
−30	−34.4	−10	−23.3	10	−12.2	30	−1.1	50	10.0	70	21.1	90	32.2	110	43.3
−29	−33.9	−9	−22.8	11	−11.7	31	−0.6	51	10.6	71	21.7	91	32.8	111	43.9
−28	−33.3	−8	−22.2	12	−11.1	32	0.0	52	11.1	72	22.2	92	33.3	112	44.4
−27	−32.8	−7	−21.7	13	−10.6	33	0.6	53	11.7	73	22.8	93	33.9	113	45.0
−26	−32.2	−6	−21.1	14	−10.0	34	1.1	54	12.2	74	23.3	94	34.4	114	45.6
−25	−31.7	−5	−20.6	15	−9.4	35	1.7	55	12.8	75	23.9	95	35.0	115	46.1
−24	−31.1	−4	−20.0	16	−8.9	36	2.2	56	13.3	76	24.4	96	35.6	116	46.7
−23	−30.6	−3	−19.4	17	−8.3	37	2.8	57	13.9	77	25.0	97	36.1	117	47.2
−22	−30.0	−2	−18.9	18	−7.8	38	3.3	58	14.4	78	25.6	98	36.7	118	47.8
−21	−29.4	−1	−18.3	19	−7.2	39	3.9	59	15.0	79	26.1	99	37.2	119	48.3

▼ Table A.2 SI Units* and Their Symbols

QUANTITY	NAME	UNITS	SYMBOL
length	meter	m	m
mass	kilogram	kg	kg
time	second	sec	sec
temperature	Kelvin	K	K
density	kilogram per cubic meter	kg/m^3	kg/m^3
speed	meter per second	m/sec	m/sec
force	newton	$m \cdot kg/sec^2$	N
pressure	pascal	N/m^2	Pa
energy	joule	$N \cdot m$	J
power	watt	J/sec	W

*SI stands for Système International, which is the international system of units and symbols.

Equations and Constants

Relative Humidity

The relative humidity of the air can be expressed as:

$$\text{RH} = \frac{e}{e_s} \times 100\%$$

To determine e and e_s, when the air temperature and dew-point temperature are known, go to the Online Appendix at www.cengagebrain.com and click on *Relative Humidity Tables*. Simply read the value adjacent to the air temperature and obtain e_s; read the value adjacent to the dew-point temperature and obtain e.

UNITS/CONSTANTS		
e	=	actual vapor pressure (millibars)
e_s	=	saturation vapor pressure (millibars)
RH	=	relative humidity (percent)

Gas Law (Equation of State)

The relationship among air pressure, air density, and air temperature can be expressed by

$$\text{Pressure} = \text{density} \times \text{temperature} \times \text{constant}.$$

This relationship, often called the gas law (or equation of state), can be expressed in symbolic form as:

$$p = \rho R T$$

where p is air pressure, ρ is air density, R is a constant, and T is air temperature.

UNITS/CONSTANTS		
p	=	pressure in N/m^2 (SI)
ρ	=	density (kg/m^3)
T	=	temperature (K)
	=	287 J/kg • K (SI)
R	=	2.87×10^6 erg/g • K

Stefan-Boltzmann Law

The Stefan-Boltzmann law is a law of radiation. It states that all objects with temperatures above absolute zero emit radiation at a rate proportional to the fourth power of their absolute temperature. It is expressed mathematically as:

$$E = \sigma T^4$$

where E is the maximum rate of radiation emitted each second per unit surface area, T is the object's surface temperature, and σ is a constant.

UNITS/CONSTANTS		
E	=	radiation emitted in W/m^2 (SI)
σ	=	5.67×10^{-8} W/m^2 • K^4 (SI)
	=	5.67×10^{-5} erg/cm^2 • K^4 • sec
T	=	temperature (K)

Wien's Law

Wien's law (or Wien's displacement law) relates an object's maximum emitted wavelength of radiation to the object's temperature. It states that the wavelength of maximum emitted radiation by an object is inversely proportional to the object's absolute temperature. In symbolic form, it is written as:

$$\lambda_{\max} = \frac{w}{T}$$

where $\lambda_{\max}$ is the wavelength at which maximum radiation emission occurs, T is the object's temperature, and w is a constant.

UNITS/CONSTANTS		
$\lambda_{\max}$	=	wavelength (micrometers)
w	=	0.2897 μm K
T	=	temperature (K)

Geostrophic Wind Equation

The geostrophic wind equation gives an approximation of the wind speed above the level of friction, where the wind blows parallel to the isobars or contours. The equation is expressed mathematically as:

$$V_g = \frac{1}{2\Omega \sin\phi \rho} \frac{\Delta p}{d}$$

where V_g is the geostrophic wind, Ω is a constant (twice the earth's angular spin), $\sin\phi$ is a trigonometric function that takes into account the variation of latitude (ϕ), ρ is the air density, Δp is the pressure difference between two places on the map some horizontal distance (d) apart.

Hydrostatic Equation

The hydrostatic equation relates to how quickly the air pressure decreases in a column of air above the surface. The equation tells us that the rate at which the air pressure decreases with height is equal to the air density times the acceleration of gravity. In symbolic form, it is written as:

$$\frac{\Delta p}{\Delta z} = -\rho g$$

where Δp is the decrease in pressure along a small change in height Δz, ρ is the air density, and g is the force of gravity.

▼ **Table B.1 Saturation Vapor Pressure Over Water For Various Air Temperatures**

AIR TEMPERATURE (°C)	(°F)	SATURATION VAPOR PRESSURE (MB)	AIR TEMPERATURE (°C)	(°F)	SATURATION VAPOR PRESSURE (MB)
−18	(0)	1.5	18	(65)	21.0
−15	(5)	1.9	21	(70)	25.0
−12	(10)	2.4	24	(75)	29.6
−9	(15)	3.0	27	(80)	35.0
−7	(20)	3.7	29	(85)	41.0
−4	(25)	4.6	32	(90)	48.1
−1	(30)	5.6	35	(95)	56.2
2	(35)	6.9	38	(100)	65.6
4	(40)	8.4	41	(105)	76.2
7	(45)	10.2	43	(110)	87.8
10	(50)	12.3	46	(115)	101.4
13	(55)	14.8	49	(120)	116.8
16	(60)	17.7	52	(125)	134.2

Weather Symbols and the Station Model

SIMPLIFIED SURFACE-STATION MODEL

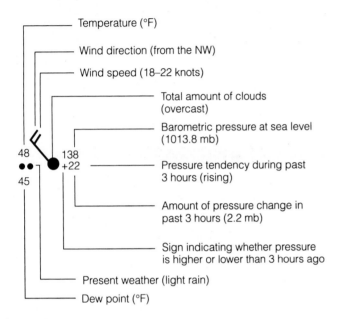

- Temperature (°F)
- Wind direction (from the NW)
- Wind speed (18–22 knots)
- Total amount of clouds (overcast)
- Barometric pressure at sea level (1013.8 mb)
- Pressure tendency during past 3 hours (rising)
- Amount of pressure change in past 3 hours (2.2 mb)
- Sign indicating whether pressure is higher or lower than 3 hours ago
- Present weather (light rain)
- Dew point (°F)

CLOUD COVERAGE

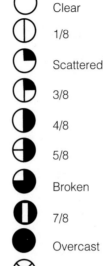

- Clear
- 1/8
- Scattered
- 3/8
- 4/8
- 5/8
- Broken
- 7/8
- Overcast
- Obscured
- Missing

UPPER-AIR MODEL (500 MB)

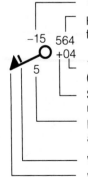

- Temperature (°C)
- Height of pressure surface in meters with first 3 digits given (5640 m)
- 12 hour height change in meters (04 equals 40 m)
- Sign indicating whether height is rising or falling
- Dew point depression (difference between air temperature and dew point, °C)
- Wind speed (58–62 knots)
- Wind direction (from the southwest)

COMMON WEATHER SYMBOLS

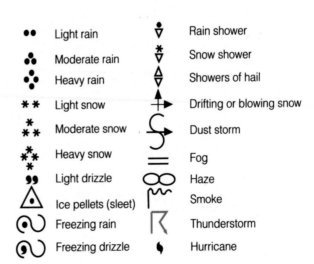

- Light rain
- Moderate rain
- Heavy rain
- Light snow
- Moderate snow
- Heavy snow
- Light drizzle
- Ice pellets (sleet)
- Freezing rain
- Freezing drizzle
- Rain shower
- Snow shower
- Showers of hail
- Drifting or blowing snow
- Dust storm
- Fog
- Haze
- Smoke
- Thunderstorm
- Hurricane

WIND ENTRIES

	MILES (STATUTE) PER HOUR	KNOTS	KILOMETERS PER HOUR
◎	Calm	Calm	Calm
———	1–2	1–2	1–3
⟍	3–8	3–7	4–13
⟍	9–14	8–12	14–19
⟍	15–20	13–17	20–32
⟍	21–25	18–22	33–40
⟍	26–31	23–27	41–50
⟍	32–37	28–32	51–60
⟍	38–43	33–37	61–69
⟍	44–49	38–42	70–79
⟍	50–54	43–47	80–87
⟍	55–60	48–52	88–96
⟍	61–66	53–57	97–106
⟍	67–71	58–62	107–114
⟍	72–77	63–67	115–124
⟍	78–83	68–72	125–134
⟍	84–89	73–77	135–143
⟍	119–123	103–107	144–198

PRESSURE TENDENCY

∧ Rising, then falling

⟋ Rising, then steady; or rising, then rising more slowly

／ Rising steadily or unsteadily

✓ Falling or steady, then rising; or rising, then rising more quickly

— Steady, same as 3 hours ago

∨ Falling, then rising, same or lower than 3 hours ago

＼ Falling, then steady; or falling, then falling more slowly

＼ Falling steadily, or unsteadily

∧ Steady or rising, then falling; or falling, then falling more quickly

Barometer now higher than 3 hours ago

Barometer now lower than 3 hours ago

FRONT SYMBOLS

▲▲▲ Cold front (surface)

●●● Warm front (surface)

▲●▲● Occluded front (surface)

▲●▼● Stationary front (surface)

——••—— Squall line

Trough (trof)   Ridge   Dryline

Average Annual Global Precipitation

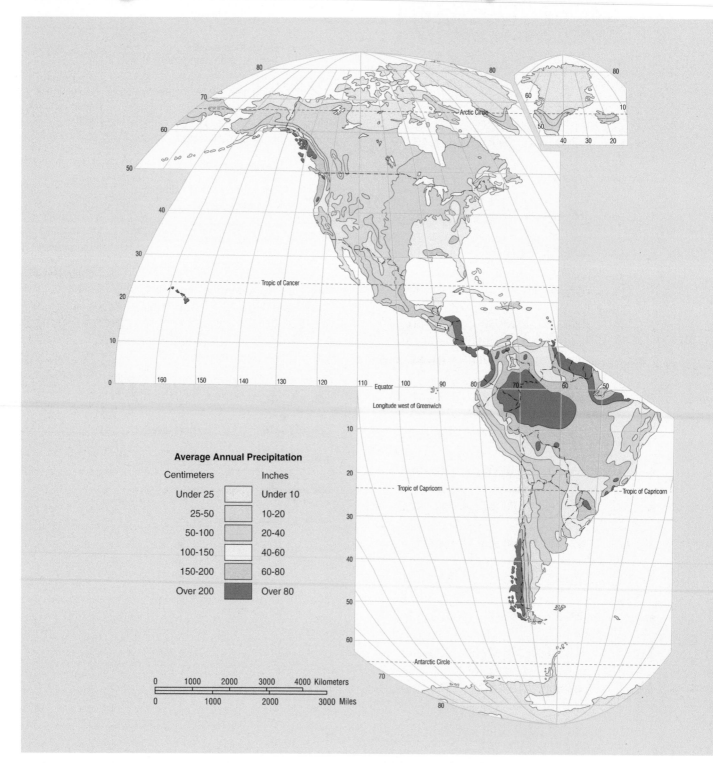

●FIGURE D.1 World map of average annual precipitation

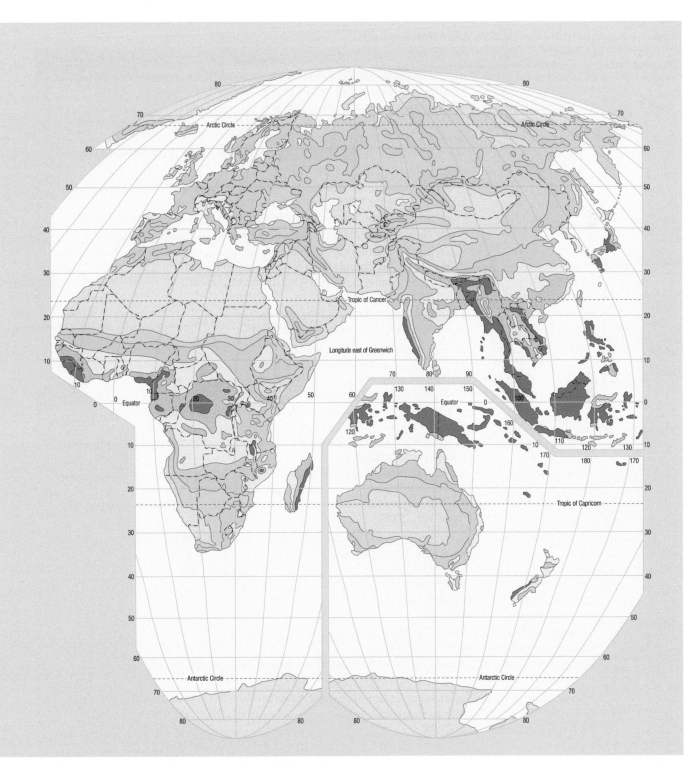

A Western Paragraphic Projection developed at Western Illinois University

Köppen's Climatic Classification System

▼Table E.1 **Köppen's Climatic Classification System**

1ST	2ND	3RD	CLIMATIC CHARACTERISTICS	CRITERIA
A			Humid tropical	All months have an average temperature of 18°C (64°F) or higher
	f		Tropical wet (rain forest)	Wet all seasons; all months have at least 6 cm (2.4 in.) of rainfall
	w		Tropical wet and dry (savanna)	Winter dry season; rainfall in driest month is less than 6 cm (2.4 in.) and less than 10 – P/25 (P is mean annual rainfall in cm)
	m		Tropical monsoon	Short dry season; rainfall in driest month is less than 6 cm (2.4 in.) but equal to or greater than 10 – P/25.
B			Dry	Potential evaporation and transpiration exceed precipitation. The dry/humid boundary is defined by the following formulas: $p = 2t + 28$ when 70% or more of rain falls in warmer 6 months (dry winter) $p = 2t$ when 70% or more of rain falls in cooler 6 months (dry summer) $p = 2t + 14$ when neither half year has 70% or more of rain (p is the mean annual precipitation in cm and t is the mean annual temperature in °C)*
	S		Semi-arid (steppe)	The BS/BW boundary is exactly ½ the dry/humid boundary
	W		Arid (desert)	
		h	Hot and dry	Mean annual temperature is 18°C (64°F) or higher
		k	Cool and dry	Mean annual temperature is below 18°C (64°F)
C			Moist with mild winters	Average temperature of coolest month is below 18°C (64°F) and above –3°C (27°F)
	w		Dry winters	Average rainfall of wettest summer month at least 10 times as much as in driest winter month
	s		Dry summers	Average rainfall of driest summer month less than 4 cm (1.6 in.); average rainfall of wettest winter month at least 3 times as much as in driest summer month
	f		Wet all seasons	Criteria for w and s cannot be met
		a	Summers long and hot	Average temperature of warmest month above 22°C (72°F); at least 4 months with average above 10°C (50°F)
		b	Summers long and cool	Average temperature of all months below 22°C (72°F); at least 4 months with average above 10°C (50°F)
		c	Summers short and cool	Average temperature of all months below 22°C (72°F); 1 to 3 months with average above 10°C (50°F)
D			Moist with cold winters	Average temperature of coldest month is –3°C (27°F) or below; average temperature of warmest month is greater than 10°C (50°F)
	w		Dry winters	Same as under C
	s		Dry summers	Same as under C
	f		Wet all seasons	Same as under C
		a	Summers long and hot	Same as under C
		b	Summers long and cool	Same as under C
		c	Summers short and cool	Same as under C
		d	Summers short and cool; winters severe	Average temperature of coldest month is –38°C (–36°F) or below
E			Polar climates	Average temperature of warmest month is below 10°C (50°F)
	T		Tundra	Average temperature of warmest month is greater than 0°C (32°F) but less than 10°C (50°F)
	F		Ice cap	Average temperature of warmest month is 0°C (32°F) or below

*The dry/humid boundary is defined in English units as: $p = 0.44t - 3$ (dry winter); $p = 0.44t - 14$ (dry summer); and $p = 0.44t - 8.6$ (rainfall evenly distributed). Where p is mean annual rainfall in inches and t is mean annual temperature in °F.

Humidity and Dew-Point Tables [Psychrometric Tables]

To obtain the dew point (or relative humidity), simply read down the temperature column and then over to the wet-bulb depression. For example, in Table D.1, a temperature of 10°C with a wet-bulb depression of 3°C produces a dew-point temperature of 4°C. (Dew-point temperature and relative humidity readings are appropriate for pressures near 1000 mb.)

▼Table F.1 **Dew-Point Temperature (°C)**

AIR (DRY-BULB) TEMPERATURE (°C)	WET-BULB DEPRESSION (DRY-BULB TEMPERATURE MINUS WET-BULB TEMPERATURE) (°C)															
	0.5	1.0	1.5	2.0	2.5	3.0	3.5	4.0	4.5	5.0	7.5	10.0	12.5	15.0	17.5	20.0
−20	−25	−33														
−17.5	−21	−27	−38													
−15	−19	−23	−28													
−12.5	−15	−18	−22	−29												
−10	−12	−14	−18	−21	−27	−36										
−7.5	−9	−11	−14	−17	−20	−26	−34									
−5	−7	−8	−10	−13	−16	−19	−24	−31								
−2.5	−4	−6	−7	−9	−11	−14	−17	−22	−28	−41						
0	−1	−3	−4	−6	−8	−10	−12	−15	−19	−24						
2.5	1	0	−1	−3	−4	−6	−8	−10	−13	−16						
5	4	3	2	0	−1	−3	−4	−6	−8	−10	−48					
7.5	6	6	4	3	2	1	−1	−2	−4	−6	−22					
10	9	8	7	6	5	4	2	1	0	−2	−13					
12.5	12	11	10	9	8	7	6	4	3	2	−7	−28				
15	14	13	12	12	11	10	9	8	7	5	−2	−14				
17.5	17	16	15	14	13	12	12	11	10	8	2	−7	−35			
20	19	18	18	17	16	15	14	14	13	12	6	−1	−15			
22.5	22	21	20	20	19	18	17	16	16	15	10	3	−6	−38		
25	24	24	23	22	21	21	20	19	18	18	13	7	0	−14		
27.5	27	26	26	25	24	23	23	22	21	20	16	11	5	−5	−32	
30	29	29	28	27	27	26	25	25	24	23	19	14	9	2	−11	
32.5	32	31	31	30	29	29	28	27	26	26	22	18	13	7	−2	
35	34	34	33	32	32	31	31	30	29	28	25	21	16	11	4	
37.5	37	36	36	35	34	34	33	32	32	31	28	24	20	15	9	0
40	39	39	38	38	37	36	36	35	34	34	30	27	23	18	13	6
42.5	42	41	41	40	40	39	38	38	37	36	33	30	26	22	17	11
45	44	44	43	43	42	42	41	40	40	39	36	33	29	25	21	15
47.5	47	46	46	45	45	44	44	43	42	42	39	35	32	28	24	19
50	49	49	48	48	47	47	46	45	45	44	41	38	35	31	28	23

▼Table F.2 Relative Humidity (Percent)

	WET-BULB DEPRESSION (DRY-BULB TEMPERATURE MINUS WET-BULB TEMPERATURE) (°C)																	
AIR (DRY-BULB) TEMPERATURE (°C)	0.5	1.0	1.5	2.0	2.5	3.0	3.5	4.0	4.5	5.0	7.5	10.0	12.5	15.0	17.5	20.0	22.5	25.0
−20	70	41	11															
−17.5	75	51	26	2														
−15	79	58	38	18														
−12.5	82	65	47	30	13													
−10	85	69	54	39	24	10												
−7.5	87	73	60	48	35	22	10											
−5	88	77	66	54	43	32	21	11	1									
−2.5	90	80	70	60	50	42	37	22	12	3								
0	91	82	73	65	56	47	39	31	23	15								
2.5	92	84	76	68	61	53	46	38	31	24								
5	93	86	78	71	65	58	51	45	38	32	1							
7.5	93	87	80	74	68	62	56	50	44	38	11							
10	94	88	82	76	71	65	60	54	49	44	19							
12.5	94	89	84	78	73	68	63	58	53	48	25	4						
15	95	90	85	80	75	70	66	61	57	52	31	12						
17.5	95	90	86	81	77	72	68	64	60	55	36	18	2					
20	95	91	87	82	78	74	70	66	62	58	40	24	8					
22.5	96	92	87	83	80	76	72	68	64	61	44	28	14	1				
25	96	92	88	84	81	77	73	70	66	63	47	32	19	7				
27.5	96	92	89	85	82	78	75	71	68	65	50	36	23	12	1			
30	96	93	89	86	82	79	76	73	70	67	52	39	27	16	6			
32.5	97	93	90	86	83	80	77	74	71	68	54	42	30	20	11	1		
35	97	93	90	87	84	81	78	75	72	69	56	44	33	23	14	6		
37.5	97	94	91	87	85	82	79	76	73	70	58	46	36	26	18	10	3	
40	97	94	91	88	85	82	79	77	74	72	59	48	38	29	21	13	6	
42.5	97	94	91	88	86	83	80	78	75	72	61	50	40	31	23	16	9	2
45	97	94	91	89	86	83	81	78	76	73	62	51	42	33	26	18	12	6
47.5	97	94	92	89	86	84	81	79	76	74	63	53	44	35	28	21	15	9
50	97	95	92	89	87	84	82	79	77	75	64	54	45	37	30	23	17	11

▼ Table F.3 Dew-Point Temperature (°F)

AIR (DRY-BULB) TEMPERATURE (°F)	WET-BULB DEPRESSION (DRY-BULB TEMPERATURE MINUS WET-BULB TEMPERATURE) (°F)																							
	1	2	3	4	5	6	7	8	9	10	11	12	13	14	15	16	17	18	19	20	25	30	35	40
0	−7																							
5	−1	−9																						
10	5	−2	−24																					
15	11	6	−10	−27																				
20	16	12	8	2	−21																			
25	22	19	15	10	5	−3	−15	−51																
30	27	25	21	18	14	8	2	−7	−25															
35	33	30	28	25	21	17	13	7	0	−11														
40	38	35	33	30	28	25	21	18	13	7	−1	−14												
45	43	41	38	36	34	31	28	25	22	18	13	7	−1	−14										
50	48	46	44	42	40	37	34	32	29	26	22	18	13	8	0	−13								
55	53	51	50	48	45	43	41	38	36	33	30	27	24	20	15	9	1	−12						
60	58	57	55	53	51	49	47	45	43	40	38	35	32	29	25	21	17	11	4	−8				
65	63	62	60	59	57	55	53	51	49	47	45	42	40	37	34	31	27	24	19	14				
70	69	67	65	64	62	61	59	57	55	53	51	49	47	44	42	39	36	33	30	26	−11			
75	74	72	71	69	68	66	64	63	61	59	57	55	54	51	49	47	44	42	39	36	15			
80	79	77	76	74	73	72	70	68	67	65	63	62	60	58	56	54	52	50	47	44	28	−7		
85	84	82	81	80	78	77	75	74	72	71	69	68	66	64	62	61	59	57	54	52	39	19		
90	89	87	86	85	83	82	81	79	78	76	75	73	72	70	69	67	65	63	61	59	48	32	−7	
95	94	93	91	90	89	87	86	85	83	81	80	79	78	76	74	73	71	70	68	66	56	43	24	
100	99	98	96	95	94	93	91	90	89	87	86	85	83	82	80	79	77	76	74	72	63	52	37	12
105	104	103	101	100	99	98	96	95	94	93	91	90	89	87	86	84	83	82	80	78	70	61	48	30
110	109	108	106	105	104	103	102	100	99	98	97	95	94	93	91	90	89	87	86	84	77	68	57	43
115	114	113	112	110	109	108	107	106	104	103	102	101	99	98	97	96	94	93	92	90	83	75	65	54
120	119	118	117	115	114	113	112	111	110	108	107	106	105	104	102	101	100	98	97	96	89	81	73	63

▼ Table F.4 Relative Humidity (Percent)

WET-BULB DEPRESSION (DRY-BULB TEMPERATURE MINUS WET-BULB TEMPERATURE) (°F)

AIR (DRY-BULB) TEMPERATURE (°F)

Air Temp (°F)	1	2	3	4	5	6	7	8	9	10	11	12	13	14	15	16	17	18	19	20	25	30	35	40
0	67	31	1																					
5	73	46	20																					
10	78	56	34	13																				
15	82	64	46	29	11																			
20	85	70	55	40	26	12																		
25	87	74	62	49	37	25	13	1																
30	89	78	67	56	46	36	26	16	6															
35	91	81	72	63	54	45	36	27	19	10	2													
40	92	83	75	68	60	52	45	37	29	22	15	7												
45	93	86	78	71	64	57	51	44	38	31	25	18	12	6										
50	93	87	80	74	67	61	55	49	43	38	32	27	21	16	10	5								
55	94	88	82	76	70	65	59	54	49	43	38	33	28	23	19	14	9	5						
60	94	89	83	78	73	68	63	58	53	48	43	39	34	30	26	21	17	13	9	5				
65	95	90	85	80	75	70	66	61	56	52	48	44	39	35	31	27	24	20	16	12				
70	95	90	86	81	77	72	68	64	59	55	51	48	44	40	36	33	29	25	22	19	3			
75	96	91	86	82	78	74	70	66	62	58	54	51	47	44	40	37	34	30	27	24	7	1		
80	96	91	87	83	79	75	72	68	64	61	57	54	50	47	44	41	38	35	32	29	15	3		
85	96	92	88	84	80	76	73	69	66	62	59	56	52	49	46	43	41	38	35	32	20	8		
90	96	92	89	85	81	78	74	71	68	65	61	58	55	52	49	47	44	41	39	36	24	13	3	
95	96	93	89	85	82	79	75	72	69	66	63	60	57	54	51	49	46	43	41	38	27	17	7	1
100	96	93	89	86	83	80	77	73	70	68	65	62	59	56	54	51	49	46	44	41	30	21	12	4
105	97	93	90	87	83	80	77	74	71	69	66	63	60	58	55	53	50	48	46	43	33	23	15	7
110	97	93	90	87	84	81	78	75	73	70	67	65	62	60	57	55	52	50	48	46	36	26	18	11
115	97	94	91	88	85	82	79	76	74	71	68	66	63	61	58	56	54	52	49	47	37	28	21	13
120	97	94	91	88	85	82	80	77	74	72	69	67	65	62	60	58	55	53	51	49	40	31	23	17

APPENDIX G

Standard Atmosphere

▼ TABLE G.1

ALTITUDE				PRESSURE	TEMPERATURE		DENSITY
METERS	FEET	KILOMETERS	MILES	MILLIBARS	°C	°F	KG/M³
0	0	0.0	0.0	1013.25	15.0	(59.0)	1.225
500	1,640	0.5	0.3	954.61	11.8	(53.2)	1.167
1,000	3,280	1.0	0.6	898.76	8.5	(47.3)	1.112
1,500	4,921	1.5	0.9	845.59	5.3	(41.5)	1.058
2,000	6,562	2.0	1.2	795.01	2.0	(35.6)	1.007
2,500	8,202	2.5	1.5	746.91	−1.2	(29.8)	0.957
3,000	9,842	3.0	1.9	701.21	−4.5	(23.9)	0.909
3,500	11,483	3.5	2.2	657.80	−7.7	(18.1)	0.863
4,000	13,123	4.0	2.5	616.60	−11.0	(12.2)	0.819
4,500	14,764	4.5	2.8	577.52	−14.2	(6.4)	0.777
5,000	16,404	5.0	3.1	540.48	−17.5	(0.5)	0.736
5,500	18,045	5.5	3.4	505.39	−20.7	(−5.3)	0.697
6,000	19,685	6.0	3.7	472.17	−24.0	(−11.2)	0.660
6,500	21,325	6.5	4.0	440.75	−27.2	(−17.0)	0.624
7,000	22,965	7.0	4.3	411.05	−30.4	(−22.7)	0.590
7,500	24,606	7.5	4.7	382.99	−33.7	(−28.7)	0.557
8,000	26,247	8.0	5.0	356.51	−36.9	(−34.4)	0.526
8,500	27,887	8.5	5.3	331.54	−40.2	(−40.4)	0.496
9,000	29,528	9.0	5.6	308.00	−43.4	(−46.1)	0.467
9,500	31,168	9.5	5.9	285.84	−46.6	(−51.9)	0.440
10,000	32,808	10.0	6.2	264.99	−49.9	(−57.8)	0.413
11,000	36,089	11.0	6.8	226.99	−56.4	(−69.5)	0.365
12,000	39,370	12.0	7.5	193.99	−56.5	(−69.7)	0.312
13,000	42,651	13.0	8.1	165.79	−56.5	(−69.7)	0.267
14,000	45,932	14.0	8.7	141.70	−56.5	(−69.7)	0.228
15,000	49,213	15.0	9.3	121.11	−56.5	(−69.7)	0.195
16,000	52,493	16.0	9.9	103.52	−56.5	(−69.7)	0.166
17,000	55,774	17.0	10.6	88.497	−56.5	(−69.7)	0.142
18,000	59,055	18.0	11.2	75.652	−56.5	(−69.7)	0.122
19,000	62,336	19.0	11.8	64.674	−56.5	(−69.7)	0.104
20,000	65,617	20.0	12.4	55.293	−56.5	(−69.7)	0.089
25,000	82,021	25.0	15.5	25.492	−51.6	(−60.9)	0.040
30,000	98,425	30.0	18.6	11.970	−46.6	(−51.9)	0.018
35,000	114,829	35.0	21.7	5.746	−36.6	(−33.9)	0.008
40,000	131,234	40.0	24.9	2.871	−22.8	(−9.0)	0.004
45,000	147,638	45.0	28.0	1.491	−9.0	(15.8)	0.002
50,000	164,042	50.0	31.1	0.798	−2.5	(27.5)	0.001
60,000	196,850	60.0	37.3	0.220	−26.1	(−15.0)	0.0003
70,000	229,659	70.0	43.5	0.052	−53.6	(−64.5)	0.00008
80,000	262,467	80.0	49.7	0.010	−74.5	(−102.1)	0.00002

APPENDIX H

Beaufort Wind Scale [Over Land]

▼ **Table H.1 Estimating Wind Speed from Surface Observation**

BEAUFORT NUMBER	DESCRIPTION	MI/HR	WIND SPEED KNOTS	KM/HR	OBSERVATIONS
0	Calm	0–1	0–1	0–2	Smoke rises vertically
1	Light air	1–3	1–3	2–6	Direction of wind shown by drifting smoke, but not by wind vanes
2	Slight breeze	4–7	4–6	7–11	Wind felt on face; leaves rustle; wind vanes moved by wind; flags stir
3	Gentle breeze	8–12	7–10	12–19	Leaves and small twigs move; wind will extend light flag
4	Moderate breeze	13–18	11–16	20–29	Wind raises dust and loose paper; small branches move; flags flap
5	Fresh breeze	19–24	17–21	30–39	Small trees with leaves begin to sway; flags ripple
6	Strong breeze	25–31	22–27	40–50	Large tree branches in motion; whistling heard in telegraph wires; umbrellas used with difficulty
7	High wind	32–38	28–33	51–61	Whole trees in motion; inconvenience felt walking against wind; flags extend
8	Gale	39–46	34–40	62–74	Wind breaks twigs off trees; walking is difficult
9	Strong gale	47–54	41–47	75–87	Slight structural damage occurs (signs and antennas blown down)
10	Whole gale	55–63	48–55	88–101	Trees uprooted; considerable damage occurs
11	Storm	64–74	56–64	102–119	Winds produce widespread damage
12	Hurricane	≥ 75	≥ 65	≥ 120	Winds produce extensive damage

ADDITIONAL READING MATERIAL

SELECTED PERIODICALS

Bulletin of the American Meteorological Society. Monthly. American Meteorological Society. http://journals.ametsoc.org/toc/bams/current

Journal of Operational Meteorology. Online-only. (Deals mainly with weather forecasting.) National Weather Association. http://www.nwas.org/jom

Weather. Monthly. Royal Meteorological Society. http://onlinelibrary.wiley.com/journal/10.1002/%28ISSN%291477-8696

Weatherwise. Bimonthly. Taylor & Francis Group LLC. http://www.weatherwise.org

SELECTED TECHNICAL PERIODICALS

From the American Meteorological Society

http://journals.ametsoc.org
Earth Interactions
Journal of Applied Meteorology and Climatology
Journal of Atmospheric and Oceanic Technology
Journal of the Atmospheric Sciences
Journal of Climate
Journal of Hydrometeorology
Journal of Physical Oceanography
Monthly Weather Review
Weather and Forecasting
Weather, Climate, and Society

From the American Geophysical Union

http://agupubs.onlinelibrary.wiley.com/
Earth's Future
Earth and Space Science
Geophysical Research Letters
Journal of Geophysical Research—Atmospheres
EOS (https://eos.org/)

From other publishers, including frequent content from atmospheric science

Nature. Weekly. Macmillan Publishers Limited. http://www.nature.com

Nature Climate Change. Monthly. Macmillan Publishers Limited. http://www.nature.com/nclimate

Proceedings of the National Academy of Sciences of the United States of America. Weekly. NAS. http://www.pnas.org

Science. Weekly. American Association for the Advancement of Science. http://www.sciencemag.org

BOOKS

Many of the titles below are written at the introductory level. Those that are more advanced are marked with an asterisk.

Ahrens, C. Donald, and Robert Henson. *Meteorology Today* (11th ed.), Cengage Learning, Boston, MA, 2016.

Ahrens, C. Donald, and Perry Samson. *Extreme Weather and Climate*, Cengage Learning, Boston, MA, 2011.

*Andrews, David G. *An Introduction to Atmospheric Physics* (2nd ed.), Cambridge University Press, New York, 2010.

Archer, David. *Global Warming: Understanding the Forecast* (2nd ed.). Wiley, Hoboken, NJ, 2011.

Bigg, Grant R. *The Oceans and Climate* (2nd ed.) Cambridge University Press, New York, 2003.

*Bluestein, Howard B. *Synoptic-Dynamic Meteorology in Midlatitudes. Vol. 1: Principles of Kinematics and Dynamics,* Oxford University Press, New York, 1992.

*———. *Synoptic-Dynamic Meteorology in Midlatitudes. Vol. II: Observations and Theory of Weather Systems,* Oxford University Press, New York, 1993.

———. *Tornado Alley: Monster Storms of the Great Plains,* Oxford University Press, New York, 1999.

Bohren, Craig F. *Clouds in a Glass of Beer: Simple Experiments in Atmospheric Physics,* Wiley, New York, 1987.

———. *What Light Through Yonder Window Breaks?,* Wiley, New York, 1991.

Boubel, Richard W., et al., *Fundamentals of Air Pollution* (5th ed.), Academic Press, New York, 2014.

*Bradley, Raymond S. *Paleoclimatology: Reconstructing Climates of the Quarternary* (3rd ed.), Academic Press, New York, 2014.

Brunner, Ronald D., and Amanda H. Lynch. *Adaptive Governance and Climate Change,* American Meteorological Society, Boston, MA, 2010.

Burt, Christopher C. *Extreme Weather, A Guide and Record Book,* W.W. Norton & Company, New York, 2007.

Burt, Stephen. *The Weather Observer's Handbook,* Cambridge University Press, New York, 2012.

Carlson, Toby N. *An Observer's Guide to Clouds and Weather: A Northeastern Primer on Prediction,* American Meteorological Society, Boston, MA, 2014.

Changnon, Stanley A. *Railroads and Weather: From Fogs to Floods and Heat to Hurricanes,* American Meteorological Society, Boston, MA, 2006.

Climate Change 2013. The Physical Science Basis. Working Group 1 contribution to the Fifth Assessment Report of the IPCC, Cambridge University Press, New York, 2014.

*Cotton, W. R., and R. A. Anthes. *Storm and Cloud Dynamics* (2nd ed.), Academic Press, New York, 2010.

Cotton, William R., and Roger A. Pielke. *Human Impacts on Weather and Climate* (2nd ed.), Cambridge University Press, New York, 2007.

Doswell, Charles A. III, editor. *Severe Convective Storms,* American Meteorological Society, Boston, MA, 2001.

Dow, Kirstin, and Thomas E. Downing, *The Atlas of Climate Change: Mapping the World's Greatest Challenge.* University of California Press, Oakland, CA, 2011.

Edwards, Paul. *A Vast Machine: Computer Models, Climate Data, and the Politics of Global Warming.* The MIT Press, Cambridge, MA, 2013.

Emanuel, Kerry. *Divine Wind: The History and Science of Hurricanes,* Oxford University Press, New York, 2005.

Encyclopedia of Climate and Weather (2nd ed.), Stephen H. Schneider, Terry L. Root, and Michael D. Mastrandrea, Ed., Oxford University Press, New York, 2011.

The Encyclopedia of Weather and Climate Change: A Complete Visual Guide, Juliane L. Fry, et al., University of California Press, Berkeley, Los Angeles, CA, 2010.

Fabry, Frederic. *Radar Meteorology: Principles and Practice.* Cambridge University Press, New York, 2015.

Fagan, Brian. *The Great Warming: Climate Change and the Rise and Fall of Civilizations.* Bloomsbury Press, London, England, 2009.

Fleming, James Rodger. *Fixing the Sky: The Checkered History of Weather and Climate Control.* Columbia University Press, New York, 2012.

Glossary of Meteorology. Mary M. Cairns, Ed., American Meteorological Society, Boston, MA. Online-only (https://www.ametsoc.org/ams/index.cfm/publications/glossary-of-meteorology/)

Henson, Robert, *The Thinking Person's Guide to Climate Change,* American Meteorological Society, Boston, MA, 2014.

———. *Weather on the Air: A History of Broadcast Meteorology,* American Meteorological Society, Boston, MA, 2010.

*Hobbs, Peter V. *Basic Physical Chemistry for Atmospheric Sciences* (2nd ed.), Cambridge University Press, New York, 2000.

Hulme, Mike, *Why We Disagree about Climate Change: Understanding Controversy, Inaction and Opportunity.* Cambridge University Press, Cambridge, England, 2009.

International Cloud Atlas. World Meteorological Organization, Geneva, Switzerland, 1987. Available online (http://library.wmo.int/pmb_ged/wmo_407_en-v2.pdf).

Jacobson, Mark Z. *Air Pollution and Global Warming: History, Science, and Solutions* (2nd ed.). Cambridge University Press, New York, 2012.

*Karoly, David J., and Dayton G. Vincent, Eds. *Meteorology of the Southern Hemisphere,* American Meteorological Society, Boston, MA, 1998.

Kocin, Paul J., and L. W. Uccellini. *Northeast Snowstorms,* Vol. 1 and Vol. 2, American Meteorological Society, Boston, MA, 2004.

Lackmann, Gary. *Midlatitude Synoptic Meteorology: Dynamics, Analysis, and Forecasting,* American Meteorological Society, Boston, MA, 2012.

Lee, Raymond L., and Alistair B. Fraser, *The Rainbow Bridge: Rainbows in Art, Myth, and Science.* Penn State University Press, University Park, PA, 2001.

Lynch, David K., and William Livingston. *Color and Light in Nature* (2nd ed.), Cambridge University Press, New York, 2001.

Managing the Risks of Extreme Events and Disasters to Advance Climate Change Adaptation. Intergovernmental Panel on Climate Change, Cambridge University Press, New York, 2012.

Meinel, Aden, and Marjorie Meinel. *Sunsets, Twilights and Evening Skies,* Cambridge University Press, New York, 1991.

Mergen, Bernard. *Weather Matters: An American Cultural History since 1900.* University Press of Kansas, Lawrence, KS, 2008.

Mims, F. M. *Hawaii's Mauna Loa Observatory: Fifty Years of Monitoring the Atmosphere,* University of Hawaii Press, Honolulu, HI, 2011.

Monmonier, Mark. *Air Apparent: How Meteorologists Learned to Map, Predict, and Dramatize Weather.* University of Chicago Press, Chicago, IL, 2001.

Pretor-Pinney, Gavin. *The Cloudspotters' Guide: The Science, History, and Culture of Clouds,* TarcherPerigee, New York, 2007.

Randall, David. *Atmosphere, Clouds, and Climate.* Princeton University Press, Princeton, New Jersey, 2012.

Righter, Robert W. *Wind Energy in America: A History,* University of Oklahoma Press, Norman, OK, 2008.

*Rogers, R. R. *A Short Course in Cloud Physics (3rd ed.),* Pergamon Press, Oxford, England, 1989.

Ruddiman, William. *Earth's Climate: Past and Future* (3rd ed.). William Freeman, New York, 2013.

Schultz, David H. *Eloquent Science: A Practical Guide to Becoming a Better Writer, Speaker, and Atmospheric Scientist,* American Meteorological Society, Boston, MA, 2009.

Sheffield, Justin, and Eric F. Wood. *Drought: Past Problems and Future Scenarios.* Routledge, London, England. 2011.

Simmons, Kevin M., and Daniel Sutter. *Economic and Societal Impacts of Tornadoes,* American Meteorological Society, Boston, MA, 2011.

———. *Deadly Season: Analyzing the 2011 Tornado Outbreaks,* American Meteorological Society, Boston, MA, 2012.

Somerville, Richard C. *The Forgiving Air* (2nd ed.), American Meteorological Society, Boston, MA, 2008.

*Strangeways, Ian. *Precipitation: Theory, Measurement and Distribution.* Cambridge University Press, London, England, 2011.

*Stull, Roland B. *Practical Meteorology: An Algebra-based Survey of Atmospheric Science.* University of British Columbia, Vancouver, 2015.

Vallis, Geoffrey K., *Climate and the Oceans.* Princeton University Press, Princeton, New Jersey, 2011.

Vasquez, Tim. *Weather Forecasting and Analysis Handbook.* Weather Graphics Technologies, Austin, TX, 2015.

*Wallace, John M. and Peter V. Hobbs. *Atmospheric Science: An Introductory Survey* (2nd ed.), Academic Press, Burlington, MA, 2006.

Williams, Jack. *The AMS Weather Book: The Ultimate Guide to America's Weather,* American Meteorological Society, Boston, MA, 2009.

GLOSSARY

A

Absolute humidity   The mass of water vapor in a given volume of air. It represents the density of water vapor in the air.

Absolute zero   A temperature reading of 273°C, 460°F, or 0 K. Theoretically, there is no molecular motion at this temperature.

Absolutely stable atmosphere   An atmospheric condition that exists when the environmental lapse rate is less than the moist adiabatic rate. This results in a lifted parcel of air being colder than the air around it.

Absolutely unstable atmosphere   An atmospheric condition that exists when the environmental lapse rate is greater than the dry adiabatic rate. This results in a lifted parcel of air being warmer than the air around it.

Accretion   The growth of a precipitation particle by the collision of an ice crystal or snowflake with a supercooled liquid droplet that freezes upon impact.

Acid deposition   The depositing of acidic particles (usually sulfuric acid and nitric acid) at Earth's surface. Acid deposition occurs in dry form (*dry deposition*) or wet form (*wet deposition*). Acid rain and acid precipitation often denote wet deposition. (*See* Acid rain.)

Acid fog   *See* Acid rain.

Acid rain   Cloud droplets or raindrops combining with gaseous pollutants, such as oxides of sulfur and nitrogen, to make falling rain (or snow) more strongly acidic than usual—typically with a pH less than 5.0. If fog droplets combine with such pollutants, it becomes *acid fog*.

Actual vapor Pressure   *See* Vapor pressure.

Adiabatic process   A process that takes place without a transfer of heat between the system (such as an air parcel) and its surroundings. In an adiabatic process, compression always results in warming, and expansion results in cooling.

Advection   The horizontal transfer of any atmospheric property by the wind.

Advection fog   Occurs when warm, moist air moves over a cold surface and the air cools to below its dew point.

Advection-radiation fog   Fog that forms as relatively warm moist air moves over a colder surface that cooled mainly by radiational cooling.

Aerosols   Tiny suspended solid particles (dust, smoke, etc.) or liquid droplets that enter the atmosphere from either natural or human (anthropogenic) sources, such as the burning of fossil fuels. Sulfur-containing fossil fuels, such as coal, produce *sulfate aerosols*.

Aerovane   A wind instrument that indicates or records both wind speed and wind direction. Also called a *skyvane*.

Aggregation   The clustering together of ice crystals to form snowflakes.

Air density   *See* Density.

Air mass   A large body of air that has similar horizontal temperature and moisture characteristics.

Air-mass thunderstorm   *See* Ordinary thunderstorm.

Air-mass weather   A persistent type of weather that may last for several days (up to a week or more). It occurs when an area comes under the influence of a particular air mass.

Air parcel   *See* Parcel of air.

Air pollutants   Solid, liquid, or gaseous airborne substances that occur in concentrations high enough to threaten the health of people and animals, to harm vegetation and structures, or to toxify a given environment.

Air pressure (atmospheric pressure)   The pressure exerted by the mass of air above a given point, usually expressed in millibars (mb), inches of mercury (Hg) or in hectopascals (hPa).

Air Quality Index (AQI)   An index of air quality that provides daily air pollution concentrations. Intervals on the scale relate to potential health effects.

Albedo   The percent of radiation returning from a surface compared to that which strikes it.

Aleutian low   The subpolar low-pressure area that is centered near the Aleutian Islands on charts that show mean sea-level pressure.

Altimeter   An instrument that indicates the altitude of an object above a fixed level. Pressure altimeters use an aneroid barometer with a scale graduated in altitude instead of pressure.

Altocumulus   A middle cloud, usually white or gray. Often occurs in layers or patches with wavy, rounded masses or rolls.

Altostratus   A middle cloud composed of gray or bluish sheets or layers of uniform appearance. In the thinner regions, the sun or moon usually appears dimly visible.

Analogue forecasting method   A forecast made by comparison of past large-scale synoptic weather patterns that resemble a given (usually current) situation in its essential characteristics.

Analysis   The drawing and interpretation of the patterns of various weather elements on a surface or upper-air chart.

Anemometer   An instrument designed to measure wind speed.

Aneroid barometer   An instrument designed to measure atmospheric pressure. It contains no liquid.

Annual range of temperature   The difference between the warmest and coldest months at any given location.

Anticrepuscular rays   Bands of light extending across the sky that appear to converge toward the horizon opposite from the sun.

Anticyclone   An area of high atmospheric pressure around which the wind blows clockwise in the Northern Hemisphere and counterclockwise in the Southern Hemisphere. Also called a *high*.

Apparent temperature   What the air temperature "feels like" for various combinations of air temperature and relative humidity.

Arctic front   In northern latitudes, the semi-permanent boundary that separates very cold, dense arctic air from the less-cold and less-dense polar air.

Arctic Oscillation (AO)   A reversal of atmospheric pressure over the Arctic that produces changes in the upper-level westerly winds over northern latitudes. These changes in upper-level winds influence winter weather patterns over North America, Greenland, and Europe.

Arid climate   An extremely dry climate—drier than the semi-arid climate. Often referred to as a "true desert" climate.

ASOS   Acronym for *Automated Surface Observing Systems*. A system designed to provide continuous information of wind, temperature, pressure, cloud base height, and runway visibility at selected airports.

Atmosphere   The envelope of gases that surround a planet and are held to it by the planet's gravitational attraction. Earth's atmosphere is mainly nitrogen and oxygen.

Atmospheric boundary layer   The layer of air from Earth's surface usually up to about 1 km (3300 ft) where the wind is influenced by friction of Earth's surface and objects on it. Also called the *planetary boundary layer* and the *friction layer*.

Atmospheric greenhouse effect   The warming of an atmosphere by its absorbing and emitting infrared radiation while allowing shortwave radiation to pass on through. The gases mainly responsible for Earth's atmospheric greenhouse effect are water vapor and carbon dioxide. Also called the *greenhouse effect*.

Atmospheric models   Simulation of the atmosphere's behavior by mathematical equations or by physical models.

Atmospheric river   A region of upper-level flow that transports large amounts of moisture, typically from the tropics and subtropics into the midlatitudes.

Atmospheric stagnation   A condition of light winds and poor vertical mixing that can lead to a high concentration of pollutants. Air stagnations are most often associated with fair weather, an inversion, and the sinking air of a high-pressure area.

Atmospheric window   The wavelength range between 8 and 11 m in which little absorption of infrared radiation takes place.

Aurora   Glowing light display in the nighttime sky caused by excited gases in the upper atmosphere giving off light. In the Northern Hemisphere it is called the *aurora borealis* (northern lights); in the Southern Hemisphere, the *aurora australis* (southern lights).

Autumnal equinox   The equinox at which the sun approaches the Southern Hemisphere and passes directly over the equator. Occurs around September 23.

AWIPS   Acronym for *Advanced Weather Interactive Processing System*. New computerized system that integrates and processes data received at a weather forecasting office from NEXRAD, ASOS, and analysis and guidance products prepared by NMC.

B

Back-door cold front   A cold front moving south or southwest along the Atlantic seaboard of the United States.

Backing wind   A wind that changes direction in a counterclockwise sense (e.g., north to northwest to west).

Ball lightning   A rare form of lightning that may consist of a reddish, luminous ball of electricity or charged air.

Barograph   A recording barometer.

Barometer   An instrument that measures atmospheric pressure. The two most common barometers are the *mercury barometer* and the *aneroid barometer*.

Bergeron process   *See* Ice-crystal process.

Bermuda high   *See* Subtropical high.

Billow clouds   Broad, nearly parallel lines of wavelike clouds oriented at right angles to the wind.

Bimetallic thermometer   A temperature-measuring device usually consisting of two dissimilar metals that expand and contract differentially as the temperature changes.

Blackbody   A hypothetical object that absorbs all of the radiation that strikes it. It also emits radiation at a maximum rate for its given temperature.

Black ice   A thin sheet of ice that appears relatively dark and may form as supercooled droplets, drizzle, or light rain come in contact with a road surface that is below freezing. Also, thin dark-appearing ice that forms on freshwater or saltwater ponds, or lakes.

Blizzard   A severe weather condition characterized by low temperatures and strong winds (greater than 35 mi/hr) bearing a great amount of snow either falling or blowing. When these conditions continue after the falling snow has ended, it is termed a *ground blizzard*.

Boulder winds   Fast-flowing, local downslope winds that may attain speeds of 100 knots or more. They are especially strong along the eastern foothills of the Rocky Mountains near Boulder, Colorado.

Boundary layer   *See* Atmospheric boundary layer.

Bow echo   A line of thunderstorms on a radar screen that appears in the shape of a bow. Bow echoes are often associated with damaging straight-line winds and small tornadoes.

Brocken bow   A bright ring of light seen around the shadow of an observer's head as the observer peers into a cloud or fog bank. Formed by *diffraction* of light.

Buoyant force (buoyancy)   The upward force exerted upon an air parcel (or any object) by virtue of the density (mainly temperature) difference between the parcel and that of the surrounding air.

Buys-Ballot's law   A law describing the relationship between the wind direction and the pressure distribution. In the Northern Hemisphere, if you stand with your back to the surface wind, then turn clockwise about 30°, lower pressure will be to your left. In the Southern Hemisphere, stand with your back to the surface wind, then turn counterclockwise about 30°; lower pressure will be to your right.

C

California current   The ocean current that flows southward along the west coast of the United States from about Washington to Baja, California.

Cap cloud   *See* Pileus cloud.

CAPE   *See* Convective Available Potential Energy.

Carbon dioxide (CO_2)   A colorless, odorless gas whose concentration is about 0.039 percent (390 ppm) in a volume of air near sea level. It is a selective absorber of infrared radiation and, consequently, it is important in Earth's atmospheric greenhouse effect. Solid CO_2 is called *dry ice*.

Carbon monoxide (CO)   A colorless, odorless, toxic gas that forms during the incomplete combustion of carbon-containing fuels.

Celsius scale   A temperature scale where zero is assigned to the temperature where water freezes and 100 to the temperature where water boils (at sea level).

Chaos   The property describing a system that exhibits erratic behavior in that very small changes in the initial state of the system rapidly lead to large and apparently unpredictable changes sometime in the future.

Chemical weathering-CO_2 feedback   A negative feedback in earth's climate system. As chemical weathering of rocks increases, CO_2 is removed from the atmosphere more quickly, which in turn weakens the greenhouse effect, causing the atmosphere to cool. Thus, an increase in global chemical weathering tends to lower atmospheric temperatures.

Chinook wall cloud   A bank of clouds over the Rocky Mountains that signifies the approach of a chinook.

Chinook wind   A warm, dry wind on the eastern side of the Rocky Mountains. In the Alps, the wind is called a *Foehn*.

Chlorofluorocarbons (CFCs)   Compounds consisting of methane (CH_4) or ethane (C_2H_6) with some or all of the hydrogen replaced by chlorine or fluorine. Originally used in fire extinguishers, as refrigerants, as solvents for cleaning electronic microcircuits, and as propellants. CFCs contribute to the atmospheric greenhouse effect and destroy ozone in the stratosphere. Their use has been virtually eliminated through the Montreal Protocol.

Cirrocumulus   A high cloud that appears as a white patch of clouds without shadows. It consists of very small elements in the form of grains or ripples.

Cirrostratus   High, thin, sheetlike clouds, composed of ice crystals. They frequently cover the entire sky and often produce a halo.

Cirrus   A high cloud composed of ice crystals in the form of thin, white, featherlike clouds in patches, filaments, or narrow bands.

Clear air turbulence (CAT)   Turbulence encountered by aircraft flying through cloudless skies. Thermals, wind shear, and jet streams can each be a factor in producing CAT.

Clear ice   A layer of ice that appears transparent because of its homogeneous structure and small number and size of air pockets.

Climate The accumulation of daily and seasonal weather events over a long period of time.

Climate change A change in the long-term statistical average of weather elements—such as temperature or precipitation—sustained over several decades or longer. Climate change is also called *climatic change.*

Climatic controls The relatively permanent factors that govern the general nature of the climate of a region.

Climatic optimum *See* Mid-Holocene maximum.

Climatological forecast A weather forecast, usually a month or more in the future, which is based upon the climate of a region rather than upon current weather conditions.

Cloud A visible aggregate of tiny water droplets and/or ice crystals in the atmosphere above Earth's surface.

Cloudburst Any sudden and heavy rain shower.

Cloud seeding The introduction of artificial substances (usually silver iodide or dry ice) into a cloud for the purpose of either modifying its development or increasing its precipitation.

Coalescence The merging of cloud droplets into a single larger droplet.

Cold advection (*cold air advection*) The transport of cold air by the wind from a region of lower temperatures to a region of higher temperature.

Cold fog *See* Supercooled cloud.

Cold front A transition zone where a cold air mass advances and replaces a warm air mass.

Cold occlusion *See* Occluded front.

Cold wave A rapid fall in temperature within 24 hours that often requires increased protection for agriculture, industry, commerce, and human activities.

Collision-coalescence process The process of producing precipitation by liquid particles (cloud droplets and raindrops) colliding and joining (coalescing).

Comma cloud A band of organized cumuliform clouds that looks like a comma on a satellite photograph.

Computer enhancement A process where the temperatures of radiating surfaces are assigned different shades of gray (or different colors) on an infrared picture. This allows specific features to be more clearly delineated.

Condensation The process by which water vapor becomes a liquid.

Condensation level The level above the surface marking the base of a cumuliform cloud.

Condensation nuclei Also called *cloud condensation nuclei.* Tiny particles upon whose surfaces condensation of water vapor begins in the atmosphere.

Conditionally unstable atmosphere An atmospheric condition that exists when the environmental lapse rate is less than the dry adiabatic rate but greater than the moist adiabatic rate. Also called *conditional instability.*

Conduction The transfer of heat by molecular activity from one substance to another, or through a substance. Transfer is always from warmer to colder regions.

Constant-height chart (constant-level chart) A chart showing variables, such as pressure, temperature, and wind, at a specific altitude above sea level. Variation in horizontal pressure is depicted by isobars. The most common constant-height chart is the surface chart, which is also called the *sea-level chart* or *surface weather map.*

Constant-pressure chart (isobaric chart) A chart showing variables, such as temperature and wind, on a constant-pressure surface. Variations in height are usually shown by lines of equal height (contour lines).

Contact freezing The process by which contact with a nucleus such as an ice crystal causes supercooled liquid droplets to change into ice.

Continental arctic air mass An air mass characterized by extremely low temperatures and very dry air.

Continental polar air mass An air mass characterized by low temperatures and dry air. Not as cold as arctic air masses.

Continental tropical air mass An air mass characterized by high temperatures and low humidity.

Contour line A line that connects points of equal elevation above a reference level, most often sea level.

Contrail (condensation trail) A cloudlike streamer frequently seen forming behind aircraft flying in clear, cold, humid air.

Controls of temperature The main factors that cause variations in temperature from one place to another.

Convection Motions in a fluid that result in the transport and mixing of the fluid's properties. In meteorology, convection usually refers to atmospheric motions that are predominantly vertical, such as rising air currents due to surface heating. The rising of heated surface air and the sinking of cooler air aloft is often called *free convection.* (Compare with *forced convection.*)

Convective Available Potential Energy (*CAPE*) The maximum energy available to a rising air parcel. it is a measure of how fast the air parcel will rise inside a cumuliform cloud.

Convergence An atmospheric condition that exists when the winds cause a horizontal net inflow of air into a specified region.

Cooling degree day A form of degree day used in estimating the amount of energy necessary to reduce the effective temperature of warm air. A cooling degree-day is a day on which the average temperature is one degree above a desired base temperature.

Coriolis force An apparent force observed on any free-moving object in a rotating system. On Earth, this deflective force results from Earth's rotation and causes moving particles (including the wind) to deflect to the right in the Northern Hemisphere and to the left in the Southern Hemisphere.

Corona (optic) A series of colored rings concentrically surrounding the disk of the sun or moon. Smaller than the halo, the corona is often caused by the diffraction of light around small water droplets of uniform size.

Country breeze A light breeze that blows into a city from the surrounding countryside. It is best observed on clear nights when the urban heat island is most pronounced.

Crepuscular rays Alternating light and dark bands of light that appear to fan out from the sun's position, usually at twilight.

Cumulonimbus An exceptionally dense and vertically developed cloud, often with a top in the shape of an anvil. The cloud is frequently accompanied by heavy showers, lightning, thunder, and sometimes hail. It is also known as a *thunderstorm cloud.*

Cumulus A cloud in the form of individual, detached domes or towers that are usually dense and well defined. It has a flat base with a bulging upper part that often resembles cauliflower. Cumulus clouds of fair weather are called *cumulus humilis.* Those that exhibit much vertical growth are called *cumulus congestus* or *towering cumulus.*

Cumulus stage The initial stage in the development of an ordinary cell thunderstorm in which rising, warm, humid air develops into a cumulus cloud.

Cut-off low A cold upper-level low that has become displaced out of the basic westerly flow and lies to the south of this flow.

Cyclogenesis The development or strengthening of middle-latitude (extratropical) cyclones.

Cyclone An area of low pressure around which the winds blow counterclockwise in the Northern Hemisphere and clockwise in the Southern Hemisphere.

D

Daily range of temperature The difference between the maximum and minimum temperatures for any given day.

Dart leader The discharge of electrons that proceeds intermittently toward the ground along the same ionized channel taken by the initial lightning stroke.

Dendrochronology The analysis of the annual growth rings of trees as a means of interpreting past climatic conditions.

Density The ratio of the mass of a substance to the volume occupied by it. Air density is usually expressed as g/cm^3 or kg/m^3.

Deposition A process that occurs in subfreezing air when water vapor changes directly to ice without becoming a liquid first.

Derecho Strong, damaging, straight-line winds associated with a cluster of severe thunderstorms that most often form in the evening or at night.

Desertification A general increase in the desert conditions of a region.

Dew Water that has condensed onto objects near the ground when their temperatures have fallen below the dew point of the surface air.

Dew cell An instrument used to determine the dew-point temperature.

Dew point (dew-point temperature) The temperature to which air must be cooled (at constant pressure and constant water vapor content) for saturation to occur.

Dew-point hygrometer An instrument that determines the dew point temperature of the air.

Diffraction The bending of light around objects, such as cloud and fog droplets, producing fringes of light and dark or colored bands.

Dispersion (of light) The separation of white light into its different component wavelengths.

Dispersion (of pollution) The spreading out of atmospheric pollutants.

Dissipating stage The final stage in the development of an ordinary cell thunderstorm when downdrafts exist throughout the cumulonimbus cloud.

Divergence An atmospheric condition that exists when the winds cause a horizontal net outflow of air from a specific region.

Dixie Alley Region in the southern United States, typically over Mississippi and Alabama, where tornadoes often form.

Doldrums The region near the equator that is characterized by low pressure and light, shifting winds.

Doppler lidar The use of light beams to determine the velocity of objects such as dust and falling rain by taking into account the *Doppler shift*.

Doppler radar A radar that determines the velocity of falling precipitation either toward or away from the radar unit by taking into account the *Doppler shift*.

Doppler shift (effect) The change in the frequency of waves that occurs when the emitter or the observer is moving toward or away from the other.

Downburst A severe localized downdraft that can be experienced beneath a severe thunderstorm. (Compare *microburst* and *macroburst*.)

Drizzle Small water drops between 0.2 and 0.5 mm in diameter that fall slowly and reduce visibility more than light rain.

Drought A period of abnormally dry weather sufficiently long enough to cause serious effects on agriculture and other activities in the affected area.

Dry adiabatic rate The rate of change of temperature in a rising or descending unsaturated air parcel. The rate of adiabatic cooling or warming is about 10°C per 1000 m (5.5°F per 1000 ft).

Dry-bulb temperature The air temperature measured by the dry-bulb thermometer of a psychrometer.

Dry climate A climate deficient in precipitation where annual potential evaporation and transpiration exceed precipitation.

Dry haze *See* Haze.

Dry lightning Lightning that occurs with thunderstorms that produce little, if any, appreciable precipitation that reaches the surface.

Dryline A boundary that separates warm, dry air from warm, moist air. It usually represents a zone of instability along which thunderstorms form.

Dry-summer subtropical climate A climate characterized by mild, wet winters and warm to hot, dry summers. Typically located between 30 and 45 degrees latitude on the western side of continents. Also called *Mediterranean climate*.

Dust devil (whirlwind) A small but rapidly rotating wind made visible by the dust, sand, and debris it picks up from the surface. It develops best on clear, dry, hot afternoons.

E

Easterly wave A migratory wavelike disturbance in the tropical easterlies. Easterly waves occasionally intensify into tropical cyclones. They are also called *tropical waves*.

Eccentricity (of Earth's orbit) The deviation of Earth's orbit from elliptical to nearly circular.

Eddy A small volume of air (or any fluid) that behaves differently from the larger flow in which it exists.

Ekman spiral An idealized description of the way the wind-driven ocean currents vary with depth. In the atmosphere it represents the way the winds vary from the surface up through the friction layer or planetary boundary layer.

Ekman transport Net surface water transport due to the Ekman spiral. In the Northern Hemisphere the transport is 90° to the right of the surface wind direction.

Electrical hygrometer *See* Hygrometer.

Electrical thermometers Thermometers that use elements that convert energy from one form to another (transducers). Common electrical thermometers include the electrical resistance thermometer, thermocouple, and thermistor.

Electromagnetic waves *See* Radiant energy.

El Niño An extensive ocean warming that typically extends westward from the coast of Peru and Ecuador across the eastern tropical Pacific Ocean, with associated atmospheric conditions. El Niño events occur roughly once every 2 to 7 years. (*See also* ENSO.)

Energy The property of a system that generally enables it to do work. Some forms of energy are kinetic, radiant, potential, chemical, electric, and magnetic.

Enhanced Fujita (EF) scale A modification of the original *Fujita Scale* that describes tornado intensity by observing damage caused by the tornado.

Ensemble forecasting A forecasting technique that entails running several forecast models each beginning with slightly different weather information. The forecaster's level of confidence is based on how well the models agree (or disagree) at the end of some specified time.

ENSO (El Niño/Southern Oscillation)   A condition in the tropical Pacific whereby the reversal of surface air pressure at opposite ends of the Pacific Ocean induces westerly winds, a strengthening of the equatorial countercurrent, and extensive ocean warming.

Entrainment   The mixing of environmental air into a pre-existing air current or cloud so that the environmental air becomes part of the current or cloud.

Environmental lapse rate   The rate of decrease of air temperature with elevation. It is most often measured with a radiosonde.

Evaporation   The process by which a liquid changes into a gas.

Evaporation (mixing) fog   Fog produced when sufficient water vapor is added to the air by evaporation, and the moist air mixes with relatively drier air. The two common types are *steam fog*, which forms when cold air moves over warm water, and *frontal fog*, which forms as warm raindrops evaporate in a cool air mass.

Exosphere   The outermost portion of the atmosphere.

Extratropical cyclone   A cyclonic storm that most often forms along a front in middle and high latitudes. Also called a *middle-latitude cyclonic storm*, a *depression*, and a *low*. It is not a tropical storm or hurricane.

Eye (of hurricane)   A region in the center of a hurricane (tropical cyclone) where the winds are light and skies are clear to partly cloudy.

Eyewall   A wall of dense thunderstorms that surrounds the eye of a hurricane.

Eyewall replacement   A situation within a hurricane (tropical cyclone) where the storm's original eyewall dissipates and a new eyewall forms outward, farther away from the center of the storm.

F

Fahrenheit scale   A temperature scale where 32 is assigned to the temperature where water freezes and 212 to the temperature at which water boils (at sea level).

Fall streaks   Falling ice crystals that evaporate before reaching the ground.

Fall wind   A strong, cold katabatic wind that blows downslope off snow-covered plateaus.

Fata Morgana   A complex mirage that is characterized by objects being distorted in such a way as to appear as castlelike features.

Feedback mechanism   A process whereby an initial change in an atmospheric process will tend to either reinforce the process (*positive feedback*) or weaken the process (*negative feedback*).

Flash flood   A flood that rises and falls quite rapidly with little or no advance warning, usually as the result of intense rainfall over a relatively small area.

Foehn   *See* Chinook wind.

Fog   A cloud with its base at Earth's surface.

Forced convection   On a small scale, a form of mechanical stirring taking place when twisting eddies of air are able to mix hot surface air with the cooler air above. On a larger scale, it can be induced by the lifting of warm air along a front (*frontal uplift*) or along a topo graphic barrier (*orographic uplift*).

Forecast funnel   A sequence of steps used by forecasters to analyze current and projected conditions, moving from larger to smaller scales during the process.

Free convection   *See* Convection.

Freeze   A condition occurring over a widespread area when the surface air temperature remains below freezing for a sufficient time to damage certain agricultural crops. A freeze most often occurs as

cold air is advected into a region, causing freezing conditions to exist in a deep layer of surface air. Also called *advection frost*.

Freezing rain and freezing drizzle   Rain or drizzle that falls in liquid form and then freezes upon striking a cold object or ground. Both can produce a coating of ice on objects, which is called *glaze*.

Friction layer   The atmospheric layer near the surface usually extending up to about 1 km (3300 ft) where the wind is influenced by friction of Earth's surface and objects on it. Also called the *atmospheric boundary layer* and *planetary boundary layer*.

Front   The transition zone between two distinct air masses.

Frontal fog   *See* Evaporation fog.

Frontal thunderstorms   Thunderstorms that form in response to forced convection (forced lifting) along a front. Most go through a cycle similar to those of ordinary thunderstorms.

Frontal wave   A wavelike deformation along a front in the lower levels of the atmosphere. Those that develop into storms are termed *unstable waves*, while those that do not are called *stable waves*.

Frost (also called hoarfrost)   A covering of ice produced by deposition on exposed surfaces when the air temperature falls below the frost point.

Frostbite   The partial freezing of exposed parts of the body, causing injury to the skin and sometimes to deeper tissues.

Frost point   The temperature at which the air becomes saturated with respect to ice when cooled at constant pressure and constant water vapor content.

Frozen dew   The transformation of liquid dew into tiny beads of ice when the air temperature drops below freezing.

Fujita Scale   A scale developed by T. Theodore Fujita for classifying tornadoes according to the damage they cause and their rotational wind speed. (*See also* Enhanced Fujita Scale.)

Funnel cloud   A funnel-shaped cloud of condensed water, usually extending from the base of a cumuliform cloud. The rapidly rotating air of the funnel is not in contact with the ground; hence, it is not a tornado.

G

Gas law   The thermodynamic law applied to a perfect gas that relates the pressure of the gas to its density and absolute temperature.

General circulation of the atmosphere   Large-scale atmospheric motions over the entire earth.

Geoengineering   The use of global scale technology fixes to mitigate climate changes.

Geostationary satellite   A satellite that orbits Earth at the same rate that Earth rotates and thus remains over a fixed place above the equator.

Geostrophic wind   A theoretical horizontal wind blowing in a straight path, parallel to the isobars or contours, at a constant speed. The geostrophic wind results when the Coriolis force exactly balances the horizontal pressure gradient force.

Glaciated cloud   A cloud or portion of a cloud where only ice crystals exist.

Global climate   Climate of the entire globe.

Global scale   The largest scale of atmospheric motion. Also called the *planetary scale*.

Global warming   Increasing global surface air temperatures that show up in the climate record. The term *global warming* is usually attributed to human activities, such as increasing concentrations of greenhouse gases from automobiles and industrial processes, for example.

Glory   Colored rings that appear around the shadow of an object.

Gradient wind   A theoretical wind that blows parallel to curved isobars or contours.

Graupel   Ice particles between 2 and 5 mm in diameter that form in a cloud often by the process of accretion. Snowflakes that become rounded pellets due to riming are called *graupel* or *snow pellets*.

Green flash   A small green color that occasionally appears on the upper part of the sun as it rises or sets.

Greenhouse effect   *See* Atmospheric greenhouse effect.

Greenhouse gases   Gases in Earth's atmosphere, such as water vapor and carbon dioxide, that allow much of the sunlight to pass through but are strong absorbers of infrared energy emitted by Earth and the atmosphere. Other greenhouse gases include methane, nitrous oxide, fluorocarbons, and ozone.

Ground fog   *See* Radiation fog.

Growing degree day   A form of the degree day used as a guide for crop planting and for estimating crop maturity dates.

Gulf Stream   A warm, swift, narrow ocean current flowing along the east coast of the United States.

Gust front   A boundary that separates a cold downdraft of a thunderstorm from warm, humid surface air. On the surface its passage resembles that of a cold front.

Gustnado   A relatively weak tornado associated with a thunderstorm's outflow. It most often forms along the gust front.

H

Haboob   A dust or sandstorm that forms as cold downdrafts from a thunderstorm turbulently lift dust and sand into the air.

Hadley cell   A thermal circulation proposed by George Hadley to explain the movement of the trade winds. It consists of rising air near the equator and sinking air near 30° latitude.

Hailstones   Transparent or partially opaque particles of ice that range in size from that of a pea to that of golf balls.

Hailstreak   The accumulation of hail at Earth's surface along a relatively long (10 km), narrow (2 km) band.

Hair hygrometer   *See* Hygrometer.

Halos   Rings or arcs that encircle the sun or moon when seen through an ice crystal cloud or a sky filled with falling ice crystals. Halos are produced by refraction of light.

Haze   Fine dry or wet dust or salt particles dispersed through a portion of the atmosphere. Individually these are not visible but cumulatively they will diminish visibility. *Dry haze* particles are very small, on the order of 0.1 m. *Wet haze* particles are larger.

Heat   A form of energy transferred between systems by virtue of their temperature differences.

Heat burst   A sudden increase in surface air temperature often accompanied by extreme drying. A heat burst is associated with the downdraft of a thunderstorm, or a cluster of thunderstorms.

Heat capacity   The ratio of the heat absorbed (or released) by a system to the corresponding temperature rise (or fall).

Heat Index (HI)   An index that combines air temperature and relative humidity to determine an apparent temperature—how hot it actually feels.

Heating degree day   A form of the degree day used as an index for fuel consumption.

Heat lightning   Distant lightning that illuminates the sky but is too far away for its thunder to be heard.

Heat stroke   A physical condition induced by a person's over-exposure to high air temperatures, especially when accompanied by high humidity.

Hectopascal   Abbreviated hPa. One hectopascal is equal to 100 newtons/m², or 1 millibar.

Heiligenschein   A faint white ring surrounding the shadow of an observer's head on a dew-covered lawn.

Heterosphere   The region of the atmosphere above about 85 km where the composition of the air varies with height.

High   *See* Anticyclone.

High inversion fog   A fog that lifts above the surface but does not completely dissipate because of a strong inversion (usually subsidence) that exists above the fog layer.

Highland climate   The climate of high elevations. Also called *mountain climate*. There is no single climatic type but a variety of different climate zones often characterized by a rapid change in temperature and precipitation as one ascends or descends in elevation.

Homosphere   The region of the atmosphere below about 85 km where the composition of the air remains fairly constant.

Hook echo   The shape of a hook on a Doppler radar screen that indicates the possible presence of a tornado.

Horse latitudes   The belt of latitude at about 30° to 35° where winds are predominantly light and the weather is hot and dry.

Humid continental climate   A climate characterized by severe winters and mild to warm summers with adequate annual precipitation. Typically located over large continental areas in the Northern Hemisphere between about 40° and 70° latitude.

Humidity   A general term that refers to the air's water vapor content. (*See* Relative humidity.)

Humid subtropical climate   A climate characterized by hot muggy summers, cool to cold winters, and abundant precipitation throughout the year.

Hurricane   A tropical cyclone with sustained winds of at least 64 knots (74 mi/hr).

Hurricane hunters   A popular term for aircraft and/or personnel engaged in the reconnaissance of hurricanes (tropical cyclones).

Hurricane warning   A warning given when it is likely that a hurricane will strike an area within 24 hours.

Hurricane watch   A hurricane watch indicates that a hurricane poses a threat to an area (often within several days) and residents of the watch area should be prepared.

Hydrocarbons   Chemical compounds composed of only hydrogen and carbon—they are included under the general term volatile organic compounds (VOCs).

Hydrologic cycle   A model that illustrates the movement and exchange of water among Earth, atmosphere, and oceans.

Hydrostatic equation   An equation that states that the rate at which the air pressure decreases with height is equal to the air density times the acceleration of gravity. The equation relates to how quickly the air pressure decreases in a column of air.

Hydrostatic equilibrium   The state of the atmosphere when there is a balance between the vertical pressure gradient force and the downward pull of gravity.

Hygrometer   An instrument designed to measure the air's water vapor content. The sensing part of the instrument can be hair (*hair hygrometer*), a plate coated with carbon (*electrical hygrometer*), or an infrared sensor (*infrared hygrometer*).

Hygroscopic   The ability to accelerate the condensation of water vapor. Usually used to describe condensation nuclei that have an affinity for water vapor.

Hypothermia   The deterioration in one's mental and physical condition brought on by a rapid lowering of human body temperature.

Hypoxia   A condition experienced by humans when the brain does not receive sufficient oxygen.

I

Ice Age *See* Pleistocene epoch.

Ice-crystal (Bergeron) process A process that produces precipitation. The process involves tiny ice crystals in a supercooled cloud growing larger at the expense of the surrounding liquid droplets. Also called the *Bergeron process*.

Ice fog A type of fog that forms at very low temperatures, composed of tiny suspended ice particles.

Icelandic low The subpolar low-pressure area that is centered near Iceland on charts that show mean sea-level pressure.

Ice nuclei Particles that act as nuclei for the formation of ice crystals in the atmosphere.

Ice pellets *See* Sleet.

Ice storm A winter storm characterized by a substantial amount of precipitation in the form of freezing rain, freezing drizzle, or sleet.

Indian summer An unseasonably warm spell with clear skies near the middle of autumn. Usually follows a substantial period of cool weather.

Inferior mirage *See* Mirage.

Infrared radiation Electromagnetic radiation with wavelengths between about 0.7 and 1000 m. This radiation is longer than visible radiation but shorter than microwave radiation.

Infrared radiometer An instrument designed to measure the intensity of infrared radiation emitted by an object. Also called *infrared sensor*.

Insolation The *in*coming *sol*ar radi*ation* that reaches Earth and the atmosphere.

Instrument shelter A boxlike (often wooden) structure designed to protect weather instruments from direct sunshine and precipitation.

Interglacial period A time interval of relatively mild climate during the Ice Age when continental ice sheets were absent or limited in extent to Greenland and the Antarctic.

Intertropical Convergence Zone (ITCZ) The boundary zone separating the northeast trade winds of the Northern Hemisphere from the southeast trade winds of the Southern Hemisphere.

Inversion An increase in air temperature with height.

Ion An electrically charged atom, molecule, or particle.

Ionosphere An electrified region of the upper atmosphere where fairly large concentrations of ions and free electrons exist.

Iridescence Brilliant spots or borders of colors, most often red and green, observed in clouds up to about 30° from the sun.

Isobar A line connecting points of equal pressure.

Isobaric chart (map) *See* Constant-pressure chart.

Isobaric surface A surface along which the atmospheric pressure is everywhere equal.

Isotach A line connecting points of equal wind speed.

Isotherm A line connecting points of equal temperature.

Isothermal layer A layer where the air temperature is constant with increasing altitude. In an isothermal layer, the air temperature lapse rate is zero.

J

Jet maximum *See* Jet streak.

Jet streak A region of high wind speed that moves through the axis of a jet stream. Also called *jet maximum*.

Jet stream Relatively strong winds concentrated within a narrow band in the atmosphere.

K

Katabatic (fall) wind Any wind blowing downslope. It is usually cold.

Kelvin A unit of temperature. A Kelvin is denoted by K and 1 K equals 1°C. Zero Kelvin is absolute zero, or 273.15°C.

Kelvin scale A temperature scale with zero degrees equal to the theoretical temperature at which all molecular motion ceases. Also called the *absolute scale*. The units are sometimes called "degrees Kelvin"; however, the correct SI terminology is "Kelvins," abbreviated K.

Kinetic energy The energy within a body that is a result of its motion.

Kirchhoff's law A law that states: Good absorbers of a given wavelength of radiation are also good emitters of that wavelength.

Knot A unit of speed equal to 1 nautical mile per hour. One knot equals 1.15 mi/hr.

Köppen classification system A system for classifying climates that is based mainly on annual and monthly averages of temperature and precipitation.

L

Lake breeze A wind blowing onshore from the surface of a lake.

Lake-effect snows Localized snowstorms that form on the downwind side of a lake. Such storms are common in late fall and early winter near the Great Lakes as cold, dry air picks up moisture and warmth from the unfrozen bodies of water.

Land breeze A coastal breeze that blows from land to sea, usually at night.

Landspout Relatively weak nonsupercell tornado that originates with a cumuliform cloud in its growth stage and with a cloud that does not contain a mid-level mesocyclone. Its spin originates near the surface. Landspouts often look like waterspouts over land.

La Niña A condition where the surface waters of the central and eastern tropical Pacific Ocean turn cooler than normal, with associated atmospheric conditions.

Lapse rate The rate at which an atmospheric variable (usually temperature) decreases with height. (*See* Environmental lapse rate.)

Latent heat The heat that is either released or absorbed by a unit mass of a substance when it undergoes a change of state, such as during evaporation, condensation, or sublimation.

Laterite A soil formed under tropical conditions where heavy rainfall leaches soluble minerals from the soil. This leaching leaves the soil hard and poor for growing crops.

Lee-side low Storm systems (extratropical cyclones) that form on the downwind (lee) side of a mountain chain. In the United States lee-side lows frequently form on the eastern side of the Rockies and Sierra Nevada mountains.

Lenticular cloud A cloud in the shape of a lens.

Level of free convection The level in the atmosphere at which a lifted air parcel becomes warmer than its surroundings in a conditionally unstable atmosphere.

Lidar An instrument that uses a laser to generate intense pulses that are reflected from atmospheric particles of dust and smoke. Lidars have been used to determine the amount of particles in the atmosphere as well as particle movement that has been converted into wind speed. Lidar means *li*ght *d*etection *an*d *r*anging.

Lightning A visible electrical discharge produced by thunderstorms.

Liquid-in-glass thermometer *See* Thermometer.

Little Ice Age The period from about 1350 to 1850 when average temperatures over Europe were relatively low, and alpine glaciers increased in size and advanced down mountain canyons.

Local winds   Winds that tend to blow over a relatively small area; often due to regional effects, such as mountain barriers, large bodies of water, local pressure differences, and other influences.

Long-range forecast   Generally used to describe a weather forecast that extends beyond about 8.5 days into the future.

Longwave radiation   A term most often used to describe the infrared energy emitted by Earth and the atmosphere.

Longwaves in the westerlies   A wave in the upper level of the westerlies characterized by a long length (thousands of kilometers) and significant amplitude. Also called *Rossby waves*.

Low   *See* Extratropical cyclone.

Low-level jet streams   Jet streams that typically form near Earth's surface below an altitude of about 2 km and usually attain speeds of less than 60 knots.

M

Macroburst   A strong downdraft (*downburst*) greater than 4 km wide that can occur beneath thunderstorms. A downburst less than 4 km across is called a *microburst*.

Macroclimate   The general climate of a large area, such as a country.

Macroscale   The normal meteorological synoptic scale for obtaining weather information. It can cover an area ranging from the size of a continent to the entire globe.

Mammatus clouds   Clouds that look like pouches hanging from the underside of a cloud.

Marine climate   A climate controlled largely by the ocean. The ocean's influence keeps winters relatively mild and summers cool.

Maritime air   Moist air whose characteristics were developed over an extensive body of water.

Maritime polar air mass   An air mass characterized by low temperatures and high humidity.

Maritime tropical air mass   An air mass characterized by high temperatures and high humidity.

Mature thunderstorm   The second stage in the three-stage cycle of an ordinary thunderstorm. This mature stage is characterized by heavy showers, lightning, thunder, and violent vertical motions inside cumulonimbus clouds.

Maunder minimum   A period from about 1645 to 1715 when few, if any, sunspots were observed.

Maximum thermometer   A thermometer with a small constriction just above the bulb. It is designed to measure the maximum air temperature.

Mean annual temperature   The average temperature at any given location for the entire year.

Mean daily temperature   The average of the highest and lowest temperature for a 24-hour period.

Mediterranean climate   *See* Dry-summer subtropical climate.

Medium-range forecast   Generally used to describe a weather forecast that extends from about 3 to 8.5 days into the future.

Mercury barometer   A type of barometer that uses mercury to measure atmospheric pressure. The height of the mercury column is a measure of atmospheric pressure.

Meridional flow   A type of atmospheric circulation pattern in which the north-south component of the wind is pronounced.

Mesoclimate   The climate of an area ranging in size from a few acres to several square kilometers.

Mesocyclone   A vertical column of cyclonically rotating air within a supercell thunderstorm.

Mesohigh   A relatively small area of high atmospheric pressure that forms beneath a thunderstorm.

Mesopause   The top of the mesosphere. The boundary between the mesosphere and the thermosphere, usually near 85 km.

Mesoscale   The scale of meteorological phenomena that range in size from a few km to about 100 km. It includes local winds, thunderstorms, and tornadoes.

Mesoscale convective complex (MCC)   A large organized convective weather system comprised of a number of individual thunderstorms. An MCC can span 1000 times more area than an individual ordinary cell thunderstorm. An MCC is a particular type of mesoscale convective system.

Mesoscale convective system (MCS)   A large cloud system that represents an ensemble of thunderstorms that form by convection, and produce precipitation over a wide area.

Mesosphere   The atmospheric layer between the stratosphere and the thermosphere. Located at an average elevation between 50 and 80 km above Earth's surface.

Meteogram   A chart that shows how one or more weather variables has changed at a station over a given period of time or how the variables are likely to change with time.

Meteorology   The study of the atmosphere and atmospheric phenomena as well as the atmosphere's interaction with Earth's surface, oceans, and life in general.

Microburst   A strong localized downdraft (downburst) less than 4 km wide that occurs beneath thunderstorms. A strong downburst greater than 4 km across is called a *macroburst*.

Microclimate   The climate structure of the air space near the surface of Earth.

Micrometer (m)   A unit of length equal to one-millionth of a meter.

Microscale   The smallest scale of atmospheric motions.

Middle latitudes   The region of the world typically described as being between 30° and 50° latitude.

Middle-latitude cyclone   *See* Extratropical cyclone.

Mid-Holocene maximum   A warm period in geologic history about 5000 to 6000 years ago that favored the development of plants.

Milankovitch theory   A theory suggesting that changes in Earth's orbit have been responsible for variations in solar energy reaching Earth's surface and related climatic changes.

Millibar (mb)   A unit for expressing atmospheric pressure. Sea level pressure is normally close to 1013 mb.

Minimum thermometer   A thermometer designed to measure the minimum air temperature during a desired time period.

Mini-swirls   Small whirling eddies perhaps 30 to 100 m in diameter that form in a region of strong wind shear of a hurricane's eyewall. They are believed to be small tornadoes.

Mirage   A refraction phenomenon that makes an object appear to be displaced from its true position. When an object appears higher than it actually is, it is called a *superior mirage*. When an object appears lower than it actually is, it is an *inferior mirage*.

Mixing depth   The vertical extent of the mixing layer.

Mixing layer   The unstable atmospheric layer that extends from the surface up to the base of an inversion. Within this layer, the air is well stirred.

Mixing ratio   The ratio of the mass of water vapor in a given volume of air to the mass of dry air.

Moist adiabatic rate   The rate of change of temperature in a rising or descending saturated air parcel. The rate of cooling or warming varies but a common value of 6°C per 1000 m (3.3°F per 1000 ft) is used.

Molecule   A collection of atoms held together by chemical forces.

Monsoon A name given to seasonal winds that typically blow from different directions during different times of the year, most often during summer and winter.

Monsoon wind system A wind system that reverses direction between winter and summer. Usually the wind blows from land to sea in winter and from sea to land in summer.

Mountain and valley breeze A local wind system of a mountain valley that blows downhill (*mountain breeze*) at night and uphill (*valley breeze*) during the day.

Multicell thunderstorms Thunderstorms often in a line, each of which may be in a different stage of its life cycle.

N

Nacreous clouds Clouds of unknown composition that have a soft, pearly luster and that form at altitudes about 25 to 30 km above Earth's surface. They are also called *mother-of-pearl clouds*.

Negative feedback mechanism *See* Feedback mechanism.

Neutral stability (neutrally stable atmosphere) An atmospheric condition that exists in dry air when the environmental lapse rate equals the dry adiabatic rate. In saturated air the environmental lapse rate equals the moist adiabatic rate.

NEXRAD An acronym for *Nex*t Generation Weather *Rad*ar. The main component of NEXRAD is the WSR 88-D, Doppler radar.

Nimbostratus A dark, gray cloud characterized by more or less continuously falling precipitation. It is rarely accompanied by lightning, thunder, or hail.

Nitric oxide (NO) A colorless gas produced by natural bacterial action in soil and by combustion processes at high temperatures. In polluted air, nitric oxide can react with ozone and hydrocarbons to form other substances. In this manner, it acts as an agent in the production of photochemical smog.

Nitrogen (N_2) A colorless and odorless gas that occupies about 78 percent of dry air in the lower atmosphere.

Nitrogen dioxide (NO_2) A reddish-brown gas, produced by natural bacterial action in soil and by combustion processes at high temperatures. In the presence of sunlight, it breaks down into nitric oxide and atomic oxygen. In polluted air, nitrogen dioxide acts as an agent in the production of photochemical smog.

Nitrogen oxides (NO_x) Gases produced by natural processes and by combustion processes at high temperatures. In polluted air, nitric oxide (NO) and nitrogen dioxide (NO_2) are the most abundant oxides of nitrogen, and both act as agents for the production of photochemical smog.

Noctilucent clouds Wavy, thin, bluish-white clouds that are best seen at twilight in polar latitudes. They form at altitudes about 80 to 90 km above the surface.

Nocturnal inversion *See* Radiation inversion.

Nonsupercell tornado A tornado that occurs with a cloud that is often in its growing stage, and one that does not contain a mid-level mesocyclone, or wall cloud. Landspouts and gustnadoes are examples of nonsupercell tornadoes.

Nor´easter (Northeaster) A name given to a strong, steady wind from the northeast that is accompanied by rain and inclement weather. It often develops when a storm system moves northeastward along the coast of North America.

North Atlantic Oscillation (NAO) A reversal of atmospheric pressure over the Atlantic Ocean that influences the weather over Europe and over eastern North America.

Northern lights *See* Aurora.

Nowcast Short-term weather forecasts varying from minutes up to a few hours.

Nuclear winter The dark, cold, and gloomy conditions that presumably would be brought on by nuclear war.

Numerical weather prediction (NWP) Forecasting the weather based upon the solutions of mathematical equations by high-speed computers.

O

Obliquity (of Earth's axis) The tilt of Earth's axis. It represents the angle from the perpendicular to the plane of Earth's orbit.

Occluded front (occlusion) A complex frontal system that ideally forms when a cold front overtakes a warm front. When the air behind the front is colder than the air ahead of it, the front is called a *cold occlusion*. When the air behind the front is milder than the air ahead of it, it is called a *warm occlusion*.

Offshore wind A breeze that blows from the land out over the water. Opposite of an onshore wind.

Onshore wind A breeze that blows from the water onto the land. Opposite of an offshore wind.

Open wave The stage of development of a wave cyclone (midlatitude cyclonic storm) where a cold front and a warm front exist, but no occluded front. The center of lowest pressure in the wave is located at the junction of the two fronts.

Orchard heaters Oil heaters placed in orchards that generate heat and promote convective circulations to protect fruit trees from damaging low temperatures. Also called *smudge pots*.

Ordinary cell thunderstorm (also called *air-mass thunderstorm*) A thunderstorm produced by local convection within a conditionally unstable air mass. It often forms in a region of low wind shear and does not reach the intensity of a severe thunderstorm.

Orographic uplift The lifting of air over a topographic barrier. Clouds that form in this lifting process are called *orographic clouds*.

Outflow boundary A surface boundary separating cooler, more-dense air from warmer less-dense air. Outflow boundaries formed by the horizontal spreading of cool air that originated inside a thunderstorm.

Outgassing The release of gases dissolved in hot, molten rock.

Outlooks (seasonal and 90-day) An overview or projection of how certain weather elements (such as temperature and precipitation) might compare with average weather conditions.

Overrunning A condition that occurs when air moves up and over another layer of air.

Overshooting top A situation in a mature thunderstorm where rising air, associated with strong convection, penetrates into a stable layer (usually the stratosphere), forcing the upper part of the cloud to rise above its relatively flat anvil top.

Oxygen (O_2) A colorless and odorless gas that occupies about 21 percent of dry air in the lower atmosphere.

Ozone (O_3) An almost colorless gaseous form of oxygen with an odor similar to weak chlorine. The highest natural concentration is found in the stratosphere where it is known as *stratospheric ozone*. It also forms in polluted air near the surface where it is the main ingredient of photochemical smog. Here, it is called *tropospheric ozone*.

Ozone hole A sharp drop in stratospheric ozone concentration observed over the Antarctic during the spring.

P

Pacific air Cool, moist air that originates over the Pacific Ocean, moves eastward, then descends the Rocky Mountains and moves over the plains as dry, stable, relatively cool air.

Pacific Decadal Oscillation (PDO)   A reversal in ocean surface temperatures that occurs every 20 to 30 years over the northern Pacific Ocean.

Pacific high   *See* Subtropical high.

Parcel of air   An imaginary small body of air a few meters wide that is used to explain the behavior of air.

Parhelia   *See* Sundog.

Particulate matter   Solid particles or liquid droplets that are small enough to remain suspended in the air. Also called *aerosols*.

Pattern recognition   An analogue method of forecasting where the forecaster uses prior weather events (or similar weather map conditions) to make a forecast.

Permafrost   A layer of soil beneath Earth's surface that remains frozen throughout the year.

Persistence forecast   A forecast that the future weather condition will be the same as the present condition.

Photochemical smog   *See* Smog.

Photodissociation   The splitting of a molecule by a photon.

Photon   A discrete quantity of energy that can be thought of as a packet of electromagnetic radiation traveling at the speed of light.

Pileus cloud   A smooth cloud in the form of a cap. Occurs above, or is attached to, the top of a cumuliform cloud. Also called a *cap cloud*.

Planetary boundary layer   *See* Atmospheric boundary layer.

Planetary scale   The largest scale of atmospheric motion. Sometimes called the *global scale*.

Plate tectonics   The theory that Earth's surface down to about 100 km is divided into a number of plates that move relative to one another across the surface of Earth. Once referred to as continental drift.

Pleistocene Epoch (or Ice Age)   The most recent period of extensive continental glaciation that saw large portions of North America and Europe covered with ice. It began about 2 million years ago and ended about 10,000 years ago.

Polar easterlies   A shallow body of easterly winds located at high latitudes poleward of the subpolar low.

Polar front   A semipermanent, semicontinuous front that separates tropical air masses from polar air masses.

Polar front jet stream (polar jet)   The jet stream that is associated with the polar front in middle and high latitudes. It is usually located at altitudes between 9 and 12 km.

Polar front theory   A theory developed by a group of Scandinavian meteorologists that explains the formation, development, and overall life history of cyclonic storms that form along the polar front.

Polar ice cap climate   A climate characterized by extreme cold, as every month has an average temperature below freezing.

Polar low   An area of low pressure that forms over polar water behind (poleward of) the main polar front.

Polar orbiting satellite   A satellite whose orbit closely parallels Earth's meridian lines and thus crosses the polar regions on each orbit.

Polar tundra climate   A climate characterized by extremely cold winters and cool summers, as the average temperature of the warmest month climbs above freezing but remains below 10°C (50°F).

Polar vortex   The semipermanent zone of upper-level low pressure found in the polar regions that is sometimes disrupted or displaced during winter.

Pollutants   Any gaseous, chemical, or organic matter that contaminates the atmosphere, soil, or water.

Positive feedback mechanism   *See* Feedback mechanism.

Potential energy   The energy that a body possesses by virtue of its position with respect to other bodies in the field of gravity.

Precession (of Earth's axis of rotation)   The wobble of Earth's axis of rotation that traces out the path of a cone over a period of about 23,000 years.

Precipitation   Any form of water particles—liquid or solid—that falls from the atmosphere and reaches the ground.

Pressure   The force per unit area. *See also* Air pressure.

Pressure gradient   The rate of decrease of pressure per unit of horizontal distance. On the same chart, when the isobars are close together, the pressure gradient is steep. When the isobars are far apart, the pressure gradient is weak.

Pressure gradient force (PGF)   The force due to differences in pressure within the atmosphere that causes air to move and, hence, the wind to blow. It is directly proportional to the pressure gradient.

Pressure tendency   The rate of change of atmospheric pressure within a specified period of time, most often three hours. Same as *barometric tendency*.

Prevailing westerlies   The dominant westerly winds that blow in middle latitudes on the poleward side of the subtropical high-pressure areas. Also called *westerlies*.

Prevailing wind   The wind direction most frequently observed during a given period.

Primary air pollutants   Air pollutants that enter the atmosphere directly.

Probability forecast   A forecast of the probability of occurrence of one or more of a mutually exclusive set of weather conditions.

Prognostic chart (prog)   A chart showing expected or forecasted conditions, such as pressure patterns, frontal positions, contour height patterns, and so on.

Psychrometer   An instrument used to measure the water vapor content of the air. It consists of two thermometers (dry bulb and wet bulb). After whirling the instrument, the dew point and relative humidity can be obtained with the aid of tables.

R

Radar   An electronic instrument used to detect objects (such as falling precipitation) by their ability to reflect and scatter microwaves back to a receiver. (*See also* Doppler radar.)

Radiant energy (radiation)   Energy propagated in the form of electromagnetic waves. These waves do not need molecules to propagate them, and in a vacuum they travel at nearly 300,000 km per sec (186,000 mi per sec).

Radiational cooling   The process by which Earth's surface and adjacent air cool by emitting infrared radiation.

Radiation fog   Fog produced over land when radiational cooling reduces the air temperature to or below its dew point. It is also known as *ground fog* and *valley fog*.

Radiation inversion   An increase in temperature with height due to radiational cooling of Earth's surface. Also called a *nocturnal inversion*.

Radiative equilibrium temperature   The temperature achieved when an object, behaving as a blackbody, is absorbing and emitting radiation at equal rates.

Radiative forcing   An increase (positive) or a decrease (negative) in net radiant energy observed over an area at the tropopause. An increase in radiative forcing may induce surface warming, whereas a decrease may induce surface cooling.

Radiative forcing agent   Any factor (such as increasing greenhouse gases and variations in solar output) that can change the balance between incoming energy from the sun and outgoing energy from Earth and the atmosphere.

Radiometer *See* Infrared radiometer.

Radiosonde A balloon-borne instrument that measures and transmits pressure, temperature, and humidity to a ground-based receiving station.

Rain Precipitation in the form of liquid water drops that have diameters greater than that of drizzle.

Rainbow An arc of concentric colored bands that spans a section of the sky when rain is present and the sun is positioned at the observer's back.

Rain gauge An instrument designed to measure the amount of rain that falls during a given time interval.

Rain shadow The region on the leeside of a mountain where the precipitation is noticeably less than on the windward side.

Rawinsonde observation A radiosonde observation that includes wind data.

Reflected light *See* Reflection.

Reflection The process whereby a surface turns back a portion of the radiation that strikes it. When the radiation that is turned back (reflected) from the surface is visible light, the radiation is referred to as *reflected light.*

Refraction The bending of light as it passes from one medium to another.

Relative humidity The ratio of the amount of water vapor in the air compared to the amount required for saturation (at a particular temperature and pressure). The ratio of the air's actual vapor pressure to its saturation vapor pressure.

Return stroke The luminous lightning stroke that propagates upward from Earth to the base of a cloud.

Ridge An elongated area of high atmospheric pressure.

Rime A white or milky granular deposit of ice formed by the rapid freezing of supercooled water drops as they come in contact with an object in below-freezing air.

Riming *See* Accretion.

Roll cloud A dense, roll-shaped, elongated cloud that appears to slowly spin about a horizontal axis behind the leading edge of a thunderstorm's gust front.

Rotor cloud A turbulent cumuliform type of cloud that forms on the leeward side of large mountain ranges. The air in the cloud rotates about an axis parallel to the range.

Rotors Turbulent eddies that form downwind of a mountain chain, creating hazardous flying conditions.

S

Saffir-Simpson scale A scale relating a hurricane's winds to the possible damage it is capable of inflicting. It is now called the *Saffir-Simpson hurricane wind scale.*

St. Elmo's fire A bright electric discharge that is projected from objects (usually pointed) when they are in a strong electric field, such as during a thunderstorm.

Santa Ana wind A warm, dry wind that blows into southern California from the east off the elevated desert plateau. Its warmth is derived from compressional heating.

Saturation (of air) An atmospheric condition whereby the level of water vapor is the maximum possible at the existing temperature and pressure.

Saturation vapor pressure The maximum amount of water vapor necessary to keep moist air in equilibrium with a surface of pure water or ice. It represents the maximum amount of water vapor that the air can hold at any given temperature and pressure. (*See* Equilibrium vapor pressure.)

Savanna A tropical or subtropical region of grassland and drought-resistant vegetation. Typically found in tropical wet-and-dry climates.

Scales of motion The hierarchy of atmospheric circulations from tiny gusts to giant storms.

Scattering The process by which small particles in the atmosphere deflect radiation from its path into different directions.

Scientific method The systematic production of knowledge that involves posing a question, putting forth a hypothesis, predicting what the hypothesis would imply if it were true, and carrying out tests to see if the prediction is accurate.

Scintillation The apparent twinkling of a star due to its light passing through regions of differing air densities in the atmosphere.

Sea breeze A coastal local wind that blows from the ocean onto the land. The leading edge of the breeze is termed a *sea-breeze front.*

Sea-level pressure The atmospheric pressure at mean sea level.

Seasonal outlooks Outlooks that extend over three-month periods ranging out to a year or more, typically showing the odds that precipitation and temperature will be above or below average.

Secondary air pollutants Pollutants that form when a chemical reaction occurs between a primary air pollutant and some other component of air. Tropospheric ozone is a secondary air pollutant.

Selective absorbers Substances such as water vapor, carbon dioxide, clouds, and snow that absorb radiation only at particular wavelengths.

Semi-arid climate A dry climate where potential evaporation and transpiration exceed precipitation. Not as dry as the arid climate. Typical vegetation is short grass.

Semipermanent highs and lows Areas of high pressure (anticyclones) and low pressure (extratropical cyclones) that tend to persist at a particular latitude belt throughout the year. In the Northern Hemisphere, typically they shift slightly northward in summer and slightly southward in winter.

Sensible heat The heat we can feel and measure with a thermometer.

Sensible temperature The sensation of temperature that the human body feels in contrast to the actual temperature of the environment as measured with a thermometer.

Severe thunderstorms Intense thunderstorms capable of producing heavy showers, flash floods, hail, strong and gusty surface winds, and tornadoes. The United States National Weather Service describes a severe thunderstorm as having at least one of the following: hail with a diameter of at least one inch, surface wind gusts of 50 knots or greater, or production of a tornado.

Shear *See* Wind shear.

Sheet lightning Occurs when the lightning flash is not seen but the flash causes the cloud (or clouds) to appear as a diffuse luminous white sheet.

Shelf cloud A dense, arch-shaped, ominous-looking cloud that often forms along the leading edge of a thunderstorm's gust front, especially when stable air rises up and over cooler air at the surface. Also called an *arcus cloud.*

Short-range forecast Generally used to describe a weather forecast that extends from about 6 hours to a few days into the future.

Shortwave (in the atmosphere) A small wave that moves around longwaves in the same direction as the air flow in the middle and upper troposphere. Shortwaves are also called *shortwave troughs.*

Shortwave radiation A term most often used to describe the radiant energy emitted from the sun, in the visible and near ultraviolet wavelengths.

Shower Intermittent precipitation from a cumuliform cloud, usually of short duration but often heavy.

Siberian high A strong, shallow area of high pressure that forms over Siberia in winter.

Sleet A type of precipitation consisting of transparent pellets of ice 5 mm or less in diameter. Same as *ice pellets*.

Smog Originally *smog* meant a mixture of smoke and fog. Today, *smog* means air that has restricted visibility due to pollution, or pollution formed in the presence of sunlight— *photochemical smog*.

Smog front (also smoke front) The leading edge of a sea breeze that is contaminated with smoke or pollutants.

Snow A solid form of precipitation composed of ice crystals in complex hexagonal form.

Snow-albedo feedback A positive feedback whereby increasing surface air temperatures enhance the melting of snow and ice in polar latitudes. This reduces Earth's albedo and allows more sunlight to reach the surface, which causes the air temperature to rise even more.

Snowflake An aggregate of ice crystals that falls from a cloud.

Snow flurries Light showers of snow that fall intermittently.

Snow grains Precipitation in the form of very small, opaque grains of ice. The solid equivalent of drizzle.

Snow pellets White, opaque, approximately round ice particles between 2 and 5 mm in diameter that form in a cloud either from the sticking together of ice crystals or from the process of accretion. Also called *graupel*.

Snow squall (shower) An intermittent heavy shower of snow that greatly reduces visibility.

Solar constant The rate at which solar energy is received on a surface at the outer edge of the atmosphere perpendicular to the sun's rays when Earth is at a mean distance from the sun. The value of the solar constant is about two calories per square centimeter per minute or about 1361 W/m^2 in the SI system of measurement.

Solar wind An outflow of charged particles from the sun that escapes the sun's outer atmosphere at high speed.

Sonic boom A loud explosive-like sound caused by a shock wave emanating from an aircraft (or any object) traveling at or above the speed of sound.

Sounding An upper-air observation, such as a radiosonde observation. A vertical profile of an atmospheric variable such as temperature or winds.

Source regions (for air masses) Regions where air masses originate and acquire their properties of temperature and moisture.

Southern Oscillation (SO) The reversal of surface air pressure at opposite ends of the tropical Pacific Ocean that occur during major El Niño events.

Specific heat The ratio of the heat absorbed (or released) by the unit mass of the system to the corresponding temperature rise (or fall).

Specific humidity The ratio of the mass of water vapor in a given parcel to the total mass of air in the parcel.

Spin-up vortices Small whirling tornadoes perhaps 30 to 100 m in diameter that form in a region of strong wind shear in a hurricane's eyewall.

Squall line A line of thunderstorms that form along a cold front or out ahead of it.

Stable air *See* Absolutely stable atmosphere.

Standard atmosphere A hypothetical vertical distribution of atmospheric temperature, pressure, and density in which the air is assumed to obey the gas law and the hydrostatic equation. The lapse rate of temperature in the troposphere is taken as 6.5°C/1000 m or 3.6°F/1000 ft.

Standard atmospheric pressure A pressure of 1013.25 millibars (mb), 29.92 inches of mercury (Hg), 760 millimeters (mm) of mercury, 14.7 pounds per square inch (lb/in.²), or 1013.25 hectopascals (hPa).

Standard rain gauge A nonrecording rain gauge with an 8-inch diameter collector funnel and a tube that amplifies rainfall by tenfold.

Stationary front A front that is nearly stationary with winds blowing almost parallel and from opposite directions on each side of the front.

Station pressure The actual air pressure computed at the observing station.

Statistical forecast A forecast based on a mathematical/statistical examination of data that represents the past observed behavior of the forecasted weather element.

Steady-state forecast A weather prediction based on the past movement of surface weather systems. It assumes that the systems will move in the same direction and at approximately the same speed as they have been moving. Also called *trend forecasting*.

Steam fog *See* Evaporation (mixing) fog.

Stefan-Boltzmann law A law of radiation which states that the amount of radiant energy emitted from a unit surface area of an object (ideally a blackbody) is proportional to the fourth power of the object's absolute temperature.

Steppe An area of grass-covered, treeless plains that has a semi-arid climate.

Stepped leader An initial discharge of electrons that proceeds intermittently toward the ground in a series of steps in a cloud-to-ground lightning stroke.

Storm surge An abnormal rise of the sea along a shore; primarily due to the winds of a storm, especially a hurricane.

Straight-line wind Strong winds created by a thunderstorm's downdraft that flows outward, away from the storm in a straight line, more or less parallel to the ground.

Stratocumulus A low cloud, predominantly stratiform, with low, lumpy, rounded masses, often with blue sky between them.

Stratosphere The layer of the atmosphere above the troposphere and below the mesosphere (between 10 km and 50 km), generally characterized by an increase in temperature with height.

Stratospheric polar night jet A jet stream that forms near the top of the stratosphere over polar latitudes during the winter months.

Stratus A low, gray cloud layer with a rather uniform base whose precipitation is most commonly drizzle.

Streamline A line that shows the wind-flow pattern.

Sublimation The process whereby ice changes directly into water vapor without melting.

Subpolar climate A climate observed in the Northern Hemisphere that borders the polar climate. It is characterized by severely cold winters and short, cool summers. Also known as *taiga climate* and *boreal climate*.

Subpolar low A belt of low pressure located between 50° and 70° latitude. In the Northern Hemisphere, this "belt" consists of the *Aleutian low* in the North Pacific and the *Icelandic low* in the North Atlantic. In the Southern Hemisphere, it exists around the periphery of the Antarctic continent.

Subsidence The slow sinking of air, usually associated with high-pressure areas.

Subsidence inversion A temperature inversion produced by compressional warming—the adiabatic warming of a layer of sinking air.

Subtropical high   A semipermanent high in the subtropical high-pressure belt centered near 30° latitude. The *Bermuda high* is located over the Atlantic Ocean off the east coast of North America. The *Pacific high* is located off the west coast of North America.

Subtropical jet stream   The jet stream typically found between 20° and 30° latitude at altitudes between 12 and 14 km.

Suction vortices   Small, rapidly rotating whirls perhaps 10 m in diameter that are found within large tornadoes.

Sulfate aerosols   *See* Aerosols.

Sulfur dioxide (SO$_2$)   A colorless gas that forms primarily in the burning of sulfur-containing fossil fuels.

Summer solstice   The point in the year when the sun is highest in the sky. Typically it occurs around June 21 in the Northern Hemisphere, with the sun directly overhead at latitude $23\frac{1}{2}$°N, the Tropic of Cancer.

Sundog   A colored luminous spot produced by refraction of light through ice crystals that appears on either side of the sun. Also called *parhelia*.

Sun pillar   A vertical streak of light extending above (or below) the sun. It is produced by the reflection of sunlight off ice crystals.

Sunspots   Relatively cooler areas on the sun's surface. They represent regions of an extremely high magnetic field.

Supercell storm   A severe thunderstorm that consists primarily of a single rotating updraft. Its organized internal structure allows the storm to maintain itself for several hours. Supercell storms can produce large hail and dangerous tornadoes.

Supercell tornadoes   Tornadoes that occur within supercell thunderstorms that contain well-developed, mid-level mesocyclones.

Supercooled cloud (or cloud droplets)   A cloud composed of liquid droplets at temperatures below 0°C (32°F). When the cloud is on the ground it is called *supercooled fog* or *cold fog*.

Superior mirage   *See* Mirage.

Supersaturation   A condition whereby the atmosphere contains more water vapor than is needed to produce saturation with respect to a flat surface of pure water or ice, and the relative humidity is greater than 100 percent.

Super typhoon   A tropical cyclone (typhoon) in the western Pacific that has sustained winds of 130 knots or greater.

Surface inversion   *See* Radiation inversion.

Surface map   A map that shows the distribution of sea-level pressure with isobars and weather phenomena. Also called a *surface chart*.

Synoptic scale   The typical weather map scale that shows features such as high- and low-pressure areas and fronts over a distance spanning a continent. Also called the *cyclonic scale*.

T

Taiga (boreal forest)   The open northern part of the coniferous forest. Taiga also refers to subpolar climate.

Tangent arc   An arc of light tangent to a halo. It forms by refraction of light through ice crystals.

Tcu   An abbreviation sometimes used to denote a towering cumulus cloud (cumulus congestus).

Teleconnections   A linkage between weather changes occurring in widely separated regions of the world.

Temperature   The degree of hotness or coldness of a substance as measured by a thermometer. It is also a measure of the average speed or kinetic energy of the atoms and molecules in a substance.

Temperature inversion   An increase in air temperature with height, often simply called an *inversion*.

Thermal   A small, rising parcel of warm air produced when Earth's surface is heated unevenly.

Thermal belts   Horizontal zones of vegetation found along hillsides that are primarily the result of vertical temperature variations.

Thermal circulations   Air flow resulting primarily from the heating and cooling of air.

Thermal lows and thermal highs   Areas of low and high pressure that are shallow in vertical extent and are produced primarily by surface temperatures.

Thermal tides   Atmospheric pressure variations due to the uneven heating of the atmosphere by the sun.

Thermograph   An instrument that measures and records air temperature.

Thermometer   An instrument for measuring temperature. The most common is liquid-in-glass, which has a sealed glass tube attached to a glass bulb filled with liquid.

Thermosphere   The atmospheric layer above the mesosphere (above about 85 km) where the temperature increases rapidly with height.

Thunder   The sound due to rapidly expanding gases along the channel of a lightning discharge.

Thunderstorm   A convective storm (cumulonimbus cloud) with lightning and thunder. Thunderstorms can be composed of an ordinary cell, multicells, or a rapidly rotating supercell.

Thundersnow   The occurrence of thunder and lightning while snow is falling, typically as part of an intense midlatitude winter storm.

Tornado   An intense, rotating column of air that often protrudes from a cumuliform cloud in the shape of a funnel or a rope whose circulation is present on the ground. (*See* Funnel cloud.)

Tornado Alley   A region in the Great Plains of the United States extending from Texas and Oklahoma northward into Kansas and Nebraska where tornadoes are most frequent.

Tornado emergency   A type of tornado warning issued when a particularly strong tornado poses the potential for major damage or loss of life.

Tornadogenesis   The process by which a tornado forms.

Tornado outbreak   A series of tornadoes that forms within a particular region—a region that may include several states. Often associated with widespread damage and destruction.

Tornado vortex signature (TVS)   An image of a tornado on the Doppler radar screen that shows up as a small region of rapidly changing wind directions inside a mesocyclone.

Tornado warning   A warning issued when a tornado has been observed or is indicated on radar. It may also be issued when the formation of tornadoes is imminent. Tornado warnings typically cover parts of one or more counties and last from 30 to 45 minutes.

Tornado watch   A forecast issued to alert the public that tornadoes may develop within a specified area, usually a portion of one or more states, over the next few hours.

Trace (of precipitation)   An amount of precipitation less than 0.01 in. (0.025 cm).

Trade wind inversion   A temperature inversion frequently found in the subtropics over the eastern portions of the tropical oceans.

Trade winds   The winds that occupy most of the tropics and blow from the subtropical highs to the equatorial low.

Transpiration   The process by which water in plants is transferred as water vapor to the atmosphere.

Tropical cyclone   The general term for storms (cyclones) that form over warm tropical oceans.

Tropical depression   A mass of thunderstorms and clouds generally with a cyclonic wind circulation of between 20 and 34 knots.

Tropical disturbance   An organized mass of thunderstorms with a slight cyclonic wind circulation of less than 20 knots.

Tropical easterly jet   A jet stream that forms on the equatorward side of the subtropical highs near 15 km.

Tropical monsoon climate   A tropical climate with a brief dry period of perhaps one or two months.

Tropical rainforest   A type of forest consisting mainly of lofty trees and a dense undergrowth near the ground.

Tropical storm   A region of organized thunderstorms over the tropical or subtropical ocean with a closed cyclonic wind circulation between 35 and 64 knots.

Tropical wave   A migratory wavelike disturbance in the tropical easterlies. Tropical waves occasionally intensify into tropical cyclones. They are also called *easterly waves*.

Tropical wet-and-dry climate   A tropical climate poleward of the tropical wet climate where a distinct dry season occurs, often lasting for two months or more.

Tropical wet climate   A tropical climate with sufficient rainfall to produce a dense tropical rainforest.

Tropopause   The boundary between the troposphere and the stratosphere.

Troposphere   The layer of the atmosphere extending from Earth's surface up to the tropopause (about 10 km above the ground).

Trough   An elongated area of low atmospheric pressure.

Turbulence   Any irregular or disturbed flow in the atmosphere that produces gusts and eddies.

Twilight   The time at the beginning of the day immediately before sunrise and at the end of the day after sunset when the sky remains illuminated.

Typhoon   A hurricane (tropical cyclone) that forms in the western Pacific Ocean.

U

Ultraviolet (UV) radiation   Electromagnetic radiation with wavelengths longer than X-rays but shorter than visible light.

Unstable air   *See* Absolutely unstable atmosphere.

Upslope fog   Fog formed as moist, stable air flows upward over a topographic barrier.

Upslope precipitation   Precipitation that forms due to moist, stable air gradually rising along an elevated plain. Upslope precipitation is common over the western Great Plains, especially east of the Rocky Mountains.

Upwelling   The rising of water (usually cold) toward the surface from the deeper regions of a body of water.

Urban heat island   The increased air temperatures in urban areas as contrasted to the cooler surrounding rural areas.

V

Valley breeze   *See* Mountain breeze.

Valley fog   *See* Radiation fog.

Vapor pressure   The pressure exerted by the water vapor molecules in a given volume of air.

Veering wind   The wind that changes direction in a clockwise sense—north to northeast to east, and so on.

Vernal equinox   The equinox at which the sun approaches the Northern Hemisphere and passes directly over the equator. Occurs around March 20.

Very short range forecast   Generally used to describe a weather forecast that is made for up to a few hours (usually less than 6 hours) into the future.

Virga   Precipitation that falls from a cloud but evaporates before reaching the ground. (*See* Fall streaks.)

Visible radiation (light)   Radiation with a wavelength between 0.4 and 0.7 m. This region of the electromagnetic spectrum is called the *visible region*.

Visible region   *See* Visible radiation.

Visibility   The greatest distance at which an observer can see and identify prominent objects.

Volatile organic compounds (VOCs)   A class of organic compounds that are released into the atmosphere from sources such as motor vehicles, paints, and solvents. VOCs (which include hydrocarbons) contribute to the production of secondary pollutants, such as ozone.

W

Wall cloud   An area of rotating clouds that extends beneath a supercell thunderstorm and from which a funnel cloud may appear. Also called a *collar cloud* and *pedestal cloud*.

Warm advection (or *warm air advection*)   The transport of warm air by the wind from a region of higher temperatures to a region of lower temperatures.

Warm-core low   A low-pressure area that is warmer at its center than at its periphery. Tropical cyclones exhibit this temperature pattern.

Warm front   A front that moves in such a way that warm air replaces cold air.

Warm occlusion   *See* Occluded front.

Warm sector   The region of warm air within a wave cyclone that lies between a retreating warm front and an advancing cold front.

Water equivalent   The depth of water that would result from the melting of a snow sample. Typically about 10 inches of snow will melt to 1 inch of water, producing a water equivalent of 10 to 1.

Waterspout   A column of rotating wind over water that has characteristics of a dust devil and a tornado.

Water vapor   Water in a vapor (gaseous) form. Also called *moisture*.

Water vapor–greenhouse effect feedback   A positive feedback whereby increasing surface air temperatures cause an increase in the evaporation of water from the oceans. Increasing concentrations of atmospheric water vapor enhance the greenhouse effect, which causes the surface air temperature to rise even more.

Watt (W)   The unit of power in SI units where 1 watt is equivalent to 1 joule per second.

Wave cyclone   An extratropical cyclone that forms and moves along a front. The circulation of winds about the cyclone tends to produce a wavelike deformation on the front.

Wavelength   The distance between successive crests, troughs, or identical parts of a wave.

Weather   The condition of the atmosphere at any particular time and place.

Weather elements   The elements of *air temperature, air pressure, humidity, clouds, precipitation, visibility,* and *wind* that determine the present state of the atmosphere, the weather.

Weather type forecasting   A forecasting method where weather patterns are categorized into similar groups or types.

Weather types   Certain weather patterns categorized into similar groups. Used as an aid in weather prediction.

Weather warning   A forecast indicating that hazardous weather is either imminent or actually occurring within the specified forecast area.

Weather watch   A forecast indicating that atmospheric conditions are favorable for hazardous weather to occur over a particular region during a specified time period.

Westerlies   The dominant westerly winds that blow in the middle latitudes on the poleward side of the subtropical high-pressure areas.

Wet-bulb depression   The difference in degrees between the air temperature (dry-bulb temperature) and the wet-bulb temperature.

Wet-bulb temperature   The lowest temperature that can be obtained by evaporating water into the air.

Whirlwinds   *See* Dust devils.

Wien's law   A law of radiation which states that the wavelength of maximum emitted radiation by an object (ideally a blackbody) is inversely proportional to the object's absolute temperature.

Wind   Air in motion relative to Earth's surface.

Wind-chill index   The cooling effect of any combination of temperature and wind, expressed as the loss of body heat. Also called *wind-chill factor*.

Wind direction   The direction *from which* the wind is blowing.

Wind machines   Fans placed in orchards for the purpose of mixing cold surface air with warmer air above.

Wind profiler   A Doppler radar capable of measuring the turbulent eddies that move with the wind. Because of this, it is able to provide a vertical picture of wind speed and wind direction.

Wind rose   A diagram that shows the percent of time that the wind blows from different directions at a given location over a given time.

Wind shear   The rate of change of wind speed or wind direction over a given distance.

Wind speed   The rate at which the air moves by a stationary object, usually measured in statute miles per hour (mi/hr), nautical miles per hour (knots), kilometers per hour (km/hr), or meters per second (m/sec).

Wind vane   An instrument used to indicate wind direction.

Windward side   The side of an object facing into the wind.

Winter solstice   The point in the Northern Hemisphere when the sun is lowest in the sky. Typically it occurs around December 21 in the Northern Hemisphere. At this point, the sun is directly overhead at latitude $23\frac{1}{2}$°S, the Tropic of Capricorn, where the summer solstice is occurring at the same time.

X

Xerophytes   Drought-resistant vegetation.

Y

Younger Dryas event   A cold episode that took place from about 13,000 to about 11,700 years ago, during which average temperatures dropped suddenly and portions of the Northern Hemisphere reverted back to glacial conditions.

Z

Zonal wind flow   A wind that has a predominantly west-to-east component.

INDEX

A

development of, 229–238
energy for development, 231, 232–238
families, 230
and jet streams, 237–238
life cycle of, 229
mature, 230
and model of, 37
movement of, 232–238
satellite images of, 8, 108
structure, 232
and upper air support for development, 235–236
and vertical air motions, 164–165
winds around, 161–162
Cyclones, post-tropical, 341
Cyclones, tornadoes (*see* Tornadoes)
Cyclones, tropical, 320, 341 (*see also* Hurricanes)
Cyclonic flow, 162

D

Data loggers, 74
Daylight hours, for different latitudes and dates, 45
Debris ball, 313
Deforestation, 397
Degassing (*see* Outgassing)
Degree-days
cooling, 69
growing, 69
heating, 67, 69
Degree, defined, 29
Dendrite ice crystals, 135
Dendrochronology, 384
Density (*see* Air density)
Deposition, 30, 92
Depression of wet bulb (*see* Wet bulb depression)
Depressions, middle latitude (*see* Cyclones, middle latitude)
Depressions, tropical, 325
Derecho, 282
Desert climate (*see* Arid climate)
Desertification, 398
Desert winds, 184–185
Des Voeux, Harold, 416
Dew, 91–92
frozen, 91
Dew-point front (*see* Dryline)
Dew-point temperature, 85, 87
average for January and July, 86
Diffluence, 234
Diffraction, of light, 459
Diffuse light (*see* Scattering)
Dimethylsulphide (DMS), 394
Dinosaurs, extinction of, 396
Dispersion
defined, 429
of light, 455
of pollution, 429–430, 432, 435

Divergence, 233, 234
cause of, 236
effect on surface air pressure, 164–165
with high-and-low-pressure areas, 164–165
around jet stream, 237–238
speed, 234
Dixie Alley, 298
Dobson units, 19, 425, 426
Doldrums, 189
Doppler lidar, 313
Doppler radar (*see* Radar)
Doppler shift, 142, 143, 312
Downbursts, 12, 279 (*see also* Thunderstorm, downdrafts in)
Drizzle, 132, 366
Drought
Great Drought of 1930, 373
defined, 372
and dry spells, 372–373
and the Dust Bowl, 373
summer of 2012, 11
Texas drought of 2011, 373
Drought Severity Index, 372
Dropsonde, 22, 168, 326, 345
Dropwinsonde observation, 168
Dry adiabatic rate of cooling and warming, 116, 120
Dryas, 386
Dry-bulb temperature, 91
Dry climates, 288, 357
Dry ice, in cloud seeding, 97, 129, 130
Dry lightning, 294
Dryline, 226, 290
Dry slots, 237
Dual-polarization (polarimetric) radar, 152, 245, 313, 314
Dust Bowl, drought, 373
Dust clouds, 184, 280
Dust devils, 184, 184, 185, 219
Dust storms, 184
Dynamic lows, 232

E

Earth
albedo of, 41, 42, 389
annual energy balance, 42–43
atmosphere, overview of, 4
average surface temperature, 17, 34, 35, 41–42, 387–388, 400–401
changes in tilt, 391
distance from sun, 41
orbital variations, 391–394
radiation from, 37–38
radiative equilibrium temperature of, 37
tilt of, 46, 48, 391
Easterlies, polar, 191
Easterly wave, 320
Eccentricity of earth's orbit, 392

Eddies, 174, 175–176, 177 (*see also* Cyclones; Anticyclones)
and air pollution, 429
and mini whirls, 336
and turbulence, 177
Ekman spiral, 332
Ekman transport, 332
El Chichón, volcano, 395, 448, 459
Electricity
and growth of cloud droplets, 127
and lightning, 290–296
Electrification of clouds, 290–296
Electromagnetic spectrum, 34
Electromagnetic waves, 33 (*see also* Radiation)
Elevated storms, 226
Elongated highs (*see* Ridge)
Elongated lows (*see* Trough)
El Niño, 187, 200, 395
ENSO (El Niño/Southern Oscillation), 187, 201
Embryo, hailstone, 139
Energy, balance of earth and atmosphere, 42–43
Energy conversions, 463
Enhanced Fujita (EF) Scale, 302–303
Ensemble forecasting, 251–252, 346
spaghetti plot, 252
Entrainment, 276
Environmental lapse rate, 117, 118, 119, 120
Environmental Protection Agency (EPA), 35, 417, 418, 421, 422, 427
Equation of state, 150
Equatorial counter current, 198, 201
Equatorial current, 197–198, 200
Equatorial low, 190
Equinox
autumnal, 47, 48
precession of, 392
vernal, 49, 50
Evaporation, 81–82
defined, 16, 30, 80
factors that affect, 81–82
latent heat of, 29
Evaporation (mixing) fog, 95
Evelyn, John, 416
Exosphere, 23
Extended forecast, 255
Extratropical cyclone (*see* Cyclone, middle latitude)
Eye of hurricane (*see* Hurricanes)
Eyewall, 321, 322
Eyewall replacement, 325, 347

F

Faculae, 394
Fahrenheit, G. Daniel, 29
Fahrenheit temperature scale, 4, 29
Fallstreaks, 134
Fall wind, 180

Prognostic chart (prog), defined, 248
Psychrometers, 90
Pyrocumulus clouds, 119

Q

Quasi-Stationary front, 221
QuikSCAT Satellite, 327

R

Radar
    and algorithms, 142, 245, 313
    bounded weak echo region (BWER), 307
    and bow echo image, 281
    Cloud Profiling, 110
    conventional, 6, 142
    defined, 141
    determining winds, 167–168, 311–312
    Doppler, 6, 9, 10, 13, 141–142, 144, 168, 222
    dual-polarization (polarimetric), 143, 245, 313–314
    and hook echo image, 307, 312, 313
    Next Generation, 304
    phased-array, 143
    polarimetric (dual-polarization), 143, 245, 313, 314
    and precipitation, 5, 6, 16, 80, 90
    and satellites, 109–110
    and tornadoes, 311–312
    WSR-88D, 245, 313
Radiant energy (see Radiation)
Radiation (radiant energy), 33–44
    absorption and emission of, 36–44
    defined, 32
    effect on humans, 33
    infrared, 36
    intensity, 34
    longwave (terrestrial), 36
    reflection of, 41, 444
    refraction of, 449 (see also Refraction)
    scattering, 41, 444
    selective absorbers, 37–38
    shortwave, 36
    and skin cancer, 35, 423
    and sunburning, 35
    of sun and earth, 33–35, 41–42
    and temperature, 33–35, 42
    transmitted light, 449
    ultraviolet (UV), 33
    wavelength of, 33, 34
    variation with seasons, 4
Radiational cooling, 59, 95
Radiation fog, 93, 94
Radiative equilibrium temperature, 36–44, 399
Radiative forcing agents, 399
Radio
    communications and ionosphere, 23
    wavelengths, 23, 34, 35
Radiometers, 73, 107
Radiosonde, 22

Radon, 418, 432
Rain, 132–133 (see also Precipitation)
    acid, 436
    intensity, 103
    trace of, 140
Rainbows, 456–458
Raindrop
    shape, 125, 134
    size, 125, 132–133
Rainforest, tropical, 359–363
Rain-free base, 283
Rain gauges, 141
Rain shadow, 125, 355
Rainshower, 132
Rain streamer, 364
Rapid Refresh, 250
Rawinsonde observation, 168
Rayleigh scattering, 445
Reading material, additional, 479–480
Rear-flank downdraft, 307, 308
Rear-flank inflow jet, 281
Red sky, 5
Red sprite, 295
Red suns, 41, 447–448
Redwood trees and fog, 94
Reflection (see also Albedo)
    and color of objects, 444, 456
    of radiation, 36, 41, 444, 456
Refraction
    and green flash, 449–451
    and lengthening of day, 450
    of light, 454, 456
    and mirage, 451–452
    cause of twinkling, 449–450
Relative humidity, 83–90
    comparing desert air with polar air, 86
    computation of, 85
    daily variation of, 84
    defined, 83
    and dew point, 85–87
    equation for, 83, 466
    in the home, 89
    and human discomfort, 88–89
    measuring, 90–91
    oppressive, 85
Ribbon lightning, 294
Ridge
    defined, 154
    and upper-level charts, 235 (see also Anticyclones)
Rime, 136, 137, 138
Roll cloud, 279
Rossby, C. G., 235
Rossby waves, 235 (see also Longwaves in atmosphere)
Rotor cloud, 125
Rotors, 175
Rowland, F. Sherwood, 425
Ruddiman, William, 398
Runaway greenhouse effect, 389
Runaway ice age, 389

S

Saffir-Simpson Hurricane Wind Scale, 335
St. Elmo's Fire, 295
Salt particles, as condensation nuclei, 126
Sandstorms, 184
Santa Ana wind, 10, 182–184, 419
Satellite images
    of clouds, 106–110
    computer enhancement of, 108
    enhanced infrared, 109, 282
    infrared, 108, 109, 187, 217, 268
    of middle latitude cyclonic storms, 7, 217, 228
    and ozone hole, 18, 425
    three-dimensional, 110, 322, 326
    of tropical cyclones, 325, 343
    visible, 108, 110, 191, 214, 228, 326, 329, 334, 342
    water vapor, 6, 109
    wind speed and direction of surface, 168
Satellites
    Aura, 425, 426
    CloudSat, 110, 143, 327
    and forecasting, 245, 267–268, 346
    geostationary (geosynchronous), 7, 106
    GOES, 108
    and identification of clouds, 107
    polar orbiting, 107
    QuikSCAT, 327
    SOHO, 394
    TIROS 1, 107
    Tropical Rainfall Measuring Mission (TRMM), 110, 143
    and weather forecasting, 267–268
Saturation, 81–82, 83
Saturation vapor pressure, 83
    over ice, 83
    for various air temperatures, 83
Savanna grass, 362
Scales of atmospheric motion, 174
Scattering, 41, 444
    of radiation, 41
    and sky color, 41, 444, 445
Scatterometer, 168
Scientific method, 4
Schaefer, Vincent, 129
Scintillation, 450
Scud, 101
Sea breeze, 176, 178–179
Sea breeze convergence zone, 179
Sea breeze front, 179
Sea ice, 408
Sea-level pressure, 152, 153
    January and July average, 193
Sea-level pressure chart, 153
Seasons, 44–51
    astronomical and meteorological, 49
    defined, 44–45
    in Northern Hemisphere, 46–50
    in Southern Hemisphere, 50–51
    local variations, 51